Theme Cities: Solutions for Urban Problems

GeoJournal Library

Volume 112

For further volumes:
http://www.springer.com/series/6007

Wayne K.D. Davies
Editor

Theme Cities: Solutions for Urban Problems

Springer

Editor
Wayne K.D. Davies
Department of Geography
University of Calgary
Calgary
Alberta
Canada

ISSN 0924-5499 ISSN 2215-0072 (electronic)
GeoJournal Library
ISBN 978-94-017-7802-2 ISBN 978-94-017-9655-2 (eBook)
DOI 10.1007/978-94-017-9655-2

Springer Cham Heidelberg New York Dordrecht London

Softcover reprint of the hardcover 1st edition 2015

Printed on acid-free paper

Springer is part of Springer Science+Business Media (www.springer.com)

This study is dedicated to my mentor, Professor Harold Carter, to the many members of the Urban Commission of the International Geographical Union, and to the next generation: Jasmine, Evan, Asmara and Charlotte

Preface

Urban places have always been difficult to comprehend, let alone plan, given their multifaceted character, heritages and constant change. Today, their increasing size, sprawl and dominance in our habitat have made this understanding an even more challenging task. At various times in history, a combination of processes led to the emergence of cities with new characteristics that often defined their age. At other times they have been threatened with a new range of problems that threatened their survival, or at least liveability. This was the situation in the developed world in the nineteenth century when the growth of innovations, production, trade and increasing liberalism promised a more fulfilling urban future. But this progress was soon accompanied by an urban development that had such high levels of mortality, squalor and poverty that they threatened the very survival of the cities. These difficulties led to new conceptions of urban public health and the engineering and social solutions that reduced mortality rates and modified the squalor and poverty. By the end of the century it led to the new ideas and policies that created the modern urban planning movement to control or at least modify urban development. In 1988 Peter Hall published an eminently readable book that summarised the major phases of planning urban places over the previous century, which included a human focus by describing the character and contributions of the major figures in the field. This valuable book and its subsequent editions not only described the various efforts to resolve the problems created by nineteenth century urban developments, but also how the new difficulties that emerged during the twentieth century were tackled. Yet many of the issues described by Hall are still with us—especially the persistence of disadvantaged areas and the seemingly built-in injustices in many places. Also many of the policies designed to provide better urban places led to new problems.

From the late 1980s increasing dissatisfaction about the urban condition emerged. It led to a cluster of new urban concepts, each seeking to improve urban places in specific ways. They have identified key problems in urban places and suggested new ways to solve, or at least mitigate, these drawbacks and their negative impacts on the environment. Many of these new urban ideas, summarized as city themes in this study, have been the subject of individual reviews. But there have been few attempts to provide a comprehensive *guide* to the character and utility of the various new themes, one that also creates an introductory *background* for later detailed studies of contemporary urbanism, in specialized academic fields, such as urban geography, sociology, economics, etc., and also in those subjects with a more active bent that contribute to the construction and functioning of urban places. These two goals are the objectives of this book.

In many ways the text describes the major trends in improving urban development from the period where Hall ended his study, where he anticipated a new 'end of the century' moment, paralleling the late nineteenth century one that created new horizons, such as the City Beautiful, Garden City and Regional Planning concepts that provided much of the impetus for the development of planning in the twentieth century. Of course, the contexts of the new concepts that have attracted attention in the developed world of the past twenty or so years are very different. They have emerged in a more interconnected world, because of electronic systems and faster and cheaper transport systems. It is one that is increasingly dominated by large, sprawling urban areas—many of which in the developed world are viewed as having a too dominant car culture—and those with unanticipated rapid growth in the less developed world. These forces have created new urban regions of unprecedented size and number, which has led to extremes of wealth and poverty, with far too many living in conditions little better than those seen in the densely populated, squalid communities of the nineteenth century European cities. Although these new urban themes are different from one another, there is an underlying general aim to make places not only more liveable and sustainable, but also fairer or more just, although it is unfortunate that this later objective is often lost as other interests dominate.

As an Urban Geographer who also helped initiate and develop an interdisciplinary Urban Studies programme in the University of Calgary for many years, there has always been an awareness of the importance of looking at urban places in more comprehensive ways than those practiced in individual disciplinary fields. In this context three issues were important background features in writing this book. First, it is about time that more attention is being paid in the urban literature to environmental issues, through sustainable development ideas and by policies to mitigate the urban disasters inflicted by extreme natural processes. Such issues were almost ignored in most urban specialisms a generation ago, yet were still seen by many geographers as important contextual issues in urban development, although such issues rarely appeared in the standard urban geography texts. Second, spatial and global perspectives, long the bedrock of geographical interests, are vital in our closely interconnected world, and are now being pursued in many urban subfields. They provide even wider comparative contexts and also emphasize the locational divergences and connections within and between urban places, which mean that one can normally expect local and regional variations that lead to differences in urban places, even from general processes.

Third, another important consideration for somebody brought up in a British urban geography that emphasized the development of the morphology of urban places has been the need for an historical perspective in urban studies. This makes it possible to understand how towns and townscapes emerged, and through what cultural imperatives. It also makes one appreciate that many of the so-called 'new' policies of the present are in part, at least, revivals of old and often forgotten ideas, although in new situations, putting the many fads and fashions of new urban ideas into a broader context. Many of the older ideas can be found in those masterful urban surveys by authors such as Mumford (1961) and Vance (1977) that deal with the history and culture of urban places. A temporal perspective also reminds one of the utility of forgotten premises of Geddes's (1915) early twentieth century work. One is the succinct phrase: 'survey first and plan second'. The other was his emphasis upon regional analysis to survey the resources, potentials and drawbacks of the region around cities, the human responses, and the variety and complexity of the resultant cultural landscapes—ideas that came from the French geographical regional tradition. This focus upon understanding the context and the problems of existing landscapes first, is not simply to appreciate the variety of cultural townscapes which enhance the human experience. It provides the opportunity to assess the limitations, risks and possible negative effects created by new developments, rather than looking only upon the positive results—or rather assumed positive results—which all too frequently have focused upon profit maximization, rather than upon the utilities of the projects for people. It also provides understandings of previous historical phases, making it possible to provide links to the new developments, rather than only obliterating the past in the name of progress. Certainly much of my own research stayed far from these roots in trying to provide more empirical rigour to urban analysis, and more recently in how exploration narratives are constructed. Indeed the biggest challenges have always been in trying to interrelate, if not integrate, a series of what were often seen as disparate ideas to provide greater coherence to the study of various urban phenomena or ideas. The attempted integration is not simply in human terms through conceptual overviews, but also through the integrative methods of multivariate analysis, although the following chapters only provide examples of the former approach.

This project developed out of a classroom need to review the emerging new urban themes against the context of the main twentieth century ideas, problems and policies of urban development. It has benefitted from discussions with many of the members of the Urban Geography Commission of the International Geographical Union (IGU). Indeed three of its members—Professor Ivan Townshend (Lethbridge University), Dr Niamh Moore-Cherry (University College, Dublin) and Dr Susan Ball (University of Paris, 8)—were kind enough to collaborate on the project by providing chapters in this study. Their assistance was appreciated. Others helped in a more indirect way in broadening my understanding of urban places, through their willingness to host IGU conferences in their home cities. Over the past twenty years these and other meetings provided me with the opportunity to visit and lecture in many diverse urban centres around the world—in Japan, China, Korea, Australia, Mexico, United States, South Africa, Tunisia, France, Italy, Slovenia, Switzerland, Ireland, The Netherlands, Britain, Germany, Denmark, Sweden, Finland, Poland and Spain—as well as the fortune to have short-term teaching positions in universities in Cape Town, Tokyo and Dublin. These visits enabled me to learn from the local experts about the essential characters and drawbacks of their cities and what polices were being adopted to provide solutions. Although many of the formal lectures in the conferences were useful, the more effective insights often came from field excursions and informal discussions, which provide the more tacit exchange of information that has been recognized as crucial in the development and adoption of policies lying behind many of the themes dealt with in this study, although most apply to the urban places in the developed world.

It must be admitted that as this project evolved additional material was incorporated, which led to delays in its completion and far longer chapters than were originally contemplated. In this context thanks must be provided to the patience of various personnel in the publishers, especially Bernadette Deelen-Mans and Mirielle van Kan, whose constant encouragement and good humour helped in the always lonely task of an author. I also appreciate the fact that the Department of Geography in the University of Calgary allowed me to use the facilities of the department in working on this project. In particular I am grateful for the help of Robin Poitras, the department cartographer, for his skill in drawing the various diagrams and his willingness to redraft figures when I wanted modifications. Various other people have also helped by reading and commenting on various chapters or parts of them, especially Ivan Townshend and Rhiannon Shah, while my former colleagues and friends in Swansea, notably Dr Graham Humphrys, have always provided a 'home from home' that has enabled me to discuss ideas and to have a U.K. base to explore other libraries and cities, helping to provide material for this study. Of course the weaknesses others will find in the text are all my responsibility. One of the biggest difficulties faced was to select the themes to be dealt with in this work. Soon after starting it was realised that that they could not all be covered because of space limitations. So choices had to be made, the problem of all writers. I only hope that most will agree that the themes chosen are the most important ones of the past few decades, and that there are justifications for leaving out those themes that others will believe are of more value to their interests. Also it must be stressed that the emphasis of the book is primarily upon the developed world, although many of the themes discussed could be usefully adopted in cities in less developed countries, so long as they are adjusted to their cultures and circumstances and have local involvement. These restrictions seem inevitable. Urban places are just too changeable and complex to be able to deal with everything, features that make them so endlessly fascinating and a challenge to understand.

Calgary
July 2014

Wayne K.D. Davies

References

Geddes, P. (1915). *Cities in evolution*. Williams and Norgate, London.
Hall, P. (1988). *Cities of tomorrow*. Blackwell, Oxford.
Mumford, L. (1961). *The city in history*. Harcourt Brace, New York.
Vance, J. E. (1977). *This scene of man: The role and structure of the city in the geography of western civilization*. Harpers, New York.

Contents

Contributors

Susan Ball Département d'études des pays anglophones, University of Paris 8, St-Denis, France

Wayne K.D. Davies Department of Geography, University of Calgary, N.W. Calgary, AB, Canada

Niamh Moore-Cherry School of Geography, Planning and Environmental Policy, University College, Dublin, Belfield, Dublin, Ireland

Ivan J. Townshend Department of Geography, University of Lethbridge, Lethbridge, AB, Canada

List of Figures

List of Tables

Chapter 1
Introduction

Wayne K.D. Davies

1.1 The Changing Urban Context

We are living through one of the greatest of transformations of our habitat that the world has ever experienced. The driving force of the transformation is the inexorable growth of the world population, which the latest United Nations urbanization report calculates to have increased from 2.5 billion in 1950 to 7.2 billion by 2013 (United Nations 2013). It will probably reach 8.1 billion and 9.6 billion in 2025 and 2050 respectively, based on the U.N.'s middle range estimates that assume the recent decline in fertility rates continues. By 2007 urban centres became the places where most people in the world lived, when the urban proportion of the total world population reached over 50 % for the first time in human history—although it must be remembered the U.N. uses the urban definitions of the various countries, which do vary, so the figure is only approximate. Of course, this urban majority did occur earlier in particular parts of the world. For example, the 1851 census showed that England and Wales had become the first area to have more urban than rural people. In subsequent decades country after country reached the 50 % urban threshold; and increasingly in the developed world this proportion often reaches over 75 %, for many people living in what are administratively rural areas are actually functionally connected to their nearest city through commuting and other flows. Yet in most of the developing countries there are still more rural than urban dwellers, although this balance is changing very rapidly.

All these changes mean that the percentage of the world population living in urban areas has increased from 29.4 % in 1950 to 54 % in 2014, reaching an estimated 66 % by 2050 (United Nations 2014). The developed world will have urban proportions rising from the current average of 78 to 86 % at that date, while the less developed world will show increases from the current 47 % urban to 64 % of the

W. K.D. Davies (✉)
Department of Geography, University of Calgary, 2500 University Drive, N.W., Calgary, AB T2N 1N4, Canada
e-mail: wdavies@ucalgary.ca

W. K.D. Davies (ed.), *Theme Cities: Solutions for Urban Problems*, GeoJournal Library 112, DOI 10.1007/978-94-017-9655-2_1

total population. The absolute numbers of urban residents in these developing areas will almost double from its current levels to 5.1 billion in 2050. So the urban areas will absorb practically all of the world's population growth by mid-century.

This spatial concentration of the world population in urban places is not the only characteristic of current urbanization trends. It is being intensified by the increasing numbers of people in the bigger places. Places with over a million people used to be rare; there were only 83 in 1950. But by 2015 there will probably be almost 500 centres with more than a million inhabitants accounting for over 2/5ths of the total urban population and another 730 places with between a half a million and one million residents. The continued growth of this large city population will mean that in just over a decade the majority of the world's population may be living in places with over a million inhabitants. This large city concentration is further intensified by the fact that in 1951 there were only two places in the megacity category—areas greater than 10 million people. There were 28 such regions in 2014, a number that will increase to 41 by 2030, the majority in the less developed areas of the world, accounting for some 14 % of the world population (United Nations 2014). Perhaps the most dramatic changes are taking place in China where a deliberate state policy of urbanization is taking place at an unprecedented rate. It has taken only just thirty years for China to move from 20 % urban in 1980 to 54 % today. By adding 13 million new urban dwellers per year, the country will probably be 70 % urban by 2030 when China's cities will account for a billion people (TE 2014c: United Nations 2014).

These urbanization trends are unprecedented in history but are not simply matters of size, proportion and changing location. They are associated with fundamental changes in the patterns, economies and social structures of urban places, with greater interactions between them, as well as problematic relationships with their environments. The change to an urban habitat makes it even more of a necessity to find and implement effective policies to create better urban settlements. Yet despite a century or more of planning in many countries, as well as the increasing involvement of governments in so many activities that affect urban places, there is increasing dissatisfaction with the results of these changes, as well as with the processes by which urban development occur.

An important basic concern about the current urban condition comes from the relationships of towns and cities with their environments. Despite our increasing technological prowess and knowledge we have only just begun to seriously consider the extent to which the power of nature, through storms, floods, earthquakes and other natural hazards, are negatively affecting, and in some cases even destroying, many of our settlements and the life of people within them. Several reasons account for our penchant to underestimate the effects of natural hazards on urban places. Our former hubris about our apparent ability to conquer nature through technical progress is fading. Our myopia about the devastating effects of natural hazards has helped blind our views and limited our policies. Short-term perspectives, often based on single generational views because of the rapid recent growth of cities, and the increasing mobility patterns of urban residents, have led us to forget the way in which such extreme natural events have often drastically affected our settlements, especially in those parts of the world that are especially prone to major physical threats. Certainly their episodic nature means it is difficult to predict many of extreme natural processes and makes them easy to ignore. But the increasing force and frequency of many of these natural events makes it imperative for us to develop effective plans to reduce their impact, especially in the burgeoning cities of the less developed world. But to blame natural processes alone for these and other problems linked to the environment is now seen as unfair. Our sprawling, larger cities make bigger targets for extreme natural events and many of our actions within our settlements make them more vulnerable.

In addition, the last two decades has seen a greater awareness of the negative effects of so many of our activities upon our cities and our planet, as well as recognizing that we are consuming finite resources at too fast a rate. The continuation of such trends will mean that the opportunity for subsequent generations to match our progress may be reduced, which is leading to an increasing understanding of the value of introducing more effective sustainable policies and to recognize the importance of green spaces for a healthier life. In addition we are adding far too much toxic waste to our environment and are almost certainly contributing to global warming through our increasing production of greenhouse gases. This is likely to produce a series of more drastic climatic events that will have a series of negative consequences of unknown magnitude upon our rapidly growing and larger settlements (TE 2014a, b). Yet it would be inappropriate to assume that all these climate variations are of the same magnitude with a single result, for the vulnerabilities of people, societies and economies vary between regions and places, an issue discussed in Chap. 5 (IPCC 2014).

Another set of problems associated with current urban trends comes from the way that the changes seem to be negatively affecting the diversity and vitality of urban places and the opportunities they provide in so many aspects of life. The distinctive characteristics of many urban places that developed over centuries—as seen by their urban morphologies, economies and social patterns and sustainability—are under threat. The modernisation of the world based on the diffusion of western values, life-styles and technologies, and the minimization of the differences produced by local environmental character, is producing a creeping urban homogeneity in places throughout the world. This is especially true in suburban or exurban areas where the low density sprawl and over-dependence on the car reduces personal interaction, increases traffic congestion and creates heath risks and environmental damage through pollution. In social terms the new communication technologies have added to a faster pace of life, with less time for leisure and contemplation, while more rapid social and economic changes have added to the pressure and problems of life in cities. In some urban areas these problems have been intensified by a rising tide of family breakdowns, anti-social behaviour and crime, and the difficulties of absorbing unskilled immigrants from many varied cultural backgrounds. The result has been an increase in the numbers of the disadvantaged in urban places.

In economic terms a de-industrialization has occurred in parts of western countries due to a globalization of manufacturing, which has created additional problems for many western cities. This change occurred because corporations found cheaper places to produce or assemble goods, which was made possible by the quicker and less expensive transportation of goods and people. The consequence was a drastic reduction in the economic vitality of many western towns and cities, leaving many to stagnate or even decline. This trend was accelerated by the increasing mechanization of many processes which required less labour, especially of the unskilled sort, which has led to limited employment prospects for the unskilled in developed countries. Yet the negative changes in the west have led to the explosive growth of many urban places in parts of the less developed world, unfortunately with limited labour rights and safety standards, poor housing, and often high levels of poverty, poor health and also pollution because of the absence of environmental regulations.

These globalization trends, combined with more intense capitalist processes and the decline in the number of communist states, have meant that most cities have become increasingly open to a similar range of economic pressures. So there is a greater urban competiveness as well as making them less local, increasingly vying with others in a global world for economic activities. Many have lost the site advantages they once possessed, for the old industrial-age 'trinity' of localization advantages based on the incidence of raw materials, differential transport costs and industrial skills, have been replaced by the new 'gods of location'—information, amenity and image—which many small and older industrial centres do not possess. Certainly these older locational requirements are still important for some industries. But even in the places that have attracted new economic activities there is now a greater danger that these may be moved to other places by decisions from remote corporations with little concern about the effects on the original location. Also, in many other parts of the word there are simply not enough jobs, certainly meaningful and well-paying jobs, to absorb the growing population. Yet the rapid pace of innovation has brought new growth to some places and placed a premium upon knowledge creation. Unlike the spread of manufacturing, invention and innovation has become increasingly localized in a small number of centres. The change has focused attention on the knowledge-creation processes in these cities in order to understand why some have able to develop and attract new innovative activities. Better understanding of these locational imperatives may help improve the job prospects in other centres as part of their efforts to remain competitive.

Many of these changes have been lubricated by the growth of almost instant world communication via electronic systems, such as the Internet and mobile devices, and by more rapid, cheaper and regular transport connections by sea, air and land, creating a far more inter-connected world. In the case of sea, the explosion of container ships and very big bulk carriers has transformed the transport of cargoes and raw materials. Air transport expanded especially rapidly from the 1970s with the introduction of wide-bodied jets and later low cost new airlines, as well the growth of new cargo planes for freight. On land, road transport has increased due to more and better roads and more efficient vehicles, while the introduction of high speed railways has slashed the connection times between many big cities in Europe. In this context it is China, with a high speed rail network of 12,000 km, bigger than Europe's network in 2014, and with plans to double this by 2020, that has made enormous strides in the last decade. Certainly some continents and parts of others have been left behind in this transport and communication revolution. But these changes, linked to political decisions that have led to freer trade, has created an unparalleled, interconnected and competitive world. One set of centres, the global command centres, sometimes called world cities, have been the ones to profit the most. They boomed after a period of stagnation of the 1980s, due to the growth of specific economic sectors—especially producer services, such as finance—and the innovative activities produced in the new economy. They also gain impetus from their roles as centres of consumption, politics and culture. Yet within these prosperous centres there is a growing gap between those with wealth—people with high levels of education and technical skills, or with inherited or offshore money—and the rest of the population. This latter group is not simply composed of the growing numbers of the poor and impoverished, but those with limited skills and education who work in such industries as the fast food and personal service occupations for minimum wages. In the past decade it is increasingly obvious that many other people are also finding themselves on the wrong side of this wealth gap. Rising housing costs and stagnant, if not reduced, effective wages, given the cost of living increases, has meant that many of the essential service workers, in occupations such as teaching, nursing, police forces etc., can no longer afford to live in the large central cities. This means they have to engage in longer and longer commutes or accept lower quality housing.

These inequalities within cities are creating what amounts to a dualism of wealth and opportunity. Although this trend is occurring in more than in the big cities, its incidence in these places provides a stark example of the fact that the end of the twentieth century saw a reversal of the long trend of decreasing socio-economic inequality, one that was helped by so much progressive legislation in previous decades. This transformation is one of the reasons why there has been an increase in the number of calls for a greater justice in cities, in which justice is viewed in redistributive, not in retributive terms, one that addresses the unequal possessions of different people, not just of goods, but also of their limited power and influence. This injustice is not simply a matter of these inequalities. The remoteness and complexity of government, even in democratic countries, combined with the power of large corporations, has led to the increasing unease, even anger of many people, about existing urban conditions, and their inability to influence changes in these places in meaningful ways.

1.2 New Solutions for Current Urban Problems

The effect of these and other major changes affecting our urban places have led to the realization that few of the problems created are being solved in a satisfactory manner. It has led to a number of alternative, urban-based ideas that are designed to address these unresolved difficulties. In some cases urban places have actively sought new economic functions to replace those being lost or under threat; many others seek to reduce their negative environmental impacts. Some cities have made determined efforts to enhance the quality of life of residents through various social and ecological rehabilitations, as well as improving the safety and health in cities and attempting to reduce inequalities. The unique history and morphological endowment of some urban places has spawned movements to retain, enhance or re-create the historic ambiance of a place, or even to invent a new one. What is clear from these examples is that individuals from many walks of life, increasingly active city managers and officials, business leaders, as well as international agencies, are not content to see urban places as only the passive recipients of some received process of change caused by external economic, social or environmental factors. In other words urban places are not being regarded only as 'containers' in which various processes created by various agents have created particular outcomes. Instead, creative thinkers, city managers and entrepreneurs have developed, or are trying to create, alternatives to the urban container perspective, by actively searching for new sets of policies and creating new degrees of activism to improve the functioning and character of their cities. Of course this activism is not new, even in the recent past. The history of post-World War II urbanism, in the developed world in particular, was replete with interventionist strategies, such as: new towns; urban expansion containment; and various regional policies to reduce the growth potential of the biggest places and to improve the prospects of towns and regions that had lost their economic base—especially those dependent upon mineral extraction or old industrial practices. Many of these post-war policies were forgotten or reduced in scope after the stagflation of the 1970s and the growth of neo-conservative governments, creating a neo-liberal phase that has reduced regulation, especially in planning, given the belief that such rules were inimical to their pro-growth and pro-market policies. At the same time there has been greater sympathy for new intellectual ideas such as post modernism, with its stress on difference and individuality, rather than grand theories and scientific approaches (Harvey 1991; Davies 1996), as well as the realization that the 'top-down' technocratic style of policy should be replaced with more community and stakeholder consultation.

These trends have led to what amounts to a new phase of urban policy conceptualizations and implementations since the late 1980s—one that has gained momentum in subsequent decades. It is a phase with a much wider scope and integrative capacity than the ideas generated in the post-war period. Many of these new ideas and approaches to urban development have been summarised in a short distinctive word or phrase, providing an adjectival prefix to the words, 'urban, cities or towns', creating such labels as New Urbanism, Sustainable Cities, Just Cities, Winter Cities or Safe Cities etc., which summarize critiques of existing situations and provide alternatives. They can be viewed as creating a new and varied set of urban themes.

The various themes are certainly diverse in nature. Each can be thought of as focusing upon a new concept, or set of concepts, designed to critique existing problems and then advocating ways of solving them. Although these themes are descriptive, they have very different functions to the many adjectival descriptors used by urban students in the past to classify the diversity of urban centres into more manageable categories of understanding. For example, economic classifications created categories such as port cities, tourist towns, transport centres, or various types of central places, at first subjectively but later with sophisticated statistical techniques

(Davies 1984). More recently, the term 'world cities' has become popular, usually describing large cities that are the command centres of the world economy, although some other cities in the world should have that designation because their global role is as a centre of some world religion. These and other examples of adjectival descriptions were, however, primarily descriptive. The new set of adjectival descriptions, that are being called 'city or urban themes' in this study, have a different rationale. They summarize some innovative strategy, or policies, that are designed to solve some specific problems of contemporary urban places, boosting such features as the future sustainability, vitality, appearance, fairness and life-styles of centres, either through the adoption of a series of new activities, or through modification of the old. Hence, the themes are not just *passive* descriptors or ones designed to draw attention to some characteristics of cities, such as contested or unequal cities. They have been designed with an *activist* agenda in mind, to improve the quality of our current urban centres by changing certain features of urban places, such as making them healthier, or safer or more sustainable. Even themes that are also descriptive, of certain conditions, such as Winter Cities, summarize a series of policies designed to make better accommodation to the harsh conditions of cold climates.

The number and variety of these new themes does make the current period seem rather unique. However the idea of using some adjectival theme to promote a new approach to urban development is not really new. Two examples from the end of the nineteenth century will demonstrate that there were other times in the past when frustration with the trends in urban development at the time led to new ideas to solve these problems. The most well-known was the Garden City movement inspired by the work of Ebenezer Howard (1898) in England which advocated a new sort of settlement combining the best of town and country through a social co-operative process of development. Another was the 'City Beautiful' movement (Hall 1988) which was initially developed in America to make the central areas of cities more attractive. It was recognized that few American cities had the range of interesting historic buildings found in European cities where centuries of urbanization had left a heritage of structures that enhanced their image, provided local pride, and brought many visitors in to admire their morphologies and quality of life. Advocates of the City Beautiful set out to improve the look of their central areas, or at least part of it, mainly through putting classical facades or other Greek and Roman traits on buildings, and adding park spaces and avenues. However, little thought was given to the consequences of redeveloping these spaces, which often meant that the former occupants, mainly the poor and disadvantaged, had no alternative but to move elsewhere, often into more overcrowded conditions. This is the type of injustice that keeps on being repeated in many urban renewal schemes today, projects that are often based on profit maximization rather than on the needs of the people. Since cities, as Harvey (1973, 1989) has constantly reminded us, are the primary places where a social surplus is generated and sequentially reinvested, these urban projects routinely create new phases of wealth for the owners of capital, but also increase the inequalities since so many do not share in the surplus. It is an issue that still needs a solution in the eyes of many critics of current developments, which has led to increasing calls for greater fairness and justice in cities.

There can be little doubt that these older urban themes achieved some success in influencing the future development of cities throughout the world, although in practice the ideas were often drastically modified and were eventually found to create their own problems. Hence the history of their applications holds lessons for us in the context of the new urban themes, making it important to find ways of ensuring that the new ideas do not get watered down in the same way, or even debased, such as Regional Planning and Garden City concepts. Of course there were many other twentieth century themes related to urban development that blossomed and then declined, as Peter Hall (1988) showed in *Cities of Tomorrow*. They ranged from various designs, through economic and social issues or problems, to conceptual and theoretical changes. In many ways Hall's review showed the way in which urban development has been constantly influenced by new ideas that become fashionable, even faddish, but then usually faded and were replaced. But by the end of the twentieth century the results of the distortions of the older policies and the emergence of a series of new problems, have not led urban scholars and practitioners to develop one or two new themes as in previous decades. Instead, a large number of new urban ideas have emerged, each advocating solutions to what were seen as particular crises of urban development at the turn of the century. This group of new themes that have emerged within a few years of one another has surely created a distinctive new phase in the history of urban development.

1.3 The Various Themes

Each chapter in this book focuses on what are considered the major themes that have emerged in the decades around the turn of the twenty-first century. In most cases the background problems that led to the new theme are reviewed and the key elements in the theme are summarized in various sections or in conceptual models. It seemed appropriate to begin with one of the earliest and perhaps the most publicized of these ideas. New Urbanism (Chap. 2) seeks to reverse the trend to urban sprawl and to create more attractive and liveable areas, with less dependence on the car and provide a more people–orientated focus through a set of principles known as the New Urbanism Charter. This is a very useful and wide ranging set of summary principles, but the neo-traditional emphasis upon design and liveability does little to solve many of the other major problems facing towns and cities that has led to other themes. However the extension of the New Urban ideas to the Smart Growth and Transit Orientated Development concepts in particular do provide more specific solutions to some of the problems caused by sprawl. What is particularly missing is the limited attention to social problems, which has led to greater interest in the Just Cities approach (Chap. 3). This chapter draws attention to the antecedents of the question of injustice in cities and the way that many people do not share in the wealth of urban places, as well as being unfairly treated and having limited influence on how places develop. The fundamental nature of the problems of justice in cities made it important to deal with these issues early on in this study, even though the approaches are so different to the chapters on either side. Practical guides to policies that can create more just cities are an essential part of this chapter, although it does deal with incremental change, rather than the radical approaches favoured by those advocating transformational societal change.

The next four chapters deal with the different ways that urban places relate to, and affect their environment. Chapter 4 (Green Cities) picks up some of the older ideas of the Garden City movement in which cities included features of the natural biotic environment especially trees and grass, within their boundaries, and describes the new rationales for such policies. It then reviews policies that lead to urban places literally becoming greener, in which the term 'green' is *not* used as a surrogate for sustainability, but deals with environmental restoration and re-vitalization. Chapter 5 (Background to Sustainable Cities) provides the context to the issue of sustainability, showing the complex, multi-dimensional nature of the term and the problems of measuring the concept. An example of the relative sustainability of the major cities of the world through the Green Index (EIU 2012) shows the wide variations in the degree of sustainability of these places. The chapter then focuses on the various issues associated with the input-output balance of cities in terms of major resource needs, such as water and energy supply, and then waste and pollution outputs, including ways of reducing the impact of these negative effects upon human and environmental health and the warming of the planet. Particular attention is paid to the ability of various renewable energy resources to replace the use of fossil fuels, whose negative effects are causing negative atmospheric changes as well as local urban effects. Chapter 6 (Sustainable City Policies) extends the discussion by focusing on the ways that greater sustainability can be achieved through policy initiatives that are specifically urban-based, such as recycling and building construction initiatives, transport change. It also provides examples of how various types of areal developments are reducing the waste of resources and decreasing various types of pollution. Chapter 7 continues the ideas of the previous three chapters but focuses upon the way these ideas are being pursued by new citizen-based initiatives such as the Transition Town movement which is designed to create more locally resilient and sustainable places. It also describes the Eco-District approach which is a relatively new initiative developed from community activists in Portland in Oregon state, U.S.A., that seeks to improve conditions at an intra-urban community level, again with local activism, rather than by government policies.

Chapter 8 (Winter Cities) focuses on a more specific environmental topic, the problems faced by urban places that have long winters, showing how various adaptations in physical and well as human terms can improve the liveability of these places. Chapter 9 (Resilient Cities) is a topic that has received a great deal of attention in the last decade, namely the way that episodic extreme natural hazards can destroy or seriously damage our settlements. Indeed the vulnerability of the largest settlements to such hazards became part of the U.N.'s 2012 world urbanization report. The discussion describes the character of these events and the range of policies by which urban places can be made more resilient to these natural events, especially in the cities of developing countries. They have far less resources to fall back on to recover from the devastating blows that come from major storms, earthquakes, volcanic eruptions or other catastrophic events.

The next two chapters describe two economic issues that are concerned with aspects of the new economy which may help the employment prospects of urban places. Chapter 10 (Creative Cites) provides a review and critique of the ideas that give priority to the contribution of the so-called creative class to the growth of cities in the new economy of the past twenty years. Despite the superficial attraction of the idea, the process of growth in contemporary cities seems to be more complex than a product of one group of people, while some have argued that polices designed to encourage this growth has increased gentrification and contributed to the marginalization of people displaced in inner cities. Chapter 11 (Knowledge Cities) is wider in scope. It deals with the character of knowledge creation and the problem of how and where new innovative activity is created, identifying the features which seem to account for the localization of these activities, rather than only focusing in on one class of people. This review also shows how cities can engage in the process of economic change and uses the example of the Urban Capital Inventory approach to show how the positive and negative features of cities can be identified in order to help urban places attract and retain more inventions and innovative economic activities.

The next four chapters are concerned with social issues that have either caused problems in many urban places, or need to be enhanced to improve the liveability and vitality of cities. Chapter 12 (Safe Cities) describes the background to the explosive growth in crime in many western cities from the late 1970s and its decline from the 1990s. It summarizes the difficulties involved in measuring crime and reviews the range of very different policies that have been created to increase the safety of citizens from crime and how their development and co-ordination has been achieved through the adoption of Safe City Strategies. Chapter 13 (Healthy Cities) summarizes the major changes carried out in the nineteenth century cities of the developed world to reduce their high mortality rates, which could be seen as the first healthy city movement. It then focuses on the new health problems emerging, some of which are providing new challenges for the medical profession, and others that stem from the wider range of health determinants or factors affecting health, from environmental factors to those that are socially determined. The review also includes the new range of policies being advocated by the World Health Organization under their Healthy City programme, that focus as much on health promotion and organization of care as much as medical-only practices. These new policies are more integrated, involving more local government and citizen engagements and are designed to include the most vulnerable people who often lack adequate care.

Chapter 14 (Festive Cities) describes a different social issue to the ones dealt with in the previous two examples. It is one that is linked to improving the enjoyment of life, not to problems of survival or disability, and sometimes has associations with older cultural beliefs. It describes the many different types of festivals that are found in urban places and shows how they have increased in recent years. These provide not just positive economic impacts but add to the vitality and excitement of urban life. The discussion shows that festivals have varied origins and are complex in their impacts. Their multi-dimensional effects are explored to show the utility of these episodic events and their often conflictual character. Chapter 15 (Slow

Cities) is also a theme devoted to enhancing life in some centres, by describing the growth and development of a movement that was originally designed to improve the quality of life in small historic towns and cities. Advocates argue that it could be achieved through the adoption of a slower life-style, in which more time is taken to enjoy food, drink, family and companionship, as well as by promoting the need to restore the historic heritages, traditions and crafts in these centres and their regions. All are attributes that have often decayed under the creeping effects of modernization, globalization and more rapid and available transport systems. The final chapter (Chap. 16) identifies some of the key features that have emerged from these various themes and their utilities in improving what is often described as the liveability and sustainability of urban places and the possibility of more active city-learning approaches.

It must be stressed that most of the themes discussed in this book were mainly developed to solve problems that have emerged in the towns and cities of developed countries, although some apply to all parts of the world. In the less developed world there is still the need to provide adequate housing, water, sewage facilities, utilities and transport in many urban places, let alone reducing poverty, creating employment, reducing pollution, and ensuring the safety and health of urban citizens and providing them with more rights and protections. Some of the policies proposed under the various themes, from Chaps. 2–15, can help the cities of the less developed world improve their situation by providing empirical evidence of the utility of successful and tested strategies and programmes that may be applied and improve their future. But what is also vitally important is that these themes also show how many of the particular problems that have emerged in western cities of the developed world should be avoided in current and future urban developments in less prosperous areas. However there is real need to avoid assuming that developed world ideas are always useful. The growth of cities in less developed countries is often associated with different societal and other contexts. So there is almost certainly a pressing requirement for students of towns and cities in the less developed world to identify their own particular problems and create new themes to solve these difficulties. In this respect two or three issues especially deserve attention. One is that new technologies will probably be required to solve the clean water supply and sewage needs of the increasingly large urban agglomerations—where the long distance solutions of western cities may not be feasible. Also solutions to the severe traffic congestion and limited traffic regulations are needed, which is leading to such high fatalities. A third key issue is to identify more potential urban areas in advance and plan for the addition of utilities before development, rather than trying to add these after informal development has taken place. Such a development requires more commitment to urban planning and rigorous land ownership allocations, providing clear titles to lands to be occupied. These are but a few of the problems that have been largely solved or reduced in scope in western countries, but which continue to plague urban development elsewhere.

Some of the new themes described in this text have been the subject of individual books, in some cases, many books, but it is rare to see the various themes being addressed as a whole. Although there is some overlap between these themes in that somewhat similar ideas are emerging in parallel, it is curious that most seem to be developing with little reference to one another. This is probably a result of the very complexity of cities, making it difficult to provide a comprehensive approach to the study of urban places. But unless we are careful there could be yet another silo approach to urban development, with the new ideas or themes helping to solve one set of problems, but providing little insight on other challenges, even though most of the themes being discussed claim to adopt wider perspectives and provide more opportunities for community engagement. Two examples may illustrate the point. The first is that those who have lauded the merits of the New Urbanism seem to have ignored the very real problems of the inequalities in society which have stimulated the growth of interest in more Just Cities, especially the fact that the inner city poverty areas have shown remarkable persistence and have rarely been solved or effectively addressed. The second is the way that advocates for the new Healthy Cities approach have had to challenge the land use focus of urban planners, since for much of the twentieth century they have paid little attention to health issues in their evaluation of development projects, as Corburn (2009) has classically shown in the case of San Francisco. Curiously, many of the advocates for these new themes argue for a more comprehensive approach to planned urban change by involving different types of people, ideas, organizations and government. Yet their focus on one theme often ignores the wider needs of urban places. This led to the realization that it was time a more integrated, or at least inter-related, approach to the study of urban development was created, in addition to these single theme approaches that dominate urban discussions today. Perhaps these single foci approaches are inevitable since the size and complexity of towns and cities make them difficult to comprehend let alone understand the many problems that exist. However progressive urban development schemes need to consider the implications or effects of all, or at least most, of the themes described in subsequent chapters; too often they are restricted to one or two.

As usual, space constraints mean that not all of the themes that have emerged in recent years can be covered. Instead the book focuses on those that are considered the most important in affecting and improving contemporary urban developments, primarily in the developed world, although with some examples from the developing world. Although most of the chapters deal primarily with one main theme, a series of related, yet not so general topics, such as Sanctuary Cities, Charter Cities, Emancipatory Cities, and Tidy Towns to name but a few, are also discussed at relevant stages within particular chapters. However there was no room to discuss the approaches described as Phoenix Cities (Power et al 2010) or Heritage Cities (UNESCO), although these do relate to quite specific types of places rather than having the more general applicability of the themes dealt with here. In the former case the adjective has been appended to describe the way that urban places that have suffered severe economic depression, especially as manufacturing has moved out and/or mining operations have closed, have developed regenerative policies to enable them to arise from the ashes, metaphorically like the mythical phoenix. Power's (2010) book outlines the policies used in several major declining cities to successfully revive these centres, showing the utility of a mixture of economic policies to attract new industries or expanding those already present, as well as improving the urban fabric and rehabilitating the polluted and derelict sites of older industries. Closely related to this approach are the active networks devoted to the problems posed by urban places that have lost population, such as the Shrinking Cities International Research Network established in 2004 (SCiRN 2013). A very different theme is summarized by the term Heritage Cities. A world network of these centres was created by UNESCO to find ways of conserving urban places designated as having unique historical urban landscapes and buildings created by previous cultural phases of growth. But they also seek to integrate this process with socio-economic development, so they become active, lived-in places, not museum pieces. Cities in the developing world especially have used the network to get expertise from specialists from cities in the developed world to identify and maintain their historic buildings and townscapes, and to find new uses for the buildings.

It must be acknowledged that some other emerging urban themes may need more attention in the future. One is the way in which the increasing application of new technologies to urban areas is creating what some have called 'Smart Cities', Since this technology-based meaning is still very much in the formative phase in terms of applications, examples are provided in various chapters, rather giving it a chapter of its own, although the rate of technical applications make it likely that in a decade the term and the various technical spin-offs will be more widely used. However, although we are finding increasing benefits from new electronic technologies, there are, as always with innovative ways of doing things, signs of new problems emerging. The most obvious comes from the ability of hackers—whether private or governmental—to break into our control systems and either turn them off, or provide alternative instructions, which can cause untold damage, not only to personal medical devices such as pacemakers, but to large ones, such as traffic systems, electricity grid control systems, or even Email networks. Hence greater vigilance and safer systems are needed to reduce the possibility of disasters emerging in our settlements from such interferences in our intimately linked, yet still fragile world of interconnections. In a related context the increasing ability of other people, companies and governments to access large amounts of personal data, especially from internet or credit card sources, together with the growing number of surveillance devices in cities, means our personal privacy can be compromised through what has been described as a techno-creep (Keenan 2014). These information sources, could, in the wrong hands, led to a future in which others have more and more influence or even control over our actions, leading to a loss of liberty. These issues need far more attention in the usually only positive discussions over smart cities, with some worrying that the second trend in particular may start us on the road to the type of controlled society seen in George Orwell's frightening futuristic novel, *1984* (Orwell 1949). This may be an extreme conclusion for democratic societies. But what does seem appropriate to note is that we need to acknowledge that the new technologies that are helping to create technologically smarter cities may not have only positive outcomes. The negative features need to be identified and resolved.

Yet it must be noted that the term has other connotations, as seen in Smart Growth land use policies (Chap. 2) and in new City Learning strategies to make them smarter places (Chap. 16). A second emerging theme of growing importance as populations in the developed world age, is the concept of Aging Cities. Again an international organization, in this case the World Health Organization has taken the lead in establishing a global network of age-friendly cities and communities (GNAFCC 2006), issues that are briefly described in the Healthy City chapter, although if space permitted a wider description of the policies designed to create the objective in the network name could have been produced.

Other themes are also still at the stage of creating additional ways of improving the functioning of urban places, rather than something that dominates a city's development, so the trends are discussed in appropriate content areas within various chapters. In addition some may find it surprising that there is no separate chapter on the idea of Liveable Cities, which has been a popular, if very elusive concept. However many of the ideas of improving liveability can be seen in the more specific themes described here that seek to improve the quality of life in cities, especially the New Urbanism and Green discussions (Chaps. 2 and 4), and Chaps. 12 through 15, from Safe Cities through Healthy, and Festive to Slow Cities. It is also worth noting how the pejorative term Crap Towns (Jordison and Kiernan 2013) has received a lot of attention in Britain, applying to towns that residents consider to be ugly, miserable, and having little in the way of attractions, facilities, or even desirable life-styles. Their commercial centres in particular are dead at night, and characterised by chain stores that are duplicates of those found in other towns, creating what could be 'anywhere' places. However this adjectival descriptive is a criticism of the character of the centres so pilloried, not an active theme that leads to new policies, so is not be dealt with here. Nevertheless, many of the ideas expressed in several of the chapters, especially the New Urbanism, Green, Just, Festive and Slow Cities policies, which could be used to improve the prospects of the places so maligned in the negative book title and in the descriptions of the places included.

These caveats combined with limitations of space mean that the themes dealt with in this book deal what were considered the most important and general of a series of new urban development ideas that have emerged within the last two decades. Other concepts, such as the recently introduced term *tactical urbanism*, was considered too small scale and diffuse in content to be dealt with here, although some examples of these types of features are found in several chapters dealing with more specific general themes . This has become a broad descriptive term for the usually small scale, low cost, often quick and sometimes temporary additions to improve or enliven urban landscapes, such as adding benches, agricultural plots, or temporary pop-up parks in parking lots or even roads. Such bottom-up, creative practice initiatives are being promoted in internet accessible books by the New York-based Street Plans Collaborative (SPC) and certainly help to improve many inner city areas in particular. Also it is worth noting that there has been an indiscriminate use of 'city' and 'town' in some of these themes, without much attention to their size, in the sense that towns are usually regarded as smaller than cities. To take one example, the term Slow Cities is rather a misnomer in English terms as the movement is restricted to places under 50,000 population, places which most people would describe as towns. Elsewhere the Transition Town movement began with, and is still dominated by smaller centres, but now includes parts of large places that most would recognize as cities. So the descriptors 'towns or cities' used by advocates of these ideas should not be viewed as necessarily implying some size-threshold of place. In addition the themes being developed rarely apply to one place, whatever their origin or application. When these themes have been used, or seen to work successfully in a particular location or set of towns, they have often been adopted by other cities. The creators of various themes have usually created a network of like-minded individuals, or city officials, establishing an association based on the decision to resolve particular problems and develop the policies that are part of each theme. The theme is promoted and enhanced, not simply by regular meetings between adopters of a particular theme, but usually by a formal organization and a website, which has quickly spread the ideas from some initiating group, or city, to many urban places, frequently to centres around the globe. The towns or cities in the network share information about how to develop new policies to solve the problems associated with individual themes and how to effectively implement them. Sometimes the theme may come to dominate policies in the adopting city; more frequently it is an addition to, or expansion of, some existing activities. However the rapid creation of these networks via the communication medium of the Internet provides one of the other distinctive features of this new phase of new urban concepts and policies. Each theme promises to provide a new focus to solve particular urban problems and to provide better environments, both physical and social, and an improved quality of life for the residents of our towns and cities.

References

Corburn, J. (2009). *Toward the healthy city*. Cambridge: MIT Press.

Davies, W. K. D. (1984). *Factorial ecology*. Aldershot: Gower.

Davies, W. K. D. (1996). Post modernism and cities: Engaging or accepting similtaneity in forms and societal character. In R. Davies (Ed.), *Contemporary city structuring* (pp. 1–17). Cape Town: I.G.U. Commission & Society of South African Geographers.

Davies, W. K. D., & Herbert, D. T. (1994). *Communities within cities*. London/New York: Belhaven/Halstead-Wiley.

EIU: Economist Intelligence Unit. (2012). *The Green city index*. Munich: Siemens AG. http://www.siemens.com/entry/cc/en/greencityindex.htm. Accessed 8 Sept 2013.

GNAFCC. (2006). http://www.who.int/ageing/age_friendly_cities_network/en/. Geneva: World Health Organization. Accessed 14 Feb 2014.

Hall, P. (1988). *Cities of tomorrow*. Oxford: Blackwell.

Harvey, D. (1973). *Social justice and the city*. Oxford: Blackwell.

Harvey, D. (1989). *The urban experience*. Oxford: Blackwell.

Harvey, D. (1991). *The condition of postmodernity*. New York: Free Press.

Howard, E. (1898). *Tomorrow: A peaceful path to real reform*. Reprinted and edited by F.J. Osborn as *Garden Cities of Tomorrow (1945)*. London: Faber and Faber.

IPCC. (31 March 2014). Climate change 2014: Impacts, adaptations, vulnerabilities. Intergovernmental Panel on Climate Change: Working Group II. Yokohama. http://www.ipcc.ch/pdf/ar5/pr_wg2/140330_pr_wgII_spm_en.pdf. Accessed 8 June 2014. Accessed 10 June 2014.

Jordison, S., & Kiernan, D. (2013). *Crap towns returns*. London: Quercus.

Keenan, T. (2014). *Techno-Creep*. Berkeley and Vancouver: Greystone Books.

Orwell, G. (1949). *1984*. London: Secker and Warburg.

Power, A., Ploger, J., & Winkler, A. (2010). *Phoenix cities*. Bristol: Policy Press.

SCiRN. http://www.ru.uni-kl.de/en/ips/research/networks-and-cooperations/shrinking-cities-international-research-network-scirn/. Accessed 14 Feb 2013.
SPC: Street Plans Collaborative. http://tacticalurbanismsguide.com. Accessed 14 Sept 2014.
TE. (2014a). In the balance. *The Economist,* 5 April, 70–71.
TE. (2014b). Another week. Another report. *The Economist,* 19 April, 73–75.
TE (2014c). Building the dream: Chinese cities. *The Economist*: Special Report, 19 April, 1–17.
UNESCO. http://whc.unesco.org/en/cities/. Accessed 15 Feb 2013.
United Nations. (2013). *World population prospects: The 2012 revision*. New York: Economic and Social Division, United Nations.

Chapter 2
New Urbanisms: From Neo-Traditional Neighbourhoods to New Regionalism

Wayne K.D. Davies and Ivan J. Townshend

> *New urbanists have developed and propagated a formula for planning the good community, and have gained international attention in the process. Beauty is arguably a necessary condition for the good community, but is it sufficient?*
> (Grant 2006, Introduction)

2.1 Introduction

New Urbanism (NU) came from the belief that there was something drastically wrong with the way in which modern cities have developed in the past century. Many had evolved as suburban, automobile-dominated residential environments with few employment opportunities, characterized by urban sprawl, inefficiency, and placelessness, with a demise of the public realm and a failed realization of true or authentic neighbourhoods and communities. Instead of just criticizing suburbanization, and advocating the advantages of inner city life in the manner of Jane Jacobs's (1961) trenchant book, NU became a movement that sought to remedy the situation. Most accept that its basic ideas stem from the building of the new community of Seaside in Florida (Brooke 1995), and the subsequent popularization and extension of the ideas used in this town. In 1991 some of the key thinkers of New Urbanism were brought together at the Ahwahnee hotel in Yosemite National Park to identify common features in their approaches, leading to a core set of principles referred to as the Ahwanee Principles. However, the movement became more formally established when these ideas crystallized through the establishment of the Congress for New Urbanism (CNU) in 1993 and the creation of the subse-

W. K.D. Davies (✉)
Department of Geography, University of Calgary, 2500 University Drive,
N.W. Calgary, AB T2N 1N4, Canada
e-mail: wdavies@ucalgary.ca

I. J. Townshend
Department of Geography, University of Lethbridge, 4401 University Drive,
Lethbridge, AB T1K 3M4, Canada
e-mail: towni0@uleth.ca

W. K.D. Davies (ed.), *Theme Cities: Solutions for Urban Problems*,
GeoJournal Library 112, DOI 10.1007/978-94-017-9655-2_2

quent Charter of the New Urbanism in 1996 (Congress for the New Urbanism 1999; Murrain 1996).

Basic to the Charter were what were portrayed as 'new' forms of towns and neighbourhoods that paid attention to various scales of development from regional setting to streets and buildings. Many of the ideas evolved from modern critiques of suburban design, with concepts from works such as Alexander's pattern language (Alexander et al. 1977; Alexander 1979). However, students of the history of urbanism will know that many of the components are actually a return to those promulgated a century earlier. So NU frequently involves the desire to rediscover neighbourhood and a sense of community through more human-scale developments that also reclaim the public realm. In addition, the approach envisages more walkable communities and better public transit connectivity as well as a more integrated approach to metropolitan or city-regional growth. However as Hall (1988, p. 24) observed

> [its] widespread appeal emanates from its all-enabling invocation of 'community', a term that provides little practical or ideological direction, yet which is vague enough to embody everybody's hopes.

Indeed, the NU term has become more diffuse in recent years, becoming somewhat of an umbrella term to cover a set of related or extended ideas, initially mainly linked to Neo-Traditional Design (NTD) or Traditional Neighbourhood Design (TND), but which has extended into Smart Growth (SG), Urban Villages (UV), and more recently into Transit Oriented Development (TOD), while New Regionalism (NR) ideas also seem relevant, although they have developed from different sources. Hence it seemed appropriate to use the term 'New Urbanisms', in the plural, to draw attention to the many types within the movement. Space limitations mean that this review cannot provide a comprehensive review of all the details of these various approaches. Rather, it will provide an overview of the problems that led to the New Urbanism movement, its basic ideas and associated developments, using examples from many cities. Subsequent sections will review its utility and impact on current development, as well as ask questions about how new many of the ideas really are.

2.2 Background to the New Urbanism

New Urbanism has grown out of a reaction to some of the perceived ills of the contemporary industrial-commercial city. The overarching issue here is urban sprawl (Duany and Plater-Zyberk 1992; Grant 2006). New Urbanists argue that the modern metropolis, particularly the American one, is inherently dysfunctional, and its sprawl is the spatial expression of this problem (Krier 1991; Audirac and Shermyen 1994). The Fordist industrial regime and increasing affluence spawned almost universal car ownership in the developed world by the mid-twentieth century, accompanied by planning and land development practices that were biased toward the supremacy of the automobile. This resulted in cities characterized by endless expansion of various types of suburbs, often varieties of planned neighbourhood units (Perry 1929). These units were zoned as mainly residential, except for a small row of shops, so the main commercial development from the 1950s occurred in strips along arterial roads or in increasingly large suburban shopping centres, which were mainly accessed by cars. Such developments have been incrementally added on to the edge of urban places, with limited thought of the spatial consequences of these successive additions (Langdon 1994; Kunstler 1993; Knox 1992; McCann 1995; Thompson-Fawcett 1996). Hence, the resultant auto-oriented and increasingly low density suburbs of the last half of the twentieth century led to the problems of urban sprawl. It led to distinctive separation between places of work, residence, shopping, and public gathering—resulting in a spatial disjuncture between the private and the public realm, and between the geography of domesticity and that of production and consumption (Grant 2006). The sprawling suburbs that developed also required considerable infrastructure investment to access them and to supply utilities, were too far removed from the city centre, and are considered wasteful of space. The resultant sprawl produced traffic congestion, gridlock, poor air quality, and necessitates automobile usage to access every key urban function, whether employment, shopping or leisure activities, even within decentralized concentrations of employment and retail in the so-called 'edge cities' that seem like small downtowns on the outskirts of the built up area (Garreau 1991). The result has been that suburban sprawl, even of planned communities, has produced automotive-dependent lifestyles with negative health consequences because people lacked exercise—leading New Urbanists to argue for the need for new approaches, in particular to reduce the excessive use of the car (Duaney and Plater-Zyberk 1992, p. 44).

There are also problems of the morphology of suburbs. They are typically comprised of low density single family housing, with substantial lots, large setbacks for the dwelling unit, curvilinear streets and cul-de-sacs that are wider than necessary and lined with large homes with snout-like front garages, such that people move from workplace to house within their vehicles with little human contact with neighbours in their daily commutes. Their internal morphology is low-rise and expansive—characterized by excessive "horizontal" infrastructure (Duaney and Plater-Zyberk 1992, p. 44). Given their similarity, these places are seen to have little sense of identity and few architecturally redeeming qualities. As Jacobs (1961) observed, this development is the antithesis of urban, as suburbs lack the heterogeneity, interaction and facilities of urban places, creating a "geography of nowhere" (Kunstler 1993).

For New Urbanists the social problems of the suburbs are equally worrying. They are primarily relatively homogeneous places, principally defined by income or socioeconomic status characteristics. Few are socially inclusive, although Duaney and Plater-Zyberk (1992), perhaps naively in the light of past history, claim that the economic segregation found in suburbs is not the American way. In addition, they point to the problems of isolation in these areas, even though many paid lip service to Perry's 1920s ideas that his planned neighbourhoods would create cohesion and togetherness. Like Putnam (2001), who entitled his book on contemporary life as '*Bowling Alone*', New Urbanists generally argue that the social experience of community and social capital is lacking in the contemporary suburb. Some go even further to suggest that the contemporary suburb means the end of authentic civic life (Duaney and Plater-Zyberk 1992). So in such suburbs, private space is celebrated, while public space is often all but abandoned. Moreover, walkability is minimal, as pedestrian movement is secondary to automobile travel. Neighbouring and casual social interaction is limited, reducing the development of social capital, and all are discouraged by the physical structure of the community. Hence the desire to counteract the supposed demise of community in suburban developments is at the very heart of the TND variant of new urbanism. Indeed, as Talen (1999, p. 1363) has pointed out, the overarching social doctrine behind NU is the creation of a sense of community. So New Urbanists have argued that new designs—if done right—can address this issue and create more socially diverse neighbourhoods.

2.3 Principles of the New Urbanism

Table 2.1 lists the basic principles of the New Urbanism Charter. It specifies a number of different scales at which NU principles need to be implemented, from regional setting to streets and buildings, making it clear that NU is not simply concerned with small-scale neighbourhood features and community, although this is

Table 2.1 Charter of the New Urbanism. (Source: adapted from http://www.cnu.org/charter)

A. The region: metropolis. city, and town
(1) Metropolitan regions are finite places with geographic boundaries derived from topography, watersheds, coastlines, farmlands, regional parks, and river basins. The metropolis is made of multiple centers that are cities, towns, and villages, each with its own identifiable center and edges
(2) The metropolitan region is a fundamental economic unit of the contemporary world. Governmental cooperation, public policy, physical planning, and economic strategies must reflect this new reality
(3) The metropolis has a necessary and fragile relationship to its agrarian hinterland and natural landscapes. The relationship is environmental, economic, and cultural. Farmland and nature are as important to the metropolis as the garden is to the house
(4) Development patterns should not blur or eradicate the edges of the metropolis. Infill development within existing urban areas conserves environmental resources, economic investment, and social fabric, while reclaiming marginal and abandoned areas. Metropolitan regions should develop strategies to encourage such infill development over peripheral expansion
(5) Where appropriate, new development contiguous to urban boundaries should be organized as neighborhoods and districts, and be integrated with the existing urban pattern. Noncontiguous development should be organized as towns and villages with their own urban edges, and planned for a jobs/housing balance, not as bedroom suburbs
(6) The development and redevelopment of towns and cities should respect historical patterns, precedents, and boundaries
(7) Cities and towns should bring into proximity a broad spectrum of public and private uses to support a regional economy that benefits people of all incomes. Affordable housing should be distributed throughout the region to match job opportunities and to avoid concentrations of poverty

Table 2.1 (continued)

(8) The physical organization of the region should be supported by a framework of transportation alternatives. Transit, pedestrian, and bicycle systems should maximize access and mobility throughout the region while reducing dependence upon the automobile
(9) Revenues and resources can be shared more cooperatively among the municipalities and centers within regions to avoid destructive competition for tax base and to promote rational coordination of transportation, recreation, public services, housing, and community institutions
B. The neighborhood, the district, and the corridor
(10) The neighborhood, the district, and the corridor are the essential elements of development and redevelopment in the metropolis. They form identifiable areas that encourage citizens to take responsibility for their maintenance and evolution
(11) Neighborhoods should be compact, pedestrian friendly, and mixed-use. Districts generally emphasize a special single use, and should follow the principles of neighborhood design when possible. Corridors are regional connectors of neighborhoods and districts; they range from boulevards and rail lines to rivers and parkways
(12) Many activities of daily living should occur within walking distance, allowing independence to those who do not drive, especially the elderly and the young. Interconnected networks of streets should be designed to encourage walking, reduce the number and length of automobile trips, and conserve energy
(13) Within neighborhoods, a broad range of housing types and price levels can bring people of diverse ages, races, and incomes into daily interaction, strengthening the personal and civic bonds essential to an authentic community
(14) Transit corridors, when property planned and coordinated, can help organize metropolitan structure and revitalize urban centers. In contrast, highway corridors should not displace investment from existing centers
(15) Appropriate building densities and land uses should be within walking distance of transit stops, permitting public transit to become a viable alternative to the automobile
(16) Concentrations of civic, institutional and commercial activity should be embedded in neighborhoods and districts, not isolated in remote, single-use complexes. Schools should be sized and located to enable children to walk or bicycle to them
(17) The economic health and harmonious evolution of neighborhoods, districts, and corridors can be improved through graphic urban design codes that serve as predictable guides for change
(18) A range of parks, from tot-lots and village greens to ballfields and community gardens, should be distributed within neighborhoods. Conservation areas and open lands should be used to define and connect different neighborhoods and districts
C. The block, the street, and the building
(19) A primary task of all urban architecture and landscape design is the physical definition of streets and public spaces as places of shared use
(20) Individual architectural projects should be seamlessly linked to their surroundings. This issue transcends style
(21) The revitalization of urban places depends on safety and security. The design of streets and buildings should reinforce safe environments, but not at the expense of accessibility and openness
(22) In the contemporary metropolis, development must adequately accommodate automobiles. It should do so in ways that respect the pedestrian and the form of public space
(23) Streets and squares should be safe, comfortable, and interesting to the pedestrian. Properly configured, they encourage walking and enable neighbors to know each other and protect their communities

Table 2.1 (continued)

(24) Architecture and landscape design should grow from local climate, topography, history, and building practice
(25) Civic buildings and public gathering places require important sites to reinforce community identity and the culture of democracy. They deserve distinctive form, because their role is different from that of other buildings and places that constitute the fabric of the city
(26) All buildings should provide their inhabitants with a clear sense of location, weather and time. Natural methods of heating and cooling can be more resource-efficient than mechanical systems
(27) Preservation and renewal of historic buildings, districts, and landscapes affirm the continuity and evolution of urban society

often seen as its emphasis. At a macro scale, nine of these principles concern the region, including the relationships between the metropolis, the city, and the town. Nine others specify meso-scale principles associated with the neighbourhood, the district, and the corridor—the essential building blocks of new urbanism. The last nine provide additional micro-scale specificity for individual building and other architectural and landscape design features. What is clear from the principles is the need to plan for the metropolitan region not just little parts, in which governance and economies are based on the region, although there are no guides of how to do this. Within the various urban units it is stressed that there is a need for a range of housing, social housing, park space and an emphasis upon walkability, mixed uses and local employment, as well as anticipating transit corridors to improve transport flows and the need to preserve historical heritage.

2.4 What Does New Urbanism Hope to Achieve?

Many planners have recognized that inappropriate planning tools of the past, including poor rules for land management, have generated sprawl, poor design, exclusionary zoning and a demise of authentic community (Grant 2006), but they also recognize that urban growth or expansion is inevitable (Calthorpe 2001). So the key issue for New Urbanists is the question of managing future growth without sprawl, or at least in such a way that attenuates its negative impacts. The NU approach to sprawl is to consider it within a broader urbanization and development context. It attempts to refocus our planning approaches to think about the regional contexts of sprawl, and to consider what has been called the 'urbanization of the whole, not the urbanization of the pieces' (Calthorpe 1994). It is here that TOD and SG approaches, which focus more on urban or macro scale planning than the other approaches, combine to address issues of sprawl. These approaches stress the need for more integrative public transit systems that link neighbourhoods with cities, and cities within regions, thereby reducing reliance on private automobiles. Their plans call for more transit-oriented satellite towns, pedestrian pockets, and an increase in the walkability of areas surrounding these nodes as well as more attention to sustainability.

NU also hopes to reverse the twentieth century trend of poor suburban design and land use planning, to develop a new architecture of community (Scully 1994), and to bring about a complete paradigm shift in community building (Katz 1994; Al-Hindi 2001). The goals here are really about rethinking the internal design of neighbourhoods, the homes within them, and the land use mixes within them. Major proponents of TND, such as Duaney and Plater-Zyberk who are often lauded as the 'evangelists' of NU (Knox 1992), envision communities that have several key features, namely: they are more compact; have higher population densities; have narrower streets; are more pedestrianized; are designed at a more human scale; have greater social and socioeconomic diversity; have more zoning flexibility to produce a greater mix of residential, commercial, and public spaces; and provide a design foundation for a more 'authentic' urban life. A key feature of this rediscovery of a more 'authentic' urban life is the need to re-assert public space as a critical element in communities. New Urbanists believe that exemplar communities of the past provide the reference point for these types of places (Hall 2000). So this approach, which is heavily design oriented, seeks to include elements of vernacular and classical architecture from 'exemplar' older towns. This strand of NU, usually referred to as Neo-Traditional Neighbourhood Design, is about reviving those elements of historic towns that worked well, ones that provided people with a sense of identity, a human scale, and public engagement, although these ideas have been labelled as 'old fangled New Towns' (Anderson 1991), or a 'Brave Old World' (Knox 1992, p. 221). Yet Bressi (1994) cautioned that NU should not be seen as some kind of Romantic Movement; rather, it is a deceptively simple response to the problems of contemporary suburban development based on one principle, namely that community planning and design must assert the importance of public over private space.

As opposed to the contemporary suburb, in which NU argue that 'community' has met its demise, the overarching aim of NU developments is to rediscover what is called 'authentic' community, although many dispute the term 'authentic' as being too vague. Indeed there have been few attempts by New Urbanists to explore the community dimensions, which have been shown to be complex (Davies and Herbert 1993; Townshend and Davies 1999). However, the key idea for NU is that the right architecture, the right planning, the right mix of land uses, and the right integration of public space will bring about the right kinds of social interaction, neighbourliness, sense of belonging, and identity through which people will rediscover community, within appropriate socially heterogeneous neighbourhoods. In short, the desire is to create a 'geography of somewhere' (Axhausen 2000). While there are elements of social utopianism in these aims, just as the NU have drawn on architecture and design principles from old towns, in North America as well as Europe, they have also looked to older urban social commentators for inspiration. For example, Jane Jacobs (1961) in the *Death and Life of Great American Cities*, heavily criticized suburban planning practice and was instrumental in beginning the critiques that led to the NU ideas that revived the idea of community, the need for social variety and spontaneity, as well as the use of public spaces for social interaction.

As McCann (1995) noted, many New Urbanists hope to achieve the kind of 'exuberant diversity' found in cites that was lauded by Jacobs. Hence many reviewers have argued that the creation of what is often called the 'good community' and a 'sense of community' seem to be the essence of New Urbanist design theory (Katz 1994; Talen 1999, 2002; Grant 2006). Perhaps it is ironic that this search for 'community', in the sense of more local social interaction, lay behind the designs in the first Garden City and Perry's neighbourhood units, which were the inspiration behind many contemporary suburbs that actually ended up as only distorted examples of these original concepts.

2.5 Various Types of New Urbanism Ideas

These New Urban ideas have progressed considerably from their origins. It cannot be considered a singular approach today. A range of related but different NU approaches have extended many of the ideas since the original Charter, often promoted by influential and charismatic proponents. Several somewhat distinctive approaches can now be identified within the general rubric of the plural New Urbanisms and are summarized in subsequent sections, namely: Neo-Traditional Neighbourhood Design (TND); Smart Growth (SG); Urban Villages (UV); and Transit Oriented Developments (TOD), while New Regionalism (NR) seems to be a necessary complement to implement some of the principles.

2.5.1 Traditional Neighbourhood Design

Traditional Neighbourhood Design (TND) and Neo-Traditional Design (NTD) are essentially different labels applied to the same approach within NU, and the term TND will be used here to refer to either. TND is probably the approach that most people associate with NU. It is the approach that has gained the most media attention, the most celebration, and the most intense planning and academic debate. The ideas of TND developed largely through the influential work of the architects and planners, Andres Duany and Elizabith Plater-Zyberk (Krieger 1991), as well as Krier's influence on their own ideas, with heavy reference to Alexander's ideas of 'timeless ways of building' and what he called the '*pattern language*' of good urbanism (Alexander 1979; Alexander et al. 1977). TND is very much design-focused. It espouses the use of a 'design code' for each community, and strives to include strong references to local and vernacular architectural traditions. The Town Centre is a crucial central point to these designs—not as an economic centre, as in a traditional central business district, but as a focal point of mixed activity that includes commercial, civic, residential, public and recreational functions. In essence, TND attempts to design new neighbourhoods (as well as infill redevelopments) based on traditional town planning principles, to achieve an 'authentic urbanism' (Duaney and Plater-Zyberk 1992; Katz 1994). This is all about creating a distinctive 'place', as well as 'the design code'—the guiding principles and design references that are the all-important place-makers (Scully 1994, p. 227).

Lennertz (1991) has suggested that there are seven design principles in NTD. The community must be based on a comprehensive master plan which lays out the core geometry and hierarchy of places within the development. In addition the designed street network should provide appropriate connectivity and is able to accept orderly future growth. There should also be a pedestrian network design that separates pedestrians from automobiles, and which ensures flows through parks, squares, and alleys. Also, street section designs ought to ensure human scale, appropriate building proportions. There should also be orderly parking and vegetative areas. In addition, a regulating plan must outline the zoning of building types and provide for integration, rather than separation, of different uses. The TND approach also specifically includes attention to public buildings and squares to ensure they are distributed throughout the neighbourhoods and to coordinate civic and open spaces. Finally, the planning codes are a central principle, paying particular attention to both architectural controls (materials, building configurations, reference to historic style, vernacular elements, etc.), and urban regulations that control how separate developments coordinate with the broader public spaces of the design. In Lennertz's (1991, p. 22) words, the design codes should *"encourage variety while ensuring the harmony required to give character to the community."*

One of the most cited applications of this type of design framework can be seen in what is commonly recognized as the first TND in the U.S.A.—that of Seaside (Mohney and Easterling 1991). Figure 2.1 shows the plan of Seaside, which was built on an 80 acre (32 ha) site on Florida's northwest Gulf Coast between Panama City Beach and Destin, 150 km from Mobile. It was designed by Duaney and Plater-Zyberk and has just celebrated its 30th anniversary. This project quickly became the coffee-table-book icon of NU (Brooke 1995). It is considered by many to be beautiful, cute, and appealing, and gained notoriety when it was used as the set location of the film '*The Truman Show'*. The building code of Seaside established an overriding rubric of building types and functions, and laid down how they are to be integrated into the final form of the town. For example, the rubric defined eight building types, including Retail, Residential, Workshop, functions etc., and then specified how different features of the building must be coordinated and integrated into the plan, establishing detailed restrictions on the design of yards, porches, outbuildings, parking facilities, and building heights, with further details shown in Brooke (1995, p. 31). The idea was to create a 'beach town' using examples from all over the southern states, with houses of different styles, colours, and picket fences, together with porches to see the setting sun, as well as a desire to increase sociability. The settlement's motto is 'A Simple, Beautiful Life', suggesting the rationale for a slower pace of life, a concept more completely expressed in the later Italian movement called Cittaslow (Chap. 15). Yet although attractive, the area is really a resort community, and like the old neighbourhood units, contains very limited employment opportunities. Politically it is an unincorporated unit in the county of Walton, so has no formal government, and only built its first school, one of the new charter schools, in 1996. Functionally it is hardly a town, so apart from its design, it

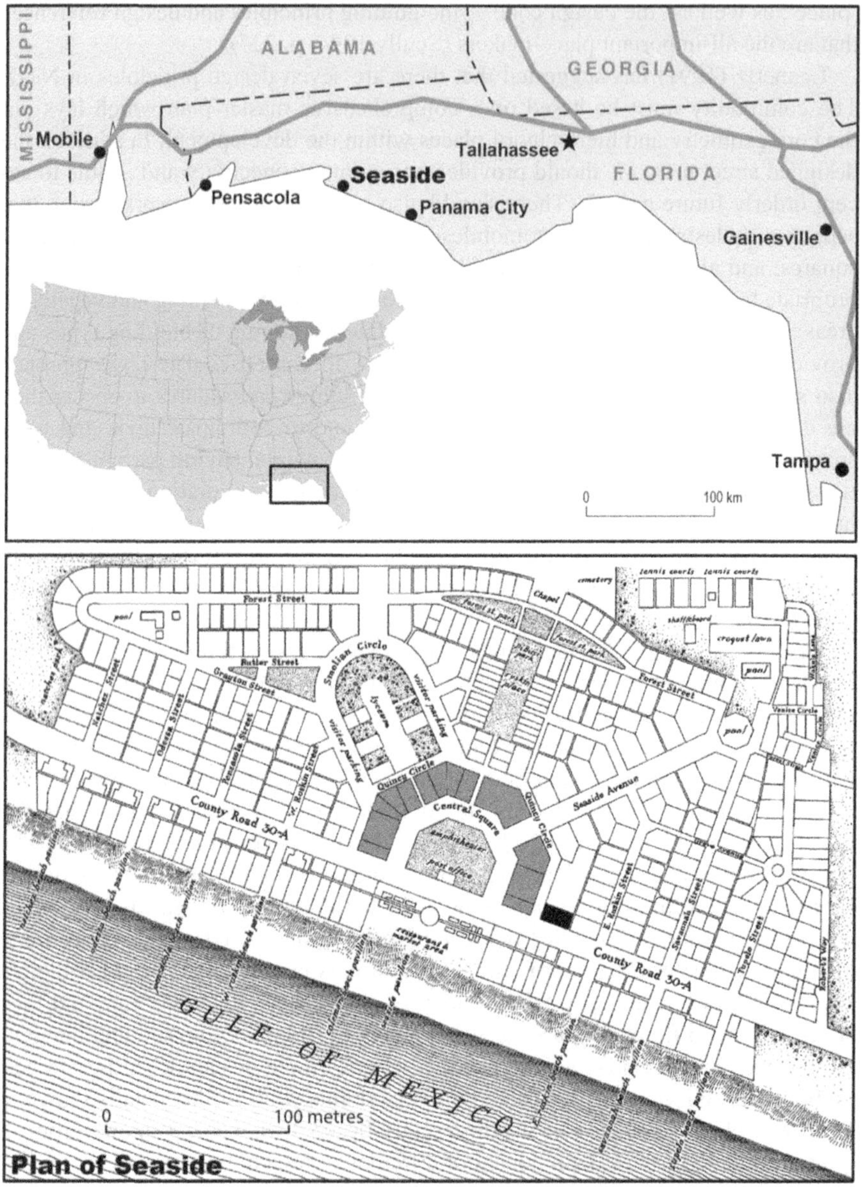

Fig. 2.1 Seaside, Florida. (Source: compiled from road maps and http://www.dpz.com/Practice/7903)

it surely contributes little that is new to the urban literature. Nevertheless it stimulated a new approach to urban ideas.

The town of Kentlands, in Gaithersburg, Maryland is another famous U.S. example of TND, and was the first application of TND principles to a real, year-round working town (Katz 1994, p. 31). The town was also designed by Duaney and Plater-Zyberk and construction started in 1989. It was built on a 143 ha site of the old Kent farm, from which the design takes its historical tradition and reference, and was designed to incorporate six distinctive neighbourhoods. Each of these combined residential, office, civic, retail, and cultural functions, while a more centralized town centre and town commons contained cultural facilities and a larger retail shopping complex on the periphery. All the distinctive neighbourhoods include a mix of residential types and sizes in order to appeal to different market segments, from families to retirees, so as to promote age and income diversity within the area. Their architectural designs place garages and domestic service functions at the rear of the property and accessible from back alleys. The community incorporates a lake, green belts, wetland reserve, and multiple small squares and park spaces of different scales. The final build-out of the project is approximately 1600 homes and a population of 5000, so in real terms the development is not much different to the average population of a census tract in a metropolitan area in Canada.

As in many other planning movements, there has been a diffusion of the ideas, and TND developments can be found in many countries, although they may be modified to reflect different planning and cultural environments (Grant 2006). In Canada, one of the first and most influential TND developments was McKenzie Towne (MT) in Calgary (Fig. 2.2), in which Duaney participated in the design with Carma developers. Built on some 365 ha (900 acres) as a new subdivision, MT was originally planned to have 12 distinctive villages, each surrounding a local square or commons area. However as development progressed it evolved into a smaller number of planned 'villages' (what are usually called neighbourhoods), themed after Scottish settlements, Prestwick, Inverness, Elgin. This was presumably to give it historical caché, although these are towns, not villages, in Scotland. Each area has (supposedly) a distinctive architectural styles, a local 'commons' (called a public park elsewhere!), connected to its commercial core through pedestrian pathways that also link residents to the central lakes within the area and which also function as storm water ponds. These pathways connect to the central 'High Street' shopping district. At the entrance to the subdivision are some buildings designed to evoke a distinctive sense of place that have heritage connections, such as a large church with a spire, a so-called 'town hall', and a fire service building built in late nineteenth century style. The community also has a logo, shown in Fig. 2.2, that identifies the area, with a deliberate addition of an 'e' to the word 'town' to stress some connection to the past, although 'the past' is obviously a mythical one. The development includes many of the principles of TND, such as: narrower streets; many houses with garages built in back alleys and not in the front of houses; higher densities than other suburban areas in the city built at the time; as well as a mixture of diverse types of houses, from various income levels of single-family houses and townhouses. However there is no social, low income housing. Despite these innovations the development was slower to complete than expected, given competition from other suburban developments, and many of the original ideals of NU design became watered down as some of parts were built according to more conventional approaches. In most cases, it is only the areas surrounding each of the squares that truly resemble the TND approach, and as one moves away from these areas the architecture and building materials are often similar to non-TND communities in

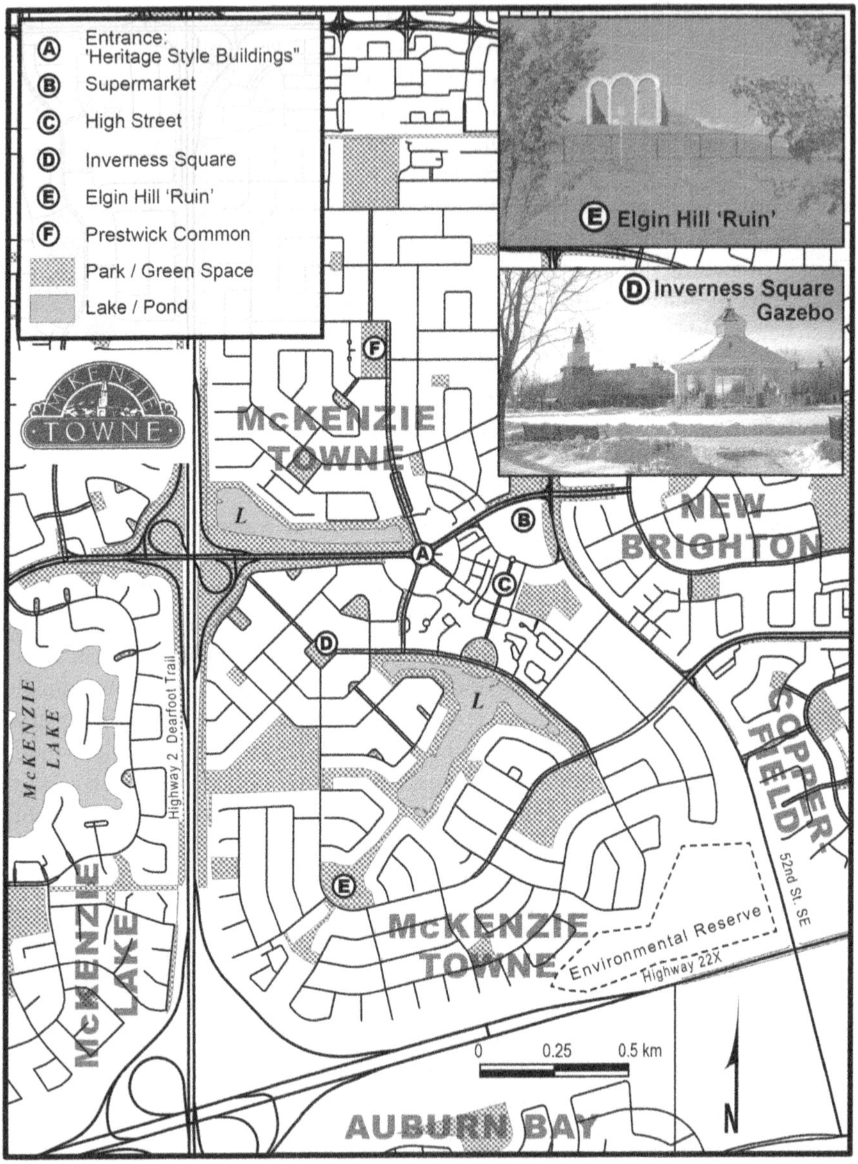

Fig. 2.2 Main features of McKenzie Towne Plan

the city that were built around the same time. Nevertheless, the developers (Carma) clearly tried to evoke the sense of history and identity in the neighbourhoods. For example, one promotional document describes the background for Elgin village:

> The historical villages that dominate much of the rural landscape of Italy, Great Britain, Greece and France have provided the inspiration for Elgin Village. The architecture in Elgin Village is influenced by the time-honored skill and craftsmanship of the finest European

designers but the neighborhood will remind you of the picturesque streets of small town Canada where every home has a unique personality of its own. (Carma n.d.)

There seems little doubt that this puffed-up description represents anything but true historical authenticity, given its excessively diverse and somewhat confused set of reference points that would be difficult to articulate in the TND code for McKenzie Towne. Indeed the description may well be more in the mind of the promoter than in any the reality. The attempt to create differences can be seen in many examples, such as in the central green with what looks like a nineteenth century gazebo in Prestwick, surrounded by town houses evoking a Georgian flair, or the many concrete garden walls in Elgin that have patches in them, presumably to give the impression of historical wear and tear. In addition, Elgin's nodal place identity is based on a created 'ruin' of what appear to be cloistered walls, built with a distinctive Prairie limestone on top of a hill constructed out of local landfill, which is this area's distinctive or symbolic centre. It is hard not to interpret the historical reference as a post-modern pastiche, since castles, not cloistered abbeys or convents, are usually on hills in Europe. Nevertheless, whatever its distorted historical reference, it does provide a distinctive local open space, with a playground alongside, and a treed 'wild' area behind that connects to a lake. Mackenzie Towne also scores over conventional suburbs in its attempt to re-create an old town street in its core, with varied western designs, and with several residences for seniors, including assisted care living units, although it is a pity that apartments were not added above the shops. However at the end of the street is a large Sobeys supermarket and car park, together with several additional commercial buildings and their parking spaces, making this part little different to older suburban developments. Yet there is much to admire in the range of housing types, green spaces, lakes and walkable areas, providing important and welcome advances on typical late twentieth century suburban designs. Also, the addition of several residential homes for senior citizens shows an attempt to make it a community for many age ranges, unlike the majority of suburban areas in the city. However there is little employment in the community, so commuting is still high; the first school only opened over a decade after its first residents moved in, which means walkability to school for children was absent for over a decade. The presence of a new large shopping centre built by other developers right on its northern boundary threatens the viability of local stores. Yet Mackenzie Towne is still an advance on previous suburban designs in the city.

Another award-winning TND development in Calgary is Garrison Woods, an infill TND development built on the site of a 65 ha redundant Canadian Forces base in the inner city (CHMC). It was developed by the federal government's Canada Lands Company in 2003 in collaboration with a local developer, architectural consultant, and the City of Calgary planners. Given its name, there has been a deliberate attempt to label streets with names of major battles in which the Canadian military fought, and with artefacts in green spaces linked to military themes. These provide actual, not invented historical references, given that military units previously based in the garrison fought in major World War I battlefields, from the Somme to Vimy. Since this development of 1600 housing units is only a few minutes' drive from the urban core, it has locational advantages compared to newer suburban TND developments on the city's fringe, while it is adjacent to a main arterial through the city and has led to the revitalization and extension of Marda Loop, a neighbouring older inner city shopping centre that was suffering from competition from modern malls before the new development was created. The innovative developments in Garrison Woods include the relocation and refurbishment of many of the former single and semi-detached military housing units, which provided residences for lower income families, while some of the former military buildings have been adapted to be community amenities. The area also contains three and four story townhouses, some with shops below, in addition to condominium blocks and single-family residences. In addition, the development included secondary suites (or 'mortgage helpers'), as small apartments above rear lane garages, thereby adding both density and social mix and varied tenure. The modified street grid pattern of the area enables integrated vehicle and pedestrian access between this new subdivision and the surrounding area. Indeed, most residents live within a 5 min walk of a bus stop, a 2 min walk to a park, and have numerous commercial and retail outlets within easy walking distance (CMHC n.d.). Although it is estimated to have cost 30 % more than a traditional subdivision, the end result was a pioneering mixed use and mixed scale development with much higher densities than usual at 25 units per hectare—which is higher than contemporary subdivision standards.

2.5.2 Smart Growth (SG)

Smart Growth (SG), which became a recognized term in the planning profession in the late 1990s (Tregoning et al. 2002; Knaap and Talen 2005; Ye et al. 2005) embraces many of the principles of the other NU approaches, such as compactness, sense of place, walkability, etc. But rather than dealing with the intricacies of design, vernacular architecture, or the minutiae of design codes and pattern language of Neo-Traditional Designs, it focuses more on the macro-scale, and particularly the interface between politics and economy in achieving a broader vision of sustainable new urbanism in economic and social terms by emphasizing local employment, community-stakeholder collaboration and directing development towards existing communities (SG 2013). The approach attempts to situate this growth within the context of an urban and metropolitan system—not just individual neighbourhoods or selective TND developments—and emphasizes the use of policies that induce the free market to pursue urban growth strategies that complement one another to develop more sustainable urban systems and metropolitan regions (Halligan 2000). According to the Smart Growth Network (smartgrowth.org), this type of sustainability can be achieved by adopting the 10 Smart Growth principles (Tregoning et al. 2002), which are shown in Table 2.2.

Portland, Oregon is often used as an exemplar of Smart Growth planning (Geller 2003; Gibson and Abbott 2002; Song 2005). Portland has a long-standing history of growth control dating back to the 1970s. This is long before the use of the term Smart Growth and included a focus on development within the existing urban envelope, the successful implementation of the Urban Growth Boundary (UGB), as well

Table 2.2 Key principles of smart growth

1.	Mixed land uses, such as residential units and employment centres
2.	Take advantage of compact building design
3.	Create a range of housing opportunities and choices
4.	Create walkable communities
5.	Foster distinctive, attractive communities with a strong sense of place
6.	Preserve open space, farmland, natural beauty and critical environmental areas
7.	Strengthen and direct development toward existing communities
8.	Provide a variety of transport choices
9.	Make development decisions predictable, fair and cost-effective
10.	Encourage community and stakeholder collaboration in development decisions

as related policies that control suburban sprawl through land use controls and coordinated transportation planning polices across 24 separate municipalities (Coroux et al. 2006). Portland also developed a multimodal regional transportation network that includes buses, light rail transit and streetcars as well as 330 km of pathways that encourage residents to walk and cycle to and from the downtown core (Geller 2003). Plans also require municipalities governed by the plan to reduce parking and driving by 10% over the next two decades (Coroux et al. 2006). Local community development has been an integral part of the overall plan that supports the revitalization of housing and infrastructure in the inner city, as well the maintenance of low-income housing stock in areas undergoing redevelopment (Gibson and Abbott 2002).

Numerous other cities are now adopting SG ideas and implementing all or at least some of the core principles. Many of these plans and policies seem to stem from growth management or environmental-sustainable development studies conducted at the local level. A recent study in Calgary, for example, a city of 1.2 million people that is the corporate centre of the Canadian oil and gas industry, has assessed the costs and consequences of sprawl, and the impediments to Smart Growth using a Smart Growth framework (Coroux et al. 2006). Ten recommendations were made to guide policy developments that support SG principles, and it is evident from these policies how closely they conform to the key principles of SG identified above. The first is to promote SG through public engagement and education about the costs of continued sprawl as well as the benefits of alternative development patterns. A second is to reform planning regulations so they advance innovative development initiatives such as SG. In terms of this, the report argues that planning regulations should be based on the following principles: (a) adopt mixed use development that combines residential land within convenient walking distance of employment, recreational and retail services, and mixed use zoning that allows for multiple uses within zones; (b) ensure compact, transit-oriented nodes in all new communities, in order to make transit more fiscally viable and efficient and therefore a more appealing transportation alternative; (c) support neighbourhood designs that prioritize the safe and convenient movement of pedestrians and cyclists, rather than the movement of cars; and (d) support redevelopment initiatives in existing neighbourhoods including more infill development patterns of development. Since SG initiatives are inherently political, the third recommendation in the Calgary report is to reform the planning process to mandate public involvement in the planning of new communities. The fourth principle is to provide incentives that encourage developers to implement Smart Growth initiatives—incentives that could range from expedited approvals of SG developments or relief from development levies. A fifth recommendation is to change municipal policies to mandate sustainable urban development initiatives such as SG. This could require amendments to higher-level legislation such as the provincial Municipal Government Act, but policies should direct city officials to ensure all regulatory frameworks are consistent with SG policies. They should also remove levy systems that may subsidize sprawl, remove barriers to SG such as inappropriate risk assessment, and mandate high energy and environmental standards for civic building, such as LEED standards (see Chap. 6) and encourage local utilities to develop and implement alternate energy sources (e.g. solar, district heating, geothermal). There is also a recommendation to develop greenbelts and Transfer of Development Rights (TDR) to maintain agricultural and wildlife habitat. The report also recognizes that regional planning is integral to the success of SG, despite the fact that the provincial government in 1995 essentially disbanded regional planning authorities throughout the province. So a sixth recommendation called for a comprehensive review of provincial legislation that affects the growth of municipalities in Alberta. However, SG approaches are not strictly macro or regional in scale. Like TOD approaches, SG also recognizes that New Urban communities need to have sensible neighbourhood initiatives. In the Calgary case, a seventh recommendation calls for the planning of sustainable communities, which include schools, transit nodes, employment centres, and a mix of housing options. Quite in keeping with the other approaches to NU, an eighth recommendation is to mandate the provision of affordable housing as an integral part of every new development. Finally, in keeping with the principles of the SG network that identify fairness and citizen input, a ninth recommendation calls for the creation of a citizens' panel to review and recommend reforms to the municipal electoral system. The final recommendation is to make monitoring and reporting of sustainability indicators an integral part of the planning process.

Unfortunately, many of these recommendations have been forgotten, as the city of Calgary continues to allow low density residential communities on its suburban periphery, so few of these suggestions have been implemented. However, a Calgary Regional Plan (CRP) has been developed in co-operation with surrounding municipalities, but like many other new regionalism initiatives discussed in a later section has only advisory powers. But this has not been as successful as expected, in large part because the adjacent rural entities fear being outvoted by Calgary and being subject to its requirements, not their own. This led to changes in the original plan advocating growth nodes in the region. Indeed, judging from new 2012 plan, sprawl in the area outside the boundary of the city now seems to be policy, rather than the type of focused growth nodes usually favoured by SG ideas. For example, the area between Calgary and the satellite town of Airdrie seems to be planned to be major growth zone, which does not seem much different to previous outward expansion, which is surely the opposite of SG principles that are against sprawl. However there are some useful initiatives related to conservation and a regional transportation plan, although the latter is still very limited.

Despite these depressing conclusions, there is a unique community in Calgary that does contain several, but not all, of the SG principles. Quarry Park is unlike the other new urbanism designs since it is a *real* mixed use development, although in some ways it is more like an office park with houses (Fig. 2.3). The area is being developed on a site of an old river gravel quarry that has been subsequently built up to a metre above grade as a flood precaution measure, together with a berm along the river bank. When complete, it is expected to contain 1.7 million sq. ft. of offices and 2200 residential units, split between single family units, town houses and apartments. Many of these are designed with what are claimed to be French countryside and Georgian influences, with steep pitched roofs, brick facades and timbered accents. In addition it has a large commercial area, with distinctive streetscapes based on the use of mansard style roofs and various pastel shades, together with symbolic centre-pieces, such as a building with a distinctive 'town' clock and a fountain enclosed by a complex sculpture imitating an old iron-work lattice design (Fig. 2.3). The colour and design co-ordination of the shopping centre buildings is in complete contrast to the usual suburban shopping centres in the city, where each unit has a different design and colour, producing a confusing and inharmonious visual mélange. The development also includes a central lake and canal that acts as a storm-water system, for in heavy storms the run-off from all the parking areas for offices and shopping centre can be considerable. So the lake has a water lift pumping system that drains the excess water into the nearby Bow river, rather than depending upon a cutting to the river which could back-up in times of flood. In addition there are many green areas with 15 km of pathways planned. Unlike neo-traditional designs it has extensive parking, needed for the offices and the large supermarket that predated the development of the rest of the commercial area. The community is bordered by suburbs built in the 1980s, which presumably means that children from the area will go to schools in these surrounding areas as there is no school planned for Quarry Park although there will be a large day care. The distinctive feature of the development is the large number of office buildings in the area. Hence, it will act as a subsidiary office employment core in the city, which is the main headquarters of oil and gas companies in Canada, enabling people to live close to work, one of the failures of most neo-traditional designs since few contain much employment. The area has attracted major tenants such as the headquarters of major companies, such as Imperial Oil and Canadian Pacific Railroads, which will move from the central city, as well as a number of regional firms.

2.5.3 *Urban Villages (UV)*

Urban Villages are a variant of NU design that Grant (2006) claims is predominantly European in origin. The roots of these ideas are mixed (Franklin and Tait 2002). The concept was first promoted in the 1980s by the Urban Villages Group in

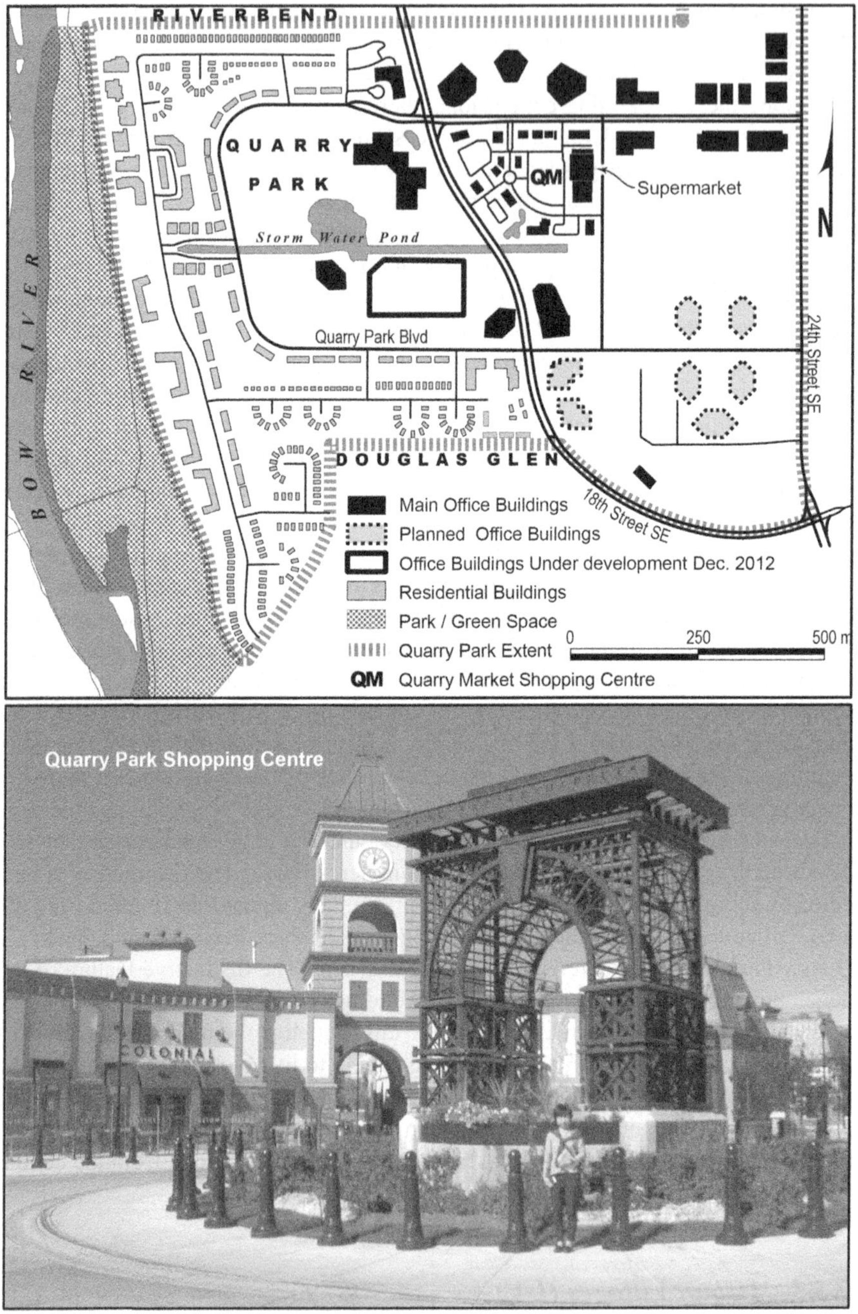

Fig. 2.3 Quarry Park, Calgary: Plan and Central area

Table 2.3 Key aims of the urban village approach. (Source: Department of the Environment, Transport and the Regions 1999)

1.	Resource consumption should be minimized
2.	Local environmental capital should be protected and enhanced
3.	Design quality should be high
4.	Residents should enjoy a high quality of life
5.	Equity and social inclusion should be increased
6.	Participation in governance should be as broad as possible
7.	The community should be commercially viable, i.e. not requiring public subsidies to maintain its performance on the other criteria
8.	Integration of environmental and quality of life objectives—a sustainable settlement would perform well on all the first seven themes, but not some at the expense of others

the United Kingdom through a publication entitled *Urban Villages: a concept for creating mixed-use urban developments on a sustainable scale* (Aldous 1992). In this publication the urban village was outlined as a means to achieve more human scale, mixed-use and well-designed places in cities, although in Europe it often also entails a focus on job creation in the development. A number of key aims can be identified in this approach, especially the stress on sustainability (Table 2.3).

The UV idea embraces the familiar NU themes of compact form, mixed land uses, and mixed housing types, but also focuses heavily on the urban village as point of employment and extends into environmental and sustainable issues. By developing these intra-urban villages for a population of between 3000 and 5000, with local shops, mixed housing (including public housing), schools, transportation networks, social and public spaces, and local employment centres, so residents should become relatively self-contained. In short, an urban village should include characteristics such as variety in social character and housing, high quality design and construction, a people-friendly environment, contain mixed land uses, high levels of activity, and flexibility. In Britain these ideas have been popular, resulting in a number of projects such as Greenwich Millennium Village (London); Allerton Bywater Millennium Community (near Leeds); New Islington Millennium Community (Manchester); South Lynn Millennium Community (King's Lynn); Telford Millennium Community; Oakgrove Millennium Community (Milton Keynes); and Hastings Millennium Community. Most reviews have argued that not all of these have been resoundingly successful in terms of the sustainability criteria identified above. For example, an external assessment of five of these communities on all of the dimensions above showed high levels of variations in outcomes, and many with poor results in particular areas (Department of the Environment, Transport and the Regions, March 1999).

Implementation of the UV ideas in Britain was also helped by the Prince of Wales's well-known criticism of British architecture and community design (HRH

Prince of Wales 1989). His views were themselves very much influenced by Krier (1998), who advocated a remedy for the modern city by a return to pre-industrial urban forms, the abolishment of zoning, and the development of urban quarters (Audirac and Shermyen 1994). Certainly the Prince was instrumental in advancing public awareness of the need for changes in ideas around city-building and has implemented many of the ideas in the new development of Poundbury outside Dorchester in Southern England. Poundbury, developed through the Prince of Wales's Duchy of Cornwall estate, was started in 1987 as an urban extension to the small historic county town of Dorchester (Dorset, England) and designed by Leon Krier. So far 1400 houses have been built, with another 1000 planned, the whole development being two-thirds privately owned and the remainder rented via housing associations to create a social mix. In some ways it is far sighted, using energy efficient measures with condensing boilers, ground heat recovery systems and photovoltaic tiles, an electric local bus system and using designated spaces at the rear of the houses to avoid street parking, with main traffic thoroughfares to ensure local streets are pedestrian friendly. But its designs, mainly in tree-lined streets or crescents and housing styles, are mainly varieties of Georgian and Victorian forms, which, although providing a distinctive sense of place, hark back to older ideas. Two problems that have emerged relate to house prices and employment patterns. The cost of an average terraced house is a third higher than equivalent sized units in nearby Dorchester, which reduces the opportunity for young couples, and although there is a range of larger housing units their prices mean they can only appeal to mature and higher income groups. In addition, many of the inhabitants commute to nearby towns rather than working in the adjoining industrial estate. This means that employment self-containment is limited, a feature reinforced by the fact that many of the shop units scattered throughout the area have not attracted much attention from retailers. There is no doubt this is a stylish area, if one appreciates Georgian ideas, and there is a sense of both community and place. So although it is an advance from the typical monotonous low density suburban extensions, it does seem to appeal mainly to higher income and mature groups, despite the mix of housing types within the streets.

2.5.4 Transit-Oriented Development (TOD)

This is the fourth variety of ideas stimulated by NU, and draws heavily on the work of Calthorpe (1993, 1994), Van der Ryn and Calthorpe (1986) and Kelbaugh (1989). While also concerned with developing mixed-use neighbourhoods and compact urban forms, the TOD approach focuses on development around mass transit stations, to reduce automobile use. Hence it is very much concerned with ideas of environmental responsibility and sustainable urban development—particularly as they pertain to transportation infrastructure, costs, and behaviour. Calthorpe (1993) has identified the key features of TOD, which are summarized in Table 2.4.

Like SG, the TOD approach also generally acknowledges that growth is inevitable, but can be better planned for. TOD seeks future growth without sprawl, or at

Table 2.4 Key features of Transit Oriented Development (TOD). (Source: Calthorpe 1993)

1.	Growth is organized to be compact and transit-supportive
2.	Commercial development, housing, jobs, parks, and civic uses are within walking distance of transit stops
3.	Street networks are pedestrian-friendly and directly connect local destinations
4.	Housing is of intermixed types, densities, and costs
5.	Sensitive habitat and high-quality open space are preserved
6.	Public spaces are the focus of building orientation and neighborhood activity
7.	Infill and redevelopment are encouraged along transit corridors within existing neighborhoods

least attenuating the negative impacts of sprawl and is more focused on inner city developments. TOD believes that transportation networks affect land use and vice versa, and so argues that public transportation nodes and their surrounding neighbourhood should become the basic unit of design organization. The nodes themselves take on a particular identity and human-scale architecture; they are pedestrian-scaled, contain a diversity of land uses, activities, and populations, and should have well-defined public spaces (Calthorpe 1993, 1994). In essence these transportation nodes are to have the qualities of small towns within the urban fabric, while at a regional scale many plans seek to identify such nodes as the growth centres in the region, attempting to build them up as mini-downtowns. Some research seems to indicate that there are links between TOD forms of development and quality of life, personal health and fitness, economic sustainability, municipal efficiency, and the production of better places to live and work (DeCoursey and Athey 2007). For example, Orenco Station, in the town of Hillsboro near Portland (Oregon) is often cited as a good example of the implementation of these initial TOD ideas. Here, the developers created a distinctive 'village' of 1800 homes, a town centre, office, retail and nearby employment surrounding the light rail transit node, as one of a number of 'town centres' that were part of Portland's 2040 regional plan. The development included a mix of residential, office, and retail spaces, together with a distinctive town centre and live-work townhouses. Many other cities have adopted similar principles and moulded the requirements to the local circumstance. Indeed DeCoursey and Athey (2007) argue that the idea of 'complete communities' is what defines a TOD, and a good TOD implementation has five traits: (a) quality transit facilities and services; (b) it is walkable with a high quality pedestrian environment; (c) it has unique destinations—complete communities with a community centre and the right mix of uses; (d) is compact, with highest densities closest to transit; and (e) parking is carefully located, designed, and managed. Most of the key principles of TOD in a locally specific context can be seen in the award-winning design of 'The Bridges' in Calgary, a TOD development in the old inner city community of Bridgeland, less than a mile from the city centre. Claimed to be one of the best examples of TOD in Canada, The Bridges was developed by the City of Calgary on an old hospital site close to an existing Light Rail Transit stop. It will be home to approximately 2500 residents in a mixture of housing types and densities, including affordable housing. The design incorporates pedestrian friendly streetscapes, regional pathways systems, local 'main street' shops, and direct access to the rail station. It includes a strict urban design code that emphasizes energy efficiency, and controls building orientation and entrances, building material and finishing, balconies, vertical and horizontal proportions. All developments in the area must be reviewed and passed by a special architectural review committee. However both in the TOD principles, and the practice, it is noticeable that recreational space, apart from pathways to such areas, is largely ignored, at a time when the value of green space is increasingly recognized (see Chap. 3).

2.6 Positive Aspects of New Urbanism

There is little doubt that the various types of NU communities, such as those described above, have attracted significant public attention, and in many cases have won urban planning or design awards. These communities are certainly appealing to particular segments of the housing market and offer advantages in three types of features often lacking in the conventional suburb, namely those found in the physical fabric, their identity and in community relations.

2.6.1 Physical Fabric, Land Use and Morphology

Compared to many new suburban tracts, NU developments have been relatively successful in creating communities with higher housing densities and smaller lot sizes, although some commentators note that this may not necessarily equate with environmental sustainability because in many situations the houses on the lots are getting larger, just as they are in conventional suburbs and with little garden space (Banai 1996, 1998; Fader 2000; Grant 2006). They are also seen to be relatively successful in developing communities with a more human scale, containing: narrower streets; houses closer to the street—often with front porches; low-rise civic and commercial buildings; public squares and gazebos; small parks and other green spaces. In addition there is usually a mixture of housing designs, although this is far from a new principle. Letchworth, the first Garden City, was planned by Parker and Unwin who had established their reputation with new village projects. They did not only emphasize the need for what we would call a 'green town' today, but planned houses for different income levels to achieve the type of mixed community advocated by Howard (Unwin 1969). This equity principle was ignored by Perry's (1929) small, residential-only neighbourhood units that were planned with a central focus on a school, with shops on the main roads that formed the edges. Moreover the Garden City approach was to create morphological variety by deliberately using vernacular-style housing. Many were in different sized blocks, set at angles to one another and facing green areas to create variety and access to nature. Together with a distinctive town retail centre near the main railway, and an adjoining avenue for public buildings, there can be little doubt that they were creating a distinctive place, although they did not elevate this and other ideas to explicit principles of form seen in NU design codes. In addition, their interest in the local environment can be seen by the fact that they built according to local site conditions, allowing views of nearby villages and physical features. The mixed land use characteristics of NU communities are generally seen to be a positive step forward (or look backward) in terms of community design (Ford 1999). Rather than the strict separation of land uses in conventional suburbs, NU communities typically succeed in integrating a variety of land uses in close proximity, blending residential and commercial functions on the same lots (e.g. apartments above shops), civic and commercial spaces, and providing more variety than what is found in previous suburban developments.

Another appealing feature of NU developments is the focus on the design and integration of so-called 'public space' and also the renewed attention to the concept of a small 'town centre' with civic functions and retail-service activities, although these are usually relatively few in number, given the limited size of the communities. This lack of retail diversity is frequently cited as one of the problems in these areas (Schleimer 1993). Most have had limited success at reducing either automobile use and ownership or commuting patterns beyond the community, although walkability is often lauded as one of the most positive aspects of NU, given its benefits in promoting health and sociability, through more physical activity and in meeting people, especially in terms of internal movement within the development (Frantz and Collins 1999). Southworth (2005) has suggested that six criteria underlie the design of a successful pedestrian network: connectivity; linkage with other transport modes; a fine grained land use pattern; path quality, path context; and safety. Many developments integrate extensive networks of pedestrian pathways (connectors) throughout the development, and are typically designed on a '5 min rule', namely that all housing within the development should be within a 5 min walk of shops and civic spaces. Schleimer (1993) found that 75 % of residents in such communities claim that these design approaches do encourage walking, leaving Ford (1999) to suggest that to some extent this type of design determinism, such as forcing people to park and walk, has been successful. Yet these NU views are not new; they seem to have forgotten that the Radburn and British New Town modifications of neighbourhood ideas of the mid twentieth century, where pathways throughout the community to schools and shops were a major characteristic.

Many have noted that NU communities are very attractive, and have architectural features that are inherently appealing to many. Some of this appeal is linked to the 'neo-traditional', vernacular, or retro elements of many of the designs, such as dormers, bell towers, porches, or style characteristics such as Victorian motifs or Georgian townhouses (Grant 2006). These communities also photograph well (Katz 1994; Brooke 1995). For some, the architectural appeal is also based on the knowledge that there will be little alteration of the design standards and architectural appearance through time, given the strict development codes and restrictive covenants that control building materials, colours, and even exterior uses (Frantz and Collins

1999). So most agree that the NU approach has been successful in designing more attractive and well-functioning spaces. Even its critics reluctantly note that the often 'cute' architecture in these places is better than the standard, front garage-dominated architecture of the late twentieth century suburbs (Ford 1999).

2.6.2 *Identity and Community*

Identity and community are foundational principles of NU that are very much linked to the desire to overcome the sterility and placeless character of the suburbs. Identity is reinforced in NU through design codes, and through reference to a historical heritage, even if it is artificially constructed or imagined. The design features which may be unique to the area may not create an immediate community identity or 'sense of place', but as Grant (2006) suggests, these developments seem to provide the residents with a head start in creating such feelings in their new surroundings. This identity, and the aesthetic character of the area, is stabilized through restrictive covenants that owners have to agree to when purchasing property. These restrictions offer a kind of guaranteed identity conformity for the future, which may help the development of a sense of cohesiveness among residents (Frantz and Collins 1999). A moot question is whether the NU designs alone have been successful at developing identity. Some evidence suggests that they do (Grant 2006), but others have maintained that the identity creation of NU seems to be associated with their niche market character (Gyourko and Rybczynski 2000) and with the fact they are aimed at a limited consumer segment, upper middle class people, often baby boomers. These people seek out such communities because they may offer an identity based on a particular constellation of lifestyle choices—what Ford (1999) called 'lifestyle zones' in the city. This socio-economic restriction of the NU developments means that there are few signs in these communities of the complete range of social types envisaged by many NU advocates.

The creation of a sense of community—what Putnam (2001) viewed as 'bowling together'—lies behind much of the design rationale of NU developments (McCann 1995; Hall 1998; Kim and Kaplan 2004). Most commentators suggest that NU may have succeeded (Brown and Cropper 2001) in creating such identities, and at least a stimulus for a sense of place. Yet so far there have been few rigorous empirical studies of the extent to which these NU designs generate real identities of community cohesion. Moreover, there is a design determinism behind the assumption of such feelings and behaviours, which by no means guarantees the successful creation of a sense of community. This is the same problem found in the British new towns of the 1950s and has been repeated in more recent work (Brindley 2003). Indeed it has been suggested that while there may be claims for a better sense of community in NU designs:

> the connection between new urbanist form and the affective dimensions of sense of community becomes more and more untenable as the complexity of meaning involved is evaluated. (Talen 1999, p. 1370)

2.7 Criticisms of New Urbanism

The NU approach is often lauded as the solution to the ills of the modern metropolis, and planners and developers have embraced many of its ideas. Yet despite the appeal of many of the ideas, NU has also met with a number of criticisms (Ellis 2002), related to five main issues: sprawl, mixed functions, nostalgia, social relations, and the extent to which they create good communities and authenticity.

2.7.1 Sprawl

Since NU aimed to remedy urbanism by addressing the suburban sprawl of the modern metropolis, one of the first and most obvious criticisms is the ability of NU as a movement to solve this problem. Critics argue that since many NU projects are built on the suburban periphery as discrete subdivision developments, often disconnected from the existing urban fabric (Ford 1999), these represent little more than a prettier form of planned sprawl or simply suburbs in disguise (Thomspon-Fawcett 1996). Furthermore, one of the key problems of sprawl that NU hopes to remedy is the reliance on the automobile, for travel within the development, and for commuting to the central city or other employment centres. Many critics argue that there is little evidence for reduced car use in NU developments and that automobile commuting patterns may be little different to that of conventional suburbs (Crane 1996; Pogharian 1996; Lund 2003; Khattak and Rodriguez 2005). Even within the developments, where New Urbanists deliberately design for increased pedestrianization and pedestrian safety, increased levels of safety from traffic remain questionable (Southworth 1997). So it is difficult to see this handful of new NU developments, especially the NTD form, as any panacea for suburban sprawl (Schleimer 1993), especially since so few of the new suburbs in current cities fully adhere to NU principles.

2.7.2 Mixed Functions

One reason why NU communities may not demonstrate significantly reduced automobile usage may be linked to an additional criticism. This is the limited success of providing enough mixed functions in these communities, especially in terms of workplace and retail functions. To some extent, critics argue that these developments have not been successful at creating communities where people live and work (or telecommute), and have typically not integrated sufficient or diverse enough retail functions (Grant 2006; Crane 1996; Steiner 1998). Indeed, most are far too small to be self-contained entities in shopping or other activities, a problem of previous neighbourhood designs that go as far back as Perry (1926). In some NU developments, even the limited retail space that does exist struggle to maintain tenants, or has high business turnover. In McKenzie Towne (Calgary), for example, the small main street shopping district has seen considerable turnover and vacant premises. In some cases, formerly upscale stores have disappeared to be replaced by cheap 'dollar stores'. So critics suggest that NU in North America has failed to understand the continental consumer mind set, where shopping behaviours are more regionally oriented and preferences centre around large malls and increasingly around big box store power centres than local community shops (Jones and Simmons 1987; Steiner 1998). However in the rather unique Quarry Park example there are considerable employment opportunities, but since the development is still under construction it is not yet known whether a large proportion of the workers in the adjacent offices will choose to live in the community. Preliminary queries about the housing construction in the area suggests that some entry-level detached single-family homes are priced at well above the city average, so it is quite likely that there will be affordability barriers to a large proportion of the office workers. It remains to be seen whether multi-family apartments and condominiums, which will eventually comprise the majority of the housing stock in Quarry Park, will increase affordability relative to other areas.

2.7.3 Nostalgia

New Urbanists insist that the way forward for more liveable areas is to look to the past for inspiration, to historical European towns and small towns in North America to create what amount to nostalgic town scapes. In these places the urban community was at a more human scale, environments were highly pedestrianized, residential and retail/commercial functions as well as land uses and houses were mixed, and the social fabric was more diverse and more authentic. This vocabulary of a 'timeless ways of building', the pattern language of urbanism (Alexander et al. 1977), and authentic urbanism, lies behind much of the way in which TND attempts, through designs and codes, to emulate the past. However, critics argue that the NU approaches have a distorted and misplaced view of the past, and rather than a timeless way of building, the planners and developers have adopted a selective look at history, choosing only those features that seem to be saleable or commodifiable in 'catchy packaging' (Grant 2006; McCann 1995; Till 1993; Thompson-Fawcett 1996; Knox 1992). However New Urbanists generally insist that the aesthetic and architectural codes of such developments should be sensitive to local tradition and vernacular architecture. Yet many of these references to tradition or the vernacular are either selective, or fabrications. In some cases these developments begin with a thematic history that is artificial, with invented family histories and regional narratives (Till 1993)—what Frantz and Collins call a form of 'Disneyesque imagineering' (1999)—and represent little more than a nostalgic imagination based on what amounts to retrogressive sentimentality. Some also criticize the historical or reference points as being elitist; traditions and design traits are often defined by historical communities populated by those of higher incomes. The same arguments have been applied to the architecture within many NU developments, with some observers claiming it to be a shallow development of style, inauthentic, cookie-cutter architecture, or excessively cute (Grant 2006; Till 1993). Others point to the artificial character of the buildings, which often include fake dormers, windows, water towers, bell towers, and so forth, that create a type of make-believe quality, what Frantz and Collins (1999) called an 'exercise in studied nostalgia' in Celebration, U.S.A., a NU style community developed by Disney.

2.7.4 Privatopia and Social Cleavage

The public realm emphasis is a cornerstone of NU design principles. Advocates of NU see contemporary suburban developments as being dominated by private space. They regret the demise of the older view of the importance of creating a public realm in communities, and see great value in its return. Developments designed to act as a public spaces appear as the centre-pieces of the planning and design of most NU communities, so public squares, benches, pedestrian paths, civic centres, parks, gazebos, bandstands, and so on abound. As Frantz and Collins (1999) demonstrate in the development of Celebration U.S.A., these public spaces are seen as integral to the lifestyle of new urban community residents. However, while these are touted as public space, they are often far from it. Many NU developments are private community developments, where all homeowners must belong to homeowner associations, and are bound to comply with numerous restrictive covenants that may cover everything from the colour of houses, to curtain styles, to exterior adornments and uniformity of appearance, and even to behavioural controls within the supposed public realm (Frantz and Collins 1999; Bartling 2004). These communities are similar to many other kinds of developments that fall within the more general rubric of 'common interest developments' (CIDs), which include the rapid proliferation of gated communities since the 1980s. They represent a form of 'privatopia' (McKenzie 1994), in which the so-called public space is really a commodified form of 'club good' (Webster 2002). The privatization of urban space in American cities in particular is gaining momentum and in these areas it is not uncommon for access to these communities to be controlled. So the nominally 'public' areas are actually restricted to residents and guests, creating an insider-outside dichotomy that underlies their conceptualization of public space. Hence it has been argued that this generates what amounts to a 'geography of otherness', one that enhances, rather than diminishes the social distance between groups. In extreme cases, such as the town of Celebration, residents must also submit to high levels of security and surveillance, such as via closed circuit cameras, and even behavioural controls and sanctions in their use of the area (Frantz and Collins 1999; Bartling 2004). Hence some critics see these essentially private communities as providing only a fake sense of the public realm, one integrally linked to lifestyles of stylish materialism, even though they claim to emphasize public space (Knox 1992; Till 1993).

This privatization is also at odds with other NU principles, since NU has criticized the contemporary suburb as being socially homogeneous, segregated according to socio-economic status, socially exclusive, and divisive. Although New Urbanists espouse more socially diverse and integrated neighbourhoods, and believe urban design can achieve this, critics frequently point to the failure to achieve this aim. Indeed these developments appeal typically to a relatively affluent class looking for lifestyle appeal, namely self-selected residents, and are typically relatively expensive. The communities rarely have affordable housing stock, or a diversity of housing options. The result is that these NU developments, especially the NTD and UV types, often become socially homogeneous affluent enclaves in which the poor and visible minorities are excluded and are not the equitable places idealized by proponents of NU. So despite the rhetoric of social inclusion, diversity, and social justice, critics suggest that such features rarely occur (Dowling 1998). Indeed many critics have suggested that NU proponents do not fully appreciate the power of real estate interests and their profit motive in shaping the urban fabric, or the ways in which these interests simply contribute to the social fractionation of the postmodern world without any intent or desire to change the economic status quo (Foglesong 1986; Harvey 1994; Baxandall and Ewen 2000).

2.7.5 Social Relations and Authentic Community

Talen (1999, p. 1363) has reminded us "the essence of new urbanist design theory is the creation of a sense of community". So through planning, design codes, architectural controls, and restrictive covenants, NU hopes to achieve an urban form that fosters social interactions, thereby building communities. Many NU developments include features that are intended to do this. Parks, playgrounds, gazebos, bandstands, civic centres and main-street shops, supposedly become the nexus for the development of these relationships. Also, NU domestic architecture typically rejects the front snout-garage style, so typical of the post-1960s suburb, replacing it with street-facing features such as front porches, in anticipation of enhanced neighbouring, and eyes-on-the-street surveillance of children. Yet there are very few studies that have systematically examined whether or not social relations and levels of community experience are enhanced in NU developments, and no studies have explored the issue using comprehensive frameworks of community dimensions that recognize that there are multiple dimensions within the behaviour, cognition and affect domains that constitute urban communities (Davies 1992; Davies and Herbert 1993). Indeed critics suggest that there is little evidence that NU delivers on enhanced social relations and authentic community experiences. In part this criticism centres around the various strains of determinism, in environment, design and architectural senses, that underpins so much of the NU approach, which assumes that design can achieve social objectives (Bookout 1992; Hall 2000). This is essentially the same scepticism made in the 1950s about the assumption that new designs would 'usher in' communities in post-war British new towns. In any case many researchers since Suttles (1972) believe the community is socially, not materially constructed. Indeed Talen (1999) notes that in NU, sense of place seems to be focused on image congruity rather than on the more important features of affect and place attachment.

Others argue that the NU approach does not and cannot deliver an authentic community experience because their notions of community are based on commodified landscapes and pre-packaged ideas of community in the first place, and because their designs are inherently artificial and inauthentic. Although NU makes much of the idea of sense of place, and may design beautiful landscapes that appeal to a niche market, its development—involving both the cognitive and affective domains identified and measured by Townshend and Davies (1999) and Townshend (2002)—takes time to establish, and is typically spontaneous rather than designed. So for some critics the NU approach simply designs more beautiful places, with no heart or soul, and an exclusionary sense of social cohesion and community identity (Till 1993; Nasar 2003). Indeed, even community authenticity is perhaps imaginary, for as Scully notes, in NU projects, "*..the rich, who can choose, choose community, or at least its image*" (Scully 1994, p. 230).

2.8 New Regionalism

Most of the New Urbanism literature deals with scales below the settlement level, even though the first section of the Charter (Table 2.1) focuses on 'region, metropolitan and city and town' principles, which might be better summarized as city-region issues. Although this section of the charter describes the advantages of using these areas as a basis for coherent development, especially in economic, ecological and fairness terms, the charter provides little guidance as to how such structures may be created, administered or spatially defined. Curiously, few advocates of New Urbanism seem to have explored these issues, since even a large collection of essays on the topic (Haas 2008) does not deal with the question of the governance of city-regions and how these principles in the charter are, or could be, implemented. Nevertheless, there is a flourishing literature in what is now known as New Regionalism (NR), part of which deals with city-region creation and governance (Wheeler 2002; Scott 2005, 2009; Sancton 2001) and which can be considered as another extension of the NU development. These issues need to be reviewed as part of the more general NU developments of the last two decades. However a caveat has to be added, since the term is also used to describe the new international regionalization, in new trade or political connections, and the political devolution within states that has led to more local governance for regions or nationalities, such as in Catalonia or Scotland etc. (Soderbaum and Shaw 2003; Telo 2007). Such issues are beyond the scope of this review.

As with so many aspects of New Urbanism, the idea of using regional perspectives to understand and improve urban planning and governance is, of course, not new. In the early twentieth century Geddes (1915) and later Mumford (1922, 1939) were pioneers in arguing for the creation of regional planning units around large urban centres to better co-ordinate planning for their hinterlands, especially to protect the vulnerable natural environments around the city that were essential for its water supplies and recreation. Even Howard (1898) did not view his Garden Cities as individual entities, but as nodes within a larger metropolitan region connected by rapid transport systems, something that was soon forgotten in the garden suburbs that claimed to mimic his ideas (Davies and Herbert 1993, Chap. 8). Yet despite the growth of regional associations in many parts of the world to promote these ideas, most were more effective in conceptual terms (Hall 1988), and only a few led to new structures of governance based on regions. From the 1950s, however, concern about the spatial economic inequalities between regions within states led to a revival of regional ideas at the sub-national not city-regional scale, with concerted efforts in many countries to remove these differences by regional planning. In addition, geographers in particular used increasingly sophisticated multivariate techniques to define the spatial spread of the emerging city-regions, defined as the functional influence of urban areas over their surrounding regions based on shopping or daily commuting flows (Davies 1984). These city regions are not homogeneous regions because of some continuous urban sprawl due to the outward spread of one large city, or the coalescence of what several formerly freestanding entities, what Geddes (1915) called 'conurbations'. Instead, they were composed of a central city surrounded by a scattered series of satellite towns, out-of-town shopping centres, industrial zones and other urban-related activities in an otherwise rural landscape. These units had coherence because they were the basic functional regions of various central cities, or cities in the case of more complex urbanized areas, although other flows, such as business or trade connections, provide other types of hinterlands. Although the Charter for the New Urbanism describes these areas as 'metropolitan' regions, there is no standard definition of what a metropolis is; census bodies in different countries use figures from 50,000 to a million population as the threshold value for the central city, although the surrounding functional region, its city-region, is usually defined by some level of daily commuting flows. These metropolitan or city-regions owe their homogeneity to their regular functional connections with the central city, so they cannot be viewed as 'finite regions', presumably defined by landscape features, as described in the first principle of the NU Charter, although if physical features impede interaction patterns they may occasionally form a rigid boundary. Instead the boundaries of city-regions are 'zones in transition'; functional regions have fuzzy edges. This causes problems when attempts are made to spatially define the boundaries for administrative purposes, since political boundaries have to be precise, not transitional zones, to ensure that the government units have control over a defined area. Most city-regions that are defined as new government entities usually just group together pre-existing local government units, which either over or under-bound the functional region and only have an approximate relationship with the functional city-region, which does present problems in coherent governance.

The impetus for a new governance in city-regions came from the recognition that planning in these regions could not be carried out effectively because administrative units had only powers over their own areas and had no ability to influence what went on in neighbouring jurisdictions. They were considered too small and powerless in a legal sense to solve many emerging city-region problems, of which six in particular had the most influence.

One was the need to *develop and co-ordinate transport, land use and utility networks* within these emergent functional regions, especially to plan the location of future growth within the region. A second comes from growing interest in *sustainability issues*, especially the reduction of the ecological footprint of settlements within these city-regions and especially the negative externalities they create in pollution and waste generation. A third issue is a cost problem, recognizing that delivering services for individual units can be expensive, whereas delivery on a city region scale can lead to *economies of scale*. Fourth was the problem of creating more *employment*, given that globalization or increased mechanization reduced jobs in many western centres, which might be easier to achieve given the larger and cheaper land resources in the city region compared to high price central city land. A fifth point was the recognition of a *size bonus* in image and marketing by using the population of the city-region not the central city alone. For example, Vancouver and Stuttgart are now marketed as centres of metropolitan areas of 2.3 and 2.67 million respectively, rather than being viewed as central city municipalities of only around 600,000 people. A sixth point is associated by the way that the conservative governments elected in most western countries from the late 1970s viewed existing planning regulations as restrictive to potential growth, so more *entrepreneurial governance was recommended* in order to attract employment and create new infrastructure and amenities to support this growth. Many adopted pro-business policies by reducing regulations, modifying or replacing the older, more restrictive managerial approaches to what amounted to controlling growth in the past (Harvey 2005). This led to an increasing desire to facilitate global linkages, in additional to traditional local and national connections, by developing new infrastructures and amenities in the city-regions. Yet it is quite incorrect to see these global linkages of large cities and their regions as always being new. Many large industrial cities in western countries developed on the basis of their world connections. For example, Swansea and Cardiff in South Wales cannot be seen just as centres of city-regions serving their local areas. The growth of the former centre in the nineteenth century was based on its role as the node of the most important copper and non-ferrous smelting and manufacturing region in the world, which led one historian of the area to call it Copperopolis (Hughes 2000). Similarly, Cardiff had important global connections; by 1913 it was the largest port in tonnage in the world because of the volume of steam coal being exported from its docks, which led to its Coal Exchange setting the price for this energy source for many decades.

All these problems meant that there has been growing public acceptance of the need to plan on a regional scale, recognizing the functional interdependency of these city-regions. But the path to regional governance has not been easy. The inertia effect, the reluctance of higher levels of government to change existing local government structures, is still considerable. Also local municipal interests are still powerful among the smaller entities in the emergent city-regions, while their fear of the potential dominance of the biggest cities in any regional grouping makes them reluctant to give up powers. In addition, in many countries most higher levels of government seemed reluctant to reduce local government powers by intervening to grant exclusive new powers to these larger city-regional entities, either because they could rival existing organizations, such as provinces or states that had the sovereign right under most federal legislation to administer municipalities, or in fear of a political backlash in future elections from parochial interests. Moreover, the people in many countries of the world have attitudes that do not favour the development of powerful regional governments below the state level in federal entities. This can be seen in the United States especially, where instead of the administrative areas of big cities expanding outwards to cover the built-up area, their expansion has been restricted, allowing small separate suburban government units to emerge in their hinterlands, a product of the strong emphasis upon local autonomy and the distrust of big government initiatives.

These jurisdictional and ideological problems of governance, combined with the rapidly changing globalizing world, have created difficulties for those seeking new forms of governance to solve the emerging problems within the larger and fragmented metropolitan regions. So relatively few adopted the older annexation route used in the late nineteenth and early twentieth centuries, whereby the largest city incorporated surrounding areas to create one large all-purpose government based on the continuous urban sprawl. Instead, various types of regional governance have developed in some of these metropolitan regions, but with very different characteristics to the older top-down, single government structure. One reviewer of this type of New Regionalism observed that the city-regional governances that have emerged were mainly local and ad hoc arrangements, often based on co-operation between entities within regions. Hence they were described as:

> centred on partnership, flexible and hybrid forms of governance… and were… persuasive by appearing to support the development of social capital, the notions of trust, participation in civic associations and a sense of community (Scott 2005, p. 453).

This description does indicate the complexity of these new city-regional governances, but the statement downplays the fact that there were also active interventions by national or state authorities to create or legalize these metropolitan entities.

The complexity of the different forms of city-regional governance that have emerged make it difficult to succinctly summarize their characteristics. However, the variations between and within four main dimensions of city-regional governance shown in Table 2.5 provide an introduction to their basic differences, although more detail could be provided by adding other dimensions, such as how

Table 2.5 City-regional governance: major dimensions and types

1. Functions of Regional Unit (s)	2. Source of Functions	3. Control Exercised Over Functions	4. Democratic Legitimacy (Council and Committees)
Complete	New	Mandatory by regional agency	Directly elected
Partial functions (Lower tier units provide other functions)	Down from State	Advisory (Needs local tier unit approval)	Representatives from other councils (or agencies)
Single agencies	Up from local units	Delivery of services	Appointed members

the regional structure was spatially defined, the way voting powers are exercised, how finance is obtained, or the type of executive and administrative units that carry out the regional functions. Despite the theoretical acceptance of some city-regional governance as being a necessary idea in the contemporary world, and essentially recommended in the New Urban Charter, there are still very few practical examples, and no recommended structure or standard evolution. The simplest theoretical approach would be to vest all local government powers in a new all-purpose government unit, with mandatory powers over the region governed by a directly elected council and created by a top-down decision of the state. This solution has rarely been adopted, because it would eradicate existing local governments, which most countries are unwilling to do. The opposite alternative would be to have separate regional agencies for each particular function, as in the last row of Table 2.5. But such a structure means that there will probably be limited co-ordination between the agencies, which defeats the purpose of coherent regionalization.

Perhaps the most typical structure is some form of a co-operative two-tier system, with a regional structure lying above the local government units. In this type individual local governments below the regional level still exercise most statutory power over a range of functions, but a *regional agency* is established by the participating councils, perhaps with state approval or initiation. This carries out some strategic functions for the lower level units as a whole, typically plans for protecting the natural environment and watersheds, controlling land use and suggesting growth centres, providing transport planning for the whole area, and sometimes economic co-ordination. Apart from planning functions some of the regional agencies may have authority for delivery of certain functions. Usually the regional council that administers the agency is composed of representatives from each participating local government, but since the regional plans are usually only advisory, not mandatory, either the regional council has to agree on implementing the plans on behalf of their governments, or the plans are sent to each unit for agreement, which frequently causes tensions within the region when local governments disagree. Although this type of structure may be the most typical, most city-regional governances are hybrid regionalizations, with structures composed of more than one of the subdivisions in the dimensions of Table 2.5. So a single tier unit of regional government may include most, but not all regional functions, with the others operated by single function agencies, such as those associated with police or emergency services that are responsible to state institutions. The following examples from four different countries show the range of variation and the complexity of these city-regional governances.

One of the most basic types of regional authorities can be seen in Canada, where the Greater Vancouver Regional District (GVRD) had its origins as one of the regional planning agencies established by the province and covered 21 municipalities. Today it is still primarily an advisory body for strategic planning, mainly in land use and transport, for its municipalities exercise statutory control over most local government services and its council is composed of members from these units. Although regional attitudes are helped by the region's concern with liveability, and more recently sustainability issues, it lacks mandatory powers and two key functions are still outside its mandate; the provincial government still controls the Agricultural Districts in the region set up in the 1970s to preserve scarce agricultural land, as well as the new Translink rapid transit system. A sign of the provincial government's reluctance to give GVRD greater powers or even image is that it still insists on the original title, although the authority voted to market itself as Metro Vancouver in 2007 (Hutton 2011). There seems little doubt that the regional planning governance has had some successes but it is always going to face problems because of its advisory nature. So it has not been successful in developing the plan to create expanded town centres in the region to act as a counterweight to the central city—a problem found in most city-regional planning. Despite being activist in sustainability terms, its ability to progress in this area is limited because it is not only squeezed geographically between the sea, the mountains, and the U.S. border—which limits its growth options—but also administratively between provincial and local municipal power, as well as an often polarized electorate between left and right wing parties with very different views of planning.

In the United States the Portland Metro is the only true city-regional government in the U.S.A. This developed out of previous regional co-operation between over 30 municipal authorities since 1966 (Cotugno and Seltzer 2011). But crucially, the regional authority had a basis in the previous history of conservation in the area. For example, the renowned urban scholar, Mumford (1938, 1961), an advocate of regional planning, created an outline regional plan in the 1930s that recommended directing growth from Portland into a series of local green belt towns that could be supplied by energy from the new Bonneville dam and hydro plants (Mumford 1938). In addition, the creation of the 5000 acre reserve of Forest Park in 1948—the largest wilderness preserve in American cities—provided Portland with an exemplar of the value of conservationist ideas (Stephenson 1999). However the real spark came from an increasing concern about protecting green space and agricultural land that led to the state of Oregon making it mandatory in the early 1970s for urban centres to create Urban Growth Boundaries (UGB) to reduce sprawl. This led to the creation of the Metro area in three counties around Portland, a regional authority that was empowered to not only define the UGB line but to create plans for land use and transport, as well as running of new transit lines that will eventually connect the various dispersed settlements in the region. Democratic legitimacy was confirmed not only by a local vote to approve the creation of a regional government, but by elections for its 10 member council.

Democratic legitimacy is also a hallmark of the Stuttgart regional authority (VRS), which has an elected council of 91 members, although many councillors are also members of other governments, which spreads their contact network widely. Originally created as one of the state of Baden-Wüttenburg's regional planning agencies in 1992, it was given additional powers as a corporate entity with its elected council 2 years later. In large part this was because of a severe loss of jobs in the auto and engineering economy in the Stuttgart region, although manufacturing still accounts for almost half a million jobs in the region. VRS is still the only comprehensive metropolitan government in Germany, with a major role in regional planning. Basten (2011, p. 284) quotes its development plan as ensuring "sustainability and socially just ecological living and economically effective development that ensures sufficient housing opportunities and the protection and creation of jobs". In addition to its general regional planning it has specific responsibility for landscape and transport planning and the development of a regional park and waste management system. It also runs the regional transport system and the regional park, while there is a separate regional economic development agency (VBS) that has additional offices in Brussels and the United States. The VRS is a partner with this agency, as well as being one of the members of an association of 30 municipalities that promotes tourism in the wider region. The regional authority has also been involved with the development of a major trade fair, various cultural and sporting events and the redevelopment of the main Stuttgart rail station and its extensive inner city rail yards. It has helped establish or encourage many regional networks, formal and informal, with various groups, from business to charity groups. So the partnerships forged with other bodies have given the VRS a great deal of influence over many other regional functions in the last 2 decades, from economic development to tourism.

One of the most comprehensive, yet complex new city-regional authorities is the Greater Manchester Combined Authority (GMCA) which came into existence in April 2011, with a new constitution in June 2013. This co-operative regional authority was not a result of a completely new initiative. It had also been preceded by several other regional governances. One was the Greater Manchester County Council from 1974 to 1986, which was the top level of a two tier local government structure, responsible primarily for strategic planning and transport, with most functions still carried out by the lower tier local governments. Later the Association of Greater Manchester Authorities was created to advise on strategic affairs and to manage some common delegated services, such as county records, ecology and geology units, for the 10 unitary authorities. These offices were deliberately scattered throughout the region to deliver examples of the tangible benefits of collaboration to each area. A 2004 national proposal to create a real regional structure for this area, plus five other councils to the south, failed to get approval. But in 2009 the 10 local authorities of Greater Manchester agreed to support a new statutory city-regional structure, stimulated by a recognition of the need for more effective regional planning for common problems in an economic downtown and the increasing acceptance of the idea that city-regions were the engines of economic growth. The hope was that Manchester could act as a counterweight to the increasing power of London and the Southeast, perhaps hoping to recreate the way in which it had become a dominant city in the Industrial Revolution.

The current council is made up of a member from each of its 10 co-operating unitary local authorities and by 2014 is one of only two new city-regional statutory structures piloted in the U.K. The GMCA has strategic powers over such functions as transport, planning, skills development, housing, regeneration, and waste management. In most cases the council operates on a majority system, but for several functions, such as budgets and finance, it needs approval by a super-majority, or seven members. For functions such as the running of public transport, the council has established a separate committee—the Transport for Greater Manchester Committee (TfGMC)—where the 33 councillors elected from the 10 local councils on a proportional population basis, administer the running of the system and make policy recommendations to the main council. The development of the GMCA has been helped by national legislative acts in 2011 and 2012 that allowed increased functions for some cities. These new initiatives meant that the GMCA no longer gets investment from the national government only a project-by-project basis; instead it gets a block grant, enabling it to choose how to spend the money, with an incentive to get a proportion back if the investment can be seen to lead to employment growth. In addition, money for housing investment, and for low carbon initiatives, have been made available by the British government for the region to administer, while finance to increase the number of apprentices and skill development is administered by the region, as are local business hubs, which are designed to allow the region to promote its own economy through advice to firms and grants based on local needs. So the GMCA has absorbed new powers, by administering functions that are usually controlled by the national state. The authority also has a series of commissions responsible for making recommendations and reports for the main council, such as the New Economy, Planning and Housing, Health, Public Protection, Investment and Efficiency. Members of these commissions are not just drawn from local councillors, but also from businesses, voluntary organizations and other groups. Hence these commissions are designed to open the structure of governance to a variety of influences and advice. However, the GMCA does not control all regional services in the area; the new authority has no powers over several other agencies that operate in the region, such as Police or Fire and Rescue. However, these so-called Partner Agencies have access to all the reports of the various councils and committees and may attend meetings as observers, a process designed to create an open regional partnership of regional bodies. In addition the main co-ordinating council has an associate membership whereby representatives from neighbouring councils can attend when matters of general interest are discussed, although they cannot vote. The result is a complex and layered structure of governance at the city-region level, not a simple one-unit structure controlling all functions, or an advisory regional body.

These four examples show the range of variation that exists in city-regional governances. What is common is an emphasis upon process and change, rather than rigid structures, often with different levels of empowerment, and attempts to create more open, collaborative approaches, instead of closed systems of government, although the frequent lack of mandatory powers usually creates problems in effectiveness. It is worth noticing that in most cases there was a history of collaborative effort between government units in these areas before the current city-region structures were established, which helped to create at least a partial metropolitan or city-region consciousness that provides support for regional actions over parochialism. However, these examples are still relatively unique cases, for there are also many failures in attempts to create city-regional governance, especially effective ones. Janssen-Jansen (2011) has described how the city of Amsterdam in 1995 voted down a plan by the national government to create a new province anchored by the city, one of four previous attempts to form regional agencies within the general area, although with different numbers of units. The creation of the Amsterdam

Metropolitan Area in 2007 may be the start of a new regional phase, although this is still an informal body. It is a bottom-up initiative designed to produce and co-ordinate plans for improving accessibility, economic re-structuring, sustainability and landscape integration for the co-operating municipalities, but still has no official status. This led Janssen-Jansen (2011) to suggest it is an encouraging sign of building city-region consciousness, but it is in an area that is still without the will for an effective metro governance to emerge. Yet even when these regional government structures are created there is no guarantee that they will survive. Their potential fragility can be seen by the experience of Melbourne where a regional co-operative structure similar to the Vancouver model was created in 2002 but was abandoned in 2010. Kroen and Goodman (2012) have argued that it failed because of several problems. The first was that there was not a strong regional consciousness in the area which never developed during the regional government's existence, mainly because the lower tier local authorities became convinced they were being ignored in the decision-making. In addition, the necessary bi-partisan support among major political parties was absent, so that the structure was easily dismantled when a new state party opposed to regional governments was voted into office. Third, there were few clear guidelines over regulation. This led to the opinion that it failed

> because of a fundamental reluctance on behalf of government to act decisively and to require, rather than recommend, its policies… (further noting that)… the expectation that market behaviour affecting the shape of the city could be altered simply on recommendation without the use of regulation would appear very naïve (Kroen and Goodman 2012, p. 319).

Hence, it was concluded that more mandatory powers are usually needed to carry out key regional initiatives, incidentally still one of the ever-present difficulties of most regional governances. Finally, the strategy did not have a clear purpose and vision and lacked the support of many in government, especially the senior civil servants and the members of key financial ministries.

These failures in Melbourne and Amsterdam provide a vivid example of the problems faced by city-regional governments in general. Indeed, many have argued that most of the current examples produce unstable and often uncoordinated structures that are fragile since they depend on co-operation and usually consensus in decision-making among the partners, which is often difficult to maintain. In a wider context Lovering (1999) and Markusen (1999) have made additional criticisms, such as the fact that a focus upon the delivery of services or functions has meant that the concern for removing inequalities that lay behind many older regional approaches, especially the economic schemes, has been forgotten. Although the critique is based more upon the larger regional entities, not city-regional planning, the critique is still relevant, for the addition of a new transport system, regional shopping centre, or new plant, is usually lauded for its economic impact, not on whether they reduce social and economic inequalities within and between regions. Rossi's (2004) critique about recent changes in the Mezzogiorno in southern Italy, provides a good example of this problem. New endogenous growth in Italy in the last few decades, based on small and middle level enterprises, especially linked to exporting fashion goods, has provided a source of local dynamism in a region known for its low growth and limited modern development. But Rossi questioned whether this export-led development should be seen as the only, or even the best, winning strategy for the region. His fieldwork revealed that many of these enterprises were pre-Fordist in their operations. In other words, the enterprises were often based in poor building structures with unhealthy working conditions, characterized by low wages, irregular work, and few rights for workers. Such conditions meant that this employment growth was exploitive, paralleling the problems found in the growth of low wage textile production in many developing countries. Moreover, Rossi showed that the gap between the northern and southern regions of Italy in 2002 was still 20 % in productivity, while the employment rate in the active population was over four times higher in the north. So focusing upon the employment gains had ignored the conditions of these seemingly dynamic workplaces of the south, producing a spatial injustice in their freedoms, let alone the fundamental issue of regional inequality. Although this example deals primarily with regionalism at a scale beyond the city-region, the fact that these exploitive activities occur mainly within the cities of the Mezzogiorno region show that it is also an urban issue, one of social injustice that is the subject of Chap. 3 (Just Cities). Such problems are curiously missing from the first section in the New Urban Charter, except for some brief comments about the need for affordable housing and sharing revenues and resources among municipalities.

2.9 Conclusions

From a design perspective there is a lot to admire in the various parts of the NU movement. Most agree that the new communities built on these principles—especially the 'town' centres of these developments—provide more attractive and focused places than the typical, faceless suburbs with chaotic and heterogeneous strip malls that developed from the 1950s (Baldassare 1986). However, in practice relatively few of these NU developments have been created, so they are massively outweighed by areas built in the last quarter of a century using modifications of older suburban designs, while apart from some New Regionalism initiatives there are few signs of really effective regional planning. Since some new suburban and inner city developments have adopted a few of the NU principles, this may hold out some hope for greater use of the ideas in future urban developments. Certainly the charter gave a fillip to the derivation of Smart Growth and Transit Orientated Development, and through its principle of preservation acknowledged the need to conserve historical heritages. Yet despite the title of 'new' it can be argued (Fulton 1996) that many of the ideas in the NU movements are really revisions of older concepts—although rarely acknowledged—ranging from those seen in nineteenth century model towns (Mumford 1961), Garden Cities (Howard 1898), Perry's (1929) Neighbourhood Unit Planning (NUP) ideas and subsequent revisions in Radburn and post-war British New Towns, which is perhaps why the term 'Neo-Traditional Planning' becomes especially appropriate. However the big difference in NU is that it provides a more explicit identification of design codes, while the term 'new' is really a contrast to mid and late twentieth suburbs. Few of the NU schemes that have been built incorporate policies that relate to sustainability ideas or to the wider issues of metropolitan planning and ecological issues identified in the first part of the New Urbanism Charter, in part because of the political fragmentation of urban regions which in the absence of any new regional governance still restricts most planning to within the various political boundaries. Yet it could be argued that the NU Charter does provide an explicit framework for improving the quality and liveability of urban places, beyond the auto-suburb, and provided the basis for the development of more detailed and focused ideas such as Smart Growth.

Despite their attractiveness, this review has shown that there is little evidence that those areas developed under NU designs have really increased either community cohesion, sense of place, social mixing, or provided real public spaces for all residents of the city, or even reduced commuting and increased walkability. What seems to have been downplayed is that it is the three fold character of the people, their behaviours, and attitudes expressed in many different affective dimensions that create the distinctiveness of communities and places (Davies 1992; Davies and Herbert 1993, p. 36). They imbue them with meanings that are often contested. So design does not necessarily create meanings; to assume so is to imply a moral determinism from some physical features. Perhaps the exception in the creation of major changes in cities from NU ideas, at least in the aim of reducing auto use, comes from the TOD examples that are based on mass transit lines. But even the TOD examples are not really new. They are planned versions of the type of development, admittedly at a higher density, of the nodes that evolved around the stations on the extensions of mass transit lines in large cities in the early twentieth century (Hall 1988). The real originality among the NU types may be the SG ideas, although few have the large amounts of employment in mixed developments that would be needed to reduce commuting. Yet the initial logic behind SG schemes tends to ignore the limits on their creation by the zoning or development control principles in most cities of the developed world, which segregate residential uses from employment-creating uses. Such rigid exclusions may still be sensible in the context of industrial plants that may be dealing with hazardous materials or processes, or even with warehouses with large amounts of traffic. They may be less relevant in the case of most sorts of offices—again unless they generate lots of vehicular traffic—which could provide the type of high employment numbers to reduce commuting that the principle of 'mixed uses' implies. In addition, it is hard not to criticize most NU developments for still being too small in size to make them the type of self-contained entities that NU advocates seem to want them to be, given their use of the term 'urban' in the NU description. Most are still not much bigger than a traditional post World War II suburban size, which are too small to satisfy the threshold requirements of most retail stores, which means there is bound to be a lot of travel outside the areas for commercial needs, let alone leisure and employment. So they may simply be examples of what Scully (1994) aptly called a 'new suburbanism', for they are rarely anything like towns.

The full flowering of many of the NU principles would seem to imply some master plan whereby these are developed as individual pods in some larger development plan designed for the whole city and region, rather than these pods being viewed on their own terms. But individual developers are unlikely to have the resources to construct at a city scale. This means that the limited results of Howard's Garden City ideas may be repeated again, since his objective was to create a self-sufficient town, which he regarded as was one unit in a larger metropolitan region. He also envisaged these as being produced by social co-operation of residents not by private or state interests, perhaps an example of *real* community effort, a type of principle that is not found in NU. Unfortunately, as the discussion of city-region governance has shown, there are still few examples of effective administrations at this scale (Janssen-Jansen and Hutton 2011a, b). What is clear from the case studies of city-regional governance is that the entrenched oppositions to such regional schemes are still considerable, despite the theoretical support for the ideas. Moreover these administrative structures take many forms, contingent upon the circumstances of the region. However, where there has been a successful implementation of such schemes this does seem to have been helped by a previous history of regional understanding and co-operative effort. It also helps to have some mandatory powers to ensure compliance to regional objectives, as well as a democratic legitimacy. Perhaps a big advantage of these regional authorities has been their ability to build partnerships with various agencies, organizations and levels of government, as well as providing tangible results in all parts of their region, so that the various actors and general population in the area can see the benefits of this regional approach. In this way governance becomes wider in scope and open to more ideas. Although some progress is being made in the development of city-regional governance, relatively few areas have such structures and many still have handicaps in their operation. But without such organizations it is difficult to see how many of the principles in the first part of the NU Charter can be fulfilled. Moreover since most of NU examples at the urban or neighbourhood scale focus on design, they pay relatively limited attention to many of the other conceptual developments of the last 20 years described in subsequent chapters that are also leading to new ways of creating better urban places, such as in urban greening, sustainability and resilience against natural disasters, greater safety, health, or other lifestyle and new employment ideas. So the search for any type of 'good' or rather better community must surely involve more than just NU ideas, useful though some of the concepts may be. What also seems important is a greater need to deal with such fundamental questions as those associated with justice in the city, issues that go beyond the provision of different types and presumably costs of housing, which is why it is the focus of the next chapter.

References

Aldous, T. (Ed.). (1992). *Urban Villages: A concept for creating mixed-use urban developments on a sustainable scale*. London: Urban Village Group.

Alexander, C. (1979). *The timeless way of building*. New York: Oxford University Press.
Alexander, C., Silverstein, M., & Ishikawa, S. (1977). *A pattern language: Towns, buildings, construction*. New York: Oxford University Press.
Al-Hindi, K. F. (2001). The new urbanism: Where and for whom? Investigation of an emergent paradigm. *Urban Geography, 22*(3), 202–219.
Anderson, K. (20 May 1991). Oldfangled new towns. *Time*, 52.
Audirac, I., & Shermyen, A. (1994). An evaluation of neo-traditional designs social prescription: Postmodern placebo or remedy for suburban malaise? *Journal of Planning Education and Research, 13,* 161–173.
Axhausen, K. W. (2000). Geographies of somewhere: A review of urban literature. *Urban Studies, 37*(10), 1849–1864.
Baldassare, M. (1986). *Trouble in paradise: The suburban transformation in America*. New York: Columbia University Press.
Banai, R. (1996). A theoretical assessment of the neo-traditional settlement form by dimensions of performance. *Environment and Planning B: Planning and Design, 23,* 177–190.
Banai, R. (1998). The new urbanism: An assessment of the core commercial areas, with perspectives from (retail) location and land-use theories, and the conventional wisdom. *Environment and Planning B: Planning and Design, 25*(2), 169–185.
Bartling, H. (2004). The magic kingdom syndrome: Trials and tribulations of life in Disney's Celebration. *Contemporary Justice Review, 7*(4), 375–393.
Basten, L. (2011). Stuttgart: A metropolitan city-region in the making? *International Planning Studies, 16*(3), 273–287.
Baxandall, R., & Ewen, E. (2000). *How the suburbs happened*. New York: Basic Books.
Bookout, L. (1992). Neo-traditional town planning: A new vision for the suburbs? *Urban Land*. January 1992, pp. 20–26.
Bressi, T. (1994). Planning the American dream. In P. Katz (Ed.), *The new urbanism: Toward an architecture of community* (pp. xxv–xlii). New York: McGraw-Hill
Brindley, T. (2003). The social dimension of the urban village: A comparison of models for sustainable urban development. *Urban Design International, 8,* 53–65.
Brooke, S. (1995). *Seaside*. Gretna, LO: Pelican.
Brown, B., & Cropper, V. (2001). New urban and standard subdivisions: Evaluating psychological and social goals. *Journal of the American Planning Association, 67*(4), 402–419.
Calthorpe, P. (1993). *The next American metropolis: Ecology, community and the American dream*. New York: Princeton Architectural Press.
Calthorpe, P. (1994). The region. In P. Katz (Ed.), *The new urbanism: Toward an architecture of community* (pp. xi–xx). New York: McGraw-Hill.
Calthorpe, P. (2001). *The regional city: Planning for the end of sprawl*. New York: New Island Press.
Canada Mortgage and Housing Corporation (CMHC). (n.d.). Residential intensification case studies, built projects: Garrison Woods. http://www.cmhc-schl.gc.ca/en/inpr/su/sucopl/upload/Garrison-Woods-Calgary-Alberta.pdf. Accessed 6 Dec 2014.
Carma. (n.d.). *A towne to call home: McKenzie Towne* (24 page sales brochure)
Congress for the New Urbanism. (1999). *Charter of the new urbanism: Region/neighbourhood, district, and corridor/Block, street, and building*. New York: McGraw-Hill Books.
Coroux, D., Keough, N., Miller, B. & Row, J. (2006). *Overcoming barriers to sustainable urban development: Toward smart growth in Calgary*. A Discussion Paper Prepared for the Calgary Citizens' Forum.
Cotugno, A., & Seltzer, E. (2011). Towards a metropolitan consciousness in the Portland, Oregon metropolitan area. *International Planning Studies, 16*(3), 289–304.
Crane, R. (1996). On form versus function: Will the new urbanism reduce traffic, or increase it? *Journal of Planning Education and Research, 15,* 117–126.
Davies, W. K. D. (1984). *Factorial ecology*. Aldershot: Gower.
Davies, W. K. D. (1992). Affective dimensions of community character. Unpublished paper read at IGU Commission on Urban Systems and Development, Detroit. Subsequently extended and

published in W. K. D. Davies (1995). The power of communities. *Acta Wasaensis (Finland), 45*(6), 49–74.
Davies, W. K. D., & Herbert, D. T. (1993). *Communities within cities: An urban social geography*. London/New York: Belhaven Press/Halstead-Wiley.
DeCoursey, W., & Athey, L. (2007). *Transit-oriented design: Illustration of TOD characteristics*. Institute for Public Administration, College of Human Services, Education & Public Policy University of Delaware.
Department of Environment Transportation and Regions (1999). *Millenium villages and sustainable communities*. London: DETR.
Dowling, R. (1998). Neo-traditionalism in the suburban landscape: cultural geographies of exclusion in Vancouver, Canada. *Urban Geography, 19*(2), 105–122
Duany, A., & Plater-Zyberk, E. (1992). The second coming of the American small town. *Wilson Quarterly,* 19–50.
Ellis, C. (2002). The new urbanism: Critiques and rebuttals. *Journal of Urban Design, 7*(3), 261–291.
Fader, S. (2000). Density by design. *Urban Land, 59*(7), 55–59.
Foglesong, R. E. (1986). *Planning the capitalist city: The colonial era to the 1920s*. New York: Princeton University Press.
Ford, L. R. (1999). Lynch revisited. New urbanism and theories of good city form. *Cities, 16*(4), 247–257.
Franklin, B., & Tait, M. (2002). Constructing an image: The urban village concept in the U.K. *Planning Theory, 1*(3), 250–272.
Frantz, D., & Collins, C. (1999). *Celebration, U.S.A.: Living in Disney's brave new world*. New York: Henry Holt.
Fulton, W. (1996). *The new urbanism: Hope or hype for American communities*. Cambridge: Lincoln Institute of Land Policy.
Garreau, J. (1991). *Edge city: Life on the new frontier*. New York: Anchor Books, Doubleday.
Geddes, P. (1915). *Cities in evolution: An introduction to the town planning movement and to the study of civics*. London: Williams.
Gehl, J. (2010). *Cities are for People*. Washington D.C.: Island Press.
Geller, A. L. (2003). Smart growth: A prescription for liveable cities. *American Journal of Public Health, 93,* 1410–1415.
Gibson, K., & Abbott, C. (2002). Portland, Oregon. *Cities, 19,* 425–436.
Grant, J. (2006). *Planning the good community: New urbanism in theory and practice*. London: Routledge.
Gyourko, J., & Rybczynski, W. (2000). Financing new urbanism projects: Obstacles and solutions. *Housing Policy Debate, 11*(3), 733–750.
Haas, T. (2008). *New urbanism and beyond*. New York: Rizzoli.
Hall, P. (1988). *Cities of tomorrow*. London: Blackwell.
Hall, D. (1998). Community in the new urbanism: Design vision and symbolic crusades. *Traditional Dwellings and Settlement Review, 9*(11), 23–36.
Hall, P. (2000). Urban renaissance/new urbanism: Two sides of the same coin? *Journal of the American Planning Association, 66*(4), 359–360.
Halligan, D. (2000). Maryland's smart growth and neighborhood conservation initiative. Institute of Transportation Engineers. *ITE Journal, 70,* 36.
Harvey, D. (1994). Flexible accumulation through urbanization: reflections on 'Post-Modernism' in the American City. In A. Amin (Ed.), *Post-Fordism: A Reader* (pp. 361–386). Oxford: Blackwell.
Harvey, D. (2005). *A brief history of neoliberalism*. New York: Oxford University Press.
Howard, E. (1898). *Garden cities of tomorrow* (Introduction by Lewis Mumford). Cambridge: The MIT Press, 1965. (Originally published as Tomorrow: A peaceful path to real reform, 1898).
H.R.H. The Prince of Wales. (1989). *A vision of Britain*. New York: Doubleday.
Hughes, S. (2000). *Copperopolis: Landscapes of the early industrial period in Swansea*. Royal Commission on Ancient Monuments in Wales, Cardiff.

Hutton, T. A. (2011). Re-thinking metropolis: From the 'liveable region' to the 'sustainable metropolis' in Vancouver. *International Planning Studies, 16*(3), 237–255.

Jacobs, J. (1961). *The death and life of great American cities*. New York: Random House.

Janssen-Jansen, L. B. (2011). From Amsterdam to Amsterdam metropolitan area: A paradigm shift. *International Planning Studies, 16*(3), 237–255.

Janssen-Jansen, L.B. & Hutton, T.A. (2011a). Re-thinking the metropolis. *International Planning Studies, 16(3), 201–215.*

Janssen-Jansen, L. B., & Hutton, T. A. (2011b). Reconfiguring the governance structures of the twenty-first century city-region: Observations and conclusions. *International Planning Studies, 16*(3), 305–312.

Jones, K., & Simmons, J. (1987). *Location, location, location: Analyzing the retail environment*. New York: Methuen.

Katz, P. (1994). *The new urbanism: Toward an architecture of community*. New York: McGraw-Hill.

Kelbaugh, D. (1989). *The pedestrian pocket book: A new urban design strategy*. New York: Princeton Architectural Press.

Khattak, A. J., & Rodriguez, D. (2005). Travel behaviour in neo-traditional neighbourhood developments: A case study in U.S.A. *Transportation Research Part A: Policy and Practice, 39*(6), 481–500.

Kim, J. A., & Kaplan, R. (2004). Physical and psychological factors in sense of community: New urbanist Kentlands and nearby Orchard Village. *Environment and Behaviour, 36*(3), 313–340.

Knapp, G., & Talen, E. (2005). New urbanism and smart growth: A few words from the academy. *International Regional Science Review, 28*(2), 107 – 118.

Knox, P. (1992). The packaged landscapes of post-suburban America. In P. J. Larkham & J. W. R. Whitehand (Eds.), *Urban Landscapes: International Perspectives* (9th ed., pp. 207–233). London: Routledge.

Krieger, A. (Ed.). (1991). *Andres Duany and Elizabeth Plater-Zyberk: Towns and town-making principles*. New York: Rizzoli.

Krier, L. (1991). Afterword. In A. Krieger (Ed.), *Andres Duany and Elizabeth Plater-Zyberk: Towns and town-making principles* (pp. 117–119). New York: Rizzoli.

Krier, L. (1998). *Architecture: Choice or fate*. Great Britain: Andraes Papadakis Publisher.

Kroen, A., & Goodman, R. (2012). Implementing metropolitan strategies: Lessons from Melbourne. *International Planning Studies, 17*(3), 303–321.

Kunstler, J. H. (1993). *The geography of nowhere: The rise and decline of America's man-made landscape*. New York: Simon & Schuster.

Langdon, P. (1994). *A better place to live: Reshaping the American suburbs*. Amherst: The University of Massachusetts Press

Lennertz, W. (1991). Town-making fundamentals. In A. Krieger (Ed.), *Towns and town-making principles* (pp. 21–24). New York: Rizzoli.

Lovering, J. (1999). Theory led by policy: The inadequacies of the new regionalism. *International Journal of Urban and Regional Research, 23*(2), 379 – 395.

Lund, H. (2003). Testing the claims of new urbanism: Local access, pedestrian travel, and neighbouring behaviour. *Journal of the American Planning Association, 69*(4), 414–429.

Markusen, A. (1999). Fuzzy concepts, scanty evidence, policy distance: The case for rigour and policy relevance in critical regional studies. *Regional Studies, 33*(9), 575–605.

McCann, E. J. (1995). Neo-traditional developments: The anatomy of a new urban form. *Urban Geography, 16,* 210–233.

McKenzie, E. (1994). *Privatopia: Homeowner associations and the rise of residential private government*. New Haven: Yale University Press.

Mohney, D., & Easterling, K. (Eds.) (1991). *Seaside: Making a town in America*. New York: Princeton Architectural Press.

Mumford, L. (1922). *The story of utopia*. New York: Boni and Liveright.

Mumford, L. (1938). *The culture of cities*. New York: Harcourt Brace.

Mumford, L. (1939). Regional planning in the north-west. *Portland City Club Bulletin, 18*(July), 1–20.

Mumford, L. (1961). *The city in history*. New York: Harcourt Brace Jovanovich.

Murrain, P. (1996). Congress for the new urbanism charter: Developing an agenda for action. *Urban Design International, 1*(2), 183–187.

Nasar, J. (2003). Does neotraditional development build community? *Journal of Planning Education and Research, 23*(1), 58–68.

Perry, C. (1929). The neighbourhood unit. Monograph 1, Neighborhood and Community Planning, *Regional Survey of New York and Environs* (Vol. VII). New York: Russell Sage Foundation.

Pogharian, S. (1996). Street design: Learning from suburbia. *Plan Canada, 36*(5), 41–42.

Putnam, R. (2001). *Bowling alone: The collapse and revival of American community*. New York: Simon & Schuster.

Rossi, U. (2004). New regionalism contested: Some remarks in light of the case of the Mezzogiorno of Italy. *International Journal of Urban and Regional Research, 28*(2), 466–476.

Sancton, A. (2001). Canadian cities and the new regionalism. *Journal of Urban Affairs, 23*(5), 543–555.

Schleimer, J. (1993). Buyers of homes in neo-traditional communities voice their opinions. *Land Development, Spring-Summer*, 4–6.

Scott, J. W. (2005). The new regionalism as a contingent governance paradigm: European and North American perspectives. In Y. Murayma & G. Du (Eds.), *Cities in global perspective* (pp. 452–461). Tokyo: College of Tourism, Rikkyo University and I.G.U. Urban Commission.

Scott, J. W. (2009). *Decoding the new regionalism*. Aldershot: Ashgate.

Scully, V. (1994). The Architecture of Community. In P. Katz (Ed.), *The New Urbanism: Toward an Architecture of community* (pp. 221–230). New York: McGraw-Hill.

SG: Smart Growth Network. (2013). http://www.smartgrowth.org. Accessed 15 Aug 2013.

Soderbaum, F., & Shaw, T. M. (2003). *Theories of new regionalism*. Basingstoke: Palgrave, Macmillan.

Song, Y. (2005). Smart growth and urban development pattern: A comparative study. *International Regional Science Review, 28*, 239–265.

Southworth, M. (1997). Walkable suburbs: an evaluation of neo-traditional communities at the urban edge. *Annals American Planning Association, 63*(1), 28–44.

Southworth, M. (2005). Designing the walkable city. *Journal of Urban Planning Development, 131*, 246–257.

Steiner, R. (1998). Traditional shopping centres. *Access, 12*, 8–13.

Stephenson, R. B. (1999). A vision of green: Lewis Mumford's legacy in Portland, Oregon. *Journal of the American Planning Association, 65*(3), 259–269.

Suttles, G. (1972). *The social construction of communities*. Chicago: University of Chicago Press.

Talen, E. (1999). Sense of community and neighbourhood form: An assessment of the social doctrine of new urbanism. *Urban Studies, 36*(8), 1361–1379

Talen, E. (2002). The social goals of new urbanism. *Housing Policy Debate, 13*(1), 165–188.

Telo, M. (Ed.). (2007). *European Union and new regionalism*. Aldershot: Ashgate.

Thompson-Fawcett, M. (1996). The urbanist revision of development. *Urban Design International, 1*(4), 301–322.

Till, K. (1993). Neo-traditional towns and urban villages: The cultural production of a geography of otherness. *Environment and Planning D: Society and Space, 11*, 709–732.

Townshend, I. J. (2002). Monitoring community dimensions: City-wide characteristics and differentiation by social region. In W. K. D. Davies & I. J. Townshend (Eds.), *Monitoring cities: International perspectives* (pp. 435–459). Calgary: International Geographical Union and Urban Commission.

Townshend, I. J., & Davies, W. K. D. (1999). Identifying the elements of community character: A case study of community dimensionality in old age residential areas. *Research in Community Sociology, 9*, 219–251.

Tregoning, H., Agyeman, J., & Shenot, C. (2002). Sprawl, smart growth and sustainability. *Local Environment, 7*, 341–347.

Unwin, R. (1969). *Town planning in practice*. New York: Ayer Company Publications (Reprint of 1934 edition).
Van der Ryn, S., & Calthorpe, P. (1986). *Sustainable communities: A new design synthesis for cities, suburbs, and towns*. San Francisco: Sierra Books Clubs.
Webster, C. (2002). Property rights and the public realm: Gates, green belts, and gemeinschaft. *Environment and Planning B: Planning and Design, 29*(3), 397–412.
Wheeler, S. (2002). The new regionalism: Key characteristics of an emerging movement. *Journal of the American Planning Association, 68*(3), 267–278.
Ye, L., Mandpe, S., & Meyer, P. B. (2005). What is "smart growth?"-Really? *Journal of Planning Literature, 19,* 301–315.

Chapter 3
Progress Towards Just Cities

Wayne K.D. Davies

If the discourse surrounding policy-making focuses on the justice of the decision, rather than simply its contribution to competitiveness, much will have been accomplished.
Fainstein 2010, p. 184

3.1 Introduction

Most people in the contemporary world see towns and cities as centres of opportunity, wealth, socialization, tolerance and culture. However, far too many others experience the same urban centres as places of low income, exploitation, insecurity and exclusion with both limited opportunities and influence on decision-making. These features create an impoverished life-style that is a depressing and often multigenerational fact of life for increasing numbers. Yet this dualism in urban places is not new; it has been a feature of urban life since cities were first created. Since the time of the Golden Age of Greek city-states, many theorists, governments and compassionate people have sought ways of resolving the problems of the urban disadvantaged, but with only limited success. Today these problems are increasingly described as injustices in cities. This is a *redistributive* justice, not to be confused with the more usual conception, which is a *retributive* justice,which addresses the way that those who transgress against the laws or customs of a society are judged and punished. By contrast, redistributive justice relates to the problems of an unequal possession of goods, in which the term 'goods' applies not simply to material goods, but also to such issues as influence, power and status. The questions associated with justice in a redistributive sense—whether it is needed, what goods, and in what quantities should be provided—are issues that have not been easy to solve, despite many attempts throughout history. Perhaps the most substantial progress made in dealing with this type of injustice in the history of western cities was initiated by the ideas of the European Enlightenment and the material growth that started with the

W. K.D. Davies (✉)
Department of Geography, University of Calgary, 2500 University Drive, N.W., Calgary, AB T2N 1N4, Canada
e-mail: wdavies@ucalgary.ca

W. K.D. Davies (ed.), *Theme Cities: Solutions for Urban Problems*,
GeoJournal Library 112, DOI 10.1007/978-94-017-9655-2_3

Industrial Revolution. Despite the negative effects at first, the latter subsequently led to huge progress in reducing the inequalities in society and in cities, while new ideas about the rights of man led to the eventual creation of democracy in the western world.

For much of the twentieth century the trend in inequality reduction continued in developed countries, often under the label of progressive policies. But by the late 1960s it was realised that many individuals and groups had not shared in this progress and, sadly, were unlikely to do so. Indeed the condition of the most disadvantaged deteriorated in the next three decades, in part because of the results of deindustrialisation, globalisation and technological change, which has led to increasing, not decreasing levels of income inequality, a situation exacerbated by neo-liberal policies favouring the market. The resultant growth in inequalities is seen in many developed countries but is increasingly stark in Britain (Mount 2012), a country that was a pioneer in so much progressive legislation. By the fourth quarter of the twentieth century the post-World War II trend of rapidly decreasing inequality in Britain ended, as shown by several studies (PG 2009; IFS 2011; Portes and Riley 2011). The Gini coefficient based on disposable household income in Britain was 0.26 in 1979, but under the new Conservative governments reached 0.35 a decade later. Despite a small reduction during the 13 years of the Labour government due to new tax and benefits policies which prevented major increases it resumed this high level, reaching 0.39 by 2009 which was the highest for 30 years. Focusing upon the amount of wealth obtained by the lowest and highest income groups provides an even starker picture of inequality. In 2009 the poorest tenth of the population had only 1.3 % of the nation's income and the second poorest only 4 %. In contrast the richest tenth have 31 % and the next richest tenth 15 % of the national income. In a more personal sense for employees it has been calculated that the average pay ratio of a chief executive officer to the average employee in Britain has risen from 47 to 128 in the last decade (Mount 2005). Recently the French economist Thomas Piketty (2014) in his book *Capital in the Twenty First Century* has tried to explain this inequality by arguing that it is a consequence of the fact that wealth rises faster than economic output, a view based on his detailed analyses of returns to owners of wealth, wage earners and economic growth. Like Marx, Piketty sees the concentration of wealth as an inevitable consequence of capitalism, except in conditions of either great technological progress, or when governments intervene, which has happened since the post World War II period, but has now been reduced by neo-liberal policies. This structural inequality in global capitalism led Piketty (2014) to the view that there should be a global tax on wealth, a recommendation that has sparked controversy, although this view is mild when set against the belief of Marx and Engels (1845) that only a transformational change in society can alter the inequality.

Fortunately, the persistent problems of the disadvantaged in this trend of greater inequality have not gone unnoticed. A series of pioneering books have gone beyond the specifically poverty issues and stressed the need for greater social justice, often focusing on the role of societal influences in maintaining the difficulties of the disadvantaged (Harvey 1973, 1989; Lefebvre 1968; Claval 1978; Wilson 1997 and

Berry 2005). Instead of arguing for efforts to alter the situation at the national scale, as seen in most welfare policies, these authors, and especially proponents of the emerging Just City approach (Fainstein 2010; Marcuse et al. 2009) and Harvey's (1973, 2013) more radical ideas, have stressed that social justice should be delivered at the urban scale. This focus seems obvious. Urban places contain the majority of people in the developed world and are the centres of consumption, exchange and production, while their governments control the development decisions that affect the location and renewal of buildings, land use policy and the delivery of many services. But in addition, pressure for change in society has come from the increasing numbers of protest movements, which are also primarily situated in urban places. These protests, urban social justice books, and the many articles using the term, Just City, are not simply descriptive of the problems of the disadvantaged—issues that have already spawned a vast literature—or of the many redistributive efforts at the national state scale. Rather, the emerging Just City advocates wish to create a broad normative framework of ideas on how this concept can be developed, not only to address the problems of the poorest people, but also to relate more broadly to those with unequal access, influence and opportunity in an urban context. In addition the idea needs to be extended to those with different lifestyles, whose hope for the ability to live openly has usually been squashed by ill-liberal, repressive laws. The term Emancipatory City was used by Lees (2004) as the title for a set of pioneering essays to explore the way in which cities and the attitudes and life-styles within them, can be centres of release from societal restrictions, something hoped for over centuries. Although this goal can be linked to problems of economic disadvantage, it also applies to the degree to which people with diverse orientations, not only gays and lesbians, but many other groups, can live freely, without prejudice and are able to find new opportunities in urban places (Roszak 1969). Hence issues such as a tolerance for diversity, not just reducing inequalities, should also be seen as part of the progress towards more just cities. However one of the problems of focusing on the urban scale is that the powers of most towns and cities are actually quite limited; it is the state that still exercises the major controls in creating laws and conditions in society that discriminate against many people.

A key element in this new approach to redistributive justice is a return to the use of values as a rationale for reducing the perceived injustices, especially the conception of 'fairness' in redistributive justice, as a fundamental principle justifying intervention. This was promoted by the philosopher Rawls (1971) in his path-breaking work *Theory of Justice*, which lies behind the quotation from Fainstein used at the beginning of this essay. It argues for using justice, not simply competitiveness—and one could add maximization of profit or returns from development—as a basis for urban policy making. This chapter summarizes the background to the development of the Just City approach, as well as its emerging elements and problems. Since the idea of redistributive justice at an urban scale has a heritage as far back as the classical Greek city-states, it does seem appropriate to set the discussion within a broad historical context to provide a background to the current debates.

3.2 Antecedents of Urban Redistributive Justice

The first known attempts in the development of western thought to describe and justify the principles upon which a city and its politics should operate were created four centuries before the Christian era in Greece, primarily in Plato's *The Republic* (1963) and subsequent works by Aristotle (Barker 1969; Hardie 1980; Crisp 2000). These early works were designed to create a more harmonious, sociable and good society in the Greek city states. These principles have been built upon, criticized and extended by innumerable subsequent scholars, providing the multifaceted themes of western political theory and governance. One of these themes deals with the delivery of justice, but since the terms 'just' and 'justice' have meant different things to various peoples through time, no universal, trans-cultural meaning has been accepted as underpinning the idea of 'just', especially in a redistributive sense. For example, many early societies accepted that justice came from god-given commands, interpreted by rulers to create a harmonious society. In Plato's discussions of the best way to organize a city-state, Thrasymachus argued that justice should be seen as being whatever is in the ruler's interest, which became a way of protecting the existing interests of those in power. However, the final recommended approach in 'The Republic' was for a rule of wise philosopher-kings. They would dispense justice evenly throughout the three class system in his ideal state, since only they had the wisdom to understand the long term interests of the state, whereas democratic rule was described as leading to the corrupt and short-term interests of the majority. Other Greek scholars thought of justice in a different way. For example, followers of the sixth century B.C scholar Pythagoras primarily viewed justice in a retributive sense. So what a person has done to another should be done to him or her, a version of the biblical 'an eye for an eye and a tooth for a tooth'. However, Aristotle (Crisp 2000) in noting that there are many forms of justice, made an important clarification in his distinction between what is usually translated as 'general or universal justice' and 'particular justice'. But one must be aware of the problem of translation, for Barker (1969, p. 362) argued that the Greek word for justice meant something more than legal concepts, for it included ethnical ideas of our word 'righteousness'. Aristotle's general justice related to goodness, especially to one's neighbour and to what is lawful in a state, including what punishments should be given if transgressions occurred. This is what is called a retributive justice today, one administered through courts. Particular justice dealt with what is fair and equal, describing many situations of injustice, such as when a man who takes more than his share should be seen as grasping and therefore unjust. This has been interpreted as a kind of redistributive justice, one designed to achieve some equilibrium in a democratic city state, although another form was related to rectificatory justice, which the "state seeks to maintain….between one member and other" (Barker 1969, 363) Aristotle also argued that there should be a roughly equal division in the allocation of goods and honours in a city-state among the citizens—although it must be remembered that these were property-owning males, and did not include those who were seriously disadvantaged, namely women and slaves. Without such equal division, Aristotle argued that some citizens will assume they are being unfairly treated, leading many to feel cheated, which could increase antagonisms, and in extreme cases could lead to civil war within the city state. So fairness in distribution, one part of Aristotle's 'particular justice', was regarded as a necessity to prevent societal conflict in a city-state. However, we must be careful of interpreting this in contemporary terms. There seems little doubt that Aristotle was referring primarily to commercial and labour exchange, and honour transactions at his time. Indeed, Hardie (1980, p. 191) has convincingly dismissed interpretations that Aristotle's redistribution principle was seen at the time as being related to the use of the common funds of the state to reduce wealth inequalities, similar to modern ideas of redistribution on welfare grounds. Certainly, during the classical Greek era there were occasional examples of the distribution of foreign grain, as well as the allocation of lands in new colonies, to the underprivileged citizens of the city-states. But it is difficult to argue that these were ever related to Aristotle's principle. Moreover, Ravallion's (2013) review of historical attitudes to poverty has shown that throughout history most people believed that redistribution was inappropriate, since poverty was assumed to be a person's own fault. This view was buttressed by the mercantilist views of the sixteenth to eighteenth centuries which argued that poverty was socially useful since it guaranteed supplies of cheap labour to owners of land and factories. In Britain the various Poor Laws were only implemented to provide minimum relief for the poorest and to reduce the shock of unusual harvest failure or other major events, not to raise their condition. Despite Adam Smith's mild suggestion that redistributive taxation was appropriate to relieve poverty, few took up the case for redistribution; indeed many in the growing field of economics argued that since the rich saved and invested the most, any reductions in their wealth would reduce aggregate growth, which was assumed to be the motor for growth. It was the increasing humanitarian views of the Enlightenment combined with public realization of the squalor of the impoverished in cities, revealed in the novels of Dickens (Tomalin 2013), and the late nineteenth century cartographic and statistical works of Charles Booth (1902) and Seebohm Rowntree's study of poverty (1901), that led to changes of opinion about the condition of the poor and the need for redistributive measures through state action.

The social and economic arguments for a redistributive form of justice must be complemented by the changing religious and political beliefs that increasingly supported such policies. The foundational belief of many religions was based on the idea that there is a moral imperative to share goods, to distribute alms to those in need, and to treat neighbours fairly. Unfortunately the very hierarchical nature of most urban societies in the past marginalised the application of these ideas in most cases, except within some religious orders. Also in political terms the innate reciprocity of relationships in most hunting and gathering groups was lost and the emergence of status levels created different amounts of possessions and power in classes. So an inequality in the distribution of goods and power was perpetuated through time by the strength, traditions and religious beliefs of the powerful elite that were increasingly based in cities. Certainly, many of these societies provided relief for the poor and the weak, especially through religious organizations, but the overall belief in the right of the elite to control both society and goods was rarely questioned, at least not effectively. Occasionally, the elite in such societies were overthrown by internal revolt or external conquest, but some form of the hierarchical pattern, if not the same families, remerged and repeated itself through time, with a belief that such inequalities of possessions and power were some sort of natural order.

The growth of Enlightenment ideas in Europe from the late eighteenth century began the process of dissolving the assumptions of the inevitability of these status differences as some sort of inevitable, foundational nature of society, which led to the emergence of new political and human rights ideas. Older state and religious beliefs were increasingly questioned and the focus of debate returned to discovering and justifying principles with which to order society, principles that were based on human, rational thought. Increasingly, people began to believe that status differences in society were man-made, not ordained by God, or by previous traditions. So relationships between people began to be regarded as social contracts based on our common humanity, in which each individual had inalienable rights. Yet as Shklar (1986, p. 25) has reminded us:

> rights have never been demands only for more shares of whatever pie was available, nor are they hostile to social conscience. Rights are demanded first and foremost out of fear of cruelty and injury from agents of governments, but also from private magnates.......rights are asserted against power abused.

The rights that were sought initially were workplace and political rights, which led to emerging democracies, although it took time to include all adult males, and even longer to apply to women. Only relatively recently have these human rights become accepted as a basis for the redistribution of goods and opportunity, especially at a state and municipal level, to improve the life of the impoverished majority in order to reduce inequality and increase opportunity. Fraser (2003) has suggested that the various movements designed to achieve these objectives can be grouped into two major alternatives: 'transformational and affirmative' approaches to redistribution. The first requires a complete transformation of societal relationships to achieve its goals. The later focuses on creating small changes that produce enough alterations through time to resolve the problems of the disadvantaged. This leads to an incremental progress, such as providing greater democracy and economic equality in western societies, in order to create a redistribution of political influence, opportunity and wealth.

One of the most fundamental and influential examples of *transformationalist* policies are seen in the communist ideas of Marx and Engels (1848) who developed a materialist critique of capitalist society. They argued that it would eventually collapse under its own contradictions, leading to a societal struggle which would phase in the dictatorship of the proletariat, who would take over the means of production and eradicate private enterprise. This change would completely alter the distribution of work and rewards through the creation of a new society, one that would lead to a distribution of goods under the principle of 'each according to his need', an egalitarian aim which would obviate the need for redistributive justice, since all would be equal. There is little doubt that Marxist views provided a penetrating view of the societal imperatives that lay behind the hidden workings of capitalism and raised the hopes of many that it would create a better and more just future for the downtrodden majority (Marx 1975; Carver 1991). However its materialist assumptions that the unequal power relationships are only due to class is flawed, for differences also come from many other factors, such as gender and race. More specifically, attempts to put Marxist ideas and the Leninist revisions into practice after the 1917 revolution in Russia, and its imposition on eastern Europe and other countries after World War II, were unsuccessful and brutal, despite their early material successes. Despite the leveling of class differences, dictatorships with coercive regimes emerged in these states resulting in millions of deaths from deliberate or misguided policies. They also led to the loss of political freedom, personal property, a limited production of goods—except in a few specialised cases—all of which contributed to the collapse of communist rule in most countries by the late 1980s. Although China remains communist in principle it has managed to adapt to the new industrial realities by becoming a major production centre, essentially via state capitalism, and the 'factory of the world' for many products. However it still maintains a centralised political regime in an increasingly wealthy, but still tightly controlled society, with limited personal rights and with rapidly increasing inequalities. Modern Marxists still argue all that these examples were distorted versions of communist principles in the various attempts to create a new society based on a redistributive principle of equality for all. However, knowledge of the experience of these societal experiments, especially the absence of personal and political freedom and the coercion of those who disagreed, led to the collapse of support in most developed countries for such transformative approaches to create a more just society.

By contrast, most western states adopted what amounted to incremental or *affirmative policies* from the mid-nineteenth century. These involved a series of new programmes, often called progressive policies, since they aimed to improve the conditions of life and work of the population. Certainly many similar policies were also used in communist regime, but often inefficiently because of the dead hand of government provision of all economic activity and with no incentives for working hard. These redistributive programmes in most western societies were implemented without losing either democratic freedoms or the basic capitalist system, with the state becoming a provider, or an implementer, of the provision of some goods, although the proportions vary drastically between states. At a national level increasing growth and prosperity from the late nineteenth century enabled governments to finance many progressive social programmes, ranging from universal education, pensions, various poverty and unemployment benefits, increased police protection, and in some countries, social housing, health care, or at least health care subsidies, for the old and poor. In addition to these state actions, municipalities continued to create safe water supplies and sewage disposal, enacted rigorous building standards and land use policies that banned noxious factories from residential areas, as well as providing a range of utilities and amenities such as parks, libraries etc. The size and scope of these measures varied in different nations, while the democratic rights that emerged took centuries, and often revolutions, to achieve. It was not until after World War II that the measures had enough impact to create a major redistributive impact, through what is now known as a welfare system. One of the blockages to the development of this system came from a lingering attachment to the idea, especially among conservatives, that since the wealthy saved and invested more, it was aggregative savings that led to growth. However Keynes's (1936) argument that it was aggregative consumption, not savings, that was the key to growth, gained political support. This helped to justify income redistribution measures, which stimulated economies through the resultant spending of this extra state-provided income. Yet in the developing world progress in redistributive policies was slow. Ravallion (2013) has argued that by the 1990s, the development of new economic models revealed that high levels of poverty reduced subsequent growth in developing countries, since it led to limited spending and a waste of potential labour who could not find jobs. This evidence, along with the growth of redistributionist ideas, has helped justify the development of many more poverty alleviation measures in developing countries, such as Brazil's Bolsa Familia policy of conditional cash transfers to poor families, so long as their children go to school and are vaccinated.

The political freedom of liberties and the right to participate in the decision-making that created representative democracies did not lead to equality in economic terms, or in decision-making power beyond the initial vote. Vestiges of the old hierarchical orders remained because of their control over capital and land. Moreover, the spread of the capitalist system, especially with the industrial and commercial revolutions of the late twentieth century, created even more inequalities, since the new relationships of production were dominated by the owners of capital and the new technologies, adding to the numbers and dominance of a relatively small and wealthy elite. Despite the advances in social justice provided by the essentially progressive measures in western democratic countries, the 1960s saw the beginning of a new questioning of the results of the policies. The civil rights and feminist movements in particular showed that people of colour and women had not shared in many of the advances, which eventually led to new legislation to improve their prospects. Young adults led counter-culture protests (Koszak 1969) about the dominance of a consumer society, with its over-emphasis on work and the unequal structure of rewards in organizations. Dissatisfaction was also expressed about the top-down nature of so many of the public planning decisions, traditionally based on utilitarian grounds in which the policies were based primarily on the maximization of benefits. This approach favoured developments that created the highest subsequent tax returns to the city or profits to developers, and ignored the poor and those most affected by development schemes, who were often left worse off. In addition, questions were increasingly asked about the planning and implementation of policies, traditionally carried out by professionals with little effective input from the public at large—specialists who were assumed to act in the public interest. However, it was increasingly recognized that these so-called experts were not unbiased; they favoured the wealthy or the powerful, and, as in the case of the many disastrous high rise social housing projects, made flawed decisions. Peter Hall, in his classic survey of planning movements in western cities, also noted that these changes were carried out by what were assumed to be benevolent states and municipal bureaucrats, a consequence that the many anarchists and communitarians who pioneered many early planning ideas "*would have hated*" (Hall

1988, p. 3). The reason is that most of the changes they envisaged were based on a society of voluntary co-operation in self-governing entities. In addition, the planners focused on new physical designs, rather than on the procedures and methods used to achieve these ends. Recognition of these problems in planning has led to increasing demands for greater transparency in decision-making, as well as for greater input from those excluded and affected by the planning decisions—especially from the disadvantaged in cities.

These dissatisfactions with the current state of affairs developed at a time in which the western world became described as an affluent society (Galbraith 1958). The fact that the majority of the population became more wealthy than before meant that the old income pyramid of pre-industrial and industrial society—with a few at the top and the majority at the bottom—had changed in shape to a spinning top-like pattern in the late twentieth century, with an increasingly disadvantaged, but still small group at the bottom, paralleled by a small elite at the top, and a very large middle. Unfortunately, the affluence of the middle led to what has been described as a 'culture of contentment' (Galbraith 1992). Popular support for progressive redistributive policies declined as a more individualistic, atomist society emerged, in which increasing numbers of people adhered to what Taylor (1986, 1991) described as 'contributionist' ideas—namely, that it was only fair for people to get more returns based on their expertise, enterprise, and ability to work harder. Hence increased redistributive measures were seen as unfair, or even unjust, by such people. Indeed, the idea of the meritocracy that emerged in the late 1950s focused on these contributionist qualities as being necessary for future progress, whereas others drew the implicit assumption that those who did not have these aspirations—the poor—would remain in their situation because of a lack of drive, laziness or lack of knowledge, showing the persistence of the older historical belief that the reason for poverty was a personal failing. Interest in justice among these growing middle income groups that probably became a majority in society, increasingly focused on individual rights, which led to a justification of differential incomes. It led one prominent Canadian philosopher to observe the following:

> outside of society….the rules of justice that prevail are not those of distributive justice but those of independent possession. Consequently, an atomist view gives one the basis to argue that what we have a right to under these original rules cannot be abrogated, since the purpose of entering society cannot be to jeopardize these rules but rather to protect them. (Taylor 1986, p. 38)

Obviously, these contributionist views had always been present in capitalist society, and were influential in the conservative movements that tried to restrict redistributive policies from the nineteenth century onwards. Yet the grossly unequal rewards in early industrial society and the squalor in which so many people lived, the poverty experienced in the Depression of the 1930s, as well as the deaths and deprivations that occurred during two world wars, provided the impetus for policies creating greater equality, at least until the late 1960s.

The societal tendencies that reduced interest in new progressive policies were enhanced by the ideological implementation of the so-called neo-liberal policies

(Harvey 2005) by governments from the late 1970s. This was a response to the economic stagnation of the period which was followed by the deindustrialisation of so many western cities and the loss of many jobs and industries, with a decline in welfare benefits and increasing competition for jobs. The election of new conservative governments led to the belief that government had to be reduced because of the high debt levels incurred through expanding many public services. In addition, conservatives argued that regaining prosperity in western cities would only occur if several approaches were followed: facilitating market-led processes to attract innovative knowledge industries and creative people; encouraging entrepreneurialism; developing new amenities within cities to provide a further draw; and removing, or at least reducing, government regulation, especially the many public planning procedures which were blamed for cutting out growth opportunities. The result was a phase of deregulation in planning and governance in general and a belief that a new phase of competitive cities would emerge, where urban centres would have to compete with one another to attract the limited growth opportunities available in the new competitive global world. Such free-enterprise and pro-growth policies became the mantra of most politicians, including many of the centre and left. It meant another source of decreased interest, this time by politicians and decision-makers, in the increasing inequalities in societies that these neo-liberal policies created.

These changes have led to four trends. One has been a *social deterioration*, affecting the position of the worse-off in cities, especially in the U.S.A. (Wilson 1987). This was not only the result of economic change, especially in the reduction of the number of unskilled jobs in cities in manufacturing and transport, but was also due to several additional social changes, such as: family breakdown, increasing crime rates and drug use, systemic ethnic discrimination, fewer effective welfare measures and especially a decline in the quality of education system in poor areas. The resultant increase in inequality has severe social consequences, including a negative effective on growth, as seen in many studies (Mount 2005; Wilkinson and Pickett 2009). A second trend is the way that the middle income groups have seen their income reduced over the past 20 years, *hollowing out the middle class*. A third trend relates to the fact that the rich are getting richer. A fourth comes from an *unemployment increase* after the financial crises from 2008, especially among the young adults. These features seem to be making people more selfish and with individualist attitudes. Indeed many developed countries have seen reduced support for welfare policies. Figure 3.1 provides an example of this trend, showing the changing attitudes of various age groups to welfare provision as measured in the British Social Attitudes Survey (BSA 2013), which has been tracking opinions for over 30 years. Although all age groups show a significant decline in willingness to fund more benefits since the late 1980s, the younger age groups now have agreement levels of only 20%, with even the people over 65 years down to 40% from 60%. Although this is only one indicator in a single country it seems that the trends are similar in other western countries, suggesting that populations are becoming less collectivist and more sympathetic to contributionist values.

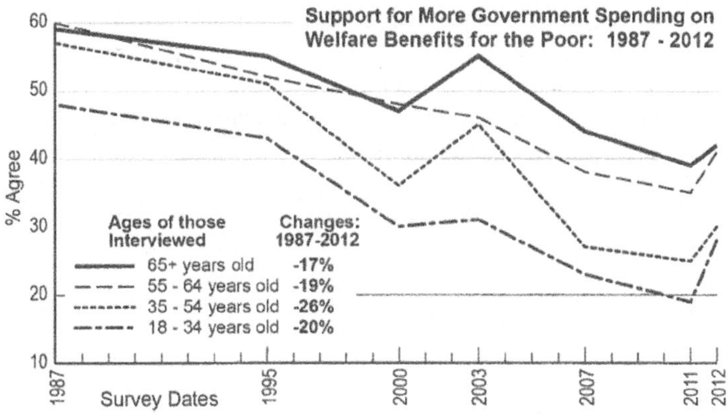

Fig. 3.1 Changing attitudes to welfare provision. (Source: Revised from British Social Attitudes Survey, 2013, Table 2-7)

3.3 New Reasons for Redistributive Justice

Despite the increasing dissatisfaction with the results and implementation of the progressive policies from the 1970s it must not be forgotten that most western governments continued to commit more money towards what can be broadly called social programmes. They use tax dollars to finance education, health, unemployment programmes, although these budgets are increasingly under threat from escalating costs, cut-cutting neo-liberal agendas and increases in user fees. Nevertheless, all western states have extensive laws that are designed to create greater equality and civil rights in society, as well as extensive social security measures in service provision and welfare etc., although the proportion of G.N.P. devoted to these measures varies considerably in different nations. In addition, the deteriorating position of a growing underclass, especially in the poorer inner city areas and social housing estates, have led to increased efforts to solve the problems in many cities. Many have been initiated by successful grass roots community movements against specific problems that affect their areas, or stimulated by community organizers, especially following Alinsky's ideas (1971). Others came from various policies by governments designed to mitigate the problems of either disadvantaged groups, such as the homeless, or by special measures targeted at areas of high crime, poverty and dysfunctional families. These areas are labelled in various ways as areas of disadvantage or priority areas, in which policies such as additional welfare measures, new schooling initiatives, and tougher drug measures, are designed to reduce their generational persistence. More generally, many studies have shown the negative social effects of poverty inequalities, even in the developed world (Mount 2005; Wilkinson and Pickett 2009).

Advocates of the need for a new approach to societal injustice in the city have argued that these essentially ad hoc successes can be matched by many examples of failures and have done little to prevent the increasing inequality in society, or the general reduction in the aims and effectiveness of planning in the public interest. For example, Harvey in his pioneering book, *Social Justice and the City* (1973), and in many subsequent works, has sought to go beyond the specific problems of disadvantaged and has stressed the need to investigate societal explanations for the problems. He argued that only transformative policies will promote greater redistribution and more societal equality, and recently suggested in a recent work, *Rebel Cities* (2013), that such policies need to be derived from the city and its social movements. In other words, Harvey and fellow thinkers believe that affirmative actions are only able to modify the margins of society and will never achieve the goal of effective equality for all. This is a consequence of their view that capitalist societies are inherently prone to inequality because only a few control the means of production and the rewards from the creation of capitalist social value (Harvey and Potter 2009), which in Piketty's (2014) view are a consequence of the fact that wealth normally grows faster than economic output in history.

Previously Lefebvre (1968, 1995) had proposed a new inclusionary principle for critical urban theorists, suggesting that all city inhabitants had what he called 'le droite à la ville' (the right to the city). This leads to a revived view of the importance of creating a new civic consciousness, accepting the fact that *all* urban dwellers are part of the same collective society, and need to act in the interest of all, without losing their individual rights to deal with their lives and the changes that occur in their neighbourhoods. Claval (1986) also argued for more progressive policies that involved participation of the disadvantaged in decision-making, instead of only maximizing economic utility in new development projects.

These pioneering arguments outlining the need for more effective interventionist policies to reduce inequality and the problems of the marginalized in cities were buttressed by recognition of the importance of three other features. One is humanitarian, stemming from the view that we should not allow such conditions to persist since we are all members of the same society. Another is that increasing numbers of studies have shown that persistent and increasing levels of poverty and inequality have many negative social effects (Wilkinson and Picket 2009) and, as the Healthy City discussion (Chap. 13) will show, leads to biological changes that increases propensity to ill-health. In addition such features stifle the investment and innovation that leads to higher growth (Ravallion 2013). Another comes from what might be called defensive reasons. This is the fear that the growth of an underclass, mired in poverty, may give up on the host society and turn to crime or even revolution to obtain the goods they require, creating an unsafe society for the majority. This justification for interventionist policies is almost a revival of the fears of many eighteenth and nineteenth century governments about being overthrown by the lower classes and city mobs that was one of the reasons for the growth of progressive policies.

These attempts to identify and explain the growth of urban problems related to the maldistribution of goods and opportunities have been complemented by the re-emergence of interest in the more general topic of redistributive justice by a new generation of philosophers. Much of the late twentieth century interest can be traced back to John Rawls's seminal work (1971, 2001). He was a pioneer in the revival of the philosophical study of the theory of redistributive justice, applying the skills of analytical philosophy to the social contract tradition seen in the Enlightenment works of Hobbes and Locke. At the core of Rawls's argument was his justification for the use of fairness as a basic principle for redistributive justice, not equality. This was derived from a hypothesized initial position, in which individuals were considered to be free and had equal opportunity, but were in a state of ignorance, in the sense that they were unaware of either their status or their goods, relative to others. He maintained that individuals would act rationally in this situation to choose an equal amount of primary goods to others, to ensure they would not be disadvantaged by being inferior to others. This argument and its subsequent detailed extensions provided a new theoretical basis for the concept of fairness as a basis for distributional justice, one that had a rational, logical basis and did not have to depend upon other justifications for legitimacy, such as commands from some gods, or some assumed natural law. Among his extensive series of deductions and recommendations, Rawls argued that political and legal institutions should be established to prevent excessive concentration of wealth, to ensure that the least advantaged members of society are not discriminated against. One of the key arguments to justify this position is his view that the distribution of natural talents in a population, and presumably the education that develops these talents, should be regarded as a common asset, which should be used to benefit society as a whole. Of course, the contributionists argue the opposite, that differential income extraction is the prize for those with particular talents or skills. In addition, Rawls attacked the utilitarian concept that lay behind many of the progressive policies of the past. The weakness is that it assumes the needs of each person are the same, thereby ignoring "*how the sum of satisfactions is distributed among individuals*" (Rawls 1971, p. 260). This focus on the sum of the utilities, not the satisfaction of the individuals or groups affected, means that a utilitarian approach fails to address their problems, which explains why those without power or interest, especially the most needy, are increasingly left behind.

The work of Rawls stimulated an outpouring of articles and books on the topic of justice (Wellbank et al. 1982; Lucash and Shklar 1986; Sen 2009). This led to: attempts to show how the distribution of various goods could be justified (Walzer 1984, 1986); the standards or criteria by which redistributive justice can be judged; the strains that exist between exponents of this approach and those derived from our increasingly atomist society (Taylor 1986); as well as to major critiques of his ideas (Nozick 1974). These and other studies have placed the issue of justice in a redistributive sense back on the philosophical agenda, and created new principles for justifying policies of redistributive justice, instead of the older dependence on moral or other beliefs. Although Rawls has provided new theoretical perspectives on the nature of justice, its basic rationale has been criticized as creating an initial a 'make-believe' situation (Jackson 1986). In other words, the initial hypothesised state that justifies 'fairness' is far removed from the complex real world situations today, in which people are aware of their goods and influence, and those with wealth and power still seek to maintain and improve their advantage. This has led to additional attempts to buttress the rationale for adopting the fairness criterion as a guiding principle. For example, it has been argued that fairness was a basic binding value in many social groups, as seen in its extensive use as a basis for redistribution by traditional hunting and gathering groups, or in the close relations found in many social animals. These examples suggest that it is important to see fairness as a foundational function of redistribution among social groups. Others have based their support for the idea on the traditional values developed in many religions, in which sharing with people in need has always been a key moral principle.

3.4 More Focused Redistributive Justice

Those who advocate a new emphasis upon societal injustices in the city argue that the essentially ad hoc successes of some recent welfare policy initiatives have done little to prevent the increasing inequality in society, or the general reduction in the aims and effectiveness of planning in the public interest. It is argued that these problems cannot be solved only by three types of policy: the type of affirmative actions used in the past, such as by more effective national welfare policies and social security nets; the pressure from various urban protest groups; or thirdly through the type of transformative societal process advocated by neo-Marxists, in which either the state dominates all sectors and directs change, or some new form of urban-based change occurs to create socialist cities.

The first approach is seen to be as too weak and haphazard in its spatial impacts and not focused enough on urban settlements. The second is often too incoherent and lacks wide-spread political support. The third solution, the transformational one, may be beloved of some intellectuals and still provides cutting criticisms of many of the problems in contemporary society. But its practice is still considered by most people in the democratic west to be tainted by the disasters of the experience of communist states with their extreme limitation in personal freedoms, as well as the corruption that created a ruling class with privileges that the majority could not share in. The result is that many believe that all three of these approaches have proved ineffective solutions against an increasingly dominant neo-liberal culture buttressed by contributionist beliefs. This has led to new ways of creating greater fairness in towns and cities, and upon the decision-making that occurs within them, which are dealt with in subsequent sections. One is based on the application of new Communicative approaches. Another deals with extensions of the Capabilities concepts. A third is upon the identification of basic principles that should be used to remove injustices. The fourth seeks to focus on specific guides to public policy in urban places, so as to delivering policies that are more just in the redistributive sense.

3.4.1 Communicative Approaches

An early suggestion to solve the problems of inequality and lack of influence of many groups in cities has focused on reforming the decision-making processes, especially in public planning in towns and cities. The approach tries to reduce the extent to which the results of most decisions have been biased by those with powerful interests, in which the voices of not only the disadvantaged, but the majority of the population, have largely been ignored. Hence new ways of approaching the decision-making process have been suggested. One approach is to *create greater transparency* in the process of planning negotiation to expose and remove biases from powerful groups. Another is to *involve stakeholders* with interests in the issues, not simply developers, planners and politicians, but including all groups affected, especially the marginalised. Their concerns could be better articulated if they were able to benefit from the use of advocates, such as specialists in law or planning, a method known as advocacy planning (Davidoff 1965). These advocates could use their expertise to develop and argue their case in public forums where technical issues and complex rules of procedure and evidence are involved. A third approach is by adopting more *communicative approaches* between all the affected parties to try and produce greater consensus between the groups. There are many alternative types of models, such as those based on Habermas's (1989) theoretical ideas of Communicative Action, or those derived from more practical experiences from actual planning (Healey 1997). They all stress the need for discussions to be far reaching, should include all affected people, and should aim to expose the reasons why different opinions are held by participants. It is suggested that this type of discursive process will mean that fair-minded people will eventually modify their initial opinions when they understand the opinions of others and the constraints involved. Hence, through good-will and compromise the discussions will lead to a reconciliation of their differences and achieve a consensus that will be fair or just.

Although there can be little doubt that these approaches have had some useful results in moving the process of planning away from its old top-down approach, towards more open and democratic procedures, critics have argued that its weakness lies in its focus upon the *process of decision-making*—the discursive element—rather than the ends or the *products*. Indeed, it is far from likely that the ends of most discursive process will produce a more just outcome. For example, it underestimates the power of those with wealth to get their way, or the probability that the disadvantaged and independents will be co-opted by the powerful. Also it largely downplays how some government bureaucracies, such as planning departments, manipulate community forums for their own ends. On the other hand it may underestimate the self-serving, even illiberal, nature of some community groups who argue for their own interests, not for the city as a whole, as seen in so many cases of developments proposed in residential areas which meets recalcitrant Not-In-My-Back Yard (NIMBY) attitudes. Moreover, there is always the problem of the inability of many of the poorest to effectively argue their case, or to consider their alternatives. More generally, of course, it pays little attention to the societal context in which the ideas have been proposed, especially the neo-liberal agendas of development projects that are so dominated by new growth and market facilitation considerations. Hence critics of these approaches maintain that although greater transparency and fairness in decision-making has been achieved, they are still handicapped by the many problems identified, so success is not guaranteed. So other approaches in the search for redistributive justice are needed, although acknowledging that some of these communicative ideas may still be useful in some circumstances.

3.4.2 The Capabilities Approach

The Capabilities approach arose from dissatisfaction with the mid and late twentieth century literature that viewed human development either as a function of growth in an economic measure such as GNP, or from utilitarian perspectives that arrive at some summary of total satisfaction or benefits. Both use one metric for what is a complex and multifaceted issue, ignoring the distributional and personal differences in societies that may see total wealth increase but ignore the persisting, and often greater inequality between rich and poor. Although the Basic Needs approach (Streeten et al. 1981) provided an initial counterweight to these economic views of development, it has often been criticized for over-emphasizing the *delivery of various commodities*, rather than upon the wider problem of the well-being of individuals, especially the poorest. Recognition of these and other issues led to the development of the Capabilities approach, pioneered by Sen (1983, 1999) and subsequently developed by many researchers (Clark 2005, 2006). This new approach has strong links to the Rawlian ideas of justice and the human rights literature (Rawls 1971, 2001), as well as incorporating ideas from many classical philosophers, ranging from Aristotle, to Adam Smith and Marx (Clark 2005). In essence the Capabilities Approach seeks to answer some deceptively simple questions.

What are individuals able to do, and to be?

What are the enabling factors that help them achieve these aims and live a full life?

Rather than adding to the debate in general terms, Nussbaum (2011) provided a set of specific answers to the questions, proposing that ten basic capabilities are fundamental to the search for a human development that improves the quality of life of persons—for the approach is based on individuals and their problems, not groups.

Nussbaum's list of ten basic capabilities is re-arranged into a diagram in Fig. 3.2, showing that they can be summarized under five basic categories which may be easier to remember, namely Life, Physical Body, Cognitive, Linkages and Control functions. Each of these dimensions consists of one or more of the capabilities, many of which are inter-related. Nussbaum stresses that the problem of improving life is a pluralist one and that all ten features which improve capabilities are basic requirements, so rather than focusing on one or two, they need to be considered as a whole to ensure human dignity. The list shows that many of the features identified consist of what are normally considered individual freedoms in developed

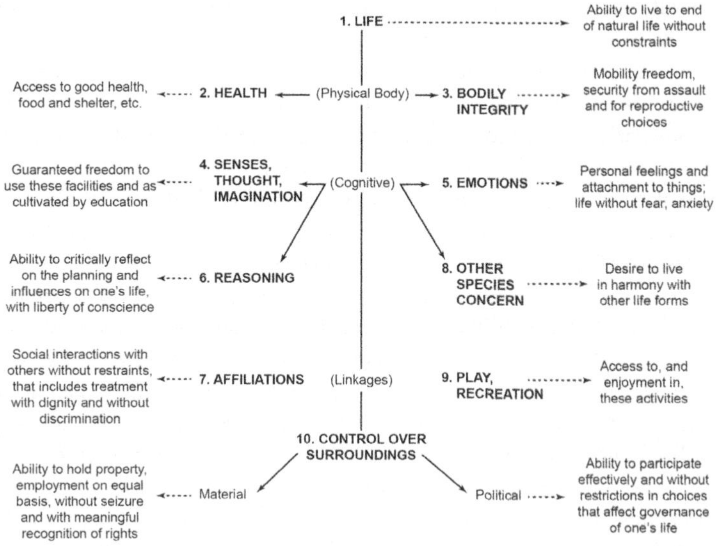

Fig. 3.2 The ten basic capabilities. (Source: The descriptions are re-organized from Nussbaum's ideas (2011, pp. 33–34))

countries. Many of these took centuries to achieve through protest and progressive legislation that enshrined the ideas in law. Sadly, in many countries, there is often a difference between what rights are described in a constitution and what happens in reality, while people in many states still do have any legal or even practical access to these rights. Moreover, the impoverished, even in democratic countries, often face what is called a *tragic choice in needs,* for many individuals have to focus on one need to achieve another, such as a woman paying for good day care or education for a child, while going without food or leisure for herself.

Advocates for this approach also stress that these capabilities come from several sources: the *innate* qualities of each person; the additional *traits* that are developed from such features as each person's upbringing and education; the *enabling environments* around, which have mores and laws that enable the capabilities to be blocked or developed; and the extent to which these are *realised*, or put into action rather than existing in theory, such as when a child may be free to be educated, but opts not to do so, perhaps under family pressure to get a job. This shows that many actions and *preferences are malleable,* for people learn to do without, and then just accept the situation. It is also worth noting that the realisation of these features also depends on the *longevity* of these features, the extent to which they will still be present in the future.

As in all approaches to human development, many critiques of the Capability approach have been made (Clark 2005). One of the most fundamental, apart from arguments about the completeness of the list, has been the suggestion that these capabilities need to be developed *by* the impoverished people who are most affected, *not* by middle class western or outside researchers (Sen 1999, p. 31). Yet Clark's (2005) work in impoverished areas of South Africa showed that the majority of people interviewed shared a common view of what was good and required to achieve a better life. Moreover placing reliance for the choice of capability indicators only on the impoverished does run the danger of corrupt leaders distorting choices, while adherence to older customs, and especially paternalist dominance, may disguise or even prevent the extent to which the capabilities can be identified, let alone realised. In an urban context the major problem of the Capabilities approach is the fact that it is the state that enacts laws and policies that allow them to develop. Most urban municipalities have restricted powers, so at first sight their fields of action to deliver policies that increase capability are circumscribed. But this is not absolute. In the fourth section of this review, the discussion on guides to the just city, as well as in many other chapters in this book, there are examples where action at an urban level can lead to the realization of many of these capabilities. For example, effective and honest policing provides one way that safety from crime can be secured and this is often locally administered, while the provision of leisure facilities or festive events are other capabilities that are very much areas of municipal responsibility. The provision of decent housing for the less well-off is also usually municipally controlled, even though most funding usually comes from national sources these days, showing that co-partnership schemes may be needed to improve such capabilities. Also, the idea of the public good, not necessarily maximization of site value or profits, has often influenced particular planning decisions, although the growth of neo-liberal ideas has often led to maximizing utility being the dominant principle.

These examples show that there are areas where the quality of life of people can be improved by actions at an urban level, although accepting that the main motors of change—especially for those linked to personal freedoms and controls—lie at the nation state level, although recent inputs from international actions and pressure groups, such as the Human Development and Capabilities Association (HDCA), are helping the growth of these capabilities in poor countries. Two of the most successful approaches to improving capabilities in employment and health contexts have been the work of micro-credit lending schemes and cash transfers. The former was pioneered by the economist Muhammad Younis who showed that providing credit to the poor without collateral, by creating a system based on trust, yet with supervision and group participation, has helped many escape from the clutches of rapacious lenders and enabled those who received small amounts of credit to create or expand their own employment, The approach which was developed in the late 1970s in Bangladesh, rapidly expanded with the creation of the Grameen Bank (Unesco 2013) in 1983 and been copied around the world. A second approach to helping the poorest has been to provide conditional cash transfers (CCTS), on condition that the recipients send their children to school and accept vaccinations, the basis of policies such as the Bolsa Familia in Brazil or Mexico's Oportunidades. These help to improve the education and health of future generations and create pressure to raise the standards of both. However unconditional cash transfer schemes (UCTS) to the poorest have also been successful, if less popular with donors in rich countries who feel that making conditions provides assurance that the money is not given for nothing. Surveys of these schemes have shown various advantages of the two types (CS). But they also reveal that the typical development aid approach of western countries has been flawed. Many of the projects provide aid in technical projects that mainly involve developed world technicians or materials. Certainly they have improved roads, dams and other infrastructure, but have often led to corruption and bypassed the needs of the poorest. Capability advocates argue that more locals need to be employed in such projects. Also when cash schemes are evaluated, researchers (Clark 2005) have shown that that the poor often know what to do with the money, spending it on improvements to houses, on education for their children and creating the limited capital needed for improving output in self-employment schemes, creating a dignity of work and self-reliance. Yet it must be accepted that most of these CCTS schemes originate in the national state, although there may be some local and city-based administration. What is clear is that towns and cities could play a bigger part in creating and delivering more of the capabilities shown in Table 3.2, although the limited tax basis of most urban places does limit what they are able to do.

3.4.3 Principles for Greater Justice

A third major approach to identifying and resolving the problems of the disadvantaged in the city has been to identify a list of principles that should be followed by government agencies in the attempt to provide greater justice. Table 3.1 summarizes the principles proposed by several prominent writers and committees. Certainly each source has its own emphasis, such as: the need to remove particular oppressions experienced by the disadvantaged, as shown in the case of Harvey's (2002) modification of Young's ideas (1990); the emphasis upon equality by the Habitat International Coalition (HIC); or the focus on Basic Needs used in Friedman's (2000) discussion of what he called the 'four pillars for the good city'. Yet despite the different content, emphases, and the terms used, there is in fact a great deal of similarity in the ideas expressed as seen in Table 3.1 which has re-arranged the individual principles of these authors and summarized them in the last column. The table demonstrates that the ideas fall into three basic categories of principles, summarized as Participation and Influence (issues that are more carefully identified by Nussbaum's capabilities), Basic Needs, and Environment. These categories are useful in providing a general context for questions of injustice, within which issues that need to be addressed are explicitly identified.

A related, but more specific approach to the problem of injustice can be seen in the 12 principles created by the 'Rights to the City Alliance' that were adopted by the United States Social Forum in Atlanta in June 2007 (Table 3.2). This alliance started as a pressure group against gentrification and the frequent removal of the poor from the re-developed areas. It has developed into a movement devoted to the assertion of human rights, democracy, racial and urban justice. Rather than using the word 'injustice' to identify the problems of the disadvantaged, the movement used Lefebvre's (1968) ideas about the various 'rights to the city' that need to be

Table 3.1 Principles of governance for greater justice in cities: selected examples

Oppressions to be removed (propositions of governance)	Equality of:	Four pillars of 'good city'	Key themes
No access to political power	Opportunity for participation in public decision—making		A. Political participation and influence
No exploitation of labour	Opportunity for productive, freely—chosen livelihood. Access to economic resources (incl. right to inheritance, land/ property ownership, credit, natural resources and appropriate technologies)	Wages	B. NEEDS (i) Employment and returns (ii) Personal rights
	Opportunity for personal spiritual, religious. cultural and social development		(iii) Social & cultural development
Elimination of marginalization of social groups			
Removal of cultural imperialism			
		Safety	(iv) Safety and protection
Humane forms of social control			
		Health Care	(v) Health
		Social welfare	(vi) Welfare
			(vii) Food and water
Mitigation of adverse ecological effects	Rights and obligations with regard to the use and conservation of natural and cultural resources		C. ENVIRONMENT
(I. Young 1990; D. Harvey 2002)	(Habitat international coalition)	(Friedman 2000)	

granted to urban residents if the problems of the urban underprivileged were to be removed. Although expressed in different words and ways there are similarities with Table 3.1 and also capability ideas. However the Alliance statement has a longer and more specific set of principles. Many will be accepted by those with liberal views who are concerned about injustice in the city. But others are far more radical, such as principles 1, 2 and 9 that call for: rights to land and housing that are free from market speculation; permanent public ownership of urban territories for public use; and the right of community control and decision-making over the planning and governance of cities; Many of these principles conflict with aspects of the current capitalist system and the legal rights or practices of most urban governments. So perhaps it is not surprising that not all of the radical principles shown in Table 3.2

Table 3.2 Platform of the 'Rights to the City' alliance (RCA). (Source:http://www.righttothecity.org/index.php/about)

1	*Land for people vs. land for speculation*	The right lo land and housing that is free from market speculation and that serves the interests of community building, sustainable economies, and cultural and political space
2	*Land ownership*	The right to permanent ownership of urban territories for public use
3	*Economic justice*	The right of working class communities of color, women, queer and transgender people to an economy that serves their interests
4	*Indigenous justice*	The right of first nation indigenous people to their ancestral lands that have historical or spiritual significance, regardless of state borders and urban or rural settings
5	*Environmental justice*	The right to sustainable and healthy neighborhoods & workplaces, healing, quality health care, and reparations for the legacy of toxic abuses such as brown fields, cancer clusters, and superfund sites
6	*Freedom from police & state harassment*	The right to safe neighborhoods and protection from police, INS/ICE, and vigilante repression, which has historically targeted communities of color, women, queer and transgender people
7	*Immigrant justice*	The right of equal access to housing, employment, and public services regardless of race, ethnicity, and immigration status and without the threat of deportation by landlords, ICE, or employers
8	*Services and community institutions*	The right of working class communities of color to transportation, infrastructure and services that reflect and support their cultural and social integrity
9	*Democracy and participation*	The right of community control and decision making over the planning and governance of the cities where we live and work, with full transparency and accountability, including the right to public information without interrogation
10	*Reparations*	The right of working class communities of color to economic reciprocity and restoration from all local, nation and transnational institutions that have exploited and/or displaced the local economy
11	*Internationalism*	The right to support and build solidarity between cities across national boundaries, without state intervention
12	*Rural justice*	The right of rural people to economically healthy and stable communities that are protected from environmental degradation and economic pressures that force migration to urban areas

have gained sufficient support for implementation in the current American political climate. So the Alliance remains a minority pressure group at the present. Yet its regular meetings and a web site that calls for members to join in building this movement does help explicitly identify many problems faced by the disadvantaged people. This in itself is progress, and provides a basis for information, debate and action, for many of the issues must to be solved if greater urban justice is to be achieved.

3.5 Guides for Just Decision-Making

The identification of the various principles outlined in the previous section has focused attention on key issues that need to be addressed to create greater justice in the city. However, the problem of most of these approaches is that they do not always provide rationales for the various principles, or rights that are demanded, which would help gain the support of the population at large. More importantly, they fail to provide specific policy guidance on how to achieve the ends required, namely to resolve, or at least mitigate, the problems or injustices felt by the disadvantaged in cities, which does not only include the poorest people. These issues led Fainstein (2010) to suggest a more policy-focused affirmative approach within the current system of global capitalism, adopting Rawls's rationale (1971) for 'fairness' as the basic value upon which redistributive justice should be based, although noting it has strong links with the liberal philosophies of natural rights proposed by many Enlightenment writers. In her book *The Just City*, Fainstein (2010) discussed the extent to which recent planning policies in London, Amsterdam and New York have adhered to the standard of fairness. Instead of only using these empirical examples to make her point about the extent to which they have helped progress towards a more Just City, she generalised her ideas by suggesting that there are three basic values of urban justice, namely, *equity, democracy*, and *diversity*, and all need to be maximized to achieve greater fairness within cities. Within each of these categories she provided succinct examples of policy areas that could be used to promote greater fairness, with the intent of resolving existing 'injustices' at a city scale, effectively creating a set of just city policy guides (Table 3.3).

Fainstein was careful to identify several caveats that circumscribe her approach. For example, she emphasized that her approach is affirmative, in the sense that it represents an incremental, not any type of transformative approach to reduce levels of injustice (Harvey 1989, 2013), although presumably many of the examples could be part of this revolutionary process. In addition, she accepts that decisions in one category may sometimes conflict with others, leading to the need for resolution between them. Another restriction is that the approach is designed to apply only to cities in states with a dominantly market economy and with a political tradition based on liberal democratic principles that has also embraced egalitarianism, some of the key issues in the capabilities approach. Finally, she accepts that these components are only part of the range of problems facing contemporary cities, so she explicitly

Table 3.3 Summary of guiding principles for just cities. (Source: Summary of Fainstein's guiding principles (2010), with suggested additions in italics.)

Guiding principles	Examples of policies
Democracy (3)	Advocates. Broad consultation. Consult with affected people
Additions (+5)	*Decision-making. Encouraging greater partnership. Fairness to others. Greater community powers. Hypocrisies of franchise and protection*
Equity (5)	Housing. Economic development. Active survey of megaprojects. Relocations. Transit
Additions (+2)	*Environmental issues. Needs*
Diversity (6)	Mixed land uses. Group needs. Inclusion. Community boundaries.Segregation-diversity
Additions (+2)	*Adding community contacts and spaces. Limitations of behaviour*
Problems (+5)	*Restricted urban powers. Inter-generational equity. Territorial issues. Equalities and freedoms. Societal contexts*

leaves issues, such as environmental problems, for other researchers to deal with. Despite these limitations Fainstein's work is one of the most useful and practical set of policy guides to create greater justice in the decision-making process within cities. These are summarized in Table 3.3 along with extensions to these ideas, although there can be little doubt that there is a need for additions to the list to provide a more comprehensive guide to just city policies.

3.5.1 Democracy and Just Cities

Although Fainstein made this criterion her second category, it may be the easiest principle to justify as a basis for redistributive policies, at least in western societies where democracy is an accepted fundamental value in society and each adult citizen is equal in voting power. Indeed the very idea of western democracy is traced back to the Greek city states of the sixth century BCE, although we often forget the citizens involved in their self-governance were only a small group of property owning males. Moreover in medieval society the aphorism that 'city air makes one free' was a consequence of the rule that being able to live in a city, at least one with a charter giving privileges, made one free from the restrictive obligations of rural feudal society, meaning the places could be seen as emancipatory cities in some respects. Today, except in a few cases, our urban governments are not direct democracies, but representative democracies, in which voters elect politicians, who use the bureaucracies that have been created to manage the system. As has already been noted, although they are supposed to serve the public interest this is not always the case, and certainly does not guarantee fairness, especially to minority interests or the needy. In any case the vast majority of cities are subject to the laws of the national state in which they are located, meaning that their powers are limited. Despite this constraint, Fainstein suggested three policies that would increase the level of justice within the public planning approach.

3.5.1.1 Advocates

The first of these is the use of *advocates*. These should be provided by municipalities for groups who do not have the skills or ability to participate in the decision-making processes in urban areas, perhaps by seconding planners or other specialists to help communities with problems.

3.5.1.2 Broad Consultation

Fainstein suggests this is especially needed when new or sparsely occupied land is to be redeveloped, although it should apply to any area being redeveloped that affect people. This should include representatives of people from the affected area, and those who may be impacted who live outside the affected area.

3.5.1.3 Consultation with Affected People

This is needed when occupied land is to be developed, although city-wide considerations mean that the inhabitants should not be the only arbiter. The ideas of communicative action discussed above may improve negotiations to create more just outcomes. All too often, the typical approaches used by city officials to consult with communities are really top-down information sessions rather than true engagement and problem-resolving sessions, in which various tactics are used to deflect attention from the main issues or decisions already made.

Although Fainstein's three guides provide a very useful beginning to creating greater justice, they need to be extended to include some more specific links to ideas that have been discussed in preceding sections. The additions are designed to improve the process of communication between affected parties and also remove the bias from existing approaches in the public planning process.

3.5.1.4 Transparent Decision-Making

All levels of decision-making should be reviewed to make them as transparent as possible. This will enable all affected groups to review and comment on various decisions. It also means that the decision processes of the representatives involved should be fairly run and monitored by a city ombudsman, or outside reviewers, to prevent local corruption occurring.

3.5.1.5 Increasing Participation

Even when new consultative processes have been created, low participation rates in meetings and elections mean that many, particularly the poor, are effectively left out, especially in existing democratic processes. This can be seen in the fact that municipal voting rates are often well below a third of the eligible voters. There is little doubt that too much complacency exists in modern urban society, although it may be argued that family, work and leisure commitments leave many with neither the time, nor the inclination to involve themselves in these activities, unless the problem directly affects them. The result has been a decline in the participation levels in most group activities as Putnam (2000) has shown in his aptly entitled book *Bowling Alone*. Altering this situation is not easy. One useful approach has been promoted by community activists such as Alinsky (1971) who suggested that more community cohesion can often be built by initially involving people in winnable actions on relatively small matters, in order to build confidence for future, more difficult tasks.

3.5.1.6 Fairness to Others

There is little doubt that some local communities or their leaders prioritize their own interests or cultures and may be biased against outsiders. This is often seen in the well-known NIMBY (Not In MY Back Yard)attitudes, where redevelopment that may be beneficial to the city as a whole is opposed by local interests. Hence, effective and fair ways of resolving this type of problem must be created, especially in the interest of the city as a whole, but without ignoring local concerns.

3.5.1.7 More Intra-Urban Community Powers

All of the above recommendations seem to assume that the decisions are made at a city level, with local groups, such as those based on communities or neighbourhoods, having a mainly advisory or subservient role. However, greater democracy, in the sense of real local influence and functioning, may be achieved if greater powers are given to local community groups within the city that have been democratically elected. For example, they could be encouraged to engage in defined tasks, together with a share of the city finances or tax powers to carry out such functions. Some cities already practice this approach, but on an ad hoc basis and without losing their over-all control. More radical changes would involve legal decentralization of some functions to community groups, together with defined tax revenues to provide greater access and control by a local population, although there may be a loss of the benefits of economies of scale in delivery at a city level. This is the typical ‘access versus efficiency’ trade-off in service delivery. Given the current state of most municipal finances, such decentralization may be unlikely, though there have been examples of tax-sharing at a community level in the U.S.A., such as in the Neighbourhood Revitalization Programme in Minneapolis (Fainstein 2010, p. 181). Similar schemes have emerged in some seriously disadvantaged neighbourhoods, where decentralised delivery and renovation also have the advantage of training local people to provide jobs and skills needed by the community, ranging from maintenance to plumbing. U.K. Community Councils have the legal right to spend a few percent of local taxes in their own area, which are usually used to help cultural projects. In addition, there are many examples where local community groups in cities run facilities in partnership with the city, while the EcoDistrict approach pioneered in Portland and discussed in Chap. 7 shows how communities can become more effective in creating better environments. All these examples create greater local participation and more effective democracy within the city, although the ever present dangers associated with the NIMBY principle must be recognized and counteracted. Yet such changes could mean that a patchwork of very different results will emerge, with middle-class communities likely to achieve better results than poor areas because of the skills and participation of their residents and a focus to create positive change. This may increase the inequalities in facilities found in different areas, which may reduce the push to fairness. So ways of counteracting such class differences need to be found, such as the type of community action in disadvantaged areas pioneered by Alinsky (1971). Despite this problem, consideration should be given to providing decentralization in service delivery, so long as it does not result in a loss of municipal solidarity and sense of common purpose through political fragmentation of the city and the creation of antagonistic community groups.

3.5.1.8 Universal Adult Franchise and Rights

The hypocrisy of a universal adult franchise should be critiqued to help the most vulnerable. There are often large numbers of adults in cities who do not vote and participate in the urban democratic process and these are some of the most vulnerable who have no real voice in decision-making. Indeed, one of the great injustices in many cities of the developed world, especially the large world cities, is associated with the immigrants—some who are illegal—who perform many of the low wage and often casual jobs spurned by the citizens. In many western countries legal immigrants may have the right to residence, but not of citizenship, even after years of living in the country. This means they cannot vote in city or state elections. Effectively, therefore, they are excluded from the democratic process, despite their contributions to the economy, increasing the likelihood of their resentment and alienation from the society in which they live. Certainly this is a complex problem, relating to the extent to which immigrants can aspire to nationality if certain requirements are fulfilled, such as acquiring the state language. However, a case can be made for giving municipal voting privileges to all residents of a city, or at least those legally resident, but who are not national citizens, to further democracy at this level, if not at national scale. One of the few examples where this is being practiced is in Denmark's municipal elections where legal immigrants who have lived in the country for 3 years can vote, although it is difficult for many immigrants to obtain citizenship. Although this discussion on voting rights is largely based on western world examples it is worth noting that this type of problem is even greater in the cities of many oil-rich kingdoms in the Gulf States where non-residents are often a majority, do much of the work, but who are dis-enfranchised and often exploited in the conditions under which they work.

A related problem associated with the franchise concerns the right of access to city services by immigrants. Even in western cities mistreatment of immigrants is widespread and they often feel they will be penalised if they expose their conditions to the authorities. Illegal immigrants, many with decades of residence, are in an especially parlous situation, always fearful that they will be deported if they come into contact with city officials or the police. This makes it easy for unscrupulous employers to exploit their labour and for criminals to threaten them if they expose illegal practices. In the United States a number of cities as well as smaller municipalities have enacted local ordinances that forbid the police or city employees to ask for proof of immigrant status before providing welfare, declaring themselves to be 'Sanctuary Cities'.

The objective of the Sanctuary City concept is to allow any resident, irrespective of citizenship status, to have full access to city services—especially social housing, food banks, medical care, education—and should be encouraged to see police as protectors, rather than agents of federal authorities with the power to deport. The Sanctuary City idea was formally implemented by written ordinance in Los Angeles in 1979 and subsequently spread to many cities. Other urban areas have chosen to implement the policy by asking municipal employees to adopt the approach informally, rather than creating legislative backing. Despite the laudable humanitarian aim of providing access to what amounts to redistributive justice to some of the most vulnerable, there has been a reaction against the policies from many states and even the U.S. federal government, especially in the case of those apprehended for even minor crimes by the police, such as driving without a current licence or littering, who have then been deported. Although a 1996 federal law requires local governments to co-operate with the federal department that has jurisdiction over immigration (currently the Department of Homeland Security), many Sanctuary Cities have chosen to ignore the law as being an intrusion into state or local rights. This has led the federal government to threaten to withhold tax dollars for various social programmes. The debate continues over the sanctuary policy. Supporters of sanctuary cities deplore the federal intrusion into state and individual rights and argue that existing community policing approaches already enable convicted criminals to be deported. So the approach provides a crucial protection and access to needed services to a very vulnerable group in cities who are not criminals. Opponents, however, maintain that illegal migrants should not be allowed to stay.

In simple justice terms there seems little doubt that the Sanctuary City idea is a laudable one, that of protecting the weak and certainly vulnerable, but the eventual outcome of the ideas remains uncertain. What is worth noting is the fact that the idea, like so many urban initiatives, is not really new. As noted earlier, in medieval Europe many towns and cities provided sanctuary for those who fled their obligations in the rural feudal society, so long as they were able to stay for a year and a day, protected by the municipal charters that gave rights to urban citizens.

3.5.2 Towards Equity for Just Cities

Fainstein suggested that 'equity' was a more useful principle in the search for a just city than the more complex term 'equality'. This implies an aim of levelling out differences in income and leads to questions of equality for whom, of what goods, and how many goods are to be distributed etc. It could also be argued that it does not cause so many concerns for those who believe in the primacy of the contributionist principle. Fainstein identified several types of policies in cities which would create more equity, all of which are designed to ensure that public policy does not favour those who were better-off at the beginning of the decision-making (Fainstein 2010, p. 36). These ideas are re-organized and described in more detail than in her initial list in order to clarify the goals to be achieved.

3.5.2.1 Housing

The old progressive goal of providing decent, affordable homes, and a suitable environment for the poor as well as for key workers in cities, should be re-asserted, since the provision of shelter is one of the basic human needs. However, Fainstein's argument needs to be modified by recognizing the many problems that occurred with the direct municipal provision and control of public housing in the past, especially a lack of safety and slow maintenance. Hence the housing objective might be better achieved by encouraging non-profit sector provision, or by municipalities taking equity in homes, especially for vital public workers, such as nurses, teachers and police, to ensure that the latter are still able to live within what are becoming increasingly expensive inner cities. In addition, she argued that all new housing developments should always contain provision for households with incomes below the median, which might mean building certain areas at higher densities. It is also stressed that the stock of affordable housing that is created should never be diminished, either by ensuring that such buildings always remain in this public or non-profit sector, or by making at least one-to-one replacements. Unfortunately, policies in many states, such as the public housing privatization in countries such as Britain from the 1970s, did not lead to the income from sales being reinvested in new social housing. This reduced the number of houses available for the working class and poor, creating a huge backlog of housing need, not only among the most disadvantaged, but among the working classes who have been priced out of many areas in the most prosperous cities. It is also worth noting that there are often hidden housing subsidies to certain groups in some countries, subsidies that ought to be made more transparent in terms of their effects. For example, in the United States the money spent on housing for the poor is relatively small given overall budgets, and is less than the subsidies given to the house-owning classes through the tax policies that allow mortgage interest payments to be set against tax. This means the owners of houses benefit more than the poor from government policies in this area, which is less than just.

3.5.2.2 Economic Development Programmes

It is suggested that programmes created by governments, or with their support, should not only focus on their capacity to increase absolute employment and provide more tax returns. Instead, they should encourage developments that are attractive and career-enhancing for workers, instead of projects based on businesses with minimum wages or part-time employment with few benefits and poor conditions of work. Also, since it is well known that more of the money that local and independent businesses receive is more likely to be spent locally, instead of being sent to the remote head offices of national chains, there should be a bias to local businesses, creating greater income circulation both locally and regionally. In addition, these ownerships usually provide a greater contribution to the development of local managerial and entrepreneurial skills, and contribute far more to local activities through sponsorships and community involvement. It is also recommended that such projects should be subject to incremental growth, with several developers involved to avoid monopolies, and with space being made available for any existing businesses that are displaced or even disadvantaged by such developments.

3.5.2.3 Active Scrutiny of Megaprojects

These projects are usually developed by outside international interests whose principal concern is profit. Their impacts should be more carefully scrutinized to reveal not only the benefits they provide in jobs and conditions of service to low income workers, but also the amenities provided to the public, such as additional, accessible open space. All too often public money has been used to assist private corporations, whether shopping centres or sports stadiums, with few returns to the public sector, even when huge profits are made by these enterprises. Fainstein (2010) has provided a very good summary of the redevelopment of the Yankees' major baseball stadium in New York, which not only cost tax payers a great deal of money but also led to the loss of valuable park land for the local population.

3.5.2.4 Relocations

Fainstein suggests that no housing or commercial development should be involuntarily relocated for economic development, or to promote diversity. If people or businesses have to be moved, then adequate compensation for renters, as well as the owners of such properties is needed, so that they are able to obtain equivalent premises elsewhere. These problems are often ignored in gentrification projects that displace the poor, but have been brought back into the planning agenda in San Francisco by the Public Health Department, supported by community groups, issues that are discussed in the discussion on Healthy Cities (Chap. 13).

3.5.2.5 Transit

The addition of new public transit systems have become an important part of transport policies in most cities—given road congestion and the environmental effects of car pollution. But the need to be fair in the pricing of transit trips is often forgotten. The poorest in cities are the most dependent on public transit and are the most affected by high fares that restrict their mobility in job searches. Hence transit fares within the cities should be kept low, while car owners could be more highly taxed, given their contributions to congestion and pollution, perhaps with the returns helping to subsidize the transit fares. In this area it is notable how the big cities of Russia still have very low public transit fares, compared to western cities, perhaps one successful heritage of the former communist system.

These Equity principles provide a useful set of ideas to guide public decision-making towards fairer solutions, although Fainstein's additional guideline about 'planners being local advocates' is assumed to be already dealt with in the Democracy category. However, it can be argued that this set needs to be supplemented by additions from several other policy areas, some of which were found in Table 3.1. The following additions are designed to provide a wider context that helps the objective of providing more equity and therefore justice in the city.

3.5.2.6 Environmental Issues

At a time in which sustainability and environmental problems have attracted a great deal of public interest, if still limited action, there is no doubt that these should also be part of the agenda that seeks to establish more just cities, although Fainstein noted that such issues were beyond her brief. Yet it does seem important to specifically include such concerns within the Just City framework. Many of these are described in other chapters, especially the Green and Sustainable City discussions (Chaps. 4–7). Here it is worth noting that important specific questions concern the environmental effects of all new developments in and around the city, such as: whether the functioning of the local physical and biotic environment will be compromised by any changes; how they can be preserved or modified to reduce negative impacts; and whether justice is also applied to the non-human species that are affected. Increasing municipal protection of rare or at risk species or existing environments, show the extent to which such policies have already became part of the agenda in many cities—a major change in the last 20 years. Some of these questions can be related to environmental hazards, such as the need to avoid developments in floodplains, since these need land use permissions that normally lie within the jurisdiction of cities, issues dealt with in the discussion on Resilient Cities (Chap. 9). However, it is still worth noting that many urban development projects still take place in hazardous floodplains, while many other environmental factors have been ignored in recent years. For example, it is almost inconceivable that many nuclear plants in Japan in particular were built near major faults, or on shores within reach of tsunamis, even though the country has a world-leading record in developing standards for buildings that will withstand all but the severest earthquakes. The devastating March 2011 earthquake and tsunami which affected such a large area of Honshu and destroyed a nuclear plant may well have been a transformative event that has changed perceptions and public policy, ensuring that far more attention will be paid to the effects of environmental hazards in future urban developments, issues discussed in Chap. 9.

3.5.2.7 Needs

Fainstein's emphasis on housing is only one of the requirements needed to create better urban places. A more vigorous and explicit redistributive approach at a municipal level should ensure that everyone has access to what are usually called Basic Needs, but extended to deal with the more specific ten principles involved in Capability Building shown in Fig. 3.2 and including: employment; assuring non-exploitive labour practices; health care; education; safety, including non-coercive social protection and control by police services; food and water supplies; housing. Also, there is usually a need for better welfare policies for the really disadvantaged, as well as freedom from coercive practices, not only by police, to allow individuals to reach their full potential. Certainly western democracies provide many of these services. But the question is often what sort of approach should be adopted to provide these needs to people who are the most disadvantaged at an urban level, for many of these services are primarily dealt with by national policies or perhaps mainly financed by the state and administered locally. A welcome sign of increasing municipal activism in this area can be seen by the May 2104 decision of Seattle to mandate a minimum wage of $ 15 an hour in the city which contrasts with the federal minimum of $ 7.25. However even this has several exceptions and will only be implemented over a 3 year period. Opponents argue it will decrease employment and hasten a move to mechanisation. The impetus for this change came from the recent election of an avowed socialist to city council who supported the many protests by service workers in corporate fast-food businesses in many parts of America in recent years. The $ 15 minimum was used as a battle cry to draw attention to the fact that existing wages kept people in poverty, whereas the CEO of McDonald's, for example, is paid $ 9 million a year and Walmart, which refuses unionization, earns around $ 27 billion annually in profit.

It is also worth noting that cash-strapped governments at all levels have reduced their levels of support in many areas since the financial and debt crises of 2008, creating additional problems for the poor and weak. In some areas the resultant decreases in provision have been resolved by volunteer or philanthropic organisations that provide needed services without the often cumbersome bureaucracies of governments. Obviously there have been many of these organizations in the past, such as 'Meals on Wheels' in many countries, and 'Food Banks' in Canada, which have provided invaluable services to the elderly and infirm in the former case, and in the second case to regular, although not daily food supplies to those who cannot afford basic foodstuffs. These and related services need to be encouraged and helped. Despite the growth of such organizations, there seems little doubt that there is often a requirement for the direct government provision of many basic needs, whether directly or through subsidies to non-profit organizations in many urban areas.

Associated with this Basic Needs approach is the provision of equity in the opportunities for personal development, not simply in economic terms, but in the sense of individuals and groups being able to maximize their personal fulfilment, whether in gendered, social (including life-styles), cultural, and spiritual areas. This means providing adequate services in such as areas as education, libraries and recreation—services that are open to all—as well as targeting groups-at-risk, such as troubled teens. In an aging western society an increasingly important consideration may be the need to ensure equitable care of the elderly population, through programmes that target loneliness or vulnerability, and if not directly via care homes or day-care, then indirectly through regulations and reviews that effectively monitor these organizations.

In some cities of the developing world determined efforts are being made through 'social urbanism projects'that have the redistributive aim to reduce the isolation of the most disadvantaged areas and provide them with needed services. For example, in Medellín in Columbia an overground metro line was opened in 1995 to connect the north and south parts of the city, which has effective police presence on every station and rapid cleaning to make the system safe and attractive. In addition, two cable car systems have been installed to link major slum areas on mountain slopes to the metro which now provides far more people with fast and cheap access to the rest of the city. Around major stations clusters of publically financed investment provide schools, libraries, public spaces and buildings for social services, which deliver new facilities for these disadvantaged areas. The social urbanism approach is a justice policy based on the view that the worse-off areas need the most public attention. However it was very much helped by funds from Medellín's profitable publically-owned utility Empresas Publicas de Medellín (EPM), which in last 10 years has added $ 3.2 billion to City Hall, enabling it to finance its social urbanism policies.

These developments were helped by mayors with pioneering agendas and preceded by a determined effort to improve security through better and more community-orientated policing, for Medellín had some of the high murder rates in the world at 381/100,000 people in 1991. This figure dropped to 50/100,000 in 2013 as drug-related gangs have been controlled. Such targeted social urban projects in slum areas are also being implemented in Rio de Janeiro where the first stage in rehabilitation has been to send in tactical police squads to root out the local gangs, to essentially pacify areas that were controlled by the drug cartels that degraded the life of residents. Once this was achieved, regular policing was followed by the addition of many social services. These examples show how far more urban centres in middle income and less developed countries could apply similar initiatives to improve the quality of urban life for their poorest residents.

3.5.3 Promoting Diversity for Just Cities

Fainstein's third category of policies to create more justice in the city is that of Diversity. She notes that policies designed to create a more just redistribution in this component in the name of fairness are often more difficult to defend, a problem that is exacerbated by the fact that there are two main senses in which diversity is normally used in the context of urban planning. One relates to land uses, the other meaning applies to different social groups, whether based on class, ethnicity, interests or culture. Although most people will agree that social diversity adds to the variety of life in cities, by adding vibrancy, new experiences and ways of doing things, there is often incompatibility between the different social groups or even users of land uses in cities which can lead to tension. Fainstein described six types of policies that illustrate how greater fairness can be achieved in the name of diversity. Again the ideas are re-organized and extended from her original discussion to clarify the issues involved.

3.5.3.1 Mixed Land Uses

In the past 20 years the promotion of mixed land uses in cities has been encouraged, especially by smart growth ideas, as a general planning principle, in order to get away from the single use functionality that dominates most western cities under planning regulations developed since the early 1900s. A century ago there was a logical and pro-health reason for these regulations, given the noxious, polluting nature of many industries, the noise and lack of space in commercial areas, and the desire to create peaceful residential areas. Therefore, the obvious solution was to separate such incompatible and unhealthy uses into distinct areas. Most now believe this land use segregation has gone too far, creating sprawling residential-only and income–band suburbs. It has led to the loss of the excitement and stimulation of street life in urban areas, especially in many post World War II suburbs and the 'dead' downtowns of many office and commercial centres at night, the type of situation so roundly condemned by Jane Jacobs (1961). Yet one must accept that many people—especially those with young families—still prefer the peace, space and greenery of exclusively residential areas, unlike the high density, diversified inner city areas full of different uses producing the street life so favoured by Jacobs.

Some compromise between these two attitudes may be seen in the Neo-Traditional or New Urbanism approaches of the late twentieth century described in Chap. 2. They have attempted to recreate the ambiance of urban areas a century earlier, by creating areas with mixtures of housing styles and designs, more emphasis on pedestrian movement, the addition of suburban employment areas, shopping centres and providing better transport services to these dominantly car-orientated areas. In addition, there have been greater attempts to use buildings more effectively by encouraging mixed uses within them, such as libraries and community halls in the same building, or making some school areas, whether playgrounds or libraries, open to community groups in the evenings. However, policies to create such land-use these mixtures ought to be supported by the voters before being implemented.

3.5.3.2 Accessible and Varied Public Spaces

They should be provided to the population at large, whether in the form of parks for recreation or open spaces that can be used for many activities, from festivals to markets and concerts. Many cities already have precise guidelines providing minimum standards for open spaces, and various types of parks and recreation facilities in cities, part of the progressive amenity provision policies created in the last century—policies which have gained greater support from the rise of the conservation and environmental movements and which will be described in Chaps. 5 and 6. In the last decade some of these open space areas have been under threat as municipalities have sold public lands, which have then been commercialised or privatised. Hence by explicitly identifying the need for various types of space, the pressure for its privatisation and loss to the public sector may be reduced. In deprived inner cities the private ownership of land means that greater funding is needed to buy land to provide more green space for the residents.

3.5.3.3 Group Needs

Those who suffer systemic discrimination should be assisted in obtaining access to housing, education and employment opportunities. Many believe that the needs of these groups should be implemented without the creation of large segregated areas and the top-down management of so many post World War II social housing projects, perhaps by non-profit groups that involve the locals in decision-making and in the provision of services—from shops to maintenance—as well as safety, which were often ignored in many of these schemes. Also, after the feminist critiques of the patriarchal nature of western society from 1970s, there can be little doubt that there is need for explicit attention to problems faced by women, such as safety, also and policies that create more affordable day-care spaces. These may be best addressed at a city level. Once gender differences are scrutinized it is but a small step to ask questions about solving discriminations faced by others, such as those suffered by people of a different race, those who are disabled, elderly or have alternative sexual orientations, which takes one into the question of whether the settlement does provide an emancipation city role for those with different life-styles (Lees 2004).

3.5.3.4 Inclusion

Related to the previous policy is the recommendation that inclusion, not discrimination, should be the aim of various zoning policies, development controls, and coping with immigrants, but without creating further segregation.

3.5.3.5 Community Boundaries

These should be porous, not rigid, as occurred in so many late twentieth century income-specific developments and especially in gated communities. Such rigid boundaries do not allow the social mixing and knowledge of people of different ideas and backgrounds, which have been two of the important experiences of city life throughout the centuries.

3.5.3.6 Segregation and Diversity

New communities or developments should not be built if they result in increasing segregation. Also, households should not be moved simply to facilitate diversity in urban areas since this disrupts social ties and generational linkages and helps to create delinquency.

These six policy guides provided by Fainstein represent important steps in explicitly showing how additional diversity in cities can be created, in order to remove many of the discriminatory practices of the past public planning, whether in private or public developments. However, they avoid addressing some of the practical tensions that exist where diverse peoples, whether in ethnicity or life-styles, are in close proximity, namely uneasiness and distrust, perhaps even violence between groups. Although Keating's (1994) work in Ohio showed that increased contacts reduced distrust between people, alternative views have been expressed by Putnam (2000) who has shown that ethnic diversity in areas, and especially increased immigration, reduced both local social solidarity and social capital. Hence at least two additional types of policy guides are required to ensure that diversity creates positive social relationships within cities.

3.5.3.7 Adding Community Contacts and Spaces

Policies designed to create more of these features are needed to allow different types of people in a city or in an area to get to know one another. These policies are almost always needed in areas of rapid social change, especially with influxes of new immigrants, in order to create more informed and trusting relations between diverse groups. This avoids or at least reduces the development of misinformed views about neighbours who are socially different, which can lead to hostility and even conflict. In this context some of the discursive ideas of communication theorists may help ease the path to understanding, especially the idea of the marginal publics, or counter-publics sphere. Essentially this involves the creation of spaces of communication and argument in which the hegemonic views of the dominant power in areas, whether market or government institutions, are challenged by alternative voices. For example, Wolf-Powers (2009) showed how the development of alternative discourses of different groups in the Bedford-Stuyvesant area of New York from the late 1960s, helped by sensitive TV programmes, and later by websites and blogs, altered the way that neighbourhood residents felt about their area and spread information about the effects of redevelopment proposals. This gave residents the confidence to raise issues and expose the relative benefits and dis-utilities of new plans. In addition, it led to different perceptions of the area by many outsiders, who realised that their prejudices of the area being only a black, dysfunctional ghetto was unfair. Even though there were high crime rates in the area, it contained many people who wanted to improve their life-prospects and who were the main victims of the local crime. This example demonstrates the utility of a multi-faceted approach in creating community spaces that has led to a great deal of success. However, the ability of grass roots organizations alone to generate positive change must not be exaggerated, especially in the most disadvantaged areas. Community or grass roots organizations suffer from many problems, such as motivating people to act to improve their area, finding effective leaders, having the ability to maintain pressure for positive change and to raise sufficient funds to continue their work (Davies and Herbert 1993, Chap. 7). Given such problems, municipalities ought to accept the need to help all of their residents, rather than automatically favouring market-led developments, such as by allowing planners to use their expertise to play a more active role as advocates or facilitators for the community groups in the interest of the city as a whole, a point that Fainstein noted in one of her Equity guides. In areas of high immigration, the creation of these greater contacts between diverse groups, and improved knowledge of one another, may also lead to the recognition that incoming groups may have specific needs that are not always understood by the original host population. Although such issues are often seen as the responsibilities of national governments and their immigration policies, some metropolitan areas, such as Minneapolis–St Paul, have been particularly active in this field. They have provided language lessons and business development organizations for new immigrants in an attempt to speed up the process of adjustment. These policies allow new residents to contribute to the local economy more quickly, enabling them to become productive citizens who bring new skills and diversity to a city, rather than being mired in low wage, dead-end occupations. Elsewhere more cities now allow marches or demonstrations in support not simply of different ethnic groups who wish to celebrate their identity and perhaps grievances but also those of different life-styles or orientations. This is seen in the blossoming of parades or festivals such as those linked to feminist, gay and lesbian pride in western democratic cities, although they usually need city permits and sometimes police presence to deter bigots who refuse to accept such diversity. Such events have led to greater emancipation in some cities, trends that would have been almost unthinkable a generation ago. But they are still banned or even criminalized in many cities in countries of totalitarian or religious rule. In such countries the hope for the emergence of more emancipatory cities is still unfulfilled, due to repressions of the moves to social diversity, equality and tolerance that have been experienced in many developed countries.

3.5.3.8 Limitation of Some Behaviours

The development of more contacts between different groups may not be enough to avoid potential conflict between diverse peoples, whether of age, life styles or ethnicity. Since the behaviours of some individuals and groups may upset others, there is often a need to create policies that have the aim of modifying the behaviours of one group if they obviously violate the rights of others. This can only be achieved by the careful management of behaviour where potential clashes may occur, which involves people with skills in resolving conflict. Perhaps the simplest example of restriction can be seen in the context of age differences where teens are not allowed to play in playgrounds designed for smaller children, since they may scare the younger age groups and their mothers, illustrating the wider problem of needing spaces for different groups. More generally, the issue shows the need for properly trained land-use officers, community wardens, or even police, to monitor and deal with the problems that emerge between groups with different cultures and behaviours. They need the power to intervene in the name of fairness if one group becomes a nuisance for others, without removing the civil rights of the criticized group. Indeed, real equality of opportunity may only be possible for the disadvantaged if some limits are placed on the behaviour of others, to avoid clashes in life-styles breaking out into conflict. Policies to create respect for all parties involved are required, so that the solutions are not seen to be coercive for any group. The most difficult problems occur when one group refuses to adjust their behaviours in the interest of creating greater harmony, claiming it has the right, perhaps on religious grounds, for its approach. Diversity may then become a problem, not an advantage for an area.

3.6 Critiques of Just City Approaches

There can be little doubt that this new emphasis upon specific redistributive justice measures at a city scale is long overdue. If these guides are adopted they will produce useful results, not simply in relation to improving the quality of life of the most disadvantaged, but in ways that can create more opportunities and fairness for the majority of urban residents. However, despite the promise of the approach, the current Just City literature does not normally deal with several broad categories of issues that affect the extent to which more just cities can be created. These relate to the limited powers of urban governments and territorial, inter-generational and societal context features. Recognizing and solving these issues will increase the effectiveness of the new approaches to a just city.

3.6.1 Restricted Urban Powers

Most cities, unless they are city-states like Singapore, have far less legislative and taxing powers than the national states within which they are located. Hence their ability to enact broad redistributive measures, such as the welfare policies found in most western democracies, are, in practice, relatively limited. However city governments often act as partners with the state in welfare policies, or as agents of the state for policy delivery, while some, as seen in the discussion on New Regionalism in Chap. 2, have acquired more powers from the state, sometimes by a new charter that defines their new rights. In addition, most cities in the developed world today are financially challenged, given their limited taxing powers, obligations to provide many services and the increasing pressure for infra-structure renewal and maintenance. Indeed, they depend on higher levels of government—such as national, state or provincial jurisdictions—for the money to provide many services. So despite the potential utility of the policies designed to create more just decision-making in the public sector, and perhaps providing some basic needs, the major effective redistributive justice measures still operate at the national state level in most countries. Hence there is a need to integrate, or at least to relate, more of the Just City literature to the redistributive justice policies at a state level with those at an urban level.

3.6.2 Territorial Issues

3.6.2.1 Localism

Most Just City policies are also limited because of the spatial restriction of their powers, which apply only to the area of control by the municipal government and even this is usually under the jurisdiction of a higher authority, such as a province, state or national state. Even if one city designs and implements some new progressive and just policy there is no guarantee that the policy will apply to other jurisdictions, since the administrative authority of a city stops at its borders. Certainly new administrative units, such as the various types created under New Regionalism (Wheeler 2002; Scott 2005) described in Chap. 2, may create policies for larger spatial entities, such as those composed of a central city and its surrounding region of separate municipalities. However, even these will be spatially restricted. Nevertheless, there is a pressing need to recognize the common interests of governments within urbanized regions and to ensure that problems are dealt with at a regional level. Historically, towns and cities often banded together, designing policies for mutual benefit, such as the medieval Hanseatic League, although these were mainly for defence or commercial advantage. Although there are relatively few examples of such entities today, the increased ease of communication and recognition of common problems has encouraged some cities or citizen groups to form networks, sharing and learning from one another, as seen in the examples described in other chapters, ranging from Winter Cities, through Resilient to Slow Cities. A more comprehensive spread of Just City ideas from one city comes from a national state mandating or encouraging other centres to adopt successful or proven policies that were pioneered in one place for all cities, or from imitative behaviour by other jurisdictions once they see that a policy in one urban place has proved effective.

3.6.2.2 Scale Effects

A second spatial restriction of most current Just City approaches is associated with scale effects. This means that the consequences of actions to reduce spatial injustice in one city may have unjust affects an international context. For example, a move by one city in the developed world, or even the states involved, to provide workers with higher wages, better conditions and lower taxes, may reduce their wage exploitation. This could lead owners to move their factories, leaving the local workers unemployed. Although the re-location provides new jobs elsewhere, there are poor conditions of work in the low wage sweat shops and factories of Asia that produce cheap goods for city workers in the affluent world. The increased demands from the developed world city may exacerbate their situation by increasing their hours of work to make the extra goods. Indeed, the practices of many first world companies in outsourcing production in the search for lower costs has increased the problems since these companies have often claimed they are not morally responsible for such conditions. The companies maintain that they simply buy the products of other firms, who set wage levels etc. Increasing awareness of the need to reduce the exploitation of the workers in underdeveloped world has led increasing numbers of developed world companies to insist that their producing partners set at least minimum standards for employment and building structures in the production centres in the developing world, which has probably marginally improved factory conditions for many workers, creating a type of off-shore justice. Others such as Puma, the sports clothing firm, have gone further. In 2011 they began to show the monetary flows at all stages of their production train, not simply to show transparency in their operation, but to encourage their suppliers to be both fairer to workers and more sustainable in their practices. An older approach can be seen in the growth of Fair Trade movements, especially in retail Co-operative Societies in the U.K., and even by some municipalities, which ensures that many products sold in stores are the result of purchases from producing units in developing countries that provide fair wages and work practices. These policies indicate some limited progress in recognizing the need to take inter-territorial consequences of global trade into account in the search for greater urban justice. The tragic and unnecessary collapse of an unsafe building in which many garment workers in Dhaka (Bangladesh) toiled, on May 10th 2013, led to 1,129 deaths, and has focused attention on this problem. The disaster seems to be leading to new policies to mitigate, if not solve, such safety problems. Some of the clothes companies that used the factory have combined to set up a fund to compensate workers who died, and also to create codes of practice to ensure the workers in factories producing their goods are employed in safe buildings. There is also pressure to allow unions to bargain for better conditions. However the scale of compensation is still relatively limited and some companies have still not contributed. Moreover progress on providing better working conditions is still limited, and will only be realized if effective factory inspections are carried out. This requires enough qualified inspectors, whose current limited numbers are still far from enough to monitor the various companies. Also it requires a national government prepared to enact appropriate laws to create these policies and to intervene if they are not implemented.

Elsewhere, some companies and even urban places have begun to implement Fair Trade policies to ensure workers in distant lands are compensated adequately, rather than simply choosing products on the basis of their lowest cost. Such movements have a long way to go. However, they do indicate greater awareness of the need to take inter-territorial factors into account in the search for practices that are more just, and that can be applied at an urban level.

3.6.2.3 Spatial Dialectics of Justice

A third issue concerns the need to investigate what can be summarized as the spatial dialectics of justice. So far the discussion has focused on the spatiality of potential injustice, in the sense of the way that urban scholars and practitioners have identified the reasons for the persistence of disadvantage in urban areas and the location of the affected areas. However, Dikeç (2009) has argued that there is also a need to examine the injustices produced by spatiality, for, space, as in a city, cannot be viewed simply as a container where injustice is found. Instead, it needs to be seen as part of a social process, in sense that the social construction of places subsequently affects people located there, or who use, or are unable to use, the space created. So the spatial dialectics approach investigates the 'site spatiality' of potential injustices. For example, it has been clear from over a century that inner city ghettos have negative generational effects, by reinforcing poverty for the majority of residents because of limited employment, poor schooling, high crime rates, dysfunctional families or lifestyles, even though some manage to escape, issues that are reviewed in discussion on Healthy Cities (Chap. 13), where such social determinants of health produce higher levels of sickness. Other injustices may be the unforeseen and unfortunate consequences of policies that actually attempted to create greater equity. For example, many European countries built many large public housing estates on raw land on the edges of existing cities. The idea of providing adequate housing for the least advantaged may have been praiseworthy, but the practice was often flawed since buildings were poorly constructed, badly maintained, and had few security features or effective on-site administration, which helped many to descend into crime-ridden slums (Davies and Herbert 1993). A forgotten spatial feature was that the remote location of many of these areas was bound to handicap the inhabitants, since they were often too far from employment sites, and often had inadequate public transit facilities to connect them to shops, employment and other services, let alone to their family networks that may have helped modify aberrant behaviour. Recent studies of public sector tower blocks that were located in more central locations showed that they were more valued and attracted a wider range of people because of their proximity to jobs and services (Murie 2005). To take another example, studies of the inner areas of large world cities with strong financial services and administrative functions have shown that these areas have become too expensive for working class households, since they cannot compete with the increasing affluence of the quaternary and quinary workers, or those with inherited or off-shore wealth who also desire locations in these areas for social, as much as economic reasons (Sassen 2001). This results in the effective exclusion of working class residences from the inner city of prosperous centres, whether office cleaners, domestic servants, or the social service workers, in teaching, in security, in health care etc. Although these workers are still needed by the affluent classes and businesses, they face huge problems, due to their relatively low wages in an expensive metropolis, with long and often expensive commutes to their places of work, some at anti-social hours. This type of spatial injustice is not due to the unanticipated effects of deliberate policy as in public housing estates, but is a consequence of letting only the market control the distribution of housing and its price. This current problem may be compared to the situation in eighteenth century London when the rich created large houses around residential squares in west London, but developers also made sure there were also smaller housing developments where the local shopkeepers and servants who serviced these areas lived and worked (Summerson 1962).

A particularly egregious spatial injustice exists in Chinese cities where as many as 250 million urban residents have no rights to the education and health facilities in cities because they or their parents are rural migrants, still classified as rural dwellers under the *hukou* system of household registration. New policies announced in 2013 will give urban residence rights to perhaps 100 million of these residents, so long as they have a stable job and legal residence and live in cities of 5 million people or less (TE 2014). This will solve the problem of less than half of these so-called illegal residents, many of whom have helped build the cities which deny their rights, although some do have the right to return to their rural locales and hence farm plots on retirement. This and other examples show the need to investigate the spatial dialectics of new developments or the consequences of old policies. The creation of new spaces often reproduces injustices, because they limit the future opportunities of people who are affected by the spatiality involved.

3.6.3 Inter-Generational Equities

In a similar vein it can be argued that the focus of most Just City guides or policies at present relates to the effects of decisions in one particular time. However, the attainment of equity in particular also needs to be investigated in a temporal sense, especially in the effects that occur between generations. The obvious example of an inter-generational problem may be the consumption of some *finite resource* without regard to the needs of a future generation for that resource, issues that are increasingly being dealt with under sustainability concepts (Chaps. 5–7). However, in the specific context of any urban area this type of problem can be seen in many examples: the loss of a public park space to private development; the exhaustion of some fossil energy source, leaving none for future generations; the long term effects of polluting some waterway or land in and around a city by industrial processes in the name of economic progress. The latter will ensure that future generations will have the expense of clearing up the mess, or suffering the health effects of the pollution. One solution has been the creation of land use approval policies that insist that companies, especially in mining or in waste disposal, should set aside a fixed proportion of operating costs each year to create a fund that will be used to remove industrial buildings and waste dumps when the production unit closes, and to restore land to a use as close to its original state as is possible, or even to improve its quality.

There is also increasing awareness of the problem that stems from the aging of western societies and settlements. Given the decreasing numbers of working people, and the increasing numbers of old people who are also living longer, it is increasingly apparent that the existing welfare nets and pension rights that support the elderly may be difficult if impossible to maintain, which creates health problems as discussed in Chap. 12 (Healthy Cities). Although governments are trying to increase the age of retirement, these policies look to be timid solutions to a major problem for future generations. In any case it is surely unfair to treat all the same: few manual workers can be expected to continue working until 70 instead of 65 years old, given decline in physical capabilities, whereas many office workers would not suffer the same disabilities. As yet these issues are being conceptualized on a national, not a local urban level. Nevertheless, municipalities will be affected, since these deliver many of the services on which the elderly depend. Moreover, since urban budgets are being reduced in many countries there are increasing doubts about the ability of most cities to continue to provide effective coverage for people in need, especially the elderly. In Britain, reduced grants from the central government since the financial crash of 2008 have meant that many municipal facilities, such as leisure and community centres, are being cut on the basis of financial exigency. Similar reductions in American cities are a consequence of the fact that pensions for former employees are often still the responsibility of municipalities rather than the state or federal government. Many urban places are struggling to pay these liabilities, which have led to bankruptcies, pension cuts and drastic reductions in police and other services. These examples show the need to be increasingly aware of such generational issues. But there is still a long way to go in ensuring that such issues are always included in reviews of policies designed to ensure greater social justice in a city.

3.6.4 Equalities and Freedoms

It must also be noted that the guidelines shown in Table 3.3 are not very comprehensive on the freedoms and equalities described in the capabilities approach (Fig. 3.1).

Such features need to be added for completeness, although it must be remembered that Fainstein's guiding principles were explicitly designed for democratic developed countries, where many of these principles are part of the rights of citizens, whereas they are absent in poor and totalitarian states.

3.6.5 Societal Contexts

Simply proposing a set of guiding principles for decision-making, or even basing policies upon 'rights to the city' in order to create a fairer and more just city, will not be enough to create progress in a practical sense. There is also a pressing need to recognize the importance of the effect of the social context in influencing effective change towards the long term goal of greater social justice (Dear and Scott 1981). Fainstein (2010) and Marcuse (2009) both note the need for *leadership* to promote the ideas that will initiate progress to justice, as well as the *mobilization of support* in democratic societies to get decision-makers to accept the utility of these approaches. As we can see from the influence of novels, such as the books Dickens produced about nineteenth century London and Britain, fictional narratives describing injustice can be as effective, if not more effective, than dry statistical accounts of problems. But drawing attention to problems is only the first step. As the whole history of the development of progressive policies in the past has shown, a great deal of political struggle is needed before action is taken to implement the ideas of greater justice. Moreover, the people involved in this movement cannot simply work on the basis of what they assume to be the utility of their ideas—whether based on the principle of fairness in the distribution of wealth and influence, distress at the condition of the needy, or a moral duty to help the disadvantaged. They must also counteract existing values and beliefs that work against acceptance of the Just City and Rights to the City approaches. Two issues in particular need to be resolved.

3.6.5.1 Counteracting Neo-Liberalism

The first problem is to effectively counteract the neo-liberal arguments about what advocates see as the necessity of creating policies that primarily serve the market (Harvey 2005). One of the key arguments for many redevelopments is that they create more amenities in an urban area, whether housing or shopping areas, making the centre more competitive in attracting and catering for the creative workers needed for the new economies. But the reverse is true. If a town or city has large numbers of poor, as well as squalid environments and high crime rates, new growth is not likely to be attracted, for these negative features will repel potential investors. So it must be made clear that such problems need to be resolved if an urban area is to find a better economic future. In addition, there is a practical reason for removing these problems, namely that these individuals will become likely more productive and better consumers, as well as being less likely to turn to crime or political violence to achieve their aims. There is also the moral benefit of improving a sector of humanity and increasing their dignity and self-worth. It is also worth noting that high levels of inequality have been shown to lead to slower recovery from recessions, in part because of lower spending by the increasingly economically challenged middle classes. So increasing inequality ought to be challenged on this practical ground as well (Wilkinson and Pickett 2009).

3.6.5.2 Countering Contributionist Views

A second problem to be addressed in the search for a more Just City is countering the atomist or contributionist views of those who question the need for redistributive policies, or at least reducing the dominance of this value in policies (Taylor 1991). This is not to deny that some element of return to individuals is justified on the basis of personal contribution. After all, if this is removed, much of the initiative for new ideas and applications in our essentially capitalist society is lost, to the detriment of progress. Rather it must be emphasized that any personal contributions must be set within the context of society as a whole, for many of its features—whether education, policing or good transportation—provide the background, support and even protection for individuals. In the words of the eminent philosopher Charles Taylor (1986, p. 60): "society is valuable as a collaborative enterprise whereby the contribution of each can be multiplied through coordinated activity". He further argued that the aims of our association in society should not only be for personal instrumental reasons, using the various organizations and services solely for our own advantage. Instead, we should accept that our associations with people of different skills allow us to be more productive, if only because we cannot do everything for ourselves and have to depend on others for certain goods and services—an obvious situation in the specializations that operate in any workplace. Indeed, the contributions that people make do not only depend on their own actions. In contemporary society many jobs involve years of education or training, which are frequently heavily subsidized by state contributions to educational institutions. In addition, the goods or services that many individuals produce can only operate in a larger institution where people depend on each other. For example, in safe and well equipped hospital buildings, cleaning staff, administrators, nurses and many skilled technicians are needed before surgeons can operate. Also, many of the skills for which some individuals wish to take sole credit, are only useful in certain types of society. For example, a mathematician is more valued and creative in a computer-based society than in a traditional rural society. Moreover, many of us live in societies in which valued qualities, ranging from personal liberties, the rule of law, respect for others, and certain cultural practices, have developed to protect and enhance our lives, with various organizations, not simply police, providing the services that guarantee these features.

Finally, it is argued that it is not enough to consider these practices and institutions as protecting our liberties and our ability to carry on our lives without retribution; we also need those that "sustain a sense of liberty" (Taylor 1986, p. 60). By this Taylor means we need to pass on the goods that we have received to future generations, to sustain and improve our society. Taylor's arguments for considering us all as common citizens can be extended to specifically address the case of the underprivileged. In a democratic society it is difficult to argue against Galton's (1986) view that the full development of an individual should be morally equal to any other. Also, in a practical, progressive sense the poor, unemployed and weak represent a waste of potential resources, for if they were able to leave these conditions the society would benefit from greater production, spending power etc. This would replace the current drain on revenues devoted to maintaining, not improving them. Perhaps many of the current needy may not achieve much, because of the limited opportunities they have received in the past. But this is not true of their children, whose potential could be released through education and opportunity, resulting in advances for the benefit of society as a whole. All one has to do is to envisage a developed world today that still had the same number of educational institutions and attitudes to women that existed in the 1930s. The expansion of both educational institutions, and the removal of many, if not all, of the discriminations against women since that time, gave many women, as well as working class people, better educations and eventually jobs. The move to gender equality in particular, although still not complete, has been to significantly increase productivity, and in many cases life satisfaction. On economic grounds alone these changes have made a huge contribution to gross national product.

Recognition of the societal contribution to what seem to be our own personal achievements, may also be extended to the argument that it is necessary to stress our *common citizenship* as a basis for action. In many ways we do this at the national level. However, we should accept the fact that most of us live in cities and their effective functioning does depend upon a common sense of identity and purpose to improve existing conditions. Our society is too complex to return to the city-state principles of the classical Greeks polis where membership and participation in the polis was an important part of collective social life, at least for the very small minority of male property owners who dominated society. However, the sense of common purpose, and the idea that we are all assisted and enhanced in our life by our common residence in a city or urban place, are issues that need greater emphasis if we are to create more Just Cities.

3.7 Conclusions

The search for social justice in urban places has a long history but with very different aims because of the varying meaning and contexts of what justice means. The focus of this review on the essentially *incremental approaches* to Just Cities will not satisfy those who believe that far more transformative approaches are need to break the injustices that stem from the dominance of the capitalist system in a neo-liberal era, given their belief that it intrinsically creates inequality. But political support for such a far reaching approach is still very limited. In a book entitled *Rebel Cities*, Harvey (2013) suggested, or rather hoped to see, meaningful change altering the capitalist value system coming from coalitions between the many movements seeking greater justice that have developed in and between cities. He observed they all too frequently flourish, then fade, failing to connect with one another. Hence he argued the case for urban solutions: "reclaiming and organizing cities for anti-capitalist struggles is a great place to begin" (Harvey 2013, p. 153). Such a view may imply that there is little hope of effective socialist change coming from state-based actions, given the built-in conservatism of the political class and the power of corporations in democratic countries, and where even the supposedly egalitarian Chinese Communist leadership have placed a new emphasis on market-based guidance in the resolutions of the 2013 Party Congress. Despite this aspiration for future actions based on cities there are few guides about how to achieve this change in a transformative sense, especially given the limited powers of most urban governments. In any case there is always the problem that success in one city may not be copied by the many other cities in which injustice occurs, despite Harvey's suggestion for some kind of association developing between cities sharing in this task, a kind of radical Hanseatic League of co-operating socialist cites. This role is a long way from the merchant capitalism that created and maintained this original medieval league, but does point to the recognition of the utility of creating networks between cities to promote some future goal, an idea that lies behind many of the themes discussed in other chapters. This view of combating injustice is still a matter of hope, with only limited examples of places where such change is occurring. Hence most would still argue there seems little alternative to developing the type of incremental and achievable policy aims discussed above that will lead to more just cities.

The types of policy initiatives to reduce injustice that have been described in previous sections are all based on two features. The first is upon *'fairness' as a redistributive principle*, one upon which decision-making in planning and redevelopment can be evaluated, and which exposes and replaces the current emphasis upon competitiveness and maximizing utilities. The fairness should not be seen only in the income or inequality sense that is stressed here; it should also be seen as a basis for ensuring the tolerance for diversity, whether in ethnicity or in life-styles etc., part of the general search for more justice. This means incorporating objectives to create what has become known as emancipatory cities (Lees 2004). The second is upon *urban-based policies*, seeing them as arenas in which more effective changes are needed, instead of only relying upon national state policies as the basis for redistributive justice. Yet some caveats must be made. The Just City literature has not yet effectively linked national, regional and urban policies together. Moreover it is obvious that many city budgets are incapable of financing many of the capabilities needed. So they, as well as local communities within their area, would need a greater share of the national tax income and additional powers to enable them to implement such additional functions. In addition, the problem that Aristotle noted millennia ago needs to be appreciated, namely that there cannot be only one form of justice. This truism is even more relevant today, given the greater complexity of contemporary cities and society, as well as the different problems that exist in various urban places.

Some may see the explicit adoption of the principle of fairness as an aspirational goal. Inertia in city governments frequently leads to a reluctance to change existing approaches and continued support for the continuation of policies based on the maximization of financial profit. The recent retrenchment of many governments—especially given the debt and unemployment levels created by the post 2008 financial crises—has not helped the progressive agenda. Yet this crisis has meant there is an even greater need to deliver many of the basic capabilities shown in Table 3.2 in cities, even in the developed world, and especially in the cities of the developing world. Nevertheless, there may be reasons for believing that changes in reducing injustices in the provision of goods and power could occur, even relatively quickly, once sufficient support for the idea is created among the general public and the decision-makers in particular, as two examples in related fields show. For example, the practice of carrying out archaeological surveys before development occurs, at least in historic places, is now standard in western urban centres. They have ensured that significant historical artefacts and evidence of the past are not lost. Similarly, environmental reviews to avoid, or at least mitigate, the negative impacts of development on the wildlife or physical environment of urban areas are also usually mandatory—at least in many parts of the developed world—although it is obvious that some impacts are still being ignored. These two examples show that the adoption of new principles in the decision-making processes that surround development or redevelopment in urban places have occurred relatively quickly in contemporary urban places. Yet it must be admitted that the attainment of greater justice, let alone the many goals of a Just City, may be more difficult to achieve than in these two examples, given the forces of neo-liberal ideology and the accumulated injustices already built into both the urban fabric and societal mores. Nevertheless the emerging Just City literature and related, more radical movements such as the Rights to the City Alliance are proving successful in focusing attention upon practical issues, showing how change could occur. Given the way that the older progressive approaches in urban areas that had achieved so much in the previous century had floundered by the end of the twentieth century, this progress is overdue. Moreover, the Just City approach, instead of just criticizing existing policies and addressing each problem of injustice as a single issue, now seeks to develop a more coherent body of policy-related literature in order to more effectively tackle the serious injustices that still exist in our towns and cities. But we must be cautious. Real progress will also require addressing the wider issue of persuading public opinion that seems increasingly contributionist to support the need for more just policies, rather than only formulating policy examples. In this regard it is worth ending by remembering the words of a prominent justice philosopher who has succinctly noted the problems of the past and identified the requirements of the future.

> We have…..made our citizens more unequal than they ought to be. We have not sustained the social infrastructure that our social life requires. We have not made a sufficient commitment to communal provision. We have not provided a wide enough range of opportunities. We have not challenged the power of private governments. Justice requires that we do all these things, but it also requires that we do them democratically. Hence the burden that

egalitarian philosophers must accept is to provide a persuasive interpretation of democratic citizenship and then of the goods and opportunities that citizens distribute to one another. (Walzer 1986, p.149)

References

Alinsky, S. (1971). *Rules for radicals*. New York: Random House.

Barker, E. (1969). *The politics of Aristotle*. Oxford: Oxford University Press.

Berry, B. J. L. (2005). *Why social justice matters*. Cambridge: Polity Press.

Booth, C. (1902–3). *Life and labour of the people of London* (17 volumes). New York: Reprinted AMS Press. (1970).

B.S.A. (2013). Spending and welfare. *British Social Attitudes Survey*, Annual Report (30), Table 2-7 and Fig. 2-3.

Carver, T. (Ed.). (1991). *The Cambridge companion to Marx*. Cambridge: Cambridge University Press.

Clark, D. A. (2005). The capability approach: Its development, critiques and recent advances. Economic and Social Research Council, for Global Poverty Research Group, London.

Clark, D. A. (2006). *The Edgar companion to development studies*. Cheltenham: E. Edgar.

Claval, P. (1978). *Espace et pouvoir* Paris: P.U.F.

Crisp, R. (2000). *Nicomachean ethics: Book V* (trans. from Aristotle). Cambridge: Cambridge University Press.

CS: Cash strings. http://www.economist.com/cashstrings. Accessed 12 Oct 2013.

Davidoff, P. (1965). Advocacy and pluralism in planning. *Journal of the American Institute of Planners, 31*, 277–296.

Davies, W. K. D., & Herbert, D. T. (1993). *Communities within cities*. Belhaven: Wiley.

Dear, M., & Scott, A. J. (1981). *Urbanization and urban planning in capitalist society*. London: Methuen.

Dikeç, M. (2009). Justice and the spatial imagination. In P. Marcuse et al. (Eds.), (pp. 72–88).

Engels, F. (1845). The condition of the English working class (trans. from Aristotle). Reprinted with introduction by E. Hobsbawn (2005), Academy Chicago Press, Chicago.

Fainstein, S. S. (2010). *The just city*. Ithaca: Cornell University Press.

Fraser, N. (2003). Social justice in the age of identity politics. In N. Fraser & A. Honeth (Eds.), *Redistribution or recognition: A political-philosophical exchange*. New York: Verso Press.

Friedman, J. (2000). The good city: In defence of utopian thinking. International Journal of *Urban and Regional Research, 24*(2), 460–472.

Galbraith, J. (1958). *The affluent society*. Boston: Houghton Mifflin.

Galbraith, J. (1992). *The culture of contentment*. Boston: Houghton Mifflin.

Habermas, J. (1989). *Theory of communicative action* (trans. by T. McCarthy). Cambridge: Polity Press.

Hall, P. (1988). *Cities of tomorrow*. Oxford: Blackwell.

Hardie, W. F. R. (1980). *Aristotle's ethical theory*. Oxford: Clarendon Press.

Harvey, D. (1973). *Social justice and the city*. London: Ed. Arnold.

Harvey, D. (1989). *The urban experience*. Oxford: Blackwell.

Harvey, D. (1992). Social justice, postmodernism and the city. *International Journal of Urban and Regional Research, 16*(4), 588–601.

Harvey, D. (2005). *A brief history of neoliberalism*. Oxford: Oxford University Press.

Harvey, D. (2013). *Rebel cities: From the right to the city to the urban revolution*. London: Verso.

Harvey, D., & Potter, C. (2009). *The right to the just city*. In P. Marcuse et al. (Eds.), (pp. 40–51).

HDCA: Human Development & Capabilities Association. http://www.hd-ca.org. Accessed 10 Nov 2012.

Healey, P. (1997). *Collaborative planning*. Vancouver: University of British Columbia Press.

HIC: Habitat International Coalition. http://www.hic-net.org/indepth.asp?PID=18. Accessed 8 Nov 2012.
IFS. (2011). *Poverty and inequality in the U.K.* London: Institute for Fiscal Studies.
Jackson, M. W. (1986). *Matters of justice*. London: Croom Helm.
Keating, W. D. (1994). *The suburban racial dilemma*. Philadelphia: Temple University Press.
Kelly, P. J. (1990). *Utilitarianism and distributive justice: Jeremy Bentham and the civil law*. Oxford: Clarendon Press.
Keynes, J. M. (1936). *The general theory of employment, interest and money*. Basingstoke: Palgrave Macmillan. Republished 2007.
Lees, L. (Ed.). (2004). *The emancipatory city*. London: Sage.
Lefebvre, H. (1968). The right to the city. In E. Kofman & E. Lebas (Eds.) (1995), *Writings on cities*: Henri Lefebvre (pp. 63–184). Oxford: Blackwell (Trans. and republished).
Lucash, F. S., & Shklar, J. N. (Eds.) (1986). *Justice and equality here and now*. Ithaca: Cornel University Press.
Marcuse, P. (2009). Postscript: Beyond the just city to the right to the city. In P. Marcuse et al. (Eds.), *Searching for the just city*. London: Routledge (pp. 240–254).
Marcuse, P., Connolly, J., Novy, J., Olivio, I., Potter, C., & Steil, J. (Eds.). (2009). *Searching for the just city: Debates in urban theory and practice*. London: Routledge.
Marx, K. (1975). *Marx's collected works*. London: International Publishers.
Marx, K., & Engels, F. (1848). *The communist manifesto*. (New edition with introduction by M. Malin). New York: Penguin.
Mount, F. (2005). *Mind the gap*. London: Short Books.
Mount, F. (2012). *The new few: or a very British oligarchy*. London: Simon & Schuster.
Mumford, L. (1961). *The city in history*. London: Secker and Warberg.
Murie, A. (2005). Lessons for development. Paper presented at RESTATE conference on Restructuring Large Housing Estates in European Cities, Ljubljana, Slovenia, May 19–21.
Nozick, R. (1974). *Anarchy, state and utopia*. New York: Basic Books.
Nussbaum, M. C. (2011). *Creating capabilities: The human development approach*. Cambridge: Belknap Press.
Orwell, G. (1937). *The road to Wigan pier*. London: Gollanz.
PG: Poverty Group. (2009). http://www.poverty.org.uk/09. Accessed 12 Oct 2012.
Piketty, T. (2014). *Capital in the twenty first century. (English translation)*. Cambridge: Harvard University Press.
Plato. (1968). *The republic* (trans. by A. Bloom). New York: Basic Books.
Portes, J., & Riley, R. (2011). *Poverty, inequality and social justice*. London: NIESR.
Putnam, R. D. (2000). *Bowling alone: The collapse and revival of American community*. New York: Simon & Schuster.
Ravallion, M. (2013). The idea of antipoverty policy. National Bureau for Economic Research, Research Paper 19210, Cambridge, Mass.
Rawls, J. (1971). *A theory of justice*. Cambridge: Harvard University Press.
Rawls, J. (2001). *Justice as fairness: A restatement*. In E. Kelly (Ed.). Cambridge: Harvard University Press.
RCA: Rights to the City Alliance. http://www.righttothecity.org/index.php/about. Accessed 8 Oct 2012.
Roszak, T. (1969). *The making of a counter-culture*. Garden City: Doubleday.
Rowntree, S. (1901). *Poverty: A study of town life*. London: Macmillan.
Sassen, S. (2001). *The global city* (3rd ed.). Princeton: Princeton University Press.
Scott, J. W. (2005). The new regionalism as a contingent governance paradigm: European and North American perspectives. In Y. Murayma & G. Du (Eds.), *Cities in global perspective* (pp. 452–461). Tokyo: College of Tourism, Rikkyo University, and I.G.U. Urban Commission.
Sen, A. K. (1999). *Development and freedom*. Oxford: Oxford University Press.
Sen, A. K. (2009). *The idea of justice*. Cambridge: Harvard University Press.
Sen, A. K., & Williams, B. (1983). *Beyond utilitarianism*. Cambridge: Cambridge University Press.

Shklar, J. N. (1986). Injustice, injury and inequality: An introduction. In F. S. Lucash (Ed.), (pp. 13–33).

Smith, A. (1776). *An inquiry into the nature and causes of the wealth of nations*. (Reprinted 1976). Chicago: University of Chicago Press.

Streeten, P., Burke, S. J., ul Huq, M., Hills, N., & Steele, F. (1981). *First things first, meeting human needs in developing countries*. New York: Oxford University Press.

Summerson, J. (1962). *Georgian London*. London: Penguin.

Taylor, C. (1986). The nature and scope of distributive justice. In F. S. Lucash & J.N. Shklar (Eds.), *Justice and equality here and now*. (pp. 34–67) Ithaca: Cornel University Press.

Taylor, C. (1991). *The malaise of modernity*. Concord, Ontario: Anansi Press.

TE. (2014). Building the dream: Chinese cities. *The Economist,* April 19, Special Report, 1–16.

Tomalin, C. (2011). *Charles Dickens: A life*. London: Viking.

UNESCO. (2013). http://www.unesco.org/education/poverty/grameen.shtml. Accessed 20 Oct 2013.

Vance, J. (1977). *This scene of man*. New York: Harpers.

Walzer, M. (1984). *Spheres of justice: A defence of pluralism and equality*. New York: Basic Books.

Walzer, M. (1986). Justice here and now. In F. S. Lucash & J.N. Shklar (Eds.), *Justice and equality here and now*. (pp. 136–150) Ithaca: Cornel University Press.

Wellbank, J. H., Snook, D., & Mason, D. T. (Eds.). (1982). *John Rawls and his critics: An annotated bibliography*. New York: Garland Publishing.

Wilkinson, R., & Pickett, K. (2009). *The spirit level*. London: Penguin.

Wilson, W.J. (1997). *The truly disadvantaged*. Chicago: Chicago University Press.

Wolf-Powers, L. (2009). Keeping counterpublics alive in planning. In P. Marcuse et al. (Eds.), (pp. 162–172).

Young, I. (1990). *Justice and the poltics of difference*. Princeton University Press

Chapter 4
Green Towns and Cities

Wayne K.D. Davies

Our old views of cities, towns, and communities are incomplete and must be substantially expanded to incorporate ecology and more ecologically responsible forms of living and settlement.
T. Beatley 2000, p. 5

4.1 Introduction

The words 'green' and 'greening' are often used today as a synonym for policies stressing environmental issues and sustainability in economies and life-styles, rather than its original, literal use referring to the addition of grass, trees and other natural features to urban places. Although there can be no doubt that this wider use of the term 'greening' is now part of the sustainability agenda, and is a short-hand for these ideas, there is a sufficient coherence of the features associated with this older literal meaning of the term to deal with it separately in this chapter. Of course, the explicit idea of developing greener towns is not new. It was popularized by Howard's (1898) Garden City ideas as well as the Broadacre City suggestions of Lloyd-Wright (1932) and to the Regional Planning of Geddes (1915). All were designed to seek new and greener forms of urban centres by incorporating natural features into the built-up areas (Mumford 1961, Bacon 1974, Hall 1988, Stephenson 1999). Unfortunately, many of these ideas were watered down, distorted or even lost in the development of twentieth century planning practices. Certainly these pioneering examples led to increasing amounts of green space in the suburbs of western cities that adopted some of their ecological principles. But the resultant low density developments, especially when based on widespread car use from the mid-twentieth century, led to a new set of problems, such as sprawl, road congestion, limited social contact with neighbours, increased energy and resource use, and increased pollution. By the late 1980s criticism of this type of development led to pleas by planners for greater concern for the environment (Steeley 1991), the growth of the

W. K.D. Davies (✉)
Department of Geography, University of Calgary,
2500 University Drive, N.W., Calgary, AB T2N 1N4, Canada
e-mail: wdavies@ucalgary.ca

W. K.D. Davies (ed.), *Theme Cities: Solutions for Urban Problems*,
GeoJournal Library 112, DOI 10.1007/978-94-017-9655-2_4

New Urbanism movement in the U.S.A. (Chap. 2), and suggestions about how new greening policies pioneered in many European cities could be applied to American cities in order to include ecological principles and ways of living (Beatley 2000), issues that have been consolidated in a recent encyclopaedia of green practice in American settlements (White and Duram 2012). However, Register (2006) and other members of what is now known as the Ecocity movement, have gone further, suggesting that cities need to be radically re-organized on ecological principles so that they function in harmony with the biosphere and ecosystems. They have promoted the idea that urban settlements should be in balance with nature, rather than the more typical case of its destroyer. This would lead to towns and cities becoming unique oases in an otherwise natural environment, rather than the green areas being isolated oases in the dominant urban sprawl that characterises so many urbanized regions. More recently, the term Biophilic Cities has been used in a recent book by Beatley (2011) to describe a new emphasis on designing and planning cities, and the people within them, so as to reconnect with the natural world in their daily life.Although far from the ideals of the Ecocity or Biophilic City principles, many towns and cities throughout the world are increasingly focused on what can be called 'green agendas', not simply literally green in the old sense, but one part of the increasing trend to what has become known as urban sustainability. To deal with all these green and sustainability issues in one chapter would make it too long. So the wider sustainability issues are described in following chapters (Chaps. 5, 6 and 7), leaving this one to deal with the more *literal* green and natural agendas. The first section deals with the changing motivations for greening. Since it is important to stress that the current interest in the greening of cities is not new, the second part deals with the main historical trends in the creation of open green space in western cities, followed by sections dealing with more recent innovations: green infra-structure concepts; the growth of environmental restoration and revitalization; the move from green conservation to production; the green roofs and walls movements; the growth of cleanliness or tidy town approaches; and finally with the principles of the biophilic city movement.

4.2 Motives for Greening

The modern drive to create open areas of green space to be used by the public, not simply the elite, came from nineteenth century municipalities bent on civic improvement. Their main motivation was to provide an antidote to the increasing squalor of industrial cities; parks were seen as providing *refuges* from the noise, bustle and dirt of streets, buildings and factories. Parks advocates were helped by the eighteenth century Romantic Movement of poets and artists who glorified the experience of nature and the need to observe and participate in it (Summerson 1962). So the new parks were also seen as providing local substitutes for nature in the city, providing places for *relaxation and contemplation*—even aesthetic pleasure. Later, parks added more functions, because they were increasingly seen as places of general recreation, initially providing walking and sometimes riding paths, but subsequently as places of more *active recreational and leisure pursuits*, containing a range of additional features, such as swimming pools, even sports fields and tennis courts, lakes for boating, and areas with cooking facilities designed for family gathering. In addition, separate specialized areas in cities were set aside for various recreational pursuits as leisure time increased and schools developed their own playing fields (Mumford 1961, Vance 1977). Later, the larger park areas became important gathering places for open-air concerts and other celebratory activities. But the growing belief that parks and open spaces were also able to *improve the social life and health* of urban residents, became a fourth justification for parks, not simply by providing places where people could escape from the built-up city and engage in physical activity, but because the presence and use of these areas seems to reduce stress, improve sociability and generally made people happier, which can be an additional part of the Healthy City agenda (Chap. 13). This old subjective 'assumption' of the health benefits of green areas has been proved by new and rigorous research. For example, a recent report (EEA 2011b) quantified some of the advantages, arguing that a 10% increase in green space in European cities leads to notable disease reduction, adding five years to life expectancy. More generally, researchers such as Richard Louv (2008, 2011), are using the term 'nature deficit' to draw attention to the fact that people in general, but especially children, are now far less exposed to nature and green areas than in the past and have much more restricted ranges in their activity patterns. So the free-roaming children of the past who learned to live in natural surroundings and to play with many others are now a rarity, a trend not helped by the fact that most children are now bussed or driven to school, rather than walking by themselves or with friends. Louv argued that our species lived in natural areas for millennia and our brains were hard-wired to assume this nature association. Contemporary children are less exposed to natural conditions because of their increasing concentration in urban areas, many parts of which have few green areas, and are often prevented from playing outside by parents fearful of crime, or they prefer the recent attractions of computer and media devices. Parks, however, provide a small way of reducing the nature deficit. Another indication of this alienation from nature can be seen from one of Louv's (2011) surveys that showed that only a fifth of American children now live within a mile of their school, or ever ride a bike to school, compared to almost three quarters for their parents.

This rapid alienation from nature in the past thirty years is having human costs—costs that are being documented by an increasing number of research reports summarized in a new network (NCN 2012) and books (Kahn and Kellert 2002). These studies have shown many important results, such as: the finding that access to green space, even views of green space, can enhance self-control, peacefulness and self-esteem in children; the discovery that exposure to natural settings increases cognitive capacity in children, as well as their curiosity, problem-solving ability and intellectual development in general; even small amounts of contact with nature can reduce the severity of the growing number of children with Attention Deficit Disorder. Moreover, if children are in schools that have gardens which children can help to cultivate, they are more likely to be interested in healthy diets and become more curious about food sources and processing. However the health benefits do not simply come from being *in* green areas. The growing evidence on this issue has led to some health care centres, such as one in Waterloo (Iowa), creating over fifty types of gardens for patients to walk or work in. Of course this is a major extension of older ideas. Howard's Garden Cities had a Green Belt which contained hospitals.

Hospital research has also shown that views of green areas have helped adult hospital patients recover more quickly from illness (Ulrich 1984, Pretty et al 2005). Experiments are also under way to create computer generated images of natural landscapes being shown to patients who are surrounded by the alien and perhaps frightening equipment of modern medicine. So patients are not simply provided with images of natural environments, which have been shown to have therapeutic value, but are able to create virtual realities in which patients can add themselves into to natural landscapes and take tours in such areas. In addition, there has been increasing interest in the value what is now called the 'walkability' of various areas in cities. This has led to attempts to measure the opportunities for walking, to improve these opportunities through the introduction of more green areas and pathways, as well as the encouragement of people to walk more, for fitness, or to obtain basic needs from local shops, rather than driving. All these health and social benefits are providing increasingly strong motivations for policies designed to increase the greening of cities, especially given the increasing obesity of people in many countries. For example, one third of Americans adults are obese, which costs $ 147 billion in 2008 in health care costs because this condition increases the risk of other diseases (McMahon 2012). Hence health care practitioners are becoming increasingly aware of issues that used to be the preserve of urban designers. People who live in areas with parks and trails have been shown to be twice as healthy as those who live in areas without such features.

In addition, a link has been shown to exist between crime and greening. Recent research by crime and safety specialists has shown the presence of even small amounts of green space and trees helps to reduce tensions and crimes in cities, especially in the high density, over-concreted inner city environments of public housing projects, as seen in the work of researchers in the Landscape and Human Health Laboratory in the University of Illinois's Urbana campus (LHHL 2012, Kuo and Sullivan, 2001). Other works have shown the human value of trees in the city (Donovan and Butry 2010)) and even claimed that areas of large trees have lower crime rates, although this may be a casual association with high income areas not a causal connection (Donovan and Prestemon 2012).

A new source of support for adding green space in urban places comes from the recognition of a fifth set of advantages, this time in *environmental terms*, for greening provides physical advantages. This is not simply from the therapeutic advantages but from real economic benefits derived from natural processes that reduce problems, whereas alternative actions to solve them involve very high costs. For example, trees, especially forests, absorb significant amounts of the carbon dioxide and particles produced by human activity, especially from fossil fuel emissions, and also generate increased oxygen supply to an environment, thereby improving the local atmosphere. Wetlands also allow plants to absorb many pollutants in run-off waters, reducing the costs in sewage systems. In addition, since trees act as noise barriers, their use along roads can reduce one of the other unwelcome aspects of urban life. Green spaces also help the absorption of rain water into the soil, whereas the asphalt or concrete cover of so many city surfaces adds to the rapidity of runoff, which in heavy storms can lead to excessive flooding. So cities with higher surface permeability due to more green areas are more likely to cope with floods, as a series of interactive maps in a recent report dealing with climate change and urban adaptation has also shown (EEA 2012). Urban areas are also well-known heat islands, since the roads and buildings absorb and keep heat, ensuring that they have higher temperatures than rural areas. This led the European Environment Agency's (EEA 2011b) climate modelling group to estimate the effect of green space in reducing temperatures and improving health. They calculated that a 10% increase in green space in urban areas reduced average temperatures in European urban areas by 1 °C, whereas a 10% decrease can lead to an 8.2% increase in average temperature. In a world in which global warming is known to be increasing, adding green space reduces the negative effect of these heat island effects in summer. These effects are not trivial, for in the summer of 2003 over 70,000 extra deaths were recorded in 12 European countries due to a heat wave.

Although these five sets of advantages provide the main positive benefits for adding green space and parks in cities, they have been complemented by the recognition that green areas in cities can also provide *environmental restoration*, and also help add to the *food supplies* of cities, issues that will be discussed in subsequent sections. As always with urban changes, the addition of green areas has many problems, even negatives. For example, there is always the problem of reconciling conflicting, or even incompatible, uses in these areas. More generally, the costs of implementation and the maintenance of parks in particular may be an important line item in urban municipality budgets, especially in areas where increasing vandalism has destroyed plants and other property, leading to extra costs—although attempts to involve schools in educating youths about respect for property and green spaces has helped reduce some vandalism problems. The increase in urban budget problems since the financial crisis from 2007 has led some towns to reduce maintenance, which decreases their attractiveness. Others have charged entrance fees to some parks, in addition to the more usual payment for various recreational facilities, which reduces their public accessibility. But a more periodic difficulty is associated with the night-time problems of parks or forested areas in cities, a long standing problem. Many become places where anti-social activities or criminal behaviour occur after dark, making them dangerous, or at least unpleasant places, for the majority of the population who avoid such areas at night, though some with alternative life-styles may be attracted to them and accept the dangers. Although regular park patrols cut out some of these problems for the majority, the usual response of municipal authorities is to lock the parks at night and to discourage night usage—which again restricts their potential use—except in areas which can have easily patrolled and well-lit access routes or which contain functions such as entertainment and food services that attract people.

4.3 The Development of Urban Green Space Policies

Throughout history the development of urban places has gone hand-in-hand with the despoliation of the natural environment and the loss of green spaces. The very pressures of growth in urban places, their increasing size—and walling in some cases—led to increasing densities and a reduction in the open areas in these settlements. But from the fifteenth century onwards, one of the new key features of European Renaissance planning was the deliberate creation of new open green areas, although organized and manicured places, in and around cities for at least part of the public, namely the elite (Mumford 1961, Vance 1977). Many began as royal gardens or parks; others were new creations. But by the eighteenth century many of these areas were opened to the public and were formally laid out as parks (Summerson 1962). A century later the addition of parks became a general trend in most municipalities, with many adding a variety of recreational facilities in addition to green spaces and gardens. The slow greening of cities was also helped by the removal of city walls and the defences outside, areas acquired by the municipality and turned into parks, boulevards and lakes as seen in Copenhagen and Vienna. In the early nineteenth century the pioneering design of Regent's Park in London added another impetus to green space in cities, as plots for 26 large houses for the elite were planned to be in the park, while the big housing blocks overlooking the park contained big apartments or houses for the wealthy. The streets around contained either large lots for middle class residents or small terrace houses suitable for servants or artisans who would serve the elite (Davies and Herbert 1993). This trend was modified in the U.S.A. by the creation of new upper class suburbs from the 1860s where houses for the wealthy were scattered in a wooded and grassy setting and connected by curvilinear roads, such as in the case of Llewellyn Park in Chicago (Reps 1965). This naturalistic feature was copied in many upper income areas of cities throughout the western world and filtered down to middle class areas, but with far few open spaces and higher densities. In addition, from the mid-nineteenth century, most European countries enacted new municipal regulations over housing to avoid health problems due to overcrowding; these ensured that all working class residences had open space, whether gardens at the back of dwellings, or a common interior courtyard area for apartment blocks, although the amount of space varied considerably.

The desire to create much more green space and especially trees throughout urban areas, rather than only in segregated parks or suburbs, received a major stimulus from Howard's Garden Cities movement (1898) discussed in Chap. 2, that aimed to create healthy towns that combined the best of town and country with lots of green space as an antidote to the unhealthy, dense industrial cities of the day. A key feature was the concept of a Green Belt to surround the planned new town, thereby providing the town with a girdle of agricultural land to help feed the urban residents, as well as space for recreation and rehabilitation functions, such as hospitals and playing fields. The ideas inspired similar examples in many countries, although many imitated the morphology and greenery, but not the social ideals. For example, Krupp's sponsorship of Margarethenhöhe near Essen, has a praiseworthy design that has similarities to Garden City ideas but with a more focused centre.

However it had carefully selected residents from the supportive middle class and from loyal factory workers, excluding those with radical views, and, like most so-called Garden 'City' projects, was really a suburban addition, not a new town. But as was shown in Chap. 2 Howard was not simply advocating what today would be called a green agenda, there were social and economic reasons behind the new form. Unfortunately, when the Garden City concept was applied to other towns it frequently degenerated into what amounted to the creation of garden suburbs, low density areas with more green space simply added on to existing areas, and usually without many services. What was also forgotten about Howard's plans was the key social concept behind the ideas, namely that groups of people got together with architects to plan their own town, incorporating community facilities for a better co-operative social life, and was progressive in encouraging recreational and educational classes of self-improvement.

Howard's ideas of community co-operation and the need for planned community facilities were, however, fundamental to the development of the Neighbourhood Planning approach in the 1920s, pioneered by Perry (1929) and later modified by Stein (1951) and the British new towns of the post-war period. For Perry, new suburban areas should be developed under an overall plan, with a park in the centre, shops on the main roads outside, with the unit being big enough to service a neighbourhood school. In addition, houses were provided with gardens, used for recreation or for gardening, as in Howard's scheme. But after World War II, greater affluence and the demand for bigger houses—and hence lot sizes—as well as the increasing use of the car and the resultant need for garages and wider roads, modified the neighbourhood approach in which inhabitants could walk to the local facilities. Cities kept on adding these planned units to the edge of urban areas, creating what amounted to a planned sprawl. Together with the addition of new high-speed four lane highways, they contributed to the increasing sprawl of cities, while developments on larger lots outside the city led to even lower density, residential-only areas that could only be served by cars. The restriction of commercial and other services in these neighbourhood unit areas led to the development of highway-orientated ribbons and shopping centres to serve the auto-dependent suburbs, which meant that even the facilities provided in the neighbourhood units could not attract sufficient threshold population from the local area to compete with these developments, which led to the decline of many of their local shops.

An alternative to the developing urban sprawl of the twentieth century was proposed by Le Corbusier (1929), with his advocacy of cities composed of high rise towers. Although much maligned because of the disasters of so many public sector high rises that were built after World War II (Coleman 1985), Le Corbusier ought to be recognized for suggesting that his tower blocks should contain many social features, such as community facilities, shops, schools and security. In environmental terms he emphasized that the towers should be set in a park setting so nature could be enjoyed, not jammed together, as well as stressing the importance of light in the apartments, with many windows and balconies at both sides of the tower blocks to catch the morning and evening sun. Most would agree that Le Corbusier over-stated the case for both high density and the too-uniform modernist styles of stark concrete and steel. Yet his desire to increase urban densities in place of the sprawling suburbs had a point, as did his advocacy of links with nature. Sadly, many of the buildings that purported to encompass his high density visions in the public sector after World War II also completely ignored the shopping and community facilities he believed should be incorporated within the buildings, as well as the need to have a mix of sizes of apartments and social class of tenants. Also, the key question of providing security was also forgotten. The result was that most of the public sector high rise buildings were badly built, poorly maintained, had no facilities and limited security, so many quickly become havens of crime and anti-social behaviour, with the limited green areas around vandalised and full of trash.

Although the key social and environmental ideas of Howard and Le Corbusier were forgotten or drastically modified in most twentieth century urban projects, the desire to avoid overcrowded slum developments and to consider the health of the population did lead most western governments to create comprehensive rules for the development of new city areas. Specific rules were enacted to control features such as construction standards, road widths, lot sizes, utilities, sanitation, etc. But in many western cities they also led to minimum requirements of amounts of open space in every new residential area were established, which in many western cities increased from the typical 10% proportion adopted in the 1960s, with larger amounts in affluent developments. Some of this space was used in parks, often with a hierarchy of different types, from small sub-neighbourhood areas for pre-school children to community then district and regional parks (Davies and Herbert 1993).

All of the above measures relate to areas within the urban settlement. However the regional aspects of Howard's scheme helped stimulate interest in Regional Planning, with advocates from Geddes and Adams, to Mumford and Abercrombie (Geddes 1915, Hall 1988). They developed new ideas about the need to create regional plans to protect the green hinterland of big cities in order to secure its water supply, recreation land and agricultural areas or simply to protect unique natural areas. Of course, previous attempts have been made in the past to control the area immediately outside the urban area, such as Elizabeth 1's designation of an agricultural zone to provide food for London, or those that restricted development on defence grounds, such as the need for an open extra-mural glacis area. After World War II Britain pioneered regional conservation measures using three major policy initiatives (Hall and Clawson 1973, Hall et al. 1973). One was the creation of formal development plans to control the growth of existing urban areas through area plans that specified the future use of land. Another was the creation of National Parks, Areas of Outstanding Beauty, and Countryside Parks, which have protected large unique landscapes areas from development, although most of these are in remoter areas of the country. The other British policy that had a more specific effect on urban growth was the creation of Green Belts around some of the major cities, in which only agricultural or recreational development is allowed (Thomas 1970, Elson 1993, p. 58). These were primarily designed to prevent sprawl, especially the coalescence of existing large urban areas, as well as to provide recreational land for the city, although increasingly they are also seen as forest and wildlife habitats. It has been estimated that 40% of the land in England is directly preserved from any large-scale development of settlements and industry by the combination of these conservation policies (Hall 1992, p. 121). Indeed the 14 designated Green Belts around the major cities and conurbations alone account for 13 % of the total area of the country (RSA s1990, p. 59). Although most of the population still support the Green Belt policy, there have been a lot of criticisms of their utility, for example: development has often leapfrogged over the belt creating sprawl in surrounding settlements; many former industrial and mineral extraction sites in these areas remain vacant; farmland deteriorates because owners are unwilling to invest in improving land, especially those close to city boundaries, not simply because intruders vandalise land and livestock, but because owners hope for windfall profits if the land is turned into urban use; house prices in the green belt increase, so that locals cannot afford to buy the properties and only the richer classes benefit; while relatively few recreational uses that would benefit the city, apart from golf courses and riding stables linked to the more affluent, were actually developed (Smith 1984, Elson 1993). This led Elson (1993) in a major report on Green Belts in England to suggest they had an important 'effect', but this did not mean they were 'effective'.

Other countries adopted different approaches to the maintenance of open land in the face of urban sprawl, such as policies that create growth areas around cities that are separated by zones of restricted development dominated by green space. The form of these green areas varies widely. Some are green sectors or wedges with urbanized corridors in between, as around Paris or Melbourne; others show green fingers, separated by urban development along main transport routes, open lands that run from close to the city centre to the periphery. This type of alternative to a complete green belt was pioneered in Copenhagen's famous 1947 finger plan, with similar ideas adopted by neighbouring countries. For example Helsinki now provides one of the most famous examples of this concept, having a green zone, Keskuspuisto park, that extends 11 km miles from the city centre to a forested periphery. In Holland, concern over the possibility of the major urban areas between Amsterdam and Rotterdam growing together to form a new conurbation led to the creation of a no-growth zone around and between the major cities, often described as the Randstadt—an area seen as the Green Heart of the country.

In the late 1970s and 1980s restrictive land use controls became criticised in many western countries that were suffering economic depression because of the higher oil prices consequent upon the creation of OPEC, and then the de-industrialization as many manufacturing jobs moved to Asia. This led to higher unemployment levels in most large cities and the introduction of many new policies to attract new businesses with a relaxation of planning regulations whose rigidity was blamed for preventing new growth and causing house prices to rise on the limited land left. So the land use policies, such as Green Belts or Green Sectors or Wedges, imposed by national authorities were widely criticized. Many cities, as different as Dublin and Seoul, abandoned their green belts, although many have survived in other parts of Europe, with relatively few areas taken out of the zones over the past decades, helped by the affluent home owners in the areas who support the policy of restricted development.

Despite these pressures to lessen policies of containment, the recent growth of interest in environmental polices has led many jurisdictions to return to policies that try to conserve areas around growing urban places, and to provide more open or green space. Many different approaches to land preservation have been used, some by fiat from a state or national government, others by various land use polices, such as: the purchase of development rights, buying the land, or zoning areas for agricultural use only. After decades of criticism even the Green Belt concept has been revived in the last twenty years. For example a Green Belt Biosphere zone around Sao Paulo was designated in 1994—although this is as much about protecting local habitat as sprawl—while a European Green Belt was designated along the line of the old Cold War barrier between the democratic west and the Communist countries, a zone that was originally heavily fortified. In urban terms the revival of interest in this approach is best seen by the implementation of the world's biggest urban green belt in southern Ontario in 2005, an area encompassing 1.8 million acres (720,000 ha). Figure 4.1 shows that it surrounds the conurbation known as the Golden Horseshoe, centred by Toronto, an urbanized area of 8.4 million people at its time of designation, one that is expected to increase to over 11 million by 2031. Most of the land within the belt is protected from development, with existing settlements in the area required to produce development plans and what amount to growth boundaries. Although heavily criticized by conservative forces at the time (Cox 2004), the plan did gain wide public support. One of the reasons for its acceptance is that the regional population was already familiar with the utility of previous conservation measures that had protected parts of the zone, namely: the Niagara escarpment, formed by

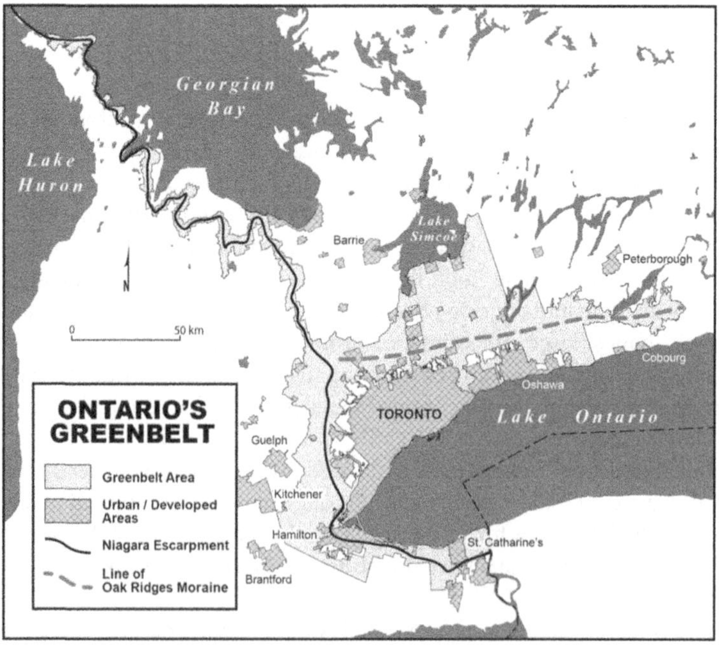

Fig. 4.1 Ontario's Green Belt

a Silurian dolomite cap over shale, which accounts for the famous falls; the Oak Ridges moraine, whose sands and gravels filtered much of the rainfall in the area before it enters the rivers and reservoirs that form the region's water supply, as well as being a popular recreation area; and the Niagara fruit belt, with its vineyards, peach and apple orchards, which together with the agricultural crops from rich soils in other parts of the belt, provide a major source of unique agricultural products in this climatically favoured part of Canada. Support has also come from the fact that the government created a Green Belt Foundation with $ 25 million seed money, a non-profit organization that has since attracted many other funds and is designed to promote the benefits of the Green Belt (OGB 2012). This source has calculated that the belt's non-market ecological services alone are worth $ 2.6 billion annually, derived from the third of the area that is composed of lakes, wetlands, river valleys and forests. In addition, produce from the area is heavily promoted and sold in a variety of local markets; farmers are helped to develop and market new crops; the benefits of tourism and recreational usage are emphasized; while the foundation helped organize the first world conference on green belts in March 2011, with delegates and speakers from around the world. This pro-active approach to publicizing the Green Belt's benefits has led to overwhelming public support to the area, one that contrasts with the way that the first green belts were just designated by top-down fiat and essentially left as agricultural areas, with few attempts to make them effective, or to create public support for the policy.

However, it must be admitted that this Canadian example provides a rare new example of comprehensive regional planning around a big western city in the last few decades, for national and state governments, especially under conservative regimes, have been reluctant to use their powers to impose such solutions. This has been especially true in the U.S.A. where national and state governments, particularly under conservative regimes, have rarely intervened to control the development around cities. Throughout the country there is greater opposition to federal intrusion, support for states' rights, and a high degree of localism with municipalities having most rights over planning. These administrative issues and increased car ownership have helped create greater amounts of sprawl in the U.S.A. than in most countries of the world, often in new local government units that operate independently from the central cities. The result was to hollow-out cities, as residents and business fled to cheaper, more accessible sites outside the central cities, with inner cities becoming increasingly obsolescent or even abandoned as residents fled and no new investment was available to counteract the decay, which was made worse by the increasing concentration of low income and often ethnic minorities in the areas. Detroit is perhaps the exemplar of these trends, a city with 90,000 abandoned homes, that has dropped in population from 2 million in 1960 to 700,000 in 2013, and which declared itself bankrupt in the summer of 2013 (Binelli 2012, LeDuff 2013). Elsewhere, the value of urban growth boundaries to define the limits of a spread of a municipality, and a preference for infill development, rather than continued urban sprawl, in order to help revitalize the central city, was pioneered in Oregon through the 1973 Oregon Land Use Act. This reduced local government powers by creating state-wide policies for land use and transportation by insisting that all urban areas establish urban growth boundaries around a zone containing the built-up area and provision for a 20 year supply of land, to prevent future sprawl. An elected seven person Land Conservation and Development Commission reviews the plans, with disputes being dealt with by a Land Use Board of Appeals. This strong direction from above, and a democratic mandate from regional elections from below, has provided a rare example of a comprehensive system of regional planning in the U.S.A. (Stephenson 1999). It has not been without opposition, for there have been several propositions on state elections to get rid of these bodies, which have been defeated with help from the Friends of Oregon lobbying group. Since the early 1970s, Portland, in particular, has emerged as an exemplar of the utility of such growth control and infill policies, helped by the addition of an elected Portland Metro council covering surrounding municipalities that is responsible for several regional-wide functions. Some of these ideas have been adopted by other American jurisdictions, although most are still plagued by a localism in planning. Yet the obvious need for planning around big cities to produce co-ordination and co-operation between neighbouring municipalities in the location of such features as new urban developments, water supply provision, recreation areas and transportation routes has led to the growth of what is called New Regionalism (Scott 2006), an issue reviewed in Chap. 2, which produces solutions,

> not through top-down regional government, but through incremental development of social capital, institutions, ad hoc partnerships and frameworks of incentives and mandates between existing levels of government. (Wheeler 2002, p. 267)

4.4 Green Infrastructure

Until recently, the approaches to the greening of towns and cities have been rather piecemeal, in the sense that different agencies, both public and private were responsible for the different types of green developments found in these jurisdictions, while there are huge differences between places in the amount of green coverage. Increasingly, however, urban governments have started to review the whole pattern of green as well as open areas in their jurisdictions to provide a more comprehensive view of its various parts. This more holistic approach has led to the term 'green infrastructure' being applied to the disparate green and open space elements within municipalities and their surroundings, a term popularized by a recent report by the European Environmental Agency (EEA 2011a). A summary of the potential assets making up the green infrastructure is shown in Table 4.1, although the various elements have been re-ordered from the original source to make the connections at different scales more obvious.

Although the EEA report noted that no widely accepted definition of the term green infrastructure exists, it does not use the term simply as a summary for the inventory of green areas. Rather it was proposed that the term was a

> concept addressing the connectivity of ecosystems, their protection and the provision of ecosystem services, while also addressing mitigation and adaptation to climate change… helping to ensure the sustainable provision of ecosystem goods while increasing the

Table 4.1 Green infrastructure: potential assets at three scales. (Source: Re-ordered from table in: Green Infra-Structure and Territorial Cohesion. EEA Technical Report, 18, 2011, p. 7)

Local, neighbourhood, village	Town, city, district	City-region to national
Local Urban Plazas	Municipal plazas Business settings	Road & rail networks
Street trees, verges, hedges		
Green roofs and walls		
Private gardens		
Greens and commons	Urban commons	Common lands
Allotments (Community gardens)		
Cemeteries, churchyards		
Institutional open spaces		
Play areas		
School grounds		
Sport pitches	Major recreational spaces	
Pedestrian & cycle routes		Long distance trials
Local rights of way	Community woodlands	
Small woodlands	Forest parks	Woodlands, Forests
Pocket parks	City or district parks	
	Country parks	Regional parks
	Urban canals	Canals
Swales (grassed), ditches		
Ponds and streams	Rivers and flood plains	Rivers and flood plains
	Continuous waterfronts	Shorelines
	Lakes	Reservoirs
Local nature reserves		
		Green belts/strategic gaps/ national parks
Vacant and derelict land	Old extraction sites Landfills Brownfield land	
	Agricultural land	Ogricultural land open countryside

> resilience of ecosystems…(one that is) …central to the overall objective of ecosystem restoration. (EEA 2011a, p. 6)

This means that the objective of the approach, like recent parallel research from the U.S. Environmental Protection Agency, is not simply to inventory the green assets. Rather it is multi-functional, emphasising the need not only for connections between the various elements, but their strengthening and potential uses in order to develop the multiple environmental benefits of these spaces in support of greater sustainability. In doing so it recognizes the various policies, services, tools and techniques used on the elements may vary at different scales.

4.5 Environmental Restoration and Revitalization

Another important recent trend in the move to greater adoption of green principles has been the cluster of policies that restore and revitalize the natural environment in urban areas. This a result of urban municipalities adopting greater environmental responsibility in their actions. Until very recently the creation of most new urban growth areas was preceded by decisions to level and clear the land, unless parks were developed, or where the local topography was too difficult to build upon, because of its steep slopes, ravines, or areas prone to flooding. So natural woodland was lost, marshy areas were drained, rocks were removed, rivers and streams diverted or modified, while the effect of the developments on local climates and local fauna and fauna were ignored. Also, roads and parking areas were sealed by the use of concrete or asphalt to make it easier for vehicular traffic, which has the negative effect of increasing run-off rates in rain storms, resulting in the need for extensive storm pond areas to prevent flash-flood problems. Hence the original natural environment was transformed into artificial urban areas, places that obliterated, or at best had little respect for, nature. However, in the last two decades the adoption of policies based on ecological principles has led to several major types of change that are designed to ensure that more of the natural environment is not simply preserved, but respected, and even regenerated.

One of the first changes came from the *modification of existing harmful municipal practices*. Many urban governments have now recognized that they adopted environmentally negative practices in the past, such as the heavy use of pesticides on green, open spaces. Such approaches have been replaced by the use of more ecologically friendly products to prevent indiscriminate and unnecessary poisoning of existing green areas and the wildlife. In addition, new attempts have been made by many cities to use tree conservation methods to maintain existing mature trees and to replace them when they die. Many also increase the arboreal stock by adding trees throughout the urban area, by providing free trees to householders in select communities on condition they look after them. The adoption of these *new vegetative policies*, often called Neighbour Wood programmes, has also helped environmental improvement in many cities, improving the look of areas that did not have many trees or other vegetation. Moreover, most arboreal policies today ensure that a variety of species is planted—unlike the almost mono-cultures of tree-planting in the past—to ensure that there will be greater diversity in urban areas. Even greater changes have taken place in areas about to be converted to urban uses. Instead of the past practices of clear-cutting and levelling, many cities have adopted new regulations to *preserve as many of the existing features of the natural environment as possible,* even though as much as 80 % of the land may eventually be developed. For example, existing woodland areas, however small, are incorporated into the green space allocation of newly developed areas; ponds and marshy areas are often kept as small ecological reserves with walkways and signage to identify local birdlife and plant species. Previously, areas difficult to develop, such as steep hills or ravines and small valleys, were often left behind by developers and became derelict lands. Now, city regulations in most western cities ensure that developments conserve and maintain as many as possible of these natural features, adding paths and family picnic areas in the natural areas to allow walking and biking—usually with tarmac pathways in high volume areas to reduce erosion on these routes and to encourage pedestrian travel. In some cases lakes or golf courses have been created within the new residential units to provide a core of additional recreational facilities, although they are more frequent in the higher value areas and use may be restricted to the local residents (Davies and Herbert 1993).

Some cities have also deliberately encouraged the preservation and even the addition of wooded areas within the urban boundary, not only for recreational use, but also in recognition of the ability of woods to help absorb carbon dioxide emissions and also to help mitigate the urban heat island effect—a product of the heat generation and retention in all the urban concrete and brick structures and roads. Other cities have been fortunate to have large areas of woodland because these were originally areas of common land preserved from the land market and eventually taken over by the municipality who took responsibility for the area, such as in the Common in what is now part of the inner city of Southampton in England. Advocates of a strong ecosystem policy also argue that particular types of areas around towns should be preserved from development and kept for public use, namely: prime agricultural land because of their value in producing food crops; ridges because of their panoramic views; coastal strips, river banks and flood plains to avoid the probability of flooding that will occur in most of these areas at some time in the future, unless, in the case of rivers, adequate reservoirs for flood prevention schemes have been built further upstream. Chapter 9 (Resilient Cities) will describe the way that the experience of earthquakes, typhoons, tornadoes, volcanoes, tsunamis and floods have led to more support for policies to protect settlements from these natural events. In the case of flood plains in particular, it must be admitted that local developer pressure, and even the bribery of public officials in some countries, has led to a great deal of building in such areas as well as vulnerable coastal areas, with disastrous consequences when the inevitable inundation occurs.

Urban development, even when green areas are provided, also effectively slice-up pre-existing habitats, which isolate these areas. So the development of parks or protected areas has usually created separate environmental niches. Ecologically they become like islands in oceans; ecocity advocates have noted that such areas will eventually have far lower numbers and diversity of species (Register 2006). So there is increasing support for what amounts to *green connectivity* policies that connect formerly separate green areas throughout a city. This not only improves human recreational access and use, but enables animals and birds to more easily move throughout the area, ensuring the species do not get isolated and suffer the island syndrome problem. To some extent the green finger or wedge development described above in Copenhagen and Helsinki has long provided a variation of this policy. This connectivity principle has been an objective of parks policy in Calgary (Canada), although designed initially to improve human recreational use, rather than for ecological reasons. The city has one of the most extensive green parkway systems in western cities, with almost 800 km of pathways and 10,000 ha of parkland occupying 12 % of the city (City of Calgary 2012). Although praiseworthy, it is worth remembering that much of the land now in park use could not be developed for houses because of the presence of steep gullies, marshy depressions, plateaus where water could not initially be delivered, or along rivers that flooded, although too much development has been allowed in river valleys which suffered the consequences in a major flood in June 2013. However in another sense the city is a leader, for by 2016 the city will complete a Greenway (GW 2014) around the city, a 140 km cycle and pedestrian pathway system encircling the city that has amenities at various points. This will connect 55 communities and link to other parks and paths in the city. It is billed by the city as the largest green urban pathway in the world.

The concept of connecting up as many of the green and recreational areas in a city as possible has been taken a stage further by the creation of *ecoducts* across, or sometimes under, major transport arteries, with Utrecht and other cities in the Netherlands, in particular, providing many examples. Some ecoducts are designed primarily to allow all fauna to cross from one area to another, with a construction above a road providing better sight lines and protection from predators; others are species-specific, such as the dormouse passages over roads near Llantrisant in South Wales. Some ecoducts are controversial as they are expensive, such as those which cross the four lane highways in Canada's Banff park and cost up to C$ 3 million each. Less open to criticism are the human-orientated ecoducts formed by pedestrian bridges or bikeways over major roads or rails, to prevent them acting as barriers to movement between different parts of urbanized areas. In a few cases the roadways have even been covered over, so that the formerly open road, rail or canal route is transformed into a short tunnel. One of the best examples of this can be seen in Vienna where 2 miles of freeway from the inner city is buried, with a treed area above that contains a series of roof vents to provide the necessary ventilation for the road.

Another general trend in the environmental rehabilitation of towns and cities has been the conscious decision to encourage the *rehabilitation of contaminated areas, including water courses*. Governments have increasingly tried to direct development on to brownfield sites—the old industrial or residential sites that have been abandoned—rather than on greenfield sites on the edge of towns and cities in order to increase urban density and decrease sprawl. Until the creation of new environmental laws in most western countries from the 1970s, material hazardous to human health was often dumped by industrial activity in rivers, ponds or nearby vacant ground. Increasingly governments try to ensure that such activity is prohibited, with large fines for offenders, while in many countries new mines or industrial plants have to set aside money for clean-up operations once the activity closes. Although there has been a great deal of progress in this area, the numbers of polluted sites in many countries is still staggering. For example, in June 2012 a Canadian federal government agency reported that there were 22,000 federal contaminated sites in the country, some were properties that were a direct result of government activities, but many others were inherited from old industrial activities that had been abandoned. Although a 15 year programme of rehabilitation from 2005 budgeted for $ 3.5 billion to be spent over 15 years to resolve these problems, the 2012 report estimated it that would take at least $ 7.7 billion to clean up these sites, showing the enormous scale of the problem, one that is not dissimilar in many countries of the world (CESD 2012). In brownfield sites within cities government funding is usually needed to provide the finance and even expertise to remove the old buildings, polluted soil and other contaminants, since their removal is expensive. This usually deters private developers who are attracted to cheaper sites on the city edge. But once brownfield sites are cleaned with government help or control, the land is usually sold to developers to recoup the costs of the rehabilitation.

Another rehabilitation trend in cities with particularly active environmental policies has been to find *ways of reducing the excessive run-off and flooding* that often occurs after extreme rain events because water rapidly runs off the developed surfaces, instead of soaking into the ground. The presence of large grassy or forest areas in urban areas helps to mitigate this problem, but several new procedures have also been adopted by cities. For example, the development of *storm ponds* is increasingly mandatory in many urban areas. These are large hollows preserved from development that can hold water from heavy rain events, reducing the rate of run-off and the risk of floods in the storm water system. Such areas can also be used as recreational space for most parts of the year since storms events are limited. *Water gardens* are also being created in some residential areas, essentially areas at the centre of drainage ditches, which may have gardens at the surface but underneath contain large sandy or gravel deposits that can hold large volumes of water. This means that rain will drain more slowly into the sewer or river system. These provide an attractive environmental feature in addition to reducing flooding.Finally, some cities have adopted *de-sealing measures* on minor roads and parking areas, reducing the amount of asphalt and concrete surfaces by replacing them with permeable bricks and allowing grass to grow between the bricks as a new surface. This allows rain water to percolate into the ground. The policy helps to replenish local groundwater stocks which are usually diminished by urban development and, by allowing water to soak into the ground, reduces the rate of runoff into storm sewers and the severity of flash floods in urban areas that had previously effectively sealed their surfaces.

Although not entirely restricted to Europe, the continent has seen far more examples of attempts to *restore the natural streams or gullies and their environment* that were either covered over, or ignored by previous surrounding developments, which meant they became derelict spaces in the city. Many waterways that were essentially ignored by surrounding developments are being returned to something like their natural state, using water filtration systems—natural ones like marshes or artificial ones—to ensure the water is clean. The addition of walkways and green areas alongside the watercourses ensures that many people are able to use and appreciate these renewed natural environments. For example, Zurich is engaged in the restoration of 40 km of these former waterways, using filtering techniques to improve the water quality and also adding recreational facilities along their banks. Old canals, abandoned for years, have also been getting the same treatment. This has meant that in many countries once derelict canals have become valued additional recreational spaces, with new pathways along their route and with boats for hire if the canal is sufficiently long and connected to other systems. Good examples of the rehabilitation of old river-ways can be seen in Britain. One of the most comprehensive areas can be seen in the city of Leicester which has created a 12 mile, 2,400 acre park by reclaiming a formerly derelict river corridor running through the city. Similarly, some of the formerly polluted and industrially abandoned Lea valley in east London has also been renovated—with part used for the 2012 Olympics site. Portland in Oregon, even removed a freeway along its principal river and replaced it with a promenade and parks, a valuable addition of recreational space for the inner city. Another recent trend in some central areas of cities that have little recreational space is to create temporary or *'pop-up' recreational areas*. This was pioneered in Paris, where some river-side roads and parking areas are closed to traffic in the summer months or on weekends, creating temporary beach-like areas on these areas by adding sand. This provides recreational spaces for local residents, another example of the way that cities are using spaces for temporary occupation, as they do with festive events (Chap. 14).

A very different new environmental initiative is particularly associated with German towns and cities. Many have been pioneers in studying the climate to maintain the *local climate advantages* and minimize deleterious effects. For example, Freiberg created new planning regulations that either prevent buildings acting as obstructions to the cool local winds that flow from the nearby Black Forest, or ensure that the design and orientation of large new buildings allows winds to 'flow through' the structure, as seen in the city's redeveloped soccer stadium. Rather more cities have attempted to avoid the shadow effects from large buildings which can worsen the effect of winter conditions. For example, in the 1980s Calgary's new City Hall was deliberately built with its higher floors set back, creating a stepped structure to reduce the shadow effect on the neighbouring central city plaza. Yet the dangers of flooding from its location in a flood plain was ignored, meaning that it suffered devastation in the June 2013 floods, which was made worse because the main heating and electrical devices were in the basements of flooded buildings. Attempts are now being made to rewrite city codes to ensure such vital utilities are located on higher levels. Other settlements have deliberately orientated their road patterns so that prevailing winds cannot funnel down straight roads, which cause particular difficulties for pedestrians and traffic in storm conditions. Similar examples of the need to relate buildings and open spaces to the local climate are also discussed in Chap. 8 on Winter Cities. It is also worth remembering that many ancient Greek and Roman towns used arcades around squares to provide cooler, shady areas for businesses, socialization and political debate, a morphological feature copied in many medieval new towns built in south-west France. In hot climates there has been a tradition of orientating buildings to cut out the direct rays of the sun, often using screens and shades to do so, with the addition of verandas around houses to create cooler areas. Fountains and other water features were also added, not just to provide a local water supply, but as cooling focal points of public squares. In many of the oldest urban areas around the Persian Gulf large funnel-shaped towers orientated to sea breezes were built, bringing cooling winds into the interior of buildings. In Augsberg a rather unique feature was the medieval development that canalized streams from the nearby Alps, so that they flowed through this prosperous medieval banking town, bringing cooling waters into the city in the summer. These and many other examples show that many urban places in older civilizations adapted features of their local environment to improve the liveability of places. Often forgotten in modernist styles and approaches, these modifications of traditional ideas are increasingly being revived and brought back into new urban developments.

Ecocity advocates also stress the need to *conserve the local biotic environment* by planting appropriate native species, which are likely to be more sustainable and able to cope with the local environmental conditions. A related important trend has been the movement to remove invasive non-native species since those indigenous to the area have a better survival rate and may cause less environmental damage. For example, in Britain, invasive species such as Japanese knot-weed rapidly colonized rivers, swamped other species and spread into surrounding areas. They have proved difficult to eradicate, given the rapid spread of their long root systems. In other parts of Europe eucalyptus trees from Australia were planted because they grew quickly and could be harvested well before native species matured. But they are now being removed because it is now recognized that fires spread more quickly through such trees, making them a potential hazard. Even in jurisdictions without strong Ecocity ideas the general concern for environmental protection has led many states to legislate to protect unique local flora and fauna, which means that new developments in cities have to preserve the existing biotic heritage. In most cases these policies are carried out independently, species by species. In some cities detailed environmental surveys have been carried out to classify the various types of habitat in cities as a prelude to subsequent protective legislation. For example, Leicester was one of the first English cities to develop a detailed *habitat survey* and *protection* strategy in the late 1980s, which led to city policies that protected, conserved and managed various parts of the city in order to protect their biodiversity. In increasing numbers of cases these habitat preservation strategies are enhancing some local distinctiveness of the ecology, which may them become a major attraction for residents and visitors. One rather unique example is the way that Austin in Texas has protected the million or more bats that roost under the bridges over its central lake. When they quickly swarm out at dusk they form a spectacular, swooping and wheeling display that not only encourages locals and visitors to view the event, but have the additional value of keeping the local mosquito population under control.

In some ways these changes contribute to what Register (2006), the Eco-city advocate, described as the need for an element of 'wildness in urban areas', of a return to some original pre-human condition. This is controversial, for in most developed countries the fauna dangerous to people, such as bears and wolves in Europe, have long been removed, while forests, so long regarded as hazardous places because of the presence of dangerous animals, have been reduced in size by agricultural developments. The result is that the local nature has been effectively sanitized. By contrast, many American cities in particular lie close to large wilderness areas where the original feral animals still survive. Since some conservationists argue that native fauna have their own rights, it has led to the development of protected corridors near urban places, where animals of whatever type can migrate through the area. Inevitably, human-carnivore encounters do occur, which usually results in the killing or removal of the threatening animals involved. This is considered part of the trade-off to maintain more natural conditions than are found in most European regions—although some would prefer to see such policies abolished on human safely grounds.

4.6 From Green Conservation to Green Production

Some policy changes in urban areas that return the environment to more natural conditions are not, however, only carried out only or conservation, recreational or health reasons. Increasingly, it is recognized that the land in and around cities should also be used more productively, for the growth of the supermarkets that dominate food sales have spatially extensive food supply chains, which means that few products come from the local area, while the food stocks carried only last a few days. So any breakdown in the transport system preventing new supplies arriving means that food stocks will diminish rapidly in cities, usually less than a week. This potential food supply problem is one that the Transition Town movement (Chap. 7) is particularly concerned with, leading them to stress the need for more local food supply sources. In the past century edible species were usually not planted or were removed in parks and open areas because their fruit leaves a mess, or attracts vandals. However more and more cities are developing what amounts to new *harvesting policies*. For example, some cities now harvest trees with fruits, as seen in Rosario's park system in Argentina, separating these areas off at harvest time to avoid vandalism. The interest in organic crops has also encouraged many more people with gardens to grow food. New urban bye-laws have allowed private urban gardeners to use the unwanted gardens of willing homeowners to grow vegetables, often sharing the crop with the owners—while volunteer citizen groups have emerged to harvest unwanted fruit from trees in private gardens. More generally, some cities have developed what amount to *ecological reserves*, either small areas within the city or in their hinterland that are open to visitors who enjoy the open space and ecological variety. In addition school gardens have been added to school yards in an attempt to help children understand how crops grow, part of the increasing adoption of more green environmentally aware practices in their curriculum that is the subject of geography education books by Williams and Lee (2006). A few western cities also have *farms* within their municipal area, incidentally one of the often forgotten features of Howard's Garden City idea. One of the best examples of this urban farm trend can be seen in Göteburg, where 60 farms cover 2700 ha. These farms are working, profitable enterprises, but also carry out functions useful for the surrounding urban residents, such as allowing school children to visit and study farming methods and animals, providing pick-your-own berry and vegetable areas, as well as riding schools. In Göteburg much of this land was bought up by the city decades earlier under a land banking scheme, so it did not have to pay high prices. Land around cities is often held for years by speculators, hoping for windfall profits when rural land is changed into urban.

A new appreciation of the need for agriculture within and around urban areas has also emerged, with many authors stressing the freshness of local supplies from farms (Mougeot 2005; Darren 2009; Mogk 2010). Some represent survivals of older patterns, but others represent new developments. Examples of the former case can be seen in countries such as Japan with its limited flat land area, where it is rare to find vacant lots or waste areas; instead many towns and cities have highly cultivated plots between the buildings, a feature that is also found in many towns in Holland. Such examples of a surviving agricultural presence in cities is also increasingly seen and promoted in many tropical and sub-tropical countries. Indeed *urban agriculture*, often on quite small plots in and around cities in the underdeveloped world, has always flourished, helped by the possibility of multiple cropping given the length of growing seasons in warm climates. Agriculture within cities is now being encouraged by many aid agencies in the last decade, especially to supply fresh vegetables and to provide more local employment and cash returns (Smit et al. 1996; Mougeot 2005). In cities such as Accra in Ghana a high proportion of the vegetables on sale in local markets come from the informal plots bordering what used to be waste land along the city's watercourses or drains. Unfortunately the polluted nature of these watercourses casts doubt upon the health of the crops, although now the city has recognized the value of these plots it is now making big efforts to reduce pollution levels in the streams.

At a household level in developing countries there is an increasing use of bags full of fertilized earth to grow fresh food, since they need little space outside the house and provide useful crops of such products as tomatoes, herbs and legumes. In western countries, of course, pressures for growth from the industrial period onwards led to the loss of most agricultural land in older cities. Yet it is worth remembering that agriculture in western cities is not new, for historically urban gardens produced many food crops. Even in the late nineteenth century dairy cows and chickens could be found in small enclosures in the big centres, fed from feed brought into the city and with a ready market for such products within the local area, thereby saving the time and transport costs of milk and eggs and, by reducing delivery time, ensuring that the products were fresh. But these animals were gradually excluded by local health bye-laws, often due to failures to effectively dispose of the waste from these animals, which led to noxious smells and breeding areas for flies and diseases. However, there are many signs that chickens, at least, will be increasingly tolerated in cities as a source of fresh eggs in particular. For example, several Canadian municipalities have changed their local bye-laws to allow limited numbers of these birds in the back yards of the houses, so long as they are large enough for the birds and neighbours do not raise objections.

More generally, the presence of increasing agricultural activity in urban areas within western towns and cities, other than in the private gardens of houses, can be seen by the revival of interest in what are called *allotments* in the U.K., with different names in other countries, such as kolonihave in Denmark or *community gardens* in the U.S.A. (Crouch and Ward 1997). These small plots, rarely over 250 sq metres in the case of Britain, produce a variety of vegetables and flowers, are usually more intensively cultivated than back or front gardens. Websites designed to offer advice to allotment users have also grown up (A-UK 2012). Most of the plots are rented from municipalities at a small annual fee, although some historic ones are owned by the allotment association. This trend is not a new one. Allotments have always been present in Britain, frequently outside the old medieval cores of cities, but the numbers expanded, especially in high density urban areas in the nineteenth century where people did not have garden space. Municipalities are required under successive national acts of parliament to ensure developers set aside land for allotments, so that the urban poor could supplement their food supply, as their ancestors had done in rural areas. Indeed the numbers exploded to an estimated 1.5 million plots during World War I and II in response to a national policy to grow more food. They were also prized in the former communist countries of eastern Europe, since they were the only areas in which people could grow, and in some countries sell, the products, providing the one example of private production in the otherwise dominant state-supply system. The numbers have often fluctuated during the last century but have usually declined. Denmark is an exception, for the numbers of urban allotments have risen from some 20,000 in 1904 to over 60,000 by 2001, and many plots have a small building that can be lived in during summer months. Britain, however, has seen great fluctuations in numbers with a decline from the 1950s as people became more affluent, had more leisure variety in their life to do other things, often away from their homes, and food products became cheaper. The decrease in interest led many municipalities to sell off the land to developers. In 2008 a review of allotments by The Guardian newspaper (Vidal 2008) estimated that there were only about a third of a million plots left in Britain. However, the rise of ecological concerns has led to a new interest in the allotments and it was estimated that there were over 100 thousand people on allotment waiting lists in 2008, with great variations in numbers between the councils (GN 2011). This situation is leading to mounting criticism of many municipalities for not providing enough allotment land for people seeking to improve local food sourcing, self-sufficiency and organic production.

A parallel urban agricultural trend, but with rather different objectives, can be seen in some inner city areas, especially in impoverished and ethnic parts of U.S. inner cities where *community gardens* have been created on derelict lots in the past decade. In many declining cities the increasing number of these vacant lots has led many cities to encourage agriculture on these plots, not only by the owners, but also by taking the land into public ownership and leasing the land out to gardeners. This policy provides not only employment and a new source of fresh crops, but also makes once abandoned areas more attractive. One of the most publicized examples is the 'Bed-Stuy Farm' in the predominantly black and hispanic Bedford-Stuyvesant area of New York, established by the Bedford Rescue Mission in 2005 that uses charity donations to provide help and food to the poor. Although barely larger than a basketball court, this 'farm' produces many thousands pounds weight of produce every year, and is a source of neighbourhood pride. It has also stimulated the creation of a local farmer's market as well as a revival of interest in 'black soul' food crops—a traditional part of the diet of Afro-American people. Another approach is the Small Plot Intensive (SPIN) ideas used in the Brick City urban farms in Newark (New Jersey) which has many obsolescent areas. Plastic crates containing soil (earthboxes) are used to grow crops, an approach that has the advantage of mobility, by being able to move the crates around to quickly colonize vacant spaces. This is another version of the growing practice of using specially filled bags of fertilized soil to grow tomatoes and other plants. On a larger scale some parts of Detroit that have large numbers of vacant lots because of abandoned houses are being transformed into productive agricultural land. Mogk's (2010) survey of urban agriculture reported estimates that the city could generate $ 200 million in sales and over 5000 jobs, providing not only a local source of fresh vegetables but new skills for a disadvantaged population.

These and other examples illustrate the possibilities of creating real change in the health of inner cities by providing fresh food in areas where local greengrocers and butchers have long disappeared and which were replaced by fast-food outlets that mainly sell highly processed products full of fat and sugar products. In the developed world San Francisco in particular promotes itself as a legislative leader in urban agriculture projects (CA 2012). Elsewhere, Lim (2010) has described a project in Guangming, China, where stepped housing blocks provide terraces which can be used for urban agriculture, by using a soil layer above a thick, impermeable membrane to protect the building. Although applied to buildings the stepped approach has echoes of the terraces used by the Incas for food production on the steep slopes around settlements such as Machu Picchu in Peru where innovative drainage systems and the use of different textures of soils at various levels ensured that the terraces were sustainable for farming. However, despite these encouraging examples in the contemporary world, the growth of community gardens alone cannot provide all the healthy food needed for their local areas. Nevertheless they can add to the amount of nutritious foods, provide work, new skills and job satisfaction, as well as physical exercise and often increased community interaction, creating a real benefit to parts of cities that were plagued by ill-health, making a contribution to creating more Healthy Cities (see Chap. 13). However if the policies are to flourish, many of the ideas, especially in less developed countries, can only be applied if there are clearer land tenure laws and implementation as well as better security.

The development of interest in community gardens has also focused on the types of food products sold in poor income areas, leading to the use of the term *'food deserts'* for areas where people do not have easy access to affordable, healthy, fresh food products, unlike the ready availability of fatty, sugar-saturated, fast-foods (Wrigley 2002; ver Ploeg et al. 2009; Whiteacre et al. 2009). Definitions of what is a food desert vary drastically, with a recent comparative survey claiming that there was little evidence of the phenomenon in the United Kingdom, although there were many examples in the U.S.A. (Beaulac et al 2009). The U.S. Department of Agriculture attempted to define these areas nation-wide, identifying a food desert as those census tracts that had a third of the population below the poverty line, and were a mile from nearest supermarket or large grocery store (a distance increased to ten miles in rural ones). The research also provided on-line maps of the distribution of these defined areas (USDA 2012). This led to the claim that 10 % of the census tracts in the country, accounting for 13.5 million people, lived in food deserts. This still seems an excessive claim due to the definitions used and the coarseness of the census tract grid. But whatever the dispute over the actual areas involved, is clear is that such conditions do exist and provide another disadvantage for many already impoverished areas. A recent report on the concept to Congress described one of the consequences of these food deserts, namely a three-fold increase in child obesity in the U.S.A. since 1980 (ver Ploeg et al 2009). So the concept of food deserts has provided a stimulus for several changes: the need for a drastic rethinking of diets among an already vulnerable population; the encouragement of community gardens; and education in schools about the need for healthy diets. In some American cities such as Chicago over 100 *'learning gardens'*, essentially special school gardens, have been created by 2013 to provide students with the experience of growing plants and learning basic botany. There has also been the encouragement of better food provision in poor neighbourhoods, such as on-line deliveries from large supermarkets, and what are now called *'pop-up grocery outlets'*, those selling fresh and healthy food. Of course, these are modern equivalents—with better technology by air conditioned trucks—to the old mobile shops or carts of the horse-drawn age, perhaps illustrating again the utility of older systems and ideas that has fallen out of use. Yet even the provision of such outlets does not guarantee that they will be used, given varying price, quality etc. What is really needed is better education and acceptance of the benefits of healthy food. In this respect more urban places need to adopt the approach used for decades in French schools or in the Slow Cities organization (Chap. 15), where students are taught from an early age of the benefits of different food types, which is then reflected in the types of school meals provided, which are carefully reviewed by parent groups.

Another growing movement that is related to the cultivation of more food in, but more usually around, cities has been the explosive increase in *farmers' markets,* which is also linked to the desire to consume healthy food, especially that which is locally sourced, fresher, or organically produced from farms in the urban hinterland. For example, the 2010 annual report of the U.S. Department of Agriculture estimated that the number of farmers' markets in the country had increased from 1,555 in 1994 to 5,274 in 2009. Similar growth has occurred in other countries. Some have been revivals of older markets that declined during industrialization and especially with the growth of supermarkets—revivals which have been stimulated by the Transition Town and Slow City developments described in Chaps. 7 and 15. Others are new developments, such as the recent opening in 2010 of a farmers' market in a new location at Pier 20 on Halifax's waterfront in Canada. This is a relocation from previous sites and maintains a 262 year tradition of a market in the city, which is claimed to be the oldest functioning in North America. This new market houses 200 vendors in a 55,000 sq. feet building which has been given a high building sustainability status (the L.E.E.D. platinum level as described in Chap. 6) because of its use of wind turbines, solar panels, geothermal heating and use of non-toxic building materials. It also includes a Green Roof and Wall. This is one of 40 such markets across Nova Scotia which are heavily promoted through an active website and provide many special events, or focus on particular crops at various times of the year (FMNS 2012). Most of the stalls in these markets provide products from the local area or from organically certified farms. Certainly the food is often more expensive in these markets, but it is fresher and the origin of the products is not only more local, but more easily discovered. In addition, these markets seem to provide a more exciting experience for consumers, adding to the liveability of places, not simply because of the products, but because there is usually greater socialization between shopper and local stall holder, an interaction that has been lost in the anonymity of big supermarkets. In the U.S.A. some states have set up schemes to boost the amount of local food consumed to stimulate agricultural production around cities, such as North Carolina's policy of encouraging firms and individuals to source at least ten per cent of their food supply from within the state (NC 2012). Although the sales of food from these approaches is still dwarfed by those from regular retail outlets in the western world, these developments add to the vibrancy of local retailing. This not only means that part—although still a small part—of the long-distance trade in food products to major supermarkets is reduced, but adds to the prosperity of local farming and hence open space in the urban hinterland. This trend has led some supermarkets to source, and to publicize the use of more food from their region, rather than typically ignoring local products, although it is still a fraction of their total supplies. Unfortunately the power of supermarkets to drastically drive down prices for local farm producers to almost the cost of production has become a problem for many small, local suppliers, unless they operate in a co-operative structure and can bargain effectively.

Of course, local markets did not disappear from all urban areas as mobility increased in the twentieth century. Some countries, especially France and Italy, maintained many of their historic markets, even in small towns. These markets developed in times of low mobility when most towns had to depend on local products for most of their food supply. Their survival and revival is itself a testament to the emphasis placed on local, healthy food products in these countries, as well as the presence of local food specialities and cuisines based on the distinctive crops from the local region, which has become a vital part of the Slow Town movement (Chap. 15). This type of distinctive food, based on the particular combination of the physical conditions and cultural heritage which create distinctive *terroirs*, is being encouraged by agricultural and economic planners in many countries, since it produces higher value goods and encourages gastronomic tourism to local restaurants. However, as researchers such as McWilliams (2009) have noted, one must not assume that local sustainability is necessarily increased by this focus on locally sourced food, especially by those locavores who have advocated a 100 mile limit for food supply. Many crops cannot be grown locally and taking only 'miles from the production to sales site' as a measure of sustainability ignores the many inputs—in energy and fertilizer especially—needed to produce crops, issues that are discussed further in Chap. 7 (Transition Towns). This often means the carbon footprint may be actually higher for the local crops, especially in areas with long winters, where higher fossil energy inputs may be needed to counteract short growing seasons.

All of the above practices are becoming more important and are providing useful increases in the amount of productive green space in urban areas or in their vicinity, as well as adding to the degree of food sustainability in many regions. Yet the trends must not be exaggerated. These developments in western cities especially are far from fulfilling all the food needs of urban residents. Moreover, given the higher cost of the local and organic products it is only the ecologically committed or wealthy people in most western cities who participate in the trends, although these are a small and growing minority. It is always difficult to predict the future, but if transport costs for food increase significantly, these local developments will become more important, especially if promoted by citizen groups such as those in the Transition Town movement discussed in Chap. 7. What also seems likely is the much greater development of more high intensity agricultural activities in, or near cities.

Greenhouse developments around cities, producing vegetables, fruits and flowers, were a common feature around western European cities by the early and mid-twentieth century. In countries like the U.K. many of these enterprises closed in the 1960s and 70 s, due to their failure to withstand competition from cheaper and earlier products transported cheaply by refrigerated trucks and by air from warmer climates. In northern Europe, these sources were initially from the Mediterranean, but now seasonal foods and flowers are often obtained from African and southern hemisphere countries that are in a different season. Yet there are increasing signs of a revival in local production near cities. Some come from farmers finding new opportunities by supplying local markets; others from direct-buy schemes in which urban residents contract to buy a certain amount of vegetables each week from farms who then deliver their products. Another trend is associated with the development of more technological greenhouses within commuting range of big cities. These produce a range of vegetable crops and flowers via hydroponic means and more sustainable energy sources, often using waste heat from district heating or industrial plants. Such enterprises reduce the need for so many goods to be transported across continents.

A greater emphasis upon vegetables and grains may also be needed to ensure future food supplies, given the rapid growth of the world population and recognition of the time and costs involved in animal protein production. For example, to produce a kilogram of meat takes the equivalent of at least ten times that amount of grain, as well as months or even years of pastoral effort in the case of cattle. Given the limited supply of suitable land for this type of production, increases in animal products to supply the growing population may not be possible; it may be time to use rapidly breeding insects for protein.

Another trend in the green movement has been the growth in the support for more *ethical practices in food production and processing*. This is best seen by the increasing number of people who abhor large-scale battery chicken and pig farms—where the animals never leave their small cages—calling it unethical treatment, which will cause problems for advocates of more intensive animal husbandry. In addition, many agricultural practices are being shown to have serious negative side-products affecting human health, apart from the transport costs and added pollution of getting so many of our foods from very distant areas, such as the routine use of antibiotics in cattle, an issue discussed in the Healthy City review in Chap. 13, which is leading to the growth of drug resistant bacteria. However some large fast-food and beverage chains are also adopting more ethical practices in their supply chains, once a concern only of ecologically-minded people. For example, Starbucks has long practiced what they call the sustainable sourcing of their coffee, paying their growers in developing countries higher rates for their products. In early 2012 the American fast food hamburger chain, Burger King, announced a five-year phase-out plan to replace the use of eggs and pork products from caged animals, which if successful in new sales growth, will probably lead their competitors to follow a similar policy. These examples should lead to a significant improvement in ethical practices and will improve key ingredients in the diet of many people in urban areas, even though costs may be marginally higher. However there are also other problems associated with food supplies that developed in the rush to create cheaper and more modified food over the past few decades. Increasing concern about the *over-use of chemicals* in so many products has resulted in some governments, but far from all, providing more explicit labelling of the origin of food products, while others insist on labels showing the chemical composition of processed food and the calorie count, although such advances are far from universal. Fears of health problems from processed food also relates to other consumption practices and have led to new prohibitions, some at a national level, others only in individual cities, such as the bans on smoking in public places and other issues discussed in attempts to create a more healthy city (Chap. 13).

4.7 Three Dimensional Green Developments

Until recently, the concept of greening urban places was restricted to horizontal, ground-level activity. In the last decade serious attention is being paid to seeing how the upper level areas or vertical structures in buildings can also be made more green. The best known approach can be seen in the Green Roofs and Green Walls movements that are reviewed in publications such as the *Living Architecture Monitor* (LAM 2012) and the annual American conference, *Cities Alive*, that promotes these ideas, which celebrated its 10th anniversary in Chicago in October, 2012 (CA 2012). Until recently roofs and building walls were uni-functional, with the former protecting buildings from the elements, while the latter acts as a frame to hold up or contain the buildings. The *Green Roof* movement seeks to add multi-use green developments to what adherents to the cause see as the wasted space of a roof, changing its singular function of only providing protection to the floors below from rain and weather. However, it must be remembered that in many hot climates with low rainfall, flat-topped houses were the norm and the roof, protected by surrounding walls, was used to collect rainfall, funnelling into a cistern, or to dry crops, or were even places to sleep at night since the area would be cooler than the rooms below which had stored heat from the day. The contemporary Green Roof movement goes beyond such traditional practices. The simplest example involves the use of grass on roofs, again to provide insulation and reduce rain run-off, to the more complex approach involving the development of gardens on the flat tops of large buildings in particular. Green Roofs are usually classified into two types, intensive and extensive, which simply describe whether the soil is deep, or is merely a thin layer. The Green Roof movement is seen to have many benefits, such as: reducing the rapid run-off of rain and heat loss in winter; providing a cooling effect in summer that helps mitigate the heat-island effect of the city; providing more recreational space; and scenic views for surrounding buildings. In addition, gardens in these areas often provide a new source of local products, especially fresh vegetables, herbs and flowers, and in some cases bee colonies to help the general process of fertilization of trees or crops in surrounding areas. Certainly there are often additional engineering costs in ensuring that a proper seamless membrane is used to protect the building from water seepage, and the building structure has to be able to bear the extra weight of the roofs, while the maintenance and gardening costs are additional expenses, although over a ten year period the returns often outweigh the costs.

North America has seen a rapid growth in the movement, as can be seen by the fact that only 2 million sq. feet of green roofs had been developed by 2005, but this grew to 8 million in 2010 and more than doubled to over 16 million in 2011, with over 500 green roof programmes on the continent, with the cities of Washington D.C., Chicago, New York, Toronto, Nashville and Philadelphia, leaders in this trend (LAM, Summer 2012). Toronto even adopted a green-roof bye-law in 2009 in its official plan, which states that new buildings over 2,000 sq m. should install a green roof, although it seems there are ways to avoid the requirement. All of these developments have been particular useful to large hotels, where chefs can access these various fresh products from their roofs. For example, the Fairmont Royal York hotel in downtown Toronto developed one of the first green roof gardens in 1998 on its 14th floor and later added bee-hives. Certainly the production is still a very small input to the overall purchasing bill, but since diners can take tours of the gardens and see the various crops, it provides a useful reminder of the hotel's commitment to fresh, healthy food.

Advocates of greening ideas have also begun to see other parts of buildings that could be more effectively used to develop a green and sustainable agenda. One has been to re-use the old Roman idea of *green atriums* in buildings, adding gardens or trees in a ground floor area that is open to the sky, or covered with clear glass, and surrounded by the rest of the building. Also there are increasing numbers of *interior gardens* in buildings, especially in areas with long winters. These places are designed to provide green recreational space for the workers in the building, or the general public, such as the Devonian Gardens on the fourth floor of a high rise complex in downtown Calgary, which is described in the Winter City discussion (Chap. 8).

An associated trend to the Green Roofs can be seen in the *Green Wall* movement, which advocates the use of climbing plants to cover the outside of buildings. In origin, the idea goes back to the historic 6 C B.C. Hanging Gardens of Babylon, although with a more recent stimulus from the ivy or plant covered exteriors of many historic European buildings. The green wall idea has recently been revived, using plants that either grow from the ground onto the building, or increasingly on hydroponic frames attached to the walls, frames that contain cells with soil, aeration and irrigation channels. This approach means that green walls can also be developed in indoor atriums, so long as there is sufficient light. The aesthetic advantages of the green walls come not simply from its greenery, but from the changing colours of some climbing plants at different seasons, providing variety to the often dominant, yet boring, grey or brown sides of most buildings of the modernist period. Such developments can attract birds and insects, thereby adding to the ecological diversity of the area. However, buildings adopting this approach must be free of cracks and have a high resistance to water penetration, otherwise the plants will enlarge the gaps and allow moisture to seep in, which will reduce the life of a building. Associated with this approach is the use of particular plants on or around the green walls on buildings that are known to attract insects, butterflies and birds, all of which add to the local biodiversity. In addition the plants that comprise a green wall provide greater insulation for buildings on which they grow and create a cooling effect in summer, in addition to acting as an air filter for the area around.

A more general type of three dimensional green practice can be seen in an increasing interest in the *Vertical Farms* approach promoted by Despommier (2010), which he claims will help feed the 2.4 billion increase expected in the world population by 2050. These are really farms in high-rise buildings, for animals or for crops, and are not just green production additions to buildings such as Green Roofs or Green Walls. He has argued that the technology exists to grow crops in special high-rise buildings, using hydroponic means which recycle most of the water used and through the use of natural lighting systems can operate on 24 h growing cycles. One of his ideas was to create an agricultural tower, a 30 storey building occupying a typical city block in U.S.A. producing crops year round, using waste energy and water inputs from the city and providing the food needs for 10,000 people (Despommier 2008). This may be a new, but compact version of Lloyd-Wright's (1932 ideas of self-sufficiency in his concept of a new suburban city called Broadacre City, although in that case the food independence was for each family, not for the city, since it was based on a plan to give them land to grow crops. At first sight the ideas of vertical farms seem futuristic and uneconomic but examples are cropping up in many cities. For example in densely populated Hong Kong, the permissive land uses regulations, very limited land, and a great local demand has led to a great increase in agricultural products from mainly informal production on roofs, or balconies of flat extensions on high rise buildings. Yet even fish farms have also been developed in high rises, such as the Oceanethix project on the fifteenth storey of one building which sells 2 t of fish each week from eleven big tanks, a market helped by the fact that Hong Kong has a high per capita annual fish consumption of 70 kg. Indeed it is estimated that these vertical food sources account for 2 % of the total food intake in Hong Kong (Shadbolt 2014). Other vertical farms exist or are being promoted in other Asian cities, but a project in a suburb of Scranton in the U.S.A. that opened in March 2014 claims to be the world's largest, with 4–5 levels of growing spaces on a 3.25 ha site that expects to be able to accommodate 17 million plants. However these vertical farm projects seem more likely to become part of the urban scene in Asian urban centres that are densely populated places. In most cities, and under current price regimes, the prices that need to be charged to cover costs of these expensive projects mean that the crops will be higher in price, at least initially, which will restrict the spread of this type of project. But vertical farms may well be an increasingly familiar part of the urban scene in the future.

4.8 Improving Urban Cleanliness and Attractiveness

One of the important trends in nineteenth century western cities was the creation of municipal public health services which led to the effective sanitation systems, regular street cleaning and rubbish removal that is found in these centres, although many urban places around the world still lack such facilities. However decreased financing for these services led to a decline in their effectiveness by the 1970s, which was made worse by other factors, leading many urban areas to become unsightly and dirty. Buildings and infra-structure had aged as development slowed down, so they were not maintained or replaced. In addition, a more careless attitude to the disposal of packaging meant that urban areas were plagued by the increasing amounts of litter from fast food outlets and from cigarette stubs. In addition, graffiti on walls, rising levels of vandalism, as well as the increase in the number of weed-infested, abandoned areas created by the economic downturns and service reductions in so many cities created not only unsightly areas, but often unhealthy conditions if organic rubbish was dumped. There can be little doubt that national cultural conditions account for many of the differences in what amounts to the tidiness of urban areas. For example, most cities in Japan are remarkably clean by western standards, not simply because of frequent street cleaning but because of the ingrained habit of the Japanese to avoid littering. Singapore is also largely free from litter, in part because of similar cultural habits, but also because of draconian regulations that give large, 'on the spot' fines people who litter, dump cigarette butts, spit or create eyesores. Such cities form a remarkable contrast with many Indian cities in particular where littering and even defecation in streets is still widespread, for few places have effective municipal sanitary services.

In most western cities a series of local laws (bye-laws or ordinances) have been enacted over the decades to control unsightly conditions or behaviour that irritates others, such as littering, graffiti, vandalism or noise. Recently there have been determined attempts to improve the effectiveness of these local bye-laws as well as upgrading existing street cleaning and garbage disposal services, including the targeting of special problems, such as the rapid removal of graffiti and pursuit of those responsible. The big problem is not the presence of local laws, but the way that compliance is achieved. Most western cities have urban officials who monitor and reduce these nuisances, giving out fines for people who transgress. However, since fines are often resented, conflict resolution methods have been adopted in some cities to try to understand why behaviours occur and to convince first or even second time offenders to change their habits. In such situations on-the-spot penalties are only applied as a last resort. In addition, there have been attempts to encourage citizens and educate children to take a more responsible approach to their environment and to involve them in protecting local areas. Although there are still huge variations in these attempts to improve the cleanliness or tidiness of cities, two new, but different, policies can be recognized, in addition to the attempts to make areas more attractive through the New Urbanism approaches. The first is essentially national, yet is still a voluntary approach, one that stems from the encouragement of local action groups to keep the settlement tidier, and make it more beautiful. The other approach comes largely from initiatives in individual cities.

One of the oldest examples of the first trend can be seen in the *Tidy Towns* campaign in Ireland that has inspired similar actions in other countries. Initially established by the Irish Tourist Board in 1958, but becoming part of the Department of the Environment and Local Government in 1995, the campaign encouraged local community groups to create a 'tidy town' committee and to find volunteers and businesses to help reduce the litter, beautify their environment, and enter an annual competition which awarded prizes to the most attractive town or villages in Ireland. A total of 52 settlements took part in the first competition, but in recent years an average of 700 were participating in what has become Ireland's best known local environmental initiative (IDE 2008). The initial approach was informal and designed to help the tourist industry, but it is now recognized that environmental improvement creates better places to live and work in, not simply to visit. Moreover, community co-operation usually creates a civic pride and sense of responsibility in reducing littering and finding ways to beautify the local environment. The informal approach of former years has also gone. A booklet produced by the Tidy Towns Unit within the national government's environment department describes the importance of creating a formal committee with a legal constitutional structure to guide improvement actions with a plan, once the initial group has created interest in the need for action (IDE 2002). The publication also shows how to go about surveying local areas to identify their uniqueness, advantages, problems, eyesores, and opportunities to improve the settlement. In this context, the built-environments, especially the various architectural heritages, are seen as important as purely physical environmental issues. The booklet also identifies the ten criteria used by judges to assess various entries, namely: overall development, built environment, landscaping, wildlife and natural amenities, litter control, tidiness, waste minimization, residential areas, the roads, streets and back lanes, and finally the general impression. Making these criteria explicit provides a focus for volunteer groups involved in the improvements, rather than depending only upon municipal action.

A different approach occurs in the places that have adopted a *cleaner city programme*. The aim is to make a settlement cleaner, not by local government action alone, but with citizen involvement, although general goals are usually established by the local council and monitored by a local sustainability officer. For example a 'Cleaner, Greener Baltimore' initiative was established in March 2009 to engage people, organizations and businesses in collaborative efforts to make this rather run-down and neglected American city a cleaner and greener place, as well as educating its citizens about the things that can make a difference in tidiness. This initiative was not just city-hall driven. It involved the participation of over a thousand volunteers who established the principles behind the initial report. A regular annual report on the achievements to date, and the recognition of the individuals or organizations who were leading by example, profiles the progress. Although the initial plan identified 131 strategies and 29 goals, six goals were chosen as being important for immediate action. Four of these, namely litter elimination, improving the tree canopy, environmental awareness and green schools, can be

Table 4.2 Policies for a greener, cleaner Baltimore. (Source: Derived from the annual reports on Cleaner, Greener Baltimore)

1.	An extensive media campaign attracted attention, based on the slogan 'Don't Make Excuses, Make A Difference'
2.	A new garbage collection service with more efficient routing and re-cycling collections, as well as a mixed garbage collection increased the re-cycling by 55 % and reduced the garbage by a quarter within 2 years
3.	Fifty community clean-up groups have been developed to improve their local areas. More rapid alley clean-ups have also been created
4.	By 2011, over 10 thousand vacant or abandoned lots, out of three times that many in the city, were deeded to the city, and over 200 were being developed as community gardens by local groups through the 'Adopt a Lot' policy
5.	An organic farm on a 33 acre site has been developed to show school groups how safe food can be produced
6.	Schools are encouraged to become green by developing a nature garden and increasing environment awareness in their curriculum
7.	The existing tree canopy covering 27 % of the city area is planned to double by 2037, by encouraging community groups and households to plant trees and providing them with the resources to do so
8.	A Community Greening Resource network has been established to provide materials and resources to encourage individuals and organizations to improve the greening and cleaning of their areas
9.	New city codes have been enacted to allow urban agriculture, while 'food deserts' in the city have been identified in which people in these areas can access a free delivery of fresh food from two virtual supermarket sites
10.	Policies to upgrade old buildings to improve the interior health of homes have been enacted
11.	Energy reduction policies have been implemented to reduce greenhouse gases by 15 % by 2015
12.	A new and free eco-friendly bus service connects many downtown and inner city locations

considered part of the specifically green agenda, while the food systems and energy reduction goals are better seen as part of the attempt to make the city healthier and more sustainable. Table 4.2 summarizes a dozen of the key policies being applied to make the city cleaner, although sustainable objectives were also plan of the plan. Despite the limited time the city initiative has been running, there are clear signs of real progress in improving the city's appearance. For example, policies to upgrade old buildings to improve the interior health of homes have been enacted, while energy reduction policies have been implemented to reduce greenhouse gases by 15 % by 2015. Also a new and free eco-friendly bus service connects many downtown and inner city locations in Baltimore and acts as a publicity site for green initiatives. What is also apparent in Baltimore's policy is the way that a variety of features discussed earlier—from garbage collection and cleaning up vacant lots, to school gardens, urban agriculture and food deserts—have been identified as issues that also need to be addressed if the city is to improve its environmental credentials. By developing these individual polices within the broader strategic framework a more comprehensive approach to improving Baltimore's appearance and sustainability has been created.

4.9 Towards Biophilic Cities

The various ideas discussed in previous sections are being implemented in a rather piecemeal manner in most cities. But the 'green agenda' is being consolidated and extended by the development of a new adjectival addition to the range of prefixes to urban places, namely biophilic cities. The term is derived from E.O. Wilson's (1993) popularization of the concept of biophilia, which deals with the extent to which humans have become hard-wired to connect to nature and other forms of life over our evolutionary history, and need this contact to ensure their health, let alone the maintenance of other species. Beatley (2011) adopted this concept into the title of his latest book, 'Biophilic Cities: Integrating Nature into Urban Design and Planning'. Based in the University of Virginia in Charlottesville (Virginia), his ideas have led to the development of another network of cities around the world seeking to share and promote the biophilic ideas, with the earliest adopters in Singapore, Wellington, San Francisco, Portland, Phoenix, Perth, Oslo, Vitoria, Wellington, Birmingham and Phoenix (BC 2013).

There can be little doubt that the concept is still an open one, subject to on-going development in which the goal is to increase the role of nature at various scales, in buildings, workplaces, areas and in urban places. The commitment to the core development goal of an abundant and rich bio-diversity in settlements led Beatley to outline a list of basic principles to achieve this aim. These have been re-arranged in Fig. 4.2 to create a summary diagram in the form of a cyclical wheel of biophilic leadership, or bio-leads. Its focus is on summarizing the development of biological or green issues in urban places and identifying their utility, not simply on renovating an environment, but creating such features as ecological knowledge and experiences within the area as well as future stewardship to maintain the improvements. Yet it is worth emphasizing that there are still few guidelines for ensuring that the addition of these ideas does not negatively impact on other species or upon vulnerable people in inner cities who may be in danger of being ignored or displaced by green revitalization projects.

More fundamental association with nature, not only in terms of functional benefits, but with what Beatley (2011) views as the deeper concept of integrating nature into everyday living, comes from recognizing how this adds to health, reduces stress, helps recovery from illness, and enhances cognitive skills, thereby helping to create more healthy cities (Chap. 13), not just more ecological ones. In many ways these ideas overlap with the Eco-Cities concept (Register 2006) described in Chap. 5, although this new approach may be more biologically based and less focused on design to achieve their ends. It is a pity that their many common aims do

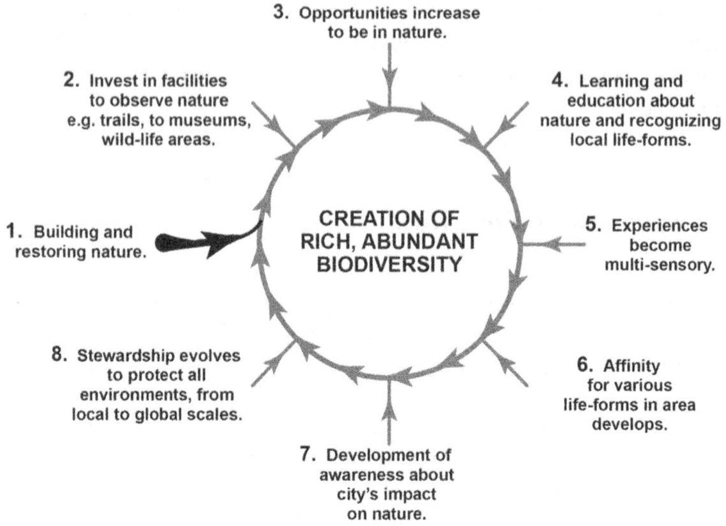

Fig. 4.2 Basic principles of biophilic cities

not lead to a more consolidated movement. Nevertheless, it is apparent is that the Eco-cities and Biophilic approaches are both helping to move this literally 'green' agenda to new heights, although most cities still have a long way to conform to the ideal of a truly green city.

4.10 Conclusions

The last two decades has seen an explosion of interest in making cities greener and cleaner, part of a more general movement to sustainability and a return to more natural conditions. Yet it is often forgotten that many of the ideas being promoted today can be traced back to the pioneering ideas of nineteenth century reformers, especially those responsible for the growth of parks in cities, and later the Garden City and Regional Planning ideas of individuals such as Howard and Geddes. Among the variety of alternative green policies that have been implemented in urban places over the past few decades, four key principles can be discerned. One is that development should try to work *within the framework of the natural ecology*, not destroy it, in order to recreate and rehabilitate areas and return them to more natural conditions. This would then improve the recreational and health opportunities of the area, as well as improve local knowledge and stewardship of the land, as promoted by the biophilia movement. Some green advocates, such as those in the Ecocity movement, which held its ninth global meeting in Montreal in 2011, are more radical, wishing to see urban places as the minority areas within a large natural landscape, rather than the opposite, which is the pattern in most urban regions today. Their principles go beyond the literally green practices discussed here. Although progress towards what advocates see as the search for urban places having greater balance or harmony with nature is slow, these ideas have helped stimulate the green agenda and continue to attract attention.

A second principle is that advocates should avoid thinking only in the top-down policy terms of the past, where only governments or city officials make all the decisions. It is important to find ways of creating *citizen interest and pressure* to persuade governments to develop green spaces, educating and involving people in the movement to make cities cleaner and more attractive as well as persuading donors to contribute to the greening of urban places, rather than leaving it only to governments. In this way the citizens will have a stake in maintaining and protecting their biotic and physical environment, rather than outsourcing responsibility to city officials and perhaps police, which may help to a return to the old ideas of stewardship and responsibility over the land.

Third, it is important for green advocates not too be too negative about development, as many environmentalists have been in the past. Porritt (2005), a leading Green campaigner, altered his previous views by observing that since capitalism is likely to continue as the primary economic system in the world, green advocates should search for policies that will appeal to market solutions without compromising their basic objectives. In this way they will *operate within the economic system*, just as the ecologically minded people stress working within the natural system.

Finally, it is also worth re-emphasizing the *benefits of nature and green spaces*. The old notions of green areas being useful as places of relaxation and recreation are only part of the many reasons for developing green policies. Many other advantages, including the improvement of the health of people, the aesthetic values, agricultural potentials, and the protection of the local and even the world environment—especially its climate and atmosphere—are being recognized. After all, nature creates its own economic benefits, from trees absorbing CO2, to wetlands removing pollutants. Yet these benefits do not only improve the life of residents or the local ecology. It is recognized that a large and varied supply of green space is likely to enhance the image of a settlement and can help attract a qualified and educated workforce and new businesses, adding to its future potential and viability. However, adding to the supply of these green spaces is not enough. There must also be ways of encouraging the *use* of these green areas in urban settlements, whether for sport or recreation. Many urban places around the world do not have enough green space, especially in the less developed world. Even in Britain where there is an extensive system of public parks in towns, one of the important pioneering heritages from the Victorian era of municipal activism, problems are emerging. Since 1996 the Heritage Lottery Fund has provided over £ 700 m to British parks which led to major improvements to a system that has an estimated 2.6 billion visitors per year (HLF 2014). But since 2010 government cutbacks to local councils has led to a serious deterioration, with 86 % of park managers reporting cuts that have led to cafes and toilets being closed, or with reduced hours, and untended flower beds etc. The result is that 74 % of respondents to a survey (HLF 2014) maintained their parks are in poor condition, a situation that is likely to get worse. Indeed 45 % of local authorities contacted are seriously contemplating selling off parks or green areas, a deplorable possibility, given the increasing evidence of the value of green space to human health. So despite progress in the greening in many settlements and improving the tidiness of places there are signs of regression. This is unfortunate, for green areas should be recognized as an essential part of urban places, improving many human experiences and providing health-giving properties, especially at a time when there are problems associated with too much indoor living, as shown in Chap. 13 (Healthy Cities).

References

A-UK. (2012). http://www.allotments-uk.com. Accessed 20 Aug 2012.

Bacon, E. N. (1974). *Design of cities*. New York: Penguin.

Baltimore, C. (2011). Cleaner, greener Baltimore : Annual reports. City of Baltimore, Baltimore.

BC: Biophilic cities.(2013). http://www.biophiliccities.org. Accessed 12 April 2013.

Beatley, T. (2000). *Green cities: Learning from European cities*. Washington D.C.: Island Press.

Beatley, T. (2011). *Biophilic cities: Integrating nature into urban design and planning*. Washington D.C.: Island Press.

Beaulac, J., Kristjansson, E., & Cummins, S. A. (2009). A systematic review of food deserts: 1966–2007. *Preventing Chronic Disease, 6*(3), A105. http://www.cdc.gov/pcd/issues/2009/jul/08_0163.htm. Accessed 12 June 2013.

Binelli, M. (2012). *Detroit is the place to be: The after-life of an American metropolis*. New York: H. Holt.

CA: Cities Alive. (2012). http://citiesalive.org/index.php/tours. Accessed 12 April 2012.

CESD: Commissioner of Environment and Sustainable Development, Canada. (2012). http://www.oagbvg.gc.ca/internet/English/parl_cesd_201205_03_e_36775.html. Accessed 8 Sept 2013, Accessed 20 Nov 2013.

City of Calgary. (2012). Annual Report of the Parks Department. Calgary City, Calgary.

Coleman, A. (1985). *Utopia on trial*. London: H Shipman.

Cox, W. (2004). *Myths about urban growth and the Toronto green belt*. Vancouver: Fraser Institute Digital Publication.

Crouch, D., & Ward, C. (1997). *The allotment: Its heritage and culture*. Nottingham: Five Leaves.

Darren, D. (2009). *Public produce: The new urban agriculture*. Washington D.C.: The Island Press.

Davies, W. K. D., & Herbert, D. T. (1993). *Communities within cities*. London/New York: Belhaven/Halstead Press-John Wiley.

Despommier, D. (2008). Vertical farm essay 2: Reducing the impact of agriculture on ecosystem functions and services. http://www.verticalfar.com/essay2. Accessed 12 Feb 2014.

Despommier, D. (2010). *Vertical farms: Feeding the world in the twenty first century. Thomas Dunne*. New York: Macmillan.

Donovan, G. H., & Butry, D. T. (2010). Trees in the city: Valuing trees in Portland. *Landscape and Urban Planning, 94,* 77–83.

Donovan, G. H., & Prestemon, J. P. (2012). The effect of trees on crime in Portland. *Environment and Behaviour, 44*(1), 3–30.

EEA: European Environment Agency. (2011a). *Green infrastructure and territorial cohesion*. Copenhagen: EEA.

EEA: European Environment Agency. (2011b). *Forests, health and climate change: Urban green space forests for cooler cities and healthier people*. Copenhagen: EEA.

EEA: European Environment Agency. (2012). *Urban adaptation to climate change*. Copenhagen: EEA.

Elson, M. J. (1993). *The effectiveness of green belts*. London: Dept. of Environment, HMSO.

FMNS: Farmers' markets in Nova Scotia. (2012). http://www.farmersmarketsnovascotia.ca. Accessed 12 Sept 2012.
Geddes, P. (1915). *Cities in evolution*. London: Williams & Norgate.
GN: The Guardian Newspaper. (2011). The English allotment mapped. Nov. 10.
GW: Greenway, Calgary. (2014). http://www.parksfdn.com/greenwaymap.pdf. Accessed 10 May 2014.
Hall, P. (1988). *Cities of Tomorrow.* Oxford: Blackwell.
Hall, P. (1992). *Urban and regional planning*. London: Routledge.
Hall, P., & Clawson, M. (1973). *Planning and urban growth: An Anglo-American comparison*. Baltimore: Johns Hopkins Press.
Hall, P., Thomas, R., Gracey, H., & Drewett, R. (1973). *The containment of urban England*. London: Allen and Unwin.
HLF. (2014). State of UK Public Parks: From Renaissance to Risk. Heritage Lottery Fund, London. http://www.hlf.org.uk/aboutus/howwework/Documents/StateofUKParks2014_advocacy.pdf.
Howard, E. (1898). Garden cities of to-morrow (Introduction by Lewis Mumford). M.I.T. Press, Cambridge, Mass. 1965. Originally publish as Tomorrow: A peaceful path to real reform, 1898.
IDE. (2002). Tidy towns handbook. Irish Dept. of Environment, Heritage and Local Government, Dublin.
IDE. (2008). Tidy towns of Ireland, 1958–2008. Irish Dept. of Environment, Heritage and Local Government, Dublin.
Kahn, P. H., & Kellert, S. R. (2002). *Children and nature.* Cambridge: M.I.T. Press.
Kuo, F., & Sullivan, W. (2001). *Environment and crime in the inner city. Environment and Behaviour, 33*(2), 343–367.
LAM (LAM). (2012). Living Architecture Monitor, Monitor: 10th Anniversary Edition, Summer.
Le, C. (1929). *Cities of tomorrow*. London: J. Rodker.
LeDuff, C. (2013). *Detroit: An American autopy*. New York: Penguin Press.
LHHL: Landscape Human Health Laboratory. (2012). http://lhhl.illinois.edu/media/thepoweroftrees.htm. Accessed 10 Aug 2012.
Lim, C. J. (2010). *Smart Cities and Eco-warriors*. London: Routledge.
Louv, R. (2008). *Last child in the woods: Saving our children from nature-deficit disorder*. Chapel Hill: Algonquin Books.
Louv, R. (2011). *The nature principle*. Chapel Hill: Algonquin Books.
McMahon, E. T. (2012). Healthy communities: New directions in development. Urban L and Magazine, Nov. 26. http://urbanland.uli.org/economy-markets-trends/healthy-communities-a-new-direction-in-development/. Accessed 2 March 2013.
McWilliams, J. (2009). *Just food: Where locavores got in wrong and how we can eat responsibly*. New York: Little Brown.
Mogk, J. (2010). *Urban agriculture: Good food, good money, good idea*. Paris: OECD.
Mougeot, L. J. A. (Ed.). (2005). *Agropolis: Social, political and environmental dimensions of urban agriculture*. London: Earthscan.
Mumford, L. (1961). *The city in history*. London: Secker & Warburg.
NC.: North Carolina. http://www.nc.10percent.com. Accessed 12 Sept 2012.
NCN.: Nature and Children network. (2012). http://www.natureandchildren.org. Accessed 20 Sept 2012.
OGB: Ontario Green Belt. (2012). http://www.greenbelt.ca. Accessed 20 Oct 2012.
Perry, C. (1929). The neighbourhood unit. In The New York Regional Plan and its environs, Vol. 7, R. New York: Sage Foundation.
Poritt, J. (2005). *Capitalism as if the world matters*. London: Earthscan.
Pretty, J., Peacock, J., Sellens, M., & Griffin, M. (2005). The mental and physical health outcomes of green exercise. *International Journal of Environmental Health Research, 15*(5), 319–337.
Register, R. (2006). *Ecocities: Building cities in balance with nature*. Berkeley: New Society Publishers.
Reps, J. W. (1965). *The making of urban America*. New Jersey: Princeton University Press.

RSA: Regional Studies Association. (1990). *Beyond green belts*. London: Jessica Kingsley Publishers.
Scott, J. W. (2006). The new regionalism as a contingent governance paradigm. In R. C. Lois Gonzalez (Ed.), *Urban changes in different scales: Systems and structure* (pp 107–118). Santiago de Compostela: I.G.U. Urban Commission.
Steeley, G. (1991). Green planning for new communities. *Town and Country Planning, 60*(March), 77–79.
Shadbolt, P. (2014). Hong Kong's fish farms in the sky. http://www.bbc.com/news/business-26627408.
Smit, J., Rattu, A., & Nass, J. (1996). Urban agriculture: Food, jobs and sustainable cities. U.N. Habitat II series, New York.
Smith, A. (1984). The green belt: Agent for regeneration or stagnation. *Town and County Planning, 53*(5), 156–157.
Stein, C. (1951). *Towards new towns of America*. Liverpool: Liverpool University Press (reprint).
Stephenson, R. B. (1999). A vision of green: Lewis Mumford's legacy in Portland, Oregon. *American Planning Association Journal, 65*(3), 259–267.
Summerson, J. (1962). *Georgian London*. London: Penguin.
Thomas, D. (1970). *London's green belt*. London: Faber and Faber.
Ulrich, R. S. (1984). View through a window may influence recovery from surgery. *Science, 224*(April), 420–421.
USDA.: U.S. Dept. of Agriculture. (2012). http://www.ers.usda.org/food-desert-locator. Accessed 12 June 2012.
Vance, J. (1977). *This scene of man*. New York: Harper.
Ver Ploeg, M. M., Breneman, V., Farrigan, T, Hamrick, K., Hopkins, D., Kaufman, P., Lin, B. H., Nord, M., Smith, T. A., Williams, R., Kinnison, K., Olander, C., Singh, A., & Tuckermanty, E. (2009). *Access to affordable and nutritious food:* Report to Congress. U.S. Dept. of Agriculture (Economic Research Service), Washington D.C. http://www.ers.usda.gov/publications/ap-administrative-publication/ap-036.aspx#.Uwk5UPldUZ4. Accessed 12 Sept 2012.
Videl, J. (22 March 2008). Coming up roses. *The Guardian*.
Wheeler, S. M. (2002). The new regionalism. *American Planning Association Journal, 68*(3), 267–277.
White, K. K., & Duram, L. L. (2012). *America goes green*: *Encyclopaedia of eco-friendly cultures in the United States*. Santa Barbara: ABC-CLIO.
Whiteacre, P. T., Tsai, P., & Mulligan, J. (2009). *Public effects of food deserts*. Washington, D.C.: National Academies Press.
Williams, M., & Lee, J. C.-K. (2006). *Environmental and geographical education for sustainability: Cultural contexts*. New York: Nova Science.
Wilson, E. O. (1993). *Biophilia*. Cambridge: Harvard University Press.
Wright, F. L. (1932). *The disappearing city*. New York: W. F. Payson.
Wrigley, N. (2002). Food deserts in British cities. *Urban Studies, 30*(10), 2029–2040.

Chapter 5
Background to Sustainable Cities

Wayne K.D. Davies

Climate change is the biggest crisis our civilization faces.
(A. Gore, March 31, 2014)

5.1 Introduction

Towns and cities have always been unsustainable places, dependent on many inputs, such as food, water and energy, and producing noxious, if not toxic outputs. Most urban places were small in the past, so many of these inputs could be supplied, and outputs disposed of, without major problems, although the larger imperial capitals always had difficulties. In the last two centuries the explosive growth in the number and sizes of urban places led to new technological solutions to solve many of these input-output issues, but others were ignored. Towards the end of the twentieth century more and more people became sceptical that continued innovation is possible while the 1960s and early 1970s saw early warnings of impending problems associated with current development practices. For example, The *Limits of Growth* (Meadows, Randers and Meadows 1972, 2004) predicted growth constraints because of the finite nature of many resources. *The Population Bomb* (Ehrlich 1968) warned about future food shortages because of explosive population growth. Rachel Carson's (1962) path-breaking book *Silent Spring* highlighted the harmful effects of so many of our industrial processes and the products they produced that poisoned flora and fauna, especially the loss of many birds and their songs, which led Carson to her title. Indeed, Carson's book, followed by many others on related themes, did much to stimulate the environmental movement. Certainly criticisms of the methods used by these pioneering authors, the slow-down in population growth rates from the huge increases of the 1960s, as well as increasing agricultural and technological

W. K.D. Davies (✉)
Department of Geography, University of Calgary,
2500 University Drive, N.W., Calgary, AB T2N 1N4, Canada
e-mail: wdavies@ucalgary.ca

W. K.D. Davies (ed.), *Theme Cities: Solutions for Urban Problems*,
GeoJournal Library 112, DOI 10.1007/978-94-017-9655-2_5

advances in the past 30 years, have tended to disguise the extent of the harmful effects that these and other authors predicted. But many still believe that the problems revealed by these critics of current practices are real ones, even if some may be in a longer term context than initially feared and may be modified by the discovery of new resources and the hope for new technological innovations to solve problems.

These early worries have now been extended and have crystallized around the need to practice greater sustainability, which has become one of the most used concepts in the past decade. More recently the need for sustainability has been fuelled by additional concerns, especially the harmful results of fossil fuel burning and the trend of global warming, all of which has increased the level of pessimism about the effects of our current progress. As the quotation above shows, former U.S. Vice President Gore, known for his popular film *The Inconvenient Truth* in 2006 and his subsequent Nobel Prize, has described the problems of global warming and climate change—one, and only one, of the key issues driving sustainability—in apocalyptic terms. Although many have accused Gore of exaggeration there can be little doubt that we should be preventing, or at least drastically reducing, carbon emissions and all the other negative externalities of many economic and social processes that are altering the environment in harmful ways, not simply in urban areas but regionally and globally. In addition we should be reducing the over-consumption of resources, many of which are finite, and should also embrace the need to solve the problem of growing enough food for the world population, although this issue is not primarily an issue that can be solved primarily by urban actions.

In the increasing concern about our limited degree of sustainability it is the prospect of climate change that looms the largest, which is linked to the consequences of our actions in urban places. An international organization like the World Wildlife Fund, which is devoted to the scientific study of the health of the planet and the impact of humans on animals and habitat, has emphasized the need for drastic policy changes to reduce carbon dioxide emissions which is contributing to climate change. In their report entitled *Reinventing the City* they claim that the majority of these emissions come from cities and suggest new policies are needed to drastically reduce these levels, based on: aggressive energy reduction using best practices; innovative funding strategies to counteract the traditional limitations of city budgets; and the use of the latest technical advances (WWF 2010). Elsewhere the more activist Greenpeace movement is more prepared to engage in confrontations to draw attention to the negative impacts of various economic activities upon polluting activities and the destruction of habitat or various species. In many developed countries concern for the environment has led to progress in many areas, especially in reducing levels of pollution from the horrors that were created in earlier phases of industrialization, but there is still a long way to go before these issues are resolved. Unfortunately the pollution problems have intensified in some areas of the less developed and rapidly industrializing world, not only in cities, as a recent review of the situation in China by The Economist noted.

> In January 2013 the air in Beijing hit a level of toxicity 40 times above what the World Health Organization deems safe. A tenth of the country's farmland is poisoned with chemicals and heavy metals. Half of China's water supplies are unfit even to wash in, let alone drink

in. In the northern half of the country air pollution lops five and a half years off the average life. (TE, Aug.10, 2013b, p. 9)

These are some of the appalling consequences of China's recent rapid industrialization, a country that is also consuming over 40 % of many of the major metals and now creating twice as much carbon emissions as the United States and a quarter of the world total. Certainly there are many signs that China, with 19 % of the world population of 7.2 billion in 2013, is now engaging in major efforts to clean up its environmental degradation with new carbon reduction policies proclaimed in 2013. For example, new carbon trading policies are designed to reduce emissions in its largest cities (Qiu 2013); new monitoring and, importantly, publication of air pollution levels in 179 cities through its Institute for Public and Environmental Affairs has been instituted; there are also promises of real time publication of the pollution, waste water and heavy metal outputs of 15 thousand large enterprises, as well as exhortations to public officials to reduce pollution levels. Such policies will take years to achieve results, although hopefully at a far faster rate than the western countries achieved in their main industrial growth period. Unfortunately many other countries are doing little to alter their development path. For example, The WHO latest report of air pollution estimates a rising death toll from 100 to 600,000 deaths in India from air pollution alone from 2000 to 2010 (WHO 2014). Many of the problems of a lack of sustainability are starkly seen in our burgeoning cities, and have led some municipalities to devise policies to at least mitigate the difficulties. But cities cannot do this alone; the problems apply to regions and countries in general, so national governments have to get involved. So to focus only on urban decision-making is to ignore the effects of the wider context of the sustainability debate, which will be dealt with in this chapter, leaving the next to deal with the specifically urban policies on sustainability.

Given the vast and expanding scope of this field this chapter can only provide signposts on the way to a more sustainable future. The first section deals with some general problems posed by limited sustainability, especially the negative externalities of our development processes. The second focuses on the difficulties of defining the scope of sustainability, especially the development of useful indicators to measure the degree of sustainability in urban places, with an example of a comparative study of sustainability in major world cities. Subsequent sections deal with the problems associated with major resources and outputs, initially water and sewage, and then the crucial resource of energy, including the prospects for renewable sources. This is followed by a review of sustainable progress in two major sectors, namely industrial processes and transport, which are essential components of the functioning of the urban economy. Space limitations mean that other vital resources, such as food supply, cannot be dealt with in this essentially urban-based discussion. It must also be admitted that many of the problems and changes taking place in the these resources and sectors involve decisions at a national level, but they need to be reviewed since they have huge effects upon the sustainability of individual urban places, although there is often room for local modification through municipal actions.

5.2 Problems of Unsustainable Practices

Urban places are the sites of many of the most significant unstainable practices. One of the most important contributors to this characteristic is the burning of fossil fuels, which has been the energy source for our industrialization and urban functioning. In the past the negative externalities of fossil fuel use were usually ignored, they were part of the price of progress. Even when a problem was obvious, such as the infamous London smogs from the nineteenth century, it was considered to be city-specific and it took over a century before serious action was taken to mitigate the problem. This intermittent type of pollution reached a new peak in four days in December 1953 (MO). A fog over the city became toxic when the emissions from the coal-burning fires and furnaces in the city became trapped because of a temperature inversion with warmer area above. The fog not only meant that visibility was less than a few yards, but created respiratory problems in large numbers of people that contributed to thousands of deaths. This crisis did lead to a clean air act which involved the banning of the use of coal in household fires in London.

Although many other cities have also suffered from similar pollution caused by fossil fuels in the last two decades, it has finally been realized that the burning of this resource is having more than urban effects. The addition of carbon dioxide to the world's atmosphere, mainly from fossil fuel use, reached 393 ppm in 2013, a major increase on the 280 ppm at the dawn of industrialization in the late eighteenth century. Along with other gases it is creating what amounts to a blanket or greenhouse effect by absorbing more of the sun's rays, thereby warming up the earth. Of course carbon dioxide is not the only gas that increases climate warming, although it accounts for over two-thirds of the total. Methane, which has even greater heat absorbent ability, accounts for another fifth, while water vapour, nitrous oxide, halocarbons and ozone also contribute to the greenhouse effect. Although these gases only account for 1 % of the atmosphere, which is primarily nitrogen (78 %) and oxygen (21 %), this change in the atmospheric balance increases warming and hence the amount of water vapour that the air can hold.

Most climate scientists believe the current warming trend is different to the other oscillations that have caused climate changes in the human past, as seen by the reports of the Intergovernmental Panel on Climate Change (IPCC 2013, 2014) that was established by the World Meteorological Organization and the United Nations in 1988. Their regular reports and the research of the vast majority of climate scientists has led to the conclusion that there will be many negative effects of climate warming. This human-induced trend will increase the intensity of extreme weather events, especially violent tropical storms, and perhaps increase their number. There is likely to be more heat waves, whose effects in cities will be exacerbated by the heat-island effect of all the buildings. In addition, droughts in many already marginal areas may reduce agricultural production. There has been a marked increase in the number of droughts in Australia and California in recent years which is leading to a greater incidence of forest fires than experienced over the past half century, while forest fires in the Canada are also well above average trends. Drought will also decrease the water supply of many cities throughout the world. Large numbers are already running short of water, a function not only of climate change and the over-utilization of artesian wells, but because nearby rivers are not being replenished. An even greater effect of the warming is the more rapid melting of ice in polar-regions which will raise sea-levels, threatening the viability of many low-lying islands, and having negative effects on coastal cities and communities. These settlements, that include many of the world's greatest cities, will not simply be at danger from the new sea levels but will also be more prone to storm surges. This means that more preventative measures are needed to increase their resilience to these hazards, issues discussed in the Resilient Cities (Chap. 9). There will also be an increased acidification of oceans, for they have absorbed at least a third of our fossil carbon emissions in recent decades (IPCC 2013). Although this ocean absorption has reduced the warming effect on land, the increased acidification of the oceans has accelerated and is at levels that are the highest for 300 million years according to a recent report from the International Programme on the State of Ocean (IPSO Oct. 2013). A major negative effect of this change will be to gradually dissolve the carbonate skeletons that form the structure of coral reefs. This will help destroy a major fish habitat, let alone a tourist resource, within the next few decades. It will add to the existing crisis in fish stocks and hence food supply, for it is estimated at least 70 % of fish stocks over currently over-exploited (Roberts 2012).

Yet the problems associated with climate change are not universal. Cities and regions may be differentially affected given the local conditions. Indeed three broad types of global warming effects on urban places can be recognized (IPCC 2014. TE 2014c). One category is when the climate change has a major effect, such as rising sea levels due to polar icecap melting, perhaps 0.2 to 0.4 metres by 2050. Hallegatte and colleagues (2013) estimated the annual costs of flooding in the largest 136 coastal cities was $5 billion in 2005 and could increase to $1 trillion by 2050, well above the $62 billion expected then due to growth and subsidence alone. In other cases the changes may be moderate, so that warming trends may be mitigated by environmental improvements to city designs, or by anticipating the likelihood of the poleward spread of malaria-carrying mosquitoes which will, without mitigating health policies, increase mortality and morbidity rates. These and other possible consequences are primarily public health and nutrition issues that are more easily modifiable. In the third case the climate change may be less influential, except in cases of already marginal environments, being related to such issues as the differential ability of various species, including humans, to survive in a location that becomes hotter or drier. For settlements in these areas there will be need for improved water supplies and better ways of reducing the effect of heat. In terms of agriculture, which may supply local or regional food needs, the changes mean that either new varieties will need to be bred, or new land use practices implemented, with help given to those affected by the transition. More generally it is important to realize that many different effects can be produced by these warming trends in various regions and in settlements. This means that many different types of policies specific to cities and regions to mitigate the effect of the changes will be needed.

It is also worth observing that there has been an unexpected pause in the global warming in the last decade. This has been calculated as being as much as 0.2 °C lower than earlier predicted levels, which is attributed not to the effects of the current solar minimum, but to stronger trade winds in the eastern Pacific which have been twice as strong in the last two decades compared to most of the previous century. This has driven surface warmer water to the depths, thereby reducing ocean warming (England et al. 2014). This seems to be only a temporary reprieve, which will be followed by the type of rapid warming seen in previous decades once this ocean effect wears off.

There is still a lot of debate about the rate of greenhouse gas build-up and whether there is a critical level above which a series of feedback loops in what is a complex atmospheric system will create irreversible changes in climate. Despite regular international conferences and the impact of Gore's film, *An Inconvenient Truth*, that publicized the problem, albeit with limitations, little progress has been reached on reducing the levels of greenhouse gas build-up through international agreements. In addition, there are still some climate deniers—individuals who deny that climate change is taking place as a result of human-created carbon dioxide emissions. This led the Canadian National Post newspaper to run a series of articles supporting these views from Nov. 2006 through 2007, which led to a summary and extension of the ideas in a book (Solomon 2008). Some of these climate deniers are in influential posts, such as those in the Republican ranks in the U.S. Congress. But many of the scientists whose opinions have been quoted by the climate deniers have rigorously argued that they were misquoted (EEEC 2007). Other initial well known sceptics, such as the physicist R.A. Muller (2012) at the University of California at Berkeley, created his own climate group to look into climate warming, based on detailed measurements of the changes in earth temperatures from stations around the world, rather than depending on climate modelling. He has confirmed the warming trend and link to atmospheric carbon concentrations. So it is becoming harder and harder to accept the opinions of climate change deniers, given the accumulating evidence from so many studies and especially the warming seen in glacial and Arctic ice cap retreat, although it must be noted that the climate system is too complex to attribute one storm or climate event to this general conclusion. The evidence from hundreds of studies led the IPCC panel (Sept. 2013, p. 2) to state their conclusion in blunt terms in a recent report:

> warming of the climate system is unequivocal, many of the observed changes are unprecedented over decades to millennia.

The combination of the problems created by climate warming with many other negative changes that affect sustainability mean that humans seem to have created a scale and type of development that is not just affecting cities and regions but is threatening to negatively transform the planet. This is the very environment that sustains us, as well as other life forms. Of course, our activity is not the first life-form to have altered the planet. The Gaia movement, pioneered by the atmospheric scientist James Lovelock (1979, 2009) and the microbiologist Lynn Margulis (1999), showed that our planet is not the simple, inert result of the endowment of the lithosphere and the physical processes upon it. Instead, it has been the generation of life itself which has altered the physical basis of the planet by changing the gas balance in the atmosphere, especially by plants that create the very oxygen we breathe and which absorb and store carbon dioxide. The result is that Gaia advocates have shown that the earth is a living ecological system composed of a series of mutually interacting and self-regulating systems, with complex inputs and outputs that help to maintain life. These increasingly large impacts of human activity on the planet led the climate scientist Paul Crutzen and ecologist Eugene Stoermer (Crutzen and Stoermer 2000) to suggest that our post-glacial Holocene period may be giving way to a new Anthropocene era, in which human activity has become one of the most dominant ecological processes in the landscape. Unfortunately many of our ecological effects are destructive, altering these interacting systems in negative ways. This has convinced many people that humans should create new developmental practices, ones that *sustain* life on earth in all its forms, rather than degrading or even raping the planet through current practices. Without such changes many believe there is a danger of our civilization being degraded, or even collapsing, paralleling the fate of some societies in the past that destroyed their environment and over-used their resources, although Diamond (2005) noted that there are many other reasons for such collapses. Even without such apocalyptic beliefs there is a pressing common-sense need to find ways of at least reducing the negative economic and life-style processes that our current prosperity is built upon, by adopting new policies to achieve a more sustainable future, based on practices that use more renewable resources and which have less deleterious effects on our environment and other life-forms.

Many of the new policies designed to attain this type of more sustainable future will come from national governments and corporations. But some will come from actions at an urban level, given the huge resource demands and the negative externalities that urban centres produce, and from individual choices to be more sustainable. Indeed, it is worth noting that even though the world urban population accounts for just over 50 % of the total population, these centres currently consume over two-thirds of the estimated world energy production and most of the carbon dioxide emissions: 70 % according to the influential World Energy Outlook (IEA 2008, p. 179), with a WWF report (2010) claiming it is as high as 80 %. The International Energy Agency estimated the urban emissions proportions will increase to 73 and 76 % respectively by 2030 if we pursue the current development path.

Many signs of a change towards greater sustainability, by individuals, businesses, national and urban governments, surround us. A recent polemical book has described the various actions as contributing to a fundamental 'sustainability leap', arguing, in a pungent phrase, that such change is vital as *"the norms of twentieth-century prosperity have become the instruments of twenty-first century collapse"* (Turner 2011, p. 1). Yet despite the encouraging examples in Turner's book of the way that many new policies and technologies are creating more sustainable practices, we need to be cautious about the extent of change that has actually taken place to date, especially given the great increase in pollution in China and other industrializing countries, and an expected future world population growth based on medium variant estimates to 9.6 billion by 2050, compared to 7.2 billion in 2013 and only 2.5 billion in 1950 according to U.N. World Population forecasts (UN 2013). To achieve a real change in the character of settlements and the sources of energy on which they function will involve the development of a new 'sustainability culture' in our lives Davies and Brown (2006), in which culture is used in the anthropological meaning as 'a way of life', not the typical meaning of high artistic endeavour. It is also worth noting that changes in new buildings and practices will not be enough, there is an under-estimated inertia effect stemming from the fact that most of the world's existing urban infrastructure and stock of buildings and life-styles are hardly sustainable, while it will not be easy to radically reduce the use of fossil fuels for energy, especially given the growth expected in energy use over the next 20 years. So despite very significant progress towards sustainability in some countries, especially renewable energy in Denmark and Germany, there is still a general failure of political will by most governments to grasp the seriousness of the current and future situation to rapidly implement sustainable policies. However there are signs of change. The European Union's policy of encouraging its members to attain a target of 20% of energy from renewable sources and a 20% reduction in carbon emissions by 2020, was raised in 2014 to 27% and 40% by 2030 respectively, with a 27% optional increase in energy efficiency. In late 2014 China also promised to cap its emissions by 2030, while the U.S. President proposed a 27% cut in emissions from 2005–2025. Yet these are enough to solve the growing problem, and too many countries do not have sustainable policies. Perhaps the 2015 Paris Climate Conference will lead to real progress, although the signs are far from positive.

5.3 Defining and Measuring Sustainability

The important breakthrough in understanding the negative effects of human development and especially cities upon the planet, at least in conceptual terms, came from the World Conference on Economic Development (WCED 1987) held in Rio de Janeiro. The resulting Bruntland report, named after the former Norwegian Prime Minister who chaired the meeting, popularized the concept of sustainability in development as a focus for altering the current trends and their negative impacts. The origin of sustainability came from the change that took place in mid-twentieth century Scandinavian forestry policies. Instead of just cutting down as many trees as the market required, foresters were required by governments to plant as many as were harvested, thereby creating practices in modern forestry management that ensured the same number of trees, although at different stages of development, survived. Of course, similar ideas had been used historically in many societies, especially in pre-capitalist days, as part of a belief in stewardship of the land and its resources, of nurturing it and passing it on to future generations—although one must remember the limitations of their technology which put limits on the extent of their resource extraction. From the Industrial Revolution onwards, the concept of stewardship decayed under the impact of the growth mechanisms within capitalism and the new technologies which enabled the greater and greater use of resources. To take one example, in the mid-twentieth century the increasing use of power-saws enabled foresters to cut down acres of trees in a day—instead of the past capacity of one or two trees a day by hand axes—trees that took generations to grow. So the amount of deforestation increased beyond nature's ability to replenish the stocks destroyed. Since similar levels of destruction were occurring in other areas of resource extraction and development, the Bruntland report argued the case for sustainable practices in all forms of development. In O'Riordan's (2004) succinct phrase this moved the debate about the type of impact that humans were having upon the world, which went "*beyond environmentalism towards sustainability*". More specific and localized plans of action to create sustainable practices came from another U.N. conference in Rio de Janeiro in 1992 attended by most world leaders. Although a series of high minded resolutions were passed, few countries have kept to their promises to protect environments, to seek more renewable energy sources, or deal with the growing human contribution to the carbon dioxide growth in the atmosphere. A subsequent meeting on climate change in Kyoto led to the agreement in 1997 of most countries to reduce the greenhouse gases build-up (GHG) under the protocol named after the host city in two periods, first to 2012 and the from 2013 to 20. This would aim to prevent these gases reaching 450 ppm which was initially considered by the IPCC panel and their climate models to be the level that would ensure world average temps would not exceed 2 °C over pre-industrial levels when the GHG were 280 ppm. After the fifth report was published in late 2013 (IPCC 2013) the chairman of the panel of climate specialists reported that the latest research showed that to restrict a global temperature increase of 2 to 2.4 °C above the levels in the 1850–1900 period would need a carbon dioxide stabilization in the 445–490 ppm range, and assumed that a peak of emissions would occur by 2015 at the latest. However the climate models also predict that an increase of 2.4 to 2.8 °C would mean a carbon dioxide stabilization level of 490–535 ppm, with peak emissions by 2020, whereas a 2.8 to 3.2 °C increase would occur if stabilization occurred at 535–590 ppm with an emissions peak by 2030 (Pachauri 2013). Despite this and other evidence of the impending crisis and regular international conferences, limited progress has been reached on an international scale, although a few countries and regional groupings are reducing the levels of greenhouse gas build up. Unfortunately the U.S.A. did not ratify the Kyoto treaty and since then many countries have abandoned their initial commitment. Only 37 countries out of 191 now have binding commitments to reduce emissions in the second phase and these only account for 13 % of global greenhouse gas emissions, although this group does include the European Union. Follow-up conferences in 2011 and 2013 produced a series of general intentions, not firm, definable plans for concerted action by nation states at an international level, although some individual countries are making major progress.

Much of the problem of a lack of agreement on emissions reductions comes from governments either sceptical of the greenhouse gas-warming connection or of their refusal to take action that would lead to higher energy prices, therefore putting themselves at a competitive disadvantage. In addition many developing countries believe that it is the industrialized world that created the problem so they should be the ones to reduce emissions, rather than expecting developing countries to do so and decrease their industrialization. Ethically they have a point, which is reinforced by the fact their emissions were 2.9 t per person year in 2010 compared to 10.4 t from the developed world (Raupach et al. 2007). But its value is reduced when one notes that it is the developing world and especially China which will soon account for more than half of the annual increases of carbon emissions into the atmosphere. Despite the progress in individual countries there is a depressing lack of international action and the continued build-up of greenhouse gases which on current progress means the attempt to limit the climate increase to an average 2 °C increase seems unlikely.

Despite the paucity of enforceable practical outcomes on this part of the sustainability debate at an international level from the Rio conference, it did help to promote the general concept of sustainability and led to increasing public concern about the issues that built upon the alarms raided by previous researchers. Moreover, of specific relevance to urban issues is that it led to the Local Agenda 21 report(U.N. 1993), which was designed to show how sustainable policies could be created at the local, urban level. The report argued that municipalities were closest to the people and could involve them to create more effective practices. These ideas certainly led to a flurry of policy initiatives in many cities. For example, 2 years after the report was published an informal network of cities in Europe was established by the European Sustainable Cities and Towns Campaign and held an inaugural meeting in Aalborg in 1994 and now has regular meetings and by 2013 2,700 communities were participants in the programme (EU 2013a). More generally The United Nations Centre for Human Settlements based in Nairobi created a Sustainable Cities programme with its first newsletter in September 1995 (UNHABITAT). In addition, a group of large world cities was created in 2005, calling themselves the C40 Climate Leadership Group (C40). This is a network committed to addressing climate change through municipal policies and which have implemented a series of pioneering policies.

Although these urban networks and parallel developments elsewhere, have been useful, progress in defining sustainability is still limited and is not helped by the complexity of the concept. It has moved a long way from its environmental origins, for it now include ideas of community participation as well as social justice and economic needs, meaning that it has become a multi-dimensional concept. Unfortunately, there is no general agreement on the individual elements that comprise sustainability. Hence the literature consists of varied, yet often overlapping descriptions of sustainability, lists of different dimensions and variables that are considered part of the sustainability concept by various international organizations, nations and citizen groups. The result is a confusing literature with many different ways of identifying and measuring sustainability.

Figure 5.1 tries to clarify the scope of the term by inter-relating the sustainability concepts used in four of the major and generally used schemes that have been suggested as summarizing sustainability. The resultant domains and dimensions of sustainability, are divided into two major categories. There are six general categories or Basic Domains, each of which can be subdivided into a series of subsidiary categories or dimensions, which are often identified as specific indicators. The first group of the Basic Domains, the most general category of all, is based on one of the most succinct summaries of sustainability ideas, that provided by the Natural Step Foundation, created by the Swedish oncologist Karl-Henrik Robert and his co-workers (NSF). As Fig. 5.1 shows, it identified four basic principles of sustainability and argued that their violations must be stopped if sustainable development is to be created. The Natural Step Foundation has complemented the conceptual scheme by the development of a curriculum to teach these ideas and a consulting

(1) Resource Over-Use.

Too many resources are being used in our growth and most are beyond their level of creation by earth processes. Too many are being concentrated or produced from by-products that create high levels of toxicity. Hence it is argued that extractions should not be used at a faster rate than their creation or dispersal.

Indicators from IEFS.

Clean and Renewable Energy. This requirement for the city should not have negative impacts on ecosystems, human health, or climate change and should be generated within the local bioregion.
Responsible Resources/Materials Use. All renewable/non renewable resources need to be sourced, allocated, managed and re-cycled responsibly without adversely affecting health of ecosystem resilience.
Healthy, Nutritious Food. It should be accessible, available and provided by maintaining ecosystems, without affecting climate change, as well as being obtained primarily within the bio-region.
Earth's Carrying Capacity Limits. City demands on the eco-system should be kept within Earth's bio-capacity limits, converting resources restoratively and supporting regional ecological integrity. (ECI: Sustainable Land Use)

(2) Environmental Degradation.

The eco-system has been systematically degraded by over-harvesting and disrupting the natural cycles that maintain resources leading to a serious reduction in biodiversity. Hence products should not be extracted at a faster rate than either they can be replenished or the ecological system can be renewed.

Indicators from IEFS.

Ecological Integrity. This should be provided by linkages between and within eco-systems to provide a contiguous habitat and corridors throughout the city.
Healthy Biodiversity. This must be sustained in species, ecosystems and genetic diversity, and natural habitat is restored.
Clean, Safe, Affordable Water. This type of water to drink should be sourced from within bioregion, while waterways need to be healthy and not negative to local eco-systems.
Healthy Soil. It should have its fertility maintained or improved and also meet the ranges of healthy ecosystem functions.
Clean Air. This is conducive to health and must be present in buildings, city and atmosphere.
(E.C.Indicators: Air Quality; Local Contribution to Climate Change; Noise Pollution)

(3) Created Compounds.

Many new products that have been created, from plastics to paints to chemicals and noxious by-products of our industrial and transport systems, take a long time to be broken down by natural processes and often accumulate as waste or pollution. So created compounds and by-products should not be produced at a faster rate than they can be re-used, rendered harmless, or broken down by natural processes.
No specific IEFS indicators.
(E.C.Indicators: Products supporting Sustainability)

(4) Human Capacity.

People should not be subject to conditions that undermine their capacity to meet their needs within the framework of the other principles. Repression, poverty, lack of education and knowledge, and opportunity, are among the main factors that reduce capacities. Hence equity and social justice should be guiding principles in government systems to allow people to achieve their potentials

From IEFS.

Healthy Culture. Urban facilities are needed to strengthen human knowledge, social learning, creativity and eco-literacy.
Individual Rights* and Community Capacity-Building. Towns and cities should support equitable participation in decision-making and provide support for all levels and scales of organizations and neighbourhoods.
Lifelong Education. All residents should have access to education, formal and informal, including vocational and social institutions, as well as access to information about all aspects of the city and its traditions. Also, environmental education should be enhanced.
Well-Being and Quality of Life. Strong citizen satisfaction is needed, based on quality of life indicators, including employment, the environment, health, education, recreation and social belonging.
Healthy-Equitable Economy. The economy should favour activities that reduce harm, benefits the environment and human health and supports many local and equitable employment options.
Proximity. There should be walkable access to basic goods - housing, shops and services, employment.
(E.C. Indicators: Local Mobility; Availability of Local Public Services and Open Areas; Childrens'Journey to Schools; Citizen Satisfaction with Local Community).

Fig. 5.1 Inter-Relationships of major sustainability principles

(5) Trans-Frontier Responsibility (from H+H).

This deals with the need to understand and resolve the spatial and global linkages that cause problems in areas removed from their source, such as acid rain from manufacturing regions that destroys northern lakes. In addition, there must be understanding of the temporal rhythms of development, which mean the adoption of long-term strategies and policy maintenance.

(6) Management (H+H).

Concern about the problems is not enough; policies have to be identified and implemented by new approaches. New management procedures. These are often needed to apply specific sustainability policies, recognizing that these are themselves the product of political decisions by governments.

Accounting. The procedures used in all development projects need improvement, such as including the environmental costs of development
- especially the negative effects of fossil fuel pollutants – should be viewed as internalized costs, not externalized, or just ignored as in the past.
Co-ordination. Packages of co-ordinated policies, instead of stand-alone approaches, are the basis of success.
Co-operation. This is needed between different levels of government, and linkages between urban areas, help implementation and enable the best practices to be identified and copied.
Subsidiarity. The lowest feasible level of government should apply policies, since these are closest to the people.
Resolution. Procedures for reconciling the different social, economic and environmental values of people is important.
(E.C.Indicator: Sustainable Management of Local Authority/Businesses)

(Sources: Categories (1) to (4) are summaries of Natural Step Principles which can be considered as Basic Domains or Categories. The additional Basic Domains (5) and (6) come from Haughton and Hunter's (H+H) principles. The IEFS standards are also summarized and located within these broad groups, as Subsidiary Principles, as are the European Common Urban Indicators (ECI). The term 'Individual Rights' has been added before the word 'Community' in this original category title, to stress its importance, since this will cover such neglected issues as women's reproductive choices and educational opportunities).

Fig. 5.1 (continued)

agency that shows businesses and urban places how to redesign their operations to achieve more sustainable practices. Despite the utility of this approach, it could be argued that the rigorous adoption of the first three of these principles would mean that most development would grind to a halt, given current practices. So the principles may be best seen as a general context within which progress to greater sustainability can be identified and subsequently measured.

The second source comes from a more detailed and specifically ecological approach adopted by the emerging Ecocity movement (Register 2006), which focuses upon the need to make urban places function in harmony with the biosphere and ecosystems. The International Ecocity Framework and Standards (IEFS) movement, which held its 9th international meeting in Montreal in 2011, proposed 15 system categories that can be used by cities or regions to measure their progress in sustainability. Since these ideas extend most of the four Natural Step Principles, their usual order is reorganized in Fig. 5.1 and placed into the most appropriate of the four basic categories prosed by the Natural Step scheme.

A third major source of sustainability indicators, namely those proposed by the European Commission in their the European Commission Indicators (ECI 2003), which were designed to be used by local authorities to measure the quality of their urban environments, are also allocated to the various groupings shown in the table. A fourth source of dimensions comes from Haughton and Hunter's (1994) pioneering book on *Urban Sustainability*. This identified what they called the three fundamental principles and nineteen guiding principles underlying sustainability, which were derived from socio-economic, ecological and management categories. An application of their ideas to a city trying to implement sustainability has shown that there is some inconsistency between the various indicators (Davies 1996). Many of their principles in the Inter-Generational Equity and Social Justice, are also covered by the Natural Step and Ecocity ideas. But their third category, Trans-Frontier Responsibility, does seem sufficiently different and important enough to be added as a fifth Basic Dimension in Fig. 5.1, since it focuses not simply on the *content* of sustainability, but upon the *spatial and temporal effects* of development that lie beyond the places where unsustainable practices are generated. Since the emphasis of Haughton and Hunter's work is on urban issues it is not surprising that the ecological issues of water, air, energy, food in the IEFS movement were not effectively covered in their list of sustainability principles. However, their interest in *management* issues, although primarily described in a set of their guiding principle categories, does seem worth elevating to a sixth Basic Dimension. This means that the summary in Fig. 5.1 does not just deal with the content dimensions and their spatial-temporal impacts but includes a management dimension, summaries of how to implement and manage policies that will create sustainable development.

The sustainable principles, standards and indicators derived from the various sources described in Fig. 5.1 provide a useful first stage in understanding the range of features that are now considered to be part of the sustainability concept, as well as showing its multi-dimensional character. However these dimensions do have different emphases and degrees of generality. For example, the ECI set provides quite specific indicators in some areas, such as the school journeys of children, and noise pollution, but are often less precise because they use the term 'sustainability' in various indicators to define part of the concept they are trying to measure, which is confusing. In addition, it is worth noting that not all of the dimensions are neutral or easily measurable entities, especially in the human and management categories. In the Human Capacity Domain, the Ecocity movement in particular identifies a series of concepts not found in other sustainability schemes. These concepts are more general than those used in the Just City and Capabilities literature discussed in Chap. 3 and do not lead to precise and easily measurable indicators, or at least measures that are universally accepted, unlike the more easily defined physical indicators. In any case different groups and governments of various political persuasions may take exception to the inclusion of particular dimensions. For example, it is well known that improvement in the less developed countries is enhanced if women are educated and emancipated. Yet none of the four sources shown in Fig. 5.1 initially included a dimension that measures the degree of 'individual choice'—particularly in the case of women's rights and their control over reproduction. Many delegates to the Rio+20 conference in June 2012 wanted this principle in their final communiqué, but it was blocked by conservative governments, especially those from Roman Catholic and Muslim countries because of their negative views on abortion and contraception. Given its importance, it has been added in the Human Capacity domain. In addition the focus on urban issues in this chapter means that the issues linked to the accelerating decrease in bio-diversity and number of life forms described in a recent review in Nature (Bradley et al 2013) is not dealt with. Even in our food supplies the U.N. Food and Agriculture Agency has estimated that 75 % of our crop biodiversity has been lost from our fields within a century. For example, India had over 10,000 varieties of rice, a figure now reduced to a few thousand, while a few favoured varietals have been replacing Italy's traditional use of 400 types of vines, many of which were restricted to one area and provided part of the distinctive local cuisine, the support of which is one of the reasons for the growth of the Slow Food and Slow Cities movement (Chap. 14). Moreover, this increasing focus on particular seeds and crop varieties developed over centuries, in favour of new hybrids, has reduced genetic diversity. It leaves food supplies derived from a few specific seed types susceptible to diseases or pests that only strike one variety, as seen in such disasters as the Irish potato famine in the 1840s or the Phlyloxera root aphid that devastated European vines from the mid-1870s, or the banana disease which wiped out the so-called Big Mike species that was the dominant cultivated type in the 1950s. Certainly there has been recovery from these problems by finding new resistant species, but the way that new diseases occur can be seen in the threat now posed to the Cavenish banana variety that has been dominant for a half century but is itself now threatened by new infections. Yet recognition of the problem of genetic loss has led to some progress. The world's first seed bank has been created in a secure vault sunk deep below the permafrost near Longyearbyen in Norway's arctic, while many countries have developed or are developing similar national depositories. The obvious problem is whether the seeds in such depositaries will be easily accessible if disaster affects particular food crops. Nevertheless, the influential work of the Global Crop Diversity Trust, in addition to the growth of the plant genetics field in breeding new species, provides hope for the future, first by finding new strains resistant to particular diseases, or giving them the ability to survive in new environments, such as drier or saline soils.

Although Fig. 5.1 can be viewed as an attempt to integrate, or at least interrelate, the key ideas associated with sustainability, there remains the problem of deriving precise and standardized measurable indicators from what are mainly a set of concepts. As yet, no quantified scheme that covers the range of variation shown in Fig. 5.1 from these six key sustainability sources has been created. Instead there have been many separate attempts to measure sustainability by various governments and private organizations that derive indicators from various concepts seen as relating to sustainability, which are then applied to countries, regions and cities. One of the most basic is the idea of *ecological footprints* developed by Rees and Wackernagel (1992) in Vancouver. Essentially this estimates how much land is needed to produce the resources used and absorb the waste for any population, leading to the dramatic conclusion that the average resource demands in the U.S.A requires 9.0 earth equivalents whereas China is only at 1.8, showing the varying extent to which resources are being consumed at a faster rate than they can be regenerated. The idea has also been frequently applied to cities. The earliest examples used values for the per capita amount of farmland needed for food, forests for wood, and land for carbon absorption, although subsequently more detailed inputs have been developed. Although a useful initial tool there are differences between various sources in the ways that the calculations are made which leads to variable results. One of the more comprehensive reviews of the approach has been made by the European Union in its thematic strategy on the sustainable use of natural resources through ecological footprints (EU-EF). The popularity of the idea has been extended to the concept of *carbon footprints* which measures the greenhouse gas emissions per person, household, city or country, and has been developed in the University of California in Berkeley (CF).

Although the methods used by these footprint calculations vary and hence produce different results, they do provide an initial and dramatic summary of the extent of the sustainability problem. This has led many people, organizations and cities to take stock of their values and seek ways of reducing the footprint through various methods, many of which are discussed in the rest of this chapter and the next. Other approaches are more comprehensive in using various variables to measure the degree of sustainability in cities, although most only use some of the range of concepts outlined in Fig. 5.1 and are not consistent with one another. However there have been major efforts by the European Commission to arrive at common indicators, such as those produced in 2003 to allow local municipalities to define their level of sustainability (ECI 2003). In addition there are sign of progress in attempts to produce more consistent and hence comparable indicators, through the global protocol measuring greenhouse gas emissions at a community scale, created by the World Bank and the United National Environmental Programme and announced at the 5th World Bank's Urban Forum, March 2010 (WB 2010). But there is still a long way to go before consistent and useful indicators that are universally accepted are derived. Perhaps the most comprehensive of recent schemes to measure urban sustainability in terms of dimensional coverage and in world-wide urban application, was developed by the Economist Intelligence Unit (EIU 2012) through sponsorship by the Siemens Engineering Group. In Europe 30 indicators derived from 8 general content categories were used to construct what is called a *Green Index* for the major cities; a similar approach was used for other continents. These indicators are shown in Fig. 5.2 but have been further generalized to four higher order descriptive domains to clarify the range and utility of the concepts used, which can be seen to relate to: Critical Resource Inputs (of Water and Energy), Damaging Outputs (such as Air Quality, CO_2, and Waste), Infrastructures (Buildings and Transport), and Governance (relating to Environmental Management).

It obvious that these indicators are dominated by variables that relate to physical characteristics, not human capacities, and do not cover the range of issues identified in Table 5.1, although they do contain management dimensions. Nevertheless it does provide a useful, mainly physical snapshot at the comparative sustainability of major cities in Europe and other continents. Once the data on the individual variables were obtained for major cities these values were arranged on the same scale so they could be summarized to create an index score for each city, running on a scale from 0 to 100. The same approach was used create the same type of index for major cities in other continents. However the lack of quantitative data for many variables in the South American, African and Asian large cities meant that the results for these continents were only able to only show the relative position of the cities chosen in five categories and are only relative within each continent and arranged in alphabetical order within each grade.

The Green Index Study of major European cities shows that Copenhagen, with a score of 87.3, tops the list in terms of the environmental sustainability as measured

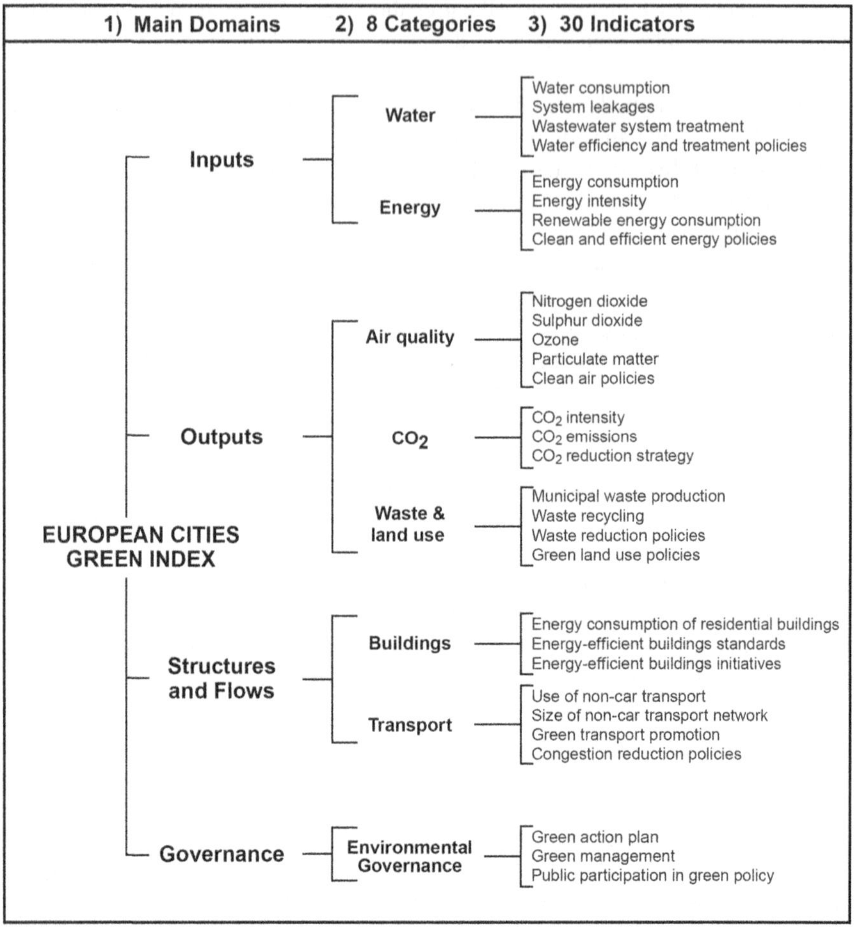

Fig. 5.2 Composition of the Green City Index for European cities. (Source: Re-arranged from The Green City Index: Economist Intelligence Unit 2012)

by the indicators, whereas Kiev is the bottom of the scale with 32.3 (Fig. 5.3). One of the striking features of plotting the scores on a map is the way it shows how the cities of the former communist countries of Eastern Europe rank far lower than those of the rest of Europe, although most of the cities of southern Europe also have relatively low scores. Theoretical isopleth lines are added to the map just to draw attention to this pattern. They also show that the large cites of Scandinavia and the Netherlands, as well as those in Switzerland and Austria have uniformly high scores, indicating relatively high levels of sustainability. This is a consequence of the fact that these mainly northern centres embraced sustainable principles at an earlier time than others and with more rigour. The lower values for Eastern European centres is a result of their heritage of development under communist totalitarian regimes that often ignored environmental problems and where criticism of these practices was

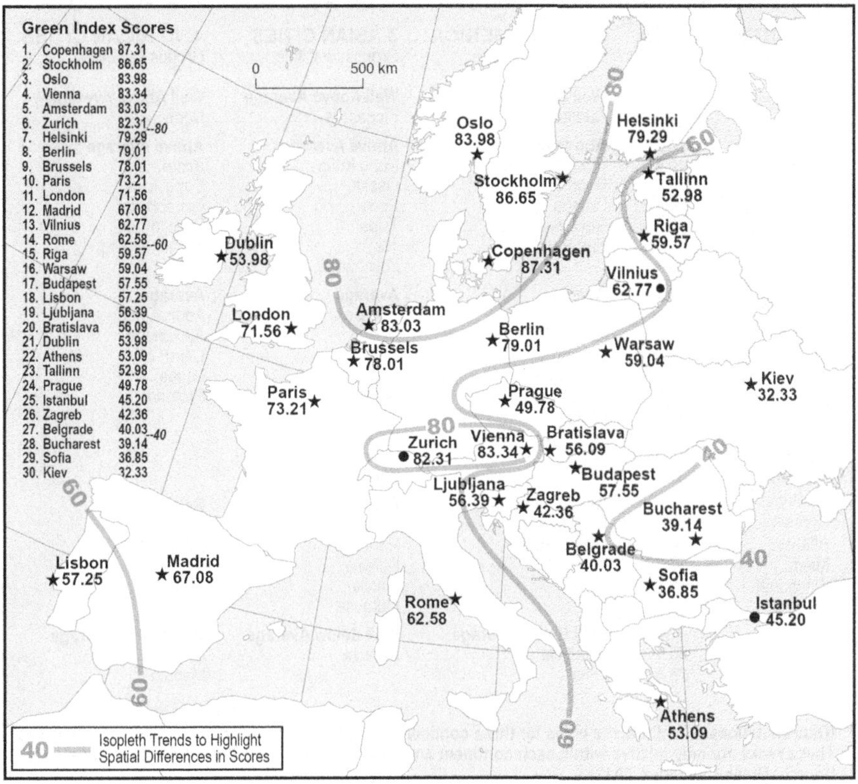

Fig. 5.3 Map of Green Index scores for major cities in Europe. (Source: Compiled by author from Green Index for European Cities, Economist Intelligence Unit 2012)

rarely allowed. It also reflects the lower adoption of sustainable practices in recent years, a function of halting economic growth and less commitment to these issues.

San Francisco tops the chart of North American centres with other west coast centres also scoring highly, while New York and Boston are both in the top six (Fig. 5.4). Detroit, a city in major decline, has the lowest scores, with Cleveland and St. Louis also scoring poorly. It is also worth noting that many of the cities of the southern Sunbelt states also have low scores, largely a consequence of their sprawl, low density and excessive dependence upon cars.

In the major cities on the other continents, precise scores could not be calculated because of the absence of accurate data, so only ranking into five broad categories could be obtained. Still the results are informative in showing how most of the cities in the developing world are in the lowest categories. Only the city state of Singapore and Curitiba in Brazil registered in the top category, with the later city receiving the 2010 Global Sustainable City award, a consequence of the innovative master plan that guided the city from its 1950s boom. At the other end of the scale are five large cites located in Latin America (Guadalahara and Lima), Pakistan (Karachi), Tanzania (Dar es Salaam) and Mozambique (Maputo), places that have

1. NORTH AMERICA
Green Index Scores

San Francisco	83.8
Vancouver	81.3
New York City	79.2
Seattle	79.1
Denver	73.5
Boston	72.6
Los Angeles	72.5
Washington DC	71.4
Toronto	68.4
Minneapolis	67.7
Chicago	66.9
Ottawa	66.8
Philadelphia	66.7
Calgary	64.8
Sacramento	63.7
Houston	62.6
Dallas	62.3
Orlando	61.1
Montreal	59.8
Charlotte	59.0
Atlanta	57.8
Miami	57.3
Pittsburgh	56.6
Phoenix	55.4
Cleveland	39.7
St Louis	35.1
Detroit	28.4

	2.LATIN AMERICA Groups for Cities	3.ASIAN CITIES Groups for Cities	4. AFRICAN CITIES Groups for Cities
Well Above Average	Curitiba	Singapore	None
Above Average	Belo Horizonte, Bogota, Brasilia, Rio de Janeiro, Sao Paulo	Hong Kong, Osaka, Seoul, Taipei, Tokyo, Yokohama	Accra, Cape Town, Casablanca, Durban, Johannesburg, Tunis
Average	Medellin, Mexico City, Monterey, Porto Alegre, Puebla, Quito, Santiago	Bangkok, Beijing, Delhi, Guangzhou, Jakarta, Kuala Lumpur, Nanjing, Shanghai, Wuhan	Addis Ababa, Alexandria, Cairo, Lagos, Pretoria
Below Average	Buenos Aires, Montevideo	Bengaluru, Hanoi, Kolkata, Manila, Mumbai	Luanda, Nairobi
Well Below Average	Guadalahara, Lima	Karachi	Dar es Salaam, Maputo

(Data limitations meant that the cities for three continents could only be ranked in five general categories. These ranks are only relative within each continent and are in alphabetical order within each grade. Source: Green City Index, 2012)

Fig. 5.4 Green Index Ranking: Major cities in other continents. (Source: Compiled by author from Green Index for European Cities, 2012 Siemens AG, Munich, 2012)

grown with uncontrolled sprawl and contain high levels of poverty and pollution, which have led to the worst sustainability indexes in the world.

Apart from the way the index provides a general summary showing how far individual cities are behind others in their sustainability, the study can also be used to isolate their progress on individual indicators. This reveals some fascinating differences on particular indicators. For example, Oslo, among the cities studied in the developed world, emits the lowest amount of carbon dioxide at 2.2 t per capita, which is far lower than the European average of 5 t, whereas the mean value in the North America cities is 15 t, a clear indication of the lower density, car use and less sustainable practices in its major cities. Also the average level of concentration of other air pollutants (SO2, N02, PM 10) in the European cities is 25 % lower than the average for Latin American cities, and 50 % lower than the average of Asian cities, which indicates how far the major cities in countries with medium and low levels of development world have to go before achieving even European levels of air quality.

The environmental performance of the cities chosen for the Green Index survey is not simply of descriptive value. It is also useful in a practical sense since it will alert some cities that they are well behind the standards being achieved by other centres. This will hopefully stimulate grass-roots groups to demand political action to improve their conditions. By comparing scores through time cities will also be able to judge the extent to which progress in sustainability is being achieved, not only in a total sense, but also through the individual indicators.

Yet despite its utility there are limitations of the Green Index method. One is that it is mainly restricted to environmental indicators. So it does not deal with many of the human dimensions that some people see as part of sustainability, although many of these issues are being dealt with in the Just City and Human Capability research described in Chap. 3. Another issue is that only some of the large range of dimensions identified with the sustainability conceptual summary shown in Fig. 5.1, are incorporated in this the Green Index. Hence it needs to be interpreted as a partial view of urban sustainability, although it is important to stress that this limitation is largely associated with the difficulty of finding accurate data for the cities chosen. A more critical technical issue comes from the fact that not all of the indicators are based on quantitative data. Some involve rank order judgements, which results in the questionable practice of integrating hard and soft data in what is then presented as a final quantitative scale. In addition, some may argue that many of the quantitative variables are measuring quite different things, so questions can be raised about whether these variables should be added together, even when they are scaled, although this seems to be a common practice in the field. Another note of caution comes from the fact that some of the spatial units used are based on political jurisdictions, not on continuous built-up areas. If the suburban areas around these centres are taken into account, part of the high degree of sustainability in some cities would be reduced, and some of the low scores in places that are primarily central cities would be increased. Nevertheless, the Green Index does make an attempt to provide exact scores rather than only the rank order listing found in many other sustainability indices, which often disguises the actual value differences between the centres. So despite the caveats that have been raised about its construction, the Green Index provides the most useful first stage comparison of the extent of the environmental part of sustainability between cities at a world scale. This is a big advance on the various regional or city-specific scales that have been produced in many centres, for these results cannot be generalised to other cities.

5.4 Creating More Sustainable Inputs and Outputs

It follows from the conceptual and practical schemes discussed above that the creation of a more sustainable future will come from using less resources, removing—or at least reducing—pollution and waste, especially man's created compounds, in addition to improving human capabilities. Many individual policies in various sectors of the environment and in society are being adopted to achieve this aim. However, a more useful way of understanding the ways that this can be achieved is not to look at these as separate variables and solutions, but to see them in terms of an urban input-output approach that tries to understand what amounts to the metabolism of urban places. For example, every urban area and the buildings, population, and functions within them, including the transport that connects them, require a series of inputs—such as water, energy etc., in addition to food and a range of created products—which are consumed in the city. They usually produce not only useful products through manufacturing or services, but also unwanted waste and noxious pollutants as outputs, some of which create regional and global damage not simply local urban ones. As cities got bigger and more technological devices were created after the Industrial Revolution, the addition of the inputs and disposal of these waste outputs were largely solved by separate, yet creative engineering solutions for each product, following what amounts to an largely linear approach. To take one example, energy in the form of coal was brought into cities in the Industrial Revolution, burned for heating and for powering various plants, and the waste was dumped, while the noxious pollutants produced from the burning of coal were initially ignored because it was an externality to the linear raw material-energy output solution. But this type of approach ignores the consequences of the linearity described earlier, such as the London smogs, or the current high air pollution levels in Beijing and other major Chinese cities of today. In similar fashion, first gas and later electricity derived from coal were used for power and lighting etc., but given the difficulty at the time of transmitting them over long distances the generating plants had to be located close to the cities.

From the early twentieth century the technical innovations in creating long-distance electrical transmission lines meant that cities could be supplied from large-scale generating plants located near raw material energy sites, or in places that used imported fuel. The next century saw an explosion in the development of bigger generating plants and long transmissions lines spreading over hundreds of miles, controlled by complex interlocking electronic systems. In addition, greater amounts of oil and natural gas as well as other sources of energy were used, all with their own supply sources and processing and transport systems. By using a more comprehensive input-output approach the effect of the materials brought in and out of cities is not only related to their immediate effect in the cities, where they are used and created, but to the whole supply and output chain. For example, this makes it possible to consider all the environmental and health effects of mining and transporting the coal, as well as the local, regional and world effects created by its processing and the final waste products. The result is to reduce the understanding of all these issues to two fundamental questions in sustainability: What are the current sizes and sources of inputs and outputs? What are the problems that affect policies designed to change the inputs and outputs so they are less wasteful of resources and create less harm to the environment?

5.4.1 Eco-Cycles

An important approach to the input-output approach to sustainability can be seen in the eco-cycle approach. This describes situations in which circular and more holistic approaches replace the traditional separate and linear solutions to inputs and outputs in towns or in industrial units, which may lead to the creation of symbiotic relationships among them (KS). One of the earliest, and perhaps best-known large-scale examples of this approach can be found in Kalundborg (Denmark) where a so-called eco-industrial complex has been created from the symbiotic relationships between three main industrial units—the Asnaes power plant, the Statoil refinery, the Novo-Nordisk chemical company—and other local activities, such as a plasterboard company and the local municipality (Ehrenfeld and Gertler 1997). This example provides a variation on most industrial clusters where local industries have grown up to produce goods that are either input to other local industries, or use outputs from them.

Figure 5.5 shows that some of the waste outputs from the three main plants are not just dumped or carted away as in typical linear solutions. They now provide valuable inputs to the other main plants and industries in the vicinity. These local inputs have substantially reduced new material inputs from other sources and their transport costs, as well as providing significant environmental benefits, since former wastes, such as excess heat and much of the sulphur dioxide pollutants from the largest coal-fired power plant and oil refinery in Denmark, are now used in the local industrial plants. This eco-industrial complex, 110 km west of Copenhagen, did grow up spontaneously, with no co-ordinated planning in the decades after the oil refinery was established in 1961. The managers of the various firms lived locally and seem to have gradually recognized how they could help each other by developing the symbiotic processes shown in Fig. 5.5. They calculated the various

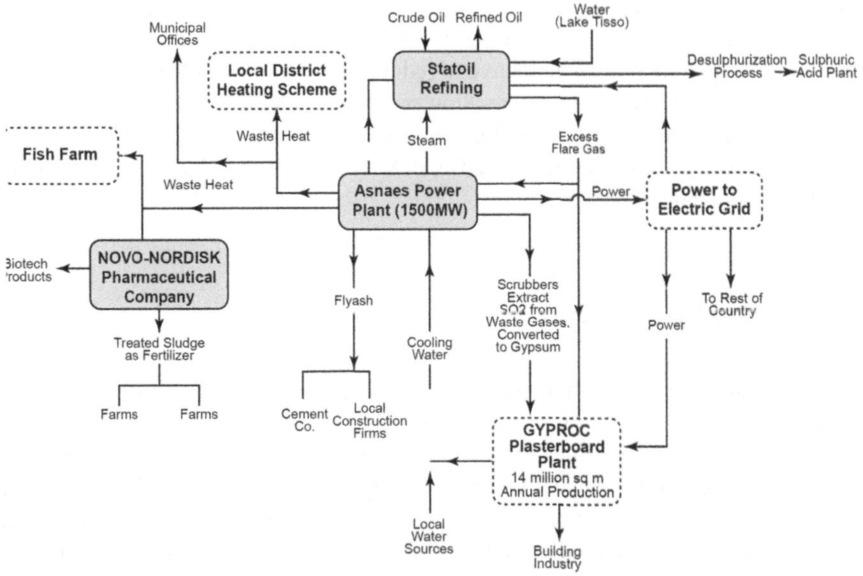

Fig. 5.5 Eco-cycle linkages of Kalundborg industrial plants. (Source: Revised from Kalundborg Symbiosis Institute and Ehrenfeld and Gertler 1997)

environmental and cost benefits of using products that were initially considered waste by others, and given the savings, decided to adopt the eco-cycle approach. Apart from the interactions between the three main companies, formerly waste products derived from the main plants are now also used by other activities. For example, the extraction of SO2 from scrubbers installed by the power plant is a major input into the creation of gypsum used in a plasterboard factory; the town created a district heating scheme using waste heat from the power plant, which replaced 3500 largely oil-burning residential furnaces, reducing heating costs and carbon emissions; and sludge from the pharmaceutical company is treated and turned into fertiliser for local farmers. There seems little doubt that part of the reason for the initial development and success of this complex was due to the close personal relationships that developed between the decision-makers in these companies. But it is also a product of the strong historical co-operative tradition in Denmark that emerged in the agricultural sector in the mid nineteenth century, as well as the rapidly developing environmental consciousness of its people. It is also worth noting that one of the key firms in the area is Nova Nordisk, not simply a world leader in biotechnology products, producing 40% of the world's insulin. It was one of the first major companies to provide an annual environmental report to shareholders in 1993 that described its impact on the environment and its determined attempts to reduce pollution and increase its sustainable practices (NN). Another, Statoil, the government-owned Norwegian oil company, has become an innovator on carbon sequestration, namely removing carbon from the emissions of its refineries and storing it underground. More recently, it has also become a major player in the offshore wind industry, adding wind turbines to many of its old production sites in the North Sea, and therefore diversifying its energy output. Although there may be only a few examples of such a successful eco-industrial complex as Kalundborg in the world, there is little doubt that eco-cycle approaches are increasingly being planned, as polluting companies realise they have to reduce emissions to conform to increasing national standards. So many are seeking ways of reducing and using their waste products, which can also lead to long-run cost benefits by the sale of the products generated from the waste, once the costs of new devices to reduce or clean up the waste are covered.

5.5 Water and Sewage

Crucial requirements in urban settlements are the need for safe and cheap water supplies and the disposal of waste matter. In the nineteenth century the explosion in the urban populations in cities of the developed world was only made possible by the engineering advances that created reservoirs in distant areas and then piped the water through long distance pipes via water treatment plants, into the growing cities and into the networks that delivered drinkable and safe water to individual houses. Although these systems have served cities well, they have not been developed to the same degree in the cities of the undeveloped world, which often means that an inadequate supply, limited use, and variable quality, are some of the major problems of sustainability in developing world. Even in the developed world increasing problems are emerging to threaten urban water supplies. One is that global warming is predicted to reduce the rainfall in some areas, or will lead to major downpours, so existing reservoirs may not be able to cope and fulfil future demand. Unless more local systems are developed and per capita water use can be reduced, future investments will be required in large storage sites, especially in sites nearer existing cities, to reduce transport costs. However, the creation of such sites is likely to run into opposition from farmers and environmentalists in the areas affected. Second, many systems are less efficient that they appear, given high levels of leakage due to aging and poorly maintained pipelines. For example, there was an average loss of approximately 25 % of the water carried between reservoirs to households in southern England in 2012 (WT 2012). This means that one way of obtaining significant urban water consumption savings, without the huge costs of new reservoir capacity, is by improving the efficiency of the existing water delivery system. Third, it is clear that there are huge variations in the per capita consumption of water. The Green Index research described above showed that Amsterdam consumed the least water in the cities surveyed in the developed countries, at 146 L per person day, compared to 262 L in New York, and 300 L in Yokohama, the leading consumers in the American and Asian centres respectively. This evidence helps to provide the background to questioning why people in different cities have such different consumption levels and about how such cultural variations in water use can be reduced. More specifically, many new policies are being introduced to reduce current water use levels, such as by: installing household water meters to encourage conservation among cost-conscious consumers; using low-flush toilets, and more effective shower-heads and taps which reduce per capita water use. Indeed, increasing numbers of urban jurisdictions now make such water-saving devices mandatory in new buildings, or provide subsidies for consumers who install such devices. Another growing trend which needs wider adoption has been the development of two different water systems in houses. One is for drinking water; the other uses recycled or 'grey' water. This is derived from water previously used for washing, or from rain water collected in separate systems, which is then re-used to flush human waste away, or to water lawns, for it does not make sense to use expensively filtered drinking water to flush toilets. A return to the greater use of water barrels to store rainwater for use on gardens also reduces the use of household water for this purpose. However, despite the growing use of these conservation measures, many cities in the world are still facing a long-term water shortage. Although local artesian sources have long been exploited, it is recognized that many have been over-used and are not being replenished, so the long-term ability to extract more water from such sources is limited in many regions. A big problem in many countries is the increasing competition for declining water supplies from rivers, aquifers and reservoirs between agricultural usage and city needs. There seems little doubt that many cities need to find new water supplies, and to reduce water use, while it is very possible that decisions have to be made to phase out crops in some agricultural areas that need large amounts of water. These problems mean that some coastal cities have turned to the sea as a source of water by using desalination plants to take the salt out of seawater and to make it acceptable for human use. Although Britain has more than its share of rainfall in most areas, southeast England is a rapidly growing and drier zone that will soon outstrip its current reservoir capacity. This led to the decision to build Britain's first major desalination plant at Beckton (east London). Operational since 2012, it produces 150 million L of water a day from the Thames estuary, supplying 1.4 million people out of London's total of almost 9 million inhabitants. Despite its £ 270 million cost it will provide a useful alternative to the construction of expensive and land-consuming new reservoirs and pipelines to serve London's growing water needs. Hopefully this type of solution for water supply and the other policies discussed above will be adopted in other cities in the world that are running low in water, especially in developing countries where clean and cheap water availability and delivery is still very limited.

Perhaps the biggest direct outflow problem, other than emissions into the air, comes from the faecal matter disposal from urban areas. In the early industrial world this was not simply from humans, but from the horses used in transportation and the many animals kept for food in cities. Historically, the so-called 'night-soil' of humans was often collected by a special class of people and used as a valuable fertilizer on crops outside the city, or disposed of in the nearest water sources. The system still exists in too many third world cities, while in countries such as India many people who lack access to toilets relieve themselves on pieces of waste ground, creating a major health hazard. The growing size of cities in early industrial Britain meant that waste disposal became a huge health problems and Edwin Chadwick's (1843) report on *The Sanitary Condition of the Labouring Population* publicized the need for new management procedures and technologies to remove waste to improve the well-being and health of the growing urban population. This eventually led to a 1846 act, The Nuisance Removal and Disease Prevention Act, which began the process of regulating waste in London, in which a new city-wide authority, the Metropolitan Board of Works became a centralized sanitation authority responsible for waste management, which was followed by similar schemes in other cities in Britain. Among the many provisions of the significant Public Health Act 1875 was a requirement that made it compulsory for every household to deposit their weekly waste in receptacles for disposal that would be regularly emptied by the municipal workers.

In the developed world the widespread use of water-based toilets from the mid-nineteenth century meant that extensive, connected systems of sewage pipes sending the outflow into sewage processing plants were built in cities. These helped solve the cholera outbreaks that devastated so many urban populations in the growing industrial-commercial cities of the early nineteenth century, where the untreated human waste was just dumped into the local rivers, contaminating the ground water and local water supplies. Although it took time to establish the link between outbreaks of disease and the faecal-contaminated water supplies, most cities in the developed world created extensive systems of water supply from reservoirs and built separate sewer systems to take the flow from the increasing numbers of toilets in buildings, which led to the development of sewage treatment systems to filter out the harmful material. However, far too many cities still dump a large proportion of their largely untreated sewage into the nearest water body. For example, in Canada the third national sewage report of the Sierra Club (SC 2004) on the 22 largest cities revealed that five cities (Victoria, St. John, Halifax, St John's, and Dawson City) annually released 140 billion L of mainly raw sewage into their local water bodies, with three others (Montreal, Vancouver and Charlottetown) using only primary treatment, which is essentially the skimming off the solids. These eight centres account for 40,000 L of effluent per second being dumped, which at best is barely treated. By contrast Calgary and Edmonton treat all of their sewage in tertiary treatment plants, which remove all of the waste. Edmonton's plants are described in more detail in the next chapter as part of a more general recycling system(Chap. 6, Table 6.1). These are examples of the way that new treatments for sewage can now re-cycle a great deal of the treated product in a safe, if costly, manner through the technological progress which converts the noxious solid matter from the sewage into useable products and safely returns water to the natural system. Yet most cities still lag behind in the adoption of such new processes, which compromises the health of their citizens and drastically reduces their degree of sustainability. Also, despite this technical progress, or at least possible progress, at the end of the human waste system only limited changes have taken place at the household level. The flush toilets have not really altered since being invented by Thomas Crapper in the mid nineteenth century, and still consume large amounts of water even with low flush systems, while an alternative on an old idea, namely modern earth closet systems, has only been adopted by a very tiny minority. However there may be signs of new advances in toilet development. For example, the Gates Foundation provided over $ 6 million for what was called 'Reinventing the Toilet Challenge'; winners were announced in 2012, with first prize going to Hoffman's electro-chemical system. Although still in the developmental stage this system, or related alternatives, may be of particular use in settlements of the developing world, for it is powered by solar panels and creates useful products from human excreta, such as hydrogen for cooking and the safe residue for fertiliser. Without the need for large inputs of water it could bring sustainable and affordable toilets to the 40 % of the world population that lacks access to sanitation, with estimated savings of over 1.8 million deaths from diarrhoea (WHO 2012). Although this technical progress shows what could be done, the general adoption of such systems will still require major investments.

Related water disposal solutions that affect sustainability levels come from the way that the run-off from roads and parking lots are treated. In the past the usual approach was to channel it into the sewers, which, in storm conditions often leads to system overloads and flooding. Many cities are now adding storm ponds into their subdivisions; they store water from major storms and even out the flow into the sewer system. Interest is also developing in reducing the size of the run-off through the use of permeable pavements, especially in large parking areas which means that rain water simply sinks into the ground through the gaps in the paving. But the problem of run-off is not simply a matter of flow. The run-off often contains large amounts of pollutants, such as oil, chemicals, PCB's and other hazardous products that come from the cars and trucks using the roads, thereby adding to the levels of pollution in the sewer system. Although expensive treatment centres had been added in some cities to remove these toxins from the water, there is increasing interest in adopting these new methods as seen by research at North Carolina State University (NCSU) and a review of the issues by the city of Victoria in Canada (CRD). In addition, there are more biological solutions to the problems. For example, the use of phyto-remediation measures are increasing. This involves diverting the run-off into vegetated areas where the microbes near the roots filter out many harmful substances. Examples of this process range from the creation of wetlands to fulfil this role, to local bio-swales, which are shallow channels containing vegetation that filter the run-off, as well as slowing it down. An early example of the bio-swale approach was developed in the Willamette River Park in Portland in 1997 to reduce pollutants running into the river. Other waste water run-off systems end in wetlands or in depressions planted with trees. Enkoping in Sweden has taken the approach further by pumping sewage with high nitrogen content into a 190 acre area of fast growing willows which filter out and convert the pollutants. These trees are sequentially harvested and turned into wood chip that is then used to generate electricity.

The examples illustrate the many initiatives in dealing with waste outputs. Despite this progress, few sewage plants remove the endocrine-disrupting chemicals from pharmaceutical products in waste flows which negatively affect sexual development, especially in male species (Wood 2014). Cities also need to reduce of replace many of the other *inputs* that have major polluting effects. The next section deals with one of the most important of these inputs in the supply of energy, given the domination of finite fossil fuels and their negative externalities.

5.6 Changing Energy Sources

Our towns and cities depend on fossil fuel sources to generate the energy that keeps them functioning; they would never have grown to their current sizes and complexity without this resource. However, one of the most significant sources of unsustainable activities in urban places comes from this extensive dependence on fossil energy and its negative effects, not simply from their direct harmful effects on the health of people in cities, but from all the negative effects they produce at previous stages in their production, namely: from environmental damage at production sites, spillage in transportation, their burning to extract energy, which creates noxious gaseous outputs, from CO2 and NO2 to a variety of particulates. These are simply deleterious for human health and have caused many air pollution problems in cities. In a human context the World Health Organization (WHO) report on the Global Burden of Disease, published in December 2013 estimated that 1.2 million premature deaths per year occur in China alone because of outdoor air pollution (Wong 2013). In addition a more detailed study by Chen et al (2013) reviewed the effect of China's free coal policy that applied from 1950 to 80 in areas north of the Huai river which approximates to the average January temperature isopleth of 0 °C. Although carried out for humanitarian grounds to keep people warm in winter, the researchers estimated that the life expectancy of people in this part of northern China was 5.5

years less than people to the south due to respiratory diseases. The cumulative effect over 30 years meant there were 2.5 billion years of lowered life expectancy in the affected area, a tragic testimony to the damaging effects of coal burning that added so many particles in the air, creating mortality rates which were 55 % on average higher than in the south.

These negative health effects must be added to the problems that the build-up of carbon dioxide and other greenhouse gases is having on atmospheric change—with coal, especially the soft bituminous types—having greater problems than the others in this regard, given its output of particulate matter as well as many greenhouse gases when it is burned for energy. In addition about a third of the greenhouse gases build-up comes from natural sources, from animals, forest fires, land use changes, or the methane released from the melting of permafrost areas, which do provide significant additional sources, although they are outside the remit of this chapter. In addition, fossil fuel use ought to be reduced because they are finite, although there is a great deal of dispute about the extent of the reserves of each fossil fuel resource.

Many of these disadvantages have been ignored until the last 20 years, or just coped with, through local policies to eradicate or reduce specific problems in specific areas. Although each city varies in the extent of these problems, depending on the mix of fossil fuels used, Fig. 5.6 (OICA 2012) shows that on a world scale the biggest

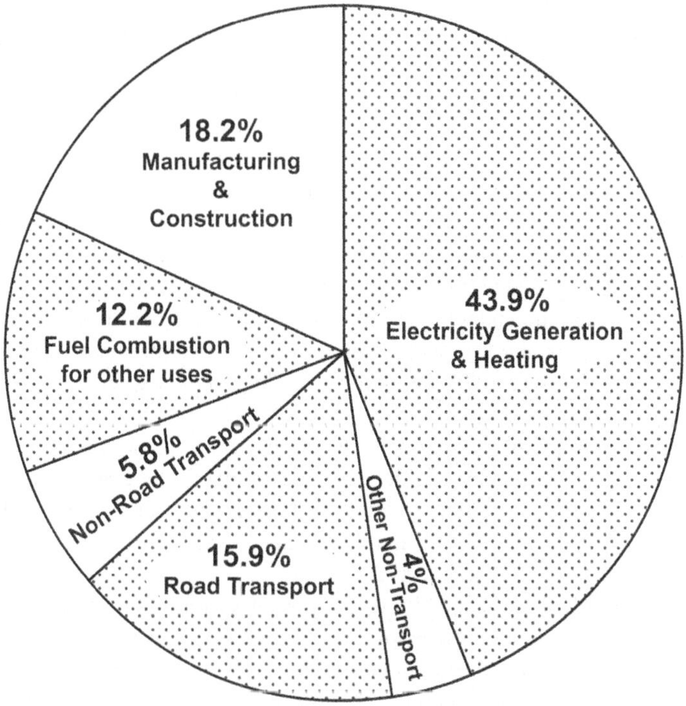

Fig. 5.6 Carbon dioxide emissions and human activities: Proportions. (Source: Revised from information in OICA, Paris, 2012)

source of greenhouse gas emissions from human activity comes from electricity generation and heating (43 %), with the next two largest proportions generated by manufacturing and construction (18.2 %) and road transport (15.9 %). Hence it is necessary to understand the growth of these fossil fuel sources and the prospects for replacing them with renewable sources if the current build-up of greenhouse gases is halted in order to reduce the current unsustainable levels of the use of these resources.

Most issues involved in reducing the negative effects of the use of fossil fuels go beyond the ability of urban areas alone to influence the situation. The crucial regulatory or incentive decisions to reduce dependence on fossil fuels are usually taken at a state or corporate level, rather than in urban municipalities, although in democracies the large number of urban voters can influence the political decisions to alter policy. Although it is popular environmental opinion to deplore the use of these fossil-fuel sources, it is salutary to remember that our industrial progress, and especially the growth of cities, has been based on the ability to exploit these sources, principally coal, oil and gas, in which subsequent technological advances have aided the extraction, transport and burning of these sources, and in doing so have created huge infra-structures and employment levels to support the generation and transmission of power. Hence the functioning of our cities is highly dependent upon the energy obtained from these fossil fuels, not simply for electricity, heating, and powering machinery, even though the generating plants may be distant from the cities, but also for transport.

This dependence stems from the fossil fuel advantages over other energy sources, namely: having very high calorific output per unit; portability, given the technological improvements in transport in the last century, though some fuels are expensive and bulky; most are easily storable; their use in the energy-producing systems developed from the late nineteenth century onwards have made it relatively easy to switch production on and off; and most have lower costs per output of energy than renewable sources at this time. Certainly the various fuels vary in their utility within these categories; for instance, oil is more portable than coal, although it takes more processing to produce the product. But overall, the advantages of fossil fuels remain, even though increasing environmental concerns about their negative effects have altered attitudes about their continued use and the amount of their reserves. The result has been a search for alternatives, leading to interest in the use of alternative renewable energy sources.

Despite the remarkable increase in the proportional growth of renewable energy sources in the last decade, all the evidence points to the continued dominance of fossil fuels for the next 20 years and the foreseeable future. This is illustrated in Table 5.1, which shows the estimates of energy sources published in the most respected and comprehensive international source of data on the topic, the annual World Energy Outlook published by the authoritative International Energy Agency based in Paris which shows figures for 2011 (IEA 2013). The second and third rows of the table show that energy production increased by 81.1 % in the 30 years since 1980 and are likely to show another increase of 42.6 % by 2035 if current energy policies and trends in growth continue. However it is expected that in the light of

Table 5.1 World energy sources and proportions: 1980, 2011, 2035. (Source: Calculated from data in World Energy Outlook, 2013, p. 71.)

1	2	3	4	5	6
Energy sources	% 1980	% 2011	% Estimates a) Current policies	For 2035 b) New policies	Based on: c) 450 scenario
Coal	24.9	28.9	29.1	25.5	17.0
Oil	42.9	31.4	27.2	26.8	24.0
Natural gas	17.1	21.3	23.4	23.7	22.5
Nuclear	2.6	5.2	5.5	6.4	10.2
Hydro	2.1	2.3	2.5	2.9	3.7
Biomass-Waste	10.4	9.9	9.3	10.6	14.8
Other renewables	0.2	1.0	2.8	4.1	7.8
(All Renewables)	12.7	13.2	14.6	17.6	26.2
Fossil fuel share		81.6	79.7	76.0	63.5
Total Mtoe	7219	13,070	18, 646	17,387	14,908
Increases over 1980		81.1			
Increases over 2011			+42.6	+33.0	+14.1
CO_2 emissions (Gt)		31.2	43.1	37.2	21.6

Key: a) Current Policies. Projections based on current policies
b) New Polices. Projections based on policies already proposed by governments up to 2013
c) 450 Scenario. Assumes new policies will lessen warming to +2 C % over pre-industrial levels by not allowing CO_2 levels to reach 450 ppm, the stage the IPCC panel predicts that major climate change effects could take effect

current policies there will be changes in the proportions of each fossil fuel energy source to the total output, with natural gas substantially increasing its proportion, with coal at a lower increase, but with a decrease in the use of oil for energy generation. A second important point about Table 5.1 is that it shows that almost 80 % of energy supplies at a world level still come from the three main fossil fuel sources of oil, coal and gas, which is not much of a decline from the 84.9 % in 1980. Nevertheless, this expected growth in energy output and the probable continued dominance of three sources shows the scale of the problem of replacing these fossil fuels with renewable energy sources. However it must be remembered that this table only deals with world figures; individual countries vary considerably in their mix of energy sources. So despite many signs of progress in the development of renewable energy sources, Table 5.1 shows such sources are only estimated to account for 13 % of the world total at the present rising to 14.6 % in 2035 *if* current policies continue. Also it is worth noting that at least half of this current renewable energy portion is mainly biomass burning, especially wood, that has its own pollution problems, especially when used in household fires, let alone the potential destruction of forests to produce the resource. Although hydro-electric power is an important renewable source in some countries and regions, it only accounts for just over 2.3 % of total world energy supply. Moreover, in world terms, nuclear power, which in the 1960s was promoted as the major new energy source, accounts for only 5.2 % of the world energy total from over 440 nuclear generating plants in 2013 and will rise to 5.5% on current trends, although countries such as France heavily depend on this source, with 80% of its electricity produced in this way (IEA 2013). Unresolved questions about the higher costs of the nuclear generating plants that produce power, the problems of safety and the difficulties of disposing of radio-active nuclear waste that takes centuries to decay have always led many to question an increase in this source of power. But negative attitudes to nuclear power have increased substantially since the Chernobyl accident in 1986 and especially after the tsunami that destroyed much of Japan's Fukushima Daiichi plants in March 2011. In both cases the release of radiation has led to the creation of extensive exclusion zones around the plants. This has led many countries to reconsider their use of nuclear power, with Germany planning to close its reactors by 2022, Italy to ban nuclear plants and Japan to close all its plants, although some are likely to reopen given Japan's limited fossil fuel supplies despite the fact that the majority of the public opinion is against them. However India has plans to double its number of nuclear plants. Nevertheless, unless political opinions change it is unlikely that nuclear power will be an increasing contributor—at least at a world scale—to future energy production in absolute terms. Also the prospect of fusion as a power source, instead of nuclear fission, is still an unresolved dream, without a major technical breakthrough. Yet there may be another alternative. In recent years there has been increasing calls for the use of thorium in nuclear power generation. This is a more abundant source than uranium and many regard it as safer, since the generating process is easier to turn off if accidents occur (Hargreaves 2012; Martin 2012). But it must be admitted that there are still many technical problems to solve before this becomes a significant option in power generation, for even if they are solved, the building of many plants will take decades.

The IEA also calculated the changing energy supplies and sources if governments pursued different policies. A slightly more optimistic figure of 17.6% from renewable sources by 2035 is predicted if the future supply estimate is based on policies actually promised by various governments by the end of 2011, but were not yet implemented. By contrast, column 5 of the table shows that the increase in energy would only be 14.1% and the renewable proportion would be 26.2% if governments enacted carbon reduction policies, especially in power and industrial plants, to limit the C02 proportion in the world atmosphere to 450 ppm, a level that many climate scientists have long held to be one beyond which major increases in global warming may occur. However, given current trends and the reluctance of many governments to drastically change their sources of energy production, this 450 scenario is looking to be increasingly unlikely.

For those advocating sustainability, these projections are depressing, since even with such drastic new measures as the 450 scenario shown in Column 6 the proportion of the three main fossil fuels still amounts to just over 63.5% of the total energy consumption by 2035, although with different energy source mixes, with coal declining by 32.7% in absolute terms, oil by 12.9% and natural gas increasing by 20.5%, again in totals, not proportions of the energy mix. Perhaps the most optimistic feature of the projections is the reduction in the proportion of coal, since the increasing use of natural gas only emits, on average, half the pollutants of coal. Yet it must be admitted that there are still huge coal deposits in the world, and it is still cheaper to use than other fossil fuels, although externality costs are rarely taken into account, so many countries will be tempted to use such sources. Others argue that the case for drastically reducing the use of oil in energy production is enhanced because it is used for many other things in the modern world, so burning such a scarce resource is wasteful. More tellingly, advocates of fossil fuel use argue these supplies are not as finite as advocates of sustainable development suggest, so that 'business as usual' can be practised. Certainly coal is still present in abundance, though greater use of coal will mean far more pollution and carbon dioxide emissions, unless major investment is made in carbon reduction or sequestration processes. The case of oil is far more problematic, for it has been argued that the stage of Peak Oil will soon be upon us. This concept was developed by Hubbert (1962) to predict that US oil supplies would reach a peak, decline and become more expensive. Although its prediction was subsequently accurate in the U.S.A. for a few decades, its application to the world scale and to current conditions is far more problematic. In the long term there are limited supplies of oil in the world and the newly discovered sources will be more expensive to develop and operate. But there is no consensus about when, or if, this Peak Oil stage will occur, with lower production levels and making transport more expensive. Early advocates of the idea placed it in the early parts of the second decade of this century, whereas more cautious commentators place it 15 to 30 years away.

Many disagree with the Peak Oil scenario, arguing that the increasing demand for energy will be met by the discovery of new sources of the fossil fuels, as in the past, which we should continue to use since they are cheaper and more flexible in their use than available renewable energy sources. Until recently even the most optimistic oil supply advocates used to admit that new sources will only be found in areas that are unstable politically, or in areas far more difficult to exploit, such as in the Arctic or in deep-sea locations, which will raise extraction costs and hence oil prices, as well as running the risks of environmental damage. This view was given support by the Deepwater Horizon explosion and subsequent oil spill in the Gulf of Mexico 2011, which was made worse because of the weeks it took to cap the well, making it the largest marine spill of 5 million barrels in history. By mid-2013 BP had already paid out an estimated \$ 42.2 billion in environmental clean-up costs, compensation to fishing and tourism in the area, and fines to the U.S. government. Yet despite these negative features, the anti-Peak Oil and 'business as usual energy approach' seems to have gained support in recent years—at least for the next few decades—given enormous increases in newly discovered, or rather exploitable, sources of natural gas since 2008. This has helped gas prices to plummet to a 16 year low in the U.S.A. by 2012, to \$ 4 per million BTU (British Thermal Units). Together with the associated release of light oil and other condensates from extensive shale beds it means that the United States will again became the world's biggest oil producer in 2014. Much of this increase comes from the recent development of hydraulic fracturing (fracking) and then horizontal drilling procedures—which involve pumping water, sand and chemicals under pressure to fracture shale beds—releasing gas and oil that would otherwise would not be able to flow from the tight shale formations that have very small pores. Certainly there are still many worries about the negatives associated with this process, such as contamination of water supplies, even fear of earthquakes in some unstable areas. Fracking also requires far more wells, as many as 40 times more in any area to maintain production compared to traditional oil wells, as they cannot draw from such a wide area as from traditional oil-bearing rocks, so they decline as much as five times more quickly a year (TE 2014b). These problems have led many jurisdictions to ban the procedure. But despite these negative features these new techniques have drastically altered the ability to produce more of the two major fossil fuels. Indeed it is estimated that by the end of this decade the U.S.A. will be producing close to 9.6 million barrels of oil per year—compared to under 5 million in 2008—a level last reached in the early 1970s. However conventional oil will only be around 40 % of this total.

The ability to transport more gas from underdeveloped gas fields via liquefied natural gas (LNG) ships is also increasing its availability and decreasing its price to many energy-deficient countries, although the costs of transport mean that natural gas prices in Japan in 2013 were four times higher than in America. What is important about this new source of natural gas for cities is its cost advantage over other sources, which mean that it is being used in more and more power stations and vehicles, with the additional advantage of being less polluting than other sources, leading to lower pollution levels. Indeed on current trends the use of more natural gas in power generation could mean that by 2020 it could be providing a third of American electricity, compared to a fifth in 2008 (TE 2013b).

The development of major new oil reserves, such as the deep-water oil discoveries off the coast of Brazil and new discoveries in many parts of the world seem to indicate increasing oil supplies for the next few decades at least (IEA 2013). So the case for immediate reductions in fossil fuel use, given the fact that the Peak Oil theory of decreasing oil supply and increased cost has not occurred, has been weakened. But it not been destroyed. Even the new sources of oil and gas only extend the horizon of relatively cheap fossil fuels for a few decades. Ultimately they *are* finite, while there remain the negative effects of using these sources. This means that the increasing use of gas will reduce emissions and could provide the world with a 'breathing space' to implement more rigorous renewable energy and conservation policies.

Table 5.2 provides examples of the way that four countries (Brazil, Denmark, Germany and Iceland) have created progressive solutions to their development of renewable energy sources, initially stimulated by the OPEC oil embargo in 1973–1974 and the subsequent rise in the cost of oil and later by environmental concerns. These examples provide grounds for real optimism in the progress to greater sustainability and can be extended by examples from other countries. Yet progress must not be measured only in new energy output and lower pollution; there are other advantages of renewable energy. For example, there are increasing tax returns from the growth of renewable energy sites and many new jobs in the expanding renewable energy sector, in both innovation and production of renewables. Also solar panels and wind turbines are often located in rural areas where unemployment levels have often been high. However we must be cautious. Although a small country like Denmark and a large prosperous one such as Germany are showing substantial shifts in energy supplies, the absolute extent of the change in most other countries is still relatively small when compared to fossil energy sources. Certainly there has been significant individual growth in particular renewables, and in locations other than those described in Table 5.2. For example, a quarter of the electricity generation in the state of Iowa in 2012 came from wind energy, with additional increases planned in the next few years. Yet satisfaction at this increase must be tempered by

Table 5.2 Examples of renewable energy developments

a) One of the positive decisions of the military government that ruled Brazil in the 1980s was to reduce the dependence upon foreign oil supplies by encouraging the development of ethanol, mainly from products of the country's extensive sugar cane industry. This was complemented by making it mandatory for cars built in Brazil to run on a mixture of at least 25 % ethanol and regular gas. Despite the initial opposition of the manufacturers they eventually capitulated. Later, so-called flexi-cars were created which have the ability to adjust the engine operation to take account of the gas-ethanol proportion that is input. The success of the scheme has led other countries to subsidize ethanol production from plants, such as using corn in the United States, but there are disadvantages. Moreover, world-wide, the proportion of bio-gas used in cars is still very low.
b) A second example comes from Denmark which in 2011 obtained 21 % of its energy from wind. The promotion of wind energy came from the distress experienced by many households in winter given the substantial rise in the cost of oil after the 1970s OPEC crisis. The government embarked on a strategy of subsidizing wind power, helped by the strong rural-co-operative history of Danish farmers who combined to finance local wind farms. Companies such as Vestas, formerly a maker of agricultural machinery, became one of the biggest turbine manufacturers in the world, making major advances in turbine efficiency and size. Although there has been some reduction in the subsidies for renewables, the growth of district heating and other schemes, such as ensuring that all new buildings in the future are designed to use 75 % less energy than in 2009, means that the government's plan to become carbon-free on balance by 2050 seems to be feasible.
c) A third example comes from the expansion of renewable energy in Germany in the past decade, which has increased from under 4 % in 1990 to 22 % by 2012 through its Energiewende (Energy Transformation) policy created in 2000, when the Green party was a junior partner in the Social Democratic government. The original policy seeks to reach a 35 % level of renewables by 2020, but on current trends this is likely to be exceeded and could reach 80 % by 2050, especially as nuclear reactors producing 16 % of all electricity are due to be shut down by 2022, a reaction to the Fukushima disaster. Wind turbines have increased in number, while Germany has seen major annual increases in its solar capacity to over 25 GW, through household roof panels, as well as by solar generating parks. Some German cities have become hubs for solar developments and technology and it has been estimated that by 2012 there were almost 40,000 people directly employed in the solar power industry. But an even more important reason for this German success came from the 2000 decision to adopt a generous 'feed-in' tariff for households who added solar panels to their roofs or installed wind turbines, paid for by a surcharge on electricity prices. This meant that any excess energy produced by households had to be bought by the electricity supply companies, even though it was a higher rate than the retail price of electricity, the difference being subsidized by the government. The rate was guaranteed for 20 years and set at a level to allow the typical household to recover the cost of installation within approximately a decade. The rate of take-up far exceeded expectations and gradually the feed-in tariff has been reduced as the costs of the subsidies mount. Yet despite the German promotion of solar power these advances still mean that only 3 % of total energy consumption in the country came from this renewable source by 2012. A greater proportion of the renewable energy so far comes from extensive wind turbine operations, many by locally organized units, such as local energy firms or co-operatives. But these developments mean that extensive new transmissions grids are needed to link supply and demand sites, and these, like wind turbines, are generating opposition from areas affected.

Table 5.2 (continued)

d) The greatest advance in utilizing geo-thermal energy sources has been made in Iceland where geo-thermal power is now used extensively, with over 80 % of dwellings in urban areas heated from this singular geological area. Until recently it was assumed that only limited areas on earth had the ability to use geothermal resources, places located where tectonic plates collide, or in areas of thin crust, places that had the capability of exploiting the heat from the earth's interior. However research by a group at M.I.T. led by J. W. Tester (Tester and Anderson 2007) has claimed that there are far more economically feasible geothermal sources available in the United States This is due to the use of new drilling techniques and rock fracturing at depths well below 5000 ft., so the source could supply even high demand time or base-load electricity, at least in larger settlements. Despite this potential, the development of this resource is still in infancy, except in a few areas.

the recognition that wind still only supplies 4.4 % of the electricity energy in the United States as a whole, which is a long way below Denmark's level. So although it is undeniable that there are major increases in the growth of renewables, they have a long way to go in most countries before catching up with fossil fuel sources, a problem made more difficult because of the predicted large expansion in energy demands shown in Table 5.1.

The prospect of increasing sustainability in energy sources at a world level has also become more difficult given the rapid industrialization and hence energy growth of such a huge country as China, which has already been shown to have major pollution problems. So even though China and other countries may be installing solar panels and wind turbines at an great rate, the fastest pace in the world, its huge population and increasing energy demands means it will still be heavily dependent on fossil fuels, especially coal, for the next 20 years. Moreover many of the wind farms in the west of the country need long transmission lines to the main consumers of electricity in cities. They are also limited because they have lower capacity levels at 23 % due to longer periods without wind than in most countries (Cyranoski 2009). All these issues mean that the proportion of world energy from renewables at a world scale is still predicted to be relatively small compared to other sources, ranging from 14 to 26 % in 2035 dependent on the scenarios adopted, as shown by the International Energy Authority estimates in Table 5.1. This does not deny that considerable progress is being achieved in some places, rather that the rate of growth does not seem sufficient to even make a world with 50 % energy levels from sustainable sources by mid-century.

The limited predicted contribution of renewables relative to fossil fuel sources leads to the question of why such renewable sources are not likely to make larger contribution. Eight general *problems of renewable energy expansion* need to be addressed before renewable sources can grow to outpace fossil fuel sources at a world scale. The first is the *higher costs* of all renewable options, although some of these are coming down as technology improves and with larger production runs to create economies of scale. Indeed, power generated from solar cells and wind powerare becoming close to parity in price with fossil sources in some countries because of technological breakthroughs and high volume production in solar cells and wind turbines. They would be equivalent if the costs of the negative externalities of the fossil fuel sources, such as costs of pollution, were taken into account. Indeed, the retail price of electricity from rooftop PV solar power in Germany in 2012 was comparable with that from fossil fuel sources, whereas it was two and a half times as expensive even in 2006 (TE 2013a). Yet all the subsidies and cost of new plants and transmission lines has meant that industrial electricity prices in Germany have doubled from 2000 to 2013, whereas those in the U.S.A. have remained similar, in part because the shale gas revolution has substantially reduced gas prices, whereas they were at parity a decade earlier (TE 2013b). So the success in renewable energy has created problems for the German grid, leading to suggestions that large industrial plants be encouraged to create their own renewable energy plants to reduce major transfers of power via the grid, while more energy conservation methods should be adopted to reduce demand.

In Britain the costs of wind generation are also still high. There has been a concentration on large turbine development in off-shore wind farms, currently accounting for a fifth of its renewable energy but only 3 % of total energy. However this has come at a high cost, for the $ 250 subsidy provided in 2013 is three times the current wholesale price of electricity, and 60 % more than for the onshore wind farms, although there is an expectation that costs will decrease as the technology of building large units in difficult environments improves. At the end of 2013 the government was continuing to expand the system, so that Britain should have 10 GW from this source by 2020 (TE 2014a). Further growth could lead to the country having a third of its total energy, from a possible 40 GW offshore capacity by 2030, but only if large subsidies continue.

Second, there are still *technological difficulties* to be faced in improving efficiencies—especially in wind and solar—although substantial progress has been made in the last decade. Few solar panels are able to achieve more than 15 % capacity, although recent advances are increasing this to 25 %. The limits and lack of flexibility in the current silicon-based panels is leading to research to find better compounds for solar generation and ones more flexible than the rigidity of silicon. In the case of wind there is often a great difference in generation rates, making it essential to choose sites with string and regular wind flows. Many turbines operate at about 25 % capacity, given the number of calm days, although offshore sites usually have much higher rates. These two problems are unlike the situation for fossil fuel use, where many initial technological difficulties have been solved by a century or more of advances in the extraction and use of these sources, while they have costs advantages as many existing plants have covered many of their initial capital expenses. So although the cost of the 'fuel' in solar and wind plants is relatively free, they have high initial capital costs to construct and install the panels and turbines, so many still need subsidies or long-term output contracts to make them worthy of investment or use.

A third problem is the need for large amounts of *costly new and more complex transmission lines,* not only to deal with the projected increases in energy demand, but also to take power from the sunny or windy areas which are often in the peripheries of countries to the points of consumption in cities. In Germany it has been estimated that at least 4,000 km of new lines are needed (TE 2013a). Many of the routes of these lines, as well as the sites of the new converter stations needed to transform high voltage direct current into alternating current, are being opposed by local residents, so progress is likely to be impeded. Also, more of these lines need to be able to cope with two-way flows, rather than the one-way flow from generating plants to customer, especially if more households wish to sell their excess solar energy.

A fourth difficulty lies in *intermittency* problems, especially of solar and wind, for solar does not work at night and winds do not always blow. So both the sources need back-ups from other power sources, or transmission via long distant lines from places where the sun is shining or winds are blowing. In addition, the intermittency leads to power surges on windy, sunny days, reducing the wholesale price of electricity, not simply to the detriment of renewable sources but to the decreasing profitability of many large generating companies in Germany. One result of these price depressions is that companies claim that the installation of modern, efficient and less polluting gas-fired power plants is uneconomic since they do not operate full time. So they keep the old, polluting brown coal stations that have covered their capital costs to produce cheaper electricity in the back-up periods.

The fifth problem is that of *storage*. Although energy from wind and solar can be used directly, it is still difficult to store the excess amounts generated at peak periods. The technology to create the very large and powerful batteries to store this excess still do not exist, although in a few locations the excess is used to pump water to upland reservoirs that can then be released to generate hydro power on demand, or to pump water underground and retrieve it later for use if deep reservoirs exist.

A sixth major problem comes when a full evaluation of all the *costs and results of renewables* is made, for many disadvantages may be concealed For example, more electric cars may reduce pollution in cities compared to oil–powered vehicles, but will increase electricity demand, which may be from fossil-fuel burning power generators that produce more pollutants, unless distributed-generation takes off, through the development of cheap batteries in residences to store power from wind or solar sources, enabling consumers to 'go off the main electricity grid'. Another example comes from ethanol production for fuel. If farm land is used to grow food crops that are turned into ethanol, food production is reduced. It will lead to higher food costs, especially given projections that show global warming will reduce crop yields in many areas. Hence high subsidies for ethanol production can be questioned, while the necessary use of fuel for tractors and fertiliser production will increase energy use and pollutants. Similarly the demand for wood pellets to fuel electricity generating plants that depend on this fuel may well lead to the destruction of wilderness areas in hotter parts of the world where trees grow more quickly than in the northern climes and which are likely to be the sources of this increased demand. Indeed, the amount of biomass and hence land area needed to create large amounts of bio-fuel, even from non-food crops, is very considerable.

The seventh problem, one of *environmental damage* comes from the very movement that is calling for the growth of renewables, those interested in environmental protection. It is rather ironic that the expansion of wind turbines in particular is being increasingly opposed by local environmental groups that see wind turbines as blots on the landscape and dangerous to birds who get caught in the blades. They also produce noise that is irritating, if not dangerous, to residents located close by, although this can be solved by having turbines placed at greater distances from existing houses.

Finally there has been criticism of the often rapid *changes in policies*. In Germany the high feed-in tariffs provided to initial solar power installations proved popular and led to a surge in demand. In recent years this has been scaled back for new customers, which has reduced the number of new adopters. Other countries who have experimented with similar generous feed-in tariffs, such as Japan and Spain, also found consumers were willing to adopt the solar option. But confusion over changing tariffs and recent reductions in subsidies, given the costs to the government in the recent recession, has delayed the potential increase in solar power in many countries. Even in Denmark it has been argued that progress is being compromised by less political commitment in recent years, such as in Maegaard's (2010) critique of conservative government policies in Denmark in the previous decade.

The limitations apply primarily to existing sources of renewable energy. In addition, each particular source has its own drawbacks, as shown in Table 5.3. The restrictions of renewable energy sources and the specific problems of individual renewable types do mean there are still daunting challenges if they are to provide similar proportions to those of the main fossil energy sources. Among their many difficulties the question of costs looms large. There is no doubt that most of these renewable sources are still more expensive than fossil-fuels and require subsidies or long term contracts, although costs have plunged dramatically in recent years, holding out the hope for comparable generating costs to fossil fuel sources in some areas, especially wind sources, in the near future. Of course critics argue that the various subsidies provided to renewable producers are uneconomic and wasteful. Yet it must be stressed that the fossil fuel industries and users are directly and indirectly subsidized by governments in a range of measures, from tax breaks to subsi-

Table 5.3 Problems of specific renewable energy sources

Source	Specific problems to be overcome
Hydro-Electric	High construction costs. Need mountainous/hilly sites for the fall of water and extensive land requirements for reservoirs. There are problems of long-term silting of reservoirs. Sites are often far from population centres and require long transmission lines. Most developed countries have already used the most suitable river basins for this power source, while there is increasing opposition to new plants along some rivers, either from down stream countries fearful of losing water supply, or from environmentalists critical of the potential destruction of habitat by reservoirs, and having negative effects on downstream water supplies, fishing and riparian vegetation.
Tidal	Good long term prospects once new turbine technologies using two-way tidal flows are developed. But they have limitations because they are restricted to coastal sites and can be hazardous for shipping and fisheries.
Geothermal	Uses heat from the earth's core to drive turbines. High construction costs for large-scale generation. Still site-limited, because it is only feasible in areas close to a thin crust, in volcanic areas, or places with underground reservoirs.

Table 5.3 (continued)

Source	Specific problems to be overcome
Biofuels	Based on deriving these products from crops rich in sugar or starch or oil. Mixed with regular gas or petrol, a result of national regulations not initially commercial viability, it contributes about 10% of the fuel used in U.S. vehicles in 2012. The big problem is that many use food crops for the feed-stock, for example creating ethanol from corn, resulting in a large proportion of the American crop devoted to ethanol production because of the subsidies available for the processing. But this has helped increase the cost of com and produced a 'food versus fuel' argument. If more and more corn or other food crops are used, the cost of food will increase, which will cause problems for the poor in a growing world, while the approach does produce a low energy output, given all input costs in growing and processing. Small amounts of second general bio fuels are being manufactured, using various techniques; these break down plant waste or woody crops into sugars to produce a cellulosic ethanol. The product is still more expensive than the previous approach but initially seems more sustainable since waste, such as sugar cane and grain stalks, leaves and wood chips are used. Yet some of these are a traditional ingredient for fertiliser, while there is increasing concern that plantations of imported plants to provide the biomass will upset local ecological balances, which may lead to a 'vegetation versus fuel' debate. Moreover the biomass feedstock required for large scale plants is also very big, which means high collection costs and often the consumption of vast acres of land that may soon be needed for food, if production is to be drastically increased. So although a useful addition to total liquid fuel supplies, there are clearly some negatives when the indirect and effects are added to the direct effects. Research in the use of algae to create a biofuel is promising, but is still a long term solution.
Bio-gas	Derived from two sources with the advantage of being available near cities creating low-transmission costs. One is by anaerobic decomposition of rubbish that creates methane that can be used to generate electricity, as in the 60 year old Los Angeles 500 ft high landfill that supplies 70 thousand homes. Another is by combusting household waste, or waste from forestry plants to produce gas to drive turbines. This involves the expense of obtaining waste and the development of conversion plants. Pollution from waste-generating plants in the past led to a reduction in this approach. But new Waste-To-Energy technologies show much promise and offer an alternative to landfills or recycling rubbish. They are based on high energy burning at over 850 °C, which breaks down the molecular bonds in toxins such as dioxin and furans, making them harmless, while the flue gases can be used to produce stream to drive turbines that creates electricity. This approach is likely to increase in use.
Wind	At first sight a free and available source to drive turbines on windmills that produce electricity. But not all areas have regular winds. Generation is often intermittent and greater at night, but high electricity demands are in the day. Generating capacity, given varied wind supply, averages only 25% of capacity of the turbines under current technology, though offshore sites are often 50% more efficient because of stronger winds. Local noise and aesthetic intrusion, as well as hazards for bird life, has led lo increasing opposition to land wind farms. Extra costs come from adding transmission lines from rural production sites to distant urban places. Technological advances in turbine design mean costs are increasingly competitive with fossil fuels. But storage of excess wind power is still a problem, although it can be used to pump water to hydro reservoirs in hilly country or to geo-thermal stores.

Table 5.3 (continued)

Source	Specific problems to be overcome
Solar	a) Photo-voltaic (PV) panels. Can be applied at individual house, community level, or in large industrial scale with fields of panels on large sites. Current panels are only 15 % efficient in capturing potential sunlight energy of 1,366 W/m^2, but research has now developed second generation panels which achieve 25 % efficiency, with greater prospects for the future. b) Solar Thermal Power plants. Focuses sunlight from 'fields' of mirrors that track sunlight either upon a tower or pipe that heats up water or oil to power a turbine. High start-up costs. Still expensive, but new technologies promise cost reductions. Both these approaches need sunny areas or south exposures, otherwise generation is less efficient. But large scale batteries to store energy generated in daylight are not yet available, so part of the utility of this source is lost.
Batteries	Electric batteries are viable for cars but have limited capacity and range (approximately 150 km) and continued higher relative costs. Few plug-in renewal outlets exist in cities, although some countries are rapidly adding them. Recent technological progress is reducing battery size and increasing energy generation. The problem of range is being reduced with the growing use of hybrid cars, and the prospect of easier night-time recharging of batteries.

dies for vehicle gas. The IEA (2013) estimated that these fossil fuel subsidies have grown substantially in recent years and reached $ 544 billion a year in 2012, over five times the subsidies to renewables at $ 101 billion, although the later need to double to $ 220 billion if the New Policy Scenario in Table 5.1 is to be achieved. In addition, lobbyists for the fossil fuel industries spend large amounts in advertising, persuading politicians to maintain these subsidies and often finance negative opinions about renewal sources and sponsor climate change deniers. Moreover, the externalised costs of fossil fuel use—whether problems from global warming or the pollution effect on human health—are rarely considered in the cost comparisons between fossil and renewable energy sources (IEA 2012). The case of the large numbers of premature deaths due to the free coal policy in north China has already been described. A report on the effects of coal mining in Kentucky is also salutary. It showed an increase in mountaintop mining which involves stripping off the tops of mountains to get at the coal beneath, then dumping the fill into the valleys (Palmer et al 2010). Apart from the loss of forest and biodiversity that this development creates, the water that emerges from the buried streams is full of toxic materials, leading to the poisoning of fish and humans. In addition, the extra dust in the air has been shown to increase rates of lung cancer and other related respiratory diseases. Clearly the damage to ecosystems, threats to human health and the lack of effective mitigation require new approaches to mining regulation.

Even in cities the externalities associated with vehicle use are obvious but vary with the density and sprawl of settlements. In a study of large American cities Glaeser and Kahn (2010) calculated that the cost of greenhouse emissions varied from $ 1150 per household in San Diego to $ 2015 in Memphis, and suggested that land use regulations should take into account these externalized costs, since the price of gas does not take into account the full cost of energy consumption. So if full accounting of fossil fuel use is made, the case for renewable energy becomes more and persuasive, even if we discount the finite supply argument.

Despite these important caveats, Table 5.1 shows that the hoped-for dominance of renewable over fossil fuel energy is still in the realm of hope, rather than being a realistic reality, at least by the mid-twentieth century unless drastic policy changes are made. Yet it cannot be denied that substantial progress has been made in the last decade compared to the previous one. However, this pessimistic conclusion is more a consequence of two political issues than only technological ability. One is the *absence of political will* in many countries to create more rapid change in renewables, for many governments fear that major investment in renewables and in reductions of pollution from generating plants will lead to cost rises in most goods, which will put them at a disadvantage compared to countries without controls. Hence international agreements seem essential to prevent such comparative advantages by countries without renewable energy policies. Despite these problems many countries have set targets to increase the level of renewable fuels, while an international agreement exists in the European Union where the 2014 revised aim is that the area as a whole should attempt to set a 27 % level of renewable energy use by 2030. In addition a few cities have also set their own targets, such as Copenhagen's plan to be carbon-neutral on balance by 2025 through a variety of policies that will be discussed in the next chapter and summarized in Fig. 6.2.

Second, new *spending priorities to favour renewables are needed.* It has been argued that if these two problems could be resolved the world could be fossil-free by 2030, as seen in the plans proposed by two influential Californian scientists. Jacobson and Delucci (2009) claim that the technology exits for a major change to develop a 100% renewable energy solution creating 11.5 TWT world energy capacity by only using wind, water and solar sources (WWS). They maintain that such sources provide the best options for reducing global warming, pollution and water use, as well as minimizing the negative effects of fossil fuel use on land use, wild-life and human health. Certainly the scale of the required plants are enormous: wind, contributing 51% of the total, will require 3.8 million large wind turbines and 720,000 wave converters; solar sources will produce 40% of needs with 89,000 solar plants, while water-based solutions will contribute 9% from 490,000 tidal, 900 hydro-electric and 5,350 geothermal facilities. The authors estimated that geothermal and hydro-electric sources have 2% and 70% of potential capacity respectively already installed, but the rest have less than 1% of the plants as of their base year of 2009. The total cost at $ 100 trillion over 20 years (or $ 5 trillion a year) makes the project seem unobtainable. But it is worth remembering that the recent annual costs of the Afghan and Iraq wars for the U.S.A. alone were estimated by the National Priorities Project to be US$ 539 and $ 804 billion, as shown in the Military and Non-Military spending allocated by Congress 2011-12, while the Defence budget was $ 525 billion/year (COW). These calculations mean that even with existing technology for renewables and current finances, the United States, let alone the world, has the capacity to alter its energy balance IF the political will existed. Yet changing the energy mix from its fossil-fuel dominance to more renewable sources is not the only way that countries, regions and urban places can become more sustainable. The following sections show that progress in curbing emissions is also occurring in two of the major sectors of the economy that contribute substantially to non-sustainable activities, first, industry and construction activities, and second in the transport sector, where the resources consumed, and waste produced, lead to low levels of sustainability.

5.7 Curbing Emissions from Industrial Activities

Figure 5.6 has shown that the second largest source of human-generated greenhouse gases and related pollutants come from manufacturing plants and construction sites—including those that extract and process raw materials—which have also released or dumped lots of other toxic materials into the environment, while those from transport is the third. In the nineteenth century there was often an assumption that active smoke stacks showed the prosperity of a town, but through time attitudes have changed. Various initiatives have led to the reduction of some of the pollutants, at least in the developed world, although there is still a long way to go before industry can be seen to be environmentally benign, while industrial cities in the developing world often have higher levels of pollution because of the greater toxicity of modern industrial processes and a lack of concern by many governments of the problems. Many of the policy initiatives are carried out at a national scale—sometimes internationally with recognition of the trans-border effect of some polluting sources. But there are often opportunities for urban municipalities, or the local citizens affected by pollution, to influence policy. For example, Vancouver has long had an active environmental movement and in 2012 the city council adopted new plans to improve its sustainability, namely by: a reduction of greenhouse gases by a third between 2007 and 2020; trying to ensure all residents live within 5 min of a green space or park; as well as a policy to plant 150,000 trees by 2020. In general most nations and cities engaged in sustainable practices tend to focus on a limited number of policy areas, but more comprehensive strategies are emerging, of which several approaches are the most used to curb emissions: environmental regulations; industrial initiatives, via new technologies at a plant level; new technologies for both vehicle propulsion systems and efficiencies; as well as those involving taxes, or cap and trade policies to reduce emissions.

5.7.1 Regulations to Reduce Emissions

New environmental regulations, at least in the developed world, enacted over the last two decades have drastically reduced not only the development of plants with toxic outputs in many areas, but have also restricted their ability to release such pollutants. However, as the chapter on Green Cities (Chap. 4) has shown, it often takes an industrial disaster, or the work of dedicated critics who have exposed the negative effect of pollution on human health, before serious attempts are made to alter environmental impacts, in this case to reduce pollution levels. Although such laws are usually made by national governments rather than local municipalities, they are often monitored locally, while the need for local development permits means that local governments are often able to influence the decision to allow industrial developments to take place, at least if they are independent of state influence or control, which is not the case in totalitarian regimes. Despite opposition from business groups concerned at the extra costs of complying with environmental regulations, they are likely to become tougher through time, so long as concern about environmental damage remains a priority among politicians. For example, the province of Ontario has plans to phase out all its coal-powered electricity stations by the end of 2014. However, in many parts of the world the absence of regulations, the limited effective monitoring of these plants, or an ability to punish offenders, means that toxic waste and emissions continue to pollute the environment. Yet there are signs of change, although progress is still slow. For example, recognition of major health problems in the past, due to the extensive use of lead, paints, asbestos and some plastic compounds, led to bans on their use through legislation, although older buildings may still contain such materials. At an international level the discovery that various halogenated hydrocarbons, such as those used in refrigerators, was destroying the ozone layer that filtered out ultraviolet light in particular and was leading to increases in skin cancer rates, led to international action to ban these products by the Montreal Protocol, which took effect in 1987. Now adopted by most of the countries in the world it has been lauded as one of the most successful of all international agreements, a success based on clear scientific evidence of the damage that is linked to a specific cause. Yet in other cases progress has floundered, as described earlier in the case of the Kyoto in 1997. So with a few regional exceptions it is left to individual countries or cities to implement policies.

5.7.2 Industrial Initiatives to Reduce Emissions

There seems little doubt that substantial reductions in pollution levels from industrial plants can be made from existing technologies, let alone future ones. As always, many large polluters are reluctant to invest in pollution-reducing initiatives on their own, because of the extra costs involved, which is particularly the case when aging plants need heavy upgrades. Increasingly more rigorous government regulations restricting high levels of emissions is leading to new technological initiatives. One of the most innovative examples in recent years, reflective of greater concern for the negative externalities of industrial plants, has been the changes that have taken place in the Inco smelting plant outside the town of Sudbury in Northern Ontario. A sign of the problems of the past can be seen by the 380 m smoke stack erected at the plant in 1972, designed to carry smoke away from the town more quickly and at higher altitudes. Dominating the skyline, it annually emitted 2 million kilotonnes (kt) of SO2 in addition to large amounts of other gases, continuing the industrial practice of just emitting waste gases. These emissions, along with similar practices by other companies, had devastated the vegetation in the area around Sudbury for over a century. Increasing environmental concerns about the effects of these emissions on human health as well as the local ecology led to various gas-capture procedures by the company, which had reduced this level to 150,000 kt. by 2011. However a new project costing over $ 100 million by the new owners (Vale Inc.) will be completed by 2015. At its heart is a new plant to convert most of the formerly flared waste gas to sulphuric acid which will reduce annual emissions to 45 kt., well below Ontario's regulatory limit of 66 kt. and with dust and metals emissions also down by 40 %. The sale of the formerly waste product will enable the company to recoup its expenditures, since can be used in many industrial processes and also converted into a valuable fertilizer. This plant is claimed to be the most ambitious project of its kind in Canada, one that is designed to help the more general environmental initiatives to re-green Sudbury.

In a more general context several new technologies are under development to reduce the emissions of carbon dioxide in particular from industrial plants (EC-Climate 2013). Recently there has been a lot of interest in a sequestration processfor carbon emitted from coal-burning electricity generating plants, such as storing it underground, or using it in oil or gas reservoirs to improve outflows from existing oil fields. But the expense of such additions to most existing coal-generating or oil-refining plants has deterred most companies from applying such a procedure, except in a few pilot projects. Yet there are exceptions. Statoil, the national oil company of Norway has successfully added carbon sequestration to several of their production systems, while the Canadian province of Saskatchewan announced in 2013 that it would invest $ 1 billion in carbon sequestration to allow it to develop more coal burning power plants. By contrast, in the same year the province of Ontario decided to close down the last of its coal burning plants in the interest of sustainability, although it is not without significance that this province has no useful coal reserves, whereas Saskatchewan does.

5.7.3 Vehicle Changes and Emission Reductions

In the transport sector the growth in the number of cars with new propulsive systems is reducing the low levels of sustainability. The simplest approach to reducing emissions is that of changing the fuel input, such as by running on natural gas, or a mixture of gas and bio-fuel. There are some signs of increasing gas use, but the need for larger and thicker fuel tanks and adjusting engines remains a problem. So the most significant progress to date is taking place by the conversion of the bus fleets in some progressive cities to gas power. Since large vehicles are major generators of greenhouse gases, this substantially reduces the emissions from oil based fuels, although there are conversion costs in altering the fuel input. The growth of bio-fuel has been considerable in the past decade, but the use of this fuel alone, or mixed with regular oil-based fuel, lags far behind the use of traditional gasoline. A more complex approach is by changing the vehicle engine, either in halfway form by hybrid cars that have both internal combustion and also electric motors, or by electric-only vehicles. Optimists believe that these two types of car will soon lead to a revolution in car types, with resultant fuel reduction and lower emissions. But the extra price of these cars, due to the still high costs of the electric motors, has meant that their growth is still limited, despite government subsidies in many countries, and lower running costs. Nevertheless most major car companies have versions of these vehicles. The international organization of car manufacturers has calculated that there were 84 million passenger and commercial vehicles produced in the world in 2012, a substantial increase from the 59 million in 2002 (OICA, 2012). But electric cars only accounted for 0.02 % of the total, and hybrid cars another 5 %, by the end of 2012. Hopefully improved electric batteries will increase the range of these vehicles beyond the current 150 km level, while the addition of more electric changing stations, will also improve their numbers and use. But despite the addition of such cars to the vehicle fleet it is difficult to see these cars as anything more than a minor, although still needed addition to the world's vehicle fleet during the next two decades at current rates of progress—an evolutionally not revolutionary change—although they may have more localized impacts within large cities where small car-sharing schemes are emerging. However the recent success of the Tesla Model S cars which uses standard lithium-ion cells rather than the batteries used by other companies which are specially designed, heavier and almost twice as costly as the Tesla approach is worth noting. Also in 2014 Tesla announced plans to develop a major battery factory with an eventual 500,000 units capacity in the USA, which it maintains will reduce costs by a further 30 % due to returns to scale, and could bring the cost of electric cars down to competitive levels without the need for subsidies, which could lead to a major increase in electric cars.

Unless the cost and size of batteries decreases rapidly, the growth of this type of vehicle may suffer because of the probable savings from next generation of clean diesel engines. Current diesels are a third more efficient than petrol engines and the most fuel efficient European car, the Volkwagen Polo with a 1.4 L diesel engine, has a better fuel consumption than the Toyota Prius hybrid at 3.8 L /100 km, compared to 4.3 L 100 km (TE 2013c). The next generation diesels from 2014 will be lighter because of the use of aluminium, not heavy iron engine blocks, with lower compression ratios and more complete combustion, which is projected to deliver 20 % better fuel efficiency and lower emissions. This may reduce current criticism that diesels produce higher pollution levels. In an experimental context, cars powered by solar cells do exist, as do a few solar planes, but the number of the cells needed in these vehicles mean that passenger space is extremely limited, so their ability to alter the transport system seems decades away. A more promising trend lies in the development of fuel-cell vehicles, which unlike the electricity carried in lithium-ion batteries, makes electricity through a chemical reaction between hydrogen, carried as a fuel, and oxygen, with only water vapour as a by-product. Again the development of such cars will take decades to charge the current transport system, but the technology is sufficiently advanced for Toyota, Hyundai and Honda to announce that they will be selling sedans using hydrogen power in 2015. Details of prices and volumes are still to be announced but early in 2014 Toyota announced its FCV model, similar to a Camry, is likely to sell for $ 50,000, which is close to double the cost of a similar sized sedan. Unlike the range limits on electric battery cars these vehicles will have a range close to 500 km but will of course need refuelling stations. Although California is spending $ 200 million to create a network of 100 hydrogen stations by 2024, few other areas are adopting such policies. So despite this technological progress these alternatives to the internal combustion engine still have many limitations, from price, to refuelling issues, and it will take some decades before these technological advances in powering vehicles provide major changes in vehicle numbers and hence pollution reductions.

Although these are the main programmes being used, there are many experimental policies being pursued. For example, tests on the island of Bornholm in 2011, which has the wind capacity to supply 2/5ths of its power needs, tried to solve the problem of storing wind energy and recharging batteries for electric cars. Essentially the project uses the possibility of cheaper pricing of electricity at night from the stronger winds at that time by re-charging the batteries of a fleet of electric cars at night from home stations, which explains the long title of the project: Electric Vehicles in a Distributed and Integrated Market using Sustainable Energy and Open Networks—EDISON for short, after the famous pioneer of electricity. The approach also means the power in the batteries could be fed back into the system by day to the profit of the owner if the car is not used. Of course, this approach demands the use of a new type of grid, one that can send power both ways. In Israel experiments have been carried out in simply replacing batteries once they are almost run down, which means having lighter batteries that can be quickly replaced and enough recharging and replacement stations. So far these ideas are still in the experimental stage.

5.7.4 *Efficiency Improvements to Reduce Emissions*

Another trend in reducing the negative effect of vehicle engines comes from increasing the input-output efficiencies of the internal combustion engines, both in terms of reducing the fuel consumption and reducing emissions. In the first case there have been major increases in the average kilometre/litre fuel consumption figures, which along with the development of many different types of smaller cars seemed likely to reduce the demand for fuel. In the latter context the growth of the compact two-seater Smart Cars is a new trend, but again, despite the increase of their numbers and the adoption of car-sharing schemes using the cars, this is only producing a marginal effect. Indeed, since the growth in the numbers of cars has increased, and in western countries there has been a trend towards the purchase of large SUVs, so many of these gains have been eradicated. More progress has been made in the last two decades by the reduction of exhaust gases and other particulate matter, especially through greater fuel purity and the addition of catalytic converters in exhaust systems and the banning of lead in the fuel, which has reduced its deleterious health effects. Although these changes have been made through technical improvements by manufacturers, it is salutary to remember the importance of government regulations in leading car manufacturers to adopt these new engine efficiency and emissions standards. For example in the EU, fuel consumption targets for 2015 have been set at 5.6 L/100 km, with lower levels of 4.9 for diesel. These are the average levels for a manufacturers' fleet, with further reductions planned for the future. Credits can be obtained for hybrid and electric cars with emissions under 50 g/km, by counting each vehicle as a double in production figures, providing an incentive for them to develop cars that do not only depend on internal combustion engines. In terms of carbon dioxide emissions the recognition that 12 % of its total output in the EU came from cars, led to legislation to reduce this pollutant. Also, an average emissions level of 130 gms/km has to be attained for the total numbers of car produced by each manufacturer by 2015, although very small manufacturers are exempted. A phase-in period to reach this level was added, so that 65 % of the fleet must comply to this level by 2012 and 80 % by 2014, with further reductions to 95 g/km for 2020 (EC-climate 2013). In North America there are also agreements on similar progressive targets for engine efficiency and carbon dioxide reductions, but these were not implemented without opposition. In the 1970s the major car companies opposed the emissions standards proposed by the state of California, but lost the resultant court case, which meant they had to conform to the regulations.

5.7.5 Taxes and Emission Reductions

An alternative approach to reducing greenhouse gas emissions involves a government imposing a carbon tax on all large industrial plants, a tax that varies with the size of their carbon emissions. This is designed to persuade companies to use new processes to reduce their emissions, which will then lead to a lower tax. Some governments use the tax to fund new environmental initiatives. One of the weaknesses of these taxes is that it is not always easy to set a tax level which will have the desired effect; too low will not change behaviour, too high and it could penalize competitiveness and led to economic losses, which is the reason why it has provoked a lot of opposition from the large resource companies. So far, opinions about the use of such taxes in various jurisdictions have been mixed, yet there are cases where the tax approach has worked. For example, the province of British Columbia in Canada introduced a carbon tax in 2008, admittedly at a relatively low level initially, on a revenue-neutral basis, meaning that the money raised was used to reduce other taxes. This seems to have garnered widespread public support. The tax increased from its initial $ 5 base to $ 30 in 2012, and is predicted to remove 3 million t of annual carbon dioxide emissions by 2020, the equivalent of the emissions from 787,000 cars per year (Gass and Sawyer 2012). But elsewhere, this type of tax has led to a great deal of opposition. For example the conservative government in Australia elected in September 2013 vowed to repeal a carbon tax imposed on the large industrial polluters by the preceding Labour administration and quickly did so.

5.7.6 'Cap and Trade' System of Emission Reductions

This avoids the command and control approach by governments in favour of a more flexible system to reduce pollution. It begins by governments putting a legal limit, or cap, on the size of emissions from industrial plants or a region, calculated from some base year, and then allocating permits or allowances to companies for various levels of emissions. The permits may be allocated for free in the first instance just to start the scheme off, but more usually they have to be bought but can be traded through a market system. Companies who reduce their emissions can sell their permits for their defined higher level to those who exceed their cap. In addition, offsets to increase emissions above the cap may be obtained by some project, such as planting as many trees in the tropics as are considered to be the equivalent of the extra gases emitted. Also the numbers of permits are usually reduced through time, to encourage companies to reduce emissions by investing in equipment to lower their emissions, or companies may buy extra permits from those who have spare capacity. Companies have to pay compensation to the government for exceeding their limits.

The obvious problem with the approach is that if the cost of the permits is too low then there is little pressure for reducing emission levels, while offsets have to be monitored carefully to ensure that compliance is being achieved. In addition, lower standards or permit costs in one country could give them comparative advantages. One of the most successful examples of this system to date was implemented by the United States Environmental Protection Agency (EPA) under the Acid Rain Programme of the 1990 Clean Air Act that applied to the industrial north-east and mid-west of the country. This created substantial drops in S02 and N02 emissions, and within a decade of the beginning of the programme the numbers of acidic lakes dropped by a quarter to a third. In terms of the consequences for humans it is estimated that annual health care benefits of between $ 85–100 billion will result by 2015 from the reduction in polluted hazy air, and with 20–50,000 fewer premature deaths a year, largely from respiratory problems (EPA 2009, 2003).

A more extensive cap and trade, the biggest in the world was implemented in the EU to reduce a basket of six greenhouse gas emissions in the EU through the European Trading Scheme (ETS). By 2012 it covered over 11,000 power stations and large plants in 31 countries and with plans to extend the system to airlines in 2014. Initially the permits were given out free to introduce the scheme, but through time more and more are being auctioned for a price, with 40 % allocated in this way in 2012. The caps are also being reduced through time and in the third phase, starting in 2015, national emission levels will be phased out, to be replaced by standard measures. Although some critics have argued that the price of the permits is too low to create major changes, it has been claimed that some progress has already occurred, with companies reducing emission levels through new technologies, resulting in carbon emission levels by 2020 that will be 21 % lower than in 1990, which is better than the reductions agreed to under the Kyoto protocol (EU: Climate 2013). But there is increasing doubt about the viability of European scheme with critics arguing that too many permits were given out initially, there are too many exceptions, and their market value, which have collapsed to around € 5 per tonne of carbon from € 15 in mid-2011 are too low to encourage effective change in many companies. Indeed when emissions are cut by many polluting companies they sell their excess ETS allowances to others who can the emit more carbon, so that emissions are not necessarily reduced. In addition doubts have been cast on the effectiveness of the monitoring of some opt-out projects, such as planting trees in tropical areas as compensation for the carbon emissions, where some schemes have been shown to be fraudulent. In any case in a world context it is worth noting that the EU countries only account for 13 % of the greenhouse gases in the world, so globally this will have a limited effect. This slow progress in dealing with greenhouse gases at an international and national level has started to convince some cities that they need to take action in reducing their emissions, recognizing that their urban environments are some of the major sources of greenhouse gases. Such actions are being actively considered in the C40 network of large metropolitan cities. For example, in 2010 Tokyo became the first major urban areas to apply the cap and trade system to 1300 of its major commercial buildings and 300 industrial facilities for its major polluters, accounting for 40 % of the area's CO_2 emissions (UNCSD 2012). New York also hopes to reduce its carbon emissions from municipal buildings by 30 % by 2017 (C40 2013). The beginning of an even greater change started in early 2013 when the Chinese government began a cap and trade scheme for the major industrial city of Shenzhen, covering the largest 630 plants with emissions (Qiu 2013), a scheme that is being extended to six other large centres and eventually the rest of the country in a determined effort to reduce the life-threatening pollution levels. The big question is whether these Chinese schemes will really have the appropriate implementation strategies to be effective as there has been a long history of local areas avoiding policies pronounced nationally.

5.8 Geo-Engineering

The costs and complexity of dealing with all these alternatives to reducing the build-up of greenhouse gases and reducing sustainability has led to the development of what is being described as geo-engineering, essentially ways of cooling the earth down by engineering solutions on land and in the air (Hamilton 2013). Some similar engineering is already practiced in areas subject to serious hail damage, where storm clouds that threaten to produce heavy hail are seeded with iodine particles which reduce the chance of large hailstones developing, but this is only applied to small areas. Among the many solutions being proposed, the simplest but very costly, is for large machines to extract carbon dioxide from the atmosphere or from major polluting sources and sequester it underground, a technology that has been shown to work in pilot projects, but would require a huge investment in many plants to make a difference. Another popular suggestion is seeding clouds with sulphate particles which would have the effect of increasing the sunlight reflected back to space. This would mimic the effect of major volcanic eruptions which add large amounts of dust into the upper atmosphere and which travel around the world. Historically this has led to lower temperatures and poor harvests even on other continents far from the eruption. But reflecting sunlight only works in the day, so this could lead to cooler days but warmer nights since the greenhouse gases are still present. A simpler idea is to 'fertilize' the oceans with iron or other nutrients which will increase the number of carbon-absorbing plankton that would later sink to the ocean floor, although this may cause other problems (Shephard et al 2007). However these and other geo-engineering solutions are likely to lead to large regional variations in weather patterns. So far there is no known way of ensuring the effects will be restricted to specific areas that might benefit. Their application could cause all sorts of political disputes if one country applied the technology despite opposition from neighbouring states that may suffer unknown consequences. Of course supporters of the idea maintain that it is important to experiment with these techniques, even on a small scale to see what works and to avoid negative effects (Keith 2013). To proponents these approaches are especially important to decrease inequalities since increased global warming will have its greatest effects on the poorest, not only because of poorer harvests in marginal areas. More droughts will occur if climate warming continues, but will also lessen their ability to use the more expensive energy. But despite passionate advocates (Keith (2013), the ethical arguments against these geo-engineering experiments do appear to be stronger, because there is no guarantee that unforeseen negative side effects will occur. It is worth remembering that fossil fuels seemed to be the solution to man's need for portable energy, with their pollution effects being ignored. But in the last 20 years their side-effects are proving to be of planet-threatening significance. So most people today believe that we cannot take the chance of using geo-engineering given our current knowledge, and especially because of the difficulty of spatial targeting and side-effect problems. This led the 2010 international Convention on Biological Diversity (CBD) in Nagoya (Japan) to propose a moratorium on geo-engineering to take effect 2 years later "*until there is an adequate scientific basis on which to justify such activities and appropriate consideration of the associated risks*" (quoted in Tollefson 2010, p. 14).

5.9 Conclusions

It is only in the last generation that serious attention has been paid to the problems caused by the negative externalities of our economic development and to the reckless over-use of our finite resources. The development of the sustainable movement has focused attention on these problems and gained world-wide attention. However the term 'sustainable' is not an easy one to define, as has been shown by a comparison of some of the major attempts to summarize its nature. By inter-relating these definitions a more comprehensive understanding of its multi-dimensional nature has been revealed, showing that it now encompasses human dimensions as well as the physical features that dominated its discussion in its early years. A related problem to the conceptual confusion associated with the term is the difficulty of finding objective indicators to measure its various dimensions that are accepted in jurisdictions throughout the world and can be used as the basis for explicit empirical comparisons. Far too many of the measurements are specific to individual studies. There is a need for much greater agreement on the choice and measurement of indictors, although the problem is being solved by action from international bodies in at least some of the physical indicators.

Despite these problems there have been attempts to provide comparable measurements of the degree of sustainability of the biggest urban settlements at a global scale, with Siemens Green Index being one of the most useful to date, although this is primarily an index of physical characteristics. Although exact scores could not be obtained for cities outside Europe and North America, the relative ranking of large cities in the recently industrialized and developing world show their limited progress on the sustainability measures. In Europe and North America there are major variations between the larger cities, but the low scores of the cities of the former eastern European communist bloc stand out, whereas the Northern Europe cities and some of the west coast of America cities show high scores which indicate the progress that these cities are making.

Given the huge scope of sustainability policies this chapter has focused on the issues and policies found at a global scale or carried out by national governments and corporations, leaving those that operate at a more specifically urban scale for Chaps. 6 and 7. What is apparent is that there are real signs of technological progress in dealing with the water and sewage problems of cities but that there are still too many urban places that do not have effective policies. However, one of the biggest problems in achieving greater sustainability comes from our dependence on fossil energy sources which not only produce many noxious emissions but are threatening to alter the world's climate, adding another dimension to the range of natural hazards to our settlements, effects that are reviewed in Chap. 9 (Resilient Cities).

Despite considerable progress in the development of renewable energy sources, the growth of the world population and its increased energy needs mean that the most informed international sources (IEA 2013) show that only a few countries are likely to achieve a 50% level of energy from renewable sources by mid-century judging by current progress. This does not deny that there have been many breakthroughs in the technology of renewable energy resources and in their use, largely from a very low base a generation ago, but in most cases the higher costs still reduce their wider adoption. In addition it has been shown that all of the renewable sources have specific problems that reduce their effectiveness. Although there are many signs of new technical breakthroughs, the political will to *radically* alter the proportion of energy from renewable sources is still limited, except in a few countries. This is not to deny that progress is occurring. Reduced emissions schemes are being pioneered in some manufacturing plants and by more efficient propulsion units on cars, let alone by increasing numbers of electric and hybrid vehicles. Also, new government regulations and standards on the largest emission sources and approaches, such as the adoption of carbon taxes or cap and trade policies, are showing how emissions can be reduced, while a range of other policies at an urban level reviewed in the next chapter are having the same effect. Although some countries are making emissions reduction a national policy and have produced impressive results, many are only taking what amounts to a slow evolutionary approach to change. Given the current trends, the total pace of change in renewable energy creation and carbon reductions is not fast enough to reduce carbon emissions to a level that will restrict the warming of the world to an average of 2 °C, the level at which the influential IPCC panel believes will trigger a series of major changes. Certainly these predictions are still only best estimates from the majority of climate scientists. But despite recent increases in the amount of new gas and oil from shale beds it does seem that to adopt the 'business as usual approach', assuming new sources and technologies will always be found, is foolhardy in the long term sense because fossil fuels, especially oil, are finite, and their use has so many negative effects despite their relative price advantage over most other energy sources. Yet the situation is made more complex by the frequent price volatility in the oil and gas markets in particular. The recent decrease in gas prices makes it possible to use more gas for energy with less polluting effects. But the almost 40% drop in oil prices from June to December 2014 may make some assume that it will be permanent, thereby decreasing the attraction of renewable energy sources. But this decline is the result of a sluggish world economy, a glut of oil from new sources such as shale beds, and the decision of the OPEC oil cartel to keep producing, perhaps in the hope of squeezing out their high cost oil-producing competitors, which includes shale oil, whose potential in the world is considerable, although at high costs. Given projected world population growth and energy demands the low oil prices are unlikely to continue, at least in the medium and long term. But this oil price decline may convince some of the inadequacy of the prediction of a Peak Oil-Limited Supplies-High Price future. However, given the finite nature of supply, all these current changes do is to push this scenario into the future. Moreover, a continued dependence on coal and oil in particular will continue to pollute the planet with real negative consequences for climate warming. As a major report for the British government argued some years ago, strong national and international action to reduce emissions should be seen as an *insurance* against greater losses due to the impact of more extreme weather events. The costs for stabilizing emissions—mainly renewable energy subsidies and charging more for carbon intensive goods—to prevent a carbon dioxide build-up of over 500-550ppm (higher than the IPCC suggestions) were estimated to be between 1–2% of GDP annually, compared to at least 5% and up to 20% of GDP annually if no action is taken (Stern 2007). Obviously these were estimates, figures that have provoked controversy (Barker 2008). But even modifications of these numbers have led most to accept that to take no action, or action on too small, is an unjustifiable risk. Many hope that an international conference on climate change in Paris in 2015 win finally produce firm action to resolve this real problem of sustainability, although if past efforts are a guide the signs for binding commitments are not encouraging. Fortunately there are other approaches to reducing emissions and increasing our sustainability which involve actions at an urban scale, instead of only international ones, issues dealt with in Chaps. 6 and 7.

References

Barker, T. (Ed.). (2008). Special issue on The Stem review debate. *Climatic Change, 89*(3–4), 173–449. doi:10.1007/s10584-008-9433-x. Accessed 30 Sept 2012.
Beatley, T. (2000). *Green urbanism*. Washington DC: Island Press.
Bradley, J., et al. (2013). Biodiversity loss and its impact on humanity. *Nature,* 486(7 June), 59–67. doi:10.1038/nature11148.
Button, K. (2002). City management and urban environmental indicators. *Ecological Economics, 40*(2), 217–233.
Carson, R. (1962). *Silent spring*. New York: Houghton Mifflin.
CF: Carbon Footprint. http://coolclimate.berkeley.edu/carboncalculator. Accessed 5 Oct 2012.
Chadwick, E. (1843). Report on the sanitary condition of the labouring population of Great Britain. Clowes and Co. for HMSO, London. New edition (1965), V. Kiernan (Ed.). London: Penguin.
Chen, Y., Ebenstein, A., Greenstone, M., & Hongbin, L. H. (2013). Evidence on the impact of sustained exposure to air pollution on life expectancy from China's Huai River policy. *Proceedings of National Academy of Science, 110*(32), 12936–12941. doi:10.1073/pnas.1300018110.
CRD: Capital Regional District. https://www.crd.bc.ca/education/at-home/low-impact-development/bioswales. Accessed 2 April 2014.
Crutzen, P. E., & Stoermer, E. (2000). The anthropocene. *Global Change Newsletter, 41,* 17–18.
CW: Costs of War. http://www.costsofwar.com. Accessed 12 Feb 2012.
Cyranoski, D. (2009). Beijing's windy bet. *Nature, 457*(Jan 21), 372–374.
C40. http://www.c40cities.org. Accessed 5 Feb 2014.
Davies, W. K. D. (1996). Sustainable development and urban policy. *GeoJournal, 43*(4), 359–369.
Davies, W. K. D., & Brown, D. F. (2006). Culturing sustainability. In J. Lee & M. Williams (Eds.), *Environmental and geographic education* (pp. 23–38). New York: Nova Science.
Davies, W. K. D., & Herbert, D. T. (1993). *Communities within cities*. New York: Wiley.
Diamond, J. M. (2005). Collapse: How societies choose to fail or succeed. New York: Viking.
ECI: European Common Indicators. (2003). Towards a local sustainable profile. Commissioned by the European Community. Ambiente Italia Research Institute, Milan. http://ec.europa.eu/environment/urban/pdf/eci_final_report.pdf. Accessed 3 July 2012.
Ehrenfeld, J., & Gertler, N. (1997). Industrial ecology in practice: The evolution of interdependence at Kalundborg. *Journal of Industrial Ecology, 1*(1), 67–80.
Ehrlich, P. R. (1968). *The population bomb*. New York: Ballantine Books.
EIU: Economist Intelligence Unit. (2012). The Green City Index. Siemens AG, Munich, http://www.siemens.com/entry/cc/en/greencityindex.htm). Accessed 8 July 2013.
England, M. H., McGregor, S., Spence, P., Meehl, G. A., Timmermann, A., et al. (2014). Recent intensification of wind driven circulation in the Pacific and the ongoing warming hiatus. *Nature Climate Change, 4*(Feb 9), 222–227. doi:10.1038/nclimate2016.
EPA: Environmental Protection Agency. (2003). *Response of surface water chemistry to Clean Air Act of 1990*. http://www.epa.gov/nheerl/download_files/publications/acid_rain_report.pdf. Accessed 8 May 2013.
EPA: Environmental Protection Agency. (2009). http://www.epa/gov/captrade. Accessed 8 May 2013.
EPI: Environmental Performance Index. (2014). http://epi.yale.edu/epi. Accessed 4 April 2014.
EU-Climate. (2013). http://eu.europa/clima.polices/july2013update. Accessed 8 Aug 2013.
EU-EF (Ecological Footprint). http://ec.europa.eu/environment/natres/studies.htm. Accessed 8 Aug 2013.
EU: European Union. (2013a). http://www.sustainablecites.eu. Accessed 8 Aug 2013.
Gass, P., & Sawyer, D. (2012). *IISD submission to the B.C. carbon tax review*. International Institute for Sustainable Development, Vancouver.
Glaeser, E. L., & Kahn, M. E. (2010). The greenness of cities: Carbon dioxide emissions and urban development. *Journal of Urban Economics, 67,* 404–418.
Gore, A. (2014). Interview. http://www.bbc.com/news/magazine-28199131. Accessed 9 July 2014.

Hallegatte, S., Green, C., Nichols, R., & Corfee-Morlot, J. (2013). Future flood losses in major coastal cities. *Nature Climate Change, 3,* 802–806. doi:10.1038/nclimate1979. Accessed 25 Nov 2013.
Hamilton, C. (2013). *Earthmasters: The dawn of the age of climate engineering*. New Haven: Yale University Press.
Hargreaves, R. (2012). *Thorium: Energy cheaper than coal.* http://www.thoriumenergycheaper-thancoal.com. Accessed 8 Aug 2013.
Haughton, G., & Hunter, C. (1994). *Sustainable cities*. London: Jessica Kingsley.
Hubbert, M. K. (1962). Energy resources. National Academy of Sciences, Publication 1000-D, Washington DC. http://www.hubbertpeak.com/hubbert/energyresources.pdf. Accessed 3 Sept 2103.
IEA. (2008). *World energy outlook*. Paris: International Energy Agency.
IEA. (2012). *World energy outlook*. Paris: International Energy Agency.
IEA. (2013). *World energy outlook*. Paris: International Energy Agency. http://www.worldenergy-outlook.org/resources/energysubsidies/. Accessed 3 Dec 2013.
IEFS: The International Ecocity Framework and Standards (IEFS) movement. http://www.ecoci-tybuilders.org. Accessed 3 Sept 2013.
IPCC. (2014). Climate change 2014: Impacts, adaptations, vulnerabilities. Intergovernmental Pan-el on Climate Change: Working Group II. Yokohama, March 31. http://www.ipcc.ch/pdf/ar5/pr_wg2/140330_pr_wgII_spm_en.pdf. Accessed 8 June 2014.
IPPC. (2013). http://www.ipcc.ch/. Accessed 15 Aug 2013.
IPCC (2014). Climate Change 2014: Fifth Assessment Report. http://www.ipcc.ch/report/ar5/wg2. Accessed 10 May 2014.
IPSO. (2013). State of the ocean report. International Programme on the State of Ocean. http://www.stateoftheocean.org/pdf_ipso_report_O51208web.pdf. Accessed 9 Sept 2013.
Jacobson, M. Z., & Delucci, M. A. (2009). A path to sustainable energy by 2030. *Scientific American, 301*(5), 58–65.
Keith, D. (2013). *A case for climate engineering*. Mass: M.I.T Press.
KS: Kalundborg Symbiosis. http://symbiosis.dk/en. Accessed 2 Feb 2013.
Lovelock, J. (1979). *Gaia: A new look at life on earth*. Oxford: Oxford University Paperback.
Lovelock, J. (2009). *The various faces of Gaia*. Houston: Basic Books.
MO: Meterological Office, U.K. http://www.metoffice.gov.uk/education/teens/case-studies/great-smog. Accessed 8 Oct 2013.
Maegaard, P. (2010). Denmark: Politically-induced paralysis in wind power's homeland and in-dustrial hub. In K. Abramsky (Ed.), *Sparking a worldwide energy revolution* (pp. 489–495). Oakland: AK Press. (Also in: European Sustainable Energy, Feb. 2008).
Mann, M. E. (2012). *The hockey stick and the climate wars: Dispatches from the front lines*. New York: Columbia University Press.
Margulis, L. (1999). *Symbiotic planet: A new look at evolution*. Houston: Basic Books.
Martin, R. (2012). Super fuel: Thorium, the green energy source for the future. Basingstoke: Pal-grave Macmillan.
Meadows, D., Randers, J., Behrens, W. & Meadows, D. (1972). *Limits of growth*. New York: Universe Books.
Meadows, D., Randers, J., & Meadows, D. (2004). *Limits of growth: The 30 years update*. London: Earthscan.
Mulle, R. A. (28 July 2012). The conversion of a climate change skeptic. *New York Times*. http//:www.BerkeleyEarth.org.
NCSU: North Carolina State University. http://www.bae.ncsu.edu/info/permeable-pavement/pub-lications.html. Accessed 12 Oct 2012.
NN: Novo Nordisk. http://www.novonordisk.com. Accessed 14 Oct 2012.
NP: National Post. (2007). Climate change: The deniers. The National Post's series on scientists who buck the conventional wisdom on climate science. http://www.canada.com/nationalpost/news/story.html?id=c6a32614-f906-4597-993d-f181196a6d71. Accessed 12 Oct 2012.

NSF: Natural Step Programme. http://www.thenaturalstep.org/en/toolkits-around-world. Accessed 12 Oct 2012.

OICA. (2012). Annual production statistics of vehicles. http://www.OICA.net. Accessed 6 March 2013.

O'Riordan, T. (2004). Beyond environmentalism, towards sustainability. In J. A. Matthews & D. T. Herbert (Ed.), *Unifying geography: Common heritage, shared practice* (pp. 117–142). Abingdon: Routledge.

Pachauri, R. K. (2013). Conclusions of the IPCC working group: 5th assessment report. *Warsaw*, November 11. http://www.ipcc.ch/news_and_events/doc/COP19.

Palmer, M. A., et al. (2010). Mountaintop mining consequences. *Science, 372*(5962), 148–149.

Porrit, J. (2005). *Capitalism: As if the world matters*. London: Earthscan.

Qiu, J. (2013). China gets tough with carbon. *Nature, 498*(June 12), 145–147.

Raupach, M. R., et al. (2007). Global and regional drivers of accelerated CO_2 emissions. *PNAC U.S. National Academy of Sciences, 104*(24). doi:10.1073/pnas/.0700609104.

Rees, W. E. (1992). Ecological footprints and appropriated carrying capacity: What urban economics leaves out. *Environment and Urbanisation, 4*(2), 121–130.

Register, R. (2006). *Ecocities: Building cities in balance with nature*. Berkeley: New Society Publishers.

Roberts, C. (2012). *The ocean of life: The fate of man and the sea*. London: Penguin.

SC: Sierra Club. (2004). *National sewage report card*. http://www.ecojustice.ca/publications/reports/national-sewage-report-card-iii/?searchterm=%20National%20Sewage%20Report%20Card. Accessed 3 April 2013.

Shepherd, J., Inglesias-Rodriguez, J., & Yooi, A. (2007). Geo-engineering may cause, not cure problems. *Nature, 449*(781). doi:10.1038/449781a. Published online 17 Oct 2007.

Solomon, L. (2008). *The deniers*. Minneapolis: Richard Viglante Books.

Stern, N. (2007). *The economics of climate change*. Cambridge: Cambridge University Press.

TE. (2013a). Special Report on Germany. *The Economist,* June 15, 13.

TE. (2013b). Can China clean up fast enough. *The Economist,* Aug. 10, 2013, 9.

TE. (2013c). American Industry and Fracking. *The Economist,* Nov.16, 68.

TE. (2014a). Rueing the waves. *The Economist,* Jan.4, 42–43.

TE. (2014b). The economics of shale oil. *The Economist,* Feb. 15, 23–24.

TE. (2014c). In the balance. *The Economist,* April 5, 70–71.

Tester, J. W., & Anderson, B. J. et al. (2007). The future of geothermal energy: Impact of enhanced geothermal systems in the United Sates in the 21st century. *Philosophical Transactions of Royal Society A, 365,* 1053–1094.

Tollefson, J. (2010). Geoengineering faces ban. *Nature, 468*(2 Nov), 13–14.

Turner, C. (2011). *The leap: How to survive and thrive in the sustainable economy*. Toronto: Random House.

UN. (1993). *Agenda 21: Programme of action for sustainable development*. New York: United Nations.

UN. (2013). World population prospects: The 2012 Revision. Economic and Social Division, United Nations. http://www.un.org/en/development/desa/population/theme/trends/index.shtml. Accessed 10 Nov 2013.

UNH: UNHABITAT. http://www.unhabitat.org/content.asp?typeid=19&catid=540&cid=5025. Accessed 5 Sept 2013.

UNSCD: United Nations Conference on Sustainable Development. (2012). http://www.uncsd2012.org/index.php?page=view&nr=297&type=99&menu=20.

WB-Blog. (2013). http://blogs.worldbank.org/sustainablecities/ghg-protocol. Accessed 9 Sept 2013.

WB: World Bank. (2009). *Systems of cities: Integrating national and local policies, connecting Institutions and Infrastructure*. Washington DC: World Bank.

WB: World Bank. (2010). http://www.worldbank.org/en/news/press-release/2010/03/23/cities-get-common-standard-for-measuring-greenhouse-gas-emissions. Accessed 8 Sept 2013.

WB: World Bank. (2011). *Making room for a planet of cities*. Washington DC.

WB: World Bank. (2012a). Global protocol for community scale GHG emissions measurement: Version 0.9. http://siteresources.worldbank.org/INTUWM/Resources/340232133225977846 6/GPC_v9_20120320.pdf. Accessed 9 July 2013.

WECD: World Conference on Economic Development. (1987). *Our common future*. Oxford: Oxford University Press.

WHO. (2012). Projections of mortality and causes of death: 2015-2030. World Health Organization, Geneva. http://www.who.int/healthinfo/mortality_data/en/. Accessed 15 March 2014.

WHO: World Health Authority. (2013). Global burden of disease. WHO, Geneva. http://www.who.int/healthinfo/global_burden_disease/en/. Accessed 10 Dec 2013.

WHO: World Health Organization. (2014). Ambient air quality and health. http://www.who.int/phe/health_topics/outdoorair/databases/en/.

Wood, C. (2014). Gender bending waters. *Albertaviews,* Oct. 34–43.

Wong, E. (2013). Air pollution in China. *New York Times,* April 2.

WT: Weekly Telegraph. (2012). News report in U.K. *Weekly Telegraph,* March 21–27.

WWF: World Wild Life Fund. (2010). http://www.wwf.se/source.php/1285816/Reinventing%20the%20City_FINAL_WWF-rapport_2010.pdf. Accessed 10 July 2013.

Chapter 6
Sustainable City Policies

Wayne K.D. Davies

Sustainability is genuinely revolutionary, striking at all aspects of our society, our social purpose, and our future life styles. Sustainability seeks to unite the planet and human family in one supportive and inextricable embrace.
O'Riordan 2004, p. 118

6.1 Introduction

The rapid growth of interest in the sustainability of society since the late 1980s described in the previous chapter has led to many new policies designed to reduce the negative externalities of settlements, especially waste and pollution, and their over-consumption of fossil energy and other resources, since many of the biggest problems come from these sources. The major background issues and problems associated with defining and achieving greater sustainability, and the ways that some places lag behind on various sustainability measures, have been described in the previous chapter. Many of the policies discussed involve state policy but since the 2008 financial and economic crisis, many national governments—although not all—have been reluctant to pursue comprehensive sustainable policies, while it has proved difficult to find international agreement to co-ordinate or stimulate new efforts. Increasingly, therefore, it is left to individuals, developers and municipalities to create and implement policies designed to increase sustainability. Some have adopted sectoral policies designed to reduce waste and recycle noxious outputs into useable resources, others have focused on innovative sustainable area developments, sometimes using New Urbanism concepts, but usually without reference to this movement. This chapter reviews the various sustainable policies that are being applied at a local or urban level. First is the issue of reduction and re-cycling. Second is the way that sustainability is being practiced at several different scales: in building construction and transport; at a community level in both new and revitalized

W. K.D. Davies (✉)
Department of Geography, University of Calgary, 2500 University Drive, N.W., Calgary, AB T2N 1N4, Canada
e-mail: wdavies@ucalgary.ca

W. K.D. Davies (ed.), *Theme Cities: Solutions for Urban Problems*,
GeoJournal Library 112, DOI 10.1007/978-94-017-9655-2_6

areas; and lastly, for whole settlements such as Eco-towns or cities. Although these polices may be useful on their own, they become far more effective when they are bundled into comprehensive strategies to improve sustainability, especially those that are designed to reduce greenhouse gas emissions, for effective mitigation of climate changes cannot be achieved by individual actions. Indeed as the quotation from O'Riordan (2004) shows, sustainability should be seen as a revolutionary goal embracing all aspects of life. Its advocates argue that we need a new form of development that embraces sustainable ideals.

6.2 Trends in Reduction, Recycling and Re-Use

Until the Industrial Revolution the amount of waste generated in cities was relatively small and easily handled. But the increasing growth of cities from the eighteenth century led to increasing worries about the amount of waste, especially from human and animal faeces, and its effect on human health. As the previous chapter has shown, the disposal of faecal matter and household garbage seemed to be solved by the development of municipal waste disposal systems from the mid-nineteenth century in the developed world; a century later a new crisis emerged. The amounts of household and business waste in towns increased exponentially, a consequence of the increasing affluence of citizens, the acquisition of more products, a throw-away mentality and the increasing amounts of packaging, especially plastics, used to wrap new goods.

Concern about this growing amount of waste led to the one of the earliest trends in sustainable policies in the 1980s through the adoption of the so-called 'three R's of sustainability', namely reduction, re-cycling and re-use. This phrase became a mantra that was used to promote a reduction in the demand for new resources, to re-cycle more, and to re-use as many of the goods and waste as possible. Although individual citizens were often the first to make the commitment to engage in these more sustainable practices, increasing numbers of companies and governments at national and municipal levels have adopted such measures in the last few decades. Some have used advertising campaigns to promote such initiatives, while in Sweden the Göteborg Eco-handbook raised public awareness and knowledge when the city provided it to over 400 thousand households in 1992. The publication described how residents could become more environmentally aware and provided tips in how to engage in more sustainable practices. Elsewhere, the new emphasis upon environmental awareness and sustainability has led to increasing research on the topic as well as courses in universities. In addition it has become a growing part of the school curriculum in many countries (Lee and Williams 2006). This has led some parents to follow the leads advocated by their more environmentally aware children;

The initial solutions for dealing with urban waste after the Industrial Revolution, other than faecal matter which was disposed of through sewers, was by *incineration plants* and landfills, although these have their own environmental problems. In the case of the former, the first incinerator in the U.K. was built in Nottingham in 1874, followed 11 years later by the first example in the U.S.A. on Governors Island, New

York. Many other cities throughout the world followed these examples. But burning rubbish led to complaints about an increase in ash from the smoke stacks falling on neighbouring communities as well as noxious fumes. In the mid-twentieth century it was discovered that dioxins, a major health hazard, were also produced. This led most cities in the developed world to abandon their burning of waste. It seemed less hazardous, as well as easier and cheaper, in most areas to simply bury the waste in what started out as inexpensive *landfills*, grassing them over when filled. However most urban areas have run into the problem of a lack of space for new dumps, which has led many to pay to simply export their rubbish to other areas where landfills are cheaper, or increasingly to incinerators in other countries. It was also soon realized that landfills also generated noxious by-products, especially methane from the rotting garbage, which created its own health hazard. Indeed almost 20 % of the U.S.'s methane comes from landfills (EPA 2013). Increasingly many cities are developing methane recovery systems by boring holes into the dumps to collect the gases, which are then used to generate electricity, rather than letting it add to greenhouse gas build-up.

The scarcity of local land for new landfills and increasing concerns about sustainability led to new waste treatment solutions from the 1990s. Most cities in the developed world started to adopt *multiple-separation approaches* to reduce the amount of waste being buried, by recycling different parts of the garbage stream. Instead of collecting unsorted garbage, many municipalities get households to separate waste into different categories, such as the threefold division into: general garbage; those products that can be recycled—such as glass, metal packaging, paper, cardboard and plastics, and organic waste. The degree to which this separation occurs varies by city and country. However, the collection of these different types of waste, as well as building waste, either by municipal workers or private contractors, is increasingly from purpose-built containers; many cities now have two or three separate household waste bins, or distinctive coloured plastic bags for different types of waste, and sometimes sequential pick-up days for the different waste products. General garbage is still mainly dumped into landfills; but more and more cities, at least in the developed world, collect, sort and re-cycle as much glass, paper and plastic products as possible. Organic waste is often collected separately and often reprocessed to be used as compost fertilizer and sent to local farms to increase their productivity—reducing the need for imported fertilizer as a result—or used to generate biogas which can be fed into the city's gas supply or used to power local buses. Cities have been to the fore in these policies. For example, an early example of this latter approach occurred in Helsinborg in Sweden which converted all of its buses to run on this fuel. Other cities encourage consumers to take various types of waste to particular disposal locations, such as in large shopping centre parking lots, where different bins are available for different products. Increasingly, however, specialized waste centres for specific types of waste have been provided, such as those for medical waste, paints and especially for electronic devices, where the valuable metals found in these products can be extracted and re-used, instead of just being dumped into landfills. Seasonal waste centres have also been created, such as leaf depots in autumn, where dead leaves can be collected and later composted, usually by the municipality.

Although there are varied methods for dealing with different types of discarded material, the problem of food waste is of particular importance. It is astonishing to realize that most developed countries throw away a quarter to a third of the food produced, either by firms discarding products in stores that are past some defined sell-by date, producers getting rid of small or misshapen products, or by consumers discarding left-over food from meals. In the U.S.A. the Environmental Protection Agency reported that 36 million t of food is thrown away, which is over 200 lbs (91 kg) per person year. This constitutes the largest share (21 %) of the waste going to municipal landfills. Indeed only 2 % of this waste is currently composted, compared to the 62 % of paper that is recycled, so there is a great opportunity to reduce or reuse this material (EPA 2013). In 2012 New York's mayor persuaded many restaurants to put their food waste in separate containers which will be collected and sent to a biogas producing plant, a scheme that is intended to become compulsory later. A similar proportion of food produced is lost in underdeveloped countries such as India, this time before it reaches the consumer—either eaten by pests because of the absence of adequate biological control, or spoilt because of limited transportation and refrigeration capacities—a severe problem in a country where millions still go hungry. So ways of reducing this waste are needed. Better ways of storing and transporting food in the developing countries would be of immense benefit in reducing food waste in areas where some people do not get enough food. In developed countries there have been moves to reduce the amount of food waste at its source, such as using fruit or other products that might be misshapen, off-colour or small size, or finding ways of using the by-products of food processing. In addition, the American Grocery Manufacturers Association, in conjunction with other agencies, announced plans in 2013 to tackle the problem, such as better stock control, and in cases where some food may be near its sell-by date by sending more of this so-called waste to food banks for use by the poor, rather than just dumping it. They estimate that 10 million out of the 50 million American people who suffer from irregular food supplies could be fed from this source, if only a fifth of the current food waste was diverted to them (Walsh 2011). In Britain there have been similar suggestions and also to get retailers to be flexible in 'best-before' dates for food, and to reduce the number of cost reductions for buying double or triple amounts of food portions, which often get wasted rather than used. These examples can be expanded by more detailed studies of what amounts to more sustainable food systems (Marsden and Morley 2014)

A second approach is by *re-cycling* or treating waste in some way so that the material can be used for other purposes. Not all of these re-cycling policies are new. For example, in terms of re-cycling one of the most successful policies involved the creation of separate returnable beer bottle depots in Canada as far back as the 1930s after the brewers agreed on a standard bottle size. A small deposit price was added to the retail price which was given back to consumers who returned the bottles to the depots; the current rebate is ten cents a bottle. This re-cycling was implemented as a way of reducing the costs of creating new bottles for each beer. Today 97 % of beer bottles are recycled and the typical bottle is used 15–20 times before being broken down and made into a new one. It has been estimated that when the bottle recycling programme is compared to the costs of a one-use bottle system, annual savings of $ 36 million a year are made in the province of Ontario alone (BAC 2013). This very successful programme has been expanded to deal with additional products, so that many Canadian bottle depots now take all beverage containers, including those of glass, metal, plastic and cardboard, and return a fee for each unit, although the rate of return and use of the containers is far less than the bottles. However the success of the programme does depend on the fact that the majority of households have a car and can transport the containers to a depot. In other countries, such as Germany and Sweden, retailers are obliged to take back the packaging of goods that they sold, which substantially reduces the waste coming from households. In addition, many countries have tried to reduce the use of plastic bags and bottles, many of which were only used once, either ending up in the garbage, or abandoned and increasing the amount of litter in streets. Legislation in many countries, as well as individual cities, has required retailers to change for the use of a bag, or even banning them, which has led to the return of shopping baskets or cloth bags, although the need, on hygiene grounds, to frequently clean these containers is often forgotten. Some cities are starting to ban the sale of plastic bottles in their premises, returning to far older solutions by adding public water fountains in public places.

Although there are clearly vast variations in the extent to which this type of recycling occurs in urban areas, it is apparent that a huge increase in sustainability could be achieved if more places adopted the best re-cycling practices of successful cities. What has been learned from the experience of these exemplars is that if sustainable consumerism is to flourish, it cannot depend on consumers behaving on their environmental values alone. Economic incentives, such as the returnable deposit, provide an important motivation for re-cycling, although it is vital for the product to have a value as a recyclable material, especially if more private companies are to be attracted into the business. This also involves acceptance by the public of using recycled material. China is the biggest customer for waste paper, cardboard and other products, so thousands of containers containing this material are sent there weekly to be recycled, which does involve large transport emissions. In addition, city regulations are needed in many cases to promote different sorts of recycling, such as the way in which many municipal governments have introduced rules, such as requiring builders to re-cycle or re-use building materials, or have sent used products such as street lamps to recycling depots, where they are broken down into their various components and then sold to re-coup costs. In many developing countries there are lots of informal pickers who search garbage dumps or litter in streets for reusable material. Although they only receive a pittance for their travail, it does provide them with an income and helps the previously poor recycling efforts, although often at the cost of the health of the workers.

Even when these various methods are used to separate the components of the waste stream so that the various parts can be recycled or re-used, the question of what to do with the general household waste remains. In countries such as Japan, where space for landfills is at a premium, incinerators are still a popular solution and often use new technologies to prevent harmful emissions. The new incinerators being developed in several countries burn at over 850 °C, which breaks down the molecular bonds in toxins such as dioxin and furans, making them harmless, while the flue gases can be used to produce stream to drive turbines that creates electricity in what are called Waste to Energy plants (WtE) plants. Japan, given its shortage of available land for infill sites, has always persisted with incineration, while Denmark and Sweden in particular have been pioneers in developing new WtE techniques (Themelis 2003). Even in 2005 Denmark obtained almost 5 % of its electricity from such sources, a figure that rises to 14 % of domestic heat consumption through its district energy schemes. Even in the U.S.A. there were 85 WtE plants by 2012, creating 14,000 jobs, power to serve 1.3 million homes and $ 5.6 billion in gross economic activity (Berenyi 2013). Hence incinerators, as part of WtE power generation, are no longer shunned. They are increasingly being adopted around the world, helped not only by the higher costs and scarcity of landfills but by government action in providing tax credits to reduce the cost of installing the new generation of incinerators (Themelis 2003). It is estimated by Ecoprog (EP 2013) that there are 2200 WtE plans in the world and by 2017 there will be 2400 with a capacity of using over 300 million t of waste per year.

In other countries more and more municipalities are developing new garbage disposal methods in place of space-consuming landfill sites. At the beginning of the new millennium research on the problem of waste disposal in San Francisco showed that 90 % of the waste from this city of 840,000 people could be recycled, which would reduce its need for new landfills or incinerators. So in 2002 it adopted an ambitious Zero Waste policy designed to eliminate waste that could be composted or recycled by 2020 (SFW), making it the first major U.S. city to have an ordinance to separate different waste streams. In the 2001–2005 period the proportion of waste recycled increased from 42 to 60 %, but soon stuck on a 70 % level. New policies were implemented, first making all restaurants and food establishments put their food waste in organic containers for pick-up, and then extending a system of compulsory organic waste and recycling bins to all residences and offices, with non-compliance leading to first warnings, then fines. By 2013 the city achieved a 80 % diversion rate, showing the success of adopting a main target, and the gradual step-by-step introduction of new policies that allowed businesses and residents to get used to the changes, and to understand the advantages of the policies. This includes the daily production of 600 t of high value compost each day from the organic waste processed in the city's new plant at Vacaville, north of the city (Pouchard 2014). A more general policy of waste processing combined with sewage treatment can be seen in the city of Edmonton which was one of the first Canadian cities to have a separate waste stream collection from households, but also claims to have one of the world's most efficient and comprehensive new plants for waste material and sewage processing and treatment. Its key features are described in Table 6.1. This example of an inclusive new approach by an urban municipality in dealing with household waste shows how re-cycling can produce major advances in the degree of sustainability in cities. If more urban places upgraded their waste systems to comparable standards, huge advances in sustainability can be made. But there is also a need for towns and cities to reduce many of their other inputs, and also to reduce or transform additional outputs to those involving waste from households and businesses.

Table 6.1 Edmonton's Waste Management System. (Source: From publications of Edmonton's, Waste Management Centre, 2012)

Waste and Sewage Treatment Plants: Edmonton, Canada
North America's largest and most advanced Waste Management Centre (EWMC) is located on a 233 ha. site in the Clover Bar district of Edmonton, a metropolitan region of 1.2 million. In 2012 the centre processed 60 % of residential waste that would have gone into the old landfill system through various treatment centres, but this figure will rise to 90 % by the end of 2013, turning the unwanted waste into resources that can be used elsewhere. In addition the centre operates research facilities and exports its expertise to other cities. Some of the facilities are public-private operations.
a) In 2000 the Edmonton Composing Centre opened to turn organic waste and sewage sludge into compost that is sold to farmers and gardeners.
b) In 2004 a world class Material Recovery Centre opened in a 6000 sq. ft. building. This sorts and processes 40,000 t of waste a year, separating out recyclable material.
c) In 2012 the Global Electronic and Electrical Processing Centre (GEEP) opened in a 4500 sq. ft building. This recycles 30,000 t of electronic products a year. There are also 3 Eco-Centres in the city where paint, electronic goods and other hazardous products can be dropped off for processing.
d) In 2012 Greys Paper and Glass Recycling Plant opened in a purpose-built dome 26 m high. It processes the paper and cardboard from the Materials Centre into paper products and glass bricks.
e) In 2013 one of the world's largest industrial scale biogas facilities will open on the same site. It will process a further 30 % of the total household waste that cannot be recycled, first into methanol and then ethanol. It is designed to process 100,000 t of waste into 38 million L of an auto-fuel, which has far lower CO2 emissions than oil-derived gasoline. At the Gold Bar site the city has also developed one of the continent's most efficient Waste Water Treatment centres, treating up to 310 million L of sewage a year, first separating out the recyclables and grit by various barriers, second using membranes to separate water that is suitable for industrial use, and then sending the residue to various fermenters and anaerobic digesters, which creates a sludge that is used for composting, in addition to creating a biogas by-product worth $ 1 million a year. After approximately 18 h of processing the now clean water is subjected to ultraviolet light to kill remaining bacteria and sent to the North Saskatchewan river.

6.3 More Sustainable Buildings

The last two decades have seen an explosion of interest in applying sustainable principles to buildings, not simply to new structures (Baird 2010; Bokalders 2010; Yudelson and Meyer 2013), but also to retrofitting old buildings to these new standards (Carroon 2010; Yudelson 2010). Given the scope of this field, space constraints mean only a few highlights can be noted, such as the intent to: make them healthier and safer; reduce inputs; adopt new management procedures; and create new industry standards.

6.3.1 Safer Buildings

Perhaps the first stage in improving the sustainability of buildings comes from the attempts to make them safer, which means strict building codes to ensure the use of appropriate and strong building materials, as well as careful designs and constructions to prevent building collapses. In most developed world countries materials that had been used for years have been shown to have serious health effects, such as lead in pipes, guttering and paints, as well as asbestos for insulation. All have been eradicated in the past 20 years, at least in new buildings, while since 1985 the dangers of radio-active radon gas leaking from building materials, or from ground sources, have been recognized and reduced. Older buildings may still need renovating to remove these noxious materials that cause brain and lung damage in the case of lead and asbestos respectively. More recently, some plastic products have also been banned, again because of their deleterious health effects when they break down. Attempts are also being made to reduce noise levels in large buildings from outside sources, especially from local traffic, given the fact that most street level noise is funnelled upwards. One of the most recent progressive examples can be seen in the NMB bank in Amsterdam which was built with sloped walls, which help bounce the sound of roads away from buildings. However the ever present problem of unanticipated effects by innovative structures may be seen in a new aluminium-cladded office tower, 160 m high in the City of London, which was finished in 2013, the so-called 'Walkie-Talkie' building at 20 Fenchurch Street. Its curved sides have had a mirror effect, resulting in the concentration of the sun's rays and their reflection down onto a nearby street, which has melted plastic and other objects in this area in a few hours. Of more general importance in sustainability has been the mandatory addition of such features as smoke alarms and water sprinklers to new buildings in many developed countries in the last decade, which has helped to decrease the risk of fires. The danger of carbon monoxide poisoning from incomplete combustion in furnaces or motors has been reduced through the addition of monitors that detect this lethal gas.

Fires have long been a major scourge in many cities, from the Great Fire of London in 1666 which destroyed much of the city, to the many fires that regularly ravaged Yedo, the historic shogunate capital of Japan, renamed as Tokyo, due to its high density of wooden buildings in an earthquake zone, where an astonishing 94 major fires were recorded from 1601 to 1867. The risk of fires has also been reduced by internal and external means. In the former case the mandatory requirement for both larger gaps between houses and the use of fire-resistant sidings and building materials has done much to prevent flames spreading from one building to another. The addition of more fire stations in cities that are permanently manned and deliberately sited to ensure a response within a few minutes, plus the infrastructure of fire hydrants with available water supply, has also led to more rapid and effective control of fires in buildings and reduced the likelihood of spread. More generally the issue of safety in urban areas must also take into account the dangers that come from extreme events generated by physical processes in the local environment, whether by earthquakes, volcanic activity storms or floods—issues will be covered in the discussion on Resilient Cities (Chap. 8). Finally, more attention is being paid to eradicate the problem of dead air in buildings because of the use of sealed windows since the 1960s, such as by introducing more effective air circulation systems, while advocates of adding more plants in buildings as described in Chap. 4, whether by green walls, or individual plant displays, emphasize their value in drawing in carbon dioxide and giving out human-sustaining oxygen.

6.3.2 Reducing Inputs

A second major trend in building sustainability comes from the new efforts used to reduce the consumption of building inputs. Issues associated with water use and energy supply have already been discussed in Chap. 5. In terms of reducing energy use, substantial savings come from such practices as: better insulation in walls and reduction of draughts through weather-proofing doors and windows; triple glazing and vacuum seals in windows to avoid heat loss; while the use of infra-red tinting of glass allows the penetration of warmth from sunlight, yet prevents the loss of heat, is also being developed. Some municipalities also encourage the use of recycled building material in new constructions, to reduce the amount of old material just being dumped in landfills. Other trends include the greater use of LED and florescent lights, which are respectively up to 90 and 75 % more efficient by releasing less heat and consuming less electricity than the incandescent lights that have dominated the market for a century. Campaigns have also been mounted by governments to encourage switching-off electrical appliances when they are not being used, since the increase in electronic devices in households mean they now account for between 5–10 % of a typical bill. Also, energy inputs have been reduced in many large office buildings by using the warmth provided by electric lights as part of the heating, while the development of more efficient gas furnaces is also substantially reducing the energy needs of many houses. In the first decade of the twenty first century many governments subsidized such developments, especially increasing insulation and adding renewable energy supplies. In Canada, a country with long winters in many parts, this support proved especially useful, but given the retrenchment of federal government spending they were withdrawn in the budget cuts of March 2012, another example of short-term thinking. Similarly the reduction or removal of subsidies for solar power in several countries since the economic depression of 2008 is having a negative effect on the progress of sustainability.

6.3.3 New Building Management Procedures

At a municipal level some governments and organizations have made progress using new building management methods that encourage greater sustainability and adopt stricter energy conservation methods in their buildings. In the early 1980s Saarbrucken became a pioneer in instituting energy efficiency in buildings. Municipal building managers were required to monitor energy consumption and to set appropriate levels of heating in various areas, such as 12 °C for stairwells and 20 °C for offices, with adjustments for different times of the day. The city also spent the equivalent of over half a million dollars to increase insulation in its buildings, such as through better weather-stripping on doors and windows, and adding energy-efficient windows. Connections to district heating schemes replaced the individual coal and oil-burning heating systems in buildings. In the 15 years prior to 1996 carbon emissions were more than halved to 35,000 t, while the heating consumption in city buildings was reduced by over 50%. This created annual energy savings of $ 5.4 million from the $ 1 million annual investment in these various approaches (Beatley 2000, p. 264). Other municipalities have followed the same conservation trends by reducing their energy outputs in street lighting systems, such as by changing light bulbs to focus lights down rather than wastefully spreading light upwards. In the case of Calgary this led to the use of a new EnvironSmart lighting system by installing 37 thousand new more energy efficient lights in 2005 which were also designed to focus light on to roads downward, whereas previously as much light reflected upwards. This policy has been estimated to save $ 1.6 million annually in energy costs and to cut carbon dioxide emissions by 19,000 t a year. More generally, German cities such as Heidelburg created E.Teams (Energy Teams) to show households and businesses how the consumption of energy can be reduced, as well as encouraging schools and other institutions to follow the same practices. Many municipalities and even businesses now have sustainability officers charged with finding ways of increasing sustainable practices, while ranking schemes have been developed to compare the sustainability of different organizations and buildings. For example Yudelson and Meyer (2013) have rigorously evaluated the cost and energy savings of the world's greenest buildings. Such comparisons are useful in showing the extent to which the poor performances of buildings or areas need to be improved, which often provide a stimulus for action to increase sustainability levels.

6.3.4 New Building Industry Standards

Another important source of progress in the development of sustainable building practice has come from the creation of non-profit groups developing and encouraging sustainable building standards and practices. Examples include Canada's Sustainable Building Council (SBC 2013), and the development of new industry practices and standards, such as the LEED (Leadership in Energy and Environmental Design) designation. Originally developed by the American Green Buildings Council (USGBC 2011) in 1993, the LEED approach provides a rating system for assessing buildings based on indicators from six different types of criteria: Sustainable Site Development; Water Efficiency; Energy Efficiency; Materials Selection; and Indoor Environmental Quality; Innovation in Design. Independent assessors certified by the organization review buildings, seeking designation on the basis of the various indicators. The total points score out of 100 assesses the degree of sustainable practice in the building and its relation with surroundings, which either lead to certification at four different levels of LEED attainment or rejection. For example Platinum is the top category that needs over 80 points, followed by the

Gold, (60–79), Silver (50–59), Certified (40–49) levels and the rejected category. Successful certification provides prestige to a building developer, a showcase for others to learn from, and usually provides costs savings. The scheme has been adapted in other countries, such as by the Canadian Green Building Council (CaGBC), which has added two other categories of criteria to the assessment, namely, Location and Linkages, especially to transit systems, as well as Awareness and Education. Despite the praise heaped upon these new standards and the progress it shows, the number of buildings that have achieved this designation since 2000 is still very low relative to the total numbers of new premises. Despite being recognized in 132 countries, the majority are still in North America, with the 2013 LEED website listing 44,220 projects that are LEED certified in the U.S.A in the past 20 years. This accounts for 593.8 million gross sq. m of building, with 20,000 homes now having the designation, up from 8000 two years earlier. Canada with 4212 projects is second on the world list (USGBC 2013).

Although most buildings being developed or renovated in urban areas today are not trying for LEED designation, there is no doubt that many of these ideas are being adopted by progressive individual developers, such as the development of new insulation levels, building materials, and reduced energy and water use. The orientation of buildings and courtyards have also been given much more attention in recent years, especially using southern exposures with the object of maximizing sunlight and solar heat, the type of approach that is being stressed in the Winter Cities movement (Chap. 8). Of course, houses in hot climates desire the opposite objective, to create cooling, and add shade by a variety of means. Sustainability has been helped by the addition of balconies to buildings which enable people to sit outside more often, or to use the space to grow herbs, vegetables or flowers. In addition, some countries have seen the increasing use of solar panels on roofs to generate energy for the building, drastically reducing or even replacing the amount of energy input from outside. Although it is still rare to find examples of buildings with sustainable energy sources and building practices accounting for practically all their power use, an exception lies in the growing Passivhaus (PH 2013) approach. Inspired by the work of architects Bo Adamson and Wolfgang Feist, the first projects were developed in Darmstadt (Germany) in 1991. By 2013 at least 30,000 dwellings around the world had adopted the approach which has been summarized in the following way.

> A Passivhaus is a building, for which thermal comfort can be achieved solely by post-heating or post cooling of the fresh air mass, which is regulated to achieve sufficient indoor air quality conditions, without the need for additional circulation of air. (PH 2013)

Passivhaus builders claim that the use of their techniques on a 70 m^2 house with gas heating will mean residents will only spend less than the equivalent of about $ 40 a year on space heating, showing the dramatic effect such buildings have on energy consumption. Yet the costs of building such houses are still higher, which is still reducing the extent to which they are adopted. So unfortunately, such ventures still represent a very small proportion of building construction.

Perhaps the most rapid change in terms of energy sustainability in buildings is taking place in many of the new big box retailing developments in new shopping centres or in large warehouses, especially in California, which are using the essentially wasted space of their large roofs to add banks of solar panels to create their energy supplies. Some large commercial companies are going beyond solar energy and also adapting a wider range of sustainable practices in buildings to save costs. One of the most explicit policy practices linked to this theme has been developed by Walmart, the world's biggest retailer, which adopted a three pronged corporate environmental ethic in 2009 based on the themes of 'Planet, Health and Life' to improve its impact on the world (WM 2013). Its objectives are to have its stores powered completely by renewable energy, with zero waste, which in some places it is close to achieving, although the other goal of 'selling products that sustain people and the environment' seems farfetched. The lights in many stores are being converted to LED bulbs (light emitting diodes) which are brighter, do not need time to warm up and use far less energy than traditional incandescent bulbs. More specifically the wider use of sustainable practices can be seen in Walmart's 400,000 sq. ft food distribution centre for Western Canada in Balzac just outside Calgary that opened in 2010 (WMC 2013). Costing $ 115 million, the facility has solar panels and wind turbines to generate power, LED lighting that is 70 % more efficient than fluorescent lights to reduce energy costs and heat generation, more efficient docking doors to reduce refrigerated air loss, as well as 71 distribution vehicles in the store using hydrogen-cell, not lead-acid batteries. The company estimates it will save $ 4.8 million in energy costs over 5 years (WMC 2013).

Most of these examples are still the product of individual initiatives by builders or the companies that employ them. Yet their adoption is helped by municipal policies, such as regulations that demand compliance to particular sustainability standards—although these standards vary drastically from town to town. Some are prepared to lend money to pay for these ventures with returns from subsequent energy savings. The previous section, dealing with reduction and re-cycling measures, has already described some of the practices being used. Many still argue that it is the fact of increased building costs that has minimized the adoption of sustainable practices. Yet it is worth noting that the Sustainability Framework Reports produced since 2006 by the International Finance Corporation (IFC 2012), the investment arm of the World Bank, has estimated that even by using existing technology the energy and water bills of buildings could be reduced by 20 %, within only a 1 % increase in costs. This shows that substantial changes can be made by small incremental costs.

Changing practices in new buildings does nothing to solve the problems of older buildings that act as a huge drag upon the level of sustainability in urban areas. Fortunately, there is increasing interest in showing how adaptive re-use of buildings can make them more sustainable (Carroon 2010; Yudelson 2010). Recognition of this possibility has led some national and state governments to adopt policies that provide subsidies to householders that increase sustainability, as noted previously. Yet given this evidence of cost savings by these sustainable practices, it is curious how so many buildings in the subdivisions based on New Urbanism ideas do not use many of these ideas.

6.4 Transport System Changes

Our modern urban and economic progress has been built on the various transportation and communication improvements that allow us to move people and goods quickly and relatively cheaply between places. Since the Industrial Revolution these new transport systems have depended largely upon various fossil fuels, from coal in steam engines to various types of oil or gas in cars, trains, planes and ships. Although there are increasing worries about the increase in pollution at high altitudes from the growing number of aeroplanes, and at sea from the heavy bunker fuels used by ships, it is the emissions from the growing numbers of car and commercial vehicles over the past 20 years that provides the major concern for urban areas, as well as increasing congestion in cities. As Fig. 5.6 in Chap. 5 has shown, transport is estimated to be the third largest urban source of greenhouse gas emissions (OICA 2012), while in the European Union vehicles are estimated to account for 12 % of its carbon dioxide emissions (EU-clima 2013). Much of this output occurs in cities, with the daily journey-to-work a major generator of emissions, given that it is so concentrated in time, with peaks to and from work, and in many cities with a focus on downtown, although this is changing as large metropolitan areas become polycentric and employment increases in suburban areas. However there are major variations in urban areas in the level of these emissions. One is related to the extent to which transport other than cars is used to get to work. The Green Index study (EIU 2012) revealed that 63 % of the population in European cities in the study used green transport to get to work (public transport systems or bicycles), although with a huge range—from 90 to 33 % in the sample cities—while in North American cities the average is only 13 %, so pollution levels from vehicles are far higher.

An important reason for the differences between cities of the world lies in the relation between urban densities and pollution levels. Figure 6.1 shows that the sprawling southern cities in the U.S.A., such as Atlanta and Houston, that are so dependent upon automobiles for transportation, have per capita emissions of carbon dioxide that are over four times the rate of cities such as London and Barcelona, and well over twelve times those of Hong Kong and Mumbai. This has led many to argue that increasing urban density, such as the Smart Growth approaches discussed in Chap. 2, will make a major contribution to the reduction of carbon emissions. However the very size and location of the existing building stock means that it is unlikely that the existing patterns of density can change significantly in the foreseeable future. So other solutions are being found to reduce vehicle use, which in turn reduces emissions and the amount of fossil fuel used, thereby increasing sustainability levels.

6.4.1 Reducing Car Use

Four main approaches have been used to decrease the use and number of cars in cities in order to reduce pollution from emissions, congestion and accidents. The

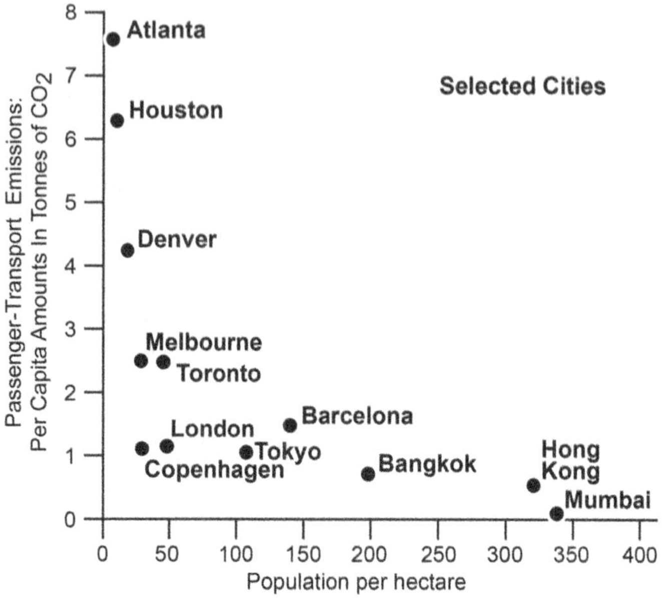

Fig. 6.1 Emissions and urban density. (Sources: Compiled from data in Kenworthy 2003; World Bank 2012; TE 2012)

simplest is by *restricting car supply*, such as discouraging ownership and use, for example by increasing taxes on new cars to reduce car sales, a policy adopted in Denmark, or as in the old USSR by keeping car production at low levels and with high prices. However, such policies are difficult to apply in democratic countries, especially given the fact that automobile production is a major economic driver. A variation on this restrictive approach was implemented in Mexico City in the 1990s, but it applied only to old cars. Since the metropolitan area was plagued with high pollution levels because the air is thinner at the altitude of over 2000 m, policies were implemented to get rid of the old, polluting cars, by offering incentives to take these vehicles off the road. The result was that the newer and more fuel-efficient vehicles with better emission standards became dominant in the vehicle fleets, which has lowered pollution levels, although the effects have been reduced by the increasing growth of the metropolitan area and numbers of cars. Seoul, a member of the C40 group of megacities concerned with climate change, has *car-free days* which it claims has reduced greenhouse gases by 10 % (C40 2014). Another form of supply restriction is to reduce the number of parking spaces for cars in central city areas or make them more expensive, which means motorists are less able to use cars if they cannot park the vehicles. Copenhagen adopted the former policy in the 1980s, restricting its parking spaces by a set amount each year.

A second approach involves *car licensing*, which has been adopted in Beijing, primarily to curb the extreme levels of air pollution. In 2008 an additional digit was added to licence plates, which indicated the day of the week when the car could *not* be used in main urban area. In 2011 a more direct licensing system was introduced, in which only a certain number of car licence plates are issued to the city's residents, and these are allocated by a monthly lottery for plates. The odds of being successful in the lottery were approximately 1 in 10 at the start of the policy; by mid 2013 they were closer to 1 in 80, a ratio that will probably double in a few years. This type of direct intervention to reduce car numbers usually needs a national government initiative and is difficult, if not impossible, in most democratic countries given current circumstances. Another approach has been to encourage *car sharing*, especially among employees of the same firm or firms in close proximity, in order to reduce the typical situation of single occupancy vehicles travelling to downtown at rush hour. Another method used to reduce car use, especially in central cities, is some kind of *congestion charge or road pricing,* such as charging more for cars to use central areas of cities. Singapore was one of the the first major cities to implement such a scheme with a cordon around the city centre. In London congestion changes of £ 10 a day have been implemented since February 2003 in London's main commercial area bounded by its inner city ring road, although this is designed to reduce congestion, and increase traffic speeds, rather than pollution. However electric and certain hybrid vehicles are exempt from the charges. Fines of £ 65–195 are levied for non-compliance, with monitoring achieved by an Automatic Number Plate Recognition (ANPR) system. Although other centres such as Stockholm and Milan have similar schemes, relatively few cities of the world have adopted the idea.

Finally a series of *land use changes to discourage car use* have been implemented in some cities. From the 1960s many changes were made in cities to improve traffic flow. In the last decade more cities are reversing these changes to reduce both the space for cars and their speed. Among the most popular policies are: widening sidewalks to improve pedestrian traffic; creating pedestrian only roads and areas; removing one-way roads in central areas to reduce speeds and to make them more pedestrian friendly; removing roadside parking in central areas; adding traffic calming restrictions on roads such as speed bumps. Another policy is to reduce speed limits below the typical 30 mph (50 kph) in most parts of the developed world, for accidents involving vehicles below 30 mph are substantially less likely to involve death. It also increases travel time, thereby reducing the car's advantage. In many parts of Canada areas around schools already have 20 kph limits. In Brighton and Hove in England, experiments with 20 mph limits around schools and a public consultation convinced the city council—which is led by Green councillors—to start implementing this speed limit from December 2013 in all but the major routes in the city.

6.4.2 Transport Systems

Some of the most important policies designed to improve mobility in cities, other than by cars, can be seen in the development of new transit systems. They attempt to reverse their mid-twentieth century decline because of increasing car ownership and limited investment in these public systems. Apart from improvements and additions to mass transit systems, underground as well as over-ground, the addition of more commuter trains into central areas of big cities has helped reduce vehicle use, again to popular areas such as downtown commercial areas. Increasing numbers of urban places, of different sizes, have added rapid bus routes, or light rail, or tram systems, which have been created or revived in many cities to encourage greater transit use, especially for commuters, often with the incentive of free or low cost car parking at the main suburban transit stops. However there is usually a need for a critical mass of population and high densities along routes for these systems to be worthwhile, so only bigger cities have light rail systems and is why Transit Orientated Developments (Chap. 2) have become popular planning policies in recent years. Although many of the early mass transit lines from the mid nineteenth century were private ventures, the high costs involved and the problem of acquiring land or getting permission to run lines underground now mean that most mass transit systems are in some kind of public ownership. Moreover, cities cannot afford these measures by themselves, so their construction is largely dependent on finance from higher levels of government, in which some justify the measures as being as necessary to urban functioning as the creation of sewer systems. In the last two decades major cities in the developing world are also adding mass transit systems, with China especially making it a priority. Although it was only started in 1993, by the end of 2013 the Shanghai Metro had over 500 km in 12 routes, with plans for another 22 lines covering 877 km by 2020, which would make it by far the biggest and most travelled system in the world. Thirty one other Chinese cities either have mass transit lines or are constructing them, part of a $ 150 billion investment by the national government. This policy is designed, along with high density apartment blocks at transit stops, to avoid the commuting congestion found in western cities, brought about by increasing car ownership among the new middle class and which is now being restricted.

Perhaps this type of Chinese progress can only be achieved by an increasingly wealthy country with a centralized decision-making structure. But it forms a great contrast to the situation in most western countries where relatively few extra Metro lines are being added, although cities such as Paris, Copenhagen, London, Madrid and Barcelona, represent exceptions, but nowhere near the scale that is really needed.

Figure 6.2 shows Copenhagen's mass transit system, which complements its extensive bus network. This consists of an underground Metro system completed in the past few years which connects major nodes in the city as well as to the airport at Kastrup. Another underground network, the Cityringen, is planned for completion by 2018 and will link more of the inner suburbs to the existing Metro system and the main suburban and rail nets. In addition, the figure summarizes the major policies adopted to increase the general level of sustainability to assist this progressive city in reaching a target of being a carbon neutral city by 2025 (WWE 2013). So the new mass transit systems are only part of a wider strategy to reduce carbon dioxide and other emissions, illustrating the way this city is using a number of policies to achieve its general aim. As Chap. 5 has shown, it is already regarded as the city with the most sustainable rating in Europe.

At the more mundane level of bus transport, many of the originally private companies that provided these services were absorbed into municipal or even state ownership in many developed countries after World War II. In some countries they have been returned to some form of private ownership. Although ownership may be

Copenhagen's Carbon Neutral Plan by 2025:
Major Components and Related Policies.

Energy Supply

Convert two main power stations from coal to wood chips/straw.

Add 100 large offshore wind turbines and geothermal plant.

Add 280,000 sq. m of solar panels (= 400 football fields area).

Obtain more biogas from organic waste.

Energy Demand

Upgrade energy efficiency of all new and renovated buildings to low energy levels (>20 kw/.m2/year), by insulated walls, windows, LED lights etc.

Reduce energy consumption in city buildings by 45%. Move to all-electric city vehicle fleet. Change street lighting to LED to cut energy in half.

Extend District Heating Schemes, already dominant in city..

Mobility

Add bike paths; create 300km of SuperCycle Highways to suburbs & local towns to improve commuting levels from 36% to 50% by cycles.

Encourage changes in engines: to c 35 % levels of biogas, from diesel, in trucks; electric cars to reach c 25% of total.

Add Cityringen (by 2018) to complement Metro system (2002-7) and older heavy rail S (suburban) system.

Persuasion

Campaigns to persuade 50,000 city workers to improve energy efficiency.

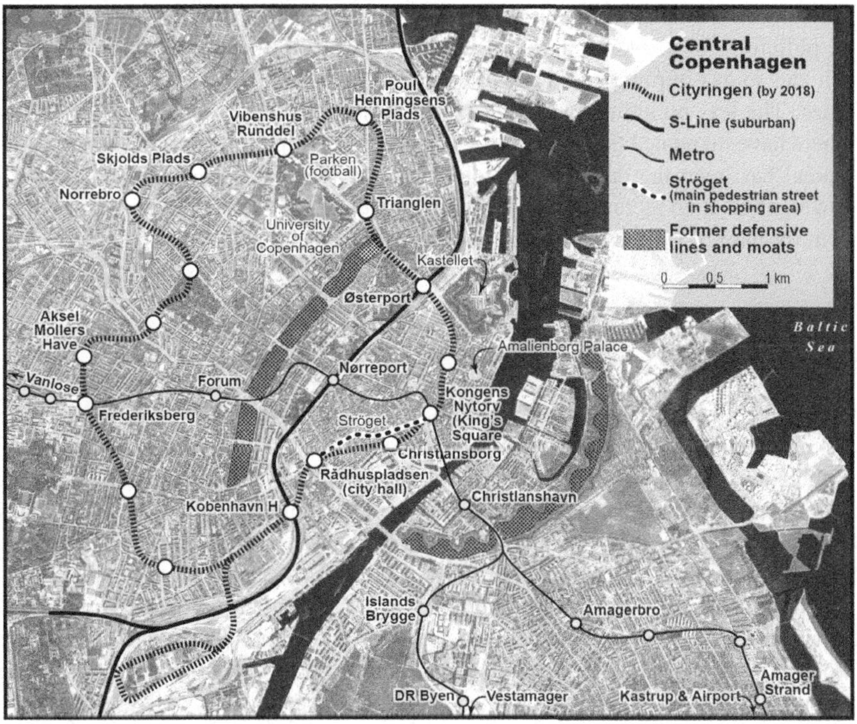

Fig. 6.2 Copenhagen's metro networks and carbon neutral plans. (Source: Compiled by author from various Copenhagen City reports)

mixed, and varies by country, even the private firms often have subsidies from the national or state governments. However the general trend in the last few decades has been one of *improving bus services* by increasing their efficiency and passenger comfort, such as: adding extensions to existing bus lines; providing better connections to the schedules of mass transit lines; putting electronic scheduling signs and better shelters at bus stops; adopting pre-paid smart cards for payment; establishing free car parks and secure bicycle racks at main transit stations in suburban areas; adding bus–only traffic lanes to speed their flow to the city centre or other destinations. In addition, many cities have implemented Park and Ride schemes, where large parking spaces are located on the edges of downtowns and free bus services are provided to the car passengers so that they can leave their car and return later, thereby reducing vehicle congestion in the city centres. Another important trend has been to add more safe and convenient walkway connections from transit stops to car parks and bus stops. In all these new or improved transport systems it is usual to find higher levels of government paying much of the cost of the new systems, and municipalities subsidizing at least part of the fare, either on each ticket or by providing reduced costs for weekly, monthly or annual travel.

Additional innovative approaches to improve people mobility and sustainability have also been adopted in some countries. In the former case, gondolas, as well as moving staircases have been used to access formerly isolated housing developments on the steep hillside shanty towns of some Latin American cities, such as Medellin and Bogota. Some cities have also tried to reduce their transport systems' reliance on polluting fuels. Increasing numbers are converting their buses and municipal vehicles to gas-burning engines, not oil-based fuel, to reduce vehicle emissions in cities. The less polluting natural gas is the usual alternative fuel source, although others have mandated the use of bio-gas generated within the city from waste. Less typical has been the deliberate buying of some renewable energy, such as wind power, often at an increased rate, to supply the electricity used by new rail systems, as in the case of Calgary's Light Rail Transit system.

6.4.3 Bicycles

At the other end of the people mobility scale from mass transit or buses is the encouragement of *bicycle use*. It is promoted because it is a healthier and environmentally friendly form of travel to work, extending its use beyond recreation. Cities with high bicycle use have usually created separate dedicated cycle lanes along many roads with a small paved barrier from car lanes, as can be seen throughout Copenhagen. This improves the safety of cyclists compared to simply painting lines on roads to designate bike lanes, which are often ignored by motorists. However the greater use of bicycles also involves creating bicycle parks in or near around major destinations such as train or bus stations, and the promotion of safe locking systems to avoid thefts, which is often reduced when people use relatively basic and inexpensive bicycles in their commutes since these are not worth stealing. It is also useful if employers provide showers and changing rooms for their commuting employees. In cities such as Copenhagen and Amsterdam over a third of people commute to work or school by bicycle, with the former city hoping to raise the amount to 50 % by 2015. Although this may seem to be an old and persisting tradition in these countries with flat terrain the number of cyclists had actually decreased drastically by the 1970s in Copenhagen as car use increased, leading to massive congestion, especially in the old city centre with narrow roads. Citizen action groups rejected the initial choice of new roads across the lakes and parks that ring the old city and are shown in Fig. 6.2. Together with the growing environmental movement they convinced city governments to create dedicated bicycle routes along major routes in the city from the early 1980s, aided by concern over the increased price and availability of vehicle fuel. In Copenhagen these lanes are constantly being extended, while an additional Green Route system of 22 segregated bicycle ways, amounting to over 110 km through parks and residential areas, was started in 2000 and was half way to achieving its goal by 2013. In addition the first of a so-called Cycle Superhighway, a 22 km segregated bike-only route to the city from Albertslund, was opened in 2013, the first of a planned series of 26 similar routes to the suburbs and surrounding towns (CPH 2012). Nearby Mälmo in Sweden has even more dedicated cycle lanes, a total of over 500 km, and all are two way lanes and separated from the car traffic by a divider. This has drastically reduced accidents, and bicycles now account for a quarter of all journeys in this city of 300 thousand people, with the city allocating $ 61 million for 2012–19 on various cycling initiatives to increase this proportion. By contrast only 10 % of vehicle trips in Stockholm are made by bicycles.

In many cites a system of low-cost or even free bicycle rentals with many drop-off and pick up points have been added, especially for tourist use. Although there are signs of progress in promoting this form of transport in the developed world it may be noted that this alternative opportunity is being lost in many cities in the developing world, where the adoption of the western-type modernizing approaches of the 1960s and 1970s has been favoured, leading to greater congestion and pollution. So the opportunity to re-plan cities on the more sustainable principles is being lost, especially in Beijing where a 'city of bicycles' in the 1990s has been replaced by congested car traffic, although the new rapid transit systems and restrictions on car use may help reduce the problem. Although there are examples of other bicycle-friendly solutions being developed in some cities—such as the Ciclo Ruta in Bogota, one of the C40 group (C40 2013)—the unruly nature of traffic behaviour in many developing countries, noted in the review of traffic accidents in the Healthy City chapter, does not hold much hope for a peaceful co-existence between vehicles and bicycles, unless more road safety measures are adopted and dedicated bicycle lanes are provided.

In addition to these main polices, a range of alternative measures have been adopted by cities to reduce car use, which are summarized in Table 6.2 in three categories other than the mainly direct intervention approaches described above, namely: land use measures; disincentives to use cars or incentives to use other transport; or persuasion by government. Many cities use combinations of these approaches to reduce the use of cars, which then decreases fuel consumption and the size of

Table 6.2 Taming the car in cities

Degree of intervention: Examples of major policies
A. Direct interventions
Expand, improve, subsidize high-quality bus systems, especially using bio-fuel or gas, with the addition or expansion of light rail and/or mass transit systems in bigger cities.
B. Controls or changes in land uses
Create pedestrian-only areas. Reduce numbers of parking spaces in city centres. Add park & ride schemes on inner city edge to access CBD to reduce shopping traffic. Designate bus-only lanes to speed their flow to city centre. Widen sidewalks to improve pedestrian traffic. Remove one-way roads in central areas to reduce speeds. Remove roadside parking in central areas. Add traffic calming measures and lower Speed Limits. Add more safe and convenient walkway connections from transit stops to car parks and bus stops. Create separate bicycle lanes—not just road marking—alongside roads or provide dedicated bicycle paths.
C. Disincentives/incentives
Restrict availability of vehicles. Add substantial taxes to car prices. Add road pricing, such as congestion charges in central cities, to discourage car use. Increase car and/or gasoline prices and taxes. Increase car parking fees. Create free or low cost bicycle rentals in central cities, with pick-up and 'drop-off at many locations. Encourage car sharing and allow such cars to use dedicated bus lanes or parking spaces. Subsidize transit fares.
D. Persuasion
Add educational programmes to promote non-car uses, e.g. advocate walking/bicycle use on health grounds.

emissions from cities. But despite all these positive examples there can be little doubt that most cities have rarely been able to reduce their vehicle emission levels.

6.5 Greater Sustainability in Area Developments

Greater sustainability can also be created by focusing upon the sustainable development of areas, not only on individual sectors, whether buildings or transport. In the developed world at least, practically all settlements have urban growth policies organized within the framework of municipal regulations or some type of area plan. The regulations, the way they are applied, and the extent of national, regional or municipal involvement, vary considerably between countries. Most deal with features such as density levels, permissible heights, usage, spacing, city infra-structure requirements and various other standards, such as green space, playground allocation etc. In European countries the compactness of old cities that were originally developed within city walls, plus a traditional preference for urban living and tolerance of smaller amounts of dwelling space, higher densities, as well as the desire to protect limited agricultural and scenic areas, led to a strong conservation element, with increasingly strict regulations over development from the mid-twentieth century. So most European settlements have been far more contained, with less sprawl onto new greenfield sites, compared to the situation in North America. This means that density levels in most European cities have remained high, as seen by the fact that Amsterdam and Stockholm have 49 and 53 persons per ha of built-up area respectively, compared to sprawling American cities, such as Phoenix at 10 persons/ha. Nevertheless, the growing environment movement and the recognition of the need to reduce the massive inputs of water and energy in particular, the build-up of wastes, congestion from car use, plus pollution from noxious emissions from cars, especially carbon dioxide and nitrous oxide and various particulates, has led to a revision of some older policy ideas and the generation of a series of new policy initiatives. These seek to increase not only the degree of sustainability of urban places, but also the attractiveness in human terms, by improving the liveability of these renovated areas of urban places and new developments.

Perhaps the most obvious trends in the development of more sustainable urban practices in both newly developed or urban renewal areas of cities in the developed world are the emphases on a dozen of the most important sustainable development principles, some of which are the familiar New Urbanism standards (Chap. 2) but applied to renewed or renovated urban areas. These involve: increasing the density and compactness of new urban developments; preferring brown-field over green-field sites for future growth; ensuring the connectivity of new developments to older areas; creating urban growth boundaries to reduce sprawl; providing greater land use mixes in place of single function areas; conserving and re-using heritage buildings; increasing transport sustainability by reducing car use; providing mobility alternatives, from new mass transit routes to separate bicycle paths; adding more green space, ecological variety and sensitivity, with more permeable surfaces to allow drainage; including carbon neutral energy generation and district-wide heating schemes; use the best conservation and waste recycling practices; create more pedestrian-friendly spaces to make places more liveable and healthier; reducing the risk from local natural hazards, by not building in vulnerable areas, such as flood plains, part of the adaptation of the settlement to local physical conditions.

In all cases the objectives are to reduce energy and other resource consumption, to re-use land, reduce pollution and the other negative externalities of resource use, and adding ecological practices. What is particularly important is the need to reduce car use in cities, using the policies described in Table 6.2. All these principles are designed to make places more liveable and healthier as well as sustainable.

6.5.1 More Effective Area Heating Solutions

One of the most important areal developments that can provide greater sustainability in energy uses comes from the creation of District Energy Schemes. Traditionally the heating and cooling of buildings has adopted an individual building approach, either through fossil fuel heating systems for the building itself, typically based on coal, peat or oil, or through connections to some grid, delivering gas or electricity, which then either needs a power furnace to burn the fuel to convert the energy into hot water or steam, or to directly power the heating and other appliances. These individual systems are very inefficient, costly, and wasteful of space. In addition, they have high installation and maintenance costs, while there are safety problems associated with features such as open flames, gas leaks and high temperatures in houses. Instead of this within-building heat provision, District Energy schemes generate heat in some District Centre and provide this via hot water or steam through insulated pipes to the end consumers, or use it to provide cooling in summer. The system has many advantages. The district heat generating centre can be based on many fuel sources and it is scalable, so can be used for small clusters of houses or large areas in cities. It reduces the problem of intermittency in renewable energy sources, because excess energy generated by the sun or wind can be converted into hot water and stored geo-thermally, to be used when demand increases. In addition, the power is generated locally so the threat of power failures from complex long distance grids is reduced. Most appealingly the central generating unit is often based on capturing waste energy from existing power generating systems, although it can also use the waste energy from large production plants, in what is often called a co-generation process. Currently most electricity power plants are only a third efficient, meaning that up to two-thirds of energy from fossil fuel sources is lost, either in the electricity production process, or through the steam or water cooling systems, which are major sources of carbon dioxide and other emissions. The heat energy that would otherwise be wasted is diverted to district heating systems which can then make these power plants almost 90% efficient in converting the calorific fossil fuel input into energy, essentially creating a re-cycling or rather energy-recovery approach. Although the district heating schemes have been available for decades, technological changes have increased their effectiveness. The industry now refers to the latest advances as Fourth Generation District Energy systems. They have several key features. They allow the use of other sources of energy, such as from municipal waste incineration or renewable energy. They typically have substations and heat exchangers to improve heat transfers, as well as smart grids that allow flexible inputs from various energy sources and two way linkages in the heating centre, to take in excess power from solar panels etc. Also the use of pre-insulated supply and return pipes allow the use of lower temperature levels of hot water (EP 2013). The EU countries, especially in Scandinavia, have the most experience with this approach, with 270 municipalities in Sweden using the new system. These are estimated to have reduced carbon dioxide emissions by 60%, and lowered dependence on fossil fuels in these plants from 80 to 3% since they have been installed. In Denmark, Copenhagen now has 90% of its buildings powered by these schemes, accounting for 60% of its heat supply needs, generating 50% less carbon dioxide emissions than the old boiler system, saving an estimated 660,000 t of this greenhouse gas annually. The plans of Munich's municipal utility company (Stadtwerke München 2013) makes it another progressive city in this context, for it is well on the way to achieving its goal of being the first city of over a million people to obtain all its energy needs, of 7.5 billion kwh, from renewable sources by 2025 (SM 2013). Although wind and solar energy is being increased, the city gets 4 billion kWh of waste energy from its power stations in its district heating schemes, saving the equivalent of 450 million L of heating oil which would have generated 1.1 million t of carbon dioxide—the same amount generated by all its vehicles in a year. Also Munich intends to be the first German city to have district heating systems wholly dependent on renewable sources by 2040, which will be helped by tapping into the Bavarian Molasse, a large reservoir 3000 ft underground that has water temperatures above 80 C.

Since many European countries, such as the U.K, and Ireland, have only a handful of such district schemes, it means that only about 9 % of the total European Union's urban heating needs is currently obtained in this way. The technology is proven, so it would be relatively easy and financial feasible to double this figure in the rich countries, for it has been estimated that the costs of the pipes and other installation pay for themselves within 3 or 4 years, with cheaper heating bills for householders and huge savings on imported fuels. As always, the development of this system depends on political will. One of the reasons why the system has proved so popular in Denmark in particular is because of early legislation to support such schemes, as well as its co-operative tradition. For example, a 1979 act made municipalities responsible for urban heating supply planning, and which enabled groups of consumers, as well as municipalities, to establish co-operatives to establish district energy companies, which were required to return any profit to consumers, and to search for, and to implement energy savings.

6.5.2 Smart Systems

The development of so-called 'smart systems' technologies are also proving useful in increasing sustainability by helping address several problems of existing practices. In the case of water supply, meters are increasingly being installed in houses to make residents aware of their water use, which frequently leads to reduced use, complementing the low flush toilets and other devices described in a previous chapter. In the case of electricity a key objective is to reduce demand fluctuations in a day, for at peak demand times more power has to be generated, which means higher prices, while the back-up generators for these peak loads are idle for most of the rest of the day. New 'smart metres' which can be installed in houses show the costs and usage of electricity at any time. They allow time-differential use of energy, so that consumers are able to reduce demand at peak time when prices are high, saving money and lessening the load on the supply system. A type of smart-grid system is being tested in Napierville (Chicago) where almost 60 thousand smart meters have been installed in buildings in recent years, which are designed to improve reliability and consumer's control over energy use, which is reducing demand at peak times, thereby lowering the costs of generation (NP). The installation of automatic heating regulation systems in cold or winter climates, or air conditioning in hot climates, is another 'smart' approach. These devices reduce heating or cooling levels when the house is empty—which is most of the day with families with two or more workers and school children—and turn on power some time before people return from work. Increasingly these systems are being adapted so that they can be activated by commands from mobile phones, which may also turn on other systems such as cookers. Among the additional smart grid developments are power lines that allow the two-way flow of electricity, rather than the usual uni-directional flow system of power flowing from generating stations to end users. With the adoption of two-way systems the addition of solar panels and other renewable energy devices in houses can send the excess power back to the system or to storage systems. Denmark is developing smart grids as a national project for the whole country, for they see their utility in solving some of the problems of intermittency and storage associated with solar and wind generating systems. Since 2000 many of these energy saving methods have been assisted by grants from governments to households improving energy efficiency in their homes. There is little doubt that new electronic devices with remote commands will increase the ability to monitor household appliances from a distance. In addition new monitoring systems of traffic flow systems are being tested which can automatically assess the level of congestion on roads and then adjust traffic lights. Researchers are developing many different types of these so-called smart systems which may help transform some of the functioning of cities in the next 20 years and reduce fuel consumption, which many believe will create smarter cities. But there is another way in which cities can become smarter. This is by the more effective use of the huge and often complex amounts of data that cities and other public organizations gather, especially if it contains locational co-ordinates to enable maps to be constructed. By making this so-called Big Data available to all departments and in some cases to citizens, better identification of developing problems and more efficient solutions to many problems can be provided, from monitoring utility lines to transport systems. The increased knowledge may improve the governance of cities and increase citizen participation, although there does exist the problem of who owns and can use data if it comes from multiple sources that include private companies. Analyses of the data allow trends to be spotted, often through corroboration from different sources, that would be difficult to detect using small, separate compilations. Hilbert and Lopez (2011) have calculated that the world's per capita capacity to store information has doubled every 40 months since the 1980s, which shows the rapidity of change. Many departments in cities can benefit from these larger and more accessible data sets, given new computing capabilities, although they almost certainly need to reorganize their data gathering and management systems. The discussion on Safe Cities (Chap. 12) will show that police forces in particular are using rapid compilations of crime figures and their locations to produce better analyses and quicker responses to crime incidents.

6.5.3 More Sustainable New Urban Areas

Some of the best examples of sustainability can be found in the late-twentieth century area developments in Northern and Western European cities, especially in the Netherlands. One is Leidsche Rijn, a growth area adjacent to Utrecht which has been designed to accommodate 30,000 houses and the same number of jobs, with a population reaching 80,000 by 2025, one of Holland's largest new developments. Instead of the typical pattern of either allowing sprawl on the city's edge, or restricting development around the city which would end up by creating spill-over growth in small towns around the city, the plan will concentrate development in this growth area, It will account for 75 % of the future regional growth of Utrecht at a 37 units/ha density level within a defined urban area and provides an example of many smart growth principles. The project also rates high on creating connectivity with the existing city of Utrecht, by ensuring that the new growth lies within 5 km of the centre of the old city which will remain as the major shopping and social heart for the whole new urban area. A potential linkage problem caused by the separation of the new area from Utrecht by a major highway and the Amsterdam-Rhine canal has been solved by the construction of three new bridges, one only for bicycles, another for public transit and bicycles, and a third for all modes of transport, as well as an ambitious plan to roof over the highway for a third of a kilometre. The new area includes both private and public sector housing, with a variety of housing types and apartments, together with pedestrian-friendly public spaces, while local shopping and industrial zones, and room for in-home offices, provide a mixed use development that will reduce commuting. The development rates highly on the sustainable use of energy and water, for buildings will be served by a district heating scheme, provided mainly by previously discarded heat from a local electricity generating plant. In terms of water use, all homes are provided with two water sources, one for clean water, another of grey-water (less clean) for toilets and non-drinking purposes, obtained from a separate major water line running through the area. Drainage channels (wadies) have been built throughout the area to collect storm water to prevent flooding, while both greening and ecological variety principles are provided by the addition of a large central park of 300 ha, which also contains allotments and sports fields, a development that connects to what is regarded as the Green Heart of Holland's urbanized region, the Randstadt. Most of the high density developments and workplaces in Leidsche Rijn are clustered near the bus and train stations to allow ease of access, while the use of bicycles is encouraged by extensive networks of bike-only lanes, to emulate the situation found in Utrecht where a third of the trips are made by bicycle. This should provide a high degree of mobility alternatives to cars. However, unlike many other new developments that aspire to sustainability, this new area does not seek to drastically reduce car use, for it still has considerable provision for cars, with 1.2 parking spaces per housing unit in the area. The fact that it is essentially a new area means there are few heritage buildings.

A second area example has more explicitly low carbon principles for its growth, and was one of Europe's first examples of this type of development, a trend summarized by Foletta and Field (2011) using 8 European cases. The GWL-Terrein project in Westerpark, Amsterdam, is one of these car-minimizing and carbon neutral developments. It is located on a brownfield site located on the edge of the nineteenth century city that since 1851 had been used as a depot and reservoir for a municipal water supply company and was the terminus of an existing tram line (Fig. 6.3). The new development area is designed for 1400 people at 230 persons/ha and is only 3 km northwest of the city centre. Development of the 600 dwelling unit complex, with social and market-priced apartment blocks and houses, began in 1995, although planned several years earlier (GWL 2013). It included the conversion of some of the heritage buildings, such as old water company's offices and warehouses, into housing, as well as keeping the water tower as a symbol of its

Fig. 6.3 GWL-Terrein Project, Westerpark, Amsterdam. (Source: From local community association, koepelvereniging Gwl-terrein)

heritage. Since the area is close to a major shopping street, schools and other social facilities, most daily needs can be satisfied locally, reducing the need for local shops within the complex. Residents have to sign declarations that they understand it is a car-minimal area and will avoid negative effects on surrounding areas by not parking on adjoining streets. However, they do have access to a 135 unit parking area, a car-sharing services and extensive storage areas for bicycles. The design included many ecological features by adding green space, allotments as well as private gardens, and includes many trees—but for safety reasons has few low bushes—while most of the roads and paths have a filtered permeable surface, with only one hard surface road into the complex to allow for fire and emergency vehicles. The project has a major interior green space, while green roofs have been developed on the larger buildings. High levels of sustainability characterizes both energy and water use. A central co-generation plant provides district heating. Rainwater is allowed to drain off naturally; underground water cisterns collect water for use in toilets; water use is reduced via low flush toilets, and kitchens have separation bins for different types of waste. It has been estimated that the carbon dioxide emissions from the development are two-thirds lower than the city average. There is little doubt that the project appealed to those who have adopted a more sustainable life-style. Like many other 'low car (bon) communities' in Europe (Foletta and Field 2011), this development was preceded by an organized grass-root support for the concept from the onset; residents were consulted at all stages of the project's evolution and often modified the original plans. To reinforce the sustainability concept, new residents are also given booklets about sustainable house-keeping to provide guides for a more sustainable life, while a fulltime 'housemaster' or custodian helps residents take care of any problems and offers advice on adoption of more sustainable practices and life-styles.

A third example of more sustainable practices in a new area development comes from North America, one which has focused on the use of renewable energy in a housing complex. This is a solar development in part of a residential community in a rapidly growing satellite town of 25,000 people in 2011 located 20 km miles from Calgary. The 52 housing unit development completed in 2007 is part of the community of Drake's Landing, Okotoks. It represented a breakthrough in the application of new solar technologies by claiming to be the first large development in the world to fulfil 90 % of the space-heating needs of its houses from solar energy (DLC 2012). As such it claims to be more efficient than the earlier and pioneering solar settlement of Freiberg's Solarsiedlung (Solar Settlement), whose objective is to reduce energy needs by trapping heat and soaking up sunlight (SSF 2012). Figure 6.4 shows the plan and operation of this development. The estate was completed with

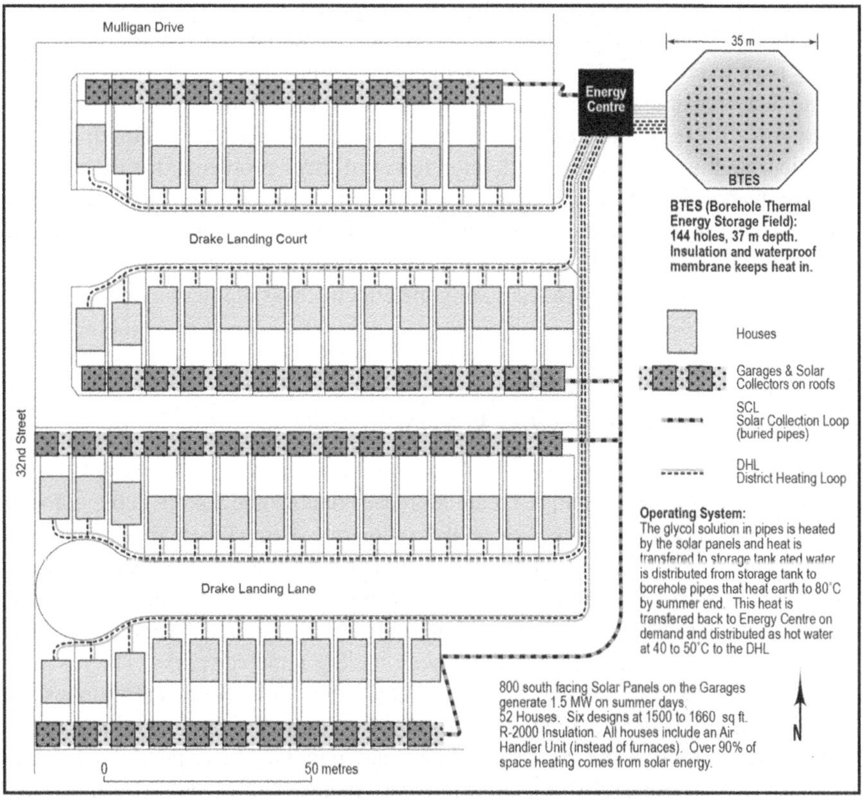

Fig. 6.4 Solar Community in Drakc's Landing, Okotoks, Alberta. (Sources: Revised from http://www.disc.ca; http://www.unitedcommunities.com)

R2000 insulation standards for all houses, with solar collectors on each south-facing roof of the line of connected garages which send power to the district heating system. This not only heats water, which is circulated back to the houses through insulated underground pipes at 40–50 C depending on the outside temperature, but also pumps the hot water underground for storage, releasing it in the winter for use in the system when the solar input is lower. Despite its real energy savings the scheme has not been copied by other developers in the area.

This pioneering example shows the feasibility of such projects, and the major reduction in heating costs, but far more houses in the town of Okotoks are still being built to conventional standards in large part because they are still cheaper per unit of space and are usually far larger than these solar powered units. It may also be a function of the reluctance in North America for developers to create, and owners to buy, houses based on some communal heating scheme, whereas, as the review of district heating schemes showed, this is not the case in many parts of Europe. This means that solar power, even if used, is still largely restricted to panels added to individual houses in most countries. But the problem of storing power at night, requiring large batteries to store energy, or getting enough power on dark winter days, still causes problems for the application of solar power to houses in many countries. Even though smart grid solutions are becoming available, when some renewable power is generated locally it really supplements regular gas or electricity supplies, unless there is some district heating scheme, or where wind power has become a major part of a region or country's energy supplies. By contrast geothermal communities, such as those in Iceland, have fewer of these problems, for the required heat for the houses is available in its unique subterranean sources. Other geothermal projects exist in several countries and the EU has three pilot projects operating in eastern European countries, part of the wider Strategic Energy Technology Information System (SETIS) that is investigating the effectiveness of alternative low carbon technologies.

6.5.4 Urban Revitalization Schemes

Another example of the trend to create more sustainable and also liveable communities can be seen in the renewal or revitalization of existing urban areas, especially in central or inner city areas. Many of the urban renewal schemes of the 1960s and 70s obtained a bad reputation for destroying both the older heritage buildings in an area and the mixed land uses in and around central city areas. Many focused on high value projects, large office towers with underground car parks. The resultant sterile, modernist concrete and steel buildings, often with windy, empty plazas at their foot, had little local distinctiveness and were empty at night. In addition, the attempt to accommodate the increasing use of cars in downtowns ruined these spaces for many people because of the pollution from their exhausts, their speed—which caused danger to pedestrians, and the requirement for large areas of surface parking, or new parking structures. In addition there were negative effects from the way that municipal authorities tried to help traffic flows by widening streets, reducing pavements, adding traffic barriers, and often a myriad of traffic lights. Since cars were privately owned and usually occupied by one person, the effect has been to priorize private cars in what had been public spaces, while the focus on high value buildings turned many downtown areas into single-use areas. Not surprisingly, a reaction set in against such developments, with attempts to apply many of the same ten principles of more sustainable practices outlined above to create new standards of liveability, especially in central city areas. Among the many different types of change has been the attempt to reclaim the major shopping streets and public places from their devastation by automobile use, with the objective of making these places more sustainable and more people-friendly.

By the late 1970s many people in western cities regretted the type of dehumanizing urban renewals that had occurred and most municipalities created at least one pedestrianized street in the central retailing area. The success of these developments and recognition of the problems of congestion caused by the car also led to the widespread use of building set-backs for new high-rise towers to create more open space at a street level. Despite the initial opposition of retailers, who felt they would lose trade if customers were unable to park outside their doors, these streets have been successful in most cities, although the attraction of homeless people and beggars to some of these areas has required careful and sensitive policing to prevent annoyances and crime. The utility of these revitalizing developments has been shown by planner-architects such as Jan Gehl in Denmark, whose ideas are summarized in his book *Cities are for People* (Gehl 2010). His emphasis on 'people, not cars or large scale projects' developed from his careful study of streets in Copenhagen in the early 1960s, observing behaviour in these areas and counting pedestrian flows. Gehl argued that safe and well-designed public spaces will attract people to sit, stroll, meet, shop, or simply to be seen and to be entertained, essentially reviving some of the functions of the old agora in Greek cities, although these were also places of religion and politics. Also he argued for the greater use of what he called 'green mobility', where most movements are by public transport, biking and walking, using the policies discussed in the previous Sect. 6.4 on transport. Not only do the latter two mobility options provide exercise and improve health, but the greater use of streets by people improves their safety, especially if transport stops are linked to well-developed and attractive public spaces. Gehl's detailed studies showed the value of pedestrianizing the main axis of narrow shopping streets in Copenhagen that ran through the old medieval city. The zone, already shown on Fig. 6.2, is now known as the Strøget and was first turned into a pedestrian zone as a temporary measure for the Christmas period in 1962. The initial success led to the pedestrian only being made permanent and its 1200 m length contains many of the most important commercial premises in the city as well as including several squares. It now runs from the nineteenth century Radhausplat (City Hall place), near the main railway station, to the Kongens Nytorv (King's New Square) on the western edge of the city's eighteenth century Renaissance developments, near the famous Nyhaven, originally a new seventeenth century quay for ships but now dominated by restaurants and related businesses. Recent surveys have shown that some 80,000 people a day regularly use the Strøget in the height of the summer, with half that number in winter. The success of this zone has led the city to pedestrianize many other streets in the old core, many of which link to the Strøget, although Gehl maintained that this type of policy is usually best done gradually, allowing people to get used to the new streets and for new functions to emerge in them. Gehl and his associates have been consultants for many cities throughout the world and adoption of these ideas in many cities have helped municipalities turn many dysfunctional, polluted, car-dominated central city spaces into vibrant, people places.

Of course, pedestrianization alone is never the only solution to improving liveability and reducing traffic congestion in various types of renewal schemes, or even in the downtown area in general. A common trend is to remove or reduce the one-way streets that were designed to speed up traffic flows and return to a two way traffic system which is likely to be slower. The concept of a 'living or shared' street has also become popular in many renewal schemes, especially in the Netherlands where the Dutch term 'woonerf' (literally a resident yard) is often used for the concept. The involves the reorganization of a key shopping street to make them more pedestrian-friendly by widening the pavements, adding benches for people to rest, as well as kiosks and street cafes, and planting trees for shade, variety and colour. Other features consist of some of the policies discussed in Sect. 6.4, such as removing parking spaces, reducing the speed of traffic by speed bumps or speed limits, narrowing and curving the traffic lanes, adding separate bike lanes and lots of crossing places for pedestrians. There has also been a revival of making city squares or other people-places more attractive, by adding striking public art, fountains, unique tiling stone or brick work—incidentally, the type of street furnishings and renovation so typical of Renaissance city squares—an approach which has been particularly successful in many Spanish towns and cities. These changes are also designed to provide uniqueness to many renovated squares or streets, which subsequently attract far more people. Many municipalities also allow local restaurants or bars to spill-over into adjacent city squares or open spaces, creating lively, people-places, especially at night, although areas with heavy rainfall, such as La Coruña in Galicia, have added glass roofs or sidings to protect their customers. The licensing of street musicians, or various artistic acts, to perform in these squares also makes these places more vibrant, attracting more people and adding pleasure. Increasing interest in the heritage of cities has also led to the renovation of old historic buildings, often changing them into new uses, such as the way many old warehouses on the edge of downtowns have been turned into offices, bars, restaurants and even theatres. This heritage approach provides more linkages to the past, creating more uniqueness.

These renovations have also helped the growth of greater night-time activities, creating a new vitality that has invigorated many central city areas, especially on the edge of the main shopping and office areas. In the bigger downtown areas attempts have also been made to encourage the creation of specialist areas to focus on certain activities, creating arts or theatre areas, or night-time entertainment zones, even areas with a concentration of other businesses, such as designers. Yet many shopping areas in the central city can still be empty at night outside the bar and restaurant areas. So many towns have attempted to bring back more residences into the old central city by encouraging higher densities or more high rise developments, whether apartments or condominiums, in and around the central areas, which is likely to add to the demand for local stores and greater night-time activity. Another trend, usually seen as emanating from New York, has seen the conversion of existing warehouse buildings into residences. Many are traditionally found near railways on the edge of main shopping streets, buildings that stored goods brought in by rail, before road transport became dominant by delivering goods, especially bulky ones, from edge-of-city locations. Elsewhere, the addition or re-establishment of residences above commercial premises to improve the vitality of shopping streets, to increase housing stocks and to re-use vacant space above stores, has been encouraged by national grants for renovation, such as the U.K. scheme called L.O.T.S. (Living Over The Shop) that started in 1989, or the similar one in the larger Irish cities between 2001–2004. Both the projects were short-lived and usually restricted to areas that were undergoing regeneration. Such schemes usually involve governments subsidizing parts of the costs of renovation and conversion of older buildings into residences, in order to encourage private development.

From the 1970s the increase in vandalism and break-ins led shopkeepers in many British town centres to use steel shutters to cover their windows at night, giving downtown areas an abandoned look and destroying their attractiveness in the evenings, which in turn reduced pedestrian flows. This has led some cities to create local bye-laws to prevent such shutter developments, while many have added barriers on the edge of pedestrianized shopping streets to prevent car access and ram-raiders. Another trend has been to install CCTV cameras in many parts of the downtown areas to deter crime or to make detection easier, while a greater police presence in the area has also aided security.

Additional efforts have been made to re-invigorate formerly unutilized spaces in city centres, such as the back lanes or alleys that lie behind the main streets and acted primarily as service lanes in the past, although many had been abandoned in recent years. Melbourne in Australia provides one of the best examples of the trend to revitalize these areas which cut between the large east-west blocks of the central area (MLaA 2012). Surveys of the Melbourne laneways in the early 1990s showed that only 8 % were actively used spaces, other than acting as delivery areas for area's offices and stores. So the city created new regulations and added incentives to encourage not only the upgrading of the often dilapidated surfaces of the lanes but the development of new or refurbished buildings along them, also allowing setbacks for large buildings to create more space. These changes dramatically opened up the former decayed and unused spaces to commercial activity, attracting small bars, clubs, cafes, restaurants and specialist shops, such as boutiques and art galleries, activities which now cover over ninety per cent of the old lanes. Over thirty of these lanes now serve the surrounding offices at lunchtime and for after-work activities, and have become a centre of attraction for night-time entertainment and socialization, making the city centre a lively area at night instead of its old deserted night time character. Other cities, such as Cardiff, are known for their covered arcades that developed during its period of Edwardian age prosperity. They cut between blocks and run from one street to another, providing extra commercial space that often attracts the smaller and local retailers that cannot afford the high rents on the main streets. However, since these are private spaces many are closed at night, reducing their effectiveness as people attractors.

Many attempts have been also made to revitalize the major focal points in and around the central city, especially the public squares that are basic elements in most historic European urban places, but were often forgotten as an essential element in city life when most North American cities were built. The renewal has been achieved by cleaning and refurbishing the squares and buildings, providing more seating areas, and making them multifunctional by day or by season—such as with local markets in the mornings, perhaps allowing parts to provide additional parking in afternoons, and by using them as concert areas at night, or as festival centres at particular times of the year. The regeneration of these areas, like pedestrianized streets, need to be complemented by efficient daily cleaning systems to remove the litter generated from both the high flows of people, and rubbish from the many commercial premises, as well as by improved security through appropriate policing. Few of these pedestrian streets can operate without some vehicular traffic. Goods deliveries to the stores have to be made, usually by providing access to trucks only in the early morning or evening. The use of these areas is also improved by having regular bus services on their edges, while underground rapid transit routes often run under them, with stations under major squares.

The problem of accommodating parking is a constant problem for central and inner city areas. In some renovated cities residential areas near city centres still have their own car spaces, but they are organized as separate traffic pods, having access from major roads outside the central area, but without through routes that would enable traffic to cut through the housing area. In many Spanish towns and cities, car parks have been developed under many historic city squares, or under promenades in coastal locations in Galicia, which preserves the ground level areas for pedestrian use. Some, like Copenhagen, instituted a system that gradually reduced the number of spaces annually to encourage greater use of green mobility solutions such as bicycles. Most others have created new park-and-ride parking areas outside the city centre, with regular and free bus services to move people into and out of the centre. Although parking structures have also been added on the edge of commercial areas, or in underground locations, many of those developed have a poor reputation, especially in British cities. They frequently have narrow car stall spaces for cars, are low ceilinged, ill-lit, and contain depressing and dark concrete interiors and minimal security. They have created a threatening, if not a crime-prone environment that inhibits use, especially by older people, and at night.

6.6 Eco-Towns

Western urban history is littered with examples of new towns being built according to new principles of the times, from Greek colonies and Roman towns to medieval bastides, through the model cities of the Renaissance and Industrial Revolution, to Howard's Garden Cities and British post-war New Towns. But despite the increasing application of many new sustainable principles being used in existing settlements, or in parts within them—as seen in the so-called 'low car (bon) communities'

(Foletta and Field 2011)—the examples are far outweighed by the continued development of traditional car-orientated suburban designs. Moreover, there are still very few examples of complete settlements being built on the sustainable ideas described above that could be classified as Eco-towns or Eco-Cities. Dongtan, outside Shanghai, promised to be one of the most pioneering eco-cities to reach 500,000 people when its plans were unveiled in 2007. However its progress stalled by 2014 because of questions of funding and corruption scandals, so it may be yet another example of a plan that was never completed (DAC 2012). Similarly a proposal for eight British eco-towns were scaled back to six in 2009, given the opposition from surrounding areas and limited government financial commitment, but the plans were eventually abandoned with a change to a conservative dominated national government. However there is one major example of a settlement that claims to being built on eco-city principles. This is the Masdar City development in Abu Dhabi, which is proceeding apace since its construction started in 2008 (MC 2013).

When completed in 2025 at a current estimated cost of just under \$ 20 billion, Masdar is projected to have a resident population of 50,000 on a 6 km^2 site, together with 1500 businesses that will probably also attract as many commuters to these organizations as its resident population. The development is south east of the centre of the main city of Abu Dhabi and is adjacent its international airport. Its name, which means the ‘source’ in Arabic, was chosen to symbolize the way the design draws upon the traditional building techniques of historic cities in the arid region, some of which were designed to mitigate the effects of the extreme heat of summer and to utilize the cooling on-shore breezes. Designed by the British firm of Foster and Partners, its technology depends entirely upon renewable energies, especially solar-generated power, and aims for an ecology without waste, thereby creating what is claimed to be the world’s first carbon-neutral, waste-free, car-less settlement. Since cars will be banned in the settlement, transport needs will be served by mass transit and personal transit vehicles. It is also designed to be a low-rise, yet high density area, with no buildings over five storeys, and with a large number of public spaces for people to mingle and enjoy the outdoors. It will also have mixed uses, so residences, workplaces and leisure facilities are in close proximity in order to encourage pedestrian travel and social interaction. Another important feature is that the streets will be narrow to ensure shade, with many patterned screens to filter breezes and sunlight, following traditional practices in the Arab and Persian world. This desire to minimize solar heat build-up and increase cooling possibilities can also be seen by the plans for streets, walls and buildings which are also orientated away from the sun and designed to catch the prevailing night breezes. Some buildings, such as the Zayed National Museum, will have elaborate funnels to bring the winds into the galleys that are buried within its palm-treed mound, again following traditional practices of adding wind towers to bring cooling breezes into buildings in the region. Buildings and roofs tops have been deliberately built to be irregular so as create turbulence, while the town will be walled, not for defensive purposes, but to reduce the penetration of the hot desert airs and storms into the settlement.

One of the first buildings constructed was a solar power plant, to be expanded to 60 MW, which will be supplemented by solar panels added throughout the settlement that will deliver twice this amount of energy. In addition, local wind farms around the city are planned to produce another 20 MW. A future geothermal development is also planned. Water needs, so crucial in a desert environment, will be provided by a desalinization plant. However conservation measures mean the town will need 60% less water than equivalent settlements in the area, with 80% of the water being re-cycled at least once and with the grey-water outflows being used in local irrigation schemes. All the wood used in exteriors will come from a hardwood developed from old palm trees; biological waste will be processed to act as a fertiliser or for biogas to fuel transit and local domestic needs; and all plastics and metals will be recycled. Economically, the town is to be based on research organizations and clean energy enterprises, including the headquarters of the International Renewable Energy Agency (IRENA), essentially a knowledge-based venture. In addition the settlement will be an economic free zone, allowing businesses to have 100% foreign ownership.

The huge advantage of this development is the financial support of the extremely rich Abu Dhabi state, whose wealth comes from its massive local oil and gas resources. Perhaps it is ironic that it is the wealth from the polluting sources of fossil fuels that is financing the development of a settlement will have an economy and energy supplies based on renewable sources. So the real 'source' of its growth comes from the wealth derived from the world's mid and late twentieth century fossil economy boom and new interest in green technologies and policies. This reverses the idea of many of its principles being derived from the older Arabic and Persian urban planning practices designed to overcome the local climate problems of heat and aridity, although supplemented by contemporary technology. The originally of the project has led the World WildLife Fund to designate the town as an example of a 'One Planet Living Community'. However, it must be emphasized that Masdar is not really an independent town; it is embedded in the urbanized region of Abu Dhabi and close to its airport. Also, in social terms there is a danger is that the development will become an enclave of the technical and wealthy, not a cross-section of the population of the region; the service workers will probably live in the cheaper, adjacent traditional urban areas. Moreover, despite the promise of this design, few countries seem able to match the funds available for such a project in gas-rich Abu Dhabi. Without a major transformation in government thinking and finance in other countries of the world, such complete eco-town examples will remain rare exceptions in national urban systems. Yet it is to be hoped that more and more of the sustainable principles being implemented in Masdar will be incorporated in additions to existing settlements and retrofitted into older areas throughout the world. The technology is known; the will and the means are lacking

If these issues are given more attention greater progress towards a more sustainable future in our urban settlements may be achieved. Yet a note of caution must be added. Given the scale of the inputs demanded by cities and the waste and pollution they generate, it is difficult to conceive of urban settlements being fully sustainable, with covered domes producing all the crops needed, although the small Eden project in Cornwall, with its two covered domes of 5.5 acres, one housing a tropical area and another a Mediterranean plant habitat, show its feasibility, albeit at major expense (EDP). However, there is little doubt that if the examples of sustainable practices already shown to be effective and efficient in many cities were copied by others, then massive changes in the extent of sustainability would occur, even without building large numbers of new settlements like Masdar. Such projects are probably best seen as model entities, demonstrating what can be done, like the Garden Cities of the past. Hopefully they encourage imitators to adopt some of the many sustainable practices being pioneered, either in new settlements or in revitalizing old ones, which will lead to a better future, through the redevelopment of exiting settlements rather than hoping for the wholesale addition of new eco-towns

6.7 Conclusions

This review has shown the range of new initiatives that have been implemented in urban settlements in the search for a more sustainable future for towns and cities, especially in the past two decades. However, there have been great variations in their application and co-ordination, let alone their efficiency and effectiveness. Most cities, even in the developed world, have only adopted a few of the possible initiatives, which is a depressing situation, given that many new proven technologies can be applied. The projected increase in urban places provides an unprecedented opportunity to restructure the way settlements are built to increase sustainability, especially mitigation of climate warming, a point made by one of the committees of the IPCC 2014:

> The next two decades present a window of opportunity for mitigation in urban areas, as a large portion of the world's urban areas will be developed during this period.... Mitigation options in urban areas vary by urbanization trajectories and are expected to be most effective when policy instruments are bundled... Effective mitigation strategies involve packages of mutually reinforcing policies, including co-locating high residential with high employment densities, achieving high diversity and integration of land uses, increasing accessibility and investing in public transport and other demand management measures. (IPCC Working Group III, April 2014, p. 26)

The initiatives described in this review represent a promising, if slow beginning in many places. Yet most experts believe the implementation of these policies need to be need to be significantly ramped up in all cities if real success is to be achieved, especially to solve the problems associated with an increasing greenhouse gas proportion and the dangers of future resource depletion in oil and other finite resources. These are not problems that can be ignored because they represent some far-off times. Many will have impacts within the lifetime of people already born. So there is a pressing need for action. Unfortunately, too many government actions are predicated on short-term considerations linked to electoral cycles, so the will to engage in comprehensive long-term sustainable planning is limited. Moreover, the low impact of these individual sustainable policies has revealed several key elements that need to be more successfully addressed if more comprehensive sustainable practices are to be created in greater numbers of towns and cities.

First, the *scale and variety of the problems* involved in making towns and cities more sustainable, through the policies described in this and the previous chapter, means that no single policy approach, or a single source of implementation, will be sufficient to achieve this goal at ether a societal or urban level. Certainly each individual policy, if effective, can make a real contribution. Yet there is a need to go beyond single policy approaches. A truly sustainable future can only be achieved if a *multi-policy* approach is pursued. This means that major gains can only come by jurisdictions pursuing a number of sustainable policies in conjunction with one another, implemented not only by governments, but also by businesses, as well as individuals. The example of Copenhagen shown in Fig. 6.2 is a good example of the multi-policy approach.

Second, a *multi-agency approach* also needs to be pursued. Urban municipalities alone cannot solve the current problems because of their limited powers and financial resources. Nevertheless, some cities, such as Copenhagen, Portland, Göteberg, Freiberg and Saarbrucken, have been particularly progressive and need praise. Yet far more municipalities need to attain the levels of sustainability achieved by these cities if real progress is to be achieved. However there are limits to urban-only policies. Most places cannot significantly improve sustainability without the involvement of higher levels of government, which will ensure the greater spatial spread and effectiveness of policies through legislation and finance. Nevertheless, urban governments can make a significant contribution to the overall effort by the direct contribution of their own extensive operations, and by setting the framework for others to act, either by regulations or by incentives.

Third, cities can gain additional momentum in their degree of sustainability by participating in another type of multi-agency approach, by joining the information and *collaborative networks* of cities often across continents that have grown up to discuss, develop and share knowledge on the best sustainable practices that have been implemented in various municipalities, as seen in the European Sustainable Cities Network or the C40 group of major cities. In addition, knowledge is increasingly disseminated by international funded programmes covering specific areas, such as transportation, energy, environment, economic development and urban regeneration initiatives.

Fourth, all too frequently, environmental activists and those who promote sustainable practices, have treated issues of growth and the domination of the market-place as the basis for decisions in a negative way, as being the source of the problems. But there are signs of changes in attitudes. For example, a well-known green activist (Porritt 2005) observed that it is time to *work within the framework of the market economy* that underpins the functioning of most economies in the world, to show the new sustainable opportunities that are available. Of course, the countries with Command and Control structures consider their approach is quicker, but such regimes have always had a bad record in democratic practices and especially on pollution because it was almost impossible to criticize waste practices, although the Chinese seem to be making big efforts to clean up their environment, but have a long way to go. Porritt's views do not only apply to the economies derived from clean technologies, but in also showing how businesses can gain from adopting sustainable practices. This is not simply to benefit their profits, but to improve the welfare of their employees and to increase their positive effects upon their surroundings. It has already been shown that increasing numbers of companies are embracing sustainable practices in their businesses, in large part to reduce their costs, but again there is still a long way to go before the majority of companies, let alone municipalities, follow suit. One of the most positive signs of progress in this context comes from the way that business organizations have also been created to encourage and rate the adoption of sustainable practices, such as the U.S. Green Building Standards Council described in a previous section.

There are still barriers associated with the extra expenses of adding many of the sustainable practices. Yet the costs of these new initiatives compared to keeping the status quo are often not properly accounted, as four main examples will illustrate. First, it is clear that many of the *full costs of current practices* are never properly taken into account, especially the externalised costs, such as already described in the case of the burning of polluting fossil fuels such as coal in Chap. 5. When these are added into the accounting system many of the existing status-quo practices are seen to be far less cost-effective than has previously been recognized. Second, although many sustainable practices do involve additional expense, the variety of polices already being practiced have very different cost implications, so there is always the possibility of choosing *cheaper but effective options* in many cases. Third, there is still insufficient research to show the *real costs* of the wholesale adoption of these practices. As described in Chap. 5, a comprehensive report for the U.K. government on sustainability argued that the annual costs would be 1–2 % of GNP (Stern 2007; Stern and Taylor 2007). Although the figure is disputed by many people, even the trebling of this figure would be within the ability of most developed countries to afford, given sufficient support from the population and governments, together with agreements at an international level to ensure countries are not disadvantaged by such policies, difficult although reaching such agreements have been. Unfortunately, the financial crises from 2008 and the subsequent focus on reducing the debt problem in most developed countries has tended to de-priorize the adoption of new sustainable initiatives at a national state level. Also the government cutbacks have led to a decline in incomes of so many of the middle and lower income classes. Hence fewer people believe they have the finances to adopt sustainable policies without government subsidies, while many governments have withdrawn or reduced such practices in an attempt to cut their expenditures and reduce debt.

A fourth cost issue involves the *long term consequences* of doing nothing, or only implementing weak practices in a limited fashion, by only looking at costs as a current or short term expenditure. There is little doubt that people in cities will be much healthier and will have more active and varied life-styles if greater progress towards sustainable practices is made. Also, given the increase in greenhouse gases produced by cities, which most climate scientists believe is leading to climate change, then the wider adoption of sustainable practices need to be promoted on the basis of being an insurance against the even greater costs that are highly likely to be incurred in the reconstruction of cities hit by the increasingly intense storms or floods that we have been experiencing, or in the higher costs of using many increasingly scarce resources. So incurring the costs of sustainable practices today, or spreading them over a near future to mitigate these effects, ought to be seen as an insurance against the far, far greater expenditures for mitigation and recovery that will be incurred in the long term. This is not only appropriate advice for governments and companies to rapidly improve their levels of sustainability by making it a real priority through all administrative levels. Increasingly, individuals, or rather groups of individuals in at least some towns are also recognizing the need to combine their efforts to develop their own sustainable policies, rather than waiting for governments to do so, as will be shown in the review of grass-roots movements, such as Transition Towns, in Chap. 7. Given the need to improve levels of sustainability in towns and cities, and the many examples of workable and efficient practices, it is rather sad to realize that we have the technical tools, but not sufficient political will, to create much greater degrees of sustainability. Far more places need to follow the example of Rhein-Hunsruck in south west Germany, a district of 102,000 inhabitants that generated over 200% if its electricity needs by 2014 from solar, wind, biomass and hydro installations, selling its surplus to the German grid (GHP). These installations, plus earlier efforts to improve building efficiencies, including district heating for public buildings, have led to 25% and 5% reductions in heating and electricity demands respectively, and a 64% decrease in CO2 emissions since 1990. The district has also ensured that energy issues and the problems of climate change are taught in the schools to ensure students are brought up to be environmentally conscious.

Acknowledgment I appreciated permission from the members of the GWL-Terrain project to reproduce a photograph of the area (koepelverenging. GWL-Terrain) as Figure 6.3.

References

BAC: Brewers Association of Canada. (2013). http://www.brewersassociationofcanada.ca. Accessed 10 Jan 2013.

Baird, G. (2010). *Sustainable buildings in practice: What the users think?* New York: Routledge.

Beatley, T. (2000). *Green urbanism*. Washington, D.C.: Island Press.

Berenyi, E. B. (2013). http://www.wte.org/userfiles/files/130820%20Berenyi%20Nat'l%20WTE%20Economic%20Benefits.pdf. Accessed 10 Nov 2013.

Bokalders, V., & Block, M. (2010). *The whole building handbook*. London: Earthscan.

CaGBC: Canadian Green Building Council. http://CaGBC.org. Accessed 11 Aug 2013.

Carroon, J. (2010). *Sustainable preservation: Greening existing buildings*. New Jersey: Wiley.

Chadwick, E. (1843). Report on the sanitary condition of the labouring population of Great Britain. Clowes and Co. for HMSO, London. New edition. V. Kiernan (Ed.), (1965). London: Penguin.

CPH: Copenhagen. (2012). http://denmark.dk/en/green-living/bicycle-culture/. Accessed 12 April 2012.

C40. (2014). http://www.c40cities.org. Accessed 5 Feb 2014.

DAC: Danish Architectural Council. (2012). http://www.dac.dk/en/dac-cities/sustainable-cities/all-cases/energy/dongtan-the-worlds-first-large-scale-eco-city/?bbredirect=true. Accessed 15 March 2012.

DLC: Drake's Landing Community. (2012). http://www.disc.ca. http://www.unitedcommunities.com. Accessed 15 March 2012.

EDP: EdenProject. http://www.edenproject.com. Accessed 20 March 2012.

EIU-Economist Intelligence Unit. (2012). The Green City index. Siemens AG, Munich. http://www.siemens.com/entry/cc/en/greencityindex.htm. Accessed 8 July 2013.

EP. (2013). Waste to energy, 2013–14. Ecoprog, Cologne. http://www.ecoprog.com/fileadmin/user_upload/leseproben/ext_market_report_waste_incineration_ecoprog.pdf. Accessed 10 Aug 2013.
EPA: Environmental Protection Agency. (2013). http://www.epa.gov/waste/conserve/foodwaste/. Accessed 24 Sept 2013.
EP: Ecopolis. (2013). http://www.ecopolis.danfoss.com. Accessed 10 Sept 2013.
EU-Clima:European Union, Climate Action. (2013). htttp://eu.europa-eu/climate action July 2013 update. Accessed 10 Aug 2013.
Foletta, N., & Field, S. (2011). *Europe's vibrant new low car(bon) communities*. London: Institute for Transport and Development Policy.
Gehl, J. (2010). *Cities are for people*. Washington D.C.: Island Press.
GHP: http://www.go100percent.org/cms/index.php?id=77&tx_ttnews[tt_news]=258&cHash=81261a7fdf5436a56620c595d7f531c9. Accessed 10 Oct 2014.
GWL. (2013). http://www.gwl-terrein.nl/files/artikelen/carfree%20housing.pdf. Accessed 12 June 2013.
Hilbert, M., & López, P. (2011). The World's technological capacity to store, communicate, and compute information. *Science, 332*(6025), 60–65.
Holmgren, D. (2005). The end of suburbia and the beginning of mainstream permaculture. *Permaculture Magazine, 46,* 7–9.
IFC: International Finance Corporation. (2012). http://www.ifc.org/wps/wcm/connect/7540778049a792dcb87efaa8c6a8312a/SP_English_2012.pdf?MOD=AJPERES. Accessed 3 Dec 2012.
IPCC. (2014). Climate change 2014: Impacts, adaptations, vulnerabilities. Intergovernmental Panel on Climate Change: Working Group II, Yokohama, March 31. http://www.ipcc.ch/pdf/ar5/pr_wg2/140330_pr_wgII_spm_en.pdf. Accessed 10 June 2014.
Kenworthy, J. R. (2003). Transport Energy Use and Greenhouse Gases in Urban Passenger Transport Systems: A Study of 84 Global Cities. Conference paper. Republished in: http://cst.uwinnipeg.ca/documents/Transport_Greenhouse.pdf.
Lee, J. C., & Williams, M. (2006). *Environmental and geographical education for sustainability*. New York: Nova.
Marsden, T., & Morley, A. (2014). *Sustainable food systems: Building a new paradigm*. London: Earthscan.
MC: Masdar city. (2013). http://www.masdarcity.ae. Accessed 20 March 2013.
MLaA: Melbourne Lanes and Arcades. (2012). http://www.thatsmelbourne.com.au/placestogo/LanewaysandArcades/Pages/LanewaysandArcades.aspx. Accessed 5 Dec 2012.
NP: Napierville. http://www.naperville.il.us/smartgrid.aspx. Accessed 10 March 2013.
OICA. (2012). Annual production statistics of vehicles. http://www.OICA.net. Accessed 6 March 2013.
O'Riordan, T. (2004). Beyond environmentalism: Towards sustainability. In J. A. Mathews & D. T. Herbert (Eds.), *Unifying geography: Common heritage, shared future*. London: Routledge.
PH: Passivehaus. (2013). http://www.passivehaus.org.uk. Accessed 5 Sept 2013.
Porritt, J. (2005). Capitalism: *As if the world matters. London:* Earthscan.
Pouchard, A. (2014). City's ambitious target for zero waste by 2020 is built on gradual change. *The Guardian Weekly,* 20 June, 9.
SBC: Sustainable Buildings Canada. (2013). http://www.sbcanada.org. Accessed 5 Sept 2013.
SETIS. http://setis.ec.europa.eu/setis-deliverables/technology-mapping. Accessed 10 Feb 2014.
SFW: San Francisco Waste Programme. http://www.sfenvironment.org/zero-waste. Accessed 10 Sept 2013.
SM: Stadtwerke München. (2013). http://www.swm.de/english.html. Accessed 6 Sept 2013.
SSF: Solar Settlement, Freiberg. (2012). http://www.werkstatt-stadt.de/en/projects/22/. Accessed 10 March 2012.
Stern, N. H. (2007). *The economics of climate change*. Cambridge: Cambridge University Press.
Stern, N. H., & Taylor, C. (2007). Climate change: Risks, ethics and the Stern Review. *Science, 317,* 203–204.
TE: The Economist. (2012). Shoots, greens and leaves. June 16, 69.

Themelis, N. J. (2003). An overview of the global WtE schemes. *Waste Management World*, July, 40–47. http://www.waste-management-world.com/articles/2003/07/an-overview-of-the-global-waste-to-energy-industry.html.
USGBC: U.S. Green Building Council. (2011). http://wwwusgbc.org. Accessed 12 Feb 2011.
USGBC. (2013). http://www.usgbc.org/articles/infographic-leed-world. Accessed 9 Nov 2013.
Walsh, D. (2011). A war against food waste. *New York Times,* Sept.15.
WM: Walmart. (2013). http://corporate.walmart.com/global-responsibility/environmental-sustainability. Accessed 20 Aug 2013.
WMC: Walmart. (2013). http://news.walmart.com/news-archive/2010/11/16/walmart-canada-opens-its-first-sustainable-distribution-centre. Accessed 20 Aug 2013.
WWE: World Watch Europe. (2013). http://www.worldwatch-europe.org/node/59. Copenhagen's Carbon Neutral Plan. Accessed 21 Aug 2013.
Yudelson, J. (2010). *Greening existing buildings*. New York: McGraw-Hill.
Yudelson, J., & Meyer, U. (2013). *World's greenest buildings: Promises to performance in sustainable design*. New York: Routledge.

Chapter 7
Transition Towns and EcoDistricts: Local Sustainable Initiatives

Wayne K.D. Davies

> *For all those aspects of life that this community needs in order to sustain itself and thrive, how do we significantly increase resilience, to mitigate the effects of Peak Oil, and drastically reduce carbon emissions to mitigate the effects of Climate Change....Our mission is to inspire, inform, support, network and train communities as they consider, adopt and implement a Transition Initiative—through a range of materials, training courses, events, tools and technical resources (Transition Town Primer 2006).*

7.1 Introduction

The last decade has seen the development of local and grass-roots organizations designed to implement greater sustainability at a local level. The most well-known is the movement called Transition Towns (TT) which operates primarily at a settlement level, while a recent organization called EcoDistricts focuses mainly on the inner city neighbourhood scale. These are local, citizen-inspired sustainability movements to increase local resilience in settlements. The background to both movements lies in increasing ecological concern among many people of the current human exploitation of the environment and the unsustainable nature of so many modern processes. However the Transition Town movement has been sparked by two specific environmental concerns that were discussed in Chap. 5. One is the negative consequences of global warming produced by increased carbon dioxide emissions from our expanding industrialised and consumer society and the apparent inability of governments to do much about the problem through international agreements. The second came from a belief in the impending impact of the Peak Oil theory. In origin it was a descriptive model that applied to the United States when production peaked and then declined from the early 1970s. But increasing supplies

W. K.D. Davies (✉)
Department of Geography, University of Calgary,
2500 University Drive, N.W., Calgary, AB T2N 1N4, Canada
e-mail: wdavies@ucalgary.ca

W. K.D. Davies (ed.), *Theme Cities: Solutions for Urban Problems*,
GeoJournal Library 112, DOI 10.1007/978-94-017-9655-2_7

of oil from other world areas meant that its impact on disrupting growth in America was limited. However when applied to a situation of growing world demand for oil, given finite supplies and world population growth, the prediction was for a peak of production, followed by declining oil supplies and higher prices, which will mean that in the long run many of our systems will not be able to operate or will be reduced because of increasing costs of this vital raw material (Herold 2012). Indeed the supply of most of the goods we consume in retail outlets or in production units depend upon on long distant inputs with fragile, often 'just in time' globalised supply chains—whether of food, energy, or other goods. These inputs are easily subject to disruption, since they depend upon an oil-based economy for the creation and delivery of these goods. The higher degree of urban self-sufficiency in the past has gone, along with the complex local rural economy that supported so many towns in agricultural areas.

The fear of these impending disruptions have led many to believe that it was time to start the transition to a new and more sustainable future, one which especially encourages the reduction of oil-based energy use and the re-building of more localised life-styles and economies. Adherents to this cause maintain that action is urgently needed because of four Transition Town imperatives.

First, they believe in *local action*, for towns cannot rely on governments to initiate the necessary changes to counteract future oil or climate shocks by top-down action. Certainly they accept that government policy will help mitigate many of these effects, such as by reducing pollution or encouraging energy-saving and alternative energy sources, But it is argued that the actions of governments have been too limited in these vital areas, and will be unable to transform lifestyles and economies quickly enough to create meaningful change, at least in democratic countries,. Real progress will only be achieved by grass-roots organisations that accept the need for change in their settlements and find ways of modifying their community to increase sustainability, rather than leaving everything to market forces with decisions being made in distant places without concern over the impact on local areas. Such changes will only occur by implementing these ideas through co-operative action and by persuading local councils to incorporate these new ideas in their administrative and planning processes.

A second feature of these grass-root organizations is that they should not be seen only as creating specific policies to reduce the future shocks of oil shortage and climate change. Their actions will also act as *community catalysts*, to re-vitalise community relationships and interactions through their transitioning efforts, for measures undertaken only by individuals will be too little to effect change. The result will be the creation of a new local resilience, enabling adopting communities to more effectively withstand the future shocks that TT advocates believe to be not only inevitable, but imminent.

Third, adherents to the movement believe that the creation of more local linkages and sustainable actions will *make towns more liveable*, improving life by making it more enjoyable, meaningful, fulfilling and ecological. This will provide an important complement to the national and global associations we have developed through access to contemporary forms of communication and transport, but which have ignored local linkages.

A fourth but more diffuse factor comes from the frustration of many individuals about their decreasing local influence, their ability to affect, or change the decisions made by ever larger and more distant governments and corporations whose decisions often negatively affect local communities. This is a consequence of the *social fragmentation in modern society*, due to various factors associated with the specialization, subordination, spatial flexibility and size of urban places (Davies and Herbert 1993; Table 2.1) that Transition Towns seek to counter.

These features have combined to lead to the growth of the Transition Town movement, whose origins, development, major concerns and achievements are described in the next three sections. However, the last few years have also seen the growth of other community-based movements devoted to similar ends of creating more sustainable ways of life, such as Eco-Villages (EV), but these describe projects in rural areas so are not relevant to this urban discussion. More recently, another urban-based movement, labelling itself as Eco-Districts, and primarily in inner city American neighbourhoods, has also developed, which will be briefly reviewed in the last part of this chapter.

7.2 Transition Towns Origin and Growth

There seems little doubt that many of the concepts behind the Transition Town movement, and the policies being adopted in these centres, can be seen in the sustainable initiatives carried out by individuals, corporations and various levels of government throughout the world. However the pioneers of this movement have been especially influenced by ideas from permaculture (Holgrem 2002). This movement drew its inspiration from the recognition that many agricultural practices in the world had developed a self-sustaining capacity over hundreds, if not thousands of years. These practices worked within the natural landscape, minimizing waste and ensuring self-sufficiency without exhausting the land or sea. Moreover, like the ecological approach which it resembles, it emphasised that the complex ecological relationships within nature—and between humans and nature—need to be viewed in total, not by single elements. These self-sustaining ideas spread beyond agriculture, such as from ecological design to engineering, meaning that permaculture has become an integrated approach to long-lived, sustainable development rather than just being concerned with historic agricultural practices. However the key role in initiating and consolidating ideas from permaculture to create the concept of contemporary settlements *transitioning* to a new future came from the work of Rob Hopkins and his Irish students in an energy-lifestyle project developed in Kinsale Further Education College in southern Ireland in the early years of the twenty-first century. This project led to a plan to reduce local energy use and start the process of change to a reduced energy future, a plan which was later accepted by Kinsale Council in 2005 (Hopkins 2008). Two of the students, Louise Rooney and Catherine Dunn are credited with inventing the term, 'Transition Town', one that came to define the subsequent movement of change, in which towns try to transition to a new and less dependent future. This movement has not been simply concerned with showing the consequences of peak oil and climate change, but in designing policies to reduce their effects on the provision of food, energy, transport, health, education, and waste reduction etc., in local towns, which could be too easily interrupted.

In some ways the Transition Town approach has parallels with three older urban social movements. One was Howard's (1898) proselytizing for new *Garden Cities* that would be built by citizen co-operation, not by governments or industrialists, a feature that was soon forgotten in the debasement of the ideas into what turned out to be garden suburbs by governments or developers. Another was Alinsky's (1972) *critical conflict* approach to community development that he developed in America during the late 1930s to help to improve the situation of disorganised and impoverished inner city communities. Although a key role was played by a skilled community organiser who would help initiate consultations, the work would only be successful through the subsequent activism of the community residents (Davies and Herbert 1993, Chap. 6). A third parallel can be seen in several *self-help movements*, such as the citizen-initiated improvements in third world shanty town developments, the self-build approaches advocated by Turner in the 1970s in Britain (Hall 1988), as well as some aspects of the back-to-the-land communes set up in the 1960s. All approaches stressed the importance of the need for concerned local citizens to co-operate to create a better future—although in the Transition Town approach this is not simply as some idealised or preferred new forms. Rather it is a local town re-organization, designed to produce practical responses that will begin the process of counteracting what the advocates of the movement see as the inevitability of future problems.

The diffusion of the Transition Town ideas occurred when Hopkins moved to Totnes in the county of Devon in South West England, a town of 8400 people in 2011, but a market centre of 24,000 if the surrounding parishes are included. This historic market town had grown to prosperity as a medieval woollen centre, and later as service node and a centre of local industries linked to its role as an agricultural centre and port on the lowest bridging point of the tidal reach of the River Dart. From the 1920s it was also the home of a progressive educational and art centre known as Dartington Hall. However, by the early twenty-first century Totnes, like so many small county towns, had lost many of its local industries, so that apart from the local retailing and service jobs most people were commuting to jobs in nearby larger nodes. Hopkins initiated a series of meetings designed to show the impending problems posed by climate change due to the increasing build-up of human created greenhouse gases and the dependence of current life-styles and economies upon the finite resource of oil, stressing the urgent need to seek alternatives and ways of countering the changes. Within a few years local residents interested in the problems had established a series of informal committees that discussed specific problems and sought practical solutions to the problems of global warming and an oil-based economy and transport systems. By September 2006 a major, open meeting was organized to discuss progress. This was attended by over 350 people who reviewed the issues brought up by the initial committees and set up a series of working groups to identify problems, imagine alternative futures and suggest policies that would allow the town to change or to *transition to a new future*, with new types of functioning. Unlike the community activists from Alinsky's critical conflict approach in poor and impoverished areas, a far greater cross-section of the town population was involved in Totnes, with perhaps a majority of the people who were middle income and educated individuals, including a high percentage of women. Many were individuals disillusioned with the failure of elected officials or governments to listen to their worries about the effects of ecological issues, especially climate change, environmental destruction, and the inevitable problems that they envisaged being caused by higher oil costs and shortages. What became known as Transition Town Workshops provided a focus for many previously unrelated people to share their concerns and search for ways of ensuring a better future, not simply as individuals, or within their groups, but linking with existing organizations. The Totnes town council was soon persuaded to support the principles behind the policies advocated, and voted to become the second formal Transition Town after Kinsale.

Lectures by Hopkins and experts in specific sectors affected were given to interested groups in other towns, not simply on the issues of peak oil, climate change and various sustainability initiatives, but also on the need to re-build relations with each other and the natural world, through community interaction, countering the increasing individualisation or atomization of society. One of the key ideas of the TT movement is to create more local community capability to counter or adapt to the changes taking place, so that the community would have greater resilience against these trends. The growth of the movement has been helped by the creation of a regular internet Transition Town Newsletter designed to share ideas, a promotional film about the ideas in 2007, and recommendations to other communities about various books and films that supported the idea of reducing carbon footprints and encouraging local production, re-use and consumption. In addition advice about how to build community support through individuals, other organizations, businesses and governments, have been provided. The creation of a formal Transition Town Primer or Manifesto (TTP 2006) has been regularly updated and extended, which has helped formalise the process by providing an overview of the ideas and organizational guides for the movement.

The Transition Town movement spread rapidly from its beginnings in Kinsale and Totnes. By the end of 2007 50 places had been formally recognized as Transition Towns, four of which were in New Zealand and Australia. The growth continued, with Fujino in Japan becoming the 100th member, followed by Kreuzberg, an area in Berlin long known for its alternative and artistic life-styles. The organization of the approach was helped by the establishment of a Transition Town Network which was registered as a formal charity in the U.K. in 2007 in order to obtain funds and to disseminate ideas and was soon supported by grants from the Tudor Trust and Fairbairn Foundation. This network has provided finance for speakers and promotional materials to develop and spread the principles behind the movement to other interested groups and towns. The development of this formal organization, together with the need to report on the expenditure of funds, means that interested groups within other communities do have to get approval of their work to date before they are recognized as an 'official'

Transition Town. In September 2013 the TT Network identified 462 places that had been formally approved as Transition Towns, with another 654 described as 'mullers', namely centres that had requested information about the movement and were thinking about joining the movement. Initially, most of the official TT's were relatively small towns, but the movement now includes some larger centres, such as Bristol and Eastbourne in the United Kingdom, as well as parts of big cities, such as Crouch End in London, or the separate municipality of Fredericksburg, part of the Copenhagen built-up area in Denmark. Locationally, however, most of the Transition Towns are still in the U.K (45.0 %), with another large contingent from North America (32.9 %). Places in Australia and NZ account for 11.2 %, while centres in Germany and France provide 3.9 % of the total and Ireland another 1.2 %. Although the TT network notes that there are representatives from 23 countries in the official list, its dominance by English-language countries is obvious, for only 5.7 % of the official members come from the other countries not mentioned in the list above.

7.3 Transition Town Organization

A review of the way that Transition Towns have been initiated and matured led to the creation of a 12 Step Process of Development as an idealized guide to help other communities develop their own organization (Table 7.1).

This sequence illustrates how the movement can grow from a small group of pioneers identifying some initial problems to wider action groups, and eventually to public meetings to identify and discuss the problems associated with unsustainable patterns and life-styles raised by the action groups. This, in turn, leads to decisions taken by the community on solutions designed to solve the problems. However rather than only depending upon their own actions to create change, the ninth stage shows the practical approach of the TT pioneers. It stresses the importance of persuading local governments to produce actions to increase sustainability, as well as identifying local practices that have been lost or marginalised, paralleling part of the Slow City movement (Chap. 15). The process stresses the need to incorporate elderly residents into the association, in order to make it more representative of the local community and to draw upon their skills and knowledge. What is also notable about this movement's desire to create greater local resilience and sustainability is the rather self-effacing admission in its newsletters that the strength of the movement comes *from* the participants, not from any directed source, meaning that the various groups should establish their own ways of operating. In other words, the 12 Steps are seen as only *guides* to the development sequence. Moreover, the various contributors to the TT newsletters and organizational primers stress they do not claim to have all the answers to developing the transition to a better future; rather the pioneers of the movement see themselves as individuals who are willing to help initiate change. This desire to create change by members of the community is based on the belief that it is taking too long for national or regional governments to act on the problems posed by climate change or to anticipate problems of future oil shortages.

Table 7.1 Transition Towns: the 12 steps to development. (Source: Summarized from the Transition Town Network 2006)

1. Steering Group Initiation. This starts the local discussion on the problems and need for transitioning. It subsequently creates at least four subgroups to discuss ways of achieving the needed changes, but has a built-in sunset clause to bring an end to their initial work
2. Awareness Rising. Developing discussion groups and using film shows, newspaper articles, expert speakers to discuss key problems and how local initiatives can solve them
3. Laying Foundations. Networking with other community organizations in the town to create interest and support for the ideas. It is emphasized that the process is dependent on local creativity and ideas, not on any pre-conceived agenda, or centralised control
4. The Great Unleashing. At this stage a formal, open meeting for the whole community is organized to discuss progress to date and ways forward. This meeting is designed to adopt formal organizational principles to guide future work and marks the end of the initial subgroups that simply acted as catalysts
5. Establishing Working Groups. These deal with specific issues, such as energy, food, transport, social co-operation, education, health. Development of local resilience indicators are encouraged to show existing conditions and potential changes in various areas, such as: how much food is grown locally; types of local production; degree of local business ownership; extent of commuting; amount of renewable building materials; amount of potential compostable waste etc. Essentially these indicators provide a community inventory, that can provide a basis for identifying areas of action and a measurable initial base from which change can be shown
6. Open Space. These are open meetings to encourage free-wheeling discussions and the development of visions for future based on ideas from the participants
7. Visible Demonstrations. This is the stage at which progress in various areas, such as increase in composting, tree planting, energy reduction, commuter-sharing can be identified. They demonstrate how the community is responding to the need for change, and provide encouragement to the activists, and proof to the skeptical, that change can come from within the community
8. Facilitate Re-Skilling. Development of workshops with elder citizens to highlight the various skills that have been lost in our specialist society, as well as workshops to help individuals develop self-reliance, in such projects as: adding insulation to houses, mechanical and clothes repairs, and the need for re-use and re-cycling
9. Building Bridges to Local Government. This is a necessary step to ensure that elected representatives and local officials are informed of the various work to date, and will be encouraged to formally adopt the various initiatives. It is expected that the Transition Town organization will eventually create a plan to deal with the problems that will soon emerge and will be officially approved by the district, town or community
10. Honouring Elders. This involves the development of archival work and oral histories to show how local life before the oil economy was organized and supported. Demonstrations of older practices, especially by involving children, will teach older skills and encourage intergenerational activities
11. Individualised Operational Practices. This approach can be practiced at various stages. It stresses the need to encourage all groups, discussions or projects to find their own operating processes and solutions, with initiators acting merely as catalysts to create new initiatives
12. Integration. The results of the working groups should be organized into an Energy Descent Action Plan to be presented to, and hopefully approved by, the relevant local authority

Titles are the same as in the original source, but the descriptions are rephrased

Hence it is argued that it is imperative for individuals, working together in their communities, to create a bottom-up process that will initiate meaningful change by their own actions, as well as convincing local governments of the imperative for local action before the shocks of oil shortages of higher prices and climate change take effect. However the various TT sources emphasize that there cannot be a single solution to the creation of local resilience to future shocks and the problems of sustainability faced by towns. The unique qualities of each settlement, plus their varied strengths and weaknesses, usually make it necessary to always adapt the movement to local circumstances. The 12 Step Process does omit one key feature, namely the recommendation that a charitable organization should be created in the adopting community to further the cause. Such an organizational step means that funds that are brought in to help the development of transition will not be taxed and also ensures that formal accounting procedures are adopted to make the transition development process accountable and transparent to others.

One of the important results of this open and co-operative process in creating a Transition Town has been the creation of greater interaction within the community and the fostering of a new spirit of co-operation within the adopting towns. This co-operation spreads beyond the various discussion groups. Transition Town activists stress the need to contact and involve church and various other local organizations in its developments. Teachers in the local schools have also been encouraged to share the TT ideas with their students and involve them in local projects to build environmental awareness. Additional contributions from the life-experiences and skills of senior citizens are welcomed, as they are often marginalized in contemporary western society. These initiatives have been found to re-build local contacts and interest, reduce the individualisation of society, and create resistance to the practice of unsustainable practices. A key goal of the movement is to establish groups to review all the key aspects of contemporary life in the town—such as food, energy, transport, health, community links, commuting levels, economies and livelihoods. These areas have been called *sectors of transitioning*, which search for ways to make its local area less vulnerable to external influences by becoming more resilient. This resilience is not primarily from the natural forces discussed in Chap. 8, but a local resilience against dependency, the socio-economic and ecological changes stemming from variations in the way society is currently organized and operates. This will help build a low-carbon, low oil-based future of greater sustainability that makes the town less vulnerable to many of the decisions made by outside forces that negatively affect the town.

7.4 Transitioning Sectors

The first goal of every group working on a particular sector in a Transition Town is to provide an understanding of existing conditions that relate to sustainability. Subsequently this leads to the search for ways to not only reduce carbon-based energy use, waste and environmental degradation, but to increase local production and consumption in the town and its region. Although there can be many working groups dealing with specific sectors of transitioning, space limits mean that not all can be reviewed here. Instead the most important are discussed, namely those dealing with energy, food and economic relationships, as well as a summary of key features in other sectors. Although many of the policies discussed in previous chapters on sustainable development can be used to achieve these goals, the TT approach attempts to be more comprehensive, integrative and community-based by encouraging local groups to identify and provide solutions to problems, not simply to issues of immediate ecological concern, but to those that might occur, following the closure of some local industrial plant, transport terminal, hospital or school. However the Transition Town movement is not the only one to encourage a transitioning to local sustainability and local resilience, so examples of related policies that have been practiced in other centres are also included in this brief review.

7.4.1 Energy Reliance

One of the crucial sectors of concern to the Transition Town movement relates to energy, its supply and use. So most transition towns try to produce a so-called energy descent plan, one designed to reduce the amount of energy consumed, which will reduce demand and greenhouse gases. Holmgren's (2009) book, *Future Scenarios,* outlines the four likely future energy alternatives and has often been used by TT advocates. The two extremes are represented by a *Techno-Explosion* scenario and the Collapse scenario. The former assumes a continued boom in technical progress and consumerism based on the discovery of some super new energy resources. At the other end of the spectrum is the Collapse scenario, characterised by a rapid societal breakdown due to a depletion in oil supplies and far higher prices which will lead to the catastrophic decline in economic activity and the spatial linkages that tie these activities together. It is often observed that the British lorry drivers' strike in 2000, a protest against an increase in fuel taxes, provided a glimpse into the way the collapse scenario would occur because of restricted fuel supplies. In 2 days the decline of deliveries to petrol stations and food to supermarkets were causing hardship because of the breakdown of the 'just-in-time' delivery systems and the absence of any local back-ups that could provide new supplies that would give resilience to local communities.

Holmgren also proposed two medium scenarios. The *Techno-Stability* Scenario assumes that the rapid addition of more renewable energy sources will allow us to carry on in a similar fashion to today, but with a gradual reduction. By contrast, the *Energy Descent* Scenario assumes a much larger reduction of energy needs because of local reducing actions and the slowing of economic activity and population decline, which will reduce the demand as well as the supply of energy.

It is argued that the first scenario is unlikely to occur because of the requisite technologies have not been invented and will take too long to implement, while the Techno-Stability model is limited because it assumes a massive growth in the proportion of renewable energy supplies, which seems unlikely given the projections of future energy demand from an expanding world population and the limitations of most renewable energy sources discussed in Chap. 5. So if Collapse is to be prevented, then Transition Town advocates maintain the only realistic alternative is the Energy Descent approach. Not all will agree with such conclusions, as the recent boom in gas and oil supplies from the fracking of American shale beds is rapidly increasing production to reach a new peak of production within a decade. Similar prospects exist in other countries, unless prevented by environmental activists. So there is increasing belief that there is no immediate danger of decreasing oil supplies and higher prices. Although this may reduce the threat in the next two decades environmentalists believe all this does is to postpone the peak oil date, so many believe it is prudent to take the idea of the Energy Descent Scenario seriously and that the new supplies of oil and gas give us a breathing space to create more sustainable practices.

The concept of energy descent is the opposite of the rapid ascent of energy supply since the Industrial Revolution, based on the growth of fossil fuel supplies and use. Hence many cities are engaged in polices to reduce their energy use. For example, Portland (Oregon), a city with a long list of environmental concerns, adopted a policy of reducing total oil use in the city of close to 500,000 by 2.6 % per year so that energy use would be reduced by a quarter by 2020. But Transition Town advocates have argued that instead of top-down policies, urban residents should combine to create an Energy Descent Action Plan (EDAP), which is now an essential goal in every member town in the network, although not a mandatory principle, such as those laid down in the Slow City approach. It assumes there is a real need to alter the supply and use of energy in a town and its region through gradual change from the citizens and businesses, since national government-led schemes will be too slow to take effect and will not cause enough change in the corporate culture that has altered the self-reliance of local towns. Unlike the many community or regional development plans produced by governments or their agencies, the EDAP does not assume a growth in demand and supply of energy resources. Instead it takes the view that there is a need for local communities to take the initiative in *reducing* energy use in all aspects of life, since it is believed that the problem of some future peak oil situation cannot be solved by a supply-led solution given the finite nature of fossil energy resources and their negative externalities, especially climate change. The approach is not simply to devise some empirical plan, but to use many methods, including storytelling and visioning exercises, to show the effects of some future of drastically reduced or costly energy. Such varied approaches will help raise consciousness of the need for change among the community residents. So the EDAP is not regarded as a plan for living a more sustainable life, but a plan to provide the *transition* to this low carbon future.

Totnes, as a pioneer in the Transition Town movement, was the first in England to develop an EDAP (TEDPA). Much of the report is devoted to showing the history of the town and its former energy sources and uses, in order to show how people coped in the past, as well as demonstrating the need to explain the emerging ideas of the EDAP to residents through public meetings and school presentations. These sessions are designed to elicit comments and feedback, not simply to improve the ideas, but also to publicize those being proposed. Many of the energy-saving ideas described in previous chapters are emphasized as being crucial to energy demand reduction, from cavity-wall insulations, to the use of new lighting schemes and energy use reviews. Among the latter is the promotion of oil vulnerability audits in local businesses, to show how all types of oil is being used by the firm, how savings can occur, and which uses are most vulnerable to future oil increases. Some English Transition Towns, such as Lewes in Sussex, have organized regular tours of homes with solar panels and other energy-saving devices to provide working examples of the cost-savings that can come from the adoption of such schemes or from Passivhaus concepts (Chap. 6). In addition, some Transition Towns are actively seeking ways of generating their own power through community co-operation in order to reduce their reliance upon the national grid. For example, the Totnes group has established TRESCO (Totnes Renewable Energy Society), a community-owned enterprise with shares owned by over 500 local investors. This has already established solar panel arrays to create a local power source and has tried to create two large wind turbines in the surrounding area to generate even more power. Unfortunately for the town, the local district authority refused planning permission of the venture in 2013. These examples show how Transition Towns are starting the process of transitioning to a future that is less oil-dependent for energy, but clearly there is a long way to go before most of the member towns can claim to be carbon-neutral settlements and dependent on local renewable energy sources.

7.4.2 Food Reliance

Before the railway age most towns in countries in the developed world were largely self-reliant in terms of food, in the sense that most supplies, except for exotic delicacies and wine, came from the local region, as did most building materials and textiles. Many of the people involved in the Transition Town movement are concerned about the way that food supplies have become so dependent upon non-local, and frequently international or global sources. Also there is a worry that so much food is now delivered by heavy duty vehicles from centralised supply centres. Indeed, between 1978 and 2005 the food supplied by this means in Britain increased by 23 % and the vehicles miles by 50 % (DEFRA 2005). These food delivery changes are a result of four trends: *globalization* or at least internationalization of food sources; *greater concentration* through larger food suppliers and delivery systems; large *supermarket chains* that search for cheaper sources of all products and often ignore local sources because of their bulk-buying through centralized purchasing systems; and also greater *farming specialization*, which means fewer types of crops are grown locally.

These trends have raised questions about whether it is really sensible for our society to continue a system in which the average carrot in the United States travels over 1800 miles from source to consumption, or whether it makes sense to get bottled water from Australia, or not worry that that half the vegetables in Britain and ninety percent of its fruit comes from overseas sources. These long distance sources have led many to argue that non-local sources can be easily interrupted by transport blockages, or reduced by rapidly increasing transport costs. So if this current system, with its just-in-time supply, became handicapped by increased fuel costs and reduced supplies, it seems logical to anticipate real local problems will emerge in food supply and access.

A key objective in many Transition Towns is to find ways of ensuring a greater and more varied production from the local area and also marketing these products, paralleling some of the Green City movements described in Chap. 4, creating a need for a *re-localization of food supply*. In addition, it is recognized that some of the problems of the amount of supply could be reduced by avoiding the enormous amount of food waste described in Chap. 6, for a third of food bought is thrown out in the U.K. amounting to almost £500 per household, while half of the food produced is lost in the supply chain (DEFRA 2010).

Obviously many products, especially fruit, that are found in food stores throughout the year cannot be grown in temperate climate countries such as Britain; others can only be grown in particular seasons. But even in the cases where the British climate is suitable for a particular crop there has been a great drop in local production and variety by the end of the twentieth century. For example, most of the apple and plum orchards in Britain have been removed in the last century. In addition, the large variety of different apples and plums that used to exist, even up to the early 1960s, have been reduced to a handful of types. Hence, local Transition Town activists are encouraging *new orchard planting*, using different varieties and also seek ways of ensuring that local production will be available in local stores. They also advocate the *local harvesting* of crops, such as many types of berries that grow wild in and around many British towns. It is amazing how wild local blackberries and mountain berries are now usually allowed to rot on the plant, as few bother to collect this crop, unlike the situation even 30 years ago. In addition Transition Town groups support the idea of *increasing distinctive local products* from district sources. For example, in Britain there is a new trend to create different types of cider from different apples—paralleling the differential varietal basis of different wines—rather than homogenizing the cider input from many varieties. Another is the creation of new cheese varieties from local inputs, for many distinctive cheese types originated in particular towns or regions, and like local wines some have been given exclusive rights to the label through national and later European Union regulation.

In another context Totnes has shown the way in increasing local food sources by planting large numbers of *nut trees* throughout the town and in surrounding areas, since these trees produce high energy crops. This venture has been helped by a government Agro-Forest Research Unit that has long been established near the town. Research in the centre has produced new varieties of nuts which can flourish in a British climate, such as sweet chestnuts and walnuts. The initiative of the Transition Town activists has been approved by the local council which has made land available for the nut tree planting project and in subsequent help with the harvesting and marketing of the crop, while individuals called Tree Guardians have been mandated to help ensure the trees are not vandalised.

Another focus in TT's has been to increase local land use intensity, for it is only a generation of two ago that back gardens and local allotments were full of vegetable crops. All too frequently the gardening and vegetative trends of the last three decades have been to design what has been described in Hopkins's (2008) pungent phrase as '*totally useless landscapes*' in a productive sense; this is not simply in individual house plots, but in the boring low-maintenance ground cover beloved of local councils. Clearly aesthetics and cost-conscious maintenance reduction have triumphed over potential productivity. Perhaps lessons can be drawn from Japanese agriculture in this regard. One of the fascinating things about Japan is the fact that there are few places where land is just left idle; instead cultivated land can be found everywhere that is flat enough, for the desire to produce local food and especially rice is a deeply held belief in the country. The situation is very different in most western countries, especially in cities, where there are great amounts of unused land, both private and publically owned, mainly because gardening has declined in activity as people seem to have less time to engage in such activities, and it is easier and cheaper to get food from local supermarket. So a common trend in Transition Towns is to produce an *inventory of unused land*, to encourage the expansion of community gardens and allotments, areas of land owned by the local council that is rented out in small plots to local gardeners, which was described in Chap. 4. Another is to develop *land sharing*. For example, the Totnes Gardenshare scheme persuades older people who do not use their gardens to let local volunteers grow vegetables and flowers on their land, share the harvests and, in doing so, creating intergenerational social links. Many Transition Towns also provide courses that involve *re-skilling local residents* in the practice of gardening and composting, while seed exchanges and excess crop swops also add to a sharing culture. A revival of the old Harvest Festival tradition also occurs in many places. Although this has often been kept alive in churches in autumn Thanksgiving Services, the small attendance and limited services in churches these days mean that such practices only touch a small part of the population today, unlike in the past. Hence *adding local festivals* at the end of the summer focuses attention on local food and its sale. These still thrive in France, unlike in the U.K. A deliberate effort is also being made to *involve schools in Transition concepts*, not simply through classroom lessons, but by hands-on experience of helping to create gardens and in some places to look after farm animals. In Totnes historical evidence has been used to show that the town school used to have vegetable gardens and even a sheep flock until the 1980s, well within the memory of many residents. It also seeks to encourage the revival of many traditions. After all, many people grew up helping parents on the farms or in gardens, gaining practical skills in such pursuits. Such life-skills have been lost and now need to be recreated through the new ecological focus in schools.

Many places had town markets in the past where food from local farms was available; many decayed as food was brought in from wider distances and farming became more specialised in the search for profits. So Transition Towns have been active in re-developing farmers' markets in areas where the local one has disappeared, especially in industrial countries such as Britain. Yet they survived vibrantly in France and Italy, in part because of the greater range of local crops that can be grown in many areas. However, the fact that so many local or regional cuisines have survived and been promoted in these countries has helped, whilst the greater emphasis upon the quality and freshness of food in countries that reveres meals, have all helped to maintain a very different approach. *Farmers markets*, which have been described in Chap. 4, also have the additional advantage of providing a local outlet to encourage farmers and smallholders to increase the variety of crops they can grow and which could be consumed locally. The development of a Local Food Directory in many Transition Towns also provide informative guides to local products as well as to restaurants where such local foods and preparation form a key part of the cuisine. All these trends mean that the Transition Town movement has given a stimulus to the idea that more food can and should be grown locally, and these ideas have been spreading to towns and communities that are not part of the formal movement. It also helps the development of what can be called the *niche production of local goods* that was described previously, those uniquely associated with local area and which can become part of a new local cuisine, like the older ones so famous in many rural French towns.

Support for the idea of a more localized production initially came from the idea of *food miles*, which identified the distance that food had travelled from production to sales sites. The large food miles travelled from many food sources was seen as demonstrating the increasing transport costs, energy use, transport emissions and hence lower sustainability of these long distance inputs. It led some to advocate the need for a more local diet, the so-called *locavores*, with some trying to restrict their food intake to within a 100 mile radius from their homes, a difficult feat in many climates. Despite the initial attractiveness of the indicator it has been shown to be flawed. It led James McWilliams (2009) of Texas State University in his book *Just Food: Where Locavores get it Wrong and How we can Truly eat Responsibly* to stress that a full accounting of environmental impacts must also take into account all the inputs of energy, water, fertiliser etc., in producing these goods, not only in terms of their costs, but the energy expended in the production of these inputs and the carbon dioxide generation. Transport costs in bringing the food to any location are only part of the costs involved, often less than 10 %. Such audits have shown that locally produced food may come at a higher cost in environmental impact than long distance supplies. One of the most comprehensive studies of this issue was carried out by researchers at New Zealand's Lincoln University (Saunders et al. 2006). It showed the comparative ecological advantages in food production, specifically in importing lamb from that country compared to production from Britain. The total energy used to produce an equivalent amount of lamb in New Zealand's more benign climate compared to Britain was four times less, while carbon dioxide output was lower by a similar amount because of the need for such things as heated pens and higher feed costs in Britain. The advantage was also seen in other products, as the energy expenditure figures for total production in Britain, including transit, were twice as much for dairy products, and 60 % more for apples, despite the 18,000 miles between the two countries that incur substantial transport costs. Pierre Desrochers at the University of Toronto's Geography Department has been a vigorous advocate of what amounts to the theory of comparative advantage in agricultural products, claiming that the more efficient and environmentally appropriate way of sourcing goods is to find the cheapest source in the best climatic area, subject to the need to avoid exploitation of local labour or the environment, for large capacity food transportation is often fuel-efficient. The arguments of comparative advantage for long distance sources seem especially cogent given the need to increase food production substantially, not simply for the extra 2.4 billion people by 2050, but for the hoped-for increase in living standards of the world's poor, without finding new crop land by destroying the large areas of tropical forests. In the *Locavore's Dilemma: In Praise of the 10,000 Mile Diet*, Desrochers and Shimizu (2012) argue that food subsistence farming is drudgery and that large-scale industrial farmers have made food more available, varied, affordable and nutritious than any time in history. The study maintains that if we want to save the planet and feed the growing human population we should be should be eating globally, not only locally.

Obviously there is a big difference in the conclusions of those emphasizing more local production and those stressing the route of comparative advantage which results in longer supply chains. Certainly local food may be desired by the more ecologically minded, who may be prepared to pay the premium price. Many will also argue that local sourcing produces fresher food, with fewer additives to ensure freshness, and also encourages pride in local products. These can be featured in local restaurants, providing an addition to the distinctiveness in towns and regions, in place of the creeping homogeneity in cuisine that is so often criticized and rejected in countries such as France and especially Italy where the Slow Food and Slow City movements originated (Chap. 15). In addition, of course, increasing local production leads to a greater variety in local employment and may also keep more farmers on the land, with resultant benefits to local economic opportunity and environmental stewardship. Many of these advantages are difficult to quantify and relate to the comparative cost argument. There is also the point that the rejection of foreign sources could reduce supplies from under-developed countries, which may limit their development and could lead to less employment, although there will always be crops that cannot be grown in many temperate climate countries. But there is no need to assume that one approach is chosen exclusively over another. Most Transition Town enthusiasts are not locavores; rather they wish to redress the current balance where practically all food is imported to urban areas. Adding more local production has the advantages described above. Moreover it is often observed that too many of the large-scale industrial agricultural and marketing enterprises in less developed countries provide limited returns to local farmers, who remain at almost subsistence level. This is why various Fair Trade schemes have grown up, especially visible in Co-operative stores in the U.K. in which products for sale in western countries come from producing units where a fair price is paid for the products. Hence this adds an ethical issue on the side of local versus long distance food. In addition, most research reviews do not show any energy or emissions advantage for local versus global supplies, or organic versus current agricultural techniques, although measurement problems in dealing with information from different countries do make the conclusions tentative (Foster et al. 2006). So only using the energy and emissions arguments for increasing local and organic food production may not be the only factors that need to be considered. Indeed the issue of food security for any area must be considered. This was recently acknowledged as an future issue of concern by the U.K. government, when food security was added to the goals of creating healthier and more sustainable food in its call for a more integrated policy on food supplies by 2030 (DEFRA 2010). So the Transition Town approach of supporting more local food production may be justified as an important local employment generator and supplementary supply resource that provides an important back-up which would be useful in emergency situations if transport systems are compromised, whether by energy constraints or disasters from natural hazards.

Yet some argue that it is impossible to return to the days of greater local food production. However, it is worth remembering that major changes have taken place within the life-time of many people. For example, in Britain the loss of imports in World War II due to the depredations of the German U-Boats led to a national decision to increase local food production within the country. In 1940 the government set-up 476 districts to co-ordinate agricultural re-orientation, for much of the productive land had been placed under pasture as grain and other crops were imported. Gardiner's (2004) survey of war-time Britain showed that food production increased by 91 % by 1944, with 19.8 million acres under production compared to the 12.9 million in 1938, helped by the creation of 'land armies' staffed by females who came from towns to replace the men in uniform. It was also estimated that 10 % of food crops in this period came from the expansion of production from allotments and gardens, issues discussed in Chap. 4 (Green Cities). This shows what a determined effort to increase local food supply backed by government can actually achieve. The big question is whether such an emergency will occur in the future because of higher transport costs and limited supplies. Consideration of such a question means that issues such as lower comparative food costs, total energy use and emissions on the one side, versus increasing local food supply employment, agricultural opportunities and fresher food on the other, are not the only factors that should be considered in evaluating the merits of local food supply.

7.4.3 Economy and Livelihoods

One consequence of the spread of capitalist practices and large corporations, from banking to supermarkets, is that most aspects of our local economies involve ownership, inputs and outputs from, and to, ever more distant places, with the result that the circulation of money is dominated by outward flows from smaller towns. Our integrated food economy, dominated by big supermarket chains, means that much of the money generated and spent locally soon disappears into the larger economy. However there is evidence that local ownership leads to more money remaining in the area. For example a comparative national study in the U.S.A by concluded that "independent, locally-owned businesses of all types and sizes recirculated a substantially greater proportion of revenue to the local economy than did their chain competitors" (CivicEconomics 2012, p. 14). For the retailers and restaurants as a whole, the advantage was 3.3 greater for local firms compared to the others, although the proportions varied between various categories of stores. This type of evidence is enough to encourage the growth of local businesses as a way of increasing local resilience. Support for this idea and concern with the decreasing distinctiveness of local areas and the economic effects of the loss of local control led the New Economic Foundation in Britain to publish two reports (Potts et al. 2005; Ryan-Collins et al. 2010). They investigated the extent to which the High Streets (main shopping streets) in British towns were filled by chain stores, creating a scale from 0 to 100 to measure their degree of domination and presence of independent businesses. The survey discovered that only 36 % of these main shopping streets had more than two-thirds independent stores, with another 41 % being dominated by chain stores. These were given the title of Clone Towns because they all looked the same, with 26 % in the border clone category. The problem was not only seen as a problem of the boring similarity in townscapes, but in the fact that big chains invested less in local activities, sent most of their profits away, and had often abandoned city centre locations for out-of-towns areas, less accessible for those without cars, or closed the stores if they were less profitable than the average in the chain. This has left big gaps in many downtown townscapes, often filled by not-for-profit charity stores. It was noted that the university town of Cambridge had the highest clone score of 92.1 in its main street, a feature which contrasted with the distinctiveness of its historic university colleges and churches. Whitstable in Kent was the lowest at 11.6, and was the most independent place, which may be a result of its local Oyster Festival and its growing reputation as a local food centre, although its small size and limited hinterland probably did not attract many national chain stores. The reports stressed the need to support local businesses to make them less dependent on negative decisions made elsewhere, which provides support for the Transition Town approach of finding ways to improve the local resilience of towns.

In terms of local ownership the Totnes Energy Company has already provided one example of the way that a re-localization is being pioneered in energy creation in Transition Towns. The Totnes Development Trust has a wider remit, for it is a local organization established to provide finance and managerial advice for local businesses. In addition it has helped create the Atmos project, which has taken over the site of a large Dairy operation that closed in 2007. The site originated as a power plant in 1845, part of Brunel's attempt to build what he called an 'atmospheric railway' to Newton Abbot and Exeter. Despite its investment of over £ 400,000 it failed 3 years later because of various technical problems, such as creating effective seals on the vacuum tubes. The site has been home to many businesses since, and still contains some of Brunel's original buildings, providing a valuable heritage resource connection. Financially the Atmos project relies on members of the community buying shares in the project to raise capital for the development. The site is being designed as a local entrepreneurial hub for start-up businesses, hopefully with more fortune than Brunel's project, and contains what is described as an eco-village of apartments using power from a large number of solar panels to create a carbon-free settlement. This type of local community investment obviously occurred in the past when local banks operated, but decayed as they were amalgamated into large national enterprises in Britain as well as other countries. However there are some regions where the local banking tradition is still alive. For example, in the Prairie Provinces of Canada so-called Community Chests have been created in many small towns, which are financial organizations in which local investors provide money for economic initiatives that are designed for the local town or area. In addition many Credit Unions —essentially local savings and loans associations—can also be found in many small towns in the Prairies, which provide a localized banking system that is rare in the United Kingdom and other European countries. The growth of these agencies in the Prairies is in part a heritage of the co-operative tradition that developed in the agricultural sector in the early twentieth century. It is also a product of a regional distrust of the five large banks that dominate the banking industry in Canada, which are headquartered in Toronto, and had a reputation of foreclosing too many farms and local enterprises that got into debt in the depression years of the 1930s.

A significant initiative of many Credit Unions has been to try and increase the degree of local shopping in small towns in place of the growing tendency to travel 50 or a 100 miles to larger centres where lower prices and greater selection attracted customers. A popular device in many of these organizations has been to extend loans to customers at Christmas time so long as the money was spent in the local town. The loan did not require any surety, other than the fact that the customer and credit history was known to the bank. The loans were repayable at standard interest rates over the following few months and seemed eminently successful since very low default levels (often below 3 %) were reported and these were easily covered by the interest repayments. Such initiatives are however far more difficult in countries such as Britain where local banking has almost disappeared. Certainly there were lots of local mortgage agencies, called Building Societies, in which local people saved, and the money was used to finance home purchases. They grew very rapidly in the mid-nineteenth century. But many of these demutualised from the late 1980s and either amalgamated into large units, or were bought by the big banks. So the opportunities for local creativity in financial agencies were lost. However the local banks do not always provide advantages. The financial crash in the western world from 2007 revealed that many local lending organizations were dangerously overexposed because of rash lending practices that assumed houses would continue to inflate in value, leading to bankruptcy or government take-over in subsequent years.

This problem of how to encourage the local ownership, local circulation of money and local buying has led some Transition Towns to create their own currency, such as the Totnes Pound or the Lewes Pound. These notes are exchanged for sterling and can be used in a large number of local participating stores, although it has been discovered that many are bought and kept as souvenirs by tourists. This approach encourages people to shop in local stores that accept the notes. The ideas were based on the pioneering example of the Berkshares dollar in New England (BSD) which was first issued in 2006 and now can be used in 400 organizations in a region of Massachusetts. However, they are really a revival of older practices where local banks often issued their own notes, such as the original Totnes Pound which was in circulation from 1789 to 1895.

Although separate from the Transition Movement itself, another recent variant of this desire to create more viable local financial organizations and provide a boost to the local food production and supply system can be seen in the *Slow Money* (SM) movement. This is based on two beliefs. The first is that money in the contemporary world circulates too fast and is too centralized, in the sense that money saved in local bank branches goes to national centres and is often invested in securities that are barely understood and in distant areas, resulting in a disconnection from people and place. Although this pattern is unlikely to disappear, the SM approach wants to slow at least part of this process down, and redirect the money flows. The second is the need re-develop more sustainable local food systems by connecting investors to places where they live, providing an alternative to the large industrial farming practices that are too environmentally exploitive, and depend on artificial inputs, such as fertilizers to the soil and antibiotics to animals. The aim is to connect investors and donors to small, local food systems, especially organic farms, thereby providing new sources of capital for local small food enterprises which will enhance food security and safety, increase local connectivity between customers and producing units in an area, and claims also to improve nutrition and health. This Slow Money approach is not only designed to slow down and redirect at least part of the financial flows to local activities, but also seeks to promote local ecological, economic diversity and the improvement of soil fertility by building up organic matter in the soil. Like many types of ecological and community approaches—especially the Transition Town concept—it is based on a recognition of the urgency to start the transition to a different future, as can be seen in one of the objectives within the Slow Money Principles, namely to

> accelerate the transition from an economy based on extraction and consumption to an economy based on preservation and restoration (SMP).

The pioneer of the movement is Woody Tasch (2010), whose ideas are described in his book, *Inquiries into the Nature of Slow Money*. In 2008 a non-profit organization called Slow Money (SM) was founded in Boulder (Colorado) and held the first of its now regular national meeting in Sante Fé. Its aim is mobilize both investors and farm system enterprises through the Slow Money website, publishing ventures, workshops, lectures and annual meetings. So far there has been a steady growth, for the movement claims that from 2010 to 2013 a total of at least $ 30 million has been invested in 221 small food outlets, while 17 local chapters have been founded and affiliated to the national organization, all but one in the USA (SM). Other more localized Slow Money organizations are also being created in cities well as in rural areas. The hope is that the enterprises that benefit from the loans will put half their profits back into the community, in order to develop a cumulatively growing source of local capital for additional enterprises. There is little doubt that investors in the movement do not seek large speculative capital gains, for they see merit in returns that are reasonable and based on the ethical principle of building more sustainable local practices through what is described as creating 'nurture capital'. In many ways this parallels the origin of the Building Society movement in Britain, which raised and loaned money locally or at least regionally, although these organizations were restricted to lending money for homes. The idea of nurture capital is based on connecting investors to their local area and learning to invest because as their slogan emphasizes: 'Food, Farms and Fertility do matter'. In 2013 the organization developed the Soil Trust whose objective is to get a million members to join the movement in future years, in which they will pledge to invest 1 % of their assets in local food systems. Despite the growth of the movements and the publicity it is attracting through slogans that envisage investments moving from 'Making A Killing to Making a Living' its impact is still small and is still largely a U.S. movement. However the approach does provide a parallel to at least part of the Transition Town ideas, that of developing at least a partial re-localization of the economy.

7.4.4 Other Transition Sectors

The various Transition Town Groups have also developed a large number of other ways in which the towns can improve their own future to increase sustainability and build-up resistance to shocks produced by outside forces. In terms of *transport*, efforts are made to save energy and reduce traffic flows by encouraging the greater use of bicycles as well as car-sharing, since so many people commute out of their area to larger centres, while some towns in Cornwall have set up eco-car rentals for tourists to use. Local recycling efforts that help to encourage the reduction or re-use of waste are also common. Health promotion schemes have also been prominent in many Transition Towns, not simply by ensuring that the town has enough medical facilities and personnel, but by publicizing the need for healthier diets through better nutrition and life-styles. In addition, historical surveys in many British Transition Towns have been used to reveal natural hazard dangers, especially showing areas that have been affected by floods or landslides in the past. This information is made available in libraries and schools, to make as many people as possible aware of the areas at risk. Other trends include enrolling more people in first aid training, who can help if disasters occur, as well as identifying the vulnerable people at risk, the old, very young, lonely and infirm residents who may need help in emergency situations, issues covered more fully in Chap. 8 which deals with resiliency from natural hazards. A focus upon the local bio-diversity also leads to evidence of the richness of local eco-systems, as well as showing which species are under threat and where conservation measures are needed, as well as providing ways of encouraging the greater use of local walks and paths in the region. Ecological awareness in education is seen as a key objective. It has already been noted that Transition Town advocates stress the need to involve children in many of the social and ecological projects, especially through hands-on practice in creating school gardens, as well as educating children in the merits of social interaction and sharing. Increased community support for keeping key local facilities, like schools and medical centres, also mean that attempts to close or reduce the range of these facilities by decisions made by distant governments can be more easily resisted. In addition, Transition Towns have advocated for increased support for many town activities, such as donating books and money to the local library, developing new initiatives, such as music festivals, and running World Cafe events with visits by local councillors to make them more accessible to the citizens. Emphasis is also placed on the rediscovery of old skills, from knitting to dress creation and the maintenance and re-use of old fabrics. In some Transition Towns there are examples of the revival of older building practices, such as using local building materials to create cob houses using local clay and wood, with straw used as insulation. Certainly these building activities are still few and far between, but in some parts of the world local materials are still often the material of choice, such as the continued use of local granite as a favoured building material in rural Galicia in North-West Spain which helps to maintain the distinctiveness of the local landscape.

In social terms the 12 Steps Guide to the creation of a Transition Town illustrates the way that many meetings are held to develop, discuss and publicize ideas and progress, creating greater interaction between people in the community. Also there is recognition of the importance of recreating old or developing new community social events, such as festivals, while lecture series bring residents up to date with the various schemes being promoted, as well as learning new ideas from experts in sustainable practices. All these events are designed to draw people together in the common cause of re-localizing practices and increasing sustainability. Making Transition Towns more liveable places, as well as being more resilient and able to withstand the probable changes that might occur in the short term from reduced oil supplies and from climate change are key goals.

Apart from the various content areas of the sectors of transitions there is also a strong desire in monitoring community progress. Table 7.2 shows an example of the 20 features that have been identified as being important in a local community recovery project in Port Alberni, Canada and which has often been quoted approvingly on the Totnes Transition Town website. These issues should be returned to regularly to see if progress is being made. In Table 7.2 the principles have been grouped into six broad categories from the original long list in order to improve comprehension, with their original order shown in brackets. It again illustrates how the grass-roots organization that developed this list of principles places value upon what amounts to greater community interaction, diversity and consensus, in order to build better, more liveable and sustainable futures for urban places.

7.5 EcoDistricts

In the last 2 decades there have been many examples of community organizations springing up to improve conditions in their local area. One of these, EcoDistricts, has moved beyond its home base and provides an example of more active and wider forms of local initiatives than those devoted to impoverishment and ill-health reduction described in the Healthy Cities discussion (Chap. 13). This is a new name for an organization whose ideas were initially developed in the Portland Sustainability Institute in Portland (Oregon) a city that has been at the forefront in America of measures to increase its sustainability. Its aim is to improve urban places from the neighbourhood up. In the words of its website:

> A Ecodistrict is a new model of private-public partnership that emphasizes innovation and deployment of district-scale best practices to create the neighbourhoods of the future—resilient, vibrant, resource efficient and just (ED).

Although there are parallels with the Transition Town movement in the emphasis upon grass-roots organization and sustainability, its scale is different, having a within-city focus. This is unlike the whole town emphasis of most of the Transition

Table 7.2 Key characteristics of a locally resilient community. (Source: Reorganized from Port Alberni Community Resilience Manual: CRM 2000)

A. Leadership and Involvement
Leadership is diversified and representative of age, gender and cultural composition of the community (1)
Elected community leadership is visionary, shares power and builds consensus (2)
Community members are involved in significant community decisions (3)
Citizens are involved in the creation of implementation of the community vision and goals (18) and have a Community Economic Development Plan that guides development (18)
B. Attitudes
The community feels a sense of pride (4)
People feel optimistic about the future of the community (5)
There is a spirit of mutual assistance and co-operation in the community (6)
People feel a sense of attachment to their community (7)
The community is self-reliant and looks to itself and its own resources to address major issues (8)
There is a strong belief and support for education at all levels (9)
C. Community Organizations
There are a variety of Community and Development (CED) organizations in the community, such that the key CED functions are well served (16)
Organizations in the community have developed partnerships and collaborative working relationships (11)
D. Employment
Employment in the community is diversified beyond a single large employer (12)
Major businesses in the community are locally owned (13)
The community has a strategy for increasing independent local ownership (14)
There is an openness to alternative ways of earning a living and economic activity (15)
E. Externality
The community looks outside itself to seek and secure resources (skills, expertise, finance) that will address areas of identified weakness (16)
The community is aware of its competitive position in the broader economy (17)
F. Progress
There is an on-going action towards achieving the goals in the CED plan (19)
There is a regular evaluation of progress towards the community's strategic goals (20)

Categories are added and the original numbers used in the source are shown in brackets

Towns, while there is also a more obvious private-public partnership. The organization also has less emphasis on the need to create change because of the problems caused by the climate warming-oil scarcity factors, for it is recognized that many neighbourhoods have additional problems that need to be addressed, such as concerns about facilities, health and justice. Moreover, rather than simply providing advice to other community groups about how to organize and develop policies the EcoDistrict organization is designed to be far more pro-active and outgoing than just having a website that describes its aims and publications or giving lectures to other town groups. For example, it has a partnership with Portland State University to develop courses and programmes to stimulate EcoDistrict ideas; these will soon be available in other affiliated colleges. It has also been organizing annual meetings, called 'summits' for interested people to attend, such as the 2013 Summit in Boston and the 2014 one in Washington DC, as well as sponsoring EcoDistrict incubators, essentially 3 day workshops of executive training, that are described as follows:

> the incubator includes a powerful mix of one-on-one technical assistance, expert facilitation, tours, interactive plenary presentations and working sessions with leading experts in stakeholder engagement, sustainability, urban design, real estate development, infrastructure, and civic entrepreneurism (ED).

Although the meetings include presentations, workshops and advice from experts, the description shows the way in which the movement is not only designed as a top-down exercise to improve sustainability. Instead there is an emphasis on the need to improve community participation and leadership, while in content terms it goes beyond issues of design and infrastructure to promote local entrepreneurship and real estate development, showing a private enterprise focus in addition to public governance. The ideas developed from the five initial EcoDistrict projects, parts of community areas in Portland, that led to reports about how the areas could be improved. By 2013 they were used as guides to create 16 similar projects in various American cities, in addition to one each in Mexico and Canada. Despite the emphasis upon local stakeholder engagement the EcoDistrict approach does seem to provide a more directed, assertive, even self-promotional approach than seen in the Transition Town movement which stresses the need for local groups to find their *own* way to debate and formulate ideas. This can be seen by the EcoDistrict's December 2013 web message asking for charitable contributions to their programme that pointedly states the programme will be developing a

> global standard for district scale development'… (and will)… create a world class programme for urban development professionals (ED2).

This self-defined, rather boosterism-type objective is very different to the rather self-effacing descriptions used in Transition Town publications. Moreover, despite their emphasis upon grass-roots engagement, the neighbourhood assessment reports completed to date seem to be initially created by experts in local and sustainable development, not ordinary citizens, as can be seen in the report for the neighbourhood of Gateway in Portland, where nine sectors of the area were identified as needing attention, and various recommendations for improvement were made (Table 7.3).

The Gateway community contains almost 19,000 people, with 12 % of the population Hispanic. It is mainly a mixed area, primarily middle to low income, with a medium household annual income of \$ 52,000, although 20 % are over \$ 100,000 and 10 % are below \$ 25,000, with a slight majority (54 %) in the owner-occupied housing category. Given the experience of the professionals involved, the assessment report for the area was completed quickly, within 4 months, showing that it is not the work of amateurs finding their way. It first explored the background to the nine sectors, identifying various metrics or community indicators that can be used to measure their current characteristics, ranging from physical issues, such as water use, to social issues such as defining local identity. A set of strategies for improvement to create carbon-neutral districts were recommended by the researchers,

Table 7.3 EcoDistrict strategies in a Portland (Oregon) district. (Source: Summarized from Portland's Gateway EcoDistrict assessment report: EDGP 2011)

Sectors	Summary strategies	Specific recommendations
Energy	Reduce, Generate, Offset (not locally conserved)	Energy conservation and solar retrofits District Energy schemes District dashboard; Education-Competition campaigns
Air quality	Enhance, Reduce, Sequester and Offset	Change street lights to LED Organize indoor air quality campaign Organize district carbon calculator
Water	Reduce, Reuse, Reconnect	Water conservation and reuse District greening and infrastructure strategy Connect to hydrological functions
Access and mobility	Shift, Share, Manage	Improve pedestrian and bike connections to main destinations Add and maintain safe bike parks Provide incentives to promote alternatives to cars
Place-Making	Identity, Form	Create a gateway brand and distinctiveness Create signature parks and high quality public space Add green infrastructure
Social cohesion	Instigate and Interact	Develop systems to engage community interactions Create capacity-building for social capital development Organize public social events: parties, bike rides etc. Create district dashboard to publicize events
Habitat-Ecosystem	Enhance, Connect, Restore, Functioning	Plant diversity of trees and other vegetation Add vegetation in wildlife corridors Install stewardship ethic in community
Material management	Reduce, Reuse, Recycle	Programmes to reduce material and product use Build commodity market for used materials Support Columbia bio-gas project
Equitable development	Access to Jobs, Housing; Equal Representation and Oppurtunities	Provide green quality housing Create local jobs Support pathways to opportunities and culturally-appropriate employment

initially by some catchy words to summarize the objectives, and then through specific strategies that can be linked to particular policies, as shown in Table 7.3. However, once these information, data-gathering and strategy recommendation phases have been carried out, the report, like those for other EcoDistrict projects, was given to the EcoDistrict Steering Committee composed of members from the community being studied. This is the group that will *make* decisions about what to implement, although guidance will be provided by EcoDistrict personnel. An important element in the subsequent EcoDistrict development process is a follow-up stage, namely the careful monitoring of the changes in the various sectors to see if real progress is being achieved.

The dynamism and commitment to local community or neighbourhood development of the EcoDistrict organization must be praised. But for those working in the field of urban sustainability and in green issues, there is little in the content and in the strategies that is really new. The uniqueness lies in their attempt to integrate or at least inter-relate vary different sectors of the character and functioning of neighbourhoods or community areas within cities to create meaningful change through grass roots mobilization with professional help. Yet there may be several weaknesses in subsequent stages, weaknesses that plague all grass-roots organizations. One is the fact that the implementation of the ideas from the experts that produced the report depends on community leaders who may lack the skills and the finances to ensure that policies are successfully pursued. As yet it is too soon to see how successful these EcoDistricts will be in transforming their neighbourhoods. Many cities do not have the track record of proven environmental interest and improvements that is found in Portland, nor the commitment of a solid majority of its population to these types of sustainability initiatives. But since the EcoDistrict approach claims to provide skill-training, and recommends policies with a proven record of success, there must be hope that the approach will be more successful than the alternative in most cities, mainly top-down policies that seem to be imposed on areas by city officials, with few attempts to involve the local residents in generating and maintaining the projects. Such approaches mean that there are few signs of any collective feelings of what amounts to local ownership and belonging in the projects, as when plans and implementation are produced by, and for, the local community. Also this organization, which also stresses the need for public-private partnerships, does seem to be a product of an American system that often distrusts direct government action and often prefers private enterprise and local solutions. So its applicability to other countries with more centralized planning systems may be problematic. But a general unresolved question is whether land owners will co-operate in this development process and what happens if they do not. Moreover it is not clear whether the citizens in the local districts within cities have really got the power to reject proposals that they think are detrimental to their area, or to generate their own. NIMBY-ism is often a powerful response against change and many in inner cities may resist what others see as progress. There may also be resistance from the different views held by city planners and city councils, although in some cases they may be considering the wider needs of the city. More generally there are few mentions of the many problems faced by all grass-roots organizations (Davies and Herbert 1993, Chap. 6). Certainly the desire of the EcoDistrict organization to provide professional advice to their various projects by training people to identify problems and to help the implementation of some of the strategies through local community action is a new and significant venture. But there can be no guarantee of successful conclusions. Nevertheless, the Ecodistrict approach does provide another useful way of improving urban areas, not by new developments, but by the more difficult task of retro-fitting districts within cities to create a more sustainable future, through an approach that involves local community development and involvement.

7.6 Conclusions

The two organizations described in this chapter are alike in wishing to create more sustainable, resilient and liveable urban places that involve grass-roots community actions in towns and cities, often using similar policies to those described in previous chapters, especially Chaps. 4–6. However the two examples operate from different imperatives, at different scales and with different emphases. The Transition Town movement is mainly based on complete small settlements, although with some exceptions, and has grown because of worries over the future problems that will occur from climate change and from the descent from some peak oil situation. As important has been the realization that the economies of many towns, in services and industries, are dominated by outside ownerships that makes them too dependent upon decisions made by remote governments and often even remoter faceless corporations who are more concerned about their profits and market share than the fortunes of the towns in which they are located. So the pioneers of the Transition Town movements and subsequent adopters seek ways of creating greater local interaction, and through co-operation try to find ways to counteract future problems in many sectors before change is thrust upon an unprepared settlement. So a big question is posed by The Transition Network.

> For all those aspects of life that this community needs in order to sustain itself and thrive, how do we significantly increase resilience (to mitigate the effects of Peak Oil) and drastically reduce carbon emissions (to mitigate the effects of Climate change). (TNTI)

The various sectors of transitioning are designed to find ways of coping with the new conditions by re-localizing many activities and to re-energize a local town or community to provide commonly accepted goals that increase the degree of sustainability in economies, social relationships and well as in environmental terms, creating a greater social and economic resilience or ability to counter changes that are likely to take place in the near future. This does not mean that TT enthusiasts are trying to create a locally self-sufficient economy, closed like the agricultural communes that were developed from the 1960s by groups that rejected the consumer-driven, individualistic life-styles. Rather the idea is to increase the town's ability to cope with future climate changes and reduced oil supplies, to increase local enterprise, so more money circulates in town and profits are reinvested locally not flowing out, in addition to creating healthier, more socially active and interactive communities. Many of these ideas can be found in the parallel movements that have developed to create sustainability within districts in cities, especially the emerging theme of EcoDistricts, through local community action. These, however, are clearly at a different scale to Transition Towns and as yet are primarily U.S. based and more linked to public-private partnerships. What this approach is attempting to do is to create a more integrated and even professional approach to inner city redevelopment, since professional urban development experts are involved in an early stage, although unlike most government planning approaches there is a determined effort to create community engagement.

There can be little doubt that both approaches are welcome and needed additions to the government or developer-led processes of change that are seeking more sustainable solutions, or by individual decisions. But in both cases, however, the problems of creating maintaining change through grass-roots organizations are never really answered (Davies and Herbert Chap. 6). How many people in the towns are involved in the process? How are the eternal problems of the burn-out of volunteers being solved? Can effective leadership be found and be maintained—a leadership based on commitment to the area and projects, and not to their potential use of the post as a ladder for future political office? Can sufficient money be found to sustain individual policies as well as finding long-term funding? Unlike governments, these organizations do not have taxing powers to finance their goals. What powers do the local organizations possess, or can they create, or do they mainly depend on convincing elected officials and their staff of the utility of their desires? These and other problems are always difficulties that impede the work and continuity of local community organizations.

Despite the rapid growth of the Transition Town network some critics have argued that too many of the objectives of the movement are designed to provide some sort of return to the past when there were more local linkages, supplies, employment and decision-making, which many see as impossible in the contemporary interconnected world; to do so will create higher prices and inefficiencies. However there is no real evidence of some type of return-to-the past movement emerging. Instead there are many signs that Transition Towns are managing to achieve some of their goals of greater local resilience and an ability to mould more of their own futures, rather than dependency on distant decision-making, by convincing other organizations and local governments of the need to change. Of course this is a matter of degree; the long-distance linkages in food supply and many resources still exist and are dominant, but more local connections seem to be emerging which may create a more balanced future. Other critics will argue that the types of changes being pioneered in the small number of Transition Towns will have a minimal effect on world climate change. Moreover the Peak Oil threat may be postponed for a generation given the increasing supply of natural gas and oil from frontier developments and from the fracking of shale beds. However, unless enough renewable energy sources and new technologies for energy generation and storage are found, the problem of Peak Oil may re-emerge in the next generation and there is always the threat to towns of a disruption of supplies from natural hazard disasters, whose impacts are described in Chap. 8. So it does seem sensible to use the extra time that we have available, given the temporary increase in oil and gas supplies and lower costs, to create more sustainable towns that are more locally resilient through the types of ideas being developed in the Transition Town and EcoDistrict movements. Both have provided guides to show how change can be achieved. Transition Towns in particular see themselves as producing a catalytic effect, leading many neighbouring local areas to also implement the changes they have pioneered, which may lead to a real cumulative effect that will be influential in increasing sustainability, let alone their other useful social functions. They also see themselves as shaming or convincing governments into more active policies. In addition, of course, any reductions in local energy use will lead to a less polluted local environment and therefore a healthier population in the adopting towns. Also the obvious community re-energization of the population in many of these towns and districts within cities seems to be substantially improving the social life and co-operative abilities within the participating towns, in addition to providing pathways to less dependent economies and more socially aware communities. But it must be emphasized that all the discussions and policies being implemented in both Transition Towns and EcoDistricts are not designed to produce one goal. Change is an ongoing process, so the adjustments in the Transition Town movement are really designed to promote a *transition* to a different future, in which there should be constant re-evaluation to determine whether enough progress is occurring, given the desire to build up local resilience in order to build a better, more liveable and sustainable future for urban places. These are more organized and directed movements than the individual, small scale, bottom-up initiatives being carried out by many individuals and community groups to improve particular places in cities. These are a diverse, but still useful, set of modifications that are often known as *tactical urbanism,* with examples promoted by the Street Plans Collaborative (SPC).

References

Alinsky, S. (1972). *Rules for radicals*. New York: Random House.

BSD: Berkshare Dollars. http://www.berkshares.org/whatareberkshares.htm. Accessed 12 March 2013.

CivicEconomics. (2012). Indie impact study series. A national comparative survey with the American Booksellers Association. http://www.civiceconomics.com/app/download/.../INDIE+IMPACT. Accessed 11 March 2013.

CRM. (2000). *Community resilience manual: A resource for rural recovery and renewal*. Port Alberni: The Centre for Community Enterprise

Davies, W. K. D., & Herbert, D. T. (1993). *Communities within cities*. London/New York: Belhaven/Halstead–Wiley.

DEFRA: Dept. of Environment Food and Rural Affairs. (2005). http://archive.defra.gov.uk/evidence/economics/foodfarm/reports/documents/foodmile.pdf. Accessed 10 Feb 2013.

DEFRA: Dept. of Environment Food and Rural Affairs. (2010). *Food: 2030*. London: HMSO, DEFRA.

Desrochers, P., & Shimizu, H. (2012). *Locavore's dilemma: In praise of the 10,000 mile diet*. New York: Public Affairs.

ED: EcoDistricts. http://ecodistricts.org/tools/framework/#. Accessed 10 May 2013.

ED2: EcoDistricts. http://ecodistricts.org.(us1.campaign-archive1com/?u=6b817cf1qb7.d40b611d521b75id=47fef44ef6e=e16593de3b). Accessed 10 Dec 2013.

EDAP: Energy Descent Action Plan. http://www.totnesedap.org.uk. Accessed 14 April 2013.

EDGP. (2012). Gateway Ecodistrict, Portland (Oregon). http://ecodistricts.org/wp-content/uploads/2013/05/101411_Gateway-EcoDistrict-Assessment-FINAL-Report.pdf. Accessed 10 Dec 2013.

EV: Ecovillages. http://directory.ic.org/records/ecovillages.php. Accessed 10 April 2013.

Foster, C., Green, K., Bieda, M., Dewick, P., Evans, B., & Flynn, A., et al. (2006). *Environmental impacts of food production and consumption : A report to the Dept. of Environment, Food and Rural Affairs*. Manchester Business School, London: DEFRA. http://www.ifr.ac.uk/waste/Reports/DEFRAenvironmental%20Impacts%20of20Food%20Production%20%20Consumption.pdf. Accessed 12 April 2013.

Gardiner, J. (2004). *Wartime Britain 1939–45*. London: Hodder Headline Book Publishing.

Hall, P. (1988). *Cities of tomorrow*. Oxford: Blackwells

Herold, D. M. (2012). *Peak oil*. Norderstedt: Herstellung und Verlag.

Holmgren, D. (2002). *Permaculture: Principles and practices beyond sustainability*. Hepburn: Holmgren Design Services.

Holmgren, D. (2009). *Future scenarios: How humans can adapt to peak oil and climate change*. Hepburn: Holmgren Design Services.

Hopkins, R. (2008). *The transition handbook: From oil dependency to local resilience*. Dartington: Green Books.

Howard, E. (1898). *Tomorrow: A peaceful path for reform*. London. Reprint (1965) as: *Garden cities of tomorrow*. London: Faber and Faber.

McWilliams, J. E. (2009). *Just food: Where locavores get it wrong and how we can truly eat responsibly*. New York: Little Brown.

Potts, R., Simms, A., & Kjell, P. (2005). *Clone town, Britain*. London: New Economics Foundation. http://www.neweconomics.org/gen/z_sys_publicationdetail.aspx?pid=20. Accessed 12 April 2012.

Ryan-Collins, J., Cox, E., Potts, R., & Squires, P. (2010). *Reimaging the high street*.London: New Economics Foundation. http://www.neweconomics.org/publications/entry/reimagining-the-high-street. Accessed 10 Sept 2013.

Saunders, C., Barber, A., & Taylor, G. (2006). *Food miles: Comparative energy and environmental performance of New Zealand agricultural products*. Agricultural Business Research Report No. 285. New Zealand: Lincoln University. http://www.lincoln.ac.nz/documents/2328_rr285_s13389.pdf. Accessed 15 Oct 2013.

SM: Slow Money. http://www.slowmoney.org. Accessed 10 April 2013.

SPC: Street Plans Collaborative. http://tacticalurbanismsguide.com. Accessed 24 August 2014.

Tasch, W. (2010). *Inquiries into the nature of slow money*. White River: Chelsea Green Publishing.

TNP: Transition Network Primer. (2006). http://transitionnetwork.org/resources/transition-primer. Accessed 12 April 2012.

TNTI: Transition Towns International. http://transitiontowns.org/TransitionNetwork/TransitionInitiative. Accessed 12 April 2012.

Chapter 8
Winter Cities

Wayne K.D. Davies

Winter is a resource and an asset.
Theme of World Winter Cities Association of Mayors' Meeting, 1982

8.1 Introduction

In the late twentieth century it was recognized that the development of settlements during the previous hundred years had paid less and less attention to the effects of local climates, especially extreme ones, upon either urban design, or the way that these conditions influenced the behaviour of people in these centres. This was in contrast to the situation throughout history, where the plans, building designs and construction materials of traditional towns and cities, as well as the technologies and behaviour of people, were intimately related to the local climate, whether cold, wet, hot or other extreme conditions. For example, the desire to cool buildings in areas of summer heat led to shuttered windows to keep out the sun, bungalows with verandas to catch breezes, or wind towers in ports along the Persian Gulf that funnelled cooling winds into houses. The design-natural conditions relationship can also be seen in pagoda designs in Japan that allowed lateral and vertical shifts, making them resilient to all but the largest earthquakes. Even some historic building regulations, such as the famous Spanish Laws of the Indies in 1573 (Mundigo and Crouch 1977), acknowledged the importance of local climates in its guides to the construction of towns in their colonies, such as recommending that narrow roads should be constructed in hot climates to ensure shade, while wider roads were better in cooler climates to catch the sun.

These climatic adjustments in settlements were particularly obvious in cold climates, such as increasing the slope of roofs in Alpine areas to reduce snow

W. K.D. Davies (✉)
Department of Geography, University of Calgary,
2500 University Drive, N.W., Calgary, AB T2N 1N4, Canada
e-mail: wdavies@ucalgary.ca

W. K.D. Davies (ed.), *Theme Cities: Solutions for Urban Problems*,
GeoJournal Library 112, DOI 10.1007/978-94-017-9655-2_8

accumulation, together with an emphasis on southward-facing doors and windows to increase exposure to sunlight and warmth. Indeed the latter concept was one of the recommendations in the Swedish national building code of the late nineteenth century. Similarly, historic societies in northern lands developed various aids to movement across snow and ice, such as dog sleds, snow shoes, skis and sleighs, which often meant that mobility across uneven land was easier in winter. These new inventions and the development of building designs to take account of various climatic conditions and variations evolved through long periods of trial and error.

Unfortunately the traditional practices that adapted settlements to the extremes of their local environment were increasingly ignored from the early twentieth century. Older and tested building practices were replaced with new house designs more suited to temperate environments, based on cheap fossil energy sources and materials brought from considerable distances. In northern lands these modernizing influences, especially under the impact of suburbanization, started to dominate the additions to existing towns and the many new settlements in the far north that were primarily linked to new mining, fishing, and administrative developments. As a result, many of these settlements were badly adapted to the often severe conditions of winter, making them less sustainable and more placeless, and containing many residents who viewed winter as a season to be endured, rather than welcomed. Recognition of these problems led to the creation of the concept of Winter Cities in the late 1980s, a movement that has gained in popularity and impact in the last two decades, although most of the settlements supporting the ideas were really towns, rather than cities in size. This was one of the first of many movements characterized by an adjectival prefix to 'cities' that sought to create new approaches to the development of settlements at the end of the twentieth century, in this case to the often severe climate conditions of northern lands. However, this movement was not simply a description of the *presence* of winter in these settlements. Rather, the Winter City approach attempts to reduce winter's *negative* consequences and to emphasize its *positive* features and opportunities, so as to create more sustainable and liveable settlements in both the structures and the life of the people living in these centres. This objective of treating winter as an asset is reflected in the introductory quotation that was the keynote concept of the first meeting of the Sapporo World Winter City Mayors' conference in 1982. It is the first theme city type that has emerged in an international context with an emphasis upon mitigating problems posed by local climate conditions, although it is anticipated that other movements to improve sustainability in other climate regimes will emerge in the future.

8.2 Origins of Winter City Ideas

The development of the Winter City approach comes from several sources. The pioneering work of key individuals, especially the planner-geographer, Norman Pressman in Canada (1988, 2004) and the retired Canadian journalist J. C. Royle, were especially influential in creating and publicising the idea in North America.

Together with W. C. Rogers of Minneapolis, they were leaders in creating the Liveable Winter City Association (LWCA) in 1983, which was designed to provide a focus for new ideas on winter cities. In Europe, urban academics such as Jorma Mänty in Finland (Mänty and Pressman 1988), and planner-architects, like Ralph Erskine and Boris Culjat in Sweden, as well as many individual city planners in cities with severe winters, were also seeking solutions to the problems posed by northern climates on settlements. Many of these academics and planners soon became linked to the growing international Winter City network. But despite the publication of many useful books, articles, a quarterly magazine, as well as frequent meetings on the Winter City theme, the LWCA almost floundered in 2005. This was not because of a lack of interest, but because its non-profit and volunteer structure could not sustain the work. Fortunately, the organization has been re-constituted as the Winter Cities Institute (WCI), based in Anchorage (Alaska), and which continues to promote plans for coping with winter conditions. At the same time that the international network developed, individual planners in northern towns were seeking more effective ways of solving the problems posed by winter. This has led to increasing internet exchanges and regular meetings between interested parties, such as the Nordic cities of various sizes, but including capitals, that created the Nordic Eight Network to discuss issues of sustainability and design (NE). More generally a world-wide network on the topic was created by town administrators under the organization of the World Winter Cities Association for Mayors (WWCAM). Initiated by the city of Sapporo in 1981 and with meetings organized biannually, the association now consists of 20 cities in nine countries, although non-member cities can attend as observers. The last three meetings were held in Nuuk (Greenland in 2008), Prince George (Canada in 2010) and Ulaanbaator (Mongolia 2012), with future meetings planned for Hwachen (South Korea 2014) and Sapporo (Japan 2016).

Meetings based on the Winter City theme by the above organizations and other interested groups have attracted delegates from different northern countries to share knowledge. They have produced publications with four main Winter City aims, with their work often supported by national and international organizations, such as the European Union. The first aim is essentially an *applied urban climatology*, essentially to understand the effects of winter and its changes on both settlements and the life-styles of people within them. The second is to identify, as well as disseminate, the *best practices about planning urban areas in winter climates*, especially those based on modifying traditional practices that had fallen into disuse through modernization, and also to promote energy conservation measures. The third is to search for *ways to improve the planning of settlements as well as the interactions of their residents,* in order to provide greater adaptation to winter conditions and to create more unique places through design initiatives. Finally, the intent is to recognize and *improve the life of the most vulnerable* in winter cities, especially children, the old, the poor and the infirm, as well as the homeless, who are handicapped by their limited mobility or poverty. These approaches all have the aim of increasing the sustainability and liveability of winter settlements for all people, thereby more positively relating the cities to the problems posed by winter. These Winter City ideas are summarized in this chapter by successively dealing with the problems posed by winter, the ways that designs can improve the sustainability and liveability of settlements, and the opportunities for behavioural initiatives to improve life in winter.

8.3 Problems of the Four Types of North

It has been argued that the problems of winter for the people who inhabit these areas of seasonally freezing conditions relate to two major issues, *the physical conditions of the 'actual north'*, and what is described as the '*psychological north*' (Pressman 2004, p. 4). Although these two ideas may be basic to understanding the difficulties of settlements in the high latitudes of the northern hemisphere, they need to be complemented by problems posed by the '*economic*', as well as the '*social*' effects of winter conditions upon settlements and their residents. Before dealing with these four Winter City problems some clarifications must be made. The term 'north' in the Winter City literature is used as a surrogate for areas that experience long winter conditions. Obviously settlements in many high mountainous areas further south have similar problems, as do places in the high latitudes of the southern hemisphere. so the term 'north' is not literally accurate, but may reflect the fact that most of the settlements affected by winter conditions are in the northern hemisphere. Also the term is usually applied to places where freezing conditions exist for several months, although it could be argued that places further south with long, rainy winters and also constant cloud cover experience some of the same sets of difficulties.

8.3.1 The Physical or Actual North

Students of Winter Cities have never really clarified the exact location of places to which the term applies. One of the simplest ways of defining such centres is by including places where at least two months of the year have average temperatures below 0 °C, while the idea of a 'hard or extreme north' can be identified within this zone, those with a more intense winter, perhaps separating out those places beyond the Arctic Circle that have at least several weeks without sunlight. The Canadian geographer Louis-Edmond Hamelin (1979) used ten social and physical indicators to measure the character of what he described as 'nordicity' in Canada. Essentially this defined the severity of winter in different northern Canadian regions, classifying the country into four zones (near, middle, far and extreme north) outside the main ecumene, the populated areas in the south. In general, the severity of winter increases as one goes north, but latitude is not the only variable involved; altitude and the regional effects of mitigating or enhancing features upon local climates, such as cold or warm sea currents, and the degree of exposure to cold winds have considerable effects. These variations mean that the term 'Winter City' is a vague one in a location sense, without any accepted precise definition. Yet without going into the detailed variations of the degree of the physical severity of winter in various areas, it is clear that all settlements that experience the challenges of this season—even if only occasionally—have to cope with several major problems, although these vary in their severity.

The most obvious problem of winter cities is the *increased amounts of darkness* in winter, with short days and low angle sunlight. There are few settlements in the far north of the northern hemisphere where the sun never rises for weeks or months at a time, but some idea of the conditions can be seen from the fact that at 78°N (mid way between the Arctic Circle and the North Pole) there are 94 days without sun between November 13th and February 15th. At the Arctic Circle the period between these days has only 6 and 8 h of sunlight respectively, whilst the sun dips below the horizon at the winter solstice on December 22nd. Hence Hammerfest (Norway) has no sun above the horizon for two months at 70°39 N/23°.41 E, and claims the title of the world's most northernmost town, although the small settlement of Huningss-vag at 70° 58 N/25°58 E. is actually further north, as is Barrow in Alaska. Further south, between 60–62°N, the number of urban places increases and places such as Anchorage, Oslo, Stockholm and St. Petersburg can count on over forty days with less than 8 h of sunlight, although cloud cover can exacerbate the extent of the periods of darkness.

Obviously, winter in northern latitudes results in many days with *freezing conditions.* Prolonged exposure to cold can be fatal, unless people are suitably clothed to keep warm, with winds making conditions worse. Hypothermia occurs because the body reduces blood supply to the extremities, such as fingers and toes, so they get frostbite. If pain is felt, then the nerve endings are still sending messages to the brain, but continued exposure leads to a numbness and then no feeling, which is life-threatening, However the effects are not just physical since exposure is marked by a slowing of activity and the ability to think, which is often indexed by what a recent survey of the effects of cold called the 'umbles', where unprotected people may experience the sequence of mumbling, fumbling, grumbling, stumbling, and finally tumbling over and dying from too much cold (Streever 2009).

The cold also means that precipitation comes in the form of *frost, snow and ice*, all of which reduces the longevity of buildings, given the weight of the last two, and the freeze-thaw effects between cold and warmer temperatures that widen cracks, as well as hindering the comfort and especially mobility of the modern populations in towns and cities. In addition, the threat of avalanches sliding down mountains and burying settlements, or icebergs in the sea and shorelines, create additional threats to roads, towns and shipping in such areas. In more northerly areas where the ground is below 0 °C for over two years, the cold creates permafrost conditions—essentially permanently frozen ground (Harris 1986). The areas that are affected are usually divided into four types—continuous, discontinuous, sporadic and alpine permafrost zones. They occupy approximately 22 % of the land area of the northern hemisphere, together with considerable areas under the Arctic seas. In Canada the zone of permanent permafrost starts at 65°N on the Alaska border, dipping southwards to the southern shore of Hudson Bay at 55°N, whilst the sporadic zone runs from 53°N on the Alberta-B.C border to the southern tip of James Bay and to the southern edge of Labrador (Bone 2005). The problem with permafrost is that when the water in the soil becomes frozen it expands in volume and bulges upwards, while any warming causes the frozen water in the soil to melt, reducing its volume. This creates landscapes that are called thermokarst, marked by depressions and domes produced the freezing and thawing processes. Roads and buildings in these areas are, therefore, more subject to cracks, subsidence and collapse. Even towns and cities with less extreme winters suffer from similar structural damage due to freeze-thaw effects, which mean that roads especially require more maintenance and do not last as long. An additional problem of winter is the greater presence of *shaded areas*, for the low angle rays of the winter sun are more easily blocked by many features, ranging from mountain ranges to individual buildings. These shaded areas are characterised by colder conditions and take longer to warm up. The effect of shaded and sunny areas in Alpine valleys was well known historically in Europe, identified by the German terms sonnenseit and schattenseit, or l'adret and l'ubec in French zones, which refer to the sunny versus the shaded sides of the valleys. These conditions led to the obvious response of more vegetation growth, as well as more settlements, on the sunnier valley sides.

Many winter settlements are also subject to additional problems that come from the *variability of winter conditions*. Intermittent storms and episodic periods of deep cold or snow storms, within or between years, increase the problems of coping with winter during these times, frequently bringing modern life, with its dependence on daily routinized rhythms, to a standstill. The frequency of high winds leads to greater risk of exposure due to wind-chill effects, as well as making travel more difficult. Obviously *the severity of these conditions* varies throughout the north, increasing with higher latitudes and altitudes, and in places that do not have shelter, such as from mountain areas. In addition, influences such as the degree of exposure to winds, their strength and frequency, or the presence of a cold offshore flowing current can increase the impact of winter upon continental margins, such as the southward cold Labrador current in eastern Canada. By contrast, warmer currents, such as the Gulf Stream from the Caribbean that moves north-eastward to northern Europe, reduces the severity of winter, ensuring that the two sides of the Northern Atlantic have different climates at the same latitudes. So the European settlements at the same latitude as those on the American continent are much warmer. Moreover, the effect of *modernity worsens the winter conditions*. For example: pollution from cars soon dirties the snow, creating an ugly environment; the impact of traffic compacts snow into ice, which is more difficult to remove or to travel on; the frequent ploughing of snow off roads to allow vehicles to travel, often produces mounds of snow alongside the roads that are difficult to traverse; when municipalities move snow from roads to dumps, an additional pollution hazard occurs when it melts, given the fact that it contains so many pollutants from the exhausts of cars. In addition, there are a series of additional climate hazards, such as the negative effects produced by wind funnelling along roads aligned with the prevailing winds. The Finnish architect Kuismanan (2005) has provided a comprehensive review of the winter climate effects upon buildings, showing how major buildings need to be aligned across the winds to provide a wind break. He also argued that the local topography should be taken into account in designing structures and recommends having tightly clustered smaller buildings to retain warmth and break the effect of wind, as well as ensuring that designs create positive micro-climates around these structures to encourage people to go outdoors.

Another problem comes from *isolation* effects. The density of population is far lower in northern latitudes, especially those that have limited agricultural activity, namely places beyond 56°N in the Prairie provinces of Canada, or 62°N in Northern Europe. The northern settlements that exist are mainly those of traditional peoples, or the relatively new mining, forestry, fishing, tourist or administrative centres, which have helped push the frontier of settlements further north in the last century. Although there are more places in the north than in the past, most are at considerable distances from one another, which means that many people in northern settlements often suffer from isolation, due to a lack of contact with other people, a problem exacerbated by the expense of travel to other places.

Focus on the winter conditions alone should not disguise the related problems that occur in the *'transitional periods'* to and from winter. For example, the initial effects of winter may be transitory, in the sense that the onset of winter may be delayed by reducing the initial impact of the cold and snow through sensitive environmental designs and modifying the behaviour of the people in settlements. But many negative effects in these periods also occur, such as freezing drizzle, which produce ice accumulations on trees, roads, buildings and especially power lines, which are hard to remove and often cause them to break. Although ice-storms are common in eastern Canada they are usually of short duration, and since the larger cities have equipment to deal with the ice their effect is limited. But in January 1998 the area from eastern Ontario, Quebec, New Brunswick to Nova Scotia and the United States border experienced a massive storm with ice accumulations often greater than 4–5 cm on exposed surfaces, such as the ground, trees and hydro lines, for six days. This was a result of moist air moving from the south being trapped over a very cold surface layer, which meant that as the storm progressed, the rain from the moist air was super-cooled as it passed through the cold layer and froze on all exposed surfaces. Electricity was unavailable for over 4 million people initially; over 1,000 steel electricity pylons collapsed, millions of trees were felled, roads were impassable due to ice and tree blockages for days. Insurance claims of over c$ 1.5 billion were recorded, but additional costs to government and industries, especially the dairy and maple tree industries, probably created a bill of around c$ 5 Billion, although estimates vary (Munroe 2012). The storm was seen by Environment Canada as affecting the most people ever in Canadian history, although only 28 people died, with the most severe effects being felt in the large metropolitan area of Montreal. Eventually 15,000 military personnel were brought in to help clear the roads, trees and restore power. This provides another example of the effect that a severe natural hazard can have on urban areas, to complement those described in the Resilient Cities discussion (Chap. 7) and shows the havoc that can be produced by such a large winter storm. More generally in hilly or mountainous areas near towns, both the snow accumulation, or freeze-thaw effects, often lead to avalanches of snow or rock falls respectively, problems that may increase as climate warms.

At the end of the winter, a different problem occurs. The melting of the heavy snow accumulation produces large volumes of semi-melted ice and snow, often creating slush avalanches in the far north and then a rapid water run-off which can lead to severe flooding. Even bigger problems of flooding occur in more southerly winter climates in spring. One comes from the ice-jams that block rivers and cause a build-up of water that then overflows the river banks. Another comes from the fact that snow and ice melts faster in the southern watersheds of northward flowing rivers, which may still be frozen to the north and therefore block drainage. In the Russian Spring this period also leads to a period of saturated soil that historically made travel almost impossible for two weeks on the dirt roads, which become seas of mud in the large flat plains. It is period known to the Russians as Rasputitsa (literally 'time without roads'), and to the Finns as Rospuuto. In Russia, the autumn was a similar period of travel difficulty, in this case because of the effect of heavy rains. The Finns also used the name Kelinkko for the spring period when ice started to thin and then melt, making travel on rivers and lakes dangerous. An additional problem in urban areas near areas with lots of lakes and rivers is the appearance of large numbers of mosquitoes and other flies in the late spring. This is the season that was often regarded by early explorers in northern lands (Davies 2004) as the most unpleasant. It remains a drawback today, despite frequent spraying in towns to reduce the insect numbers, which leads to problems of chemical pollution.

8.3.2 The Psychological North

The attitudes of the majority of residents in urban places that experience severe winter conditions range along a scale from finding it 'an inconvenience' to feelings of 'fear', although it must be admitted that some people, especially long-time residents, prefer winter. Humans have counteracted the worse effects of cold by using more layers of clothes, developed specialised insulated clothing, and created more effective insulated houses with better warmth, for exposure to extreme cold can lead to hypothermia and death. But it has been gradually realised that solving the problems of cold through layered clothing is not enough. Winter conditions, especially the reduction of sunlight, can create severe psychological distress in some people. This is a condition that is now known as Seasonal Affective Disorder (SAD), a mental state characterized by misery and even clinical depression. This condition seems to be directly related to the limited sunlight of winter, which is often further reduced by cloud cover. It is a condition recognised by different local or regional terms, such as the term 'Lappsjuka' (literally Lappsickness') used by Stockholmers. A related health problem due to the limited exposure to sunlight and ultra-violet rays in particular is that the body is unable to make enough Vitamin D in winter. This leads to poor bone growth that produces diseases such as rickets, and the deficiency may also be one of the reasons why influenza is more prevalent in winter. So people in northern climates need vitamin D supplement to compensate for this effect, although a diet with a high fish oil intake can also help counteract the problem. More recently the deficiency has been alleviated by the development of sunlight lamps or pads. The rapid variability of conditions in winter can also create additional difficulties for humans, especially in places close to mountains where warming winds, such as the Chinook in Alberta, or the Föhn in Alpine areas, frequently occur in winter and often within a few hours. The majority of the population view these warm, dry winds as a welcome relief from the winter cold, but some people get migraines during these events that are characterised by rapid climate changes.

When these area-specific health difficulties are added to the drab, sterile and colourless environment produced by the absence of plant and animal life in winter, it is not surprising to find that many people who are not native to these climates view winter with at least trepidation. A more general result is that many people feel less inclined to go outside the warmth of their own house, leading to lower patterns of interaction and socialization during the winter. Yet one must note that there can be compensation in these winter climates, especially at times of anti-cyclonic conditions or in the dry continental interiors where the presence of many days of open, blue skies, can be psychologically uplifting, so long as the cold is kept at bay. This means the brisk, clear and bracing air of winter is seen as a bonus by some, so long as it is not too cold. Hence one objective of the Winter City advocates has been to reduce the negative psychological effects of this season and to develop ways of creating more winter activities and better designs to embrace and enhance its positive elements.

8.3.3 The Economic North

Residents of settlements in high latitudes face increased economic costs. Compared to areas further south they have much *higher heating costs,* whether from the need to buy specialised clothes, or to warm buildings. They also have *greater costs of transportation,* since cars need chains or special winter tyres to get grips on roads covered with snow or ice, while the isolation of many settlements means that many, especially in the 'hard north', can be reached only by expensive air connections for much of the year. However the increasing use of snowmobiles—the modern equivalent of the sleigh—on snow surfaces does improve mobility within cities and certainly to places outside. In high latitude maritime areas the sea freezes, so ports have to be closed down in winter, reducing their annual level of economic activity. In Canada this gives a big advantage to ice free ports such as Halifax, compared to Montreal where the port is closed for three months or more in winter, while a port such as Churchill on Hudson Bay is typically open for only three months of the year, restricting its ability to act as an entrepôt to the agricultural Canadian Prairies to the west. Yet in the Far North of Canada many mining camps are served more cheaply and efficiently in winter, especially for the transport of heavy equipment, because ice roads have been built over the snow that covers the rocky, undulating terrain and many lakes, an environment that makes it uneconomic to build permanent roads. Nevertheless, the isolation and distance of the settlements from agricultural areas in the south results in *higher costs of basic foodstuffs*, especially in the areas beyond the zones where agriculture is possible, since far more products have to be brought into the region. Some countries, such as Canada, provide subsidies to reduce the prices on many perishable, nutritious foods for people in the far north, although they are still more expensive than in the south, as seen by the Nutrition North website in Canada (NNC).

All modern settlements in these climates also incur *increased maintenance costs* that need specialised trucks to clear snow and add gravel or salt to roads to reduce the effect of the snow and ice from impeding movement on roads as well as other pathways, and even reducing snow accumulation on roofs. Places to dump the snow in winter and remove the gravel placed on roads in winter must also be found. This can also be a problem because the additives to roads make the dumps a pollution hazard. At the end of winter they also have to remove the gravel and other products that were placed on the snowy, icy roads to increase vehicle traction in winter, whilst the increased corrosion on cars due to the use of gravel or salt is another cost, although this falls on the vehicle owner, not the municipality. All these problems mean that settlements with winter conditions need to keep a stock of specialized equipment, and the personnel to maintain them, to clear roads, while airlines need de-icing equipment and vehicles to clear runways. Places that only occasionally have snow rarely invest in enough equipment and suffer accordingly in their periodic winter episodes.

It has already been noted that many areas of the north have also to counteract the additional problems caused by *permafrost,* areas of permanently frozen ground. Utility lines, such as water, power and sewage, will heat the soil, causing the frost to melt and the utilities to sink and buckle. Hence utilities must be above ground and insulated, adding to the infra-structure costs of the settlements. Paved surfaces are also affected by the freeze and thaw, meaning they have to replaced earlier than in southern settlements. Finally, there is little doubt that people in winter cities face *greater medical costs.* Some are due the effect of the conditions, in which colds easily turn into pneumonia unless precautions are taken, or when prolonged exposure to the cold results in frostbite; others are the result of more people falling on ice and breaking limbs, or from more accidents as a result of cars skidding on ice.

In general, therefore, the economic costs of life in the settlements that experience long winter conditions are much greater, though few accurate comparative accounts have been made to quantify the differences. Yet these economic costs have been reduced by the decisions of some national governments to provide tax rebates for people who live in northern settlements, often, like Canada, on a sliding scale that increases with the isolation and severity of winter conditions, or by bigger grants to municipal and other governments in these northern climes. These advantages are justified by the need to maintain the population in northern settlements, either on equity grounds within the state, or to ensure territorial control by a permanent population. In addition, of course, most companies and governments pay higher wages in isolated settlements in the north, in part because of the recognition of greater living costs, but also to ensure they continue to maintain or increase their workforce.

8.3.4 The Social North

Traditional societies certainly found ways of adjusting to winter, for travel by dog-sled, skis or snowshoes was often easier in winter, making it possible to cross frozen lands or seas. The modern use of snowmobiles continues this tradition of easier mobility in winter. But people in modern societies have found that their major means of transport, by car or rail, is seriously handicapped by winter, since there are so few roads, at least in the far North. However, the modern routine of life, with its set work and school hours, make it difficult to take time off from work when weather conditions become really extreme. Moreover winter has its most *serious effects on the most vulnerable*, the poor and infirm, who may not be able to afford appropriate housing and heating, or negotiate snowy streets. The homeless in particular need to be catered for in winter by shelters and provided with hot food, since they are especially at risk.

Winter conditions, whether physical or psychological, often make it *more difficult to travel around.* Also there is often less desire to be exposed to these conditions. This usually leads to *lower outdoor activity and less social interaction* between people in the settlements, at least in modern communities. The exception may lie in those northern agricultural areas that have long winters, such as the Canadian Prairies, where the limited amount of work on the land in this season does lead to a greater socialization in visiting friends, news exchange and participation in winter sports activities in the small service towns that dot the region. Yet it is important not to over-generalise about the degree of winter activity and attitudes towards the season, for different people vary in their attitudes to winter depending on their traditions of living in winter climates and personal experience. In addition, age differences occur. For example, surveys in Hokkaido, and later in other northern climates (Enai et al 2002), have shown that many young children relish the onset of winter, because they are able to play in the snow and ice, whereas most adults view the change with dread, unless they are winter sports enthusiasts. However as the children age, these surveys have shown that *children's attitudes to winter* change towards greater negativity, unless they have been encouraged in school to play outdoors and to engage in winter activities. More recently, Sweden has long advocated this approach and, recently has been persuading the children of immigrants, who often fear winter, to enrol in playgroups and kindergartens from the age of 18 months. This means they will have an early start on learning Swedish and being exposed to outdoor activities, rather than arriving at full time school with their parent's negative attitudes to winter, and usually with limited ability in Swedish, which handicaps their subsequent progress.

Recognition of these problems have led advocates of the Winter City movement to search for ways of re-designing settlements or retro-fitting older ones to mitigate the negative effects of winter upon settlements, as well as searching for ways of improving the life-style of people in these places to enrich life, and create a joy in the experience and conditions of winter that *cannot* be obtained in other seasons and environments. Most of these improvements can be summarized in two broad categories, one in design changes to increase the sustainability and quality of life by Winter Cities, the other by providing opportunities for more social activities in winter.

8.4 Winter City Settlement Designs

The Winter City movement has helped to encourage a revival of interest in the modification of older traditional winter settlement designs to counteract the effect of winter on cities, while many new planning principles with the same objectives have been developed. These have either involved the addition of new structures and plans based on these principles, or have retro-fitted new buildings and designs into existing urban fabrics. Most of these schemes are still city-specific, although often based on guidance from national planning agencies. Only in planned economies such as the former Soviet Union have attempts been made to develop a national system of planning systems of cities with appropriate adaptations to northern conditions, although it has been argued that the Soviet approach failed to produce the desired results because of poor building construction methods (Bukin 1988). All these ideas have the goal of making cities more comfortable in winter, to reduce inconveniences and enhance the opportunities available, in order to create a better life, in work as well as play, during the cold, dark months.

8.4.1 Settlement Scale

A general trend in the recent planning of cities in winter climates has been to try and increase the degree of *compactness and density* of these habitations, which leads to reductions in heating costs and a benefit from the urban heat island effects. In addition, higher densities reduce overall travel times between different activity areas, reducing costs and exposure to the worse conditions of winter. The development of *multi-functional buildings* is also encouraged, again to avoid going outdoors when moving from one function to another. In some small northern towns, large town centre buildings have been constructed in place of the typical single shops or office buildings. If oriented correctly, these buildings act as *wind-breaks* against the prevailing storm tracks, creating shelter for areas behind.

Perhaps the best example of this trend can be seen in the new town of Fermont (literally iron mountain) built in the late 1970s near a new iron ore mine in North East Quebec. This is dominated by a major wind-break, a building known as Mur-Écran (wall screen). This large multi-functional structure is 1.3 km long, five and a half storeys high in the centre and tapering to three and an half on the edges. The outer edge does contain projections that house many different functions. The building contains shops, administrative buildings, schools, recreational facilities and over 300 apartments. The building is orientated to the north-northwest but with an angled construction, providing a large wind break to the prevailing winds, sheltering 300 houses within its lee and partial shelter for a similar amount (Fig. 8.1).

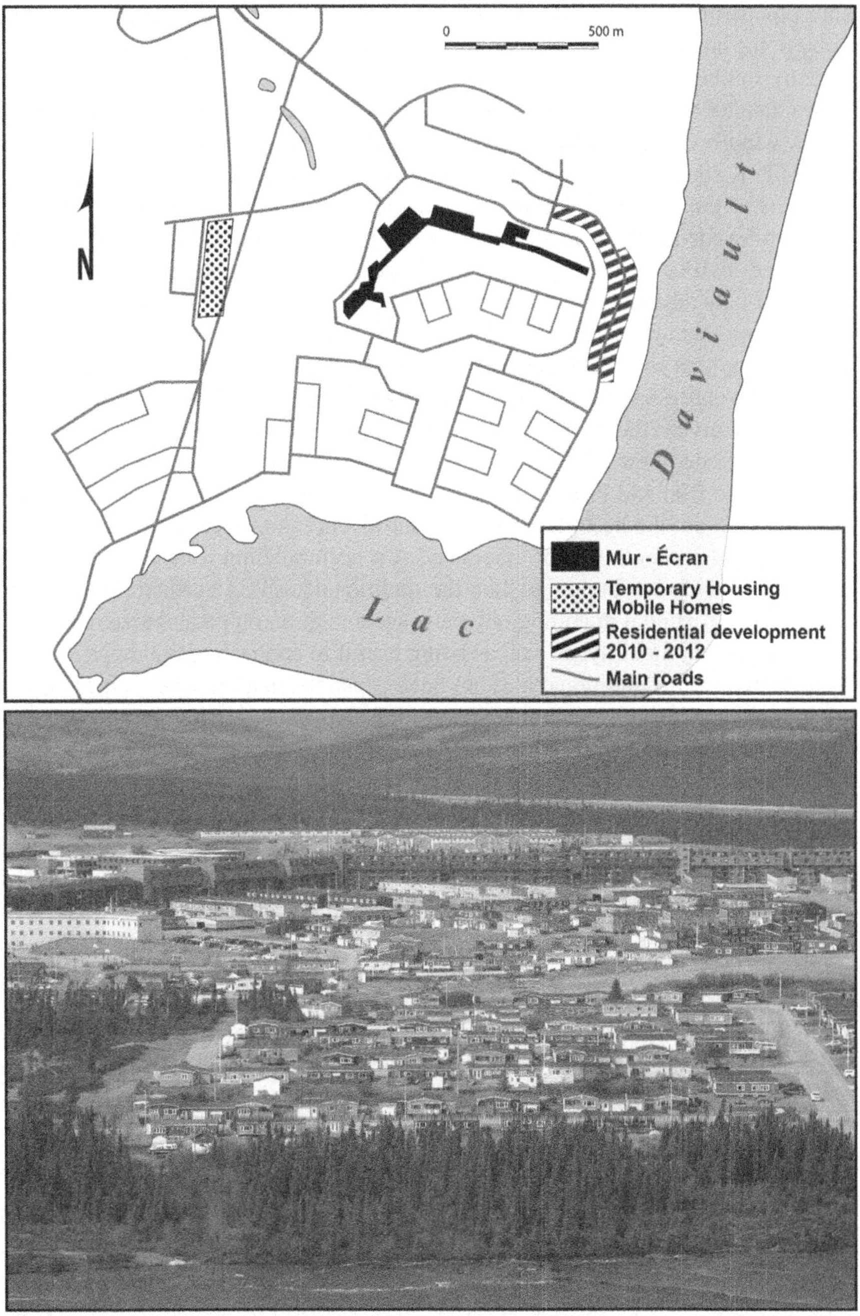

Fig. 8.1 Fermont, Quebec: Plan and Mur-Écran. (Source: Revised from Simard and Brisson 2013)

This pioneering structure was an innovative trend in northern development but many of the houses in other parts of the town do have similar designs to those found in many southern Canadian centres, so together with a recent mobile home area it shows that not all of the area is as adequately winter-proofed as its initial phase.

The wind-break design was inspired by similar, although much smaller buildings created in Kiruna and Svappavaara, also iron mine towns in northern Sweden that lie just over the Arctic Circle. In many ways the new iron ore town of Kiruna in Northern Sweden developed after the rail line to the port of Narvik in Norway was completed in 1902 contained many of the ideas that were revived a century later by Winter City advocates. The Swedish architect Per Hallman, inspired by Camillo Sitte's views that town plans should reflect existing topography, built the town on a hill, knowing that cold air drains into valleys. He also created a winding main road, thereby preventing *wind funnel effects* through the town, and ensuring other roads had similar curves. Today, developers in winter cities now use wind tunnel simulations to calculate the wind-breaking effect of buildings before they are constructed, instead of the trial and error of the past. At a smaller scale the technique also allows working out the best places to erect snow fences to prevent drifting snow, or the placement of vanes on roof to reduce snow accumulation. *Covered areas,* such as walkways between buildings, like the traditional 'roji' in northern Japan, have also proved influential in cutting out snow accumulation on paths, thereby helping to maintain local mobility, as well as being useful in connecting buildings in rainy areas. However this traditional idea has been expanded in all Japanese cities in new ways, such as connecting shopping areas to transit terminals, while local shopkeepers' associations have often financed the covering of shopping streets to provide sheltered areas with more interior retail and service outlets to make shopping easy and comfortable. These are similar to the covered arcades in European cities, although these run through building blocks from one street to another, adding to retail space and with the intent of providing shelter from rain as much as snow. In the central business districts of larger cities exposure to winter conditions has also been drastically reduced by providing *covers over some main streets* to provide protection in shopping areas against winter. For example, overhangs or canopy cover for pedestrians, were developed in the main shopping streets in Sherbrooke in Quebec in the mid 1970s. They revitalised Wellington Street in particular and enabled it to compete with suburban malls. A similar successful design change result took place in 1974 when a five block, 700 m stretch of a major shopping street, rue St. Roch in Quebec city, became the first in North America to be pedestrianized, roofed over, and provided with a climate-controlled system. The success of these ventures has led to the addition of many downtown shopping malls in cities that experience long winters. The Eaton Centre in downtown Toronto was opened in 1977 and is now one of the world's biggest, with 230 stores and 160,000 m^2. of floor space on five levels, all set around a large interior galleria illuminated with natural light. This helped revitalise the deterioration in Yonge Street, the original main shopping street in Toronto, that had started to decline due to competition for shoppers from the competing climate-controlled new suburban shopping malls.

In some cities interior gardens have also been developed, such as the Devonian Gardens located on the fourth story of a building complex in downtown Calgary and surrounded by shops and offices. Originally opened in 1977 it occupies a 10,000 sq. ft. or 1 ha area containing pools and 20,000 plants, above part of the Core Shopping Centre in downtown Calgary. The plants are organised in displays that change with the seasons. The area contains several fountains, pools with fish, small waterfalls, public spaces for events, as well as quiet spaces for contemplation, together with recorded birdsong to give the area a natural ambiance. Sixty per cent of the C$ 9 million cost of this unique area was financed by a grant from the charitable organization of the same name. A major renovation costing C$ 37 million was completed in 2012, which involved extensive replanting and a re-organization of what amounts to an indoor park, to which children's play areas, a continuous skylight and a living wall of plants were added.

A more widely used approach to providing comfortable conditions in downtown areas in winter has been through the development of either skyways or underground tunnels to *interconnect buildings at another grade*. This is not usually done as a single construction event, but by planning regulations that retrofit existing buildings and specifying that all new buildings must be connected to the above or below ground system. One of the most extensive examples of such above ground connections in winter climates can be seen in Calgary's Plus 15 system of *skywalks* approximately 15 m above ground that link individual buildings through these pedestrian walkways that provide entrances to retail facilities that line the corridors within the buildings (Fig. 8.2). Beginning in 1970, at a time in which Calgary's downtown was starting to boom with high rise office towers constructed for oil companies and their associated businesses, the system was designed to provide a comfortable system of connectivity between buildings and shops. The decision to develop connectivity above ground was made because of the high local water table in river gravels and the costs of excavation and disruption. By the end of 2013 the Calgary system

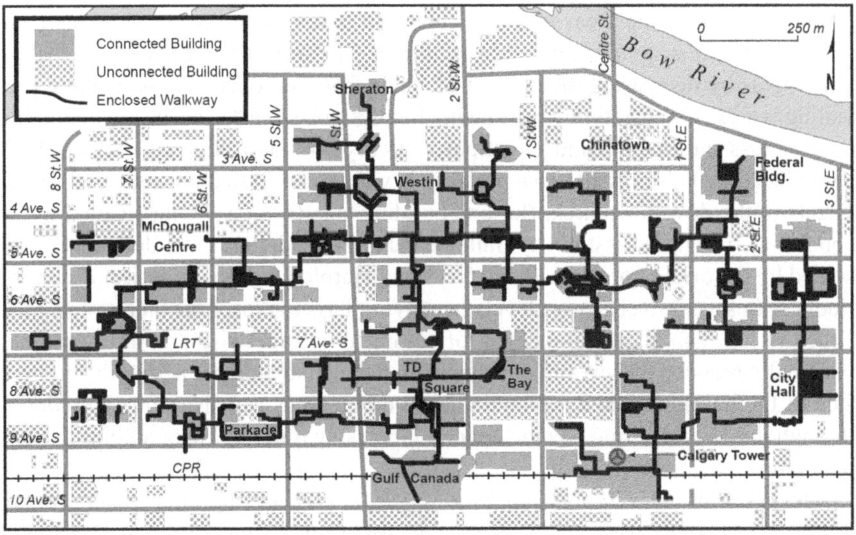

Fig. 8.2 Plus 15 Walkway System in Calgary, Canada. (Source: Revised from CP15 website)

consisted of 59 bridges and 18 km of walkways connecting the main downtown buildings which also included extensive retail areas (see Fig. 8.2), which has created a more extensive network than a similar system in Minneapolis. Building developers pay the cost of construction of the bridges and interior walkways because they are compensated with permission to include higher densities in their buildings. New buildings that are still isolated pay development costs into a fund that is used to construct the bridges when density increases in their area.

By contrast, the downtown areas of Canada's two biggest cities, Toronto and Montreal, have developed a series of *underground connections and plazas* dominated by retail facilities which makes it easier to shop or walk between many of the key downtown buildings in winter. Toronto's PATH system now connects over 50 major downtown buildings as well as subway, railways and entertainment facilities, with 27 km of paths and 1,200 shops and services in 370,000 km^2 of retail space (Fig. 8.3). A similar system in Montreal began in 1962 with a simple connection between a new office building and a major hotel, but has now expanded to 32 km of tunnels and walkways in the downtown commercial area that is used by a daily average of over a half million people and is called call La Ville Souterraine by the francophone population. Of course, as with all plans designed to prevent some problem, some unexpected negative factors often emerge in these underground or above-grade systems. These passages are essentially private spaces, except on the public bridges, so security personnel have the ability to exclude people thought to be undesirable. At ground level the loss of retail space to the interior of buildings along the walkways, and the frequent blank concrete walls of major office blocks, provide an unattractive visual experience on the streets. These problems can be counteracted, in part at least, by a more enlightened approach to the use of these spaces and by display cases at street level for the shops above or below ground level. However it must be admitted that all three of these solutions are designed to effectively exclude winter from these parts of the city, rather than accommodating to the conditions.

Within the built-up areas it is also important to design areas so as to *reduce the severity of cold as well as snow accumulation*. The most obvious ways of dealing with this is by decreasing the amount of shade from buildings and preventing prevailing winds from creating wind tunnel effects along streets, such as by orientating them across the major wind directions. Calgary's modern high rise towers built along narrow roads present a real problem in winter due to shading and wind funnel effects in the downtown area. But one building stands out as being adapted to the conditions. The modernistic City Hall opened in the mid-1980s was deliberately designed by the originator of the Plus 15 system, Harold Hanen, as a series of stepped floors with successively lower levels occupying larger areas. This blue-tinted glass structure has also added an element of colour to an otherwise drab area, and has drastically reduced the amount of shadow on the city square that lies on the western side of the building. Snow accumulation in urban areas can also often be reduced through the careful sitting of snow fences, increasing roof angles, or adding spoilers on them.

The *addition and redesign of public spaces* is another important general trend, to make them more user-friendly and adaptable to multi-purpose use. These public spaces, especially the smaller pocket areas, need to be designed to absorb as

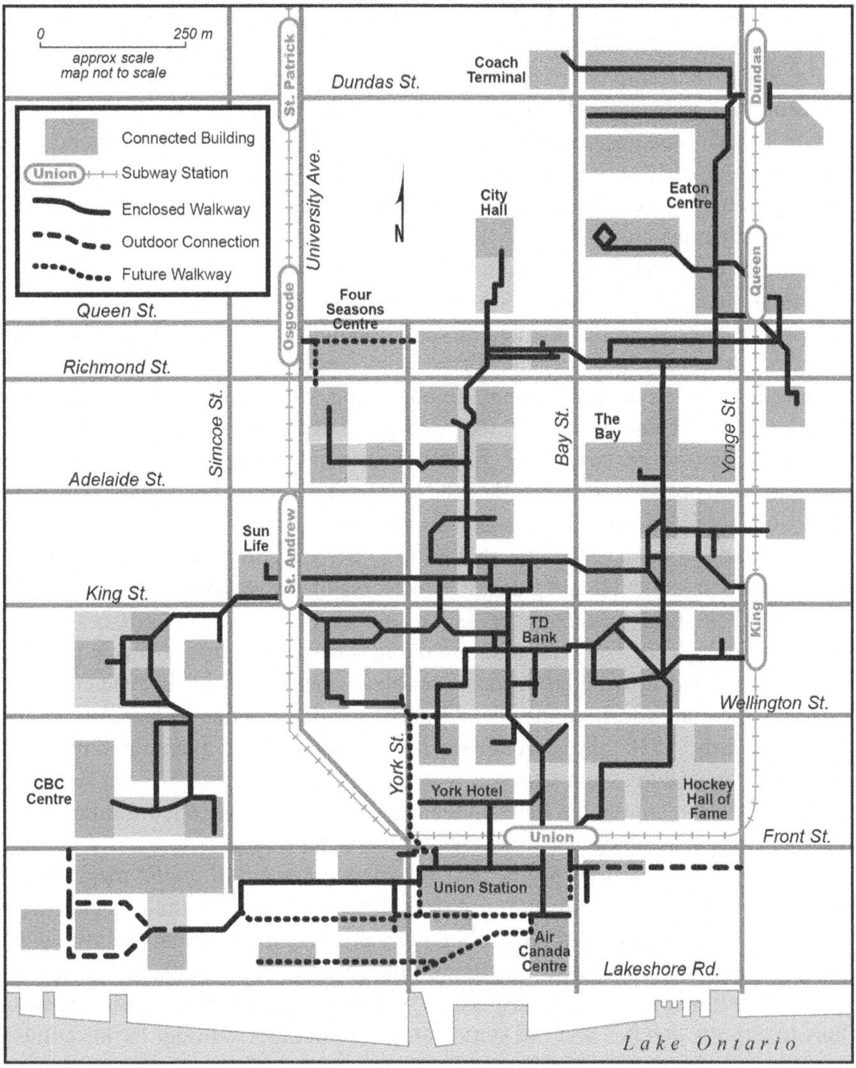

Fig. 8.3 Central Toronto's underground path system. (Source: Revised from TP website)

much of the low angle winter sun as possible, typically by ensuring they open to the south in northern regions, and with vegetation to protect them from the north. This will create solar traps in which people are more likely to congregate because they are more sheltered from the wind and shade. The addition of sculptures, skating areas, as well as street furniture improves public spaces that are designed to be used in winter, with vegetation also used to provide cover from the harsh prevailing winds. Edmonton's recent winter city plan (CE 2013) also advocates the use of firepits in parks, to act as warming stations and congregation points, as well as the encouragement of commercial enterprises such as cafes or food trucks in these public spaces, to provide additional facilities for people using the outdoors in winter.

Enclosed residential courtyards which also act as sun pockets are a popular example of this approach for they help encourage social interaction. Increasingly, winter cities also have cafes with open patio areas warmed by gas radiant heaters in the late autumn or early spring, whilst customers may also be provided with blankets, thereby extending the outdoor season of the business. These are typical features in most Scandinavian cities.

Some cities in winter climates are also *adjusting the natural environment* to provide more protection, especially against winter storms and spring floods. In the former case annual spring flooding along ice-jammed or north-flowing rivers has led to measures to reduce or break-up the blockages. Flood protection works, such as those that have a separate channel to take flood waters around the settlement in the spring run-off, can be seen by the example of Winnipeg in Chap. 9. The widespread adoption of the concept of green belts around cities popularized in Garden City plans and post-World War II British New Towns has also been adapted to assist in Winter City protection. Green Belts or Green Wedges with large belts of conifers have become a popular device for sheltering settlements, mitigating the adverse effects of wind in particular and providing an accessible recreational area for the local residents. One of the best examples is the so-called 'Green Scarf' of trees around Reykjavik in Iceland, a city that also discourages housing developments in exposed areas. Iceland had a particular exposure problem since its inhabitants had cut down most of the natural forest on the island by the twentieth century to increase pasture areas. Since the 1930s the government has encouraged tree planting—involving citizens in the task to give a sense of ownership over developments—not simply to provide shelter in winter, but to also create greater recreational opportunities in the forested areas in summer.

Another frequently used principle is the encouragement of *public transit* to reduce individual car use and its resultant pollution, as well as the associated car parks. To make these systems work effectively there is the additional need for warm bus or rail shelters, frequent service to minimize wait times, with clocks and devices to show the time of the next transport, as well as integrated public transit connections to airports and bus and rail stations to avoid outdoor waiting. In this context Scandinavian cities in particular are light years ahead of American or Canadian centres, where the result of more conservative governments, lower taxes, and often fragmented political jurisdictions, have meant that public transit systems have never been developed to the same extent. This means that taxis or infrequent bus services are the only ways that people can access their public transit terminals. In Copenhagen, by contrast, Chap. 6 has already described how a new underground Metro transit system running through the city built an extension to the airport as an early priority, and is now being complemented by a underground circular transit line that will connect the inner suburbs to the city centre by 2018.

Improving and maintaining circulation patterns in Winter Cities is also a major goal, such as by ensuring that pavements and roads are more adapted to winter conditions and with fences alongside to reduce snow drifting on to their surfaces.

Frequent snow removal is a standard now in all but the most northern places, although until the 1950's many northern cities, especially in Finland, accepted the winter snow accumulation, so that horse-drawn sleighs and sleds could be used on snow-packed roads. Today this is unacceptable in most big cities which have various policies for clearing roads. Hence, cities have a stock of expensive snow-clearing equipment that often must be quickly applied to the roads after snow. In Canada cities vary widely in the costs of snow clearance, with Calgary, a city of 1.2 million people and an average of 100 cm of snow allocating c$ 22.2 million for the task in 2008, whereas Edmonton and Winnipeg spent much higher sums of c$ 46.7 million and c$ 29.8 million respectively. The differences are not a consequence of variations in the amount of snow, but local policy in the degree of clearing that is considered necessary, linked to the reluctance to pay taxes to finance the clearing. In Calgary only the main roads and bus routes are normally cleared as soon as possible after a snow fall, which has led to a fleet of 67 sanders and snow ploughs and nine graders. In spring there is a need to clear the over 40,000 t of sand and gravel that had been placed on the road in winter to prevent skidding. Calgary's policy is based on the assumption that the typical warming Chinook winds of winter, due to its location near the mountains, will eradicate much of the snow build-up in other roads surfaces in winter. Unfortunately these winds vary in their frequency in winter, so there are many years in which the snow is not cleared by this means, which becomes a real problem in years of greater snowfall such as 2008–2009 and 2013–2014, when almost double the average snow levels were experienced. Despite frequent widespread citizen complaints City Council rejected the move to a 'bare pavement' policy of clearing all streets in September 2009, after a city report estimated it would cost c$ 100 million annually, which would have resulted in a 2% increase in city taxes to implement such a policy. However the debate over snow-clearing remains an issue of much dissention in the city. By contrast, Edmonton, the capital city of the province of Alberta, 300 km to the north, is outside the Chinook belt and not only removes snow from the main roads but also grades residential streets to maintain a snow-packed condition, which explains a budget over twice as high as Calgary. It is also adopting a system that ensures bicycle paths are cleared of snow in winter to try and make this commuting option an all-season one. The variations in snow clearing policy can be seen by the fact that Toronto, with a smaller winter period, has some of the most rigorous requirements among Canadian big cities, with city regulations requiring the clearing of all roads down to bare pavement within 15–20 h of a snowfall, with expressway ploughing after a fall of 2.5 cm, main roads after 5 cm and side roads after 8 cm of snow accumulation. Places that only occasionally have snow rarely invest in sufficient snow cleaning equipment and suffer accordingly in their periodic winter episodes, as can often been seen in the chaos affecting Britain's main airports and roads after a large snowfall occurs.

Pedestrian circulation in older settlements in winter was often helped by the creation of high connecting pavements between houses, such as in Hammerfest, where steps down from the pavements outside houses and shops to the roads allow for snow accumulation on the roads. In Newfoundland older houses often had a door in the second story to facilitate exits when snow accumulation buried the lower levels.

Such modifications are rarely seen today, but pavements are often made higher in Winter Cities to allow for snow and ice accumulation on roads and runoff in spring, but with scooped-out areas at crosswalks to allow people to cross roads easily. In Reykjavik many parts of the downtown now have *underground heating* for pavements, based on the use of waste water from the buildings that are heated from the cheap geo-thermal sources available because of Iceland's volcanic geology, due to its location between the Eurasian and North American tectonic plates.

More attention is also being paid to the *design of roads* in winter cities, such as reducing the number of steep or cambered slopes, as well as ensuring that as many cross-roads as possible are at grade, to prevent cars sliding down hills and across icy intersections. Road widening and orientation across the prevailing winter winds also reduce shadows and wind funnel effects. The use of tunnels, or sheds over roads in exposed areas also reduce the impact of snow, especially in avalanche-prone areas. In icy areas under-road heating can be used to reduce the build-up of ice in exposed locations. But the latter is only effective if the roads have a coarse surface and sufficient drains for the melted water to run off. Certainly, all of these features lead to increased installation and maintenance costs of the circulation system. Yet recognition of these costs reinforces the argument for a compact city form which reduces the number of these adaptive features. In most cases these design principles have been added to cities in an ad hoc fashion. However some northern cities, such as Umeä in north-east Sweden, have co-ordinated them, by focusing on increasing density, reducing sprawl and redesigning their city centre in recent decades using these principles to provide greater adaptation to winter.

8.4.2 Building Scale

Many new approaches to provide greater adaption to winter can be recognized in building designs and functions. The most obvious is to create *more efficient housing structures* in northern climates. These include the use of effective insulation and building materials, encouraging triple fenestration, and reducing the effect of snow accumulation on roofs, which can cause them to leak or collapse—although it must be remembered that snow accumulation can act as an insulator, as seen in the traditional igloos of the Inuit. However there is also the need to provide effective ventilation to prevent indoor pollution from household activities, which can get worse with increased insulation that does reduce heat loss but traps air and creates a dead air problem (Enai et al. 2007).

A second trend is the increasing interest in *heating schemes based on new energy sources* to reduce heating costs, such as the use of biomass, wind, solar and geothermal power. They often have the advantage of being based on local renewable sources, rather than on expensive imported fuel. It must be admitted that the use of these new energy sources in northern settlements is still in its infancy, for they are far more expensive initially, and wind and solar sources may be intermittent in winter, while the latter is non-existent in the days and months of darkness or short days, although solar radiation during the short winter days of sunlight can still be captured. But as renewable energy techniques improve, the costs of these sources are decreasing. So many national and city governments are seeing the wisdom of subsidizing such costs, especially through district heating schemes, such as those that are so popular in Scandinavian countries (see Chap. 6). Also, the use of solar panels is gradually spreading, despite their cost and their disadvantages in periods of winter darkness—although they have advantages in the long summer nights, which would be even greater if more effective ways of storing the energy generated can be created. In the new suburb of Viicks, outside of Helsinki, 40 % of the power in one area of 3,000 houses is provided by solar sources. Perhaps the greatest advance in new energy sources and district schemes has been made in Iceland where *geo-thermal power* is now used extensively, with over 80 % of dwellings in the capital heated from this singular geological resource. Until recently it was assumed that only limited areas on earth had the capability of tapping this potential resource. However research by a group at M.I.T. led by J. W. Tester and B. J. Anderson (2007) has claimed that there are far more economically feasible geothermal sources available in the United States, due to the use of new drilling techniques and rock fracturing at depths well below 5,000 ft, making them a potentially important source for even base load electricity, at least in larger settlements.

The third and far more immediate and less costly approach to adapting buildings to winter is to use more careful designs to improve the *micro-climate of spaces around buildings*, using wind tunnels to model the effects of different building sizes and placements (Kuismanan 2005). Towns in Scandinavian countries in particular are now full of examples where small courtyards or squares are being orientated to the south to maximize solar input and create suntraps with exposure to the limited northern sun and shelter from the harsher winds. Indeed, two Scandinavian planners (Culjat and Erskine 1988) estimated that the period of comfortable outdoor living could be extended by approximately six weeks in the sheltered squares through careful micro-climate policies, such as those outlined above, with a main building acting as a shelter. In Reykjavik a housing development has been built in a crenulated form. Each section has three sides of open garden-like courtyards facing south and connected by a single row to another open block to the south, which means that the alternate, corresponding open cells face north, are colder and are used as parking areas. Access to buildings in areas with extreme winter conditions is also improved if entrance ways have a three-fold sequence, a solid outer door against the elements, a lobby area to take-off/add outer clothes, and inner doors to protect the interior against cold spurts from the outside. It is also important to *reduce shading* and *increase the sunlight into buildings*, which may not be easy in low angle winter sun conditions. Again public policy in Sweden has been far in advance of most other countries. Since the 1950s the Swedish national building code requires at least half of the windows in new residential structures to face south, and developers of new apartment blocks are encouraged to have apartments that run through the buildings so that morning sun is found on one balcony and evening sun on the other. Certainly these balconies will not be used in severest winter conditions, but there are many days in winter, and certainly in the transition periods, where such exposures result in a real bonus of extra daylight and exposure to the sun, as well as being especially beneficial in summer months. Incidentally this feature was one of the key features recommended by Le Corbusier in his much reviled, but misunderstood recommendations for high rise housing blocks of the 1920s. Unfortunately, most of the environmental and community features he recommended were ignored in the rush to develop post World War II high-rise public housing structures that rapidly degenerated, in large part because of the failure to incorporate his security and community ideas and adequate maintenance (Davies and Herbert 1993). It is also worth noting that Swedish municipalities since 1971 have been required to create an energy plan for their area that shows how energy is supplied, distributed and used, an approach designed to inform the public and to publicize the development of innovative solutions to make the settlements more sustainable.

A fourth approach to help adaptation to winter can be seen in the way that northern settlements are deliberately *using bold colours and extending lighting systems* to reduce the drabness of the universal whites and greys of winter. Nuuk in Greenland has become a pioneer in the addition of vibrant colours and striking designs to as many buildings as possible. More powerful lighting, especially if coloured and in different designs, also provides an additional positive effect to counteract the darkness, particularly in city centres, at major markets and during festivals—all with the intent of bring more people outdoors. In Hammerfest an illuminated winter ski-track has helped to create more outdoor sports activity, with a design that reduces the effect of the high local winds blowing snow off the track. Even minor individual decisions by businesses can help, such as the tradition of having small oil lamps outside restaurants in cities such as Copenhagen, which provide a psychological welcome on the streets in the cold months of winter. Another unique approach to the addition of light has recently been developed in the small town of Rjuken, a small town of 3,400 inhabitants, 175 km N.W of Oslo. The town was built in the early twentieth century around Sam Eyde's innovative saltpetre (fertilizer) and hydrogen factories using hydro-electric power from plants on a major waterfall that later led to a heavy water facility. Today the town is better known for its location near a national park and for its walking and rock climbing facilities. But its location in a narrow east-west trending valley surrounded by high mountains reaching 1,500 m means that the sun does not show above the mountains from 28 September to 12 March, ensuring than the winter gloom in the valley bottom is of much longer duration than other towns in similar latitudes. Recognition of this problem led the local industrialist to construct a cable car, called Krossobanen, to a location 500 m up the valley sides so that inhabitants could ascend to a look-out and coffee shop that was bathed in sunlight in the early winter. Recently a new approach inspired by a local artist, Martin Anderson, led to the construction of four solar mirrors 450 m up the mountain. These 17 m^2 panels are computer-controlled to track the sun and reflect it back down to the town square, creating a 600 m^2 pool of bright sunlight with around 80 % of the original intensity, as well as generating solar power to operate the mirrors (Henley 2013). The project called Solspeil, is a German engineered project that cost 5 million kroner ($ 814,000), but most was covered by various grants and government subsidy leaving only a fifth to be paid for by the municipality, which is less than 1/650 of its annual budget. There is little doubt that the scheme has been a big psychological success which means that many residents take breaks to sit in sunlight in the square, instead of just seeing the sunlight in the sky above, and which has invigorated this site. A similar but smaller scheme was created in 2006 in the small village of Viganella, 120 km N.E of Turin in Italy, which was also handicapped by a location in a deep valley. In this case a 40 m^2 polished steel mirror reflects the sun and produces a more diffuse light over a wider area in the valley bottom, at a cost of $ 160,000. The success of these schemes is likely to encourage other northern towns to create similar projects to effectively extend the period of limited sunlight in winter.

Finally, it must be emphasized that there are *specific problems of construction in permafrost areas* where modern building techniques from southern climates have proved inadequate, given the freeze-thaw problems that lead to cracks and subsidence in roads and buildings. (Harris 1986, pp. 80–98). The Central Housing and Mortgage Corporation in Canada (CHMC) is one of the agencies in northern climates that have shown how many new engineering techniques have been used since the 1960s to mitigate these effects, especially for housing developments. A popular effective approach is the use of a thick gravel pad under buildings, preferably with insulated polystyrene layers to keep the sub-layers frozen, with or without rigid slabs above, upon which the building is built. The sinking of reverse T supports or straight piles deep into the unfrozen permafrost zone are two other techniques that can provide rigidity to the building structure, although it is always a essential to have the floor of the building above ground, so that a zone for air to penetrate lies immediately beneath the building, again to keep the surface layer as cold as possible. Not surprisingly utility lines between buildings are almost always above ground. Although it might be assumed that the biggest building problems lie in the coldest areas of continuous permafrost, in practice the less severe northern zones with longer periods of thaw-sensitive layers have greater maintenance costs. Since they have greater periods of warmth in the surface zones, this affects the layers below, leading to the soil heaving and collapse described in the previous section on winter conditions. It is now recognized that in all areas with permafrost it is essential to understand the specific climate, soil and moisture conditions before designing structures that are adapted to the local situation.

Similar difficulties exist with the construction of transport routes and airfields where the use of insulated pads in a gravel surface above ground level is designed to keep freeze-thaw at bay. Some idea of the scale of the learning-on-the-job that is often involved in new northern construction projects can be seen in the fact that the construction of the 1,300 km Alyeska pipeline across Alaska between 1969 and 1977 cost $ 7 billion to complete. This was eight times what was expected because of the freeze-thaw problems in permafrost regions, for many parts of the pipeline had to be built on supports and above ground. Indeed, until the 1960s there were few all-weather roads or railway lines in permafrost areas. Even today in many parts of the remote Russian and Canadian north many settlements are primarily served by winter roads, either composed of compacted snow, or ice aggregates that are compressed and then sprayed to create a winter consolidated road-bed. But new roads and rail, albeit it with much shorter lives and greater maintenance, are being built into these zones, especially to serve new mining camps and existing settlements. In China and Tibet the new $ 4 billion railway line from Golmud to Lhasa opened in 2010 and needs refrigeration in the 340 miles where permafrost exists, to keep the tracks stable. Clearly this solution is very expensive and can only be justified on well-travelled routes, or strategic ones in this example, where China is determined to control Tibet.

All these examples show how settlement and building designs can be used to counteract the worse effects of winter conditions. Certainly all these policies involve higher costs at the installation stage and frequently higher maintenance costs, especially as roads and other structures are subject to the winter freeze and spring thaw which causes surfaces to buckle, ensuring more frequent repairs. However increasing interest in sustainability ideas mean that governments, taxpayers and developers now seem prepared to pay the extra money for development, given the savings in maintenance and fuel bills over the long term, the obvious increases in comfort and liveability, as well as profit for retailers and other businesses in the new commercial buildings.

8.5 Developing Winter Social and Economic Opportunities

People who have lived in northern climates for many generations have patterns of living that were adapted to the presence of winter. Traditional societies, such as the Inuit or the Lapps, learned to live effectively in these conditions, although at a subsistence level, using seasonal migrations to take advantage of the food opportunities offered at different locations, or shelter in times of severe conditions. In addition, the Inuit developed igloos as their basic winter housing habitat to take advantage of the insulating properties of snow and ice and the lack of other local building materials. Also, the invention of vessels such as the kayak for fishing and seal hunting, and the use of dog-sleds for travel over snow and ice, shows the ingenuity of these traditional societies. But they also depended upon high levels of social co-operation within family and clan groups to survive. Also some, like the Inuit, have been shown to be more physiologically adapted to cold, for blood circulation to their hands and skin is greater than in Caucasian peoples, making them less prone to frostbite (Streever 2009). However many are far more prone to problems from alcoholic drinks, which was not part of their traditional fare, so they have little resistance to them.

Such subsistence life-styles are impossible for most contemporary people in winter climates, who live and work in contemporary capitalist societies and who are organised in single family dwellings and dependent upon a routinized wage economy. These societal conditions mean that there is a need for much more *economic and political development with sustainable goals* that are adapted to winter conditions. Unfortunately, many traditional societies in northern lands, like their counterparts in tropical and temperature lands, have limited economic opportunities, having suffered from the effect of modernization that destroyed their native cultures, leading to high levels of unemployment, considerable substance abuse, limited education and feelings of despair. Yet some native peoples still manage to practice traditional hunting techniques for meat and furs, using snowmobiles rather than dog sleds in Canada, or herding reindeer as in northern Finland and Sweden. However, the prejudices of southern societies, such as the European anti-fur and seal campaigns and bans, have destroyed the market for many northern hunters. By contrast the prospects for native artists, such as the soapstone carvers of traditional Inuit themes, have improved, although most now use more modern technologies. Despite these examples there is still a need for far more native peoples to be involved in the burgeoning economic activities and in new housing or community ventures in the far north. Some companies and governments are making determined efforts to achieve these aims, but there is still a long way to go. The trend for more involvement of native peoples in political affairs has been helped by the creation of government decentralization in some areas, such as the new territory of Nunavet in the eastern Arctic of Canada. This is providing the people in some northern regions with a greater degree of self-government in internal affairs, adding to the skills and confidence of the peoples, as well as providing a greater share of the growing revenues from local resource projects which can help local development. These governments are also having a greater degree of involvement in, or decisions over, the many mining and resource developments being planned in northern lands. Since many minerals are known to occur in these remote regions, they will become more exploitable as ice levels recede and the seas become more open for longer periods of the year, thereby extending the shipping season. Fortunately, the dangers of potential oil or resource spills in these pristine northern regions is recognized, ensuring that in most jurisdictions rigorous environmental impact statements have to be produced before any industry can proceed, and there are examples of communities rejecting resource development because of insufficient controls to prevent such potential dangers. In towns and cities further south that are still affected by severe winters some governments have recognized the developmental handicaps in northern settlements and have initiated new projects to expand their economic opportunities, such as supporting tourism, to adding educational, medical and research facilities in some of the larger northern centres. One of the best examples is Umeä in North East Sweden, which has progressed from being an isolated northern regional service outpost in Sweden to an educational and medical research centre. In the 1950s the Swedish government made Umeä the fifth repository in the country for all books printed in the country, providing the basis for a research library in the university, which was followed by major investments to locate additional universities and specialized schools, such as dentistry and design, in the city. It now has almost 40,000 university students, together with a current permanent city population of 80,000 inhabitants, almost 60 % higher than its population in the early 1960s before the first university was established. The spin-offs from its educational facilities and its average age of 38 years has led to a new vibrancy reflected in its many leisure activities, inside and outside the city, perhaps why the city adopted 'Curiosity and Passion' as its motto. Such is the strength of its cultural activities and its promotion of the role of culture in sustainable development that it was designated the European City of Culture in 2014, the most northerly centre to receive such an award.

In social terms the most obvious example of an adaptation to mitigate the psychological effect of colder climates can be seen in the creation of *ceremonies* during winter to provide a celebration of life within the season, or one to mark its end. A good example of the former is the traditional Swedish Festival of Light on 13th December, approximately the date of the winter solstice in the old Julian calendar. This was probably an older pagan festival incorporated into Christian form in which a parade of young girls, carrying candles or wearing head-dresses symbolize the triumph of light over darkness a few days before Christmas. Similar light festivals on the eve of St Lucia's Day celebrate the patron saint of visions with torchlight processions in several parts of Spain. However the end of winter is also celebrated. The festival in Zurich, held on the third Monday in March since the end of the fourteenth century, is another marker of a seasonal change, this time the symbolic ending of winter. Its climax is the ceremonial burning of the Böögg, a snowman made out of wadding. Many other cities have developed modern light festivals to banish the gloom of winter. For example, in the five week Light Festival in Amsterdam from early December many light sculptures and projections on buildings illuminate the historical centre of the city and attract tourists and residents alike.

A more general winter festival, of course, is the way in which the Christian Christmas is now celebrated, even by people without spiritual beliefs, bringing life and excitement to one of the darkest times of the year. In the last century this has turned from being only a religious event into a public holiday of several days, characterised by family re-unions, the sending of greeting cards, and an orgy of present-giving and conspicuous consumption. The period has been enhanced by non-religious themes, such as the Victorian invention of so many practices that have become traditions, from the mythical figures of Father Christmas delivering gifts, to the decoration of houses with conifers, holly, ivy and even religious symbols. Such is the size of retail purchases, mainly Christmas gifts, in the weeks leading up to the event, that many retailers obtain a large proportion of their annual sales in this period. In addition, the North American tradition of decorating their houses, stores and main streets with Christmas lights and decorations to provide a sparkle to an otherwise drab time of year has spread to other countries, and lights are often left on for a month. Yet it must be remembered that this modern Christmas festival, based on the assumption of a birth date for Jesus on December 25th, has spread to non-Christian groups, and has merely taken over the celebrations of the winter solstice found in many a pre-Christian society, usually at or near the 22 December solstice, when the sun is at its lowest angle in the northern sky. From this time on, the amount of daylight slowly increases in the northern hemisphere, even though the worst climatic conditions are still to come. Historically, people assembled to take part in these winter festivals, lighting bonfires and holding feasts to celebrate this change in the sun's position and the period from which the sun would gradually return. One of the most important of these examples occurred in Rome's Festival of Saturn that lasted a week from 17th December and which included the giving of gifts and providing a holiday for slaves.

In modern society these traditional ceremonies, originally based on religious beliefs, have been complemented by the *deliberate development of contemporary winter festivals* designed to celebrate the season, in which the presence of snow and ice in particular is turned into positive not negative features. The increasing number of these festivals in most winter cities is a reflection of the desire of local residents to enjoy winter, but the largest have outgrown their local roots. They attract people from distant cities and countries, creating a major tourist industry based on the celebration of winter in the cities that have such festivals, adding substantially to the receipts of the local hotel, restaurant and retail sectors in an otherwise low sales period. The presence of these festivals, and the associated increase in tourists, has also led to the development of more exhibitions and concerts at this time, adding to the attractions of the event and increasing the enjoyment of the period. Two of the biggest world examples of these festivals are the seven day Sapporo Winter Festival in the principal city of the northern island of Hokkaido, Japan, which began on a small scale in 1950, and Quebec's City's Bon Homme Festival. The latter is now spread over 17 days and only became an annual, formal event to attract visitors in 1955, although informal local parties on the same theme can be traced back to at least 1894. Official statistics show that these events annually attract over 2.3 million visitors and 1 million people respectively to the main show grounds in Sapporo and Quebec City. These winter festivals often brand some theme to particularize their event. In the case of Quebec it is seen by Quebec's adoption of Bon Homme—a large, symbolic snowman—and his helpers, the Knuks, to personify the festival and its joie de vive. In all these examples artistic and recreational activities are developed in outdoor locations to attract city residents and tourists alike. For example, the development of huge ice sculptures as well as ice lanterns allows artistic expression to flourish out of doors. In Sapporo over 400 hundred large ice sculptures are constructed every year to provide a sparkling, crystal-like environment in the show grounds. In addition, the addition of well-lighted areas, multi-coloured light and firework displays, as well as torchlight events, also get more people outdoors, despite the darkness. More generally these festivals also advertise winter sports, such as tobogganing, skating, snowshoeing and skiing, mainly cross-country, but also downhill in hilly or mountainous areas. All provide recreational features that cannot be easily duplicated in non-winter environments, as do the presence of dog-sled races. More specifically, the town of Rovaniemi in northern Finland has claimed to be the home of Father Christmas and his elves, and thousands flock to the town with their children every winter to seek out these characters and to visit their buildings, providing a type of northern Disneyland experience for these visitors. The creation of ice-hotels and bars, actually constructed out of ice, such as those in northern Sweden, add to the unique experiences that people in other climates cannot duplicate. The most famous of these is in Jukkasjärvi outside Kiruna where a local entrepreneur developed the concept in the early 1990s, creating a hotel made of ice which has been fully booked from December to May for many years.

More generally, many cities with long winters have deliberately developed *outdoor markets* as well as entertainments in their downtown streets or central squares, especially at Christmas time. Even though these are not really winter cities a trend in the U.K. has been to open markets based on French or German themes in the downtown areas of many cities in the weeks before Christmas, focusing especially on speciality foods from these countries. Hot drinks and foods are available, as well as local entertainments, to add to the festivity of the winter events that are enhanced by coloured light displays. What seems important in the vitality of these winter activities is the inclusion of many different types of events at specific locations, not only to attract diverse people but also to maintain the interest of others. In addition it has been found that greater success occurs if they last a long time to keep people's attention, overlap with one another, and emphasize local themes—in specific activities or food and drink—to give the place a distinctive feel. Yet such events rarely occur spontaneously and need effective and creative management, either by the urban authorities or local business organizations, to ensure success.

Even without the presence of a distinct festival many of the individual elements that dominate these events are also found in many northern settlements and contribute to their *winter recreational opportunities*. This means that many residents prefer the winter season over others. It has led to such popular winter pastimes as: toboggan (glissade) runs, downhill and cross-country skiing, ice-skating along water courses, as well as snowmobile trips, ice-fishing and hunting in those towns surrounded by forests. One well-known example is the way that Ottawa's Rideau canal, which traverses the downtown area, has become a major ice-skating recreational area in winter, whilst some who live along the route use it to commute to work on skates. These uses have required city maintenance crews to regularly clear the snow from the ice to make it suitable for skating. Other winter cities have created large open-air skating rinks in winter that are turned into pools in the summer, demonstrating an all-season approach to structures that will attract people to outdoor activities. Since the late nineteenth century, in Canada and the Scandinavian countries in particular, the growth of the winter sport of ice-hockey was nurtured in local ponds and outdoor rinks created and maintained by local community groups. Today it must be admitted that most children and adults skate in indoor, purpose-built rinks that guarantee ice-time irrespective of the weather. This also allows the sport to be extended from the fall, through winter and spring, a trend that also applies to the sport of curling. Sweden in particular has developed ski or snow-shoe trails in and around most towns, each with suitable rest and comfort stops as well as warming areas based on fire-pits or shelters and food facilities. These facilities have long provided an important leisure experience in winter. However as the recent Edmonton Wintercity plan (CE 2013) has noted, there is usually a need to provide information about these opportunities, such as by developing a Wintercity information hub and clear signposts pointing the way to recreational opportunities in parks, many of which need to be restructured to provide winter activities as well as the typical summer use. In Scandinavia many of the winter trails are converted to walking or bicycle trails in the summer, although the warming effect of urban areas means that it is more difficult to keep snow on the urban trails in early or late winter. Intensive use of these facilities leads to increased levels of fitness and socialization through shared experiences among the participants. Indeed many Swedish schools provide their students with ski lessons in an attempt to encourage outdoor activity in winter. By comparison, relatively few Canadian cities have developed such local winter facilities—their emphasis is still upon golf courses and green spaces mainly for summer use—although cross-country ski routes are developed and maintained in many areas of the more popular provincial and national parks. Edmonton is now providing free snowboard and ski lessons in some river valley locations to encourage more people to use the outdoors in winter.

The development of snowmobiles has also brought more people out in winter to enjoy this new sport, although there is increasing concern among landowners at the damage caused by these machines in churning up the ground and scaring livestock. Hence there is a real need for more dedicated and controlled areas devoted to the sport. In other areas, more traditional forms of winter activities have been extended, such as hunting in areas where there are still many wild animals, or ice-fishing on frozen lakes of lakes or rivers. Travel by dog sleds or sleighs, once the only way of moving in snowbound areas, has been transformed into a winter experience for tourists. For example, the development of long races for slushers, or dog-sleds, such as the gruelling cross-country Ididarod from Anchorage to Nome in Alaska, provides a different type of winter competitive sporting event that brings many tourists into small towns along the route. More generally, of course, the last half of the twentieth century has seen a huge increase in the popularity of downhill skiing, which has led to the huge expansion of many ski-resorts and towns in Alpine environments or in hilly terrain near large cities. The best of these places have tried to ensure that the emerging new ideas from Winter City principles are used to increase the vitality and sustainability of these settlements

It is also worth noting how *competitive Winter Games* have been developed in most countries or regions with long winters in order to encourage more participation in a range of winter sports. At a world scale the Winter Olympics has moved from an essentially elite or essentially Alpine event of the mid twentieth century to one that is now a multibillion dollar enterprise, with participants from countries in most temperate and northern climates and with a worldwide television audience, even though many of the events are still dominated by athletes from Scandinavian and Alpine countries. Every four years the Winter Olympics moves to a different location that is chosen by the International Olympic Committee from the competitive bids from rival countries. Its growth and success has done much to popularize winter sports. Moreover, the successful bidders usually need to massively increase the numbers and quality of the facilities for the thousands of athletes that take part. This provides a permanent legacy for the host winter city and an opportunity during the Olympics themselves to showcase their other attractions to the world, providing a permanent boost to their tourist potential. For example, Calgary hosted the 15th Winter Olympics in 1988 after spending c$ 25 million to promote its bid. The local organizing committee spent $ 98 million on a new ice hockey and figure skating rink, c$ 25 million on a local Canada Olympic Park to host ski-jumping and luge runs, as well as c$ 103 million for an athletes' village in the university, which included a $ 40 million dome arena for speed skating. Outside the city c$ 32 million was spent on the downhill ski-racing centre at Nakiska and c$ 17 million on a Nordic centre for ski racing in the nearby town of Canmore. The result of spending over $ 300 million is a heritage of sporting venues that can be used by locals and sports teams as well as attracting many subsequent international events. The events also fostered a spirit of co-operation through the thousands of volunteers who gave up their free time to help organize and provide guides and information services throughout the games. Most of the expenses were paid for by television and company sponsorship. Similar results have occurred in other winter cities that have hosted the Olympic Games, but now the events are becoming far bigger and far more costly. For example, the province of British Columbia spent almost a c$ 1 billion on the 2010 Winter Olympics in Vancouver and vicinity, three quarters on new and upgraded venues and transportation. In addition C$ 2.1 billion was needed for three major transport upgrades, such as the Vancouver airport-downtown rapid transit, improvements in the sea-to-highway road route to the ski resort of Whistler and a new Convention Centre. Another billion dollars was probably spent on security by various levels of governments. Yet a figure of around C$ 4 billion in costs for the 2010 Winter Olympics pales beside the estimated dollar equivalent of $ 51 billion spent on the Sochi Olympics in south eastern Russia in 2014 (Lally 2013). Much of this was spent on new and major transport routes, resorts and housing, as well as new sports facilities, a figure that surpasses the $ 43 billion that Beijing was estimated to have spent on the 2008 Summer Olympics. It is clear that the hosting the Winter Games is no longer a minor venture associated with some small alpine town to highlight a series of winter sports. Hosting the Olympics has become a vehicle for transforming the city is which it is held, using television income, sponsorship from major companies as well as significant, even massive government investment to finance the event which has also become a matter of national pride and prestige (Hiller 2012). Yet the costs and disruptions are becoming too much for some centres, as the citizens of Denver in the 1990s and Krakow in 2014 showed by their rejection of the opportunity to bid for the games, a trend that is likely to increase.

Despite the success of all these examples it must be admitted that winter cities still face often challenging climates while they are more expensive to live in, given the economic costs that were discussed earlier. Governments around the world have recognized this fact by creating northern allowances for people who live in their climates, usually by reducing income taxes. However this policy is also designed to help keep an economically active population in the north for regional equity and territorial possession reasons. Companies also pay higher wages to workers to compensate for the harsher conditions and often operate fly in-fly out camps to house workers in these remote northern locations who often work 12 h days for two or three weeks and fly back home in the week off. So these settlements are full of temporary transient workers, rather than being the remote fully-fledged towns created in the past.

The Winter City movement's focus on this one season should not lead us to forget that there are some physical benefits for winter settlements in northern areas. One comes from the *long daylight hours in summer* - brief though the summer may be—which leads to the rapid growth of many vegetables. Also the profusion of wild berries in summer, responding to the long summer days, also provides an additional unique resource in these areas that Scandinavian countries are well known for. However in recent years the reluctance of the local population or students in Nordic countries to gather berries has led companies to bring in seasonal workers from overseas, which has often led to the overharvesting of berry crops and to disputes with local residents. In addition, many of these northern settlements are in more isolated areas of countries that have been extensively glaciated, leaving large numbers of lakes and bare rocky outcrops, and with almost no permanent agricultural activity, which is, therefore, very cheap land or held in state reserves. The development of second homes in these sparsely populated hinterlands of some winter cities, has created the so-called *'cottage country'* (Halseth 1998), located within a few hour's drive of the towns. Most are constructed around lakes, providing unrivalled *recreational opportunities* for summer residents or vacationers. It has been estimated (Mänty 1988) that one third of Finnish families spend part of the summer months in such houses in a rural environment, which empties the big cities at holiday time. Finally, the presence of the surrounding wild forest means many opportunities for *photography or for hunting* the often extensive wildlife resources in their hinterlands. These features add to the positive nature of life in Winter Cities in the summer months, providing opportunities not normally available in more southerly lands. The result is an additional distinctive and much valued life in the short summers for many northern residents, one that complements the activities of the long winters.

8.6 Conclusions and Opportunities

The Winter City movement is a design and behavioural approach to improve the quality of life in northern settlements, to ensure that winter is better coped with and celebrated, not just endured. Obviously the degree to which a city is affected by winter varies according to its location, but many of the features outlined in this review can also be applied to those places that only have a limited cold season. Although rapid progress in creating a better accommodation to winter is being made in some towns and cities, there seems little doubt that North American settlements still lag behind those in the Scandinavian countries in developing and implementing these new Winter City ideas. Perhaps the exception lies in the city centres of the larger, economically vibrant cities, where commercial considerations have led to the development of covered interior retail spaces. They certainly protect citizens from winter conditions but make few attempts to help citizens adapt and use the beneficial aspects of winter. The more advanced application of Winter City ideas is probably a result of the fact that residents of the Nordic countries have had far more centuries of living with, and coping with, winter, whilst Americans and even Canadians have been more prone to use modernist ideas in urban buildings, which have often ignored winter. Also, until recently they have been less prepared to plan for, or spend public money on Winter City initiatives. However there are signs of real change even in bigger North American centres, such as the city of Edmonton's recent vision and implementation plan for "*reclaiming the joy of winter in order to embrace the season that is so central to our identity"* (CE 2013, p. 2). Apart from the many initiatives being promoted, it is worth noting that the report was developed with a great deal of citizen input, and claimed that the costs of implementing these ideas are relatively limited. A key feature is the need to convince more people of the need to embrace winter activity, as it is an 'attitude, not latitude approach' that should motivate behaviour in this season. It is also noteworthy how the city is trying to encourage private businesses and other organizations to contribute to the new plans and to link the Winter City ideas to the other strategic themes of the development of the city, rather than seeing it as a separate venture. In this way it becomes part of what the city sees as Edmonton's distinctive brand.

However, the modification of settlements to be more reflective of winter is now being challenged by the rapidity of global warming which seems to be affecting the northern settlements more quickly. A negative effect of this trend could be an increase in sea levels with their threat of the erosion of man-made structures in the many northen towns alongside oceans, and an increase in the volatility of weather conditions that can increase the severity of winter storms. However the northern latitudes settlements will also need to adjust to longer and warmer summers, which means that some sheltered places may now be able to grow more of their own vegetables for local use, whilst the fact of longer ice-free periods will increase the level of economic activities such as fishing, even in small centres, and make it easier to mine the vast minerals and oil reserves that are known to exist in northern regions beyond the Arctic Circle—although opposition from environmentalist groups may well halt many developments. In addition the rapidity of ice loss in the Canadian and Russian Arctic means that the shorter North West and North East passages from Europe to Asia may well become a reality in the next decade, after centuries of explorers trying to prove such routes (Davies 2004). All these changes will probably lead to an increase in the size and number of settlements in these largely uninhabited northern lands, although the increasing creation of more fly-in and fly-out work camps will probably be a more usual trend in really remote places than creating new towns. Nevertheless any extension in settlements in the far north makes it even more important to apply Winter City principles to counteract the effects of winter, which will still occupy a major part of their year.

So far, the Winter City movement could be regarded one of the most successful of the various themes developed to solve particular urban problems, in this case cold-related ones, in order to improve the liveability and vitality of settlements. Moreover, like many of the other themes associated with new approaches to urban development, progress has been improved by the various global networks that connect administrators, planners and academics interested in Winter City ideas, where best practices can be applied elsewhere and new initiatives and problems of mutual concern discussed. Yet it will take a long time to eradicate past mistakes in settlements that experience winter, places that were badly planned to cope with these conditions, let alone find better solutions in newly constructed areas of settlements that experience severe winter conditions. It has been argued that better solutions must involve a multi-faceted approach. In the words of one well known Winter City advocate, if settlements that experience long, cold winters are to be successful then "social values, customs, building materials, orientation, climate and site must intertwine to produce a unique weave" (Pressman 1995, p. 1).

It also may be worth suggesting that the partial success of the Winter City movement in influencing public opinion as well as planning policy could be used as an exemplar for improving urban life in other climatic regimes, such as those with conditions of extreme heat in summer, and especially those that experience many wet, cold and cloudy days. In the former case traditional building designs do exist to modify the effects of heat and can be modified and extended. But few parallel ideas for wet climates have yet to emerge, other than ensuring that roofs and roads are able to carry off the burden of extra moisture and that buildings are made more proof against rain and damp penetration. There is certainly a real opportunity for cities in wet climates to co-operate to find ways of mitigating this type of climatic effect, and like winter cities, improve the quality of life in such areas. Indeed some of the Winter City ideas could be modified and used in wet, temperate environments. However far more specific policies will be needed to counteract the often very wet, grey and dismal days of winter in these environments, for they do not have the vibrancy of blue skies and the apparently cleansing aesthetics of snow found in winter cities to counteract their often depressing conditions. In general, it must be admitted that despite the advances made to date in the Winter Cities example, there is still a long way to go before our settlements become more adapted to the specific climatic problems of their environments. In this way they could create greater sustainability and liveability, ensuring they are more adapted to their environments, rather than being imposed upon them and suffering the consequences.

Acknowledgments I appreciated permission from Dr Simard and the journal Cybergeo to reproduce the picture of part of Fermont, Quebec as Figure 8.1, and also information about policies in Sweden from professor Olof Wärneryd in Lund.

References

Bone, R. M. (2005). *Regional geography of Canada*. Oxford: Oxford University Press.

Bukin, Y. K. (1988). The North 2005 program of the Soviet Union. In J. Mänty & N. Pressman (Eds.), *Cities designed for winter* (pp. 270–318). Helsinki: Building Book Ltd.

CE: City of Edmonton. (2013). For the love of winter: Winter city implementation plan. City of Edmonton, Edmonton. http://www.edmonton.ca/city_government/documents/TheLoveofWinter-ImplementationPlan.pdf. Accessed 10 Oct 2013.

CHMC: Central Housing and Mortgage Company. Northern housing designs. http://www.cmhc-schl.gc.ca/en/ab/noho/noho_035.cfm. Accessed 12 April 2012.

CP15: Calgary Plus15 Walkway. http://wcalgaryplus15.com. Accessed 10 May 2013.

Culjat, B., & Erskine, R. (1988). Climate-Response social space: A Scandinavian perspective. In J. Mänty & N. Pressman (Eds.), *Cities designed for winter* (pp. 347–363). Helsinki: Building Book Limited.

Davies, W. K. D. (2004). *Writing geographical exploration*. Calgary: University of Calgary Press.

Davies, W. K. D., & Herbert, D. T. (1993). *Communities within cities*. London/New York: Belhaven/Halstead–Wiley.

Enai, M., Pressman, N., & Lüttgen, A. (2002). Increasing children's enjoyment of winter and enhancing winter perception: Effective use of small open spaces in Winter Cities' neighbourhoods throughout the year. In: Proceedings of Winter Cities Forum: Aomori, Japan, pp. 215–220. http://www.wintercities.com/Resources/schoolchildren_adaption.pdf. Accessed 10 June 2012.

Enai, M., Okamoto, A., Kikutal, K., & Hayama, H. (2007). On a simplified method for measuring ventilation rates. http://www.inive.org/members_area/medias/pdf/Inive/IAQVEC2007/Enai.pdf. Accessed 10 June 2012.

Halseth, G. (1998). *Cottage country in transition*. Montreal: McGill-Queen's Press.

Hamelin, L.-E. (1979). *Canadian nordicity: It's your north too* (trans B. Barr). Montreal: Harvest House.

Harris, S. (1986). *The permafrost environment*. Exeter: Croom Helm.

Henley, C. (6 Nov 2013). Here comes the sun. *The Guardian*.

Hiller, H. (2012). *Host cities and the Olympics: An interactionalist approach*. London: Routledge.

Kuismanan, K. (2005). The influence of climate on the design of housing. http://www.wintercities.com/Resources/Influence_of_climate.pdf. Accessed 10 Dec 2012.

Lally, K. (17 October 2013). Sochi is ready to host the Winter Olympics. *Washington Post*. Accessed Jan 2014.

Mänty, J. (1988). The winter factor in Finnish urbanism. In J. Mänty & N. Pressman (Eds.), *Cities designed for winter*. pp. 133–144. Helsinki: Building Book Ltd.

Mänty, J., & Pressman, N. (Eds.). (1988). *Cities designed for winter*. Helsinki: Building Book Ltd.

Mundigo, A. I., & Crouch, D. P. (1977). The city planning ordinances of the laws of the indies revisited: Part I: Their philosophy and implications. *Town Planning Review, 48*(3), 247–268.

Munroe, S. (2012). The Canadian Ice Storm of 1998. http://www.canadaonline.about.com/cs/weather/p/icestorm/htm. Accessed 10 July 2013.

NE: Nordic Eight Network. Nordic solutions for sustainable cities. http://www.hel.fi/hel2/ymk/julkaisut/nordic8Catalogue_060612.pdf. Accessed 8 Oct 2013.

NNC: Nutrition North Canada. http://www.nutritionnorthcanada.ca. Accessed 19 Sept 2012.

Pressman, N. (1995). *Northern cityscape: Linking design to climate*. Yellowknife: Winter Cities Association.

Pressman, N. (2004). *Shaping cities for winter*. Prince George: Winter Cities Association.

Simard, M., & Brisson, C. (2013). L'industrie minière et le développement urbain en milieu nordique: l'exemple de Fermont au Québec. http://cybergeo.revues.org/25817. doi:10.4000/cybergeo.25817. Accessed 5 Dec 2013.

Streever, B. (2009). *Cold: Adventures in the world's frozen places*. New York: Little Brown.

Tester, J. W., Anderson, B. J., et al. (Eds.). (2007). The future of geothermal energy: Impact of enhanced geothermal systems in the United States in the 21st century. *Philosophical Transactions of Royal Society A, 365,* 1053–1094.

Thorvaldur, S., & Thorjarnarson, T. T. (1988). *Reykjavik, Iceland: A northern capital*. In J. Mänty & N. Pressman (Eds.), *Cities designed for winter*. pp. 189–204. Helsinki: Building Book Ltd.

TP: TorontoPathway. http://www.torontopath.com/uploads/PATH%20Map%202013.pdf. Accessed 12 Oct 2012.

WCI: Winter Cities Institute. http://wintercities.com/. Accessed 12 Oct 2012.

WWCAM: World Winter Cities Association for Mayors. https://www.city.sapporo.jp/somu/kokusai/wwcam/introduction_e.html. Accessed 14 Dec 2013.

Chapter 9
Resilient Cities: Coping With Natural Hazards

Wayne K.D. Davies

> *Although no city or community can ever be entirely safe from natural hazards, they can be more resilient to the destructive forces that claim lives and assets.... Community Resistance is characterized by: its capacity to withstand stress or destructive forces through resistance or adaptation; its capacity to manage or maintain certain basic functions and structure during disastrous events; and the capacity to recover or 'bounce back' from an event. Twigg 2007, p. 6.*

9.1 Introduction

Throughout civilized history many urban places have experienced losses from destructive natural events, such as violent storms, floods, fires, landslides, volcanic eruptions and earthquakes. Despite these catastrophic events, most have recovered, because of the resilience of their people and the inherent advantages of their site and situation which encouraged re-building on the same or an adjacent site, as in Lisbon which was largely destroyed after the 1755 earthquake and tsunami. Others have been annihilated and never rebuilt. Perhaps the best known example of the latter in historical times was the rapid burying of the Roman towns of Pompei and Herculaneum under 4–6 m of the explosive acidic ash and pumice thrown out by the violent eruption of Vesuvius in 79 AD. Many other settlements have been destroyed through the advance of deserts and the erosion of shorelines. The loss of human life in more recent disasters testify to the damaging force of some of these extreme natural events, such as the deaths of over 500 hundred thousand people after the Bhola cyclone in Bangladesh in 1970, to the Haiti earthquake of 2007 that claimed over 316,000 lives according to the government, around the capital city

W. K.D. Davies (✉)
Department of Geography, University of Calgary,
2500 University Drive, N.W., Calgary, AB T2N 1N4, Canada
e-mail: wdavies@ucalgary.ca

W. K.D. Davies (ed.), *Theme Cities: Solutions for Urban Problems*,
GeoJournal Library 112, DOI 10.1007/978-94-017-9655-2_9

of Port au Prince, or the 1556 Shensi earthquake in China with estimates of over 800,000 killed (USG).

Despite the scale of these disasters it could be argued that until recently we have not fully acknowledged, or even prepared for the possibility of destructive natural events striking our cities. Certainly most urban places are built to cope with what appear to be regular events, such as seasonal flooding or weather variations, but the majority are not equipped to deal with the extreme natural processes that occur irregularly and cause such extensive damage to property, livestock and frequently leads to many injuries, deaths and the destruction of livelihoods, which may take decades to recover. For most of history these events were often considered to be Acts of God, as punishment for moral misbehaviour, so they were just accepted as something that had to be endured. In the last century the growth of scientific knowledge has led to greater understanding of the natural origin of these events, but several factors led to the belief of many that these hazards can be made less threatening or safely ignored, issues described in Chap. 1 as our hubris, myopia and what amounts to environmental blindness or forgetfulness.

Fortunately the geophysical and environmental research of the last two decades has led to greater recognition of the increasing intensity and impact of many natural hazards, plus awareness of their catastrophic impacts on lives and property in urban places, as seen in many academic papers and also major texts focusing on environmental issues (Zebrowski 1997; Blaikie et al. 2004; Abbott and Sampson 2009; Smith and Petley 2009). In addition, there has been a growing literature on a more community-based approach to the effect of natural hazards, dealing with community capacity building and local risk reduction (Townshend et al. 2014; Kulig et al. 2013), initially stimulated by concepts of ecological resilience in plant ecology (Holling 1973) and the work of psychologists dealing with stress. More generally, an international effort on the problems began when the United Nations declared the 1990s as the International Decade for Natural Disaster Reduction (IDNDR) This led to major conferences (Hyogo 2005), reports from organizations at an international level (World Bank 2006; United Nations Office for Disaster Management or UNISDR 2011; Arup 2011), and new policy initiatives from governments who have recognized their need to respond to major disasters, as well as from major aid agencies, such as Oxfam and Red Cross-Red Crescent (Walter 2004; Jennings 2008). The UNISDR (Blackburn and Johnson 2012) report entitled *Making Cities Resilient: My City is Getting Ready* has popularized 'Resilient Cities' as a shorthand term for work that seeks to expose and reduce the risk of natural hazards to all sizes of settlements, which is a different sort of resiliency to that used in the Transition Town literature (Chap. 7). This review summarizes the background to the emerging Resilient Cities literature, identifying the problems posed by these natural hazards to urban places and the way that human activity often exacerbates the problems they create. It is followed by a summary of the various ways in which urban places are improving their capability to minimize the number of deaths, injuries and amount of property damage from these various hazards and to recover from the disasters they have created in so many places.

9.2 Natural Hazard Variations

9.2.1 *Types of Hazards*

We live in a world in which our regular life and activities generally function within the normal range of both seasonal climate changes and small-scale geophysical processes. But we are in a period in which the old climate averages based on records of the previous 30 years are being adjusted to new normals to take into account the effect of an average global warming of at least 0.8 °C degrees since 1880. According to the 2012 reports of the U.S. National Oceanic and Atmospheric Administration (NOAA 2012) temperatures have been rising above the long term average for the 36th year in a row, although the last decade has seen a reduction in the rate of growth, primarily because of the oceanic absorption of the extra heat. In addition, some natural events are occurring with much higher magnitudes and in new locations—effects often not seen for generations—that involve the input of sufficiently large concentrations of energy and materials into the environment to cause threats and harm to humans and their assets. For example, it has been revealed that the average latitude at which tropical cyclones reached their maximum intensity in each year over the last 30 years had moved polewards by 53 and 62 km per decade in the north and southern hemispheres respectively, paralleling the shift of the tropical zone expansion (Kossin et al. 2014). Given the coastal location of many major big cities this trend will make more liable to larger storms in the future. This means that many buildings and infrastructures will have to be moved or built stronger to withstand the emerging climate regimes. Significant losses of life and environmental degradation also come from unintended consequences of direct human actions, such as various industrial accidents, while indirect human influences can also cause disasters and may help exacerbate the effects of some of the physical hazards. Although these potential hazards to cities can be classified in many ways, they are basically of five broad types shown in Table 9.1.

It is impossible for a single chapter to cover all of these issues in any detail, so the focus of this discussion is upon the consequences of the first three of these types of natural or physical hazards and the emerging range of risk management procedures

Table 9.1 Types of hazards affecting settlements

Main origins	Major hazard examples
1. Geophysical	Earthquakes, volcanoes, landslips, avalanches and mudslides, extra-terrestrial-meteorites etc.
2. Climate	extreme heat, cold or drought, intense precipitation (rain, hail, snow, ice), fog, lightning, storms and winds (hurricanes, typhoons, tornadoes, local winds).
4. Biological	Diseases, toxic plants, landscape fires.
5. Human	*a) Direct:* transport accidents, industrial plant failures, toxic material spills, building and infrastructure decay and collapse, fires, wars. *b) Influences:* pollution, land-sea degradation, climate changes.

needed by urban places. For the sake of completeness some brief digressions are made into the hazards of the last two categories, namely biological issues and from human activity—although not personal behaviours, wars or terrorist attacks—since they may produce similar disastrous effects to life and property. However it is worth noting that many of these potential events inflicting disasters upon cities can be reduced by many of the same procedures being adopted to combat natural hazards.

9.2.2 Natural Hazard Trends

Despite extensive reporting of the effect of natural hazards there is still a dearth of reliable, global information about the exact numbers and losses, in deaths, disruption and property, even from recent events (Bouwer et al. 2007). Hence it is not surprising to find that it has been even more difficult to estimate impacts from historical environmental hazards, although Smith and Petley (2009, p. 23) estimated that there have been 35 major disasters with over 100 thousand killed since 1,000 BCE, with seven events having over a million dead. In the last half century far more sources of data have become available, often compilations by national governments, such as The Canadian National Data Base (CNDB), but between government sources there are many variations in accuracy and in types of compilation. The two most reliable world sources at the present are the EMDAT (Emergency Events Data Base) records complied in the University of Louvain's Centre for Research on the Epidemiology of Disasters, both natural and human (CRED), and the annual estimates compiled by Munich Re, a major insurance company. For inclusion in CRED a disaster must have killed more than ten persons, and affected a hundred people, with a threshold of 2000 for drought and famine cases, although those with government appeals for national and international assistance are also included. Each source has its own bias, with those stressing fatalities being more typical of reports from the less developed world where more poor people die, whereas companies that focus on insurance losses have greater reliability in the developed world, as well as much better statistics on economic losses in the past 30 years. Although these different sources show variations in the actual numbers, there is general agreement that there has been a trend towards both increasing numbers and impacts of these hazards in recent years. For example Fig. 9.1 shows the changes by type of hazard since 1980. The rising general trend line is unmistakeable, although the variations by year must be stressed. It also shows the estimated proportion of insured losses compared to total losses at 5 year intervals, which reveals that in 2011 only 28 % of losses were covered by insurance, a proportion often lower in previous years, since few of the losses in developing countries are covered by insurance.

Figure 9.1 also shows that the increase is primarily due to meteorological and hydrological events, with geophysical causes much smaller and relatively stable, even though these often have huge impacts and attract the most attention. The bigger problem is trying to explain the trends. Jennings (2011) has argued that it is unlikely that better reporting accounts for the increase, or that population increases alone is a sufficient cause, but cautiously states that there is insufficient evidence to exclude

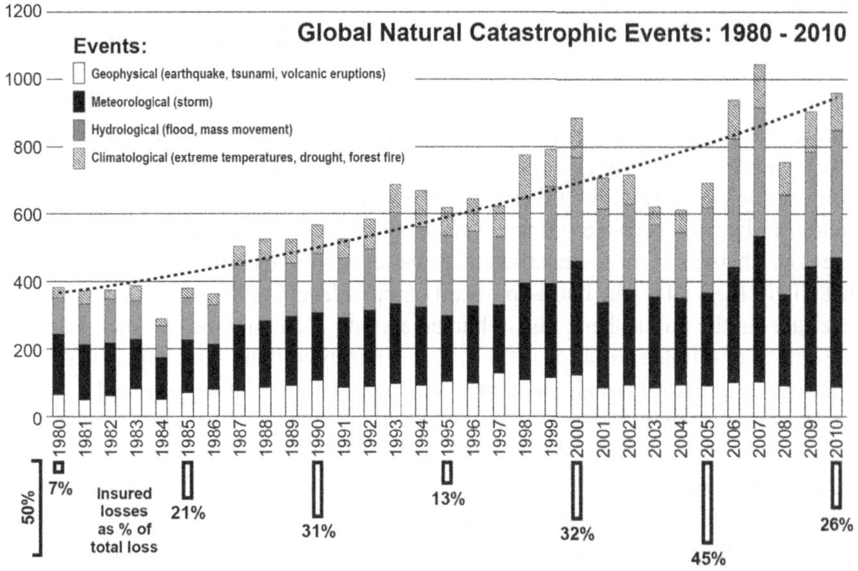

Fig. 9.1 Trends in major natural disasters, 1980–2010. (Source: revised from graphs in reports of Munich-re insurance company on natural disasters, 2011)

the possibility of climate changes causing the increases. Whatever the cause, the increase in numbers seems undeniable and developing world countries seem to be most at risk in terms of deaths, although the greater prosperity of developed countries means that financial losses are higher in these countries. Jennings (2011) shows that in a sample of 41 countries in the developing world for which comprehensive data exists since 1980, there has been an average increase of 4.1 % in the number of natural disasters from 133 in 1980, to 350 annually in recent years, a rise of 233 % over the period. The increase is mainly due to more flooding, wildfires and temperature extremes, for geophysical events have remained approximately the same. Of course, numbers alone are not the only issue. The World Bank (2006) has estimated that the global cost of natural disasters increased 15 times in real terms from the 1950s to the 1990s, and that the number of people affected in terms of displacement, injuries and death, grew to 2.6 billion in the decade from 1993, compared to 1.6 billion in the preceding decade. Two events alone, the 2004 Indian Ocean tsunami and the Kashmir earthquake in 2005, killed over 120,000 people and left millions homeless in the sub-continent. The floods of 2007 affected over 30 million in Bangladesh, India, Pakistan, Nepal and Sri Lanka, leading to estimates that the region annually loses up to six percent of its GDP to disasters from natural hazards annually (Oxfam 2008). Indeed, it has been estimated that global flood losses between 1980 and 2011 cost over $US 470 billion, with 2011 in particular experiencing a series of major floods in many countries in the world, while extreme drought affected the American Mid-West in 2012. In Canada the Public Safety & Emergency Preparedness authority calculated that in the decade 1995–2005 over

700,000 people in more than 250 communities had been threatened by wildfires, although with few deaths. Such examples of the size of the impact of natural hazards can be duplicated in many other countries.

More significant than past trends is the fact that these climate and hydrological natural hazards are likely to continue to increase for the foreseeable future, largely due to the global warming which increases average temperatures and hence moisture content and pressure differentials. This gives greater power to storms and leads to more fire hazards because of increasing temperatures in grasslands and forests. As previous chapters showed, most scientists believe that the current global warming is heavily influenced by the build-up of carbon dioxide and other greenhouse or warming gases in the atmosphere, due to human activity. Certainly the complex nature of the causes of climate variations still leads a minority to doubt that human activity is the primary cause, although the numbers of such sceptics among physical scientists is steadily decreasing. So the progress of global warming is now widely accepted and will increase the number of floods, storm surges, melting glaciers and permafrost. To take one example a report prepared for the Asia Development Bank (ADB) estimated that climate-related disasters accounted for over 70% of the frequency of the most intense natural hazard tragedies in Asia from 1971 to 2010, where the large populations and coastal or river locations of cities make them vulnerable (Thomas et al. 2013). Of course individual climate events cannot be attributed to global warming, but the trends can be. In this respect the ADB report shows the Philippines in particular is experiencing a greater frequency of large storms, which shows the importance of recognizing regional variations within the general trends. Among the other consequences, crop production is likely to suffer on balance, through increased droughts in some areas, which is likely to cause problems in providing adequate food supplies for some cities. Although there may be some regions where the warming increases productivity, there will be more losers if a global accounting of the crop-productivity–global warming association is used. For example, recent forecasts suggest a 20% drop in the productivity of spring wheat by 2050 in the developed world if current trends in greenhouse gas accumulation continue (Arnell et al. 2013). It has also been predicted that Bangladesh, located mainly in the low-lying delta of major rivers, is especially at risk (Danda & Sriskanthan 2011) and, will lose at least a third of its rice crop and a tenth of its rice yields by mid-century (Oxfam 2008). The general trend therefore is one of decreasing agricultural yields in a situation of increasing world population. This will lead to food shortages and famine in many areas, especially urban areas, unless remedial action, or some new agricultural revolution occurs, such as from genetically modified crops more able to produce greater yields and being more resistant to diseases. Yet such changes are opposed by those who see too much of a farming concentration on a limited number of crop varieties, which reduces the gene pool and increases vulnerability to disease, as well as worries about the increasing global dominance of a small number of multi-national suppliers of seed.

However much the evidence of disasters from natural hazards hits the headlines, one of the greatest threats to human life comes from an indirect effect of human activity, namely pollution, not only from manufactured or waste products, but from the burning of fossil fuels in all forms. Indeed, the annual report on the Global Burden of Disease (WHO 2013) showed air pollution in the top ten of killer diseases for the first time in 2012, estimating that 2 million deaths a year are probably due to air pollution in East and South Asia alone. Yet this focus on the danger from environmental events should not be allowed to disguise the fact that the loss of life and injuries from other human activities are often far higher—the most obvious being automobile accidents. For the world as a whole the latest U.N. report of Road Safety (United Nations 2009) calculated there were 1.3 million deaths annually due to road accidents and 20–50 million non-fatal injuries, with many not formally recorded, issues dealt with in the Healthy City discussion (Chap. 13), which also briefly deals with other threats to human safety, such as those posed by biological causes.

9.3 Natural Hazards Effects

Obviously not all extreme natural events are as catastrophic as the examples already described; they vary and have differential effects on the areas and settlements affected. They can only be considered as creating human disasters when they involve significant loss of life or destruction of property and infrastructure. One major difference comes from variations in their *incidence*, for not all of these potential hazards apply in all regions. For example, the threat of earthquakes and volcanoes is concentrated in areas where tectonic plates separate or collide, especially the so-called 'ring of fire' around the Pacific, while urban places on rivers and sea shores are under particular threat from floods. Yet settlements in some countries, especially Japan and Indonesia, are more prone than the majority to natural disasters, because of a combination of features, such as regular earthquakes, active volcanoes and long seashores frequently affected by many tropical storms. Some low-lying countries are especially at risk from floods. For example, the Netherlands has 60 % of its population and 65 % of their Gross National Product generated in areas located in flood-prone areas adjoining rivers or shorelines, for much of the inhabited area has been reclaimed from the sea or from marshes. Recent reports of the Bangladesh area suggest that one fifth of the five million delta residents will probably become refuges by 2050 due to rising sea levels, more tidal surges, increased salinity, pollution from the rivers and the loss of protective coastal mangrove swamps (PW 2012). Settlements in northern climes have to regularly cope with large winter snow accumulation and storms, while the intense downpours associated with monsoons are regular potential hazards in Asian tropical lands.

The *spatial impacts* of these natural events also show very considerable variations. Although the most obvious destructive effects are local, some may have regional or sub-continental impacts. The large tsunamis triggered by the Indian Ocean earthquake of Boxing Day 2004, that involved slippage along an estimated 1600 km north-south fault belt from just off the Sumatra coast to the Andaman islands, devastated the local shores of Sumatra and southern Thailand, destroying many coastal settlements. But these tsunami waves also spread across the ocean and obliterated many settlements in eastern Sri Lanka, so the total death rate in the region could have been as high as 100,000 people. However it is the large explosive volcanic eruptions that have really global effects. The eruptions from Mt. Asama in Japan in 1783 or Krakatau in 1883 (Winchester 2003) threw out vast amounts of dust, rock and pumice into local areas, killing thousands of people in their region and destroying the productivity of surrounding agricultural areas. Krakatau's explosion was reputedly heard over 3,000 miles away and was estimated to have had a destructive capacity 13,000 times the Hiroshima atomic bomb. But its biggest effect lay in its dust output that eventually blanketed the earth. This has been estimated to have affected the world climate for 5 years, because the dust reflected the sunlight back, reducing the temperature by an average of 1.3 % C, leading to bad harvests around the world, with similar consequences to those after Asama's eruption.

One of the most worrying things about the effect of many of these hazards upon settlements is their *speed of impact*. Most of the geological events occur rapidly, with no or little warning, and often produce major loss of life if they affect a populated area. Weather-related events take more time to develop into life or property-threatening major hazards and usually affect wider areas, which means more warnings are often provided by modern weather monitoring systems. Most of these hazards are also time-limited in their *duration*, in the sense that their most intense direct effect is measured in hours or days rather than weeks, although their effect on the disruption of life in settlements may be measured in years. However, some natural hazards, such as droughts affecting large areas, may last for years, having huge effect on the size and density of settlement patterns. This cause—aided by poor agricultural practices—created the so-called Dust Bowl in the American mid-West in the 1930s. It must also be stressed that many of these natural extreme events do not have only direct effects; many create what amount to *related and indirect effects*. For example, many large storms produce rain, high winds and floods; droughts cause vegetation to dry-out and become more vulnerable to fire; high winds help to rapidly spread these fires. Similar related effects occur with many geological events. Earthquakes and landslides into oceans or lakes can push up or replace large columns of water, leading to an increase in both the size and length of waves radiating out from the impact centre which increase in amplitude once they reach the shore, causing the flood waves known as tsunamis, the largest in recent history being around 30 m. Also there should be more recognition of what has been summarized as Na-Tech or Hybrid origins of many disasters, where a dam breaks due to seismic activity and floods an area that was developed because it was thought the dam would protect the region from river floods (Smith and Petley 2009). The list goes on and on.

These examples of indirect effects can be complemented by other historical cases that have *societal effects*, not just local ones, such as the rainfall decrease, smaller crops and political instability that led to the eventual collapse of the impressive city-states system of the Maya in Yucatan and surrounding areas from the sixth century (Kennett and Breitenbach 2012). Of course, even the most extreme of these examples pale into insignificance beside the effect of a large asteroid impact on the earth which could wipe out life as we know it. Most of the millions of meteoroids that enter the earth's atmosphere every day are small, burn up quickly and have little impact. Very occasionally a large one occurs, but few have been experienced and documented in our human memory. In 1908 a stony, iron-rich meteor of 30–50 m in diameter created a massive fireball that exploded some 8 km above the earth. Its blast was heard 1000 km away, people and animals at half the distance were knocked off their feet, and a 1000 km^2 area of forest in Tunguska, Siberia in 1908 was destroyed, luckily an uninhabited forested area. Similar sized occurrences have been estimated to occur once every 300 years, but with only ten times that incidence probability of hitting populated areas (Abbot and Sampson 2009, p. 485). The Chelyabinsk meteor that fell in southern Urals on 15 February 2013 was the largest recorded since the Tunguska example. It has been estimated to have been originally 19 m in diameter and over a million kg. (12,000 t) in weight and entered the earth's atmosphere at 67,000 km/h creating a huge fireball. However more than three-quarters burned up during its decent which climaxed in a big explosion which released half a million kilotons equivalent of TNT, 20–30 times the size of Hiroshima's atomic bomb, so that the largest fragment found was only 635 kg. The huge shock wave caused damage to 7200 buildings, mainly window and balcony damage, with 1500 people seeking medical attention mainly from cuts caused by the flying glass. One teacher who saw the fireball told her students to duck and hide under desks which saved them, although she was injured. Fortunately the area affected was relatively sparsely populated and no settlements were at the impact site (Popova et al 2013).

However, a large meteor may not just affect one place, but life in general. For example, there is now general agreement that it was large asteroid impact on what is now Yucatan some 65 million years ago, in combination with a series of volcanic eruptions, that led to a mass extinction that wiped out the dominant dinosaurs, and heralded the end of the period we call the Cretaceous. This was not because of the direct local or regional effect, but through the resultant cloud of dust that cloaked the earth for years, reflecting solar radiation back, which created colder temperatures and the loss of vegetation, followed by much warmer temperatures when the dust settled and the greenhouse gases generated remained in the atmosphere. Debate continues about whether governments should co-operate in devising measures to combat such large scale hazards. The only real progress so far has been the willingness of the U.S.A. in 1998 to fund an asteroid tracking programme, and 10 years later we know the presence and orbits of some 6 thousand large asteroids, of which almost a sixth have been classified as hazardous if they hit the earth. However extra-terrestrial impacts do not only come from space rocks. There have also been recurring episodes of large solar flares, such as the largest one recorded to date in early September 1859 (Boteler 2002; Lovett 2011). It produced spectacular auroras in most parts of the world, but more seriously it disrupted the nascent telegraph system, causing electrical shocks to operators and destroying many pylons. It has been suggested that a similar solar storm today would disrupt radio communication because of the flare's high energy output of X rays, while the slower travelling coronal mass ejection of charged particles would interact some 3 days later with the earth's magnetic field. This would create power surges in electrical grids, destroying many transformers and cutting out power to cities for weeks, if not months, as well as destroying GIS and satellite connections. Prediction of such space events is in its infancy, but it has been suggested that solar storms of this magnitude only occur every 500 years. Nevertheless, these events may be some of the most severe threats to what really are the fragile electronic linkages that connect the contemporary world in so many ways. Yet despite all these examples of destructive effects it is worth remembering that some of these extreme natural events create longer term *positives*, such as the annual river flooding that deposits silt in flood plains and increases their fertility, as seen in the Nile and other major river valleys. With the social organization that provided the catalyst to organize the surplus of the greater productivity from such lands, these areas developed many of the world's first urban civilizations.

9.4 Hazard Risks and Human Activities

At first sight it would appear that the loss of human life and destruction of property and infra-structure from one or more of these natural hazards is a product of these environmental events alone. However there can be no doubt that the scale of the disasters from natural events has been actively increased by human activity, as stressed by many texts on environmental hazards (Smith and Petley 2009) and such reports as *Rethinking Disasters* (Bray et al. 2008) or Arup's (2011) report for the Red Cross. Several factors contribute to this conclusion. The most obvious is an increasing *population size and areal spread* of settlements, which provides a bigger target area for natural hazards, given the limited spatial incidence of at least some of these events. Their *wealth* often makes losses higher. In addition, *location* factors are involved. For example, many settlements have been built or extended on flood plains or seashores, ignoring the risk of exceptional floods, tides or storm surges and the sewer back-ups that occur. Although these problems exist in the developed world especially on the low over-built South Florida coast, it is in the developing countries that some of the greatest risks occur. For example, many large cities in lowland basins, such as Jakarta, are increasingly prone to flooding, as seen by the 2002 and 2007 floods, displacing 365 and 450,000 people respectively—which in the latter case led to three-fifths of the city being under water for several days. Part of the destruction is due to greater cyclone incidence, but as much is due to the result of more intense human activity in its site. It has been estimated that the area is sinking ten times as fast as the current sea level rise, due to ground water extraction, compaction from high rise buildings, and the over-concreting of lands (Jha et al. 2012), while local wetlands have been built over. However Jakarta's failure to dredge its dozen rivers and canals—which have silted and receive a third of the garbage output—have contributed substantially to the problem, although the World Bank and the Indonesia government are now embarked on major waterway maintenance and rehabilitation projects. Similar problems are expected in many other coastal Asian megacities (Adikrai et al. 2010). The floods around Bangkok in 2011 inundated many of the area's industrial parks, resulting in $46 billion estimated losses in this 1 year .

In the context of earthquake damage it is worth noting the locational differences in destruction in areas of different material. Buildings on clay, sand or silt on shorelines suffer more from earthquakes than those on rock. The reason is that the earthquake-shaking tremors are amplified in such sediments and if the sediments are water-saturated, as they often are near shorelines because of high water tables, the bonds between the composite grains break, leading the sediments to liquefy and lose strength, ensuring that buildings become unsupported, topple and sink. Another example of increased risk comes from the use of different *construction materials* which increase the scale of devastation from some natural hazards. Zebrowski's (1997) penetrating review of the disasters produced by natural events provided a salutary example of this point in his comparison of the differences in death tolls between the 1906 San Francisco earthquake, which recorded 8.3 on the Richter scale, with that of Messina in Southern Italy in 1908 which had a rather lower intensity of 7.5. Although lower in destructive size, an estimated 83,000 people were killed in the Sicilian city, with only a 45 % survival rate, whereas 99 % of people survived in the Californian centre. The difference was in large part due to the greater density of development in Messina and its more rigid buildings—which used heavy masonry construction, granite walls and brick-tile roofs—compared to the many wooden and one-storey buildings of San Francisco, which proved more flexible and less likely to collapse, even though they were damaged. Indeed, San Francisco suffered more damage from the fires that came from the ruptured gas lines than the actual earthquake, another example of greater collateral damage from a natural disaster. Perhaps the example illustrates the old aphorism that 'it is buildings that kill more people', rather than the direct effect of the natural disaster. This problem of the greater impact of natural hazards on certain types of construction can also be illustrated by the vulnerability of shanty towns or informal settlements, which often account for between 20–40 % of dwellings in the cities of developing countries. These houses are flimsy constructions made out of bits and pieces of unwanted materials that are barely cobbled together. Not only are these easily destroyed by storms but they are often on precarious sites—usually previously unwanted land, not developed by formal plans—such as the steep slopes of Rio de Janeiro's hillside favelas, or the wetlands now occupied by slums in Lagos.

Another set of problems stemming from human activity come from *variations in the quality of services and regulatory standards* in settlements. In the developed world the industrial nineteenth century cities had unsanitary conditions, densely packed housing and limited safety and medical facilities which led to high death rates. As described in Chap. 13 (Healthy Cities) it led municipalities—usually with national government laws—to create systems that provided safe drinking water from new reservoirs and treatment plants, combined with new sewage systems and better regulation of housing construction and plans. Zoning or development control measures were also implemented to prevent incompatible land uses being built in close proximity, which meant factories with noxious products were banned from residential areas. So the release of toxic substances either from human error or from buildings destroyed by natural hazards was drastically reduced.

There have been many cases where natural hazards have exposed the *inadequacies of traditional defences against environmental problems*. For example, the embankment of rivers in many cities may have reduced the seasonal risk of regular flooding, but they are rarely capable of coping with the water flows from extreme events, as seen in New Orleans during the Hurricane Katrina storm of 2005. The *larger and more complex infrastructure* of contemporary cities also provides a more vulnerable target, for high winds destroy power lines or cause trees to fall and break. During periods of freezing rain, ice accumulation destroys many power lines since they cannot bear the additional weight of the ice. Snow accumulation causes increased traffic hazards and accidents; open air car parks or lots are vulnerable to hailstorms.

Despite these examples of failures in the developed world, the increase in the provision of services, and the regulatory environment and rules that ensure compliance to the standards, meant that cities in the developed world have become less vulnerable to natural hazards. By contrast, there is *greater vulnerability in cities of the developing world*, for most have few of the regulations and services that reduce the impact of natural hazards and, until recently, few had risk reduction schemes. So the loss of life from natural disasters is often many times worse, although the property losses are usually greater in the developed world because of the greater wealth in their cities. In addition, many aid projects in developing countries have actually reduced the capacity of many areas to withstand the effects of natural disasters. Such new transport routes as the Sheberghan and Khuzdar highways, in Afghanistan and Pakistan respectively, were hailed as exemplars of modern design and development assistance. But they have blocked natural drainage channels in many places, leading to increased floods and destroying scarce productive land (Oxfam 2008). Unthinking human activity has also increased risks through *environmental degradation*. Deforestation in many areas has increased rain run-off and hence flooding and silting downstream, as can be seen in Nepal, where 60 % of the country has been deforested in the last century, allowing the deforested slopes to become more unstable and liable to mudslides or landslips. The shock of high winds and tidal surges, especially tsunami waves, has been absorbed by coral reefs, mangrove swamps and offshore islands which provide some protection for the inland areas behind the immediate coast. Once these areas are destroyed, or the sites urbanized, it is not only the shorelines that are devastated by tropical storms; settlements inland are now less protected, thereby adding to the scale of devastation.

Finally it must be noted that the incidence and scale of devastation from natural disasters have *differential group impacts*. Those who are old, infirm, young and female suffer the worse, for they are often malnourished in less developed cities—meaning they are less able to cope with the conditions—whether cold, heat or floods—so more die from exposure to these natural events. In addition, many are too weak or do not have the ability to escape, such as being unable to swim when confronted with floods, or succumb to subsequent diseases because of their weakness. Also most have few reserves or assets to survive, let alone recover their livelihood, in the post-disaster period of deprivation without help from either national or international agencies. It is salutary to record that the Kashmir earthquake in 2005 killed approximately 75,000 people, most of whom were poor and impoverished with few resources, whereas less than a twelfth of that number were killed in the much more urbanized, but prosperous region around Kobe, Japan, by the Hanshin earthquake of similar strength in 1995. Elsewhere, authorities in Sumatra estimated that a third of the deaths in the 2004 tsunami were children, and women were four times more likely to be victims than males since they were mainly looking after children and were less mobile.

9.5 Human-Created Hazards

The threat from environmental hazards needs to be complemented, albeit briefly, by those derived directly from human activity, since disasters from such causes can also be mitigated by the risk reduction and relief schemes developed to cope with natural events. Throughout history major disasters in cities came from the collapse of poorly constructed buildings, and even more from the conflagrations due to human carelessness, combined with high densities and wooden buildings, that destroyed much of London in 1666, or the major 1657 fire in Edo (now Tokyo) that destroyed over two-thirds of the city—although this was only one of scores of major fires in that city's history. Today new building regulations and effective fire services have reduced the widespread risk of fires in developed world cities but our technological progress has created new ones, given the complexity of industrial processes and their products and the possibility that things can go wrong. Some of these come directly from various failures from our technology, such as the collapse of buildings and infrastructures, spills from toxic chemicals or oil wells, fires from broken electricity or gas lines—some of which may be triggered by a natural process. In general increased attention to safety procedures, building regulations and rapid deployment of fire engines and medical services, has drastically reduced the impact of potential disasters from human causes, although they still occur. Through time these standards have become more rigorous while recognition of new problems have been identified, such as the health hazards of lead in paint and petrol, asbestos in insulation, which led to bans of their use. More generally the pioneering work of Rachel Carson (1962), in exposing the negative effects of so many man-made chemicals in the 1960s, eventually focused attention on these deadly issues, leading to increasingly rigorous policies to reduce the problems they have caused, at least in the developed world. Unfortunately there are still many places in the world where inadequate safety procedures lead to unnecessary deaths, such as the appalling safety record in many Chinese coal mines and the pollution from many new industrial factories in developing countries where safety regulations, especially over the release of toxic chemicals, are either few or routinely ignored. However there are signs of change in some countries. After years of denial of the problem and repressing information, a Chinese government 5 year environmental report in February 2013 admitted that there may be over 3000 places in the country, so called 'cancer villages,' where cancer rates have spiked because of high levels of chemical pollution from negligent factories (Dewey 2013). Even in the developed world the aging of the city infrastructure and the lack of plans and finance for regular replacement provides another source of potential accidents leading to death and injury. So there is still a lot of room for improvement and vigilance. Moreover it must not be forgotten that it is only 30 years since the pioneering efforts of Erin Brockovich and her lawyer against the toxic emissions from a chrome plant in New York state revealed wide-spread denial and cover-ups before liability was accepted, events that led to a major film based on her name. This led to more stringent safeguards against pollution from industrial sources.

Perhaps more dangerous in their long term impact are the human mistakes that have led to explosions in nuclear plants, not simply from the direct loss of life involved, but from the subsequent abandonment of large areas around nuclear plants because of radiation dangers, and the longer term cancer victims who were exposed to radiation. The worse modern example is the 31 km exclusion zone around Chernobyl (Ukraine) following the 1986 explosion that led to the abandonment of the town of Pripyat, and rural communities in the zone, which will take centuries for radio-activity levels to drop. This disaster was due to human error alone. Yet even after this experience, safety standards are not always as strong as they could be, as seen in the limited attention paid to the site of the Fukishima nuclear plant on a vulnerable sea shore and the ill-considered location of its back-up generators. On 11th March 2011 the plant was partially destroyed when a large tsunami wave from a large offshore earthquake destroyed the generators controlling pressure and cooling which led to explosions in the reactor, although brave work by a small number of engineers managed to partially control the damage. However, enough radio-activity escaped to contaminate surrounding areas, which led to a 20 km exclusion zone around the area. Official reports on the disaster show that there were poor safety standards and monitoring at the plant, as well as subsequent cover-ups by the government and the company owners. Concern about nuclear safety has increased drastically around the world, despite the recognition that many of the problems could have been prevented. It has led several countries to review their nuclear power policies and some, such as Germany, have started to close down their nuclear plants, although other countries are still committed to the power source.

9.6 Trends in Risk Reduction Policies

Four major approaches have been identified as reducing the level of risk and impact of disasters (Smith and Petley 2009). The first approach, involves *engineering* solutions, as seen in the building of embankments and dams to reduce floods. Another, initially associated with the natural hazard school of geography (White 1945, 1974; Kates 1978), focuses upon how *human behaviour* helps cause or magnify the disaster, such when poor risk perception allows development in flood plains. The *developmental* emphasis came from the recognition that the size of many disasters was a consequence of increased human vulnerability due to uneven development processes, especially in large, unplanned and poor urban areas (Wisner et al. 2004). More recently, a *complexity* approach (Smith and Petley 2009, p. 8) has emerged, one that focuses on resolving problems by understanding the many interconnections of nature and society that underlie disasters. This complexity approach can be extended to incorporate research from a more *community-based focus* on the effect of natural hazards, which seeks to summarize the features that contribute to places being more able to recover from their effects (Cutter et al. 2003, 2008; Townshend 2014; Kulig et al. 2013). However, these risk-reduction trends need to the complemented by what has been described earlier as the addition of *security services and standards* to settlements, features that often precede many of the other approaches, which have the effect of reducing disaster impacts and provide relief and recovery. These involve the development of standards that improve the construction and zoning regulations that were noted previously, as well as the need for specialists in professions devoted to health and safety, from police, fire services, ambulance and medical services, which often provide the first line of help when disasters occur. Parts of all five approaches can be seen in current disaster-reducing policies. But it is clear few cities can cope on their own with the effects of some extreme natural hazard, either in terms of personnel and finance. This has led to greater state or national government involvement and contributions by international organizations in the greater monitoring of natural processes, such as by weather forecasting, seismic laboratories or tsunami warning centres, using earth-based and satellite systems. Additional reviews of risk come from new global organizations, such as the Global Network for the Forecasting of Earthquakes (GNFE). In 2012 it described a study that assessed the degree of risk from earthquakes in major cities in the world, leading to the prediction that Kathmandu, Istanbul, Delhi, Quito, Islamabad and Manila were the top six cities in a long list of centres likely to experience large numbers of deaths because of their seismic risk, the frailty of buildings, probabilities of fire and landslides, and the limited quality of the local medical and emergency services.

Usually, when major disasters occur, higher levels of government also get involved through their *National Emergency Organizations* and call in the army and personnel from other areas to provide assistance, as well as asking for help from international relief organizations. In the last decade there has also been the emergence of specialist *international agencies* that help in natural disasters. Some have developed from existing international bodies, such as the United Nations and the World Bank and the work of academics (Blackburn and Johnson 2012). Others have come from private relief organizations, such as Oxfam, which have recognized that they must do more than simply react after a disaster. Rather, they realise they should be *proactive* in investigating and reducing the risks before they occur. These agencies and researchers have produced valuable guidelines summarizing ways of creating more resilience in cities and regions to natural hazards. Four in particular stand out. The first comes from the Hyogo Framework for Action (HFA 2005) for the 2005–2015 period, produced in a conference in Kobe, Japan, organized by UNISDR, and which built upon on a previous conference in Yokohama in 1994. This framework identified five priorities for action, although the resolutions were not binding on governments: make disaster risk reduction a national and local priority with strong institutional basis for implementation; identify, assess, and monitor risks as well as enhancing early warning; use knowledge, innovation and education to build a culture of safety; reduce underlying risk factors in development plans and post disaster relief; and strengthen underlying disaster preparation and effective response. The second guide comes from an Oxfam report entitled *Rethinking Disasters (2008)*, which identified four key spheres for action: physical, by creating sound structures and environmental protection; social, moving from reaction to preparedness; economic, by tackling poverty so that disasters do not make the poor worse-off; and political, by reducing inequalities and protecting rights during the crisis and in the aftermath. A third guide came from a review of the characteristics of a safe and resilient community (Arup 2011) from the International Federation of the Red Cross and Red Crescent, which built on previous work by Twigg (2007). This proposed six main characteristics (Knowledge and Safety, Organizations, Connectivity, Infra-structures, Economic Opportunities and Managing Natural Assets) as the key summary elements to improve risk reduction. These categories generalized field work that had been carried out on risk reduction in 30 communities in several South Asian countries, which had identified 3000 relevant factors to aid resiliency. Finally, substantial research by the UNISDR (Blackburn and Johnson 2012) and parallel agencies have produced a global snapshot of how local governments can reduce the risk of disasters from natural events by following the *Ten Essentials for Risk Reduction.*

All these books and reports have led to substantial progress in understanding natural hazards but have different contents and emphases, and have created overlapping lists of what are considered to be the major approaches to risk reduction strategies, which does not help comprehension. To clarify the situation Fig. 9.2 combines, reorganizes, and extends previous key principles, to provide a basic framework of the major categories of ideas and policies in creating greater resiliency. This consists of four primary domains, namely: the *Context or Background* conditions of the settlement or region; the *Initial Knowledge* of the hazard risks and vulnerabilities; the *Needed Improvements*, which describes the various features, physical and human, required to reduce risk and strengthen disaster preparation; and the *Aftermath*, the basic stages in the recovery from a natural hazard disaster. A series of basic

MAIN DOMAINS	**CATEGORIES FOR IMPROVING RESILIENCY**	
	a. PHYSICAL FEATURES AND STRUCTURES	**b. HUMAN MILIEU**
A) CONTEXT	i) Environmental Conditions.	i) Culture, Economy, Knowledge, Well-Being, Governance.
B) INITIAL KNOWLEDGE: OF RISK, VULNERABILTIES (3, b)	i) Hazard Mechanisms. ii) Buildings, Infrastructures.	i) Level of Preparedness: Governments, Businesses, Households.
C) NEEDED IMPROVEMENTS: IN RISK REDUCTION (d) **AND PREPAREDNESS** (e)	i) Early Warning Systems. (9) ii) Regulation Systems: in Planning, Buildings. (6) iii) Infrastructures. (4,d) iv) Priority Facilities. (5) v) Nature: Protecting and Working With. (8)	i) Identifying Risks, Making Priorities (1,a) ii) Building Organizational Capacity. iii) Connectivity. iv) Communication-Awareness. (6) v) Education and Training for Safety. (7, c) vi) Finance and Budgeting. (2)
D) AFTERMATH-RECOVERY (10)	i) Relief. ii) Rehabilitation. iii) Recovery in Long Term.	

Numbers in brackets relate to similar issues for local government action covered in the Ten Essentials (Blackburn and Johnson, UNISDR 2012); lower case letters in brackets are similar to those from the Hyopo Framework for Action (HFA 2005).

Fig. 9.2 Major approaches to making cities more resilient

sub-components or dimensions can be identified within many of these main domains or categories, each of which involve a number of individual policies designed to help communities cope with natural disasters, although each category could be subdivided into hundreds of specific areas, as seen in more detailed schemes attempting to summarize the risk reduction needs (Twigg 2007) .

It has already been shown that the incidence and number of potential natural hazards vary widely, while citizens and governments have different abilities to understand, let alone the capacity to implement policies that will reduce the risk from natural hazards. So the priorities for action vary according to the character and needs of the urban place in question. The resiliency deficiencies may be greatest in settlements in developing countries. Most have few resources to cope with these hazards, having poor physical infra-structures that only barely cope with the regular natural rhythms of climate or geological activity—let alone extreme events—and limited management structures or finances to create new policies. Indeed a World Bank report (2006) estimated that 95 % of deaths from recent natural disasters were in developing countries and their loss in terms of the percentage of G.N.P. was 20 times as great in developed countries. The same source has shown that in some countries the loss from natural disasters has seriously reduced progress from development aid. For example, the February 2000 floods in Mozambique, due to rainfall over six times the monthly average which was followed by a cyclone, destroyed over 500 schools—more than the 487 schools financed by the World Bank over the previous 20 years—as well as over 40 health units, in addition to widespread loses of cultivated and grazing land and most of the irrigation systems. Similarly, the losses from the massive Kashmir earthquake in 2005 was equivalent to the total official development assistance received over the preceding 3 years.

However the developed countries show the largest damages from natural hazards, not in deaths, but in economic terms, due to the greater losses in personal possessions and infrastructure in their cities. So far, the insurance coverage of their businesses and inhabitants, as well as their emergency organizations and resources, mean that the people and the settlements usually recover more quickly. Yet few cities even in the developed world are safe from the devastating effect of many extreme natural events. Much of their infra-structures and many of their buildings are aging and not up to current safety standards, let alone having the ability to cope with the new climate normals of many weather regimes. In addition, human organizations that are part of the risk reduction strategies need upgrading in view of the new threats. Since the precise mix of requirements to reduce risks from natural disasters vary from place to place, each urban area will need to select very different policies according to their risk exposure and priorities as part of their overall planning for coping with, and recovering from, natural hazards, in addition to those from strictly human causes.

It is also worth noting that the old, largely reactive approach to some extreme natural event, which usually affects only one of a limited number of sectors in urban places, is being replaced by more *proactive risk reduction approaches* that are also more integrated, at least in developed countries. This means that a number of actions to cope with possible disasters are increasingly being taken *before* the hazard occurs, while they are also integrative in the sense of dealing with many different sectors. So the *physical* improvements to reduce the risk from disasters are complemented by policies to improve *socio-economic* conditions, such as future employment, empowerment and quality of life issues in general. Moreover, in accordance with the main goals of the Hyogo resolutions in 2005, disaster risk reduction policies are gradually becoming part of a more general sustainable development agenda, while local institutions are being strengthened to build resilience.

One of the most comprehensive of these new approaches is seen in The Second Delta programme in the Netherlands, inspired by a 2006 spatial plan to protect this low lying country against climate change in which sea levels are expected to rise by 2–4 ft. by 2100. This long term national priority project, with its many initiatives, was formally adopted by the Netherlands government in 2009 to provide transparent, balanced plans and designs to protect the country against floods, create urban, economic and ecological development, tourism and freshwater management, as well as building societal and political support, using € 1 billion in annual funding. The new comprehensive and adaptive approach is quite unlike the First Delta programme implemented after the devastating floods of 1953 which focused on engineering solutions. In the words of one reviewer:

> adaptive delta management is not an approach that requires a simple technological change from current practice. It demands a fundamental change in institutional capacity at multiple levels including new knowledge and skills, relationships and policy frameworks......Moreover, contexts and transitioning pathways may vary from country to country and should be taken into account. (Zevenbergen et al. 2012)

9.7 Resilient Settlement Policies

9.7.1 Context

Any study or evaluation of the likely effect of some natural hazard in any settlement or area has to begin with its context or background. Since countries and specific settlements vary in these conditions it is inevitable that the ability of places to cope with natural hazards vary considerably. The contexts consist of two sets of features. One is *physical*, consisting of the local environment and the range of potential natural hazards that can cause problems to the population, as well as the material physical constructions—the buildings and infra-structures in the settlements. The other is the *societal context*, composed of such features as: its culture and history, the economy and resources, knowledge and innovative capacity, the well-being of the population—which includes health and their social protections—and the state of governance. These characteristics provide a milieu which either helps the ability of settlements to derive and apply risk reduction strategies, or provide barriers to progress in this field.

9.7.2 *Initial Knowledge of Risks and Vulnerabilities*

9.7.2.1 Hazard Mechanisms

The first stage in any approach to risk reduction is to assess what types of natural hazards are likely to affect any community or urban place, especially by reviewing historical records of previous natural events in the area, and also to understand the mechanisms involved. In our scientific era increasing research efforts have improved our understanding of the way that these processes develop, move and decay, not simply empirically, but also by modelling techniques (Zebrowski 1997; Blaikie et al. 2004; Abbott and Sampson 2009; Smith and Petley 2009; Diacu 2009). Although we do not have the ability to predict the next incidence of many of these potential hazards, the general area of risk is usually known, and the path and speed of their development—especially in climate-related events—can now be predicted and publicised through various forecasting systems that involve national environmental research agencies, and international co-operation. This specialist research on hazard development and incidence is often beyond the capacity of individual urban places to produce such work, although some carry out monitoring functions. Usually the information from these specialist hazard agencies is forwarded to local and national governments, although there are still too many examples where the warning is delayed, reducing the time for emergency measures to be implemented.

9.7.2.2 Vulnerability of Structures to Hazards

Greater understanding of the mechanisms of various natural hazards needs to be complemented by detailed knowledge of the vulnerability of the physical structures, transport systems, utilities and food supplies. This involves an assessment of the extent to which the physical infrastructures and buildings will be able to withstand extreme natural events by reviewing the strength and flexibility of buildings and infrastructure to withstand any extreme event, such as earthquakes, storms or floods. Although expert advice is often needed to assist in the process it should be the responsibility of local government to have people capable of undertaking these assessment exercises, which not only assess the city's buildings and infrastructures but also those of private companies. For example, the severe storm codenamed Sandy in October 2012 that affected 15 American states over an area as big as Western Europe, only led to 120 deaths, a small number when one considers it left damages of $ 30–50 billion. It was particularly devastating because of the flooding it produced in and around New York and New Jersey; in the latter state almost a quarter of a million people live less than five feet above high water level. Although the main electric utility company that services New York City and nearby counties designed its facilities to withstand a sea-surge of 3.8 m, the storm created sea surges of over 4.3 m in some areas, illustrating that the risk level for key facility locations had been under-estimated, even in an affluent city at the heart of the world economy.

So some power and transport systems were inoperable for longer than they should have been. But the storm has also shown that private companies and public agencies need to pay more attention to their risk exposure and cannot just rely on the city or other levels of government to provide protection. Some companies in the flooded Wall Street area, such as the investment firm, Goldman Sachs, had enough sandbags to protect their lower floors, while their back-up generators were above flood level. By contrast, many others had few precautions, while several hospitals were without power for long periods because their back-up generators and pumps were in areas reached by the flood. It is worth noting that most of the flood affected zone in Manhattan lies on areas that have been built out over the centuries from the 1609 shoreline where the city was located. So this zone has become more at risk from the natural hazard of flooding and in recent years has been heavily overbuilt, adding to the risks. Yet what must also be considered is the way that a solution in one area affected by a storm may cause problems elsewhere because of displacement effects. For example, in the aftermath of Hurricane Sandy, the June 2013 update of the 2007 New York Plan (PlaNYC), originally developed to cope with climate change, recommended providing incentives for owners to move electrical equipment to higher floors, raising new buildings and utilities, adding flood walls and levees to protect vital infrastructure and erecting storm surge barriers on rivers and creeks, with flood gates to protect high value areas, copying the approach used in London, St Petersburg and Rotterdam. The cost of such a scheme was estimated at $20 billion. However, even if these improvements were made, the storm surge water has to go somewhere. So less valuable areas on near by shorelines—sometimes areas with poor residents—are likely to be affected instead. This means weighing up the value of protecting one area, yet accepting the destruction of another, which is clearly a question of relative public justice for city authorities.

9.7.2.3 Levels of Human Preparedness for Disasters

A human knowledge component that helps reduce the consequences of disasters consists of creating a detailed evaluation of the existing state of readiness of the population in any urban area to cope with a natural hazard. This involves reviewing the amount of preparedness by households, communities, businesses and various levels of government. Once a function only of local and national governments, the scale of many disasters now involves international organizations who have started to provide guides to local governments, especially in the developing world, about how to measure their level of preparation to combat natural hazards (UNISDR 2011). This is not done by some top-down process through commands from the national or international organizations. Instead the guides help local governments to evaluate their own risk reduction strategies by responding to an extensive questionnaire. This helps them understand their own limitations in risk awareness and reduction capabilities, as well as their strengths, while discussions with local stakeholders and community groups can identify their priorities, which may be quite different. These actions build up local knowledge of their current level of preparation and ability to cope with a disaster. It also helps create pressure from the public to change attitudes, puts pressure upon local politicians to improve readiness, and encourages local residents to identify and perhaps undertake some of their own risk reduction policies, rather than simply assuming that higher levels of government will solve all their problems

At a local level the first line of response in the case of a threatening natural hazard involves the *government agencies*, principally the basic public security and safety services—fire brigades, ambulance services and police—as well as specialist services, such as firefighting teams in forested areas, or avalanche recovery teams in alpine areas. Questions must be asked as to whether these personnel are adequate in numbers and skills to cope with extreme events. In developed countries there has been greater attention to the need to provide specialist training to cope with large scale emergencies from natural or human causes, as seen in texts such as those available for nurses (Veenema 2012). Additional trained help is often provided from within existing government departments, neighbouring authorities, armed forces—whether reservists or permanent military—and national emergency organizations. Specialists also provide advice on how to deal with disasters from human technological failures, whether toxic chemical spills or radiation leaks, and are able to assess what potential hazards exist in the local area, or from transport passing through the towns. Several key questions need to be answered. Is there an efficient and effective command and communication structure able to take the lead in organizing response, or is this delegated to an existing organization, such as the fire brigade, that has the responsibility to lead and co-ordinate the emergency response? What plans and resources exist to warn people of the degree of risk from the hazard and are they understood? For example, in flood-prone areas there is often a three stage system of Alert (possible flooding)—Warning (probable floods)—Severe Warning (threat to life), while sirens are often used as alerts in earthquake or tsunami prone areas. But are these alerts easily understood by the population and do people understand what they need to do in such situations through emergency drills? Are there effective plans and available public transport to evacuate people to shelters and safer areas if necessary? How will this be implemented to avoid panic, and has the population been alerted to the plans? Is there sufficient information on the location of the disabled and elderly who need help to be moved? Are the local supplies of food, water, mobile power and sanitation sources enough to cope with the demands from what will be a distressed and disorientated population? If local supplies are not available, are there regional storage places with supplies to cater for disasters that might occur anywhere in the region? Are these supplies frequently checked and renewed to make sure they are still useable? Are there enough safe shelters or emergency accommodation, both in temporary or permanent structures, to cope with the displaced population? Is there an effective National Emergency Organization to provide extra assistance in personnel, supplies and subsequent relief? How does this get involved and integrated with local organizations so that there is a single chain of command and help, rather than agencies duplicating or disputing with one another? How can extra relief be obtained from existing international relief operations and various NGOs and how can these work in co-ordination? Are the local airfields, road-rail systems, or ports, able to accommodate the possible inflow of national and international aid, and can they be quickly repaired if damaged?

All of these issues need to be complemented by finding out the level of preparedness in two other main areas. One relates to the *assessments needed by businesses and organizations*, for a natural disaster can affect their future viability. A summary report by UNISDR (Blackburn and Johnson 2012) reported that the global economic losses from natural disasters in the decade since 2000 amounted to $ 1.5 trillion in losses and, significantly, that a quarter of all businesses affected close down after the disaster. The report produced a summary list of the 12 key factors in reducing the impact of natural disasters. This included such features as: businesses identifying risks from various hazards, especially for critical facilities; the presence of updated and tested emergency plans and supplies; back-up facilities for computer records, updated personnel contact numbers; identifying alternative supply sources; ensuring adequate insurance schemes and using above-standard building code improvements for buildings. Many private companies only evaluate risk in their own location. This is often not enough. Companies also need to understand their exposure to natural hazards throughout their supply chain. These often encompass factories in many parts of the world—often in Asian locations that have high risks of damage by natural hazards. For example, the devastation of the eastern Honshu island settlements north of Tokyo by the March 2011 Tōhoku tsunami did not just affect this area. Many 'just in time' or specialist parts that are assembled into final products elsewhere were manufactured in the areas affected, so the destruction of plants in these areas caused factories further up the supply chain in other parts of the world to close down, or to be placed on short work weeks for months, drastically reducing the outputs of many firms outside the area. These examples show the need for private firms to improve their own level of preparededness, as well as to lobby government and utilities to increase their investment in risk reduction.

Similarly, the ability of *households* to cope with some natural hazard emergency needs to be understood, not simply by the people themselves, but also by government agencies that may help their level of preparation. Agencies, such as U.S. Federal Emergency Management Agency (FEMA), have created check lists of advisory suggestions on their website of what to do before emergencies occur. It is argued that prior preparation is the best defence from any disaster, so family meetings should discuss the dangers of the particular local natural events that can affect the household. Plans should also be developed to: share responsibilities within the family; how to respond to particular problems; and to find safe spots in the home and areas outside for each type of hazard. What are called 'disaster kits' should also be created and regularly checked. These should contain supplies of food, light and power to cover the time of the emergency when power is out and shops are closed, in both the house and in vehicles that could be used in emergencies. Key records should also be stored in a safe location. Obviously the list is aimed at people in a developed country with high resource levels. In developing countries few households have the ability to keep stores of food or other supplies that can sustain them during some natural disaster. This makes it more imperative for their municipalities and higher levels of government to have sufficient stores for poor people to rely on in emergency situations.

9.7.3 Risk Reduction Policies

This stage involves creating policies to upgrade deficiencies revealed in the physical and human capacities to cope with risks from natural hazards. Although these improvements can be derived directly from the Initial Knowledge Domain, it is often found that a *catalyst* to change comes from the impacts of a previous natural disaster, which proved to be so calamitous that a new attitude towards risk-reduction occurs in the general population and in the decision-making structures of cities or indeed nations. For example, Winnipeg, Canada, had suffered from major floods for years, given its location on a flood plain at the junction of the northward flowing Red and Assiniboine rivers. But the 1950 floods were far worse than usual with 10,000 homes destroyed, 100 thousand people evacuated and an estimated damage of $ 125 million—probably a billion dollars in contemporary money. Opposition from conservative forces meant that it took until 1959 before a decision was taken to build the Winnipeg floodway around the city to divert high flood waters. In other situations a major disaster may create public pressure that leads to policies that have effects beyond the area affected. For example, a landside from a coal spoil tip located on the steep valley slope above the mining village of Aberfan in South Wales in 1966 rapidly covered an elementary school that lay immediately below it, killing 116 children and 28 teachers. It had been known for years that valley-side tips were unstable and there were many examples of parts of tips slumping down slopes. But the exceptionally heavy rainfall of previous weeks had essentially saturated the overextended tip, compromising its stability, leading much of it to collapse quickly and without warning down the steep valley side. The resultant outcry led the government to remove all coal tips on slopes in Britain, a task completed in subsequent years—one that should have been made years earlier. So it took the sacrifice of over a hundred young lives, blighting a generation in the coal-mining community, for a long overdue reaction to prevent future natural processes working upon the careless storage of coal waste.

A more general example of creating new attitudes to the risks of natural hazards can be seen in new policies being applied in the developing world. Overseas Development experts now insist that all new projects involving outside help, whether financial or technical, should incorporate resilience to natural disasters into their plans, as well as their implementation, something that was rarely done even two decades ago. Moreover, municipalities are increasingly taking a more pro-active approach, identifying the deficiencies in the existing local level of preparedness to cope with some major natural hazard *before* a disaster occurs, often learning from other jurisdictions, or from advice provided by international organizations such as UNISDR and Oxfam. Finally, it is worth noting that although the list of requirements in subsequent sections may seem enormous, and beyond the resources of many urban places in developing countries, it has been found that relatively small improvements can produce major returns, especially in the human sphere. For example, the powerful storm, labelled Sidr in November 2007 that affected the flood and cyclone-prone delta lands of impoverished Bangladesh killed around 4000 people. Although a tragedy, it was a huge improvement over the consequences of the Bhola cyclone of similar intensity in 1970 which led to the deaths of perhaps a half a million people in the same region. The enormous difference in death tolls has been attributed to the addition of many risk-reduction policies, such as: storm warning systems which gave people time to prepare for the storm; community disaster-risk committees to identify and create local emergency responses; the maintenance, clearing and deepening of drainage ditches; the construction of elevated cyclone shelters in which approximately 2 million people took refuge; effective plans that ensured the evacuation of 3.2 million people from the most at-risk areas in 2007 (Bray et al. 2008). Although supported by the national government many of these risk reduction policies were developed with the assistance—technical, social and financial—of international advisers and organizations. Yet the cyclone still led to the loss of over 2 million acres of crop land and 1.2 million livestock, devastating the livelihoods of many farmers and leading to food shortages and price increases, especially in cities.

These examples show some of the ways in which risk reduction policies have led to improvements in the level of resilience to the threats posed by natural hazards in settlements. In more specific terms it is helpful to look at the range of possible measures in two contexts, the physical and the human domains. In the former context there are five main categories in which improvements often need to be made, namely, early warning systems, priority buildings, building codes, infrastructures and policies that work with nature.

9.7.3.1 Early Warning Systems

Increasing knowledge of the mechanisms of natural processes now provide earlier warnings of the possible onset of many extreme natural events. Some people in traditional societies had better skills than the modern population in living off the land, anticipating the arrival of major storms, either by better sensing of weather patterns or through knowledge of previous extreme events, passed on through the generations that lived in the same area. This often led them to secure their dwellings in advance of storms, or to move to higher ground before the arrival of floods or tsunamis. Today, advances in the scientific knowledge of the mechanisms that create natural hazards and the areas they will affect—combined with rapid communication of the results through modern satellite and computer linkages in the past three decades—has vastly increased our ability to show the size and places of impact of many impending natural weather events. In the case of major storms, tornadoes, excessive rain or hail, satellite coverage as well as networks of weather stations regularly identify the start and progress of these hazards. It is now routine in some countries to broadcast emergency warnings of impending storms by sirens and by breaking into all radio and television channels, while in America alerts are also made via social media sites such as Twitter and through mobile phones. Increasingly early warnings often give days of preparation to areas that will be affected by a storm, allowing residents time to stabilize or strengthen roofs, sandbag entrances, add shutters to cover windows, to take shelter, or in case of really large storms providing local authorities with time to organize evacuation from the areas under risk. However such messages may not reach people on the increasing number of tourist areas or beaches in underdeveloped lands. This has led to the installation of sirens to alert people to take precautions and to proceed to emergency refuge locations, although there is always the problem that visitors may not know these places. Yet landslides, tornadoes, hail and intense localised rainfall in particular occur with little warning. Also first world communication systems may not be present in many remote third world areas. So many governments, often with assistance from agencies such as Oxfam, have been issuing radios to families and to inaccessible local villages—especially wind-up radios that do not depend on batteries—so that their recipients have the ability to monitor emergency alerts. Similar advanced warnings occur in relation to potential hazards for human health, such as by the routine broadcasts in developed countries about extreme wind-chill conditions, or even UV indexes in summer, a result of information about the increase of skin cancer rates from overexposure to ultra-violet rays. Also there are finally signs of greater responsibility by some governments in rapidly industrializing countries to warn people of high air pollution levels in many large cities. In early 2013 China announced that 74 of its cities would now begin monitoring and publicizing all sorts of air pollutants, such as SO_2, CO_2, CO and ozone, and also including PM 2.5 (those particulates of under 2.5 μ in size. The latter has been shown to be a particular health hazard since they can penetrate to the lungs and blood stream. Such systematic monitoring and policies to reduce pollution is long overdue, especially in Beijing, for readings of PM in late 2012 reached over 1000 micrograms, a result of the burning of coal in many plants, whereas the WHO promotes 25 micrograms as an acceptable health level.

These advanced warning are less easy to provide in the case of major geophysical events which are largely still unpredictable, although some volcanoes do display build-ups to major eruptions, and small pre-shocks may precede a major earthquake. Historically, these shocks often led people in earthquake-prone areas to take the precaution of sleeping outside houses to reduce the risk of being buried in a building collapse. However, there is no guarantee that these geophysical events will display prior warning; many of the most devastating eruptions of volcanoes or earthquakes have occurred suddenly. Indeed in the case of geophysical hazards we still lack the ability to predict earthquakes and most volcanic explosions with any accuracy (USG). But the network of monitoring stations around the world is becoming more sophisticated in recording and publicizing the incidence and intensity of these events, which can lead to preventative measures by people in areas some distance from the focal points. However, the areas immediately around these epicentres may have little warning before being overwhelmed, as seen in many recent examples of earthquakes or volcanic eruptions.

Many of these new monitoring devices are beyond the ability of individual cities to pay for such procedures; hence national government funding and international co-operation has been needed to establish these stations and to monitor them permanently, often using remote devices to cover large areas to obtain extensive coverage that will transmit geophysical or weather data to collecting stations in real time and to quickly dispense the information. The problem is whether the residents of urban areas that could be affected get the information in time. Thirty years ago this was a real problem. Today our new communication devices mean there is less excuse for information not being dispensed quickly to the emergency centres in cities that should have the ability to alert residents. But even in the past decade there have been failures in communication. For example, the 2005 Indian Ocean tsunami that had its origin in a large Sumatran earthquake just off the coast, devastated the near-by shore lines and probably happened too quickly for advance knowledge of the enormous sea wave to reach the nearby shores. But this tsunami took 90 minutes to reach Sri Lanka. This should have allowed time to issue warnings on its shore lines. This did not happen. The result was thousands of unnecessary deaths. The absence of an international earthquake and tsunami warning system in the Indian Ocean did not help—something that is now being remedied—whereas the long standing Pacific warning system has proved invaluable for decades.

These immediate or short term early warning systems are being complemented by advanced alerts that come from new sophisticated modelling of natural hazards that provide knowledge of the scale and incidence of possible future events. Combined with the vulnerability data from assessment exercises, these methods provide urban authorities and governments with information about the probable scale of the increased intensity of natural hazards, which may lead far-sighted authorities to revise their estimates of possible future risks based only on historic events. For example, Copenhagen's Climate Plan has predicted that global warming will lead to greater summer precipitation from more intense storms and these levels could be 50% higher by 2050. It has led to the decision to allocate € 3 billion in procedures to cope with the greater rain intensity and flood problems in a city that is close to sea level. Despite this example of action to reduce anticipated problems, so far relatively few urban governments and their residents have paid attention to these warnings of future hazards—especially if they are seen as decades away—and do not allocate sufficient resources to counteract such potential problems.

9.7.3.2 Building Codes and Planning

The vulnerability of buildings and infrastructures can be reduced by appropriate building codes and planning policies, which define the structural integrity of buildings and the places where people can build. One of the major problems seen in third world cities devastated by natural disasters has been the absence of building codes that enable structures to withstand local natural hazards. Although urban authorities in these countries are improving their regulations, the lack of qualified engineers and trained building inspectors and ineffective penalties for non-compliance, mean that standards are often ignored. Indeed the problem is really a matter of political will, not technological knowledge, in many countries. For example, failure to comply with regulations specifying earthquake-proofing new buildings in Chile leads to heavy fines and even the removal of licences to build, whereas in neighbouring Peru the regulations are often ignored, with few penalties for non-compliance. Of course, given the variety of different types of hazards in various locations, building regulations will vary according to the local dangers. One of the most obvious requirements in areas of high snowfall follows traditional practice in Alpine areas, namely ensuring steeper roofs. In areas subject to minor floods, raising pavements can help people to move around. The ability of buildings to withstand high winds is a vital element in most places, but especially in areas subject to tornadoes. There are also a number of basic principles that should be adopted in buildings in areas prone to earthquakes, creating flexible rather than rigid structures (Abbot and Sampson 2009; Guevara-Perez 2012). Many deaths still occur because too many of the buildings are built of adobe or some insubstantial material, or of unsecured heavy masonry which falls apart and crushes people in earthquakes. Also the popularity of an open storey first floor since it was advocated by Le Corbusier, has been shown to be significantly related to building collapse during earthquakes (Guevara-Perez 2012). There are some basic requirements to reduce disasters from the human-created hazards: the need to reduce fire risk means that buildings should use non-flammable materials, and houses need to be separated with flame-proof sidings, so that fires do not spread easily to adjoining properties; land use planning needs to be more rigorous to ensure that incompatible and noxious or hazardous activities are not located next to residences; and that sites to be urbanized should not be located in areas that are subject to natural hazards, such as floods. These are only a few of the types of regulations needed to improve resilience. Many are standard in urban areas in the developed world—although the constant upgrades in both national or local regulations over the past few decades and improved engineering practices shows the increasing rigour of these codes. Such advances are still limited in the developing world, again because of the shortage of trained personal and the absence of the abilities of municipalities to ensure compliance to any regulations.

9.7.3.3 Infra-Structures and Buildings

Many urban places need to pay more attention to the location of new urban developments and to improve their physical infra-structures in order to cope with extreme events. There is no doubt that much of the infrastructure of developed world cities is aging and needs to be regularly inspected and replaced, and there are many examples where the former is not adequate, and few cities have plans or the ability to finance the replacements of many existing infrastructures. These problems are compounded by the recent development of more rigorous construction standards, which means older structures need upgrading, and recognition of problems posed by new climate trends. The higher temperatures being experienced in many countries means that: new standards are needed to withstand greater drought probabilities; buildings need to take into account the effects of higher wind shear and more precipitation, even in temperate lands. In northern climes global warming is leading to greater melting of the formerly frozen permafrost layer, which is causing buildings to sag, tilt and even collapse, while there are less possibilities of constructing the temporary ice or snow roads that used to connect settlements in winter, which means creating expensive all-weather roads across inhospitable terrain. In addition, in areas of high snow accumulation or heavy rainfall, the widespread use of flat roofs, especially in large warehouse or retailing operations may be cheaper, but run the risk of collapse under the weight of snow or rain accumulation. One frequent requirement in areas prone to earthquakes is to improve the integrity of structures by reconstructing or upgrading buildings to better withstand hazards, as seen recently in much of the central area of the city of Mashhad in Iran, or in individual buildings such as the renovation of the historic Kabukiza theatre in Ginza, Tokyo, which opened in, spring 2013. Maintenance of structures, especially infrastructures, is also important, even on such minor issues as the frequent blocking of drainage ditches by garbage or invasive species, which reduces their utility in dealing with heavy rainfall.

Desert or low rainfall areas are often prone to infrequent but intense downpours that lead to flash floods which transform natural gullies into raging torrents in the rare rains. These locations in urban areas must be preserved from development. An alternative, used in Las Vegas, was to construct a series of underground tunnels to rapidly drain off the excess rain. The city now has over 270 miles of these run-off tunnels, although unfortunately many have been occupied by homeless people, which make them especially prone to harm when heavy rainfall occurs. Too much urbanization has been allowed on flood plains, while bridges, roads and other infrastructures are often built without due consideration to extreme floods, which mean they are often destroyed, adding to the costs of reconstruction. Traditionally, a wide range of measures have been adopted to reduce the risk of floods, especially in areas subject to winter melts after high snowfall accumulations, or heavier rainfalls, such as: creating upstream reservoirs to absorb the extra flow; clearing out debris in rivers; blowing up ice obstructions in spring; straightening and dredging rivers in order to improve channel flow. Another traditional approach has been to increase the size of embankments, or to make them stronger by using concrete or stone in critical parts, especially where currents come close to the banks, to reduce the risk of collapse, but there are limits to the utility of such measures, for if they are breached major flooding will occur. An additional, but very expensive, policy adopted by coastal cities such as London and Rotterdam has been to build tidal surge barriers in their estuaries that can be raised when higher tides or storm surges are predicted. This type of solution is also being developed by the Venice MOSE system in the three entrance channels to the Venice lagoon and the vulnerable heritage city, which has been sinking for centuries. These barriers can be raised when high tides or surges over 110 cm (3.5 ft) occur which will protect the city to some extent. In addition a series of other flood accommodation measures have been implemented, such as raising the pavement levels alongside its historic canals.

An alternative practice is to divert flood waters around an affected city. Figure 9.3 shows the location of a 47 km long artificial channel to the east of the city of Winnipeg built between 1962–1968, with entry barriers and submersible gates on the main river. This carries off flood waters, usually in the spring melt from the Red River, and returns it near Lockport. When completed it was the second largest earth-moving project in the world at the time, second only to the Panama Canal. Although costing over $ 63 million at the time, it is estimated to have prevented at least $10 billion of flood damage. Recognition that its standards of preventing a 1

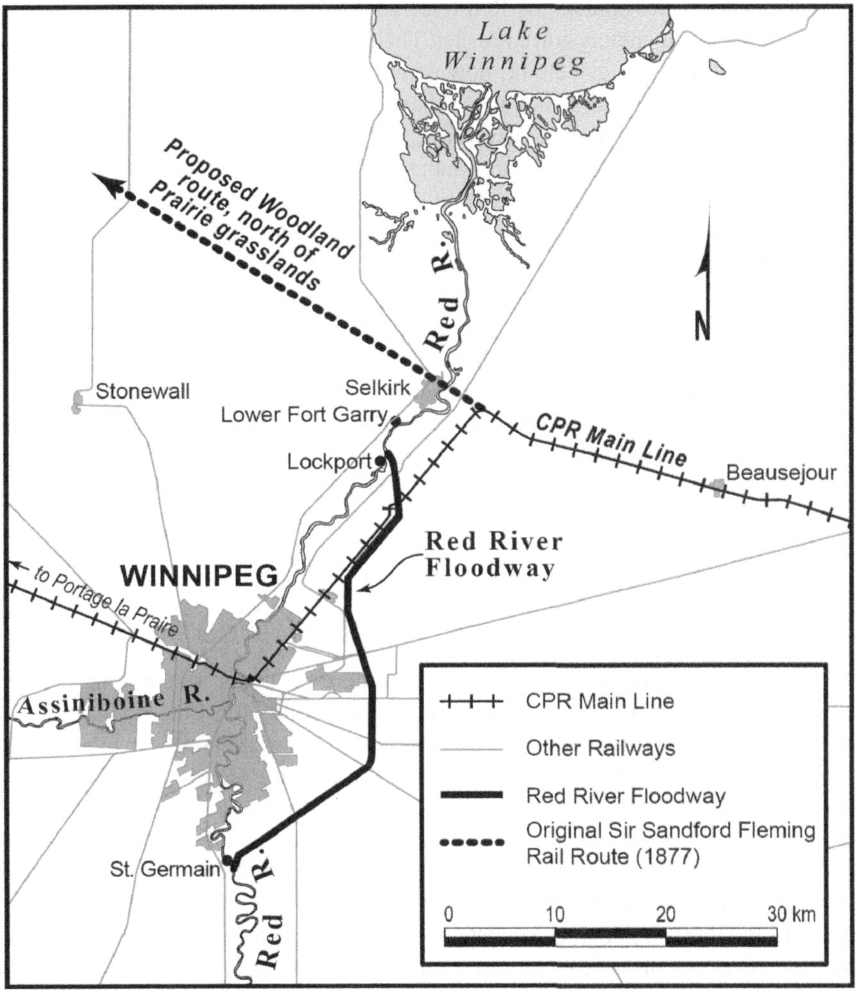

Fig. 9.3 The Winnipeg floodway

in 90 year flood was not adequate, led to a widening of the channel and upgraded control gates between 2005–2012 costing $ 665 million (FA). This increased its flow capacity ability by over 60 %, predicted to be enough to prevent a 1 in 700 year flood, although all such probabilities of flood returns are being questioned, given global warming (FA).

Such expenditure would have been unnecessary if more caution had been taken in the location of the city. In 1881, when the transcontinental Canadian Pacific Railway was being built, engineers were aware that the small town of Winnipeg and its predecessor, the fur trading fort of Ft Gary, were subject to regular high flooding in spring and also with summer flash floods. This was a result of a flat environment with the major water course, the Red river, running north, but having its source in the south, so spring melt occurred earlier. Their original plans called for the railway to cross the Red River at Selkirk, downstream from Winnipeg, where high banks protected the area around, which would have been a more suitable site for a crossing and the projected major railway yards. However, the offer of free land near the Red-Assiniboine junction and a substantial cash sum from the local business community led the railway to divert its main line south and build its railway yards and other facilities in their nascent town. This was part of a new route change in 1881 to directly cross the Prairie grasslands, rather than through the wooded area to the north that Sir Stanford Fleming had preferred and surveyed by 1877. The change stimulated the growth of what became the gateway city to the development of the Prairie Provinces, but in a location prone to disastrous floods, a problem that could have been prevented by a more sensible choice of site during the early railway era (Berton 1970). However, as subsequent sections will show, a very different approach to dealing with these types of problem is being pioneered by the Dutch, in which the emphasis has moved away from engineering solutions that attempt only to contain rivers or seas.

9.7.3.4 Priority Facilities

An essential requirement to improve urban resilience is the need to ensure that certain key or priority buildings have the structural capability to withstand major natural events, as well as being sited in areas that are as safe as possible—in other words not on exposed shore lines, flood plains, beneath unstable slopes, or on fault lines. In addition, these buildings should be on routes that are easily accessed in emergencies and can be cleared quickly if they are subject to natural disasters. *Schools* are one of these priority buildings because they contain a vulnerable population that may not be able to care for themselves in a disaster. Also schools are often designated as temporary emergency shelters to cater for a displaced population in the immediate aftermath of some event causing destruction to surrounding areas. In some countries the regularity of major cyclonic storms and floods mean that it is also appropriate to have special shelters to act as refuges when storm strikes, equipped with sufficient bedding and food to allow people to remain in safely and relative comfort, as already described in the case of Bangladesh. *Hospitals* should also be priority buildings. It imperative that these facilities are built to the highest standards so they remain open and functioning to cater for the injured during an emergency, and again should be located in a place with good access by the injured as well as by medical personnel and emergency services. Unfortunately, there have been many cases in which these priority buildings have been among the first to collapse due to shoddy construction and poor structural designs. For example, in February 2010 a massive earthquake, of 8.8 on Richter scale, centred near the populous Chilean Valle de Itata, and its related tsunami affected over 2 million people, with over a third of a million houses seriously damaged and losses of over $ 30 billion. Over half of the 73 hospitals and over 4 thousand schools were destroyed or badly damaged, reducing their ability to provide medical aid and emergency shelters respectively. Elsewhere, inappropriate sites have been chosen. International relief for the Caribbean island of Montserrat after the destruction of hurricane Hugo in 1989 led to the construction of a new hospital. Unfortunately it was built below a volcano that erupted in 1995 and destroyed the new building under a rain of pyroclastic material. The *headquarters of local emergency measures organizations* and security services should also be seen as priority buildings, simply to ensure that key personnel, as well as their secure communication facilities, are able to provide organization, support and advice in the critical hours after a disaster. This ensures a continuing, and importantly a local, organizational response to the problems, especially before national and international organizations are able to assist.

9.7.3.5 Protecting and Working with Nature

Chapters 4–6 have described the ways in which many urban places are producing greener and more sustainable practices to reduce the negative effect of our human activities, although there is still a long way to go before these effects are neutralized. Many of these new practices have the additional effect of providing greater protection against some unexpected natural event. Perhaps the most obvious is the protection of wetlands that act as storage areas in times of excessive rain and whose plants filter out noxious materials created and washed out from urban areas. In addition, it has already been noted how off-shore coral reefs, mangrove swamps or islands help absorb the impact of storms and provide protection for areas behind, so they should be re-established if they have been destroyed. Shelter belts of bushes or trees in dry or windy locations have traditionally been built around settlements to provide protection against the wind or in desert areas against the advance of sand, whose abrasive properties can quickly degrade buildings. It is also inexcusable to allow deforestation near settlements, for this increases the risk of rapid rainfall run-off and floods, apart from increasing the probability of landslides from surrounding slopes. Development should not be allowed on steep and unstable slopes, especially those formed by unconsolidated materials that are more prone to slump, while re-afforestation and measures to improve slope stability and water retention need to be undertaken to reduce the risks of disasters. A promising recent trend is for cities or governments to pay people in rural areas to manage the upstream lands from cities to preserve water quality, which reduces the need for expensive filtration plants. Yet many environmental restoration policies must also ensure that negative effects are not produced. The restoration of swamps in tropical areas can help breed mosquitoes which increase the risk of malaria; the presence of trees immediately around settlements in forested, or extensive grassy areas subject to drought can increase the probability of fire risks, unless there are adequate firebreaks outside urban places, which is still rarely the case. Also there is always the need to evaluate the positive and negative effects of natural barriers that may affect risk reduction. Yet all these policies are relatively limited approaches to this new theme of working with nature to reduce the risk of damage from natural hazards.

More comprehensive approaches to working with, and not against, nature, have been developed by the Dutch (Bezuyen et al. 1998 Saeijs 1991; Kabat et al. 2005; Scheur et al. 2011). For centuries they have reclaimed land from the sea and flood plains by ingenious reclamation techniques and through co-operative efforts. In the last two decades they have recognized that the increased size and frequency of floods during this period of global warming means that the older engineering control solutions alone are not going to be adequate to solve the problems posed by future flooding levels, solutions that are still being applied world-wide (Zevenbergen et al. 2012). Hence they have adopted new ideas. One is the '*Building with Nature*' approach, which attempts to work with nature, rather than only attempting to dominate or control it. Artificial barriers have been placed in coastal shallow seas, which allow natural processes to create salt marshes, while sand dumps have been created to provide a source of material which will be used by tides and winds to build up existing dunes that are themselves stabilized by vegetation and will form a more effective natural barrier against storms. Along water courses the new '*Room for a River*' strategy is important (Rohde et al. 2006). It recognizes that confining flows through higher and higher embankments will not cope with the predicted larger and more frequent floodwaters in rivers. Certainly strengthening embankments with rocks and concrete on bends still have a place in the fight against floods. But it is realized that a more effective risk reduction strategy is to create more room for a river, rather than only artificially confining it, by policies applied not only in one area but throughout the water-basin. Hence rivers are being given more room in which to flood, such as by: moving dykes back to widen the rivers; deepening channels to increase water flow; lowering groynes that protrude into the river and reducing flow obstacles; removing some riverside areas from current use to act as storage areas or flow channels in flood conditions. All these areas that are created to make room for floods are grassed and pathways are added to create new park areas. These add significantly to the recreational areas of cities, at least when the river is not in high flood. In cities such as Dordrecht experiments have been carried out to see how flood water affects particular housing areas, in order to create more effective flood prevention, while regulations for infra-structure and buildings are designed to minimize water damage. One of the most striking examples of the application of these ideas is taking place in the city of Nijmegen in the Netherlands. Located mainly on the south bank of the river Vaal, it grew up around major river bridges that span a river which narrows in a major bend as it passes the city. This made crossings easier, but ensured that the settlement has always been at risk from flooding. Radical new solutions are being implemented: moving the northern embankments on the river back an average of 350 m; creating an artificial channel on the north bank for the main river channel to straighten out the river to improve flow, which will create an island between the two river branches; raising the land-level in new urban developments on the north bank to reduce flood risk; adopting many of the other techniques described above to widen the river to accommodate high flood levels. The result of the changes will be to reduce the flow in the old river channel on the edge of the city. The old river channel and the new island will become a new recreational area within the city, rather than a dangerous peripheral feature that has often overwhelmed the city with major floods. At a more local level various environmental restoration trends range from creating storm ponds and water-gardens, to grassing or using permeable paving in parking areas or even roads, as described in Chap. 4. They help store more water temporarily, or allow more water to percolate into the ground so as to reduce the run-off that has been accelerated by the paving in so many urbanized areas.

9.7.4 Human Requirements in Risk Reduction

9.7.4.1 Culture of Safety

In human terms the pressing need to reduce the dangers from natural hazards is to find ways of creating a new culture of safety in urban places by improving the level of human resiliency in five main areas, namely: identifying risk and establishing priorities, building more focused organizations and institutions, increasing connectivity and communication, better education and training to combat risks, and finding ways of ensuring that new budgeting procedures will enable a better financing of new policies. Again, few urban places are able to carry out all the necessary changes by themselves, so higher levels of government, with more taxing powers and resources, are needed to be part of the process. Given the changing nature and varied incidence of many natural hazards these risk reduction policies have to be flexible to respond to new conditions and must be frequently reviewed to maintain a state of readiness, especially as new information about the utility of newer best practices become available.

An important requirement is the need to create the *political will* to identify and address the problems posed by natural hazards and convince the general public and all stakeholders—whether in existing political forums, as well as private and voluntary organizations—of the importance of actions to deal with the potential problems. All the levels of preparedness described earlier must be improved if they are not adequate. Since time and resources are needed to solve the problems, there is a need to priorize risk reduction actions, not simply within cities but by national agencies to focus attention on what can affect large areas. Unfortunately, the episodic nature of most disasters means that most politicians ignore or marginalize the threats posed by natural hazards, since their term of office is often short and dominated by day-to-day social and economic issues. Similarly, government departments are inevitably concerned with their own short term problems, not additional ones that are seen to be only in the realm of possibility. The result is that the potential risks from episodic natural hazards have usually been side-lined in the political agenda. For example, Healey and Malhotra (2009) reviewed voting patterns and natural disasters in the USA and showed that although voters rewarded presidents that responded quickly, and with financial aid to natural disasters with more votes in the next election, those who spent money on disaster preparedness did not receive any discernible increase in votes. Such inconsistencies in what is described as the degree of 'citizen competence' lead to the inertia of many politicians and public officials. This is unfortunate, since the authors show that $ 1 invested in prevention saves $ 15 in post-disaster spending. Usually it takes a strong executive or a government committed to preparedness to ensure that pre-disaster planning and implementation does occur, rather than acting only after a crisis.

9.7.4.2 Building Institutional and Organizational Capacity

The scale of the problems posed by natural hazards means that the people and their organizations in areas potentially affected need to be involved in co-operative relationships to build institutional capacity to mitigate risks. From medieval times Dutch reclamation projects have recovered land from flood plains and the sea, which was only achieved by cooperation and organization, leading to water boards to manage the reclamation process, and the subsequent monitoring and maintenance of dykes and other water control systems. Governments in most developed countries, at a national, regional and local level, have created Emergency Organizations to deal with the effects of disasters, whether from environmental or human causes, which link with existing safety services, such as police, paramedics, fire brigades, and medical facilities. In some countries and cities there is often an Emergency Information Day to remind people of the role of the agency and its importance. These organizations have been slow to emerge in cities in the developing world, for existing safety and health agencies are often overwhelmed by day-to-day problems. Hence international aid agencies such as Oxfam and various organs of the United Nations have stepped in and have encouraged the formation of local Disaster Preparation Committees to involve local communities in the risk reduction process. In the Philippines, for example, these are based on relatively new units of civil organization created by a Presidential decree in 1972. The barangay, and the smaller purok units below—a grouping of perhaps 40–50 households—were in some ways a revival of traditional family groupings, originally designed to broaden the base of citizen participation, to promote sanitary and beautification projects, as well as improving intergroup relations and links to existing government organizations (Kendall 1976). Now they are also used to discuss levels of local preparation to combat the potential losses from natural hazards, such as finding more effective ways of protecting areas and responding when disasters occur.

Apart from the discussion of needs, these committees often lead to *reducing risk by changing behaviours*, such as avoiding throwing garbage into local drainage ditches, or regularly cleaning them, which has obvious results by helping prevent or reduce local flooding in rainy seasons. The advantage of such organizations is that they increase participation, spread awareness of the risks and allow people to contribute in discussions about the type of risk procedures that are needed, rather than the more traditional top-down approach associated with government actions. To be effective, these organizations frequently require charismatic *leadership* to publicize the need to anticipate problems and to encourage action, as well as being aware of the need to listen to the local community, whose willingness to help is increased if they believe that their problems are being taken into account and not ignored. Studies of fires in communities in Canada (Kulig et al. 2013; Townshend et al. 2014) have confirmed the importance of this feature in building local community resilience. Their multivariate analysis of questionnaire data in affected communities have also shown it is intimately linked to the degree to which a local community feels *empowered*, in the sense that the local people affected believe they can help solve problems. These studies have also shown that resilience is increased in situations where a community has high levels of *engagement*, as measured by such features as local pride, similar values, engage in community events, are willing to help others through volunteerism, and, as well being open to new ideas, many of which contribute to a sense of belonging to the community. By contrast, variables that measure what might be called *negative local geographical feelings about the local environment, and feelings of isolation* from the larger political entities, proved to be a separate dimensional attribute in their Index of Perceived Community Resilience. Such negative attitudes need to be countered if the degree of resilience in local communities is to be improved.

9.7.4.3 Connectivity

These social capital and organizational issues for urban places need to be complemented by more effective connections at various levels, both horizontally within a community and vertically to higher levels of government and private organizations. This involves creating *multi-sectoral mitigation structures*, so that the emergency organizations are linked into all relevant agencies, rather than being isolated as a single sector. This means emergency organizations should also have effective contacts to key politicians and bureaucracies at higher levels of government, so that support from outside the urban area can be given in crisis conditions, as well as to civil society organizations. It is also important to acknowledge the help that can be provided at times of emergency by existing social organizations at a local level, especially church and women's groups, as well as informal social networks. They provide an invaluable source of information and community support. Links to relatives and friends in other locations are also important, for they often offer shelter to people displaced by disasters, reducing the need for public assistance. These integrated risk management structures need to be designed with clear command structures in emergency situations. Unfortunately, many of these organizations have been designed by higher level of governments as top-down structures. Studies of disasters have found that these are often too remote and lack local knowledge. Hence studies of disasters have stressed that it is usually necessary to have an effective *local base* to direct responses, in which a horizontal structure is developed. During emergency situations it is also important to plan for at least twice daily meetings with key personnel so that the often rapidly changing situation is understood by all parties. These should be carried out at a local level so that the knowledge becomes part of the process of disaster relief. These organizations should also be based on a *multi-stakeholder* structure, This means involving local governments, existing security and health services, as well as representatives of local organizations and voluntary groups to create an integrated approach, for it is recognized that local governments alone cannot bear all the responsibility for providing risk reduction plans and relief during a natural hazard event. For example, the Barcelona Resilience Board contains representatives of 37 different organizations, from utilities to various levels of government and industries. But recourse should not only be made to governments, utilities, security and medical organizations etc. Private companies should be made aware of the way they will lose the ability to operate if the potential damages from natural hazards are not anticipated and contained. So they are increasingly encouraged to help, not simply by senior personnel joining risk management committees, but through financial contributions to infrastructure improvements, as well as by seconding specialists or materials to assist with planning for resiliency or for post-disaster relief operations.

Responses to major disasters frequently require more than national level assistance. International links are increasingly important. This is not simply to draw on outside resources in times of crisis, but also to obtain information from other urban places that have suffered from similar hazards, so that a local government can draw upon the experience of others, especially about what pro-active policies are the most successful. Traditional aid agencies like Oxfam and various United Nations agencies have realised they should do more than just provide relief *after* a disaster; they have become *pro-active* in providing books and organizing conferences about risk assessment and mitigation, giving advice to local governments at risk about the best practices to reduce the scale of disasters. In addition, increasing numbers of city-to-city resilience networks have grown up. One example is the International Network for Designing Earthquake Resistant Cities (INDERC) whose members share information about earthquake issues and design practices (Guevara-Perez 2012) Another is an earthquake forecasting group (GNFE). Also, some cities in the developed world have set up partnerships with centres in poorer countries to share ideas, and specialists to help reduce the dangers from natural hazards. For example, Bonn in Germany, perhaps because of its experience of being devastated in 1993, 1995 and 2005, when the Rhine rose above the city's 10 m flood protections, has developed partnerships with several cities to help them cope with natural disasters, namely: Bukhara (Uzbekistan), Cape Coast (Ghana), La Paz (Bolivia), Minsk (Belarus), Ulaanbaatar (Mongolia) and Chengdu (China). It is also home of the United Nations International Strategy for Disaster Reduction Platform for Promotion of Early Warnings (UNISDR-PPEW).

Adoption of such integrated approaches means there is usually a need to resolve *potential conflicts* that can occur between different groups and participants who have different values and ways of operating. The interviews carried out by Reimer (2012) and his colleagues in Canadian communities affected by fires has shown the difficulties that arise between the risk mitigation approaches adopted by what are called bureaucratic, market-based, associative and community groups, all of which have different ways of operating. Meetings between the various stakeholders before a crisis can help resolve difficulties, such as by identifying problems that can hold up progress, in order to maximize resources in a co-operative fashion. For example, one of the key problems in the aftermath of the Haiti earthquake has been the way in which it has proved difficult to persuade the large landowners to release land on which to build houses for these displaced. The result is that many of the displaced people still live in temporary tent villages, 5 years after the event, very much a consequence of a conflict between the market-based private enterprise ideas of a small elite land owning group and the bureaucratic-based relief efforts from governments and NGOs attempting to re-house displaced people.

9.7.4.4 Communication-Awareness

The need for better communication to avoid misinformation and panic, before and during disasters, can be seen in two different contexts. First, it is increasingly accepted that sharing information on risks with the general public to improve their understanding of the dangers from various hazards is vital. Also they can often help in identifying past flood levels, mapping faults under sediments or recording the path of previous landslides or avalanches. The more the knowledge of the mechanisms and effects of these natural hazards are shared, and are extended through historical research on the previous impacts of extreme natural events, the greater the local awareness of potential impacts. This raises public consciousness of the need to adopt risk reduction strategies, which then empower people to help in the process of understanding and managing the risks. This can be seen in developed countries, such as the creation of a Neighbourhood Empowerment Network in San Francisco in 2007 (NEN) between residents, non-profit and faith organizations, as well as businesses and academics, to develop the capacity for neighbourhoods to help themselves, such as by increasing their resistance to potential disasters. The local disaster committees set up in many developing countries also empower people. This approach also ensures that the needs of the people affected, especially the most vulnerable, are placed at the core of resilience policies, for it is the residents in an area that know those who are limited in their ability to respond to warning alerts.

The second context is a vertical one. Since local municipalities rarely have the resources to cope with large natural hazard-induced emergencies alone, it is vital that they *develop effective connections with other organizations associated with risk mitigation* to provide more comprehensive risk reduction strategies, rather than working in isolation. In addition, during the emergency it is essential to have 24 h help phone lines to provide information and assistance to people affected, secure information places where personal enquiries can be answered, as well as ensuring that local radio stations broadcast officially sanctioned updates of the imminent threats from a natural hazard, or recommending evacuations. Some places have also installed sirens to alert people to imminent problems, such as flooding or tornadoes. In addition, communities should create safe places to assemble where affected people can go for help. However, since local broadcast systems can be destroyed by the natural hazard, back-up facilities should be constructed. Given the increasing value of social network sites, such as Facebook, Twitter and mobile phones, these communication devices should also be used to spread helpful information from reliable sources about the scale of the disaster, although official sources should try to ensure that fallacious information should not be spread by these sites, which can add to the problems caused by the disaster. Some universities already use personnel media sites to inform students and staff of potential risks on campus.

9.7.4.5 Risk Understanding and Education

Given the infrequent nature of the extreme natural events that cause loss of life and property, many in the general public are often unaware of the risks posed by these hazards, and have little experience in coping with them. This can create a dangerous passivity, which adds to the potential damages caused by such events. This lack of comprehension, or inability to act to reduce risk, is often higher among people who have only recently moved into urban areas—a growing problem given the increasing migration and mobility rates in most countries. So there is often a pressing need to inform the public, and especially children in schools, about the local environmental risks—not only from examples drawn from similar events in other areas, but from case studies drawn from local historical events. These often make a bigger impact on urban residents who learn to understand that their local area has suffered major damage from some natural hazard in the past. In many cases the required information is in handbooks provided by local governments to residents, or by lessons in schools about the problems posed by local hazards. More active preparation can come from organizing risk-reduction events in existing local organizations, or more specifically through the local Disaster Preparation Committees described above. In some countries there are national programmes that provide information about specific risks, and alerts, such as what to do when early warning sirens erected to warn of an impending extreme event are heard.

People may also be encouraged to take part in disaster drills. One of the oldest of these is in Japan. Since the Great Kanto earthquake of 1923, which directly or indirectly through fires killed 140,000 people in the Tokyo-Yokohama area, the government has encouraged schools and workplaces to run annual disaster drills on its Sept. 1st anniversary, providing advice about what to do when an earthquake strikes. The devastating Honshu earthquake and tsunami on March 11th 2011 revived interest in the need to practice for such possible disasters. In addition Japanese teachers, even in elementary schools, describe precautions that are needed in case of environmental emergencies and practice following escape routes. Local geography lessons also teach students about their local human and physical environment to ensure that students are aware of the local hazards and what to do if they strike, while cities such as Cairns in Australia have an annual cyclone awareness week. More recently, the unexpected and catastrophic earthquake in Canterbury, New Zealand, that destroyed much of the downtown area in September 2010, gave rise to a national campaign organized by the national Civil Defence and Emergency Management Office. It encouraged people to participate in an emergency drill at 9.26 am on Sept. 26 2012. After warning sirens sounded, people were told to practice their basic response under the aphorism 'Drop, Shelter and Hold-on'. This provided a short-hand reminder of key responses when an earthquake occurs, namely: to stay low so as not to be hit by flying debris; shelter under a table or heavy object that can offer protection against falling objects, ceilings or walls; and to hold on to the object, to avoid being thrown across a room by the shaking, and being exposed to objects that often fall from walls or ceilings. They were also encouraged to stay where they were until the quake stopped. An estimated 1.3 million people (over a quarter of the country's population) took part in the first drill. It is too soon to know the lasting effect of this type of national action, although it has been argued that similar drills carried out in Japan were embraced with alacrity in the early years after a major disaster, but that participation rates fell-off in later years, with decreasing responses and general indifference. One only has to look at attitudes towards fire-drills and responses to emergency procedures described on airlines to see similar complacency. In Japan, schools and workplaces regularly practice earthquake drills, as well as being made aware of safe routes to places of refuge on higher ground in the case of tsunamis. In the case of the March 2011 tsunami many children in coastal towns died because although they were ready to move to higher ground, their teachers had not been given the instruction to move, given the speed of the tsunami advance. There is also a need to be aware of safe places to shelter from major storms or wave surges. One of the few good news stories that emerged from the trail of destruction and death on the shores of the Indian Ocean after the 2005 tsunami was the way that a young English school girl on holiday on a Thai beach convinced her parents to move to higher ground as quickly as possible when she saw a rapid sea level retreat. She had learned about tsunamis in a geography lesson in her school and realised that this unusual retreat occurred before the main tsunami wave overwhelmed the shore. Her knowledge saved her family, whereas others around them moved down the now exposed beach in search of the fish that had been stranded and paid the price with their lives. Similarly in Japan many people survived because they moved to higher ground or to the fourth or higher stories of concrete buildings in March 2011 when they heard the warning sirens which rapidly alerted them to the threat of the earthquake. However some were not quick enough to move to safety, given the speed of the tsunami advance that builds up when the waves hit shallow shorelines. It is worth noting that it was not simply the wall of water, reaching over 20–30 metres high in some locations, that caused such destruction to the coastal settlements, but the fact that it contained the remains of so many buildings and boats. The result is a boiling mass of debris that acted as an abrasive, overwhelming sea walls that were too low to contain the size of the waves and utterly destroying most structures in its way.

Specific advice to reduce risk is also needed for other types of natural disasters. In the case of tornadoes it is wise to take shelter in ditches if caught in the open, or in cellars, if indoors. In the case of places that are in danger of floods survival rates increase dramatically if people can swim or at least know how to stay afloat, a particular problem for small children, females and the elderly who have been shown to suffer greater death rates in developing countries because they are less likely to have had swimming lessons. Yet many still drown because of the speed of water flow and the buffeting from debris carried by the flood water, or in the case of many tsunami by the long fetch of the waves that means the wave advance can often continue for ten or fifteen minutes, unlike the limited time of a normal wave. In addition to all this type of advice and basic training for the general public, there is also a requirement to have adequate numbers of trained first responders to natural disasters. This means encouraging local people to obtain basic first-aid training, and adequate numbers of trained rescue people and medical personnel to quickly deal with those injured.

9.7.4.6 Finance and Budgeting

No urban government has all the resources it needs to reduce risks from extreme natural events so it usually necessary to establish priorities as to what types of improvements are needed, and in what order. Once this has been settled it is necessary to create *adequate and dedicated budgets* to ensure they can be implemented. Another problem is that there are usually more funds available for post-disaster recovery than for pro-active approaches to reduce the risk of disasters from natural hazards. As noted above this is because of the budget emphasis on immediate problems—not on events that seem in the distant future. The source of funds also varies. Usually local taxes are too small for the demands of risk management policies, or for post-disaster relief, so urban centres have to depend more on funding from higher levels of government. However, there are increasing attempts to get private firms to provide specialist personnel or materials for disaster projects, and even finance—at least for projects in their immediate surroundings—for if government-run schemes break down, their business may suffer badly.

In the case of developing countries many urban places now obtain funds, or expertise in terms of advice and technical assistance, from international non-government aid agencies, although these are often for specific projects. The way that the national government money for these risk reduction schemes is allocated also varies. Some have dedicated funds supplied to a specific agency that manages various disaster strategies; others have specific line-items for such uses in the budgets of existing departments; even more have aspects of both systems. There is probably no one right way to budget for disasters, given the very different levels of both preparation and hazards faced by various urban places. However, a separate budget for the agency charged with disaster risk management is probably the most appropriate one, since it does not have to compete with demands from other functions in various departments, which may be subject to reduced funding if the anticipated risks do not materialize. It is worth remembering that many risk reduction schemes have been the subject of criticism when their facilities were not used. For example, the Winnipeg Floodway that was described earlier was often ridiculed as Duff's Ditch—a term derived from the Premier of the province, Duff Roblin, who implemented the project—and was seen by many as an expensive folly. But the relatively low flood levels of most of the 1970s and 80s did not last; in the last two decades extremely high floods have occurred and the value of the diversion channel was shown.

In the Philippines a series of bad floods in cities led to a new 2010 Disaster Management Act. This ensured that risk management budgets allocated to local governments were not used for other purposes, as it mandated local governments to spend 5 % of their budget on disaster prevention or relief. In addition, it made the rare decision to allow a larger proportion (70 %) to be available for risk reduction schemes, and the rest for post-disaster relief. In this latter case there is little doubt that if an urban place is affected by an extreme natural event the local ability to finance relief operations is far too limited to be effective. Hence higher levels of government and international organizations have to step in, although in most of the former cases there has to a declaration by the national government that a disaster has occurred before funds are released. Also it must be emphasized that the effectiveness of the allocation of funds for disaster risk reductions often depends upon the degree of corruption in government. In far too many countries, especially in the developing world, funds and resources are stolen by corrupt politicians, bureaucrats or businessmen overcharging for work or materials. This means that the money is not spent for the intended purpose, resulting in greater death rates and property damage from the natural hazards. Indeed, Escaleras and colleagues (2007) have shown that there is a direct and linear relationship between the level of losses from natural disasters and corruption. So the reduction of corruption and more transparency in government are vital components in the implementation of effective risk reduction policies.

9.7.5 Aftermath of Hazard Disasters

In the immediate aftermath of some natural disaster there are usually major requirements to rescue people, treat injuries, provide water and food, often through mobile units, and shelter, usually in temporary structures. It is also vital to provide security to prevent looting and bodily harm to the shocked and debilitated victims. Access to the areas affected via road clearance and bridge reconstruction must be re-established, along with repairs to port, rail and airport facilities to ease the flow of goods and personnel into the areas affected. In addition it is important to quickly provide alternative clean water supplies as well as sanitary facilities to prevent diseases, for water is often contaminated. It is also necessary to rapidly repair power lines or provide alternative energy sources. Increasingly, military helicopter surveillance of the areas affected is being used to quickly judge the extent of devastation so that priorities in assistance can be created. Since children especially are more at risk and may have lost their parents, they need to be especially cared for and their routines re-established. One approach is to quickly open schools, even if they have to be in the open air in warm climates initially. Help in identifying and burying the dead is also a priority, but without upsetting local spiritual beliefs. International agencies have been prominent in recent years in providing DNA testing to establish identities of the victims, using test information from survivors. The rapid creation of local bulletin boards to act as contact points is crucial, as are communication facilities to allow people to establish contacts with family and friends, even if in other parts of the world, for these can often provide moral and also financial support. In the less developed world many of these tasks need to be assisted by local members of the community, or those familiar with the culture. Such people provide what amount to 'local culture-brains' to assist outsiders in understanding local mores and making the victims feel more comfortable with the foreign assistance.

Many of these tasks are usually beyond the resources of the area affected by a major natural hazard event. The region requires outside help to provide these functions, as well as heavy equipment to move damaged structures and repair infrastructure. So it is imperative that national emergency teams have plans to quickly respond, with help from specialist personnel from surrounding areas and local militia groups or armed services. In addition the full involvement of senior political leaders is needed to cut any bureaucratic blockages that occur, and getting assistance from major international aid agencies. Such complex assistance becomes a major logistics operation, so it is vital that careful plans with recent revisions have been drafted by national agencies to fulfil these immediate needs, which means rapid access to the necessary supplies. All too often in the past such rapid response has not been forthcoming, meaning that the loss of life was often greater than it should be. However there is little doubt that more responsible governments, quicker communication facilities and the increasing skilled assistance from international organizations has been of great help in reducing the post-disaster death tolls in many areas in the last decade. The involvement of social organizations such as the Rotary Clubs in relief is also helpful. They have developed a Shelter Box programme, providing what amounts to emergency supply boxes containing a whole variety of useful products from high protein bars to lights (SB).

Despite all these helpful features, the increased level of risks from natural disasters described previously shows that a lot of work is still required. Unfortunately, there are many cases where not all victims of disasters are treated equally, such as when dalits, the lowest castes in Indian cities. have been refused access to emergency cyclone shelters by others. Elsewhere, favouritism in relief supplies has led to one ethnic group being favoured over others. Also there is always a need to protect the contents of damaged houses from looters—which means sufficient security personnel to restrict access to disaster sites—while the victims in temporary relief accommodation needed to be protected from thieves and molesters. Another problem that has been noted in relief efforts has been the failure to co-ordinate the many organizations that rush to help in any area. Duplication of effort and rivalry between relief agencies need to be avoided and resolved through some over-all co-ordinating agency in which the host government plays a major part.

Once the immediate problems of providing food, health care and shelter have been solved—all major tasks—the next duty is to clear the devastated sites. Rather than only using assistance from outside, and especially N.G.O's, it is increasingly realized that the victims of the disaster should also be encouraged to help with the tasks of clearing sites and reclamation. The may involve training local personnel, providing them with wages and basic skills in construction or in security. All too often the victims of disasters remain homeless and unemployed for too long, without hope and with limited security, in the temporary camps. By contrast, the well-meaning NGO personnel obtain the jobs. By involving as many locals as possible and paying them, the disaster can be used as a way of creating a buffer for the next big storm or hazard and in improving the economic and social life of affected people. Such approaches will provide the victims with an income and perhaps new skills. In addition, international aid agencies in particular have been stressing the need to supply victims with money for them to purchase food etc—at least in developing countries, where the people have few assets or reserves to fall back on—rather than the continuation of free relief, which leads to conditions of dependency, not individual or family resilience. This will stimulate the growth of markets and often local and regional agricultural production to supply goods, all of which help local economy to recover. Some jurisdictions, such as Makati in the Philippines, have gone further and provided loans to known residents of disasters for hospital fees, burials etc. In developed countries faster assessment of insurance claims, or assistance from government, also helps the recovery process. For example a week after devastating floods in Calgary and areas around in late June 2013 the provincial government provided newly homeless families with preloaded credit cards, enabling victims to obtain food and other supplies, some of which could be used to repair damage. However the issue of compensation for residences destroyed, and whether rebuilding on the flood plain should be allowed, are still contentious issues a year after the flood.

One factor that was often forgotten in many cases of disaster relief by international agencies is the need to treat the minds of the victims as well as their bodies and livelihood. Most natural hazard disasters create a great deal of what amounts to post-traumatic stress, and the loss of self-esteem among the victims about their ability to cope with the future, especially in less developed world situations since their meagre possessions have been lost, which seriously restricts belief in their ability to recover from the events. Hence there is a pressing need for trained psychologists to assist in the evaluation of victims, especially in the case of children. Their feelings of safety have been destroyed, not simply by the negative experiences of the disaster itself, but by being uprooted and perhaps faced with the deaths of many friends and family. In addition, regular counselling sessions and roundtable discussions must be created. This allows victims to take part in the process of recovery, as well as sharing their stories and describing how improvements can be made, based on their local expertise or even experiences in dealing with government or insurance agencies. However it has been noted that attempts to help victims over stress-related problems need to use the methods familiar to the cultures affected, not the external, person-oriented approach of western medicine. Watters (2011) has shown that western approaches, often individual and personal in one-on-one sessions, can have harmful effects on more group-orientated societies where grief is expressed and resolved collectively, for local cultures and belief systems shapes understandings of what has happened and the ability to recover. In the longer term one of the sad effects of the 2011 Japanese tsunami has been the increased number of family breakdowns due to stress, inability to find work and worries over exposure to radiation in the areas near the Fukishima nuclear plant. Many, especially women with children, want to move to other locations, which disrupts the family unit, but those who move often find little assistance in finding jobs. By contrast many of the elderly in particular, wish to remain in more familiar surroundings, but soon face a lonely and bleak existence given the slow development of new housing and facilities in affected towns.

Once the immediate problems have been solved plans are required for the *long term rehabilitation* of an area subject to disasters as well as the compensation of victims. The people in too many disaster areas remain in conditions of dependency for years. There is a pressing need for governments to help in restoring the living conditions and livelihood of victims of disasters. Certainly it is likely that many former residential areas will be considered at risk from further disasters, since they were built on flood plains or open to other hazards. Decisions on areas that can or cannot be developed should be quickly made, after thorough assessments of the areas and when people are compensated for loss of land or residences. For example, Sri Lanka provided compensation for owners of 19 thousand registered fishing boats that were lost in the tsunami waves of 2005. This has led to rapid recovery, through the ability to finance the building of new boats and then of fishing. But it did nothing for those who had rented vessels or had unregistered boats or were landless labourers. Such examples are rare. Indeed, there seems little doubt that the long term recovery of areas affected by natural hazards disasters may be the biggest problem of all, although the short term recovery efforts gain the most attention.

9.8 Conclusions

Most scientific authorities agree that the world is likely to see an increasing number of disasters from natural hazards, which will particularly impact the big cities of developing countries. Hence there is a pressing need to decrease the level of risk from these problems, as seen by the increasing attention paid to the topic by the U.N. and major international aid agencies in the last decade. Not all of these hazards can be predicted in advance, but some can be. The discussion of various problems that must be addressed to increase the resiliency capacity of settlements issues as outlined in Figure 9.2 provides an integrated overview of the key issues, with examples of successful policies from around the world. A major change in thinking about how to cope with many these problems has occurred in the last decade, in which it is recognized that large engineering solutions are far from being the most effective ways of protecting settlements. Certainly such approaches have their part to play, but the major projects that work with, rather than against nature, and with people, show a new approach to reducing the impact of natural disasters. In addition, it is recognized that the high death rates and property loss in underdeveloped countries is associated with the fragility of the settlements and the poverty of their inhabitants. Hence it is not enough to protect and rescue people; consideration should also be given to their possessions and resources, assets which may enable them to recover more quickly. So reducing poverty and increasing the economic status of people helps people to better cope with disasters and enables faster recovery. In developed countries there is likely to be insurance payments that provide some relief from loss, but such schemes are rare in the poorer countries. So despite the progress that has been made in recent years, and the greater degree of international co-operation in assistance and knowledge sharing, the problems posed by the greater incidence of natural disasters to settlements will still be a major issue on the future agendas that seek to improve the life of people in many settlements throughout the world.

It must be also stressed that few places can cope on their own. Urban governments require help from higher levels of government and international agencies to help mitigate the problems from natural hazards. Moreover, it must be accepted that the increasing risks from extreme natural events is not simply due to basic geological processes or features such as global warming. It is also a human creation. The greater size and growth of urban places, and failure to create adequate risk reduction strategies in too many centres, has increased the level of vulnerability and potential loss in human lives, infrastructures and economies. Hence far more attention needs to be paid to developing more resilient policies for urban places to cope with the dangers posed by natural environmental processes, issues that have been ignored for too often, or just accepted as one of the hazards of life, because of their irregular occurrence. Such policies should also be developed with the co-operation of residents who often provide valuable insights, rather than being imposed from above, which are the main objectives of reports by Arup (2011) and UNISDR (Blackburn and Johnson 2012). This will also ensure that the people most affected are able to understand the scale of the risks which will make them more alert to the need to be prepared, which used to come when settlements were primarily occupied by people who had historic experience of the problems posed by local natural hazards.

References

Abbott, P. L., & Sampson, C. (2009). *Natural disasters*. New York: McGraw-Hill-Ryerson.

Adikrai, Y., Osti, R., & Noro, T. (2010). Flood-related disaster vulnerability: An impeding crisis of megacities in Asia. *Journal of Flood Risk Management, 3,* 185–191.

Arnell, N., Lowe, J. A., et al. (2013). A global assessment of the effect of climate policy on the impacts of climate change. *Nature Climate Change, 3,* 512–519. doi.10.1038/nclimate1793. Accessed 10 Oct 2013.

Arup: Arup International Development. (2011). Characteristics of a safe and resilient community. Geneva: International Red Cross and Red Crescent Societies. http://www.alnap.org/resource/12165. Accessed 10 Oct 2013.

Berton, P. (1970). The national dream. Toronto: McCelland and Stewart.

Bezuyen, M., van Duish, M. J., Onderzoek, C., & Leeders, J. A. (1998). Flood management in the Netherlands. *Australian Journal of Emergency Management, 13*(2), 43–49.

Blackburn, S., & Johnson, C. (Eds). (2012). *My city is setting ready: how local governments reduce risk.* Geneva: UNISDR: United Nations Office for Disaster Risk Reduction. Second edition.

Blaikie, P., Cannon, T., Davis, I., & Wisner, B. (2004). *At risk: Natural hazards, people's vulnerabilities and disasters.* London: Routledge.

Boteler, D. (2002). The super storm of 1859. *Annals of Space Research, 38*(2), 155–158.

Bouwer, I. M., Crompton, R. P., Faust, E., Hoppe, P., & Pielke, R. A. (2007). Confronting disaster losses. *Science, 218,* 753–755.

Bray, I., Kenny, S., & Chughtai, S. (2008). *Rethinking disasters. Why death and destruction is not nature's fault but human failure.* Geneva: Red Cross.

Carson, R. (1962). *Silent spring*. New York: Houghton-Mifflin.

Cutter, S., Boruff, B., & Shirley, W. (2003). Social vulnerability to environmental hazards. *Social Science Quarterly, 84*(2), 242–261.

Cutter, S. L., Barnes, L., Berry, M., Burton, C., Evans, E., Tate, E., & Webb, J. (2008). *Community and regional resilience: Perspectives from hazards, disasters, and emergency management.* CARRI Research Report 1, Hazards and Vulnerability Research Institute, Department of Geography, University of South Carolina.

Danda, A. G., & Sriskanthan, G. (Eds). (2011). *Indian Sundarban delta: A vision.* New Delhi: World Wild Life Fund and Jadavpurta University. http//:www.deltacommissaris.nl/english/topics. Accessed 14 March 2012.

Dewey, C. (2013). Chinese state media release a map showing the spread of 'cancer villages'. *Washington Post,* Feb 22.

Diacu, F. (2009). *Megadisasters: Science of predicting the next disaster*. Princeton: Princeton University Press.

Escaleras, M., Anbarci, N., Register, C. (2007). Public sector corruption and major earthquakes: a potential deadly interaction. *Public Choice, 132,* 209–230.

FA: Floodway Authority. http://www.floodwayauthority.mb.ca. Accessed 14 Feb 2012.

GNFE: Global Network for Forecasting of Earthquakes. http://seismonet.org/page.html?id_node=182&id_file=96. Accessed 10 Jan 2014.

Guevara-Perez, L. T. (2012). "Soft story" and "weak story" in earthquake resistance design. Posted on INEDERC (International Network for the Design of Earthquake Resistant Cities). http://inderc.blogspot.ca/2012/11/soft-story-and-weak-story-in-earthquake.html. Accessed 23 Nov 2012.

Healy, A. And Malhotra, H. (2009). Mypoic voters and natural disaster policy. *American Political Science Review, 103*(30), 387–406.

HFA: Hyogo Framework for Action. (2005). Building the resilience of nation and communities to disasters. U.N., Geneva. http//:www.unisdr.org. Accessed 12 Feb 2012.

Holling, C. S. (1973). Resilience and stability of ecological systems. *Annual Review of Ecological Systems, 4,* 1–23.

Jennings, S. (2011). *Time's bitter floods: Trends in the number of reported natural disasters*. Oxford: Oxfam

Jha, A., Bloch, R., & Lamont, R. (Eds.). (2012). *Cities and flooding*. Washington D.C.: World Bank.

Kabat, P., van Vierssen, W., Veraart, J., Vellinga, A., & Aerts, P. (2005). Climate proofing the Netherlands. *Nature, 438,* 283–284.

Kates, R. W. (1978). *Risk Assessment of environmental hazard.* SCOPE report 8, New York: John Wiley.

Kendall, S. H. (1976). Barangay as community. *Ekistics, 242*(Jan.) 15–19.

Kennett, D. J., Breitenbach, S., et al. (2012). Development and disintegration of Maya political systems in response to climate change. *Science, 338,* 788–791.

Kossin, J. P., Emanuel, K. A., & Vecchi, G. A. (2014). The poleward migration of the location of tropical cyclone maximum intensity. *Nature, 509,* 349–352. doi:10.1038/nature13278.

Kulig, J., Edge, D., Townshend, I., Lightfoot, N., & Reimer, W. (2013). Community resiliency: Emerging theoretical insights. *Journal of Community Psychology, 41*(6), 757–775.

Lovett, R. A. (2011). Solar flare: What if the biggest known sun storm hit earth today: *National Geographic Daily News,* March 8. http://news.nationalgeographic.com/news/2012/03/120308-solar-flare-storm-sun-space-weather-science-aurora/. Accessed 14 Feb 2012.

NEN: Neighbourhood Empowerment Network. http://www.empowersf.ca. Accessed 14 Feb 2013.

NOAA: National Oceanic and Atmospheric Administration. (2012). http//:www.ncda.noaa.gov. Accessed 15 Jan 2013.

Popova, O. P., Jenniskens, P., et al. (2013). Chelyabinsk airburst, damage assessment, meteorite recovery, and characterization. *Science, 342*(6162), 1069–1073. doi:10.1126/science.1242642.

PW: Partners for Water. (2012). Towards a Bangladesh delta plan. http:www.bangladeshdeltaplan.org. Accessed 10 Jan 2013.

Reimer, W., Kulig, J. C. Edge, D. S., Lightfoot, N., Townshend, I. (2012). The Lost Creek Fire: An example of community governance under disaster conditions. *Disasters, 37*(2), 317–332.

Rohde, S., Hostmann, M., Peters, A., & Ewold, K. (2006). Room for rivers: Strategy for flood plain restoration. *Landscape and Urban Planning, 78,* 50–70.

Saeijs, H. (1991). Integrated water management, a new concept: From treating of symptoms towards a controlled ecosystem management in the Dutch Delta. *Landscape and Urban Planning, 20*(1), 245–255.

SB: ShelterBox. http://www.shelterboxcanada.org. Accessed 12 Feb 2012.

Scheur, S., Haase, D., & Meyer, V. (2011). Exploring multi-criteria flood vulnerability by integrating economic, social and ecological dimensions of flood risk and coping capacity: From a starting point view towards an end point view of vulnerability. *Natural Hazards, 58,* 731–751.

Smith, K. And Petley, D. (2009). *Environmental hazards: Assessing risks and reducing disasters*. Abingdon, Oxon: Routledge. Fifth edition.

Thomas, V., Albert, J. R. G., & Perez, R. T. (2013). Climate related disasters in Asia and the Pacific. Asia Development Bank Working Paper 358. http://www.pacificdisaster.net/pdnadmin/data/original/ADB_2013_WP358.pdf. Accessed 15 June 2013.

Townshend, I., Awosoga, O., King, J., Fan, Hy. (2014). Social cohesion and resilience across communities that have experienced a disaster. Natural Hazards. DOI 10.1007/s11069-014-1523-4. Accessed 20 Nov 2014.

Twigg, J. (2007). Characteristics of a disaster-resilient community: A guidance note. http://www.benfieldhrc.org. Accessed 12 April 2012.

United Nations. (2009). *Global status report on road safety*. New York: United Nations.

USG: U.S. Geological Survey. http://earthquake.usgs.gov/earthquakes/world/most_destructive.php. Accessed 15 June 2013.

Veenema, T. G. (2011). *Disaster nursing and emergency preparedness*. New York: Springer.

Walter, J. (2004). *World disaster report*. Geneva: International Federation of Red Cross and Red Crescent.

Watters, E. (2011). *Crazy like US: Globalization of the American psyche*. Free Press.

White. G. F. (1945). *Human adjustment to floods*. Research Paper No. 29. Chicago: Geography Department, University of Chicago.

Wisner, B., Blaike, P., Cannon, T., and Davis, J. (2004). *At risk: natural hazards, people's vulnerability and disaster*. London: Routledge.

WHO: World Health Organization. (2013). *Global burden of disease report*. Geneva: WHO.

Winchester, S. (2003). Krakatoa: The day the world exploded, August 27, 1883. London: HarperCollins. SB: ShelterBox. http://www.shelterboxcanada.org. Accessed 12 Feb 2012.

World Bank. (2006). Hazards of nature: Risks to development. In R. Parker & W. Hurlbut (Eds.), *Independent evaluation group for World Bank*. Washington D.C.

Zebrowski, E. (1997). *Perils of a restless planet*. Cambridge: Cambridge University Press.

Zevenbergen, S., van Herk, S., Jeroen, D., Rijke, J., Kabat, P., Bloeme, P., Ashley, R., Speers, A., Berry, A., Gersonius, B., & Veerbeek, W. (2012). Taming global flood disasters: Lessons learned from Dutch experience. *Natural Hazards, 65*, 1217–1225.

Chapter 10
Creative Cities

Niamh Moore-Cherry

The creative city, if there is such a thing, is surely an arena for contentious politics over the character of the city and for whom it works.
Markusen 2006, p. 1937

10.1 Introduction

Cities are fundamentally shaped by their social and economic life; they obtain their specific character from the people that congregate and activities that take place within, as well as their interactions with places outside, their specific spatial or functional boundaries (Massey 1994). Economic change goes hand-in-hand with urban transformation. The de-industrialisation of cities in Europe and North America and the growing importance of Asian cities in the late 1970s and early 1980s is one good example of this trend, as large proportions of global manufacturing moved from some historic industrial cities, such as Sheffield (UK) and Detroit (USA), to those such as Guangzou (China), Dacca (Bangladesh) and Bangkok (Thailand). Equally, the emergence of the so-called 'New Economy' in recent decades has resulted in significant spatial, social and economic urban change. This new economic order is indexed by the growing importance of the service sector, especially by international financial services in the biggest cities, and has been facilitated by significant advances in information and communication technologies (ICT), although stalled by the dot-com crash of the early 2000s and the financial crash from 2007. Because of these new technologies, the new economy is more footloose and capital has become highly mobile. This seems to make the nature of contemporary economic activity less place-dependent on traditional locational factors, such as land, labour and capital, which means that cities are facing significant competition to attract their share of global investment. To address this issue over the last two decades, policy-makers have used a range of tools to improve urban competitiveness, such as fiscal

N. Moore-Cherry (✉)
School of Geography, Planning and Environmental Policy, University College, Dublin, Belfield, Dublin 4, Ireland
e-mail: niamh.moore@ucd.ie

W. K.D. Davies (ed.), *Theme Cities: Solutions for Urban Problems*,
GeoJournal Library 112, DOI 10.1007/978-94-017-9655-2_10

incentives, advertising and branding, as well as the assumption of some of the risks of development through supporting public-private partnerships that are favourable to businesses. However, this is no longer considered enough. In this so-called 'age of creativity', knowledge and innovation have become key attributes in driving urban economic development (Peck 2005). The new found importance of knowledge and brain-power over other kinds of economic inputs, has given rise to the need for new approaches to understanding urban growth, where 'knowledge' is seen as a necessary addition to the classic locational factors of land, labour and capital. This has led to such new urban-economic themes as Knowledge Cities (Chap. 11), But the combination of particular types of knowledge and recognition of the importance of 'soft factors' in determining the attractiveness of urban areas, especially as related to the growing cultural industries, has also led to the idea of the creative class in generating new growth in what are described as Creative Cities. Most recently, this view has been popularized by the work of Richard Florida (2002a, 2005) in the United States, who argues for the importance of technology, talent as well as tolerance, as the key factors of future economic success. This chapter examines the lineage of the Creative City concept and its many interpretations, and then focuses substantially on the interpretations promoted by Florida. His ideas have given rise to heated intellectual debate, and attracted many critical comments, Yet the urban policymakers looking for a solution to the challenges generated by economic restructuring and resultant urban unemployment have often eagerly embraced his concepts, without considering the kinds of questions posed by Markusen (2006) in the comment above.

10.2 The Roots of the Creative City

All cities are creative places by their very nature. They are resourceful, productive, innovative, original, imaginative and, most importantly, should be dynamic. While the Creative City idea is now something to which a very large number of city governments explicitly aspire in the hope of guaranteeing economic growth, it may be understood as a new spin on a very old idea. Successful cities have always been defined by their creativity and flexibility, especially after the Middle Ages, and no city at any time has ever had any monopoly on creativity, or for that matter the good life (Hall 1999). However during the Renaissance a new phase of creativity in some cities emerged, such as the artistic ones in the city states of Italy, stimulated by the patronage of private individuals and families, like the de Medici family in Florence. Today the activities of contemporary urban and cultural policymakers have attracted creative people, such as artists, writers and musicians etc. to particular cities. Their presence has in turn raised the city profile, bestowed prestige on it and attracted even greater numbers of talented people, which in recent decades has moved beyond the traditional arts to large numbers in media, advertising and computing. This basic premise of attracting talented people, sometimes called 'creatives', underpins contemporary ideas of the Creative City. However, the terms

Creative and Knowledge Cities are often used interchangeably; indeed Musterd et al. (2007, 2010) have provided research and a policy guide combining the Creative and Knowledge City ideas into the concept of Creative-Knowledge Cities. But as the next chapter will illustrate, the Knowledge City literature is much broader in its conception of creativity, placing major emphasis on the whole learning or knowledge acquisition process, not simply individual sectors, such as patent-based innovation, as the basis for new economic growth. So this chapter adopts a very specific focus only on the cultural sector of creativity.

While it is undisputed that creativity plays a key role in urban and regional development generally (Hall 2000) and is a key requisite for any city that aspires to be successful. What creativity constitutes, how it is measured and how it relates to urban life more broadly has been, and continues to be, the subject of intense debate. The roots of the current debate on creative cities can be traced to the cultural economy work of various authors writing on urban affairs since the late 1980s, such as Bianchini and Parkinson (1993), Garnham (2005) and Landry (1995, 2000, 2006). Each of these recognised the potential for one form of creativity to promote urban employment and vitality, especially in what has been called the cultural industries (Scott 2000). In a European context, much of the research on creative cities has been undertaken within the context of a response to economic decline and the need for urban revitalisation. Cultural resources were seen as a key to the diversification of local economic bases and the promotion of growth in new economic sectors such as leisure, tourism, the arts and media (Bianchini and Parkinson 1993). Central to the economic restructuring of many cities and the development of more vibrant urban environments, the cultural industries became critical elements of regeneration strategies across North American, British and Australian cities. Scott (1997, p. 335) explains this shift by suggesting that at this particular moment:

> in contemporary capitalism, the culture-generating capabilities of cities are being harnessed to productive purposes, creating new kinds of localized competitive advantages with major employment and income-enhancing effects.

This finding emphasises the way in which arts and culture are no longer primarily regarded as being consumption activities in the contemporary city. The productive capacity of cultural activities has become one of the defining characteristics of the post-industrial age. In terms of the new post-modernist style they support what has been described as:

> the architecture of festival and spectacle, with its sense of the ephemeral, of display, and of transitory but participatory pleasure…whole built environments become centrepieces of urban spectacle and display. (Harvey 1989, p. 270)

This cultural underpinning of Creative Cities is behind major initiatives such as the UNESCO Creative Cities Network which evolved out of the *Global Alliance for Cultural Diversity initiative*, set up in 2002. The network is a way for cities to promote their local cultural scene and is a mechanism for cities to share experiences, ideas and best practices for cultural, social and economic development. The promotion of 'creative tourism', was defined as tourism that

> involves more interaction, in which the visitor has an educational, emotional, social, and participative interaction with the place, its living culture, and the people who live there. (UNESCO 2006, p. 2)

Such a key aspiration is achieved through specific marketable designations. The UNESCO designations are summarized in Table 10.1. It shows that Edinburgh is one of the cities described as a City of Literature, whereas Santa Fe is a City of Craft and Folk Art, and Shanghai is identified as one of the Cities of Design. This list indicates that some towns and cities are now being defined by an authoritative international body according to particular types of cultural activity, profiling specialisms within their urban economies. Yet the list has been criticised, as it is hard to see how Milan and Paris especially were not designated in the Design categories, and how other places get their designation over others. Nevertheless, the overall goal of this programme is to make culture more accessible through a 'less museums, more squares' approach to innovation and creativity, but one that also provides a very useful branding tool for promotional strategies. The Creative City Network of Canada (CCNC) provides another example of how culture and arts is appropriated for urban development in a country. The network provides a forum for municipal staff working on arts, culture and community development to share their expertise and experience and develop tools for the community of practice in Canada.

The overarching assumption made by both of these networks is that the arts and culture are important for urban vitality, and that cities have an internal creative capacity that needs to be harnessed to make them more interesting and attractive places. Similar ideas have underpinned the work of urban theorists such as Peter Hall (2000) who has argued that creativity is essential for urban and regional economic development. Charles Landry (2006) has also suggested that the shift in economic focus from "brawn to brain" provides new opportunities for cities to think anew, although this does undervalue the creative abilities that existed in many an industrial city that created new machines and products. Landry sees creative cities as those that have the capacity to grow through harnessing their existing cultural and social capital in new ways—a capital that goes beyond just the arts and cultural sectors. Indeed Landry's original work (2000) maintains that creativity goes beyond the economic to embrace all aspects of urban development, including its governance. Rather than providing a solution to a particular economic problem, creativity is seen as an approach to urban development. He argues that every city has a latent internal creative capacity to facilitate growth and development through inventiveness and inclusivity, so the convergence of innovation and creativity can be facilitated through enlightened governance (Landry 2006). Drawing on the earlier work of Tornqvist (1983) and Andersson (1985), Landry suggests that a creative institutional and economic context is critical to the emergence of an urban cultural milieu that enables cities to grow successfully. Bayliss (2007) has discussed this at length in relation to the city of Copenhagen where he argues that public policy has provided a favourable framework to facilitate, rather than impose or import, creativity. This is in direct contrast to policymakers in other places who have interpreted the work of writers such as Richard Florida (2002a) in a more prescriptive way. This has led to the suggestion that there are two approaches to providing the necessary

Table 10.1 Members of the UNESCO creative cities network. (Source: Adapted from UNESCO 2013)

UNESCO designation	Cities
Cities of literature	Edinburgh (UK)
	Iowa City (USA)
	Melbourne (Australia)
	Dublin (Ireland)
	Reykjavik (Iceland)
	Norwich (UK)
	Krakow (Poland)
City of film	Bradford (UK)
	Sydney (Australia)
Cities of music	Bologna (Italy)
	Ghent (Belgium)
	Glasgow (UK)
	Seville (Spain)
	Bogota (Colombia)
	Brazzaville (Congo)
Cities of craft and folk art	Aswan (Egypt)
	Kanazawa (Japan)
	Santa Fe (USA)
	Icheon (Republic of Korea)
	Hangzhou (China), Fabriano (Italy)
Cities of design	Berlin (Germany)
	Buenos Aires (Argentina)
	Kobe (Japan)
	Montreal (Canada)
	Nagoya (Japan)
	Shenzhen (China)
	Shanghai (China)
	Seoul (Republic of Korea)
	Graz (Austria)
	Saint-Etienne (France)
	Beijing (China)
City of media arts	Lyon (France)
City of gastronomy	Popayan (Columbia)
	Chengdu (China)
	Östersund (Sweden)
	Jeonju (Republic of Korea)
	Zahlé (Lebanon)

tools for achieving creative city status: first, the development of world-class physical spaces in the city—such as landmark buildings and public spaces—and second, the attraction of external human talent to the city.

As the discussion above has shown, much of the debate on creativity in cities began with a focus on the cultural economy in the 1980s, building on other more disparate bodies of work that had emerged from the 1960s onwards. While not directly identifying with the idea of creative cities, Jane Jacobs (1961), the well-known community activist and planning critic, can be credited with influencing much of the recent work in this field. For her, the urban renewal policies of the 1950s that razed city neighbourhoods for road building and other urban projects were entirely counter-productive. She advocated the need for dense mixed-use neighbourhoods with a diversity of people and activities to ensure urban vibrancy and success. This kind of diversity of people and activities and constant change as a pre-requisite for urban dynamism is a recurring theme in the creative cities literature. For example, Andersson (1985) argued that the conditions that promote urban synergy or interaction come from the instability that is crucial to the promotion of creativity. This idea can also be seen in the work of Peter Hall (2000), who suggested that innovative capacity in cities only happens in a society in extreme flux, where there is a large number of new and young arrivals mixing and merging into a new kind of society. More recent discussions of the creative city have taken these ideas as their starting point and developed them in entirely new directions to some acclaim and much criticism.

10.3 Creative Class Theory

The existence of a Creative Class, as proposed by Richard Florida, has become a key element of the Creative City idea since 2002. Adding value to the attractiveness of the city by a particular social group now underlies much the planning and development goals and urban tourism strategies which aim to attract high-spending visitors to maximise their impact across the wider economy. The city atmosphere or 'city buzz' is crucial to perceptions of vitality and economic growth and these ideas have become the starting point for the most recent work on creative cities. Rather than using these characteristics to simply market a city to visitors, they are now seen as central to attracting highly educated and capable migrants. Rather than being simply dependent on the classic location factors of physical infrastructure, schools, and low tax rates for urban competitiveness, successful cities are those that use a range of less tangible attributes to attract highly talented people who are seen as the driving forces of economic development in this creative age (Florida 2005). Grounding this in the kind of language beloved of policy-makers, it has been emphasized that

> the key dimension of economic competitiveness no longer lies in large endowments of raw materials or natural resources or even labour cost advantages. Rather, it turns on the ability to attract, cultivate and mobilize creative assets. (Florida and Tingali 2004, p. 12)

But the argument goes further. If cities are to have creative ambitions, then it is stressed that they need to invest heavily in building a high-quality urban environment rich in cultural amenities and conducive to diversity in local social life in order to attract talent. Florida (2002a) identified three components in this ability to achieve urban success, namely the 3T's of success: technology, talent and tolerance. While the first two of these assets are important components of the new urban growth, the tolerance of a range of cultural and lifestyle choices is also seen as a critical element in attracting the necessary talent to create success. Florida argued that cities cannot compete successfully without all three components and provides a number of examples to support this claim. He suggested that places like Pittsburgh have not grown because although they have 'deep reservoirs of technology and world-class universities', they lack the open, tolerant atmosphere required to retain creative talent (Florida 2002a, p. 250). Other places like New Orleans, which have open and tolerant societies, have the opposite problem, in that they lack a major technology base. He suggested that, only when the 3T's converge, can a place be truly creative. One of the major criticisms of his ideas is that, unlike Landry who argued that everyone within a city has the potential to contribute to its creativity, Florida reified one group of people—the Creative Class—and argues for urban policies that nurture the conditions to attract and retain them.

In many ways, what Florida has done is to turn traditional ideas in urban and regional economics on their head. He argues that what cities should be doing is attracting talent so that the high technology companies will follow, rather than attracting jobs for people to follow. While innovative, this approach is not entirely new and draws some inspiration from previous work. For example, Gouldner (1979) wrote about the 'rise of the new class' in line with the new economy, while Lucas (1988) suggested that productivity gains resulting from talented people clusters are the driving force behind the growth and development of cities and regions. More recently, Glaeser (2005) has provided empirical evidence of the association between regional economic growth and talent. Like Gouldner (1979), Florida provided a Creative Class definition, those employed in the upper stratum of the new economy. Within this, he identifies a number of sub-divisions: the creative professionals or knowledge-based workers such as those working in law, business and healthcare; the super-creative core "*whose economic function is to create new ideas, new technology, and/or new creative content*" (Florida 2002a, p. 8), such as artists, musicians, and architects; and bohemians. In his first book on the topic in 2002, Florida calculated that the Creative Class constituted 38 million Americans, or 30% of the workforce, which shows the importance of this sector. However this growth in the creative class numbers is not just an American phenomenon. A more recent EU-funded project comparing the Creative Class in thirteen European cities has highlighted the importance of this group in employment terms to the general economic structure. For example, in Munich (Germany) the category constitutes over 32% of total employees (Hafner et al. 2008), while in Barcelona (Spain) they represent almost one quarter of all workers, with the distribution in the various occupations shown in Table 10.2.

Hall (2000) has also contributed to the debate by suggesting that creative cities are those within which outsiders feel a sense of ambiguity, and where they are neither excluded from opportunity, nor so embraced that they lose their creative drive. However, Florida's thesis is more specific since he focuses on the attraction,

Table 10.2 Employees by Economic Sector in Barcelona. (Source: Pareja Eastaway et al. 2008)

Creative industries	% of total employees, 2006
Arts/antiques trade	3.8
Designer fashion	2.9
Architecture	1.4
Computer games, software, electronic publishing	1.1
Video, film, + radio and tv	1.0
Music, visual & performing arts	0.9
Publishing	0.7
Advertising	0.5
Knowledge industries	
Law & other business services	3.3
Finances	2.6
ICT	2.3
R+D and higher education	1.2
Rest of activities	77.8
Total	*100.0*

retention and embedding of external talent in the city to develop what he described as the Creative City. He identifies four specific creative city planning and design goals that are essential for achieving this general objective.

The first is to *maintain authenticity and create distinctiveness*. One way of achieving this is through retaining evidence of urban history in the built fabric. This is an important strategy in promoting the city as a place with an interesting lineage, rather than erasing the past. These issues need to be embraced and harnessed for future growth.

The second is the adoption of a *small-scale approach to urban development*. The retention of small independent shops, rather than the construction of chain stores and big-box retailing, will produce a more sustainable urban area. Florida draws heavily on Jacobs (1961) in suggesting that the scale of the neighbourhood is of major significance in the fortunes of the overall city. These neighbourhoods should have a clear sense of place and an attractive atmosphere.

The third is to create *attractions to appeal to younger people, their interests and lifestyles*. Facilities such as cafés, bars, restaurants, distinctive shops, live music venues and other forms of nightlife, plus other cultural activities, add to the overall 'buzz' of the urban area. Amenities that contribute to a better work-life balance will prove attractive and popular.

The fourth relates to *measures to facilitate and support a diverse population*. This will include people with a range of lifestyle preferences, as well as an ethnic, cultural or social variety. Rather than focusing on the development of strong, community ties, as advocated by a social capital approach (Putnam 1995), it is felt that strategies to support the development of weak social ties and broad contacts are more attractive to creative people.

These four goals can be seen in some of the other theme city approaches. For example, diversity is a key element in creating Just Cities (Chap. 3), while maintaining authenticity is a basic Slow City Idea (Chap. 15) and keeping local stores is an aim for several approaches, not simply Transition Town (Chap. 7). Peck (2005) has interpreted Florida's version of the Creative City as one representative of our new culture and society in which people are motivated not only by material rewards but also by a stimulating and interesting life. Moreover, it has been argued that the Creative Class want to be in a place that validates their identities and provides new experiences.

> Creative Class lifestyle comes down to a passionate quest for experience. The ideal.... is to 'live the life'—a creative life packed full of intense, high-quality, multidimensional experiences....
>
> ....[Members of the Creative Class] like indigenous street-level culture, a teeming blend of cafes, sidewalk musicians, and small galleries and bistros, where it is hard to draw the line between participant and observer, or between creativity and its creators....More than anything, they crave intense experiences in the real world. (Florida and Tingali 2004, p. 166)

The major reason why the Creative Class theory has been so widely embraced is because of its fortunate timing. At the turn of the twenty-first century, a growing realisation of the need for economic restructuring in a whole host of cities across the U.S.A. coincided with the initial publication of Florida's book *The Rise of the Creative Class* (2002). Unlike in Europe, where the Creative Cities idea evolved out of discussions on the arts and cultural economy, Landry (2006) noted that previous pioneering work by Scott (1997) on the cultural industries in the U.S.A. had failed to make the same policy impact. Also, some academics have criticised Florida's theory as an example of what has been described as an example of 'postmodern weak thought' (Ponzini and Rossi 2010, p. 1038). But this weakness may be its very strength. Because of the ease with which it can be translated across cultural and spatial boundaries, Florida's ideas for urban economic growth captured the imagination of urban policymakers, perhaps more than any other urban economic theory in the last 30 years. Part of the attractiveness of the theory has been the development of a ranking system that enables urban areas to be compared with one another, an important activity in an increasingly competitive global environment. This ranking is based on four individual Creative City indices developed by Florida (2002a), for American cities: the '*gay index*'—an indicator of tolerance; the *'technology index'*—a measure of technology; the '*bohemian index*'—the density of artists, writers and performers in a place; the '*talent index*'—a measure of those with a bachelor's degree or above in a region. A composite index, called the '*creativity index*', was derived by Florida from these initial indices and described as follows:

> the Creative Class share of the workforce; innovation, as measured by patents per capita; high tech. industry, using the Milken Institute's widely accepted Tech Pole Index...; and diversity, measured by the Gay Index, a reasonable proxy for an area's openness. (Florida 2002a, pp. 244–245)

In *The Rise of the Creative Class*, Florida (2002a) ranked San Francisco as the most creative city in the USA, followed by Austin (Texas), with Boston (Massachusetts) and San Diego (California) tied at third place.

10.4 'Business-As-Usual' in the Creative City?

Perhaps one of the most interesting things about the Creative City debate is the manner in which the ideas have become so quickly part of planning and economic development goals without much definitive evidence to back the theories. As Pratt has argued:

> Those peddling culture or creativity follow a long line of previous potions: environment, safety, liveability, hi-tech, bio-, or nano-industry. To be clear, it is not the moniker that matters for the exercise but what it is suggestive of: growth. (Pratt 2008, p. 5)

The rankings that have been produced seem to provide a statistical basis for general policy prescriptions, but these have been subject to a range of methodological and normative criticisms that have been largely ignored by policymakers, issues which will be discussed later in this chapter,. It seems reasonable to suggest that what the translation of Florida's ideas into practice actually represents is the latest phase of what has more typically been referred to as 'city boosterism' or 'place marketing'. Miles and Paddison (2005, p. 833) have suggested that culture-driven urban regeneration now has "a pivotal position in the new urban entrepreneurialism" and the Creative Cities debate is just one part of that. Some argue that the same principles and philosophies that underlay the property-led urban regeneration in the 1980s have been brought back into play through Creative City strategies. Now, rather than fiscal or other incentives to attract capital, cities are focused on implementing strategies to attract talent in order to entice growth industries. These strategies are founded on the same discourses of interurban competition, gentrification, consumption and place-marketing, as earlier rounds of regeneration but they now have a 'softer' spin. Ponzini and Rossi (2010, p. 1039) have described this *urban policy culturalization* as:

> a new stage of entrepreneurial urbanism centred on the dynamic of the immaterial (atmosphere, vibrancy, creativity) and material factors (regeneration areas, monuments etc.).

This is seen as the latest attempt for urban areas to push for a stake in the new economy. The more fine-grained locational decision-making processes now adopted by both the creative and high-technology industries means that cities have had to go beyond offering what became the standard international package of flagship projects, stadia and conference centres to gain a competitive advantage. The development of the creativity and other indices by academics such as Florida have provided a tool to identify those cities who have successfully achieved this goal and given other cities something to emulate. However, Landry (2006) disagrees with this 'business as usual' approach and argues that creativity should be about a *journey* to creating a 'better' city, rather than a strategy to become a 'winning' city.

Peck (2005, p. 760) has also demonstrated how the creative city in the U.K. is just a *"market-friendly urban placebo ... (that) ... can quite easily be bolted on to business-as-usual urban-development policies"*. This view seems to be duplicated in other countries. Lund Hansen et al. (2001) in their review of the discussion on creativity in Copenhagen (Denmark) argue similarly that no-one has questioned how the creative city is different from the property-led regeneration that it claims to replace. More generally, Rausch and Negrey (2006) argued that strategies aimed at Creative Class attraction are just another tool for those *controlling* the economic development agenda in order to distinguish the city from other, potentially competitive centres. In Ireland, for example, the newest Dublin Economic Development Strategy (2009) embraces the creativity agenda as an added-value component of a pre-existing development approach. This is explicitly articulated in the aim of this Plan:

> to further develop the Dublin City Region, the engine of Ireland's economy, as a significant hub in the European knowledge economy through a network of thriving spatial and sectoral clusters providing a magnet for creative talent and investment. (Dublin City Council 2009, p. 7)

The delivery of the Plan was designed to be aided by the Creative Dublin Alliance, which is a network led by the Dublin City Manager, with members drawn from the most senior level in Local Government, Commerce, Industry, Education, State Agencies and the Not-for-Profit Sector. The role of the Alliance is to identify challenges and issues of citywide significance that could most effectively be dealt with through the synergies created in the Alliance. The influence of Creative Class theory, concepts and approaches as shaped by Florida (2002a, 2005) on facilitating a business-as-usual planning and economic development agenda in Dublin is clear:

> Good urban quality is of central importance in attracting investment and talent. This includes the quality of the built and natural environment, the vibrancy of street life, density and intensity, café culture, arts and music, outdoor activities, public spaces, a choice of quality places to live, a child friendly environment, tolerance and social harmony. (Dublin City Council 2009, p. 18)

Unfortunately, this type of development agenda is being adopted with little regard to the evidence emerging from empirical research. For example, Lawton et al. (2009) have demonstrated in the same context that the people described as being in the creative class are not as highly mobile as Florida would suggest. Their work, undertaken in Dublin (Ireland) as part of a broader European project on transnational migrant workers and other members of the creative class, suggested that personal trajectories and family-kinship ties—things that Florida criticizes—along with the more classic hard factors of location, land availability and infrastructure, are still the most important factors in locational decision-making. This crucial evidence identifies a problem about the degree of mobility of the creative class; there is a significant disjuncture between rhetoric that assumes ease of mobility and the spatial immobility reality in creative-class decision-making.

Despite these problems there have also been some surprising Creative City benefits in the emergence of this focus. For example one significant outcome of this type of urban development strategy has been the emergence of a clear rationale for government spending on culture and the arts, which has benefitted those involved in these activities. Markusen (2006) has also argued that while the agenda of artists and neo-liberal political regimes cannot be conflated, the Creative City agenda *has* given artists a newfound visibility in social and political debates. Her work demonstrates that the artists' visions of urban space are much more in line with Jane Jacobs 'mosaic of neighbourhoods' ideal (1961) and the approaches to creativity of Hall and Landry. The irony is that because the Creative City agenda has become so intertwined with the goals of inter-city competition, urban decision-makers are increasingly trying to orchestrate this 'creative buzz' or atmosphere, but so far with mixed effects.

10.5 Creative City Critiques

While there is little disagreement that all cities have a creative capacity and function, there has been intense criticism of the Creative City concept as currently conceived. Four major types of criticism have been made, linked to: methodology and context of the study; causality between the presence of the Creative Class and urban success; widening intra-urban inequality; and whether Creative City policies are being used to legitimate questionable policy prescriptions.

10.5.1 Methodology and Contexts

One of the most attractive aspects for policymakers of Florida's approach to Creative Cities has been the emphasis on rankings, providing the ability to benchmark cities against each other in order to assess relative competitiveness. Obviously the value and validity of these rankings depend on the variables used and how they are methodologically combined, yet there has been little real justification for the choice of variables selected by Florida to create his series of indices or the rest of his Creative City methodologies. His technical approach is also limited for no attempt is made to use more sophisticated factorial multivariate methods to search for common dimensions or key drivers among the variables used, which could be measures of the same thing (Davies 1984). This is an oversight that contributes to the second critique below, on the lack of, or contested, evidence around causality. In a previous study Florida (2002a) calculated correlations between a bohemian index (measured by location quotients of what amount to creative artistic workers) and various other indices—of talent (by degrees), technologies, gayness etc. He shows high levels of concentrations of these indices in a limited number of the largest 50 metropolitan areas in the U.S.A., and finds significant associations between the bohemian index and the other indices, although the best were only in the 0.5–65 range. Since the bohemian index showed a 0.6 correlation with population, one of the higher values, it surely shows that these indices are primarily linked to population size. Also some doubt can be placed on some of the conclusions, especially his graph (Florida 2002b, Fig. 3, p. 64) which shows the relationship between several levels of education and the bohemian index. Although there is a high negative correlation between the index and low education levels, the correlation shown with graduate degrees is only +0.1, hardly convincing evidence of high positive relationships between the two! Although Florida is careful to note that the correlations between his bohemian index and other variables cannot be interpreted as causal, the discussion frequently implies that this is the case, such as by observing that: "the presence of such human capital concentrations in a region in turn attracts and generates innovative technology-based industries" (Florida 2002b, p. 56). Yet there is no evidence of any analysis which shows such temporal migrant activity. Moreover the aggregation system of Florida's approach means there is no unanimity on the choice of variables that should be used to measure aspects of creativity or their combination into various indices. Indeed, different choices of variables and scoring systems will produce different measures. This can be seen by a comparison made by Martin Prosperity Insights (MPI 2009) of Florida's early work with their later analysis of 374 North American metropolitan regions. This had different results, placing Seattle (Washington) at the top of the creative index, with San Jose (California) and Ottawa-Gatineau (Ontario) not far behind. Innovative though Florida's ideas seemed to be at first, there is a pressing need for researchers to develop more standard and widely acceptable variable sets and integrative techniques, as well as temporal change analysis to vigorously review the ideas if the approach is to gain more widespread academic credibility.

While the methodology involved in defining Creative Cities is flawed, a more general critique in the popular literature comes from the work of Malanga (2004), who has argued that the Creative City economic evidence is flawed. Using data from the U.S. Bureau of Labour Statistics, he showed that the top creative cities identified by Florida—San Francisco, Austin, Houston and San Diego—were underperformers in relation to jobs over the period 1983–2002. His least creative cities—New Orleans, Memphis, Las Vegas and Oklahoma City—were actually job powerhouses over the decade from 1993 to 2003 adding 19 % new jobs, and showing faster growth than the national economy as a whole. Florida (2005) has addressed these criticisms in part by suggesting that it is not the quantity, but quality of jobs that should be considered. So while places like Las Vegas may have had significant job growth in the decade to 2003, the *per capita* income from these jobs is very low. Nonetheless, the most recent data from the U.S. Bureau of Labour Statistics (2013) would suggest that even in Florida's top-tier creative cities the most recent greatest gains in employment have *not* been in high paid jobs. The major growth has not been in the creative sectors in a city like Austin (Texas) but in more traditional primary and secondary economic sectors—logging, mining and construction. In many other U.S. cities, the major job growth in the 12 months from July 2012–2013 has been in the leisure and hospitality sectors, traditionally low paid and low skilled jobs.

Other critiques of Florida's Creative City thesis centre on the contexts of the study, in the sense of where the ideas apply. Many have questioned how new these centres of creativity really are, given that the world's great cities have always been centres of innovation and creativity. For example, Hall (2000) had previously suggested that even if all of the type of factors identified by Florida above is taken into account, and so-called Creative City strategies implemented, creativity cannot be produced overnight. The specific geographical and historical trajectory or tradition of innovation in each city ought to be studied, but this is ignored in the 3T's cross-sectional analysis of Florida. Indeed, this is part of the issues that others, such as the ACRE (Accommodating Creative Knowledge) research team based at the University of Amsterdam, have explored in developing a better understanding of the specific contextual conditions that support and generate creativity (ACRE 2006, Musterd et al. 2010). They have reviewed the features that generate what they describe as creative-knowledge growth, identifying the key factors of path-dependency of previous history, place-making and networks. In addition, something that has been given little attention within the geographical and planning literature on Creative Cities and Classes is the developed world emphasis of this theory. It could be argued that the debate so far has simply ignored the reality of cities in the world where secondary production remains a central component to economic survival, such as in India, Brazil, and some Asian cities, although some of the countries and cities are now competing successfully in innovation with places in the developed world. Moreover it is also worth asking about creativity further down the occupational scale, for example that associated with the bottom-up, small-scale creativity of craft workers and street traders etc. They are much in evidence in the cities of developing countries, so there ought to be room for them in the current creativity and creative class debates by academics and policy-makers. All these points need to receive more attention in coming years as the contemporary critiques on the issue based only on the developed world exhaust themselves.

10.5.2 The Creative Class and Urban Development: Unproven Causality

Perhaps the most fundamental flaw with the Creative Class theory and its implications for urban development is that there is no evidence to suggest that a causal relationship exists between the creative class and economic development, as shown in early research by Montgomery (2005) and McCann (2007). Moreover, the presence of the Creative Class alone is not a sufficient driver to explain urban success or growth. While Florida's indices may pick up on factors that he believes define Creative Cities, they do not necessarily explain their growth. In a range of studies, different authors have tried to replicate the creativity index or to test other possible causal factors for economic success. The same conclusion has emerged from almost all of them: it is not the size of the Creative Class in a city that actually matters to growth, but rather high educational attainment (Glaeser 2005, Rausch and Negrey 2006, Markusen 2006). Given that educational attainment is just one aspect of the creative index and one in many cases that is given least attention by policymakers, it raises questions about the kinds of strategies and investments being driven by the Creative City agenda.

One of the other assumptions underpinning the Creative Class theory is that once this group have been brought together, their intrinsic capabilities and drive will lead to economic vibrancy (Scott 2006). While there are well-demonstrated benefits of clustering of firms and talent, this does not guarantee growth. There are so many different occupational groups under the 'umbrella' of the Creative Class that the concept would need to be significantly disaggregated to draw any meaningful conclusion about economic impacts. Markusen (2006) argued that is it simply nonsensical to try to suggest that the role and impact of artists and musicians can in any way be compared to that of scientists, engineers and other members of the Creative Class. Based on her empirical research, she questioned whether the relationship between the creative class and economic development may in fact work in the opposite direction to that suggested by Florida. Drawing on her work in Minneapolis-St-Paul, where the 3 M high-tech firm is based, she observed that artists may be attracted to high-technology, wealthy areas, as it may be relatively easier to find markets and patrons for their work. That important view takes one back to the situation found in the early modern period in Italy where artists flocked to the political and religious patrons in the city states, for these were the ones needing art for their palaces, cities and churches.

10.5.3 Intensifying Inequality

A second body of work critiquing these concepts has focused on the impacts of these ideas, in particular the relationship between Creative City strategies and intra-urban inequality. It has been suggested that inequality is an inevitable outcome of the strategies, given the emphasis placed on the needs and demands of a particular social group—the Creative Class. Yet many places have assumed the opposite. An extreme example of this can be seen by the 'Cool Cities' Initiative in the state of Michigan, which put the need to attract and keep creative people first, in the expectation that they will then attract new activities to reverse economic decline. This is an approach that has been described as:

> creating hot jobs in 'cool neighborhoods' throughout Michigan….It's about attracting and encouraging people—especially young people—to live, work and shop in the 'cool cities' we are working hard to create together. Build a 'cool city' and they—young knowledge workers and other creative class members—will come. (CC: Cool Cities)

Examples of similar strategies abound, given the uncritical acceptance of Creative City ideas. But to think that the inverse relationship between equality and creative city growth is new would be a mistake. After all, Hall (2000) has reminded us that Creative Cities throughout history have been bourgeois centres where high culture was created by an artistic minority to fill the demands of a wealthy, but small elite. These cities were characterized by a highly unequal distribution of wealth that led to high consumption by the elite, paralleling the way that creative boosters today promote the consumption experience as central to the creativity agenda. This link between inequality and creative class growth has been revived in a more contemporary context by McCann (2007), who has shown a clear correlation between creative and economically polarized cities. For example, Austin in Texas is one of North America's high-technology nodes, with companies such as Dell, IBM, Apple, 3 M, eBay, Intel and Sun Microsystems having major operations there, leading Austin to be ranked as the top performing metropolitan area in the United States by the Milken Institute in 2009 (De Vol et al. 2009). But this ability to create and sustain creative jobs has gone hand in hand with increasing polarization, particularly as a result of decreasing housing affordability, which has increased the numbers of the disadvantaged. Moreover, arguing that Austin's growth is only a matter of the Creative Class ignores the salient fact that it is the state capital, so entrepreneurs have access to decision-makers, while it is also the site of the very large, research-based University of Texas. In a European context, researchers such as Lund Hansen et al. (2001), as well as Bayliss (2007), have picked up similar themes and trends in relation to the city of Copenhagen (Denmark). The former suggest that the construction of a new built-environment to facilitate the Creative Class is nothing more than public sector-funded gentrification and the subsidization of middle class consumption. In effect, these authors see the Creative City strategies as representing nothing more than the state stepping in to 'take back' the city for the middle classes, marking the re-emergence of a highly interventionist state.

> What at first glance appears to be an unambiguously positive characteristic and goal—the creative city—becomes on closer inspection a dubious ideological smokescreen to cover up the social costs associated with compulsive adaptation to the "requirements" of the "new" flexible globalized economy, including reduced transparency in urban governance, social and geographic polarization and large scale transformation of the urban landscape involving considerable displacement. (Lund Hansen et al. 2001, p. 866)

These kinds of observations are almost a re-run of the debates on the merits and costs of 'place marketing' and 'urban imagineering' in the early 1990s. One particular case in point was the *Glasgow: European City of Culture* events held in 1990 that attracted hard-hitting criticism from a number of quarters for the amount of money spent on high culture at a time when suburban and inner-city areas were in significant need of regeneration. The same criticism has been levelled at the manner in which Creative City strategies are being enacted in many cities. Peck (2005), describing the operation of the Cool Cities programme in Michigan, illustrated how the neighbourhoods with creative potential are being targeted for financial incentives, rather than those with the greatest needs. In effect, already deprived areas are being condemned to a future of economic decline, through the particular bricks and mortar approach favouring creative activities being adopted by the State government and other agencies. This determination to follow the Creative City path seems to be leading to more gentrification and marginalisation on a class basis (Ponzini and Rossi 2010). This approach exemplifies the "dark side of the dialectic" described by Gouldner (1979) in relation to the new economy and the new class who benefit most significantly from these types of policies. The result has been increasing urban inequality as examined in the Just City review (Chap. 3). This has been acknowledged by Florida (2013) in his more recent work where he concedes that talent clustering is responsible for rising economic inequality in a U.S. context. The bohemian lifestyle promoted as part of the Creative City appeals to a very small proportion of the urban population and has limited impact on overall regeneration strategies, particularly in the former rustbelt cities. This led Florida to admit that the clustering of talented people in a few cities did not lead to trickle down benefits for the poorer people; the highly skilled professionals were the ones to gain.

10.5.4 The Politics of Urban Restructuring

One of the other powerful critiques stems from the relationship between Creative City policies and city governance. Peck (2005) suggested that the creative strategies could be described as state-sponsored gentrification because of the actions of the local state in promoting consumption lifestyles and the agenda of a particular social class. Because of the initially positive discourse around the Creative City concept, the local state has been enabled to adopt certain policies that may otherwise be more questionable, as they are enacted on the premise or promise of achieving broader economic growth. The manner in which policy is developed and implemented has also been the subject of concern, as new agents become privileged and some pre-existing ones become ever more powerful. This has been particularly true in the case of Baltimore where Ponzini and Rossi (2010) argue that the freedom from certain responsibilities brought about by neoliberalism has enabled the local state to become newly powerful and this has benefited certain political elites. They described how Creative Class strategies have been used to enact network politics within the city that is directly controlled by the Mayor. By creating a new 'macro-actor' structure within the city, made up of a range of groups brought together under the Creative Class banner, a politically manageable core has been created that supports their interests. This means that support for other urban policies has been effectively neutralised. Issues such as socio-spatial justice become buried in support for the arts and cultural agenda. They suggest that Baltimore provides a classic example of how neo-liberalism becomes legitimised through a veneer of cultural policy and has resulted in a complete political re-ordering within the city (Ponzini and Rossi 2010). The dependency that has emerged between creative and political actors has also been clearly identified by Montgomery (2005, p. 343), who noted that

> the key figures in all of this are the visionary political leaders and the artists, inventors and entrepreneurs, the former creating the conditions for the latter to invest and prosper.

Related to this point is the fact that the Creative City debate as promoted by Richard Florida has also posed new questions on the scale at which urban restructuring takes place. Given his opinion that Creative Cities cannot be planned from above, cities are given new agency, so the local municipal state becomes heavily privileged, as in the example of Baltimore above. Within the context of a wide-ranging theoretical debate within geography and other disciplines on the role and power of the state, Florida's arguments, and those of his critics, support those who suggest that state power is not declining, rather it is actually being transformed. We are increasingly witnessing more, not less, government intervention as a result of neoliberal policies (Ward 2003). Indeed this discussion has shown that the local state—be it at the state, regional or city level—has been the key agent in the development and implementation of Creative City initiatives. What Florida's concentration on a particular class has also done is give significant agency at an ever smaller scale to particular groups of individuals. The core of his argument relies on the idea that cities need to attract talented people and to do this they must guess or influence individual locational decision-making preferences. For Florida, the most creative and successful city will be the one that can cater to as many different lifestyles and choices and cluster as many talented individuals together as possible. However the biased financing of this trend, given other needs, and the dubious link between creative workers and growth, has produced a new set of urban problems that seem to have been ignored by policy-makers.

10.6 Conclusions: The Great Creative City Debate

This discussion has shown that there is little new about the concept of the city as a creative entity. Florida has applied the idea in a specific contemporary form in the United States and Canada and claimed primacy for what he calls the Creative Class in generating new employment. The influence of this particular view has been far-reaching, with European and Australian policymakers also drawing on the ideas to produce Creative City policies. By their very nature, the world's great cities have always been driven by innovation and creativity, attracting innovative migrants that have aided further growth. However, in recent decades, ideas about urban creativity have gained new momentum in academic and policy circles as they have become closely linked with an expanding cultural sector, and the new economy and urban regeneration agendas. Scott (2006, p. 15) has argued that a focus on developing creative environments within cities through transformation of the social and physical fabric provides opportunities for "*bringing the dimensions of economy, culture, and place back into some sort of practical and humanly reasonable harmony*". Perhaps one of the major benefits of the debate has been a renewed focus on the arts and cultural sectors from the perspective of both production and consumption in cities. For many working in the cultural economy, the Creative City debate has provided a platform from which they can argue for increased funding and support from government agencies.

Despite some advantages the Creative City literature has many flaws, especially in its methodology, and limited contexts, while there is research evidence that shows that the Creative Cities as identified by Florida have not necessarily been in the forefront of the turn of the century growth. Apart from this essentially urban system critique, there are intra-urban problems. One of the biggest drawbacks of the Creative City idea as embraced by policymakers is that it has failed to address, and in some cases reinforces, pre-existing social divisions. From the basic premise that a particular class are responsible for 'creativity', through to the unequal consequences of creative policy implementation, as discussed earlier, the Creative City agenda has become equated with a neoliberal agenda that favours urban elites and essentially writes-off whole sectors of the urban population and environment. Caught up in the rhetoric of hyper-competitiveness, Creative City policies increasingly highlight the agency of the individual and a specific group, rather than the idea of the city as community and a place for social justice.

The prescriptive nature of some of the literature that has been generated also brings the entire concept into question. If creativity is seen as a spontaneous process, how can urban policymakers use it in what has been seen by some as 'a recipe for success'? Moreover, if it is this simple, and every city does it, then how creative are they all really? Perhaps it is worth bearing in mind the conclusions of one prominent geographer who questioned the path to economic restructuring and growth suggested by Richard Florida and his supporters as some sort of general panacea for all cities,

> A few fortunate centers perhaps may achieve something that approaches a creative, high-quality environment across the board, but in most metropolitan areas, developments of this type will most likely continue to exist only as enclaves in an urban landscape where poverty and social deprivation still widely prevail. (Scott 2006, p. 12)

This means that not all cities can expect creative growth and that comprehensive evaluation is needed to critically assess the broader impacts of creative initiatives, especially the negative ones. Otherwise, cities may add to disadvantage and undermine their own path to creativity and to economic regeneration. It must be concluded, therefore, that although this urban economic theme has some useful qualities, it is not really new, has been overstated—with Florida's version especially being criticized—for in some cases it has created negative consequences that were rarely anticipated by cities that rushed to apply the idea.

Acknowledgements Sincere thanks to Veronica Crossa and Philip Lawton for their constructive comments on an earlier draft of this chapter.

References

ACRE: Accommodating Creative Research. (2006). http://acre.socsci.uva.nl/. Accessed 12 Oct 2012.

Andersson, A. (1985). Creativity and regional development. *Papers of the Regional Science Association, 56,* 5–20.

Bayliss, D. (2007). The rise of the creative city: Culture and creativity in Copenhagen. *European Planning Studies, 15*(7), 889–903.

Bianchini, F., & Parkinson, M. (1993). *Cultural policy and urban regeneration: The West European experience*. Manchester: Manchester University Press.

CC: Cool City. http://www.downtownbaycity.com/CoolCity.htm. Accessed 18 Nov 2013.

CCNC: Creative City Network of Canada. http://www.creativecity.ca/. Accessed 8 Aug 2012.

Davies, W. K. D. (1984). *Factorial ecology*. Aldershot: Gower.

DCD: Dublin City Council. (2009). *Economic development action plan for the Dublin City Region, July 2009*. Dublin: Economic Development Unit.

De Vol, R., Bedroussian, A., Klowden, K., & Hynek, C. F. (2009). *Best-performing cities 2009: Where America's jobs are created and sustained.* Santa Monica: Milken Institute.

Florida, R. (2002a). *The rise of the creative class*. New York: Basic Books.
Florida, R. (2002b). Bohemia and economic geography. *Journal of Economic Geography, 2,* 55–71.
Florida, R. (2005). *Cities and the creative class*. London: Routledge.
Florida, R. (2013). More losers than winners in America's new economic geography. *The Atlantic*. http://www.theatlanticcities.com/jobs-and-economy/2013/01/more-losers-winners-americas-new-economic-geography/4465/. Accessed 19 Nov 2013.
Florida, R., & Tingali, I. (2004). *Europe in the creative age*. London: Demos.
Garnham, N. (2005). From cultural to creative industries: An analysis of the implications of the creative industries' approach to arts and media policy making in the United Kingdom. *International Journal of Cultural Policy, 11,* 15–29.
Glaeser, E. (2005). Reinventing Boston: 1630–2003. *Journal of Economic Geography, 5,* 119–153.
Gouldner, A. (1979). *The future of intellectuals and the rise of the new class*. New York: Seabury.
Hafner, S., Heinritz, G., Miosga, M., & von Streit, A. (2008). Requirements and demands of Munich's creative knowledge workers: Understanding the attractiveness of the metropolitan region for creative knowledge workers. ACRE report WP 5.7. AMIDSt, Amsterdam.
Hall, P. (1999). *Cities in civilization: Culture, innovation, and urban order*. London: Phoenix Giant.
Hall, P. (2000). Creative cities and economic development. *Urban Studies, 37*(4), 639–649.
Harvey, D. (1989). *The Urban experience*. Baltimore: John Hopkins University Press.
Jacobs, J. (1961). *The death and life of great American cities*. New York: Random House.
Landry, C. (1995). *The art of regeneration: Urban renewal through cultural activity*. New York: Demos.
Landry, C. (2000). *The creative city: A toolkit for urban innovators*. London: Earthscan.
Landry, C. (2006). Lineages of the creative city. *Research Journal for Creative Cities, 1*(1), 15–23.
Lawton, P., Redmond, D., & Murphy, E. (2009). *Transnational creative knowledge migrants in the Dublin region: The view of transnational migrants*. ACRE report WP7.13. AMIDSt, Amsterdam.
Lucas, R. E. J. (1988). On the mechanics of economic development. *Journal of Monetary Economics, 22*(1), 3–42.
Lund Hansen, A., Thor, A., & Clark, H. E. (2001). Creative Copenhagen: Globalization, urban governance and social change. *European Planning Studies, 9*(7), 851–869.
Malanga, S. (2004). The curse of the creative class. *City Journal*. http://www.city-journal.org/html/14_1_the_curse.html. Accessed 18 Nov 2013.
Markusen, A. (2006). Urban development and the politics of a creative class: Evidence from a study of artists. *Environment and Planning A, 38*(10), 1921–1940.
Massey, D. (1994). A global sense of place? In D. Massey (Ed.), *Space, place and gender* (pp. 146–156). Minneapolis: University of Minnesota Press.
McCann, E. (2007). Inequality and politics in the creative city-region: Questions of liveability and state strategy. *International Journal of Urban and Regional Research, 31*(1), 188–196.
Miles, S., & Paddison, R. (2005). The rise and rise of culture-led urban regeneration. *Urban Studies, 42*(5/6), 833–839.
Montgomery, J. (2005). Beware 'the creative class': Creativity and wealth creation revisited. *Local Economy, 20*(4), 337–343.
MPI: Martin Prosperity Insights. (April 2009). *Leaders and laggards of Ontario: How our metro regions stack up*. http://martinprosperity.orgimages/stories/jmc/cache/mpi-leaders-and-laggards-of-ontario-how-our-metro-regions-stack-up.pdf. Accessed 19 Nov 2013.
Musterd, S., Bontje, M., Chapain, C., Kovacs, Z., & Murie, A. (2007). *Accommodating creative knowledge: A literature review from a European perspective*. ACRE Report 1, AMIDSt, Amsterdam.
Musterd, S., Brown, J., Lutz, J., Gibney, J., & Murie, A. (2010). *Making Creative-Knowledge cities*. Amsterdam: AMISSR:Amsterdam Institute for Social Science Research.
Pareja-Eastaway, M., Turmo Garuzm, J., García Ferrando, L., Pradel i Miquel, M., & Simó Solsona, M. (2008). *Why in Barcelona? Understanding the attractiveness of the metropolitan region for creative knowledge workers*. ACRE report WP 5.2. AMIDSt. Amsterdam.

Peck, J. (2005). Struggling with the creative class. *International Journal of Urban and Regional Research, 29*(4), 740–770.

Ponzini, D., & Rossi, U. (2010). Becoming a creative city: The entrepreneurial mayor, network politics and the promises of an urban renaissance. *Urban Studies, 47*(10), 1–21.

Pratt, A. C. (2008). Creative cities: The cultural industries and the creative class. *Geografiska Annaler: Series B—Human Geography, 90*(2), 107–117.

Putnam, R. D. (1995). Bowling alone. *Journal of Democracy, 6*(1), 65–78.

Rausch, S. E., & Negrey, C. (2006). Does the creative engine run? A consideration of the effect of creative class on economic strength and growth. *Journal of Urban Affairs, 28*(5), 473–489.

Scott, A. J. (1997). The cultural economy of cities. *International Journal of Urban and Regional Research, 21*(2), 323–339.

Scott, A. J. (2000). *The cultural economy of cities: Essays on the geography of image-producing industries*. London: Sage.

Scott, A. J. (2006). Creative cities: Conceptual issues and policy questions. *Journal of Urban Affairs, 28*(1), 1–17.

Törnqvist, G. (1983). Creativity and the renewal of regional life. In A. Buttimer (Ed.), *Creativity and context: A seminar report* (Lund studies in Geography Series B, Human Geography, No. 50, pp. 91–112). Lund: Gleerup.

UNESCO. (2006). *Towards sustainable strategies for creative tourism*. Discussion Report of the Planning Meeting for 2008 International Conference on Creative Tourism.

UNESCO. (2013). Creative cities network. http://www.unesco.org/new/en/culture/themes/creativity/creative-cities-network/. Accessed 18 Nov 2013.

Ward, K. (2003). Entrepreneurial urbanism, state restructuring and civilizing 'new' East Manchester. *Area, 35*(2), 116–127.

Chapter 11
Developing Knowledge Cities

Wayne K.D. Davies

> *A knowledge city is a city that aims at a knowledge-based development, by encouraging the continuous creation, sharing, evaluation, renewal and update of knowledge. This is achieved through the continuous interaction between its citizens themselves and at the same time between them and other cities' citizens. The citizens' knowledge-sharing culture as well as the city's appropriate design, IT networks and infrastructures. support these interactions*
> Ergazakis et al. (2004), p. 5

11.1 Introduction

The development of the Knowledge City (KC) as a distinctive theme for the economic and social growth of cities has been a rapidly evolving concept in the last two decades. Originally created from extensions to traditional urban and regional growth theory by economists and geographers, it is being increasingly expanded by contributions from management theorists. All have recognized the increasing importance of intellectual capital (IC), or knowledge-based value, in the creation of new types of growth in particular urban areas, especially in or close to large prosperous cities. The mixture of ideas from different fields of study means the concept still lacks any definite methodologies of study or any complete theoretical structure. There is also some loss of focus because the term overlaps with several related ideas: Intelligent Cities (Komninos 2002), Learning Cities (Holden and Connelly 2004), Cities of Knowledge (O'Mara 2005), Technopolis (Okubo 2012), Creative Cities (Florida 2002; Landry 2005), Creative Regions (Cooke and Schwarz 2007), or the hybrid Creative-Knowledge Cities (Musterd et al. 2010a, b). But the Knowledge City concept does seem to be emerging as a more comprehensive approach to understanding, developing a new type of urban growth which subsumes these related ideas. However, it must be remembered that the idea that knowledge is vitally

W. K.D. Davies (✉)
Department of Geography, University of Calgary, 2500 University Drive, N.W.,
Calgary, AB T2N 1N4, Canada
e-mail: wdavies@ucalgary.ca

W. K.D. Davies (ed.), *Theme Cities: Solutions for Urban Problems*,
GeoJournal Library 112, DOI 10.1007/978-94-017-9655-2_11

important in the productive process is not new, as one of the classic works on economics, emphasized.

> By Capital is meant all stored-up provision for the production of material goods....It is the main stock of wealth regarded as an agent of production rather than as a direct source of gratification...... Capital consists in a great part of knowledge and organization....Knowledge is our most powerful engine of production; it enables us to subdue Nature and force her to satisfy our wants. Organization aids knowledge. (Marshall 1916, p. 115)

Although Marshall's reference to subduing nature shows an early twentieth century view, far from the sustainable concerns of the present, the link of capital to knowledge seems to have been downplayed for decades. It has taken a century before most locational theorists began exploring the implications of the idea in the context of urban economic growth.

Advocates of the contemporary Knowledge City (KC) idea argue that the *concentration* of knowledge assets in an urban place is *not* its only key characteristic, for this could apply to any university town. Instead, the definition above shows that the advocates of a Knowledge City envisage it as a place where there is a *deliberate process of development* that involves the upgrading of the human capital, infrastructure and organizations to *create knowledge*, not simply to teach or apply it in urban places in general, although the term 'city' is often used as a shorthand term for the places where this occurs. This approach is based on a new valuation of the importance of knowledge, not simply for personal development, but to continually create new innovations within the city and its region. In the words of one of the early geography pioneers of the concept, the result is to ensure that *"city development must... become more intentional, less accidental and less determined by external forces.... (and)...shaped by endogamous processes"* (Knight 1995, p. 227). This means that KC advocates believe that this type of development should not only occur spontaneously. Rather, it requires effective management to constantly create and renew IC and its needs in cities to ensure future growth. These requirements mean that the KC concept does not simply involve an understanding of how knowledge can be turned into innovations in some kind of continuous sequence; it also requires an appreciation of the locational factors that are involved in this process, as well as understanding how cities can assist and manage this process to attract and keep knowledge-generating activity. This chapter outlines the development of these KC ideas and then deals with five key issues in the field: a clarification of the nature and applications of knowledge; the spatial patterns of knowledge activities; the factors that account for these patterns; the concept of the city as a capital resource; and the key stages that are involved in the creation of a knowledge-based economy in cities, which must be based on a knowledge-sharing culture. Within these sections specific policies developed by particular cites provide examples of the approaches that have been successful.

11.2 Origins of the Concept

The contemporary Knowledge City concept stemmed from attempts in the developed world to understand and resolve the pressing economic problems that faced many cities and societies after the stagnation of the 1970s and recessions of the 1980s, especially the loss of so many industries from the western world to competitors in less developed countries. This led many researchers to argue that developed countries should accept that they would never be able to compete with these new, mostly low-cost production locations; instead they should move up the value chain by *applying knowledge* to create new innovations in the development of goods, services and processes, rather than dealing with the routine production of goods. This would not only result in new employment in these activities, but new markets for these innovations, as well as reducing the cost of existing goods and services (Porter 1990). The evidence at the time seemed to show that the western cities that had followed this route had prospered—especially those based on newly emerging computer and science-based technologies, information processing or management. They worked on the product development of these innovations, with the actual assembly carried out in low wage-cost locations. In addition, the last half century has seen a rapid expansion in the numbers of people with tertiary and university level education, which has led to increasing numbers of people working in research and development activities which are concentrated in a small number of locations. At the same time the explosion in new communication devices produced an information revolution that allowed rapid contact between people in different locations, creating new jobs in information processing and the media, and, in the last two decades, allowing greater access to knowledge through the Internet. All these variations provided an example in economic terms of what evolutionary biologists have called 'punctuated evolution', a stage of rapid burst of evolutionary change in organisms, not the slow, incremental progress that is often assumed, incorrectly, to describe progress. They also contributed to the emergence of what has been called a new Knowledge Economy (KE), which is being regarded as a transformative phase in socio-economic development, world views and basic values (Drucker 1993a). Urban and economic growth theorists also began to appreciate the increasing importance of the application of knowledge in the productive cycle, realizing that this could not be regarded just a subset of the labour component in the three traditional factors of production: labour, land and capital (Feldman 1994). So knowledge, or intellectual capital, should be seen as a primary and distinctive factor in the development of many of the new innovative activities, both in production and services (Edvinsson and Stenfelt 1999). The growth of the KE ideas soon spread to the urban case through several key works, such as in: pioneering individual papers (e.g. Knight 1995); special issues of journals (e.g. Urban Studies 2002; Journal of Knowledge Management 2004); investigative reports on the impact of knowledge and innovation on economic growth (OECD 1999) and increasingly by new contributions from management theorists (Carillo 2004, 2006; Viedma Marti 2005). In addition, the reviews of more practical attempts to create growth based on the KE in cities (Ergazakis et al. 2006; Musterd et al. 2010) or Intellectual Capital generally (Bounfour and Edvinson 2005), as well as in less developed regions (Morgan and Nauwelaers 2003) and nations (Dahlman et al. 2007; Senor and Singer 2009), have shown how a new focus on knowledge development can transform urban economies that had suffered stagnation or even decay. The cities that have also used these ideas to create urban economic transformation have also shared their experience to help others, by creating the Knowledge City network (KCN).

11.3 Concepts and Applications of Knowledge

Although this new emphasis upon the deliberate application of knowledge has created a new type of national and urban economic growth, the meanings and applications of the word 'knowledge' are often vague and need to be clarified. Knowledge has been defined as "a dynamic framework or structure from which information can be stored, processed and understood" (Howells 2002, p. 872). This knowledge definition means it has several characteristics, namely: it is *relational*, between some person and an entity or event; it needs a *memory*, which through the process of learning can store and assimilate information; it also allows *creative actions* to be undertaken from it, which includes the development of new knowledge. In addition, the word can apply to two very different types of *knowledge sources*. 'Codified or explicit knowledge' is the knowledge that has been created, processed, written down, or even organized in services or machines that can routinely produce some product or service. These are all features which occur *after* new knowledge has been created and does not necessarily require direct personal experience. As such, it is normally available to others in printed or electronic form, either by free access or by being bought. Very different is the knowledge that cannot be communicated in any explicit, formal way through writing or other prescribed forms of communication. This knowledge is embedded in the personal, unique and qualitative attributes of people who have specialist information. These are individuals who can solve problems and also can create *new* knowledge, which usually involves original ways of creating or envisaging things, or ways of doing. These particular attributes usually depend upon a person having a relevant information base upon which to begin, as well as having the capability of extending or even overturning existing ideas and practical procedures. This is the type of knowledge called 'tacit knowledge'. It is typically passed on to others quickly by face-to-face interactions where questions can be posed and answered far more effectively. This creates a quality and speed of interchange that other communication methods cannot emulate. Usually this tacit knowledge is only shared if trusting and reciprocal relationships exist between the parties, creating what amounts to a social capital between the participants. Howells (2002) revived Polyani's (1966) original argument that they should be seen as a continuum, not opposites, in which knowledge exchanges have various degrees of codified or tacit knowledge. For example, as any reader of a text in some specialised field will attest, this codified knowledge often needs tacit knowledge for its interpretation and especially valuation. Nevertheless, the codified-tacit distinction is very useful in dealing with the process of innovation. Although many innovative processes depend upon the application of codified knowledge, the creation of original inventions and innovations frequently need tacit knowledge to acquire the new understanding to create some new product, process, or ways of doing things. This exchange of tacit knowledge usually depends upon physical proximity, and is also heavily influenced by the particular contexts and institutional structures involved, especially in the urban places that have the environments to facilitate this knowledge transfer.

Another important issue in the character of knowledge comes from the distinction between knowledge as 'being and doing' (Drucker 1993a). The acquisition of knowledge throughout most of history, in both Asian and Western societies, was related to '*knowledge as being*', essentially the personal understandings of oneself, our relationships with the world, and learning the ability, through logic and rhetoric, to know what to say, or to write effectively. Certainly there was a related concept of '*knowledge as doing*' in these early civilizations, essentially applying knowledge to create some material object or service. Such work was usually distained by leaders or educated men and was carried out by lower classes, even slaves. In any case, the application of knowledge as 'doing' to a whole range of activities, such as sailing, grinding corn or metal working, was limited until the Industrial Revolution because it was often craft-based and practiced by individuals or small groups of people, subordinate to the leaders in status-based societies. Moreover, each craft was often regarded as a skill-set on its own, with few attempts to relate the information involved in creating the goods or services in other areas of economic activity. A major change occurred in early medieval Europe when tradesmen and merchants in many towns obtained freedoms from the surrounding feudal society that was rural-based, enabling them to practice their skills on their own behalf, even though they were subject to some constraints. Along with the growth of a market economy it led to a rapid growth in trade, production and wealth in the towns and cities in which they were located. Yet there were limits on growth, not simply from constraints from technologies or transport limits, but because guild and trade organizations still controlled these occupations. Entry to these organizations was usually only available to those who had passed a long apprenticeship to acquire the skills through personal contact with craft specialists, skills that were often jealously guarded and usually passed down to their family members.

A major change in access to knowledge among the limited number of literate people occurred with the development of printing in the late 1400s, and later in the Enlightenment, when individuals such as Diderot and d'Alembert developed the Encyclopaedia project between 1751 and 1772. Their objective stimulated the process of summarizing all knowledge, including descriptions of the former craft 'mysteries'. This made the craft skills potentially available to others who were literate, although some skills still had to be acquired by practice and instruction. It also led to the recognition that many of the principles in one craft field could be applied to others to develop new processes and products. A new word, 'technology', signified the changes. These and other intellectual changes led to a greater understanding of the world, accompanied by a series of related advances: the development of new mechanical and chemical processes in factory settings; the application of fossil-fuel energy to industry and transport; in some places the replacement of government regulations, monopolies and extraction, with a more laissez-faire and entrepreneurial economy. They led to the explosion of new and modified goods production in the period now known as the Industrial Revolution. Throughout this period, knowledge was increasingly applied to 'doing'. This led to the invention and improvement of tools, processes and products, based on the new technologies that were concentrated in the many new industrial towns and cities, as well as being applied to 'work' in many industrial plants by pioneering researchers such as F. W. Taylor from the 1880s (Drucker 1993a), creating new, more efficient work practices that led to greater productivity.

The last half of the twentieth century saw the emergence of another phase in knowledge creation, the application of '*knowledge to create new knowledge*', in other words not just 'doing', but to 'knowledge' itself, in order to produce new innovations, in production as well as services and organizations (Drucker 1993). This has also led to huge increases in the numbers of people in education and research, as well as a management revolution in the 40 years after the mid-twentieth century, with large numbers in the organization of production and services, as well as generating innovations in products, processes and organizations. It has resulted in the explosive growth and availability of codified knowledge via new electronic communication sources and through unprecedented access to this storehouse of information by the Internet. These changes led to the realization that knowledge, through its acquisition, creation and applications, has become *the* key resource, if cities and economies in the developed world are to increase their productivity.

11.4 The Location of Knowledge Activities

11.4.1 Disaggregation of Productive Processes

This new focus upon knowledge as a distinctive factor of production is one of the key features of the current economic phase. However, it must be emphasized that from the Industrial Revolution onwards many industrial towns had concentrations of the specialised technological and organizational skills that enabled their enterprises to produce new machines and industrial processes, as well as new goods and services, over many years. Hence the towns and often the regions in which they were located were centres of innovation in the industrial period, not just of production but also organization, since the headquarters of most firms were also there. In the last century or more many changes have led to the spatial disaggregation of the various parts of economic activity, including sites of innovation. First, firms have got *bigger and more complex,* producing many products and in different parts of the world, which has led to many manufacturing units, whether material goods or services, in a large number of locations. Although there are some exceptions, the main HQ or regional HQs and the research centres are now mainly separate from most, if not all of the production units, because they have different locational requirements. Second, *production itself is also disaggregated* in the sense that most goods are now composed of a series of specialised parts that may be produced in a large number of locations, not simply because firms outsource their requirements and drive down costs, but because they are less dependent on bulky raw materials. This is a consequence of the development of the third feature, namely that *cheaper and faster communication* enables easier raw materials and parts transfer, as well as rapid access to markets, which may be in other parts of the world. This has allowed the move of many productive activities to low cost locations where labour is cheap, and even if initially unskilled, employees can work the machines that do the routine production, instead of firms depending on inherited industrial skills. A fourth feature has been a greater spatial flexibility in the *choice of locations* for many manufacturing processes, for they are not so tied to raw material or skilled labour sources. Fifth, the productive process is not now a simple matter of assembling raw materials, or even specialised parts. More and more intervening activities called *producer services* are required, from those that provide financial resources, legal and banking facilities, designs, specialist technical skills not covered in-house, to advertising, marketing and communication activities. Of course some of these were present in the early industrial era, but as products became more complex, and market areas have expanded, the size of sector has exploded. Few of these producer services are located on the production sites. Sixth, the need for old industries to innovate to keep their market share, and the new phase of more rapidly developing innovations creating new foods and services based on new technologies, means a *new emphasis upon knowledge* has occurred. This confirms Marshall's opinion (1916) that knowledge is the 'most powerful agent of production', although its implications were not explored. The difference in the last 30 years has been that the location of these knowledge-creating and innovative activities are not bound to the old production sites as many were in the past; many they have their own locational patterns and preferences.

11.4.2 Location of Knowledge Workers

It has not proved easy to precisely measure the numbers and locations of people involved in the knowledge industries. These range from the scientific and technological activities in Research and Development (R&D), to those in innovative health-care services, through producer services (such as finance, law, advertising, market research), to media firms and various forms of information processing, as well as those involved in organization and decision-making in corporate headquarters, and people in higher education and cultural industries. The reason is that most national statistical bodies collect and compile data on various industries, rather than the functions carried out within firms or organizations, which include many jobs that are routine and are essentially support services rather than true knowledge-based activities that create innovations and new employment growth. Yet despite the lack of precision in measuring knowledge-based activities, there seems little doubt that the last half century has seen substantial increases in the numbers of people working in these fields, especially in the bigger cities. A recent attempt to provide a quantitative evaluation of the relationships between city size and urban organization showed that many of the characteristics of cities in national urban systems could be summarized as a power relation that varied with size (Bettencourt et al. 2007). Such relations have been found in many organisms, for growth often displays a − 1/4 power function, so that devolops it has slower growth and gets bigger in a predictable way. The infrastructure aspects of cities, such as road surfaces, power cables, gas stations, displayed a similar − 1/4 power relation. However, when variables linked to inventions, patents, R&D activity etc, were measured in various urban systems, the power relations of these variables clustered around a + 1/4 scaling relationship, meaning that growth in these variables accelerated as the cities became bigger. Bettencourt and his co-workers speculated that these relationships stemmed from the fact that infrastructure growth involved economies of scale as the city got bigger, effectively slowing down the rate of growth in these variables, whereas variables of innovations and wealth creation accelerated with size. It is worth noting that similar rates of growth also occurred with crime and the spread of infectious diseases, showing that growth may have some disadvantages, as studies of urban diseconomies have shown, although most were attributed to poor urban management in resolving problems (Richardson 1973).

This finding of a general relationship between the concentration of knowledge activities and larger urban places has been shown by many empirical studies. For example, an early report on the Knowledge Economy locations in the European Community (Drewett et al 1988) showed that in the 240 major city regions in Europe with over 250,00 people, only 15 contained more than 10 headquarters of the largest 1000 industrial firms and 469 of these biggest companies had head offices in only five locations (London and South-East England (216), Paris and Ile-de-France (131), Dusseldorf and Rhine-Ruhr (51), Randstad (39), Frankfurt-Darmstadt (32)), with a clear emphasis upon the first two. Other studies in Europe have shown similar patterns (Capello 2013), while a more detailed survey of companies employed in four new technologies (biotechnology, aeronautics and space, artificial intelligence, and textiles and clothing) revealed that 80 % of all innovative R&D projects were located in ten major regions of innovation and that over three-quarters of public research funding was also concentrated in these areas, namely, the previous five centres, plus Munich, Lyon-Grenoble, Turin and Milan (Hilpert 1992). Similar concentrations have been found in the United States. Audretsch and Feldman (1996), using innovation citations from over a hundred scientific and trade journals, showed that California (41.7 %) and Massachusetts (12 %) accounted for over half of the innovations in the computer industry by 1982, a result of the well-known concentration of these activities in Silicon Valley (O'Mara 2005) and Route 128 around Boston (Rosegart and Lampe 1992; Mackun 2013). This pattern of concentration of the research side of the industry in a few locations, which are usually called research clusters, has been duplicated in many other counties (Cooke 2001; Cooke and Swartz 2007).

A more general attempt to measure the location of initial research *activity*, rather than numbers of research companies, spending, or employment, has been made by Matthieson and his colleagues at the University of Copenhagen. Their analysis of the Scientific Citation Index showed that the number of papers published in the major science journals grew from 2.79 million papers in 1996–1998 to 3.14 million in 2002–2004 (Matthieson et al. 2006). Figure 11.1 also shows that 32.9 % of the papers in the1996-8 period were written by researchers located in the top 30 locations. This proportion increased to 34.56 % by 2004, but even within this pattern the top two and top ten locations revealed even greater concentrations, whilst several Asian

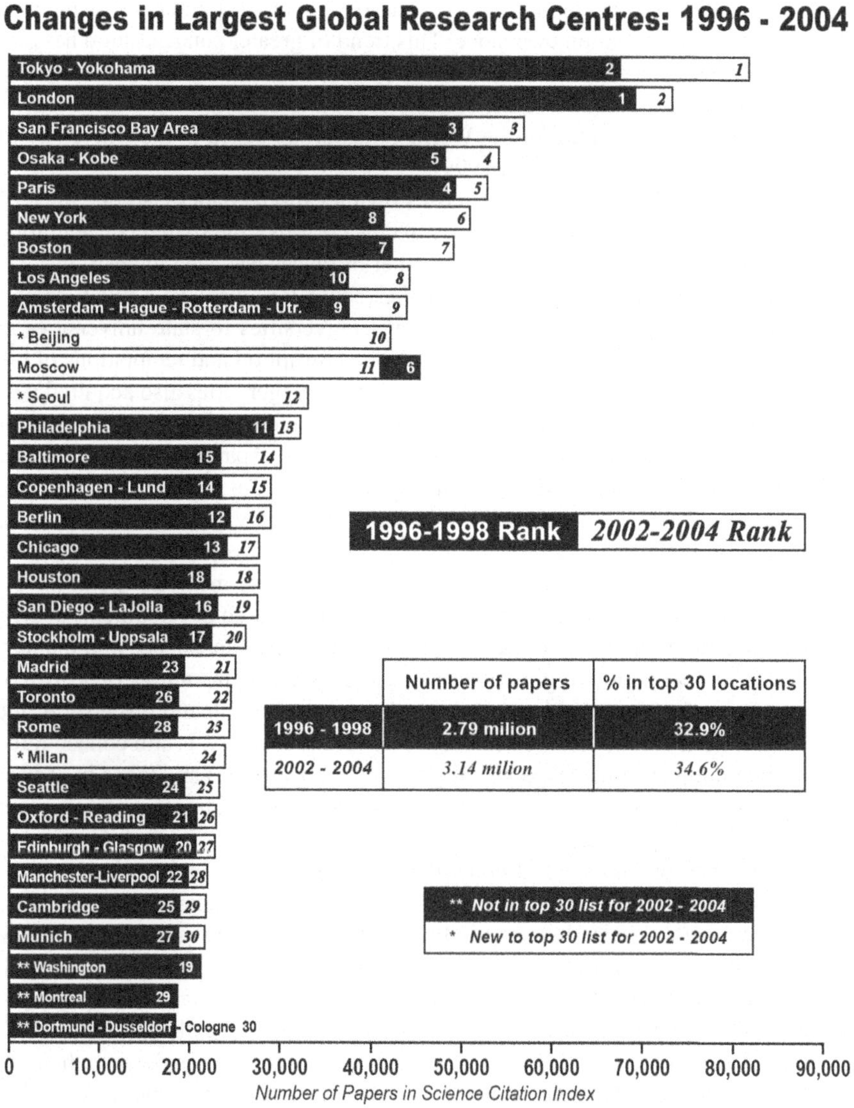

Fig. 11.1 Largest Research Centres: Measured by Scientific Papers. (Source: Revised from Matthieson et al. 2006)

centres increased in importance. Another example of this type of concentration in a different knowledge industry was revealed by a study of global financial services location (Zaheer and Manrakhan 2001). The number of cities in which firms were engaged in currency trading increased from 103 in 1973 to 154 in 1993, only 11 years after the introduction of the B2B electronic network, illustrating that the electronic changes facilitated more access to global markets and a wider dispersal of participating cities. However, even though a dispersal occurred, banks in the four major centres, London, New York, Tokyo and Hong Kong, increased their share of trading from 22 % of the total in the base year to 33 % in 1981 and have maintained that degree of concentration ever since. This trend of greater concentration has also been also revealed in the studies of financial and producer services that have been used to define what amounts to the world's economic command city locations, as shown by the various studies of the World Cities group based in the University of Loughborough (Beaverstock et al. 1999).

However, each knowledge sector has its own locational pattern, so one must be careful of over-generalisation, although it seems clear that the spatial concentration of most of these activities in a limited number of places has been intensifying. Yet one knowledge sector, the location of higher education, displays a much greater dispersal. Many early centres, from medieval universities to state universities in the U.S.A. were located away from the assumed corruption and temptations of big cities, although from the late nineteenth century the bigger cities also acquired new universities at this time, a matter of prestige and a recognition of their need for more educated workers in their new industries and the expanding professions and bigger bureaucracies in government and corporations. From the late 1950s another expansion occurred, with a veritable explosion of university and higher educational foundations in western countries, so that most centres with populations over 50,000 obtained some type of college or university. Many of these institutions still have a big emphasis upon a liberal education, rather than on research development, so their research output is limited, whereas others have specialised in science and the new technologies. This period of growth also saw an increase in the number and size of universities in the biggest cities. However, despite the growth and dispersal of the higher education sector, the main centres of *university research* is still concentrated in a limited number of places, especially the older universities and those in or near the big cities, as was seen by the list of the cities with the largest research paper outputs shown in Fig. 11.1.

Another example of dispersal rather than concentration in the knowledge sector can be seen in those research centres deliberately created by governments to act as the base for secret science-based activities, such as Oak Ridge in the U.S.A. or the originally 'closed' scientific Soviet cities in the former U.S.S.R., some of which still operate as major research nodes in Russia. These centres do represent exceptions, for most government and corporate research activities have usually been either added on to existing urban centres, primarily the largest cities, or very near to them. In 1962 something very different occurred when the Japanese government announced their intention to develop a new town based on research and learning at Tsukuba 25 miles from central Tokyo, although this is clearly within the metropolitan built-up area. During the 1980s 46 major national research laboratories were moved to this new town, two new research-based universities were created, and many major firms were encouraged to establish their research facilities in the centre. By 2012 the city contained 217 thousand people. It is not simply a traditional centre of learning. It is also one of Japan's premier research nodes, for there were 240 private research facilities and 60 national research facilities located there by 2000, when the centre absorbed half of Japan's public R&D budget. A smaller scale pattern of deliberate locating knowledge-based activities, this time in towns, can be seen in the development of *Technopolis* centres in Japan, where 18 nodes were established after an act of 1983 (Edgington 2008; Okubo 2012). But as their name implies, they were far more technologically based, and most clustered existing government research activities, rather than creating many new activities. Similar attempts to develop research clusters can be seen in the post 1970s creation of many *Research or Science Parks*, around university campuses. These are the knowledge-based equivalent of the branch factories established in industrial estates in the late 1930s in Britain, although these were designed to alleviate areas of high unemployment. Most science parks are relatively small compared to the Tsukuba example, but some, in places such as Lund and Cambridge, have a really significant research presence. These areas contain research spin-offs from the university laboratories, and in the really successful areas, have also attracted corporate and government research centres. At a smaller scale, so-called *Incubator Units* have been created in many countries to provide premises in small purpose-built buildings for 'start-up' companies in the knowledge sector, in the hope that they will expand.

11.5 The Locational Imperatives of Knowledge-Based Activities

11.5.1 Headquarter and Research Locations

If knowledge industries are to become the cornerstone of the economies of many cities of the future then the reasons for their location in a limited number of places need to be understood. This has led to studies of the geographical clusters found in knowledge-based areas (Maskell 2001), analyses of the extent to which urban regions provide effective contexts for the development of knowledge (Lambooy 2002; Musterd et al 2010) and the trends in the territorial patterns of innovation (Capello 2013). In many ways the spatial pattern of concentration of so many knowledge activities flies in the face of the popular belief in the idea of the increasing dispersal trends in contemporary society. This dispersal has been lubricated by the transport revolutions that have led to the greater sprawl of cities, the rapid movement of people and goods throughout the world, as well as the communication innovations that has enabled people with access to computers, and the infrastructure of high speed fibre optic and satellite links, to send information and interact almost instantaneously with people who possess the same communication devices. Hence the increasing concentration of so many knowledge-based activities in a limited number of cities and regions means that the process cannot be space-less, although the imperatives for the spatial intensification differ for different knowledge-based activities.

In the case of corporate headquarters and producer-services, which are the services needed to support and improve production, from finance, lawyers, advertisers, to designers etc, there seems little doubt that the need for proximity to similar businesses, power sources and the presence of agglomeration economies are some of the vital factors in accounting for what is increasingly described as their '*locational stickiness'* in a limited number of big cities (Lever 2002). The concentration of corporate headquarters is a result of the pressing need for the face-to-face contact and tacit knowledge exchanges between decision-makers in these offices and their various associates in banking and legal services. Many of these large centres are also the places where government decision-makers and regulators are located, again allowing quick and personal contact. However, Marshall's (1916) original discussion of the importance of the agglomeration economies concept is now normally split into three: *economies of scale*, relating to cost savings per unit in individual firms as size of output rises; *localization economies* that refer to the cost savings and advantages that apply to all firms in an industry in an area, such as through a specialized workforce; and in the case of *urbanization economies* to all firms in an area, providing cost savings from the presence of the overhead capital of the city or region. The latter are often publically provided goods, ranging from roads, major transit terminals, especially airport hubs, through police, leisure, education and health services to modern high speed communication facilities, as well as the private sector provision of houses, stores, etc. These are all part of what Lambooy (2002) described as the *selective city environment*—of institutions, markets and spatial structures—that complement the *entrepreneurial capabilities*, in the cognitive, innovative and organizational abilities, that help create new knowledge. If these facilities did not exist, the firms would have to provide them, adding substantially to their costs of operation. In many developing countries firms have to create or improve this missing or poor infrastructure—such as power generators, health facilities, adequate roads and transit terminals etc—which adds to their costs, since even their limited facilities have not kept pace with population growth. In addition, the largest centres in most countries do have the *information-gathering services*, including media centres, that are attracted to the hubs of transport and communication, and have far more of the *amenities*—from restaurants, theatres and leisure facilities produced by market demand—making them magnets for the wealthy and upwardly mobile. All of these features make them attractive locations for knowledge workers, a point emphasized in the Creative Cities discussion (Chap. 10).

Although it has proved relatively easy to understand the locational patterns of corporate headquarters, it has proved more difficult to explain the spatial associations of inventive and innovative activity in general. Not only is there no adequate theory to explain how innovations develop, but also it has proved difficult to predict either successful innovations, or to define the personality types of successful innovators or innovating firms. The classical economic theories of the firm, with their limiting assumptions, such as those of rationality, perfect knowledge and an emphasis upon states of equilibrium, have proved of little value in understanding innovative processes (Hodgson 1996). More relevant insights are emerging from the evolutionary approaches adopted by investigators who have studied the actual experience and patterns of people and firms involved in innovative activity. What has been increasingly obvious from these studies is that innovation depends more on the *application of knowledge* than any other economic activity. This explains why the processes involved in the creation of inventions and innovations have proved more difficult to understand than the assembly of the factors of production in any location to produce material goods. However research has revealed that four main sets of factors seem to be involved in the location of innovative processes: the characteristics of invention and innovation; the uncertainties to solve; the way that knowledge is accessed; and the degree of encouragement for innovative activity.

11.5.2 Invention and Innovation Characteristics

The processes of invention and then innovation into final goods rarely involves a simple linear route or sequence. It is quite unlike established businesses which are organized for predictability, often breaking tasks down into their various parts, as well as ensuring that employees complete these tasks. Instead, invention and innovation is a high risk, non-linear process that is uncertain and unpredictable. They usually involve complex feedbacks, as well as a lot of trial and error, have many dead-ends, and more failures than successes. So most successful inventions do not come from some sudden 'flash of insight' or Eureka moment of popular imagination—although this might start the process—but from organized and purposeful approaches that involve lots of hard work and the fortitude to overcome setbacks. There is often a compulsive behavioural trait in inventors and innovators to succeed, as well as crucial interactive contacts with other people who have special skills and knowledge. Johnson (2011) has reviewed the ways that inventions develop and showed there is no single path. Instead he identified seven main ways that new ideas—frequently the first or inventive stage—have been created, although sometimes there are overlaps between these alternatives. The first three of his paths to invention are related to the locational contiguities of major actors in the inventive process that provides the breakthrough in ideas.

The *'adjacent possible'* is the type of process by which a new idea arises as an extension from a previous one, such as the various modifications from steam pumping engines in mines that led to the development of steam powered engines on rails and roads, and subsequent applications in many different areas in factories and transport systems. *'Liquid networks'* are situations in which tacit information is derived from surrounding people, whether coffee shops in the development of London's financial markets in the seventeenth century, or in Silicon Valley today, enables an inventor to make a new breakthrough. *'Platforms'* refers to those situations in which the invention, or subsequent innovation, is based on essential inputs from the local environment, whether materials, skills or services, a situation that especially applies to the advantages provided by a big city or research environment, compared to a small isolated town. Two other types of innovation are based on discovering new relationships with different fields or processes. Thus *'exaptation'* involves borrowing from a related product or process that may be seen in a local environment to solve a new problem, such as the adaptation of the perfume spray to allow petrol to be finely dispersed in the internal combustion engine, or the way the crucial compression function of the initial printing press mimicked the wine press. In the *serendipity* case the invention comes not from borrowing, but by recognizing how a new application can be derived from some unexpected results found in an experiment or procedure that was originally created for some different reason. The final two types identified by Johnson illustrate how hard work, persistence and support are key features of the creation of another set of ways that inventions occur. The *'slow hunch'* describes those situations in which the exploration and solving of a series of problems leads to the gradual recognition of how something works, or is constructed, which then leads to a new process or product. The *'error'* type describes those cases in which false conclusions about some process are initially derived, which stimulates additional research to find out why they are incorrect, leading eventually, through many trials, to a new and viable understanding.

The study of knowledge-based development should not only focus on the initial invention but also on *subsequent developments*. Invention is only the first of many phases of development and refinement before the invention becomes an innovation that can be used or marketed. These various transitions require the application of very different types of knowledge. Also Rosenberg (1976) noted that innovation is not just about the big discontinuities in products or processes based on technological breakthroughs. There can be many innovations at various stages in production or process cycles through applying lower level knowledge, such as in revised engineering procedures. These may radically improve some productive activity, transforming existing products or processes into new product innovations, often in the downstream side, rather than in the upside level of deliberate research and development, the type of continuous improvement that Toyota pioneered (Liker 2004). Finally, there seems little doubt that it is a mistake to see the innovative process only in technical terms. Morgan and Nauwelaers (2002) have stressed that *innovation is a social process* that needs a strong collective endeavour to succeed, involving various agents, people and organizations, within and between firms, cities, and regions. It is one that must be spatially structured, since important requirements for the invention or innovation usually need to be available locally to prevent interruptions in the development process.

11.5.3 Uncertainties in Knowledge to Solve

Another reason why the invention-innovation process is so difficult to predict and replicate is that potential inventors and innovators have many uncertainties to solve. Firms have to learn to overcome these problems if they are to innovate successfully.

These problems can be generalized into four main categories. The first, and usually most important, stems from *technical problems,* a result of the limited experience and knowledge of a person or firm in carrying out the procedures needed, especially if the processes are complex. Indeed, the greater the complexity involved, the greater the uncertainties involved. So the inventor or innovator needs a good understanding and grounding in the appropriate knowledge field, or has access to it. This, in turn, is critically influenced by educational systems that stress original and critical thinking, as well as co-operative working. A second set of problems is associated with the *market*, such as whether there is a real demand for the product or process and, if the product is commercialised, whether the selling price will make sufficient profit to justify its development.

A third, but frequently crucial type of uncertainty, consists of the availability of various *producer services* to support the innovation at each development stage. This requires access to experience and knowledge from outside the firm, so that locational contiguity becomes an advantage. An especially crucial service requirement is venture finance, not simply to enable the initial invention to be translated into the first stage of innovation, but also to subsequently support what is often a long phase of product development. The availability of private venture capital was certainly crucial to the success of electronic innovations that developed in Silicon Valley (O'Mara 2005) whereas the Route 128 firms around Boston were more dependent on government support and linkages (Mackun 2013). Senor and Singer's (2009) penetrating study of the growth of Israel's high technology sector has shown it was helped by a per capita venture capital amount that was 2.5 times that of the U.S.A. and over 30 times that of Europe. However the close university-business linkages in Israel, helped by the small size of the country and many military contracts, has been influential in making it easier to develop personal contacts. Also, compulsory military service gave young adults experience with technical work. Israel's educational system also helped since it emphasises critical thinking and challenging existing issues, features that stem from the Telmudic tradition of discussion in Judaism. The ability to find adequate finance is a crucial requirement, but even that is not enough. The later stages of the innovative process require the use of quality producer services, such as designers, lawyers, advertising and marketing experts. So innovation is a process involving many stages and inputs and is not a single inventive act.

Finally, there are uncertainties due to the many *resistances or barriers* that impede innovation. One comes from a culture of negativity to inventors or innovations within some firms, due to the costs involved, ignorance of opportunities, or decisions to resist developing an invention, perhaps because of a perceived negative impact on the main product. A classic example was the decision of Kodak not to produce digital cameras, which it had initially pioneered but had not developed and marketed. It led other firms to produce the product which now dominates the camera market. Other barriers come from regulations, whether laws, planning requirements, or even local customs; each of these problems may also critically hinder one or more of the stages of innovation development, or prevent the use of the product or process. To take one example, Jaffe and Lerner (2004) have vigorously argued that the 1980 changes in the U.S. patent system created a barrier to invention and innovation, with one of the authors subsequently describing why public sector involvement has usually failed to create greater entrepreneurial activity and supplies of the crucial venture capital (Lerner 2009). Hence, it is important that a receptive milieu exists for inventions and innovations, both within the firm or circle within which the creator is working and within the eventual market.

11.5.4 Knowledge Acquisition

Overcoming the various constraints facing the creative process also involves understanding how the necessary knowledge to invent something and carry it into the innovation phase occurs, which involves a demand and supply side.

On the *supply side* the most obvious source of obtaining the knowledge to create innovations should lie within firms. The inevitable day-to-day focus on existing products and processes in a competitive environment usually takes priority within firms, so new ideas or products are usually limited, given the uncertainties that have to be overcome. Many firms often lack initiative in pursuing new ideas and have severe information deficits in their ability to create innovations. Indeed, most firms, especially small and medium sized enterprises, are still reluctant to pay the costs and engage the uncertainties of what amounts to research and development, certainly beyond their existing product range. So the 'culture' of the firms, their visions and leadership become critical variables in the extent to which they are innovative. Cooke and Morgan (1998) have argued that firms have very different capacities for knowledge creation, as well as the desire to obtain, use and apply this knowledge to innovations in a continuous learning sequence, or to share this information with other firms, for in the last resort many of these may be competitors. Although the focus of innovation is often on products from new firms, Govindarajan and Trimble (2010) have shown that large companies do play a major role in innovation. They suggested that it is the *implementation* of new ideas, rather than the initial invention, that is crucial. It is the execution that leads to new products or processes, and this involves unacknowledged and often long, grinding work behind the scenes, a phase that lacks the glamour of invention. The authors have also argued that bigger, successful firms have the resources to do both, but need to develop an effective managerial approach to achieve such results. They suggest that companies need to build what they call 'dedicated innovation machines', teams of employees recruited from within or from outside the organization on the basis of their entrepreneurial and creative skills, rather than by their rule-abiding qualities, an idea that could also apply to cities. However, these units must share staff with the main company and be able to convince other sectors within the firm of the need to support the process of translating their ideas into products, rather than being allowed to become some remote independent sector within the company, unrelated to the main organization.

The second way that knowledge is obtained, and then applied to create innovation, comes from other *formal sources outside the firm*. Examples range from: obtaining information from research laboratories—although private ones, unless non-profit, rarely share their knowledge; higher educational institutions, whose researchers not only publish their findings in available journals but may also help individual firms who seek advice on problems; and from knowledge of patents that have been published. Howells (2002) reviewed several studies that used different measures to show how knowledge in firms is acquired, revealing significant location effects even when codified knowledge is used. For example, he discovered three main effects, namely: corporate patent activity at a state level in the U.S.A. was influenced by university research spending; industries that are more knowledge-intensive were more spatially concentrated; and university R&D centres usually have a major impact upon employment in several high technology sectors in adjacent cities. This means that existing research institutions become an important locational asset for innovating firms.

Knowledge acquisition also occurs in a third way, in a more *informal context*, either from other knowledge intermediaries and local specialists outside their firm, or from interpersonal contacts between people working in similar or related industries. Such knowledge spill-overs have a strong distance-decay effect and are more likely to occur when there are clusters of similar or interlinked businesses in an area (Cooke 2001; Cooke and Swartz 2007). Such clusters create not only a knowledgeable local workforce, but a network of co-operative behaviour between individuals and firms, even when they are competitive with one another (Cooke and Morgan 1998). Although there can be an element of codified knowledge in this transfer, the really crucial knowledge spill-overs are the types of tacit knowledge exchanges described earlier. These exchanges are not only based on the specific technical or other problems being discussed. As important are the contacts with people from different specialist fields who can use their own expertise to throw new light upon, or even partially solve, the difficulties being experienced in the various phases of invention or innovation. This can be in a technical field, or in knowing where to get finance or some other essential producer service. As various areas of science and technology become more and more specialised it is almost impossible for anybody to be cognizant of all developments even in a particular field. So the presence of what is being called a 'buzz' in a local area produced by personal contacts, and the local availability of specialist firms to obtain relevant knowledge, is even more important. However, Simmie's (1992), study of innovative firms in south-east England showed that although the possibility of local knowledge spill-over may be a necessary condition for the local concentration of innovative activities, it was not a complete explanation. Two-thirds of the innovations produced by his sample of successful firms were products entering a competitive world market, and forty per cent of the firms surveyed obtained much of the knowledge necessary to develop the projects from overseas. So he found another crucially important knowledge acquisition variable was present, one that came from *international access and linkages*. Hence it was concluded that access to a major airport hub, not only to local research establishments or universities, was crucial to the smooth flow of international knowledge transfer that occurred mainly through personal contacts. What was also striking from his results is the paucity of knowledge-transfers from areas outside London and the south-east to other parts of Britain, showing that the international associations were more important that national ones. Yet one must be careful of assuming that all innovative activities are the same. Those industries that are more scientific in nature and depend on codified information are less likely to be localised, given that codified knowledge can be easily transmitted over distance, although even the exchange of this type of information often has localised patterns of acquisition (Howells 2002).

A fourth source of information acquisition comes from the presence of *knowledge networks* between firms in related businesses, such as trade organizations in an area, wider professional organizations, or those developed by informal contacts between specialists in an area in social settings or rented space (Malecki 2002). In the latter case the coffee houses of London provided the locale for interaction in the development of the capitalist trading regimes in the eighteenth century. In the past these networks have been local or regional, providing mechanisms of knowledge exchange that help innovation, often providing technical and market contacts, forums for the interchange of ideas, as well as setting standards for the practices that emerged. More recently, the growth of the Internet has led to an explosion of network associations that often share information and best practices between members, providing a positive asset for innovations. But the presence of networks alone is not enough. Morgan and Nauwelaers (2003) have shown that inter-firm networks in regions of decline or underdevelopment usually reinforce the status quo, keeping new firms out and often maintaining outmoded practices. This creates what can be called 'negative network social capital', whereas, in areas of innovation the networks are open to new ideas and opportunities, providing 'positive network social capital'.

It must also be noted that the last two decades have seen the growth of an increasing number of institutions and agencies—inspired and often financed by national governments, trade associations or even cities—which have been developed to assist in the innovative process. These innovation agencies seem especially helpful in providing advice, contacts and assistance in the crucial stages of commercializing a product, which requires skills that inventors and their firms do not always have. Finland was one of the first to adopt such methods to help support innovative activity in this way and its success in transforming its economy to a knowledge-based one in many areas from the late 1980s was helped by its strength in such agencies (Dahlman et al. 2007).

In a related context it is recognised that a key element in knowledge development is the availability of *strong and continuing financial investment sources*, especially in the initial start-up and application phases. One of the big advantages of areas such as Silicon Valley has been the number of capital sources in the area, many of which come from previously successful entrepreneurs who later become investors themselves. Different phases in this money supply chain need to be recognized. It may be easy in some technology clusters to find the initial seed money; more difficult are the subsequent phases of testing and production expansion when many multiples of million dollars may be needed.

These supply side categories of knowledge acquisition have to be complemented by *demand side* factors, although this has largely been downplayed in the research literature. Perhaps the most important demand factor comes when firms seek help in innovative activity from *specialist knowledge organizations* in their locality, where big cities have an advantage because such firms have been deliberately located in such places to serve the large potential market for their services. In addition, *suppliers to firms* may also help in innovation. For example, firms often ask suppliers to upgrade their products, or to develop replacements in response to new product development, and may provide crucial knowledge or logistical support to their suppliers to help the process of innovation. Also suppliers may suggest innovative new products or processes that help to create change in the sponsoring firm, whilst many businesses engage in joint ventures to share not only personnel but also the often large development costs associated with new ventures. These stimuli to the demand side innovation process are particularly obvious in times of war or national military rivalry when governments actively seek solutions from firms to create new types of products, as seen in such developments as atomic energy, radar and new weapons systems. Although these demand side advantages may be helped where clusters of interacting firms are found in an area, the marketing of successful products to distant places frequently adds another international dimension to the locational attributes underlying the innovation process.

11.5.5 Invention-Innovation Support

Although it is important to understand the processes by which knowledge is acquired and used in the innovative process, one of the key issues already identified in the discussion of the various uncertainties has to be resolved. This is the degree to which there is support, or barriers, from two types of governance: one from within the firm, the other from the political system. In the first case effective management within the firm is needed to provide opportunities, not barriers for invention and innovation, to develop and reward entrepreneurial initiative, as well as providing sufficient finance and other aids to support the creative process. The second comes from various levels of government within the state. It also seems that more success is achieved if the innovator lives in a society with a supportive government that encourages the creative process. Unfortunately, the history of attempts to pick winners of new innovations by governments through the world is littered with failure and huge wastes of public funds as Lerner's (2009) book, '*Boulevard of Broken Dreams*', has shown. Yet we must be careful not to be too condemning. Mazzucato's (2013) study, '*The Entrepreneurial* State', has argued that although there have been failures in industrial policies, government financing of research, directly, or indirectly through research grants, has either created many successful inventions, or been the enabler for subsequent innovative activity, especially by providing basic research, which is why she described the state as entrepreneurial. For example, parts of the consumer-electronics industry were dependent on the way that the armed forces help pioneer GPS systems, the Internet and provided a lot of early funding for the advances made in Silicon Valley. However it did take creative ability to adapt many of the ideas obtained from other sources into consumer-friendly products, such as those associated with Apple in particular. Moreover, in the case of funding, it must also be remembered that major research initiatives often involve huge expenditures that only governments can afford, such as the various military initiatives in World War II that led to radar, the atomic bomb and eventually nuclear power, or in peace, the various spin-offs from the space race. In addition, there were vast government expenditures in health studies and in supporting research into new drugs. The failures in the industrial policies of governments to finance successful innovation in particular firms, rather than in basic research, often make headlines and are regarded as outweighing the successes. This led, during the conservative revolution in politics from the early 1980s, to a decline in policies that attempted to finance new innovations. However, the economic recession from 2008 seems to be leading to a renewal of interest in sponsoring industrial projects by some governments in the developed world, in attempts to create new employment. Yet most authorities agree that the more effective government efforts come not from attempts to pick industrial winners in innovations, but from two other indirect measures. One is an approach that focuses on developing '*site assets*', such as providing excellent infrastructure in transport, premises for innovative firms, rapid broad-band computer linkages, educational upgrading, as well as research and development funds (Viedma et al. 2001). These are concentrated in cities, so such urban places are helped to become better selective environments Another comes from what are considered '*soft locational factors*' that provide a more supportive environment for the operation of an innovator or firm, such as: effective patent rights; less red-tape for start-up firms; no penalties for innovators who wish to start again after recovering from initial failures and perhaps bankruptcies. The latter is a feature of many ultimately successful Silicon Valley entrepreneurs, whereas in France and Britain the stigma of a failure often prevents people trying again, or being unable to gain new funds to support their research because of their previous financial record. Other important factors include: flexible labour markets; removing restrictions to competition; creating networks of innovation between universities and firms; creating or facilitating institutions to link inventors with those involved in the development and commercialization of products; and immigration regimes that do not restrict the supply of top-quality people from other countries. In addition, it helps to add competition to the state provision of basic research, through research councils that make scientists compete for funding and leaving the decisions to award finance to experts. All these types of policy do seem to be more effective than attempts to choose specific firms to promote innovation, although their effects are difficult to measure. In an historical context, Rosen's (2010) book on the Industrial Revolution in Britain came to the conclusion that it was not the presence of factors such as abundant raw materials, military demands, or a Protestant work ethic, that were the secret of Britain's early industrial success. Instead, he argued that the new patent laws of the eighteenth century, which gave propriety rights to the inventors, were a major influence in the rapid growth of inventions and innovations. So instead of inventions being the pursuit of academics, or the bored rich, these patents allowed creative individuals to profit in a spectacular way from the inventions they subsequently successfully developed and commercialised. Yet to identify one factor alone as the secret of success may be an exaggeration, for the other factors were surely additional influential contributory influences. Moreover there is rarely one path to success even within a knowledge industry. Superficially the concentration of the invention and development of electronic products in the Silicon Valley, California, and on Route 128 around Boston, would seem to be linked to the spin-offs and associations with major research universities in their vicinity. However, Mackun (2013) has shown that their developmental profiles are radically different. Silicon Valley has been characterised by greater risk-taking and partnerships, with open labour markets and dense social networks of informal linkages. By contrast Route 128 was more based on the growth of established firms with rigid hierarchies that internalize many operations, characterised by the region's convention and self-reliance and a greater dependency on federal research and defence contracts.

What this review has shown is that successful innovative activity comes from a complex of related factors, some of which may be present in a particular area and which are helped by the agglomerative advantages of the bigger cities. This assumes that there are no serious diseconomies, such as produced by traffic congestion, limited communication possibilities and dis-utilities such as high crime rates and limited individual rights. Nevertheless, Rosen's work does throw new light on the enduring value of what amounts to one of the intangible 'soft locational factors' in helping to create success, namely the liberation of individual creativity and the prospect of rewards for those who are successful (Brunt et al. 2011). In this sense it is hard to resist the comparison with the explosive period of medieval urban growth, where it was often said that 'city air made a man free', not only because urban residents were freed from most of the constraints of rural feudal society, but that they were able to use their freedom to carry out and expand trade for their own benefit and profit.

11.6 Conceptualizing Knowledge Cities

The previous section has shown the large number of factors that affect the development and locations of inventive and innovative activity, which explain why some locations, especially some cities with their strong infrastructures, varied networks and specialist firms, have been more successful in developing knowledge industries than others. But much of the current literature on this knowledge-based development seems a-historical, given its focus on contemporary issues. This is unfortunate. Any historical review of urban development will reveal that urban centres have always been places where the knowledge assets of society have been concentrated. The biggest centres have always attracted able and creative people; many have been centres of learning, initially in religious institutions or in the royal court, later in secular institutions. In the more practical sense of 'knowledge as doing', the political capitals of the past had the patrons who enabled art to flourish. They also contained people with the skills to conquer and manage their dependencies, whilst the centres of trade had what Vance (1966) described as an 'intelligence complex of merchants' who knew where trade opportunities existed and had the knowledge to finance and organize them, But late medieval large specialised trading cities, such as Venice or Ragusa (now Dubrovnic), were not only successful because they only had commercial acumen. Their knowledge also included the diplomatic expertise to negotiate treaties and support a network of ambassadors and contacts which provided advance knowledge of political problems as well as trading opportunities, and the ability to negotiate trade concessions. Even in the Industrial Revolution it could be argued that many industrial towns were creative, for they had concentrations of the specialised technological and organizational skills, enabling the enterprises to produce new machines and industrial processes as well as new goods and services over many years. So the towns and often the regions in which they were located were centres not just of production, but also of organization since the firm's headquarters was there, as well as being centres of nascent producer services and of knowledge-creation and innovation in the sense of knowledge as 'doing'. Yet the last century has seen the increasing disaggregation of the functions of organization, producer-services and innovative activity from many of the places where production occurs, a consequence of the fact that each of these parts of the productive cycle have their own locational imperatives.

Since the emphasis in this chapter is upon knowledge and innovation in cities, the question that must be asked concerns the conditions that help places attract and keep these activities, not knowledge as 'being' or 'doing', but as the deliberate attempt to foster knowledge-based innovative activity. This is based on the management and encouragement of continuous innovative activities and which need to be initiated and organized by co-operation between the various stakeholders in the city. Since it has been recognized from studies of the concentration of knowledge-based activities that some cities have been the vital seed-beds in which these activities have been nurtured and grown, then it seems critical to understand the assets, advantages and disadvantages of these locations, especially the problems associated with converting knowledge into innovations. This has led to new paradigms to understand the role of the city milieu, including the attitudes within it, for facilitating knowledge-based development. These models of development reject the old material-based paradigms of urban and economic growth that explained the location of activities only by the differential costs of assembling the traditional factors of production (land, labour, capital) and taking the products to market (transport). Although some of these factors are still important for many industries, the discussion above shows that they are not the only, or even the primary factors in the location of the knowledge-creation sector. The new approaches also deplore the old dichotomy between economic growth and overall social value. Instead, they focus on the utility of the assets of cities as potential seed-beds for knowledge-based growth, seeking to identify the factors that underlie and assist this development of what amounts to a selective environment. Most research on the topic tends to focus on showing the effect of a few variables for particular types of knowledge industries. But in the last decade there have been more attempts to provide more comprehensive summaries. Some focus on developing indicators that index the range of key assets that aid some sector of knowledge development, usually in particular types of knowledge-creation activity. Others, such as Florida's (2002) indexes, purport to measure the creativity of American cities, although they concentrate mainly, although not exclusively, on the characteristics of people that he considers to be creative, issues that were discussed in Chap. 10. More general approaches seek to look more widely at the features that help cities attract and retain knowledge-development activities that go beyond the old locational factors approach to look at the whole context of the city's environment. In Europe the ACRE project with researchers from several European countries represents a good example (Musterd and Murie 2010b). One of its recent reports designed for policy makers provides a list of factors that seem to be associated with what they called creative-knowledge development, together with policy examples that illustrate the utility of these features from various European cities (Musterd et al. 2010). These are grouped into three easily remembered categories: *pathways* deal with the historical legacies of the city that may influence future development; *place*, with the local characteristics that help attract and keep knowledge workers and organizations; *networks* that are the ties that enable people to obtain and share information and promote knowledge activity, a category that acknowledges the role of tacit knowledge in innovative activity. This report is eminently useful in providing a succinct summary of the key features in knowledge activity development that will appeal to policy makers. It is also valuable in stressing that there is no single path to success, for different cities with varied characteristics have chosen alternative approaches. However more comprehensive approaches that identify the many components of places that attract knowledge activities are needed, especially those that apply the type of capital-asset of cities approach, initially pioneered in studies of organizations (Edvinsson and Malone 1997), and applied to the urban case (Viedma 2005; Carillo 2006).

Essentially, this new Knowledge City approach begins with the idea that the city is a capital system, which creates social surplus products from its capital store, which can be viewed as all the value elements that can aid or repel new knowledge-based development. Carillo (2004, 2006) has provided one type of taxonomy identifying the categories or domains within which individual indicators can be grouped, but it does seem to have several weaknesses and has been revised in Fig. 11.2 to create a more direct set of domains and indicators. It shows that the capital store of a city or city-region can be disaggregated into five major groups or domains of value elements: Human, Physical, Associational, Heritage and External Capital. These five broad categories or domains summarize the large range of factors which may, or may not, provide advantages for the growth of knowledge-activities and provide leverage in value generation. Each of these domains can be subdivided into a more detailed set of sub-domains which contain specific dimensions within these five broad categories, which are summarized in Fig. 11.2. By identifying, and later measuring, the value of the various elements in the taxonomy in any city, a way of understanding their advantages and disadvantages as a basis for the presence or attraction of knowledge-based activities is created.

11.6.1 Human or Agent Capital of Cities

This domain consists of the value-generating capacity of the human resources present. These can be divided into three main categories, the attributes of people, growth trends and their attitudes. In the *attribute* category, a key feature is the character and skills of the population, such as whether they have high levels of intelligence,

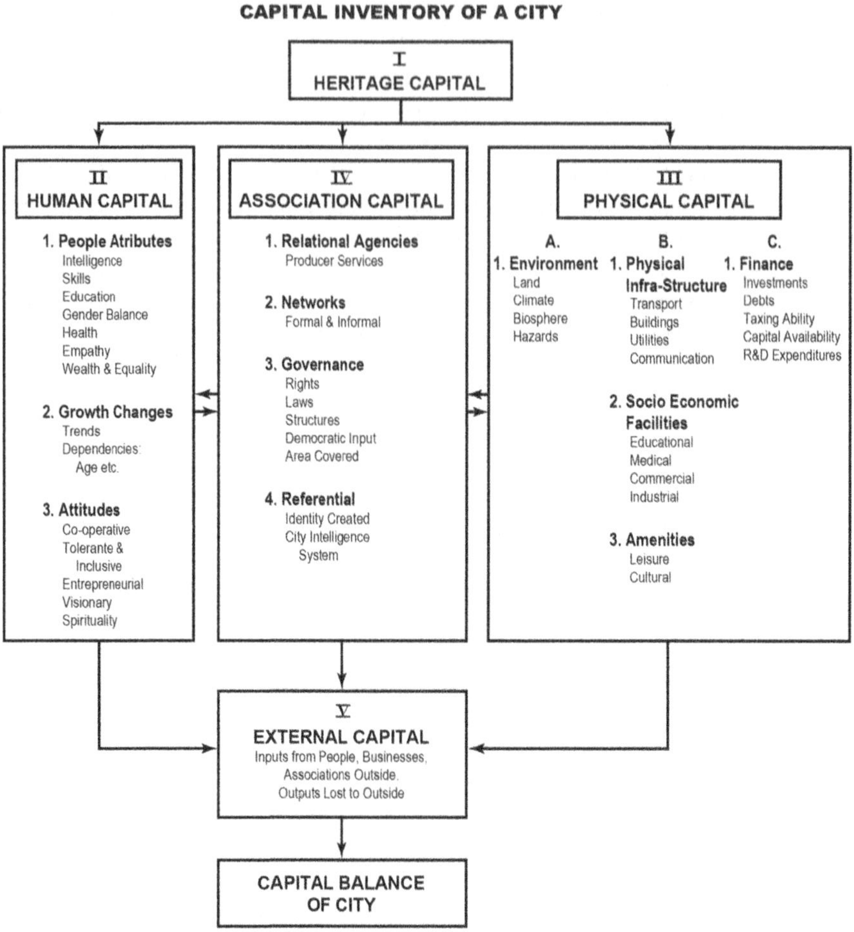

Fig. 11.2 The city as capital inventory

educational qualifications, skills—including entrepreneurial ones—and creativity in innovations of goods or ideas The importance of 'creativity' in the population has often been regarded as the crucial element in the new knowledge society in cities (Florida 2002; Landry 2005) but does seem only *part* of the range of features that affect knowledge development. In addition, it seems that the value of the population capital is enhanced by the presence of international, ethnically diverse groups. They bring different attributes, work-habits, skills and points of view to a city and its activities, as well as providing contacts and knowledge of opportunities in foreign countries. The health of the population should also be stressed as a vital factor as well as diets and physical activity levels. This is not simply an issue of reducing days lost through sickness, or the possibility of premature death, thereby increasing death rates. It is also associated with the fact that health seems intimately related to intelligence and hence knowledge capacity and future capability. It has long been assumed that the growth in average intelligence levels in rich countries is due to better and longer education (Flynn 1987; Lynn and Vanhanen 2006). But Eppig et al. (2010, 2011) have also shown by a study of 192 countries that there is a strong (0.67) inverse correlation between the average intelligence levels and those subject to a high incidence of disease, as measured by an index based on 28 infectious diseases. Although this finding is at a country scale, the same relationship probably applies at a city level, for large cities in developing countries often have lower health levels and higher levels of infectious diseases compared to those in advanced states—even though the urban conditions are often better than in the rural regions. Other explanations tested for the relationship, whether income, education, proportion of agricultural workers, or climate, proved negative. It seems that the presence of parasites and pathogens, especially malaria and diarrhoea, seriously diminish the early development of the brain and cognition in general. So Eppig concluded that the intelligence growth in developed countries may, in large part, be due to the abolition of serious infections in these countries, which had previously impeded brain development in many children. Ensuring good quality health facilities is not simply a humane policy to keep active people alive and working; it may be an essential element in raising the intelligence of a population, whether in countries or cities. Similarly the average wealth and its distribution in a population is also important. Low income inequalities mean that more people have money to spend on goods and services, whereas a highly unequal situation means that only the elite have much spending ability and they are more likely to buy luxury goods from overseas, which does not improve the local circulation of money and employment. Large numbers of poor and unemployed provide a negative element to a city in terms of productivity and wealth generation.

The human capital category should also take into account the possibility of future *changes,* such as the rate of growth and age structure of the population. An increase in old age dependency may mean that there are fewer active workers, so resources have to be spent on supporting the very old who are unable to contribute to employment. At the other end of the age distribution a very young population again means high dependency rates, but may provide a future advantage so long as job generation rates are high in the city. The third human capital sub-category involves the extent to which the population can be seen to be a capital asset because of their *attitudinal capacities*. The degree of tolerance and also the extent of co-operative ability to work in groups are important features, since it avoids any clash of languages and cultures that would hinder progress. In addition their attitudes to work, and recognition of the value of knowledge, are also seen as key elements in future progress. Related to this point is the extent to which a population is inclusive, in the sense of not denying others—such as those of different colour, culture or gender—from employment opportunities and education. One example may make the point. Despite the continued gap in wages between males and females in many occupations there seems little doubt that women have contributed hugely to the development and wealth of many societies in the last 30 years, whether in cities or states. By contrast, some other societies have often denied females the opportunity to contribute outside the household, especially in those countries where they are not considered equal partners. This has reduced progress in so many fields since half the adult population is denied the opportunity to develop and use their intelligence in the job market improve their capabilities. Finally, it must be noted that the capacity of cities, or communities within them, to develop both *visions for the future* and a recognition of the value of knowledge in creating new wealth are additional vital elements in the human capital domain. All these attributes go well beyond the limited number of features described by Florida (2002) is his Creative City and Class ideas.

11.6.2 Physical Capital of Cities

This can be divided into three distinct groups or sub-domains of features. The *environment* dimension does not only include the amount and quality of land present and its cost, but also the climatic, geological and biotic features that can either help, or act as a negative force for, development. For example: flat areas are easier and cheaper to build on but may be liable to flooding; some regions may have climates too hot, too wet or too cold for comfort, although the artificial heating or cooling of buildings now reduces these problems in the workplace, although not always for leisure times. Other places may be prone to regular natural hazards; some regions have had advantages in the past in being on, or close to raw material or energy sources; previous industrialization sites may be too polluted to redevelop, or would cost too much to reclaim. Historically these environmental features were often crucial for the development of cities. Although such factors are far less important today, they cannot be discounted completely.

A second part of the physical capital consists of those man-made objects that assist and provide leverage to the initial inputs and human agents that generate greater productivity, features that can be summarized as *infrastructure*. All of these must be viewed in terms of quality as well as quantity. In a physical context are the traditional features such as the extent and quality of various transport routes and terminals, water and energy supplies, or sewage facilities, or the more recent importance of new communication facilities such as the addition of high speed digital connections that are essential for knowledge-workers. It also includes the stock of buildings and houses etc., needed for shelter and workspaces. If these material features are already present in the city, and are of sufficient quality, they do not have to be provided separately by the knowledge-based firms, resulting in enormous costs savings. The social infra-structure is also an essential part of this type of capital stock, which include the presence of varied and good quality educational facilities at various levels, from schools to universities, libraries, health facilities as well as safety services. Such advantages should also include the availability and expense of child care; high costs and few places can deter young families from moving to such centers and make it more difficult for young mothers to return to work outside the family. Another set of capital assets in this group come from the commercial facilities, from shops to cafes, bars and restaurants. The amenities that are present in the city are also crucial, whether in leisure facilities—ranging from parks to recreation fields, centres, and stadiums for sporting events—to those in the cultural field of heritage buildings, various arts activities, and theatres, whether catering to high-brow or popular tastes. Improvements in all or some of these features may attract and retain knowledge-based workers, although the extent of these attractions is still not clear as the discussion on Creative Cities (Chap. 10) has shown.

A third set of features within the physical capital category consists of what can be called the *mobile capital or finance* features, since unlike many of the other capital assets they can be easily moved. One of the most important is the actual stock of financial assets. Those assets that are available and useable can be used to upgrade all the other physical capital stock as well as improve human capacity. In a knowledge context they include R&D expenditures and the availability of risk capital sources discussed earlier that is available to support new innovative enterprises. Within this category one must also consider the tax levels in the city, inflation rates, and the municipal debt. The latter can lead to fewer services if not resolved and can be a drag on value creation, unless it is used for investment purposes and can be paid back from existing resources or future output. Several cities in the U.S.A. are close to bankruptcy, in part because of large unfunded pension commitments to former public workers, a problem not faced by European cities because these are usually funded by national schemes. These financial exigencies have led to cuts in municipal services, meaning that citizens lose access to many services, while opportunities for new initiatives are limited, which reduces the attractiveness of the cities for incoming businesses. In addition the category also includes the range and availability of machinery, from vehicles to factory machines and technical assets.

11.6.3 Association Capital of Cities

Carillo (2006) called these features Meta Capital. Since they all involve some type of association that connects people and objects, and represent the interactions necessary for the human and physical agents to create value, the term 'association' may be more easily understood. The domain is divided into four categories. The *relational* sub-domain consists of the activities that create and lubricate the flow of the productive process, from the historical breakthroughs that saw the development of record-keeping in cities, and later currencies to act as exchange mechanisms, to the wide range of activities now usually known as producer-service activities, such as financial, legal, advertising and marketing services. These latter services are necessary for the productive process that is a result of the combination of agents, inputs and instrumental objects. Closely related to these activities are the various *networks* that were discussed earlier as crucial to inventive and innovative activity since they facilitate the exchange of tacit information. These include: formal organisational connections that link businesses or parts of businesses, websites that bring people and organizations together, to the informational sources—from newspapers and journals to media outlets and information-gathering services that provide access to specialised information required by businesses and people. It also covers the informal networks, such as social and business linkages, or the more formal associations designed to promote the interests of the group, such as trade organizations, or the many websites that now connect people interested in particular issues, such as those dealt with in other chapters from the New Urbanism to Cittaslow network sites. All of these features create a greater understanding of the functioning of urban places and the policy opportunities to solve problems. Another important Association Capital category consists of elements that can be summarized as the *governance* dimension, not only the services provided by various levels of government to the city, but also the political institutions that provide and support democracy, legality and citizen rights. These rights are not simply the words or phrases used in constitutions that are never fulfilled in many totalitarian states, but those in an actual, practical sense. They ensure that all individuals are free and fully able to fully participate in and be rewarded in a society, and are free from arbitrary coercions or sanctions, some of the basic Capability principles discussed in the Just City discussion (Chap. 3).

The final Association Capital sub-domain uses Carillo's (2006) term *referential* category. This deals with the identification of the city and its relation with all other factors, essentially its links in an affective context. It consists of two different sets of elements. One is the way that each city has a distinctive identity to its citizens and others throughout its region, country and the world, based on its inheritance from the past, its current status and what it wants to achieve in its future. Many cities have negative images—such as high crime rates or records of lengthy labour strikes—which can impede future development. These handicaps have long been known to cities and many have attempted to create positive identities for themselves by branding themselves through creative advertising. This is not new. Historically the competition between the emerging towns in the American and Canadian midwest from the end of the nineteenth century led to the so-called 'booster movement' where the settlements often produced quite exaggerated views of their prospects to gain investment and new businesses, especially railway lines (Artibise 1981). In the 1990s the deliberate advertising of cities to emphasize their strengths became a standard policy. One of the most original slogans was provided by a consulting company for Dunedin, New Zealand. Kearsley (1995) has shown how the phrase 'Dunedin: It's All Right Here' was a brilliant publicity success. The simple words expressed the double meaning of a satisfaction with life in Dunedin with the idea that the city has *all* the needed facilities for a good life, or for new businesses searching for a new location. Yet this discounted some of the real problems of the city, its old infra-structure and relatively limited hinterland, as well as a location in a peripheral part of the country. The second set of elements in this domain consists of the intelligence capacity of the city managers and the various stakeholders to understand and respond to events and changing situations that affect its current and future development, finding ways of eradicating problems and ensuring progress by the positive use of its existing assets and attracting new ones.

11.6.4 Heritage and External Capitals

These elements of value generation come from beyond the specific time and place of the city. The former deals with a temporal externality, the heritage of the city, or pathways, that have created positive or negative values for the contemporary situation, such as a reputation for excellence in particular skills, or research capability, or the negative perception from pollution spills or poor security. The latter deals with spatial externality. One example is the ability of outside sources, from governments to private organizations, to contribute to the city's stock and availability of financial assets. Many cities complain that the services they provide should be financed by a fairer sharing of general national tax revenues with cities, rather than from ad hoc, project by project transfers. In some cases it may be possible for the city to create its own taxation regime. Also in the external capital category is the extent to which governments can change the rules that affect capital formation, such as encouraging new business opportunities, or allowing more highly skilled people into the country and hence the city, by changing immigration regulations. Another element of the external capital is whether the *profits* that are generated by businesses remain in the city, and are also used effectively on its behalf. Although a city can embark on a strategy to remove the negative features of the past, it is usually dependent on outside agents to provide a positive externality in capital provision.

This Capital Inventory approach provides a comparative inventory and assessment of urban settlements. Its main advantage is in identifying the advantages as well as the problems of any town and city. But by exposing these issues it is useful in helping to build consensus among interested stakeholders by identifying areas where improvements are needed, especially when compared to other cities. This can help develop priorities among the policies for future improvement. It can also be extended, as in Carillo's scheme where a category called Output Capital identifies and measures issues such as the products of the city during various time periods, using measures such as GDP and growth forecasts as summaries of the total output of a city. However the interest here is upon identifying how the various elements in the Capital Inventory can contribute to the generation of knowledge-creating activities and upon the plans and policies to achieve this aim. Before dealing with such applied issues it does seem worth commenting on the fact that many of the components of the type of Capital Inventory model shown in Fig. 11.2 contain echoes of older ideas of the advantages of particular cities, from the nineteenth century emphasis upon the importance of site and situation factors, to agglomeration economies from physical infrastructures, as well as basic-non basic ideas and the more recent stress on the importance of producer services in urban growth. In addition, the labour element in human capital is clearly one of the traditional factors of production, although the emphasis is now upon the educational, creative and entrepreneurial qualities of the workers, as well as the additional attitudinal variables that condition productivity and creativity. So the city is regarded as the site where activities such as research may be generated and attracted, rather than as the passive container to which various factors of production are brought. These new approaches view towns and cities in an activist way; they are regarded as environments that may possess both the desire and qualities to attract new knowledge activities by reason of their own character, assets and visions.

11.7 Creating Knowledge Cities

Despite the increase in information about the characteristics of places that can be described as Knowledge Cities, the processes by which they can be created are still not entirely clear. Cities or parts of cities have very different features and problems and those that have developed a strong basis of knowledge activities have done so in different ways. Nevertheless, a broad outline of the key processes that have been proposed as favouring Knowledge City development for the city as a whole, or at least part of it, is emerging, and can be summarized in terms of several main stages. Within each of these stages it is argued that a potential Knowledge City must be an active one that evaluates itself, acts upon its existing assets, and creates future ones to achieve its own rejuvenation and sustainability, rather than being the passive recipient of policies from other decision-makers, whether in firms or national governments, as the quotation that began this chapter showed (Ergazakis 2004, p. 5). This involves creating a strategic vision and plan, that involves several different stages.

11.7.1 The Spark

There seems little doubt that the only way in which a city is to begin the process of transformation to a Knowledge City is when its decision makers and citizens *explicitly recognize the social urgency* of the need to respond to a situation of decline. In addition, a city must *possess the social and political will* to attempt re-generation in order to re-position itself for a new future, which implies an activist government able to act. Unfortunately, many western cities that experienced industrial losses from the 1960s, due to the loss of their own material-based production facilities, clung to the belief that their future would be assured by reviving these old activities through campaigns and policies to attract similar industries. Such policies were usually supported by Chambers of Commerce and city administrations. Knight's (1995) pioneering work used the case of Cleveland (Ohio) to show that such policies made the existing situation worse by not cleaning up the old industrial sites for different uses and by failing to recognize the importance of keeping the many Head Quarter offices that the city possessed from moving elsewhere. Knight argued that Cleveland wasted almost two decades before recognizing that the industrial past was dead and charting a new direction based on emerging knowledge-based activities and co-operation between the existing companies. This was pioneered by the 'Cleveland Tomorrow' vision that outlined the new course to be developed by a partnership between public, private and voluntary sectors, in which it was recognized that many of the problems faced by individual companies were the same and could be only solved by co-operative effort. This needed explicit recognition of the problems, followed by creative endogamous action by governments, businesses and citizens to solve them, rather than hoping for future development generated only by exogamous decision-makers in distant corporations or governments. Yet there is a need to be cautious about the assumption that this is the only path to success.

Morgan and Nauwelaers (2003) have stressed that the scale of unemployment in many old industrial centres and regions has been so high that exogenous developments helped by government assistance to create jobs—only some of which will be knowledge-based activities—will always be needed, since it will take time for the new sectors to grow and be self-sustaining.

11.7.2 Developing the Vision

The development of a coherent and ultimately successful new vision for the city must be integrative in several contexts. First it must involve a wide range of *stakeholders* in the city, from individuals, firms, organizations, city officials and input from higher levels of government, to ensure that the effort involves as many people as possible and to develop a consensus on the way to proceed. Second, it is important to ensure that there is widespread *understanding of the value of knowledge and knowledge resources as a wealth-creating activity* for the city, frequently a more difficult task than identifying material products created from factories. This understanding should be shared by the general public, as well as by civic politicians who are needed to provide leadership, and especially by planners, whose existing rulebooks have often been found to slow down or even restrict new ways of development. Third, the integrative process should be applied spatially, with *regional perspectives* designed to involve other municipalities around often the politically fragmented city-regions, one of the reasons for the development of New Regionalism ideas discussed in Chap. 2. This will ensure that the functional unit of the city-region as a whole is developed, rather than some areas being left behind, while others accumulate all the future successes, a depressingly familiar pattern in the administratively divided city-regions of many American cities in particular.

11.7.3 Creating the Plan

The first stage is an *evaluation of the various assets and disadvantages* of the city, such as through the various dimensions of the Capital System approach (Fig. 11.2), focusing on those which help or hinder the creation of a knowledge-based economy. The goal is to make city development less accidental, where growth occurs because of individual decisions made in various sectors of the economy—especially by employment growth from exogamous sources that can just as easily be reversed by distant corporate decision-makers. Instead it should be more intentional, with a focus upon more endogamous or internal influences that stem from, or can be influenced by, the city's own capital assets. Most reviewers of the knowledge-based plans produced by various cities stress the necessity of a *clear focus*, identifying a limited number of *niches with high growth potential*, rather than spreading effort over many different sectors. It is also recognized that the various types of knowledge activities need very different techniques and managerial capacities, so the development effort needs to be concentrated, not dispersed. At least one of the new sectors should have a high profile, not simply to symbolize the new economy or society, but to act as a catalyst for related growth. An example may be provided by the success of the Guggenheim Museum in the redevelopment of the obsolescent, polluted, heavy industrial city of Bilboa (Azua 2006). The success of the museum has not simply been in attracting visitors to its own exhibits and unique design or adding knowledgeable art experts to its workforce, or in leading to the upgrading of surrounding hotels, restaurants and stores. Rather, its success lies in the fact the museum has lots of space for other exhibitions, not only those with an international profile, but those that provide a focus for the positive elements of Basque identity and creativity, rather than the violent approach of the political separatists. Hence the museum is not only a symbol but has helped the city recreate its own regional profile, developing as a new focus for a regional identity (perhaps a 'national' identity for many Basques) within a federal Spain. This replaced its old role as an international exporter of steel and other manufactured goods that now could be produced more cheaply elsewhere. In the Capital Inventory model the success of the museum can be seen as an example of the importance of creating a specific and positive identity for the city in the referential sector, just as buildings as diverse as the Eiffel Tower, Empire State Building or Sydney Opera House grew to be internationally known and positive symbols of their cities. A related referential image can be seen in the increasingly determined efforts of many cities to brand themselves to create a distinctive profile that emphasizes their advantages and unique features, in view of the increased competition between cities for new businesses and tourists (Dinnie 2011; Baker 2012).

11.7.4 Specific Implementation Measures

Given the very different histories, sites, capital assets and problems of cities, it is not surprising that there is no single approach that provides a path to successfully implementing knowledge-based development in urban areas. In addition, it is recognized that the greatest success is usually achieved in economies that are mixed, with a range of businesses sizes, and with economic activities that spread their effect across many economic and social sectors, rather than being restricted to a single industry, because if this fails there is nothing to replace it. This mixed economy character should be a general aim of development. This approach avoids the type of separate silo focus that has characterised the production of many goods and services in cities, most of which have limited life-cycles. Hence, sustainable economic success is more likely to occur by increasing the capacity and quality for future learning and knowledge development within the city environment and beyond a single industry, in addition to infra-structure investment and to the various intangible and indirect soft factors described earlier. Although all the issues identified in the Capital Inventory System need to be addressed, reviews of knowledge-based development have identified a number of key policies that seem essential for success, which can be related to the capital system model outlined in Fig. 11.2.

In the *Heritage Capital* domain, it is important to identify and solve weaknesses in the local environment, such as clearing up pollution sites and creating a healthy and attractive environment. It is also useful to build upon and develop a city's heritage of distinctive features, such as its historical legacy of buildings or ways of life and replace negative features. This may help attract and retain skilled and innovative people.

In the *Human Capital* category the importance of having an *intelligent, educated and imaginative workforce* is essential, since this is most likely to have the creativity and skills to produce innovations. There may be innate differences in people, but a quality education is always essential to nurture and extend these skills. This does not simply apply to science and technology learning, but also requires an education that nurtures organizational, entrepreneurial and imaginative abilities, which involves imaginative arts graduates as well as applied social scientists. The importance of promoting *tolerance* is another crucial feature. In addition, it seems useful to have a workforce that includes people who have *international perspectives and come from varied ethnic backgrounds*. Unfortunately few cities have the ability to have separate immigration regulatory regimes, although cities may be powerful enough to lobby for such changes. Indeed surveys of key areas of knowledge–based development, as in Silicon Valley, have shown that almost half of the 'start-ups' are made by people of Indian and Chinese backgrounds. In this context the European Union has encouraged students to study outside their country to broaden their experience and contacts through schemes such as Erasmus, which allows university students to spend time at a university in another country. This leads to greater language skills and knowledge of other places and reduces the cultural barriers to labour mobility. In addition, it may be appropriate to *recognize and reward workers* in the knowledge industries, such as by instituting prizes or incentives to profile the worth of these individuals during their productive years, not like the Nobel Prizes which are rewards for past breakthroughs (TE 2010, Brunt 2011). Such policies will demonstrate the contribution of these individuals to the city development. In addition, policies may be needed to improve the *health of the population*, especially in less-developed countries. This is not simply because intermittent sickness reduces productivity but because an early death means the premature loss of a human capital asset which has taken a long time to educate and maintain. Problems that exist through identification of growth trends and increasing dependencies, such as an increase in old and retired people, also need to be addressed.

Within the *Physical Capital* dimension of the Capital Inventory System most reviewers emphasize the need to *remove environmental problems* and find ways of creating resilience to natural hazards, issues discussed in Chap. 8. The *specific and continuing investment in infrastructure and facilities is also vital.* At the minimum this means improving transport facilities and utilities. One of the best examples of direct investment to combine these two features can be seen in the successful 1950s French metropole d'équilibres (Rodwin 1971), although few of such bold initiatives seem attractive to policy makers today. The policy was designed to improve the prospects of the main large regional centres in France because of the belief that Paris was too dominant and was growing too fast, whereas the main peripheral nodes were too small to attract significant growth and needed strengthening. Most of these nodes achieved subsequent growth by large regional incentives and the development of high speed train connections. Toulouse has been one of the most successful of these growth poles, being transformed from a rather sleepy and stagnant large regional node of the 1950s into one of Europe's premier knowledge centres due to the addition and expansion of its universities and a focus on aerospace (Murayama 2002). It is a classic example of what directed government investment can do. More recently the city has added a new industrial cluster based on research into cancer and bio-medical activities. Cancérpôle Toulouse has been created at a cost of over a billion euros and by 2013 employed over 4.000 people on a 220 ha site. This type of development has involved the co-operation and support of city, regional and national agencies as well as from the European Union. Apart from improving all internal transport routes, good long distance connections through airports with lots of destinations are also crucially important, as few knowledge activities are found in remote and small towns, unless engaged on secret research. The importance of air transport in urban growth has led to the concept of an Aerotropolis, a city founded around an airport, rather than the other way around (Kasada and Lindsay 2011).

The necessity for cities to invest in accessible and cheap broad-band high speed computer networks, as well as excellent schools, training schemes and universities, is obvious. In the case of universities there have been few examples of the type of bold initiative practiced by Japan in the creation of the knowledge hub of Tsukuba city described earlier, but again this was a national government, not locally inspired development. However, some cities around the world have been trying to duplicate this approach to create knowledge hubs, but not on the same scale. The city state of Singapore is one of the most impressive with its focus on biotechnology and other research sectors. An ambitious city initiative is New York's knowledge hub, on Roosevelt Island in the East River, although again it is supported by state and federal funding. Formerly a hospital and low income housing site, much of the island is being redeveloped as a new science-based university campus. The campus was initiated by New York city when its development authority realised that the city had lost three-quarters of its manufacturing jobs since 1960 and that the 2008 financial collapse showed how dependent the city had become on the financial industry, with over a third of the city's private payroll coming from Wall Street-related businesses. The city invited major universities around the world to make a bid to create a new science-technology university. Cornell University, in partnership with the highly successful Technion University in Israel, won the bid. Their plans envisage spending $ 2 billion on creating a new campus on the island, one that is also sustainable in being dependent on solar and geothermal energy for its operation. The plans call for the university to have 8,000 permanent jobs and it is anticipated that various spin-off companies will generate 30,000 more jobs in 600 new technological companies within a few decades. The objective is for this new 'knowledge city' (actually within a city) to provide the basis of a new science-based industry in New York, having some similarities with the planned science–based town of Tsukuba, or the informal growth of knowledge-based hubs (Mackun 2013) that developed in Silicon Valley (Stanford, California) and around Boston's Route 128 (near Harvard and MIT).

The incremental potential of *existing higher education establishments* may also be developed, by moving beyond seeing these places as learning centres, or even research nodes, to a third phase, a focus on valorization as a major, not minor activity. This involves bringing the knowledge created in universities to market via patent development, engaging in research outsourced from large firms or government, or creating new incubator units for product development by staff or graduates. The process involves creating closer links to the local business sector so that networks of mutual benefit can be established. These new roles of universities are helping to transform their local urban economies although the process has already been important in some places in the past. Perhaps it is not without significance that Detroit did not create a major university, and focused so exclusively on the auto industry, so that when the latter declined and many production units moved to other locations, there was little ability to re-orientate the local economy. This failure, combined with its social problems, has led to the massive decline of the city and to its declaration of bankruptcy in 2013 (Martell 2012; Binelli 2012).

Studies of the new knowledge economy have drawn attention to the need to create *contact and innovation sites*, not simply by increased numbers, but also in their quality. Edvinsson (2006, p. 69) has reviewed such developments in Scandinavian cities and argued that these new knowledge spaces (KS) should provide a creative context, adding networking space and designs that help to enhance the quality of personal relationships to encourage tacit information exchanges. These spaces range from trade or service organizations for like-minded people to meet—whether open-access formal ones, such as libraries and information centres, or episodic exhibitions and conferences to disseminate expertise—to informal meeting places, such as the availability of cafes, bars and restaurants where people can meet and discuss progress and plans. They include 'umbrella offices', places rented by the hour or short periods that can be used by small and start-up companies. These examples are based on a recognition of the need to encourage ways of fostering the interchange of ideas in the tacit knowledge dimension through personal contacts, creating the 'buzz', which is now seen as being essential to invention and innovation. Adding these features reduces what Edvinsson (2006) called the 'friction cost of knowledge flow' in cities, although it is clear that the provision of these physical facilities only work if they also contain enough interactions and linkages, showing that such attributes overlap with associational capital factors.

A focus on generating new employment alone must not obscure the *housing and other needs* of knowledge and creative workers, ensuring that there are a range of housing options at reasonable prices and distances from workplaces, as well as recognizing that residential needs will vary at different stages of a worker's life-cycle. In the specific area of cultural activities, many creative artists are on low wage or intermittent incomes, so affordable housing and working space is crucial. Historically these activities have often clustered in decaying and old premises unwanted by most businesses, often in the inner city, but as redevelopment and gentrification has occurred, these cultural workers get squeezed out. Amsterdam created the Broedplaatsan scheme in 2000 (BP) to address this problem. This does not actually supply low rent premises; instead it created a pro-active forum that initiated contacts between various actors in the finance and building industries, as well as various city departments, to encourage the supply of low cost facilities. A policy based on similar reasons was developed by Leipzig to counteract the effect of suburbanization, inner city decay and low investment. In 2008 it was estimated that the city had over 38,000 empty premises in the decaying inner city, even though many were of rich architectural heritage. Few private owners were prepared to pay the costs of re-development. However, a private housing association called HausHalten (HH) was given tax breaks by the city and has brought many areas back to life by renovating property. The association was keen to attract low income creative or cultural workers and students who would help revitalise the area by their artistic skills. The new premises were called Guardian Houses since the new occupants look after the premises. This replaced the options of either continuing decay, or the development of a more typical urban renewal sequence of gentrification which would be residential-only. This example shows the wider need to improve the cultural amenities in the city in order to improve the liveability of the city, which may then attract creative workers, although unlike the emphasis on this feature by Florida (2002) this is only one of the features necessary for success in creating knowledge centres.

The third major category in the Physical Capital Domain is the availability of adequate *venture capital* to support new innovative activities. Big companies may have the resources to do so, but many inventions come from new entrepreneurs whose access to capital is limited. Hence many cities and governments have created new sources of money or agencies to help find this support. For example, in the emerging Tech City cluster centred on Old Street station in London's East End, a 'Million Pound Start–Up' competition was organised by Digital Shoreditch in 2013, while the government's 'Seed Enterprise Investment' programme reduces the risk of inventors losing all of an investment in a start-up (TE 2013). However a real problem in many high-tech clusters is not the initial start-up or seed money but the multi-millions needed for subsequent phases of development. Hence the government persuaded the London Stock Exchange to list fast-growing firms if they float 10 % of their equity, instead of the normal 25 %. Access to enough *talented people* is also a problem that needs to be solved as national immigration rules are often cumbersome to overcome or too restrictive. Hence there have been attempts in Britain to make exceptions to the normal immigration rules, with engineers and executives now being defined as 'exceptional talents', a rule that previously made it possible to import artists and scientists.

The *Associational or Meta-Capital System* category summarizes the specific relational opportunities designed to maximize the interactions within and between the various input capital, human and incremental capital domains to create what amounts to an eco-system of related activity. Given the importance of contacts in the tacit knowledge dissemination for innovative activity, ways of developing existing networks or enhancing others are essential, in addition to providing physical contact spaces. In most towns and cities organizations of businesses have always existed, but these need to be strengthened and extended, especially in specialist fields. In Munich a cluster of Audio-Visual Media (CAM) firms links together over 300 companies in this field. The Milan Network of Design (MInd) established by the municipality, includes a design museum and the largest design schools. It sponsors an annual international competition which brings in over a hundred new students to the local design schools and provides internships to those who graduate. This is another attempt to maintain the city's position as a major world centre for design. The creation of wider networks of specialist knowledge activities outside the city, in national and international terms, is also important to avoid isolation, thereby allowing people in local clusters to gain knowledge of developments elsewhere. A more general example of network creation comes from the utility of creating closer and co-operative links between the local educational establishments and the city in which they are located, while the maintenance of links with alumni also often provides access to knowledge and often finance for their former school.

Also essential are the establishment of *knowledge-based agencies*, whether public or private, that are devoted to, and are expert in, the encouragement of this type of development through the provision of services, or information on how to access the range of services necessary for this type of growth. In the Kitchener-Waterloo region the creation of Communitech (CT) in 1997 in one of Canada's major high-tech areas provides an example of support for local innovators that helps develop and commercialize new technologies. The organization moved into a 440,000 sq. ft. renovated industrial building in 2010, helped by grants from the local region, province and federal government. This provides a meeting and conference centre, technology hub, room for start-ups, as well as being the organizational centre that provides the varied types of assistance needed at different levels in the development of new technology businesses, providing peer2peer support and contacts between people with different skills. The organization claims to have helped over a thousand companies with sales of $ 30 billion by 2013.

The previous examples show that *supportive governance* in the Association Capital context is also crucial, but there are wider issues involved. In the context of rights, it has already been noted how crucial is the ability of all people to participate freely and easily in public information sites, with access to education and health guaranteed through effective legal protection, as well as tolerance—issues that are crucial parts of the Capability dimensions discussed in Chap. 3. In this context, of course, the city usually has to correspond to the laws, institutions and practices of the state. If they do not support personal freedoms and do not have effective controls on exploitation, the laws can often be a constraint on growth, as seen in many developing countries.

The fourth category within the Association Capital Domain, namely *referential capital*, includes several factors that contribute to a city's capital assets. For example, urban places need to develop a *city intelligence system* with skilled forecasters and publicists to respond to economic and social challenges. Also the presence of a clear, up-to date, *single portal metropolitan website* describing city plans, potentials and available services and assistance for knowledge industries is helpful. In too many cases the available government services are unclear and ineffective because they are provided by a multitude of overlapping and often contradictory sources. Given the increasing value of attracting skilled highly skilled immigrants it is also useful to create an office or at least a portal to help *immigrant adjustment*. By solving residential, language and social problems quickly, they will quickly become more productive. In 2006 Amsterdam created an Expatcentre which claims to have assisted over three-quarters of the skilled immigrants who moved to the city in recent years. Many centres also need to improve their image. Some have managed this by *positive advertising and branding*. Others have used the presence of some major international event to focus attention on their assets which often involves major infrastructure improvements. For example Barcelona's hosting of the Olympics in 1992 helped transform its structure and image from its role as the second city in Spain, largely based on manufacturing and service, into a more dynamic cultural, tourist and research node.

The *External Capital* of a city and its region also has potential for improving development, especially by bringing in skilled people and venture capital. Another valuable approach is to see whether the outputs of various firms in the city can be used as raw material inputs to others in the area. This is seen in the type of eco-cycle links described in Chap. 5, but has also been important in the growth of many industrial or even service clusters through *linkages* within cities. Older attempts to create new industrial bases in depressed regions, such as the various industrial estates in Britain established in high unemployment areas in the 1930s, did add new factories to these areas. But very few ever developed spin-offs, providing inputs for other businesses that would create growth between them. So there was little possibility of self-generating growth. Many were branch plants that declined when the parent company experienced problems. In addition, attention should be paid to other problems that may inhibit growth in an area, such as high tax or utility rates, or even low productivity in a city or region; such features limit innovation opportunities.

11.7.5 Charter Cities

One of the most obvious problems of cities is that they are rarely free from state control or influence in many areas, so their room to create opportunities by themselves may be restricted by state regulations, or more generally by the state culture, laws, rights and freedoms. In some places, such as California, towns have the right to set up a charter by which they can govern themselves, though this is mainly the choice of council structure, for they are still subject to state law in other areas. But the term 'charter' is also being used by cities who wish to rewrite national or state law on the powers of particular local municipalities. This would provide them with a new charter which would give them more ability to act on their own, which could involve taxing powers and local courts.

Recognition of the way that deficiencies in state culture, institutions and limited intellectual knowledge, have restricted growth in cities in underdeveloped lands led the economist, Paul Romer (2013), to advocate for what he called Charter Cities in these areas. He envisaged countries allowing the development of a new city in its lands which would adopt a charter of rights based on the legal regulations of some other country. This is assumed to be a free-enterprise, developed country that has personal freedoms and legal constraints to prevent the corruption, distorted economic policies and inertia that often plagues development in many undeveloped states. Hence the receiving country would acquire the knowledge and institutions that encourages growth and provides a charter for a defined territory. The charter supersedes the laws of the surrounding state and provides the guarantee that the people in the new city will be able to follow these new rules. The idea seems to have come from the concept of special administrative zones, such as Hong Kong's colonial status with Britain that protected it from communist rule in China and helped it achieve such spectacular post World War II growth. Romer's attempts to persuade Madagascar and Honduras to adopt the approach initially seemed successful, but by 2012 both examples had failed, the former because of a change of government from a previously supportive regime, and the latter because of political opposition from the country's constitutional court that did not want to give up jurisdiction (Melkin 2012), although the idea was being talked about again in Honduras in early 2014.

Much of the opposition came from the self-serving elites that did not want to cede powers, although some development economists, suspicious of neo-colonialism, have also been vocal in their criticism (Mallaby 2010). Although the idea seems new, to some extent the export-processing zones that have grown up in many developing countries are examples of similar practices of what amounts to a replacement of some national rules with special ones in the zones to encourage industrial growth. Moreover, it is usually hoped that the zones will create knowledge acquisition in at least the making of things, by importing skills that locals will subsequently acquire, which might lead them to create future growth.

In an historical context it can also be argued that the idea of the Charter City may be a new version of older ideas, ranging from classical era Greek city states using their own legal practices when they established their daughter cities in other lands, to the charters that gave separate powers to many new European medieval towns as well as privileges and land to tempt traders to these outposts (Beresford 1965), such as in the powerful Hansa League of commercial centres in Northern European lands. In both cases the invading peoples, traders or princes establishing the centres were more powerful than the locals and there were less obvious restrictive territories at the time compared to the national units of today. Although these historical examples may not be knowledge cities in the sense of the intellectual capital that is used here, their knowledge as 'doing', in creating new services and crafts, combined with the ability to be free from local political overlordship, provide examples of the way that societies in the past managed to create thriving urban entities in new lands based on legal and new economic principles from the old.

11.8 Conclusions

The last half century has seen a seemingly inexorable development of what are now called the knowledge industries, many of which are spatially concentrated in a small number of areas. By understanding the various factors that create these concentrations in cities, as well as the negative features that repel them, it may be able to improve their economic importance. Devising new policies built upon this knowledge could increase creativity and innovation in a wider context than just the focus upon one class, as in the Creative City approach (Chap. 10), although there is recently a tendency to combine the terms in the Creative-Knowledge City concept (Musterd and Murie 2010b). Yet it is important not to be too carried away with these ideas as the only solution to future economic success. Not all urban places can be knowledge centres, especially in an innovative high-tech sense, for the number of such centres is quite limited. There are still many other service and industrial activities that can lead to urban economic successes by employing far more people, although they can be helped by the addition of new knowledge innovations. Also, the type of globalization of the last 30 years may well be transformed. There is no guarantee that the Asian source of cheap labour will continue for ever, as Chinese workers are demanding more wages and rights, which erodes the differential from the Americas and Europe, while the rapid growth in innovation in China and other Asian countries means that many have become even more competitive in creating new products. Although there is a move to lower cost locations in places such as Thailand or Bangladesh, some industries have started to return to developed countries by what is called on-shoring, although not at a scale to replace industries lost. This might continue if transport or wage costs rise in Asia. Also, the development of new technologies, such as 3D printers, that will allow the production of many more goods in developed countries on a customized basis, may replace the need to import many goods. This is an example of the capital substitution, in the form of a new technology for labour. To take one example, much of the success of Toyota lies in their corporate and workplace culture that not only focused on innovation but values each worker and encourages them to seek ways of improving the way they do their job, searching for ways of continuous improvement and lean production systems (Liker 2004). More generally the re-connection of research centres and production places could produce what Clark (2013) has called 'working regions' in a valuable recent book. She argues that economic sustainability may be better achieved if industrial locations had the capacity for production *and* innovation, and German nodes such as such as Munich and Stuttgart seem to have achieved this association. Also it is worth stressing that despite claims about a space-less world, place does matter, for the costs of assembling the various factors of production at various places are far from irrelevant for many industries, so some cities can build on these advantages.

Recognition of the range of influences that affect knowledge-based development led to the Capital System Inventory approach to cities. This identifies and measures the many value elements that affect the ability of urban places to attract and retain existing knowledge industries in order to understand their potential for growth. The approach moves away from the silo-like approaches to development, where various policies operate mainly in one sector of the economy of the city. It emphasises that real progress has to be based on the involvement of many stakeholders and the creation of a knowledge-sharing culture and a continuous knowledge development. This relies on the social context as much as the economic, to create what are increasingly described as technological ecosystems full of linkages that provide the opportunity for self-creating growth. The addition of knowledge activities is important for future urban growth. However, the accumulation of such features alone is not seen as a justification for the label Knowledge City, as stressed by one of the leading advocates in the field.

> No urban development project, however strategic, justifies the use of the Knowledge City label if it aims primarily at economic development, or can be described in terms of available techno-economic development frameworks....for they fail to account for most of the distinctive value dimensions of knowledge-based production and overall social worth. (Carillo 2006, p. xiii)

This restriction of the Knowledge City concept to the deliberate and *continuous fostering of knowledge activity*, rather than only its concentration in area, does seem rather extreme, for even in places where this type of process occurs there are many people employed in parts of the city that are clearly not knowledge workers. The requirements suggest that the criteria that have been established for a Knowledge City covering the whole urban area may be better seen as essentially aspirational, since they will rarely be completely attained. After all, the historical record shows that centres of invention and innovation are subsequently superseded and do not last for ever. Hence it might be useful to consider one subset of the ideas, such as Knowledge Clusters, as being more practical, describing areas *within* some cities where continuous inventive and innovative activity are concentrated, nurtured and developed.

Finally, it must also be emphasized that cities lie within the jurisdictional regime of their countries, which create most of the major laws, regulations and institutions within which urban places operate. This usually makes it difficult for a city to improve all relevant aspects of its own selective environment; only city-states, such as Singapore, or special administrative regions like Hong Kong, represent real exceptions. So not all of the knowledge factors identified above can be created or influenced by the city government alone; progress will usually only occur by the involvement of multiple stakeholders and by agreement of higher levels of government, unless new powers are given to cities or learning strategies are followed (see Chap. 16). In the creation of a more purposeful and endogamous city development, the need for an active role by the state is also important, for only national governments have the resources to provide funding and encouraging basic research. Yet progress towards Knowledge Cities, or Knowledge Clusters within these cities, is more likely to be achieved by the state and individual cities acting in what European Union regional experts call an *animateur* approach, namely through evaluation, guidance, partnership, encouragement and even finance of basic knowledge activities, and improving the key parts of a city's capital systems, as outlined in Fig. 11.2. Many believe this is likely to achieve more success than the state adopting either a directed approach by choosing firms, or by the *dirigeste*, state-providing tradition, although the role of government in creating and financing basic research must not be downplayed. Yet whichever approach is used, the policies adopted must be firmly based on an understanding of the many factors that influence the location and character of inventive and innovative activities and especially the extent to which cities are able to provide a viable and supportive selective environment for knowledge development.

References

ACRE. (2006). http://acre.socsci.uva.nl/. Accessed 12 Oct 2012.

Artibise, A. F. J. (1981). *Town and city: Aspects of western urban development*. Canadian Plains Studies, 10. University of Regina, Regina.

Ashworth, G. I., & Voogd, H. (1988). Marketing the city. *Town Planning Review, 59*(1), 65–79.

Audertsch, D. B., & Feldman, M. P. (1996). R & D spill-overs and the geography of innovation and production. *The American Economic Review, 86*(3), 630–640.

Azua, J. (2006). Bilboa: From the Guggenheim to the knowledge city. In F. J. Carillo (Ed.), *Knowledge cities* (pp. 97–112). Amsterdam: Elsevier.

Baker, B. (2012). *Destination branding for small cities*. Portland: Creative Leap.

Beaverstock, J. V., Smith, A. G., & Taylor, P. J. (1999). A roster of world cities. *Cities, 16*, 34–52.

Beresford, M. (1965). *New towns of the middle ages*. London: Lutterworth.

Bettencourt, L., Lobo, J., Helbing, D., Kuhnert, C., & West, G. B. (2007). Growth, innovation and the pace of life in cities. *Proceedings of National Association of Sciences of the U.S.A., 104*(17), 7301–7306.

Binelli, M. (2012). *Detroit city is the place to be*. New York: Holt.

Bounfour, A. and Edvinson, L. (Eds). *Intellectual Capital for communities, nations, regions and cities*. Oxford and Waltham, Mass: Elsevier Butterworth-Heinemann.

BP: Broedplaatsen. http://bureaubroedplaatsen.amsterdam.nl/en/index.php. Accessed 12 July 2013.

Brunt, L., Lerner, J., & Nichols, T. (2011). *Inducement prizes and innovation*. Harvard Business School, Working Paper 11.118.

CAM: Cluster Audio-Visual Munich. http://www.cam-bayern.de. Accessed 12 July 2013.

Capello, R. (2013). Territorial patterns of innovation and economic growth in European regions. *Growth and Change, 44*(2), 195–227.

Carillo, F. J. (2004). Capital cities: A taxonomy of capital accounts for knowledge cities. *Journal of Knowledge Management, 8*(5), 28–46.

Carillo, F. J. (2006). *Knowledge cities: Approaches, experiences and perspectives*. Amsterdam: Elsevier.

Clark, J. (2013). *Working regions: Reconnecting innovation and production in the knowledge economy*. Abingdon Oxford: Routledge.

Cooke, P. (2001). Regional innovation systems: Clusters and the knowledge economy. *Industrial and Corporate Change, 10*(4), 945–974.

Cooke, P., & Morgan, K. (1998). *The associational economy: Firms, regions and innovation*. Oxford: Oxford University Press.

Cooke, P., & Schwarz, D. (Eds.). (2007). *Creative regions: Technology, culture and knowledge entrepreneurship*. Abingdon: Routledge.

CT: Communitech. http://www.communitech.ca. Accessed 10 Sept 2013.

Dahlman, C. J., Routti, J., & Ylä-Anttila, P. (2007). *Finland as a knowledge economy*. Washington DC: World Bank Institute of Development Studies.

Dinnie, K. (2011). *City branding*. London: Palgrave.

Drewett, R., Knight, R., & Schubert, U. (1988). *The future of European Cities: The role of science and technology* (FAST-Monitor. Prospective Dossier No. 4, DG XII). Brussels: European Commission.

Drucker, P. F. (1993a). *Post-capitalist society*. New York: Harper.

Drucker, P. F. (1993b). *Innovation and entrepreneurship*. New York: Harper Business.

Edgington, D. W. (2008). The Kyoto research park and innovation in Japanese cities. *Urban Geography, 29*(5), 411–451.

Edvinsson, L. (2006). K-City and society entrepreneurship for intellectual capital growth. In F. J. Carillo (Ed.), (pp. 59–73). Knowledge cities, Amsterdam: Elsevier.

Edvinsson, L., & Malone, M. S. (1997). *Intellectual capital: Realizing your company's true value by finding its hidden roots*. New York: Harper Business.

Edvinsson, L., & Stenfelt, C. (1999). Intellectual capital of nations for future wealth creation. *Journal of Human Resources, Costing and Accounting, 4*(1), 21–33.

Eppig, C., Fincher, C. L., & Thornhill, B. (2010). Disease and intelligence. Proceedings of Royal Society B. http://www.royalsocietypublishing.org/content/eob/2010/06/29/rspb2010.0973. Accessed 12 July 2012.
Eppig, C., Fincher, C. L., & Thornhill, B. (2011). Parasitic prevalence and the world-wide distribution of cognitive abilities. *Proceedings of Biological Sciences, 277*(Dec.), 3801–3808.
Ergazakis, K., Metaxiotis, K., & Psarras, J. (2004). Towards knowledge cities: Conceptual analysis and success stories. *Journal of Knowledge Management, 8*(5), 5–15.
Ergazakis, K., Metaxiotis, K., & Psarras, J. (2006). An emerging pattern of successful knowledge cities' main features. In F. J. Carillo (Ed.), *Knowledge cities* (pp. 3–26). Amsterdam: Elsevier.
Feldman, M. P. (1994). *The geography of innovation*. Boston: Kluwer Academic Publishers.
Florida, R. (2002). *The rise of the creative class*. New York: Basic Books.
Flynn, J. R. (1987). Massive IQ gains in nations. *Psychological Bulletin, 101*(2), 171–191.
Govindarajan, V., & Trimble, G. (2010). *The other side of innovation: Solving the execution challenge*. Cambridge: Harvard Business Press.
HH: Haushalten. http://www.haushalten.org/de/english_summary.asp. Accessed 14 July 2013.
Hilpert, U. (1992). *Archipelago Europe: Islands of innovation*. Brussels: European Commission.
Hodgson, G. M. (1996). Varieties of capitalism and varieties of economic theory. *Review of International Political Economy, 3*(3), 381–434.
Holden, M., & Connelly, S. (2004). *The learning city*. World Urban Forum, Simon Fraser University.
Howells, J. R. L. (2002). Tacit knowledge, innovation and economics. *Urban Studies, 39*(5), 871–884.
Jaffe, A., & Lerner, J. (2004). *Innovation and its discontents: How our patent system is endangering innovation and progress*. Princeton: Princeton University Press.
Johnson, S. (2011). *Where good ideas come from: The natural history of innovation*. New York: Penguin.
Kasada, J. D., & Lindsay, G. (2011). *Aerotropolis*. New York: Farrar, Straus and Giroux.
KCN: Knowledge City Network. http://www.knowledgecities.com. Accessed 14 July 2013.
Kearsley, G. W. (1995). Changing urban images and economic transformation in Dunedin, New Zealand. In M. Palomaki & J. A. Karunaratne (Eds.), *Urban development and urban life*. Vaasa: Acta Wasaensia.
Knight, R. V. (1995). Knowledge-based development: Policy and planning implications for cities. *Urban Studies, 32*(2), 225–260.
Komninos, N. (2002). *Intelligent cities*. London: Spion Press.
Lackhani, K. R., & Lepeeson, L. B. (2007). Getting unusual subjects to solve R & D puzzles. *Harvard Business Review, 85*(5), 30–33.
Lambooy, J. (2002). Knowledge and urban economic development: An evolutionary perspective. *Urban Studies, 39*(5–6), 1019–1035.
Landry, C. (2005). *The creative city*. London: Earth Scan.
Lerner, J. (2009). *Boulevard of broken dreams*. Princeton: Princeton University Press.
Lever, W. F. (2002). Correlating the knowledge-base of cities with economic growth. *Urban Studies, 39*(5), 859–870.
Liker, J. (2004). *The Toyota way: 14 management principles from the world's greatest manufacturer*. New York: McGraw-Hill.
Lynn, R., & Vanhanen, A. (2006). *I.Q. and global inequalities*. Augusta: Washington Summit Publications.
Mackun, P. (2013). Silicon Valley and Route 128: Two faces of American technopolis. http://www.netvalley.com/silicon_valley/Silicon_Valley_and_Route_128.html. Accessed 20 July 2013.
Malecki, E. J. (2002). Hard and soft networks for urban competitiveness. *Urban Studies, 39*(5), 929–945.
Mallaby, S. (2010). The politically incorrect guide to ending poverty. The Atlantic, July–August. http://www.theatlantic.com/magazine/archive/2010/07/the-politically-incorrectguide-to-ending-poverty/3081341. Accessed 20 June 2013.
Marshall, A. (1916). *Principles of economics* (Revised edition). London: Macmillan.
Martell, S. (2012). *Detroit: A biography*. Chicago: Chicago Review Press.

Maskell, P. (2001). Towards a knowledge-based theory of the geographical cluster. *Industrial and Corporate Change, 10*(4), 921–943.

Matthiessen, C. W., Schwarz, A. W., & Find, S. (2006). Global research centres: An analysis based on bibliographic indicators. In F. J. Carillo (Ed.), *Knowledge cites* (Chap. 3, pp. 31–41).

Mazzucato, M. (2013). *The entrepreneurial state: Debunking the private v public sector myths*. London: Anthem Press.

Melkin, E. (2012). Plan for charter city to fight Honduran poverty loses its initiator. *New York Times,* Sept. 30. http://www.theatlantic.com/magazine/archive/2010/07/the-politically-incorrect-guide-to-ending-poverty/308134/.

MInd: Milan Network for Design. http://www.milanetworkdesign.it. Accessed 10 July 2013.

Morgan, K., & Nauwelaers, C. (2002). *Regional innovation strategies*. London: Routledge.

Murayama, Y. (2002). Changing spatial structure of the Toulouse metropolitan area under European Union integration. In W. K. D. Davies & I. J. Townshend (Eds.), *Monitoring cities*. Calgary: International Geographical Union, pp 213–225.

Musterd, S., Brown, J., Lutz, J., Gibney, J., & Murie, A. (2010a). *Making creative-knowledge cities*. Amsterdam: AMISSR-Amsterdam Institute for Social Science Research.

Musterd, S., & Murie, A. (2010b). *Making cities competitive*. Oxford: Wiley-Blackwell.

O.E.C.D. (1999). *Science, technology and industry scoreboard: Benchmarking knowledge-based economies*. Paris: O.E.C.D.

Okubo, K. (2012). Technopolis: Towards realization. *Journal of Information Processing and Management, 29*(6), 471–480.

O'Mara, M. P. (2005). *Cities of knowledge: Cold war science and the search for the next Silicon Valley*. Princeton: Princeton University Press.

Polyani, M. (1966). *The tacit dimension*. Chicago: Chicago University Press (Reprinted with new introduction by A. Sen, 2009).

Porter, M. (1990). *The competitive advantage of nations*. London: Macmillan.

Richardson, H. W. (1973). *The economics of urban size*. Westmead: Saxon House.

Rodwin, L. (1971). *Nation and cities*. New York: Houghton and Mifflin.

Romer, P. http://paulromer.net/charter-cities-new-cities-more-choices-better-rules/. Accessed 12 Nov 2013.

Rosegart, S., & Lampe, D. R. (1992). *Route 128: Lessons of Boston's high technology country*. New York: Basic Books.

Rosen, W. (2010). *The most powerful idea in the world: A story of steam, industry and invention*. New York: Random House.

Rosenberg, N. (1976). *Perspectives on technology*. Cambridge: Cambridge University Press.

Senor, D., & Singer, S. (2009). *Start-up nation: The story of Israel's economic miracle*. New York: Hachette.

Simmie, J. (2002). Knowledge spill-overs and reasons for the concentration of innovative SMEs. *Urban Studies, 30*(5), 885–902.

TE. (7 August 2010). And the winner is. *The Economist*. http://www.economist.com/node/16740639. Accessed 10 Nov 2011.

TE. (2013). Start me up. *The Economist,* Oct. 5, 60–61.

Van Geenhuizen, M., & Nijkamp, P. (Eds.). (2012). *Creative knowledge cites: Myths, visions and realities*. Cheltenham: Edward Elgar.

Vance, J. E. (1966). *A geography of wholesaling*. New Jersey: Prentice-Hall.

Viedma, J. M., Bounfour, A., & Edyinsa, L. S. (2001). Remote electronic access and innovation. *Journal of International Business Studies, 32*(4), 667–686.

Viedma Marti, J. M. (2005). Cities' intellectual Capital Benchmarking System (CICBS): A methodology and a framework for measuring and managing intellectual capital of cities, a practical application in the city of Mataró. In: A. Bounfour and L. Edvinson (Eds), (pp. 317–335). *Intellectual Capital for communities, Nations, regions and cities*. Oxford and Waltham, Mass: Elsevier Butterworth-Heinemann.

Zaheer, S., & Manrakhan, S. (2001). Remote electronic access and innovation. *Journal of International Business Studies, 32*(4), 667–686.

Chapter 12
Safe Cities and Communities

Wayne K.D. Davies

The system is fractured...governments, courts and the police cannot do it alone.
Alberta Justice 2007, p. 5

12.1 Introduction

Safety is one of the vital human requirements. In Maslow's (1943) classic hierarchy of human needs it occupies the second level after hunger, thirst and housing but before the social, esteem and self-actualization needs of people. Since few people can provide or afford their own protection, one of the most important duties of governments has been to try and ensure the safety and the security of the population. Although these two terms are often used interchangeably, an often used conceptual distinction describes 'safety' as protection from threats from within the society, such as criminality that involves violence or loss of possessions, while 'security' is more commonly applied to threats to people from war, terrorism or environmental disasters, mainly, but not exclusively, from outside causes. Both are needed to maintain the quality of life of residents, their ability to move around unmolested, to generate and keep part of the wealth they have created to satisfy their own needs and enjoyment, and even to avoid death or severe injury. So safety with low crime rates and security from outside threats represent basic requirements for the continued prosperity of urban places and even their survival. Although safety considerations are important for the day-to-day life of citizens it is the problems that created insecurity that have been the main causes for the destruction of cities throughout history. This is why the decisions to wall towns goes back to the dawn of urban life, based on the need to provide security against outsiders and protection of its economic assets. Within cities, local crime was reduced by local policing and court systems. But by the eighteenth century advances in the effectiveness of cannons had rendered walls obsolete. The growth of the modern state led national governments in most

W. K.D. Davies (✉)
Department of Geography, University of Calgary, 2500 University Drive, N.W., Calgary, AB T2N 1N4, Canada
e-mail: wdavies@ucalgary.ca

W. K.D. Davies (ed.), *Theme Cities: Solutions for Urban Problems*, GeoJournal Library 112, DOI 10.1007/978-94-017-9655-2_12

countries to take on the main burden of providing security against outside threats through standing armies, while safety for the citizens was based on creating more effective laws, policing and judicial systems to fight against local crime. Although many of the larger municipalities in some countries are empowered to manage their own police forces, it is the state, not the local government, that exercises final control through legislation either passed by provinces or states in federal countries, or by the national state in unitary political regimes; nevertheless, the governments and citizens of urban places still have a role to play in keeping crime under control in their jurisdictions. This review will focus primarily upon the safety problems and range of policies used by cities of the developed world in the never-ending fight on crimes against people or possessions, which has led to the development of what have been called Safe City strategies in many countries during the past 20 years, with similar ideas applied to areas within cities, leading to Safe Community policies. Many of these policies may be applicable in other countries, but the continued existence of corruption, coercion and limited civil rights in some of the less developed states, as well as totalitarian regimes, mean that the police and judicial systems are not trusted by large numbers of the population and often contribute to, or largely ignore, crime. Hence the fundamental issue of reforming these agencies and making them effective have to be dealt with first, before many of the policies described here can be successfully implemented. Space constraints also mean that the discussion mainly focuses on the role that local governments can play in reducing crime, so only brief sections on societal-based actions to improve safety and security issues are provided. Most of these problems are primarily dealt with by national or state governments, although there may be local, urban components.

12.2 Changing Levels of Crime

Throughout history, urban places have usually had higher levels of crime than other areas and with their increasing size, density and heterogeneity even greater escalations in these rates have occurred. The reasons for high urban crime rates are straightforward. The wealth of urban places provides attractions and opportunities for criminals. The greater social disorganization, individual alienation, anonymity, reduction in community linkages and mobility of people increases the propensity to crime among some people and made it easier for those who committed criminal acts to escape detection. All of these features are reinforced by the development of subcultures in urban populations with behaviours that would be considered deviant by the majority (Turk 1997). In these groups crime becomes a way of life for those who make the decision to pray on others, or by those who cannot gain access to legitimate employment. Not surprisingly, urban citizens and governments have always worried about how to control, or at least reduce, crime, in order to maintain the safety of its citizens. But the explosion in the size of urban places, the lack of building and health standards and their inequalities after the Industrial Revolution exacerbated the safety problems. These conditions led to an increase in crime and the creation of specific crime areas in the largest nineteenth century cities as pre-existing policies to control crime proved ineffective. Areas dominated by people of criminal behaviour were vividly described as 'Rookeries' in the novels of Charles Dickens (Tomalin 2011); they were essentially 'no-go' areas that outsiders penetrated on pain of their life. From the mid-nineteenth century onwards, many cities deliberately razed these densely populated, slum areas that had high crime rates, replacing them, as in the case of Haussmann's work in Paris, with wide boulevards and attractive apartments for the middle and upper classes (Mumford 1961). Yet more generally, crime, especially petty crime, was held in check by three main changes in the countries that became the known as the developed world. First, historic crime reduction was due to the development of more comprehensive policing and judicial policies from the 1840s, increasingly using professional and technical methods. Second, new prisons were built, many based on enlightenment principles designed to get away from the brutal prison standards of the past—in which torture and mutilation, the drastic punishment of small offences, even the transport of prisons to colonies, were typical. However there are great variations between countries in the extent to which these prisons are designed for retribution—creating harsh conditions as punishment for crime—or rehabilitation, the attempt to change the behaviour of offenders before they are released. A third factor of change lay in the general socio-economic and political changes in society from the mid-nineteenth century onwards that reduced the numbers of the criminal underclass and opportunistic offenders. These involved: more employment opportunities and improved work conditions and pay; universal education; as well as greater democratic rights and legal acceptance of individual rights; and new housing standards and improved sanitation, which eradicated or at least reduced most of the old slum areas.

These changes, with increasingly professionalised and technically advanced crime-fighting methods, seemed adequate to hold the rates of urban crime in check from the mid-nineteenth century. But from the mid-1970s to the early 1990s the amount of crime and anti-social behaviour recorded by police increased dramatically in most developed countries except Japan. However, the rates and types of crime varied from country to country, and from city to city as McClain (2001) has shown in her comparison of the United States and European countries. Some of the biggest post 1970s crime increases were experienced in America, as seen in Fig. 12.1. Violent crimes and property crimes rose four times between 1960 and 1990, trends that are also present in most other categories identified in the annual crime reports of the Federal Bureau of Investigation. The growth is dramatically shown in the spike in murders in New York City, from around 500 per year before 1960 to a peak of over 2,245 in 1990. Since crimes against persons and property have many causes it is not surprising to find that many different theories have been proposed to explain the phenomena (Georges-Abeyie and Harris 1980; Gottfredson and Hirschi 1990; Graham and Clarke 2001). In general, however, the general causes of crime can be summarized in terms of five sets of explanations, namely: deep-seated societal forces, group or subculture causes, explanations based on individual behaviour; areal factors; and those linked to short-term changes in society (Davies 2005). Although various criminologists differ in the extent to which they stress one set of factors over

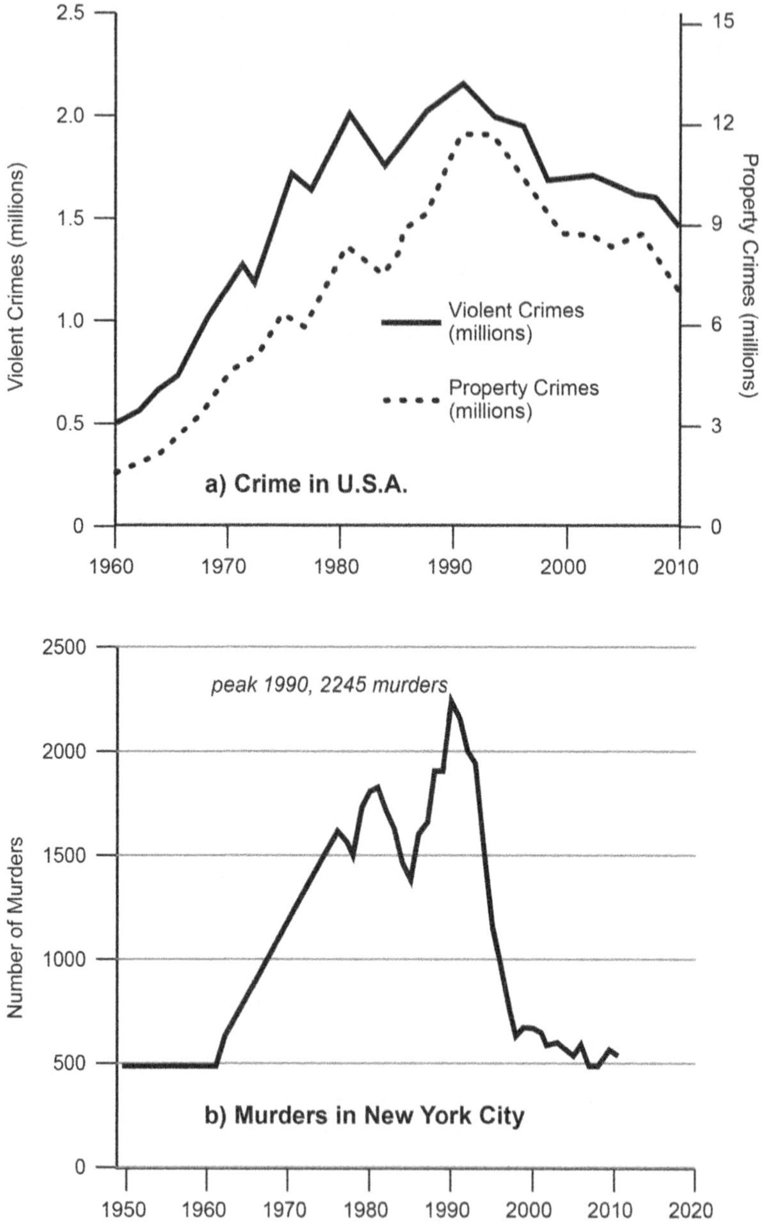

Fig. 12.1 Crime in the U.S.A. and murders in New York City. (Source: Revised from various F.B.I and City of New York Annual Crime reports)

another, and there are big differences in types of crime and crime numbers between countries (Herbert 2002; Racine 2002), there does seem a consensus that the rapid rise in crime rates was associated with several major sets of changes that affected societies and big cities in the developed world from the 1970s.

12.2.1 Reasons for Crime Growth from the 1970s

Historically the amount of crime, especially violent crime, has been disproportionally associated with societies that have unequal distributions of wealth, as meta-analyses of scores of studies have shown (Hsieh and Pugh 1993; Wilkinson 2004), meaning that countries with the most *inequalities* have higher crime rates, which are often concentrated in particular areas of cities. Many people in these unequal societies have few opportunities for employment or advancement, trust each other less and have limited community relations, leading to low levels of support or social capital among people, or even moral influence to resist crime. Yet the inequalities in western cities were reduced substantially by the progressive social policies of the late nineteenth century to late twentieth century so one might have expected crime rates to decline. From the 1970s the loss of blue collar employment in inner cities due to deindustrialization and mechanization led to their decay, creating dereliction in these areas and alienation by some from the population at large. In addition, neo-liberal policies in many countries reduced interest in increasing the amount of remedial help to the disadvantaged and those prone to crime.

Although these *social disorganization* ideas explain some of the increase in crime, the trigger for the substantial growth of crime was the corrosive effects of the accelerated use of *illegal drugs*, especially crack cocaine from mid 1980s, with supply often controlled by violent international gangs, leading many addicts to carry out crimes to obtain money to buy these addictive products. This is especially the case in the U.S.A. where half the prisoners in jail have been charged with drug offences. In addition, the increasing availability and use of *guns,* especially assault rifles in some societies, as well as a more widespread use of knives, has amplified the level of violence associated with many crimes.

The growth in the *young adult population* was another important factor, for crime rates, especially those associated with violence, are far higher in the young adult male population which increased in the two decades from the late 1960s. For example, young offenders (under 20 years of age) commit half of the violent crime in the U.K. and it has been reported that 10 % of adolescents who exhibit violent acts commit 70 % of the violent acts (Fonagy 2004, p. 181, 193), figures that are typical of many countries. Risk-taking, peer pressure for mischief and proving themselves, has always been present in the group. With some individuals these characteristics have turned into an oppositional culture that condoned criminal behaviour, not simply because of poverty and decreasing employment opportunities in inner cities due to deindustrialisation, but to the attraction of finding success and status—however temporary—through membership in gangs and in the drug trade.

Fifth, the spike in crime was probably also associated with the growth of a more *atomistic society*, with fewer positive social connections between people in local areas, creating higher levels of anonymity and less willingness to help others, which is particularly acute in unequal societies and in the disorganized, impoverished inner areas of large cities. This is a result of the greater mobility and individualization of modern life, helped by the breakdown of traditional positive socializing forces from the 1960s, forces such as family, schools, neighbourhood, and perhaps a decreased spirituality, for most religions promote moral standards. The result has been far more single parenting by poor people, fewer moral role models, and less trusting relationships between individuals. It has reduced the positive socialization between people that discouraged criminal behaviours in the past, and increased the propensity to criminality through participation in gangs, especially for youth. Gangs often encourage crime among members and in return provide a sense of allegiance, belonging, and perhaps personal esteem and a recognition that they cannot get from their family or neighbours.

A sixth factor stems from the growth in the numbers of people with *mental health problems*, either untreated or who were released after the reduction in hospitals and treatment centres in the 1980s—supposedly into the 'community' which was supposed to help them adjust. This community did not exist in practice—with the result that they swelled the ranks of the homeless, so many engaged in petty crime in cities. It is pertinent that a U.S. Department Justice study (La Vigne et al. 2006) reported that 16% of prisoners in jail had mental health problems. Another causal factor comes from the consequences of the erosion of local community linkages and the development of more *individualistic personal behaviours* that focus on the self (Taylor 1986; Fonagy et al. 2002). Although this has led many to greater self-awareness, confidence and less deference to authority, such a focus without socialization into a consideration for others often leads to increasing levels of incivilities and the breaking of existing mores, from carelessness about littering, jostling others, playing loud music, defacing buildings and environments, to a range of related anti-social behaviours. These behaviours may not be criminal initially, but if not restrained often escalate into criminal actions. Even without the escalation, they create feelings of unease or even fear to those exposed to these actions, creating a higher perception of crime. Linked to this lack of respect for, or concern about others, has been an increase in the levels of gratuitous violence which results in a lower moral consciousness about the effect of violence, or stealing. In addition, Pressdee (2000) has described what he calls the 'carnival of crime' in which many people have a rational, ordered life in the week, but some engage in anti-social or criminal behaviour on holidays or weekends to escape the constraints of their normal life, focusing upon their own needs for excitement and ignoring the consequences of their behaviour on others. An eighth factor comes from the initially *ineffective responses from police forces* from the 1960s, where the abandonment of foot patrols in favour of cars reduced links with residents of communities, especially racial minorities, and the ability of the police to understand their problems and crime potential.

Although the spike in crime from the 1970s has been enough to worry citizens and governments, there can be little doubt that the increased *perception of crime* also increased the level of insecurity in the population, especially among the elderly and minority groups (Gray et al. 2008). For example, by the early 1990s, it has been reported that over two-fifths of the population in both the U.S.A. and in Scotland were afraid to walk alone at night (Source Book of U.S. Criminal Justice Statistics

Online, McClain 2001). Among the broader background causes of increased perception of crime, apart from exposure to its increase, has been the *rising safety expectations* by the population and their lower tolerance of risk; it is assumed that others—especially police forces—will solve all problems, rather than accepting that there is also a personal responsibility to take protective measures. In addition, there was also a decrease in the degree of trust in the police and other agencies, such as local government, as well as neighbours, to provide protection or help in time of need. In some countries police forces are often regarded with fear, given the corruption of the forces and their ability to coerce local populations, which means some police are either responsible for a great deal of violations, or are in league with criminal elements, allowing the latter to flourish without intervention. The enormous increase in the number of *TV crime series* has also probably had an effect on the increased perception of crime, for some apply these images to their own area—without evidence. In a more local context it has been shown that the perception of crime, especially an individual's worry about his or her safety, varies with *areal familiarity*, the degree to which a person is familiar with and comfortable in an area. Feelings of being lost, or being in an unfamiliar place, generates at least anxiety and unease, if not outright worry about safety, especially if the local people are of distinctively different backgrounds, as seen in dress and behaviour. Hence, fear of the unknown, may come from the area itself, or the inhabitants, whose unfamiliar customs and racial characteristics may make them appear threatening. Even the behaviour of groups of young adults often evoke the same emotions. In addition, it has been shown that most people have greater fear of their safety in *badly maintained areas*, places that are dirty, unkempt and polluted, and which contain abandoned or poorly kept buildings, features that increased in many inner city areas because of deindustrialization processes. By contrast, clean, more attractive urban environments and especially public spaces, in which with plenty of people of similar origin or class are found, reduces the fear of crime (van den Berg et al. 2006).

By the 1990s all these factors increased the perception of crime and when added to the fact of higher crime rates led to increasing demands for political action to solve the problems. Indeed the scale of the problem led many to believe that the existing system of crime prevention, especially in cities, was broken and that new approaches to reducing crime needed to be sought, in which more attention should be paid to applying new policies to strike at the *reasons* for crime, rather than only dealing with the *results* of crime. Yet there was certainly a need to be more effective in solving crime by applying new methods

12.2.2 Decreases in Crime from 1990s

At the same time that greater political attention were being focused on ways of fighting the huge increase in crime and heightened perceptions of fear of crime in the 1990s, the statistics on crime in most developed countries began to show a decrease in most categories. Indeed, Fig. 12.1 showed that violent crime and property crimes in the U.S.A. have fallen to levels not seen since the early 1970s. Although there are great variations between cities, it is remarkable that the annual number of murders in New York City declined from 2245 in 1990, to 536 in 2010. More generally, crime in cities in America has fallen by almost a third since 1990 and almost two-thirds in the biggest cities, although not all have seen this drop. This trend seems common in most western countries, although one must be cognizant of variations in the timing between countries. For example the fall in the United States began around 1991, whereas it was 5 years later in Britain. Also there are differences in the rates of decline, as shown in Fig. 12.2 for three categories of crime in the G7 countries between 1995 and 2010, although the downward trend is similar (TE 2013). Such trends in the developed world contrast with the persistence of very high crime rates, especially violent ones, in many middle and less developed countries, as in many parts of Mexico due to the still unsuccessful 'war' against the drug cartels. Yet one must not be too sanguine about the declines seen in many crime categories in the past 15 years, since levels are still as high as the 1970s, and the fear of crime has not decreased in the same way. No consensus has emerged to

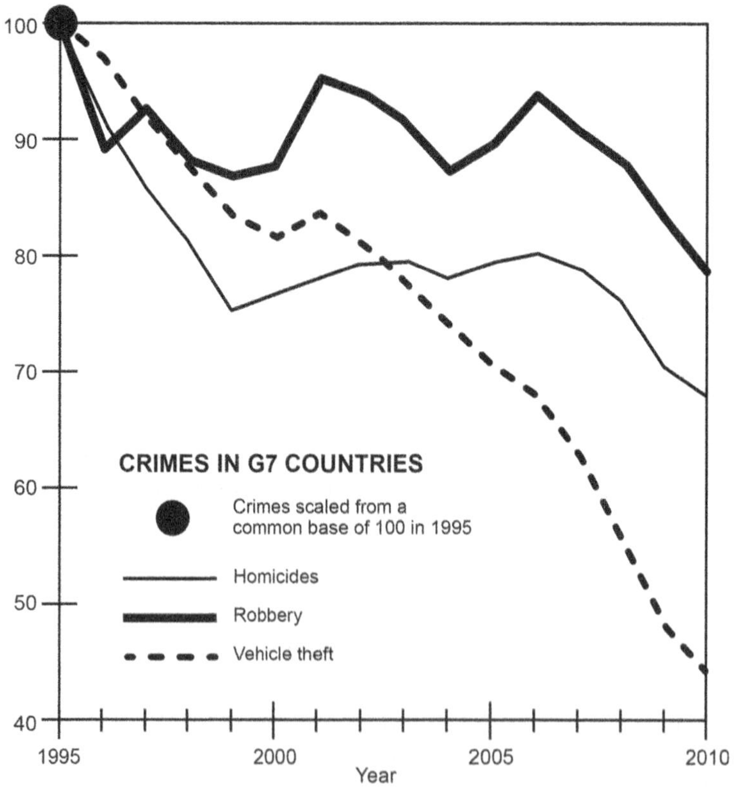

Fig. 12.2 Drop in crime: selected categories in G7 countries. (Source: Revised from TE July 20, 2013)

explain these new patterns of falling crime rates in most developed countries but a combination of eight features seem to account for the changes.

Those of a conservative persuasion have argued that the decreases are a positive consequence of the more repressive policing and judicial measures adopted from the late 1970s, in which more *offenders have been incarcerated* and therefore unable to engage in new crime. Although this may be part of the explanation, many of those being jailed are for non-violent offences, especially associated with the drug trade, and there still seems to be a large reservoir of people in areas of disadvantage and segregation, who are still prepared to engage in criminal activities since they have few good employment opportunities. A more influential reason for the decrease comes from a general decline in the *young adult* population, which is traditionally responsible for a large amount of the crime, although there are often large numbers of youths in cities. Also more of this group are likely to be in education or work and still living at home than when crime spiked, so fewer seem likely to engage in the extreme behaviours of their 1970s counterparts, while there has also been a decline in the use of crack cocaine which caused such problems two decades ago. This decline in the age group most likely to offend is primarily a result of declining fertility levels in developed countries. But an important factor in this trend has been the *ease of abortion* in most countries especially for poor, single mothers. This has reduced the number of children born into poverty and unstable childhoods, who are not always able to get the benefit of positive nurturing and socializations. In addition, the *gentrification* of many inner city areas, as a result of the increasing prosperity of larger cities in particular in the developed world, has led to a social displacement of some of the people more likely to engage in crimes against persons and property. Another explanation comes from the intriguing analysis by several researchers of the effect of what amounts to *lead poisoning*, since it has been shown that lead in gasoline correlates with the growth and then decline in crime after it was phased out from 1984 in US and in Europe from 1990s (Nevin 2000; Reyes 2007; and Markus et al. 2010). The link seems to be that lead hinders brain development, especially in growing children, by impeding the neural pathways that reduce aggression levels, so that those exposed to lead exhibit more delinquency and crime, lower IQ's and more teen pregnancies. The research has shown there was a greater effect on males, but also a higher incidence in poor black children in inner cities where there was greater exposure to traffic exhausts and to old leaded paint from crumbling buildings. This has led some to the conclusion that the environmental change in banning lead made good social policy (Reyes 2007; Monbiot 2013). Since lead has long been known to be toxic for many organs, especially the heart and kidneys, it has increasingly been phased out from its older use in paints.

Another persuasive argument is that there are *fewer easier opportunities* and rewards *for crime*, since many households, and especially businesses, have taken more precautions, installing more alarm systems, more secure safes and CCTV cameras, while cars, a big target in the 1980s, have more secure systems. In any case the rewards from household break-ins in particular are far less, given the lower relative costs of so many electronic items and their limited resale value as new models emerge. This reason accounts for the big drop in burglary, a crime that also takes time to plan. In addition, it is also likely that there is a *greater risk of being caught*. This is a result of several factors: the increasing number of private security guards who now outnumber police in many countries; the use of new police technologies and smarter tactics, issues that are described in subsequent sections. This is linked to the application of *more preventative policies*, especially with the type of Safe City strategies described below, which are also starting to have an effect, especially in the past few years. Yet these decreasing recent crime trends should be placed in perspective; they are still only declines from the very high rates of crime that peaked in the early 1990s and the costs of crime keep growing. Also one must remember the on-going safety problems caused by the actions of long lasting criminal organizations in many countries, such as the various Mafia organizations in America and Italy, Corsican gangs in France, South American drug cartels or the Yakuza in Japan—problems that have rarely been solved. So most commentators believe that crime levels need to be drastically reduced before the public begins to feel substantially safer, especially in the high crime areas, such as inner cities and many public housing estates. Despite all these changes that account for lower crime there is far less public confidence in the ability of the current police-judicial system-custodial system in developed countries to achieve the goal of reducing current crime rates, mainly because of deficiencies in the operation of its various parts.

12.3 Problems of the Current Safety System in Cities

The typical approach to crime prevention over the last two centuries has been to rely on the traditional three-fold approach represented by: the policing of urban areas, followed by the investigation and arrest of people who committed crimes; the judicial review, which involves the decision of courts about the guilt or innocence of the accused about the evidence presented about the crime; and the punishment of those found guilty, which may involve a custodial sentence. Most people in democratic countries, apart from those opposed to what they see as the oppressive apparatus of government, see the system as mainly, but not completely corruption-free. Certainly there are scandals involving corrupt or over-violent police, biased judges and political interference, as seen by a recent exposé of a rogue narcotics squad in Baltimore who robbed small grocery stores and fabricated evidence as well as stealing drugs (Ruderman and Laker 2014), or the devastating report by Judge Macpherson (HO 1999) on the inadequacies of the London Metropolitan Police in investigating the murder of a young black teenager, Stephen Lawrence, with the police accused of 'institutional racism among many other charges. But in most cases in the developed world judicial reviews of these problems leads to the implementation of new procedures to prevent their recurrence, although many minority groups, especially if coloured, still have many unresolved grievances at the way that they are treated. This is not to deny that the problems of poor police behaviour are ever completely solved; there is always the need to be vigilant against corruption, against police intimidation, aggression, the sentencing of the innocent, or the failure to find, or to convict offenders, or solve problems posed by criminal gangs. In the case of the judicial systems there are many different structures in various developed countries, but there is a general expectation that justice officers, like the police, must be held accountable for illegal actions, over-coercive behaviour, or decisions that deny people their civil rights. Sadly, this is not the case in most countries throughout the world; the police and security services are usually feared and seen as part of the problem of safely in the city, since individual officers may seek bribes, are in the pay of criminals, or act as enforcement agencies on behalf of a coercive state. Similarly, corrupt judges, or deliberate political interference, often make a mockery of the fairness of the judicial system. But even in countries where such problems have largely been reduced, by the development of what a majority may see as fairer and less biased policing and judicial systems, the last 20 years has seen increasing dissatisfaction with the crime fighting system's effectiveness. Four sets of issues account for this situation: problems in defining and measuring crime; problems in policing, deficiencies in the judicial systems; discontent about how the consequences of crime are dealt with.

12.3.1 Problems in Defining Crime

At first sight the fact of a crime, a burglary, a murder or car theft, seems easy to measure. But apart from these examples many types of what are seen by a society as criminal behaviour are less easy to conceptualise, which causes severe problems in the compilation of statistics of crime and hence the public's perception of crime levels. The reason is that crime is not a rigid and unchanging category of behaviour. We need to be aware of the *social construction* of crime, because a crime is a result of the state's decision as to what constitutes a crime (Davies 2005). What is a crime in some jurisdictions (such as abortion in the Irish republic or rape in a marriage in most western countries) is not criminalised in others, has been decriminalised in others, or reduced in the level to which punishment applies.

Figure 12.3 shows it is possible to recognize a hierarchy of approaches in the attempt to measure crime, which creates a sequence of crime reporting. It runs from the way crime is defined: the extent of criminality disposition in a population; through anti-social behaviours—those that cause unease or distress to other—but are not criminalised; to the actions that are criminalized by the laws of a nation—but which vary considerably between national jurisdictions; to crime events (as well as failures); to the fact that not all crimes are reported to police; to the even smaller numbers that are recorded; to crimes subsequently investigated; and those that are solved in the sense that people are charged; to the numbers of the people that are convicted and those that are punished.

Each of these stages involves a reduction in the amounts of crime recorded and eventually solved. This crime recognition sequence illustrates that that there is a

1. **Social Construction of Crimes:**
   Behaviours legally identified & punishable as crimes;
   they are culturally, spatially, temporally variable.
   (Incivilities or anti-social behaviour are not normally crimes).
2. **Pre-Disposition to Criminality. ------Variations by: type of crime/areas/people**
3. **Acts of Crime: a) Attempts made and failed.**
   b) Crime events.
   -------Variations by: type of crime/areas/people.
4. **Crime Reported ----------------------------------Variations as above.**
5. **Crime Recorded by Police --------------------Variations as above.**
6. **Crime Investigated ------------------------------Variations as above.**
7. **Crime Solved ---------------------------------------Variations as above.**
8. **Criminalisation of Individuals, Groups ------Variations as above.**

↑*All the above also vary by time*↑

Fig. 12.3 Stages of interpreting crime

big difference in what is considered to be the fact of a crime, and its identification and solution. Indeed, it was estimated by Radzinowicz and King (1977) that approximately 80 % of crime goes undetected and unpunished, although there are great variations across crime categories. One of the lowest offence-detection rates is seen in the case of rape. Only 5 % of those accused in court in the U.K. for rape are ever convicted, while the underreporting of the crime is seen by the fact that surveys have shown that only one in seven rape victims ever report the crime. In many countries even fewer rapes or reported or even upheld. This type of crime may be an extreme example of under-reporting, but illustrates the problem of crime measurement. Through time the problem of effective measurements becomes worse, since there are also variations in all these stages and the behaviours that are criminalised vary. Spatially, the incidence of crime also varies considerably across various jurisdictions, both within and between nations. In some cases there are alternative measurements of crime which throw light on the situation. For example, Britain, like most developed countries, publishes regular *statistics on the crimes recorded* by the police, but since different aggregations of crimes may be used, as well as differential recording by forces, they must be treated with a great deal of caution. Although they provide at least an initial approach to crime amounts, many believe that a truer picture of crime is obtained from the type of information annually published by the British Crime Survey (BCS), which records the extent to which surveyed people say they have *actually been a victim of crime*, which is further up the crime recognition sequence described above. For example, between 1981 and 2003 the amount of crime reported by the surveyed people in Britain rose from 36 to 44 % (BCS 2004). But of these crimes, the amount recorded by the police was only 62 % of the 36 % in the base year and 77 % of the 44 % in 2003. This means that only 22.1

and 33.9 % respectively of the crimes reported to the BCS surveyors were formally dealt with by the police. This does mean that there was a substantial increase in the reported-recorded crime difference, but it still leaves 78 % in 1981 and 66 % in 2003 of the crimes experienced unknown to the police and judicial system. In addition, even these figures do not record the extent of anti-social behaviour or incivilities to others, behaviours not criminalized but which are often a major cause of what makes people uneasy or even fearful of their safety in cities or city areas. In general, therefore, all comparative crime statistics, and certainly changes through time, need to be treated with prudence rather than as absolutes.

12.3.2 Policing

Initially, the response to the increase in crime in most nations and cities was to adopt more repressive crime-fighting methods, from more coercive policing, to stricter laws and longer prison sentences for offenders, resulting in major increases in the prison population. The last 30 years has also seen an increase in the number of private security companies, and the development in some countries of many new gated communities, some with guards, for middle and upper income families. In some ways this is a reversal back to pre-industrial standards of local area protection for those groups within cities who could afford it. In terms of policing there is general agreement that increasing the numbers of police does not necessarily produce a safer society, since it depends on how effective the police are. Even in democratic countries it is difficult to see a correlation between changing police numbers and crime reduction. Indeed there are major differences in the size of police forces; in Europe the numbers of police per 100,000 population for the bigger countries range from 525 and 457 in Spain and Italy to 258 and 298 in U.K. and Germany, with smaller countries showing lows of 152 and 200 in Finland and Denmark respectively (EuroStats: Crime 2013). More specifically, there has been increasing evidence of the ineffectiveness of some policing during the last quarter of the twentieth century when crime spiked. One reason is the increasing specialisation and bureaucratization of the police forces. Some of this is certainly needed to improve crime detection and evidence collection and processing. But a consequence has been that the police are far less visible on the streets, since office work takes up more and more time, rather than actual crime prevention. Not surprisingly, many argue that the police have become too separate from the public at large and they do not respond quickly enough to any but the most serious crimes, such as murder. Also the failure of most police forces in countries that have experienced immigration or have minority races to add people of different ethnic origin to their organizations has led to tensions between these communities and the police who are not representative of the changing population. However in the last decade determined efforts have been made in some countries to add more people of colour and different backgrounds to police forces. In addition, there are often limited links or even trust between the different levels of police forces in a country, or between the police and other agencies, especially the rehabilitation arms of various welfare agencies. This led to rivalry and limited sharing of information between neighbouring police forces or with other agencies, reducing the opportunity to identify potential offenders and to prevent crime, especially from repeat offenders or those that moved to other areas. Finally, as noted above, that there have been many examples in democratic countries where police have been shown to have used excessive violence or have covered up their own misdemeanours or even crimes, which has led to decreasing trust in the police force. In most western and industrialized countries there have been greater attempts to ensure these problems are eradicated through better training, supervision and independent reviews of problems. However, it seems to be an unfortunate fact that in most many states of the world the police are not trusted to be fair and are part of the problem of the absence of real safety for the population.

12.3.3 Judicial System

The judicial system also varies in different countries. But whatever the system, there has been increasing criticism of the results of the judicial system, whether in the prosecutorial stage, the defence system or the judicial review (Turk 1969). Staff shortages and limited funding in most developed countries have led to longer procedural delays in bringing a case against a person charged with crime. This creates a large gap in time between the crime and the consequences imposed by the court, ensuring that many people changed spend far too long in custody before being judged, making it difficult to provide a meaningful impact between the offence and the consequences, whether for the offenders or the victims. In addition, some argue that there is too much emphasis upon the rights of the accused, and not enough on the problems created for the victims of crime, or the provision of protection for witnesses who are often afraid to testify given fear of retribution, especially when criminal gangs are involved. A particular problem in the public consciousness comes from appears to be a 'revolving door' justice system, where some offenders are not found guilty because of technicalities in the court, or are given too short sentences, or are released and re-offend. There has also been a great increase in the use of mandatory sentences for crimes, especially for drug offences in the U.S.A., rather than allowing the judges to exercise discretion in decisions, resulting in an enormous increase in the numbers incarcerated. Yet prison often stigmatises for life those convicted of relatively minor offences. Many, especially first offenders, are socialised into criminal behaviour during their incarceration, setting them on a crime life-course, instead of prison being a period in which rehabilitation was provided.

12.3.4 Consequences of Crime

There are huge differences between countries in the use of custodial sentences to punish offenders. Although there has been a substantial increase in prison sentences in most countries, the incarceration rates in the United States are astonishing, a direct and seemingly publically supported response to the increase in crime (Useem and Piehl 2008). The prison population in the U.S.A. increased from 110 per 100,000 people in 1975, a level that had not fluctuated much for decades, to 476 prisoners per 100,000 population in 2005. Indeed the total prison population reached 2.31 million in 2008, with a slight decline since, but this is still 1 prisoner per 107 adults, about five times the world average, with prisons on average at 30% over capacity, and an estimated one in six of the inmates having some mental illness. Also the numbers of people on parole or probation rose from 1 to 5 million between 1980 and 2005. Other countries have also shown increases in the number of inmates but with much lower levels of incarceration. Few doubt the need to give prison sentences to the incorrigibly violent, the socio-paths and those who committed horrendous crimes against people, though it is dubious that such offenders should be in same place as those who carried out less serious offences. Although it has often been argued that the increase in repressive policies has worked in keeping criminals off the streets, almost half of the prisoners in U.S. federal prisons are there for drug offences, not violent behaviour. Many were guilty of small amounts of drug possession, but are in prison for life because of the high rates of re-offending and the 'three offenses and prison for life rule' adopted in many states of the U.S.A. This is a rule that is being increasingly seen to be unjust and has been a major reason for the increase in the prison population, leading to recent moves to remove the statute in some states. Similarly, problems are seen in the use of mandatory sentences, allowing judges little discretion in their decisions, or the limited use of alternatives, such as treatments not coercive punishment.

Another problem comes from the fact that prison populations are often dominated by minority groups, such as the black population in the U.S.A. and aboriginals in Canada. Without condoning the crimes to which they have been charged, there can be little doubt that the over-representation of these groups means that something has gone wrong in their socialization and education and ability to find lawful employment. This essentially repressive approach has led to the creation of a brutalizing and violence ridden society in many prisons, with limited rehabilitation of offenders. These conditions have had the effect of providing a socialising and learning environment with other offenders, one that often persists when they are released from prison, increasing the propensity to violence of many ex-prisoners and their level of alienation from the rest of society. Since there are few rehabilitation programmes it may not be surprising to find that a U.S. Department of Justice report (La Vigne et al. 2006) revealed that two-thirds of prisoners reoffended within 3 years and most went back to prison, while in the U.K. 58% were reconvicted within 2 years, with rates of 76% for young males (HMPS 2002). Although rates of recidivism do vary between countries these high rates of re-offending persist despite the presence of probationary services in most countries that are supposed to monitor people released from prison, mainly for good behaviour and remorse. Unfortunately limited money and trained-staff shortages to provide rehabilitation are typical in most countries. So if the Enlightenment's idea of prison, then subsequent rehabilitation, was to deter and re-educate, then the current system is seen by many as being ineffective in achieving such goals.

Some of the responsibility of the current problems lies with the ideological differences between those who see the consequences of crime as punishment, and those who take a more nuanced view and argue for greater rehabilitation. Most of the conservative governments from the 1980s have argued for the former approach, in which money allocated to the crime agenda has supported a law and order approach, rather than also increasing the amount of money or resources to try and re-educate offenders. As prisons have become ever more crowded and increasingly expensive there has been less opportunity for such progressive measures to be applied.

All these problems have led many to believe that new ways of dealing with the escalation of crime were needed. Also politicians, government officials and citizens in many individual cities realised that they had a major role to play in the fight against crime and its perception, rather than relying only on the existing crime fighting agencies. This explains why the last decade of the twentieth century saw the emergence of a series of new strategies and policies designed to reduce crime levels against property and persons.

12.4 Creating Safe City Strategies

There can be little doubt that individual parts of the existing criminal prevention, detection and punishment systems have implemented many new measures to become more effective, which have helped reduce many crime rates. Yet increasing numbers of criminologists believe the individual agency approach is no longer adequate to deal with the problems of crime and that more integrated and comprehensive approaches are needed, of which the various types of 'Safe Cities' or 'Safe Communities' strategies seem to be the most effective. These have been the subject of many individual reports or comparative research studies in several countries (van den Berg et al. 2006; Alberta Justice 2007). Even the adoption of this label provides a greater profile for the new strategies adopted in urban areas. Although the content and organization of these Safe City strategies vary from city to city and between countries, their success depends on the application of several key factors that underlie the creation, development, and implementation of these plans. These issues are summarized in Fig. 12.4 and have some similarity with the types of strategies that underlie many of the other new themes in urban development of the last decade.

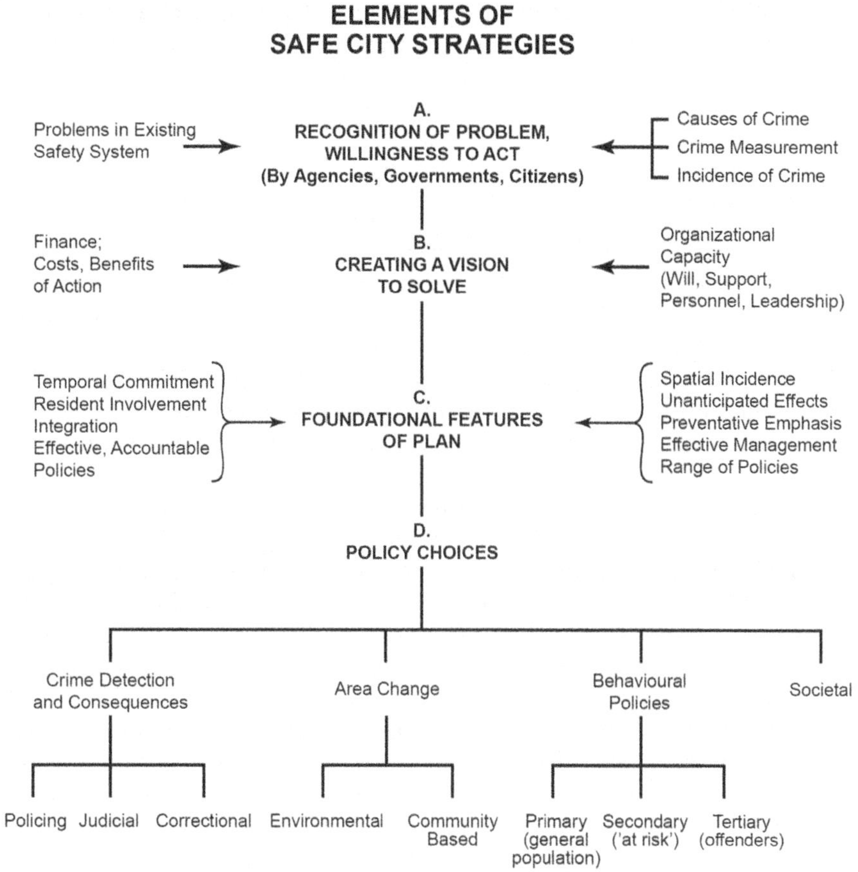

Fig. 12.4 Basic elements of safe city strategies

12.4.1 Recognition and Willingness

First, there must be an explicit *recognition* by urban citizens, agencies and governments in a city, or areas within a city, that there are problems in the existing safety system that are not being resolved. In addition there must be a *willingness* of the leaders and the population at large to do something about the situation. Figure 12.2 shows that this involves moving beyond the generality of concern about crime in general and providing specific empirical evidence about its causes and incidence. It means identifying the various types of crime, the numbers and rates of change, as well as the varied temporal and spatial incidence in crime, for crime varies by day, week and seasons. The decision to create pressure for change does not simply depend upon the assumption that crime causes problems and distress for individuals. It also comes from the understanding the effects of crime, namely that high crime rates reduces the quality of life, or liveability of cities, by increasing fear among citizens and visitors. They also increase the likelihood that the city will lose existing business and experience low inward investment, thereby reducing its future employment basis, perhaps contributing to a process of economic decline.

12.4.2 A Vision

The next stage is the creation of a strategic *vision* for the reduction and control of crime through specific measures. The development of such a strategy depends on what has been called the 'organizing capacity' of the members of the various agencies that deal with crime and its consequences, as well as various levels of governments and the residents of the city (van den Berg et al. 2006, p. 26). The crime agency organizing capacity depends on several factors that need to operate as interacting features, rather than isolated factors, namely: the political will of governments, together with the societal support from the population at large, to embark upon such measures; effective leaders in various crime and welfare agencies to develop such policies and qualified people to carry them out; the availability of sufficient resources to pay for these new initiatives, which almost certainly involves some evidence, or at least reasoned estimation, of the costs and benefits that will occur from what amounts to a new approach in the fight against crime. In addition there should be a realistic assessment of crime statistics and trends, given the problems of measuring crime.

One of the best examples of the development of a Safe City strategy can be seen in the emergence of the 'Liveable Rotterdam' political party, which proclaimed 'safety' as its main election goal in 2002 and won a third of the municipal votes. The previous decade had seen many attempts to reduce high crime rates but most initiatives failed, primarily because the policies were either mainly reactive, not based on sound information, or were too orientated to individual projects that were not integrated with each other (van den Berg et al. 2006, p. 39). A new vision was created, based on the primary goal of ensuring that the city would have no more unsafe neighbourhoods. A key measure in the new Safe City strategy was the creation of the Rotterdam Security Index, based on tax returns, socio-economic data, and actual as well as subjective perceptions of crime incidence. This grouped the 62 city neighbourhoods into five categories from 'safe to unsafe'. In 2001, 13 of these areas, mainly in central, west and south Rotterdam, had been classified as unsafe. But by 2004 only one area was still in this category, demonstrating the progress in creating a safer city. Certainly the composition of this type of index always provokes debate, but its value lies in the way such an index focuses attention on the main problem areas, and provides a base-mark from which progress in reducing crime can be measured. The danger of such indexes is that they may label inhabitants of the areas as criminal, although most residents are actually the primary victims of crime. Similar attempts at quantifying the degree of cleanliness of various areas of Rotterdam also focused attention of what needed to be done to improve specific areas of the city, with the aim of contributing to the reduction of anti-social behaviour and crime rates.

12.4.3 Foundational Features of Safe City Plans

Any effective Safe City strategy needs a series of foundational features to be built into the safe-city strategy in which various crime reduction policies are chosen and implemented. Nine separate issues can be identified as important features to be considered and built into the plan.

12.4.3.1 Temporal Commitment

It must be accepted that successful policies to reduce crime usually take time to both implement and to be effective. This does not deny that some short-term measures can lead to success. But in most cases the real solutions to major crime problems—especially to the eradication of high crime areas or those associated with particular groups – require policies that are implemented over a long time period. This means an on-going commitment in personal and resources. There are few easy, quick solutions to many crime problems; most have evolved over years and may take a similar time to resolve.

12.4.3.2 Resident and Citizen Involvement

It has been a mistake to allow crime-fighting and rehabilitation agencies, such as social services, to operate mainly in isolation from the population at large, or with only token links to the public. The promotion of citizen involvement in crime needs to ensure that citizens are active and prominent partners in the evolution and evaluations of general strategies and individual policies, not just as nominal members of committees. They need to be effective members who are able to provide advice to the respective crime agencies and to provide linkages and feedback from the people who are most affected. They should also have the power to prevent the agencies that fight crime from adopting practices that may be self-serving, and even biased by excluding ethnic or religious minorities, eroding civil rights, or be perceived as negatively affecting the population at large. Moreover it is important not to allow citizens to take the law into their own hands in some sort of vigilante action, which usually leads to either targeting the wrong people, or unnecessary violence.

12.4.3.3 Integration

Too many of the existing policies designed to combat crime are silo-like in approach, in the sense they are associated with single agencies that primarily deal with part of the system of justice, or with associated departments, and do not co-operate effectively. So crime-fighting should be treated in a co-ordinated way, by ensuring that there is a strong focus upon both the people and places most associated with crime, as well as using different agencies, not simply the police, with expertise to find, punish and rehabilitate offenders. Hence, most of the Safe City strategies that have been effective stress the necessity of adopting more multi-agency crime fighting approaches within the city. So different police jurisdictions, government departments, social and medical agencies, justice systems, and even schools in the case of juvenile crime, should co-operate to share information about potential criminal behaviour. Certainly there will always be organizational rivalries and differences of opinions as to the best approaches, even to tendencies to restrict information flows about crime or those prone to crime to people within an organization, for many believe that the rights of privacy and protection of a citizen's rights should restrict the general availability of information. But this co-operation is necessary if the battle is to be successful. For example, an important part of the Liveable Rotterdam strategy (van den Berg et al. 2006) was the decision to forge formal agreements between different municipal departments as well as with parts of surrounding government departments, in addition to police, justice and welfare agencies and seeking the co-operation of the national government. The main message of this feature is that no single part of the crime-fighting or crime prevention units can complete the task alone.

12.4.3.4 Effective and Accountable Policies

Any Safe City strategy should contain individual programmes that have been shown to be effective and accountable, crime-fighting policies that are appropriate for particular problems. In the last few decades more policies are being subject to rigorous scientific experimentation to measure effectiveness and costs, especially in relation to evidence-based policing (Sherman 2002) or violence control (Tolan and Guera 1994; Elliot 1993; Greenwood 2004). This led to U.K. officers creating a Society for Evidence-based Policing in 2010 (SEBP). The implementation of each policy involves access to sufficient resources—in qualified personnel and especially long-term funding—to solve the problems faced. The policies adopted should also have specific, measurable targets with progress being monitored to ensure that such goals are attained, again as far as possible. These need to have realistic attainments, so several monitoring periods may be required before real success is achieved. Advocates also argue that managers responsible for the policies should also be held accountable for the success or failure of policies.Finally, they should be especially sensitive to the safety problems of women, following the 'gender mainstreaming' approach to gender equity (U.N. 2002) and the policies of the UNWomen initiatives (UNW).

12.4.3.5 Spatial Incidence of Crime

Several spatial issues are involved. Many crimes vary in their spatial incidence so most Safe City Policies target what are now called the spatial 'hot spots' of crime and also accept that the police will have quite different policies for fighting crime in various parts of the city where diverse levels and types of crimes may occur. In addition, it is also necessary to be aware of diffusion or spill-over effects. When crime-fighting policy has been successful in one area, those involved in crime often move to another area and cause problems there. So it must be anticipated that a solution in one area may not result in an overall crime reduction, or the problem becomes less visible because criminals change tactics by using threatening behaviour to restrict knowledge of their activities, or turn to new methods and types of crime. Another spatial problem that needs to be faced is the fact that many crimes are committed by people who may be just visiting an area, and increasingly by groups with international connections. So crime-fighting policies must take these issues into account. Hence the sharing of information across jurisdictions, especially national ones through agencies such as Interpol, is increasingly important.

12.4.3.6 Unanticipated or Negative Effects

Many actions that are designed to improve the situation of humans and the human environment have unanticipated negative effects which may counteract the positive features. It is asking too much to be able to always anticipate all these problems. However, a careful scrutiny of the effects of particular crime policies that have been applied in other jurisdictions often provides clues to unanticipated crime results, or what may happen if a policy is applied elsewhere, or is inappropriate.

12.4.3.7 Preventative, Not Reactive Emphases

Most of the traditional policing and court practices in twentieth century western cities and society focused on apprehending criminals and punishing them for their crime, which is essentially a post-crime approach. There can be little doubt that this will continue to be an important part of the law and order agenda. A major trend has been to adopt preventative crime programmes, such as those designed to reduce the numbers of people developing criminal behaviour and to reduce the rate of recidivism of people after jail sentences. Unfortunately the financial resources for such policies are still limited.

12.4.3.8 Effective Management Structures

This may be one of the most fundamental features of an effective Safe City strategy. To some extent it could be considered part of the integrative approach, but its role is more fundamental since it refers to the need to have a crime management structure that involves the top officials in the all government agencies that are related to crime, to ensure that the strategy is considered to be a major priority in the city and that sufficient resources are devoted to the tasks. Again one of the best examples in seen in a Safe City approach in The Netherlands where a *Security Steering Group in Rotterdam* was established to co-ordinate approaches to reducing crime, one that regularly meets to discuss progress and establish priorities. This high profile committee is composed of the Mayor, Police Chief, Public Prosecutor, the two Aldermen responsible for Security and Neighbourhood Affairs, as well as additional individuals with special responsibilities for crime prevention or criminal detection. The involvement of senior figures from various types of government ensure that priority is given to the policies adopted, as well as encouraging co-ordination between departments.

12.4.3.9 Range of Policies Adopted

There has been a huge increase in the number of policies suggested and implemented in attempts to deal with crime and its prevention. What these approaches have shown is that there is no single path to success, perhaps a not unexpected conclusion because cities have different problems and capacities. Many approaches to summarising these policies have been used, each with their individual advantages, such as classifications that arrange policies along two scales according to Proactive-Reactive and Repressive-Preventative measures (van den Berg et al. 2006, p. 23), or according to Primary (dealing with the whole population), Secondary (people at risk), Tertiary (after crime) initiatives to prevent crime (Devine et al. 2004, pp. x–xi). A broader scheme is prevented here, one that addresses the major domains of crime policies that can be used to reduce crime and its perception, as well as being used to apprehend and charge criminal activity to prevent recidivism. These domains are: *Crime Detection and Consequences; Area-Based Approaches; Behaviour Modifications; and Societal Policies,* although it is inevitable that some individual policies may overlap such broad types. Each of the domains can be divided into a smaller number of categories that focus on specific policies that have been used by various Safe City initiatives, although some pre-date the formal creation of such plans. Since crime is a consequence of many factors, Safe City strategies recognize that crime will only be effectively reduced in all areas by a combination of these policies, although it must be noted that not all of these can be applied by municipal based administrations alone.

12.5 Crime Detection and Criminal Consequences

12.5.1 New Policing Policies

Despite the general criticism of the effectiveness of the existing police-judicial-corrections systems in the last decade there seems little doubt that these agencies will continue to play a major role in the attempts to contain and reduce crime. However many new policing policies have been adopted in democratic countries in recent decades, although not always under the Safe City rubic, to provide better solutions to the age-old problem of crime. Six in particular stand out.

12.5.1.1 Increased Police Visibility

The addition of more foot patrols in high crime and central city areas seem to have drastically reduced crimes (Sherman and Weisburd 1995), although even more regular car patrols may also help. They deter people from criminal actions since they might be observed, and also increase public confidence in the effectiveness of policing, especially if the same officers are used, so that the residents get to know the individual police and vice versa (Eck and Weisburd 1995). Since most of the victims in high crime areas, or those scared of being attacked, are the local residents, it is in their interest to help the police, although it must be recognized that previous police harassment, bullying, racist attitudes or even corruption, make many residents scared of co-operation. So such problems in the police must be eradicated. The addition of community wardens, such as in the U.K. has also been used to provide more police presence, creating 'eyes on the street', and may increase public-police interaction, although their limited powers mean that they are often ignored by those with high crime propensities. Neighbourhood policing, especially in ethnic and immigrant areas, has also been increased, so the police and public get to know one another and help each other's efforts to reduce crime. More controversial has been the development of local volunteer patrols, especially in low density and rural areas where the police rarely have the resources to visit, in order to add surveillance and interaction within the community. Their limited powers ensure the patrols do not dispense justice or become a vigilante force—a constant danger—for some have been known to exceed their mandate. Hence these patrols should be seen as an aid to public safety, for their presence often provides useful surveillance information to the police, deters potential criminal or even anti-social behaviour, especially by bored teens, who often congregate in particular areas. In most cases the teens will be known to local patrol residents, so inappropriate behaviour can be reported to parents and the police.

12.5.1.2 Increasing Sophistication of Staffing, Organizations

The creation of specialised units for specific crimes and for gangs has also been successful, as has greater co-operation with other police forces to share information and ideas. More training, the weeding out of unsuitable officers because of racist, sexist attitudes, or corrupt behaviours has improved many police forces. However a major breakthrough came from the pioneering CompStat (short for computer statistics) system implemented during William J. Gratton's tenure as Police Chief in New York from 1994. The success of the approach in New York helped the reclaiming of many public spaces from criminal activity, in addition to reducing crime in general, which led to its rapid diffusion around other U.S. police forces, so that over a third of forces with over a hundred officers were using it by 2000 (Graton and Knobler 1998; CMP). Although often attributed to Gratton, the idea of spatial targeting crime in New York came from Jack Marple, originally a transit police officer, who stuck pins in maps to show the location of crimes in transit stations, enabling him to spot patterns and dispatch police officers to these locations quickly. The identification and then focusing of resources at crime problem-spots led to major reductions of crime. This idea was then applied to the computerization of all crime and related data to provide a rapid analysis of crime to guide police reaction and Marple was moved to the Police Chief's office to implement his ideas (Marple and Mitchell 1999). This was only part of a new approach which was not to just respond to crime, but to *anticipate, prevent and quickly fight crime* in the areas where it took place, and to improve the morale, management and equipment of police. Essentially the ComStat system evolved into a new crime management process based on the wider collection of data obtained from the public, patrols, and other sources, using four principles. *Knowledge* on crime and disorder of all sorts is acquired, processed and disseminated quickly to local police stations through summaries of crime incidence in time and especially space through mapping techniques. *Effective policy* responses are made, with changes if initial solutions do not work. *Rapid action* is taken to prevent escalation of the problem. *Assessment* of the results is made, both weekly and at other time periods, over the whole police area, to show city-wide not just local trends with mandatory attendance at meetings by police officers, to identify and improve weaknesses and responsibilities. Usually, this approach also involves the creation of local crime officers in every station to review and publicize local crime statistics at the beginning of shifts, using computer techniques, to provide almost real-time information on the latest crime scene that needs attention. The approach has been credited with drastically reducing crime in many cities, although new emphases on problem areas, intelligence-led approaches, and evidence–based policies are also being used, together with a recognition of the need to break down older hierarchical orders in police forces and giving local stations more power, which often increases the effectiveness of responses to crime.

12.5.1.3 Adoption of New Technical Procedures

Obviously there have also been spectacular breakthroughs in specialist crime-detection techniques in the last 20 years. Part of this comes from the rapid GIS mapping and dissemination of crime statistics, which has been very influential in developing spatial awareness of crime patterns, as shown in a comprehensive book by Chainey and Radcliff (2006). The increasing provision of crime statistics and maps to the public in websites, or in local community newsletters, also provides the public with more factual information on crime areas and trends. In some police forces it has become standard to monitor social network messages to gain awareness of forthcoming events that might cause problems or the development of unruly crowds. The major breakthroughs in scientific procedures have also been important, from DNA testing to many other specialist forensic techniques. In addition, profiling, to identify criminal tendencies, and the use of other psychological techniques, has also helped solve crimes. The use of CCTV cameras to monitor areas, especially in city centres, may not prevent crime, but certainly help in the identification and often prosecution of those engaging in criminal activity. The use of body cameras by police may also resolve subsequent disagreements about police-suspect interactions.

12.5.1.4 Hot Spots and Differential Policing

An important trend has been a greater recognition that crime is concentrated in particular areas and is often predictable (Eck and Weisburd 1995). Sherman (1995, 2007) found that half the crime reported in some cities came from 3 % of the addresses. These so-called 'crime hot spots', therefore, are increasingly targeted through a large police presence, and with the use of daily up-to-date crime statistics to guide officer assignment. Yet the roots of the criminality and violence shown by many in these areas lie not only in their material conditions and early upbringing, but in their attitudes and feelings, especially of alienation and rejection, leading Davies (2004a) to hypothesize that ten separate 'terrains of community character' characterized areas of high crime rates, such as: exclusion-discrimination of the population, despair and limited goals in its population, social inadequacy, etc. Although the new policing approach has been to ensure that all infringements are dealt with in these areas as part of a zero tolerance approach, when demonstrations by large crowds in festive events occur, it is often useful to adopt an initial tolerance of minor misdemeanours to avoid provoking locals and participants for minor crimes. The use of 'stop and search' procedures in areas of high crime is often adopted to reduce criminal activity. But there is increasing concern that it is used by some police to target particular visible ethnic groups, which can lead to alienation of these groups. Hence it needs to be used with caution. More controversial, but increasingly used in some high crime cities, has been the criminal targeting of known offenders—especially criminal gang members—with close surveillance, an irritant that monitors their movements and hence their opportunity for criminal activity. For example, police in Rotterdam compiled of a 'Criminal List' of the 700 most well-known criminals and used a special unit to intrusively monitor the location and behaviour of these individuals, cracking down on any evidence of criminal activity (van den Berg et al. 2006). A rather wider approach called 'Weed and Seed' (WS) was developed in the 1990s by the U.S. Department of Justice in the U.S.A. which spread to over 300 high crime areas a decade after. The idea is to target high crime areas with a multi-level strategy to reduce drug use and crime by 'weeding' out major criminals and gangs through rapid law enforcement, then 'seeding' a neighbourhood with human services to prevent and reduce crime, together with a Safe Haven located in a school or community centre to provide anti-crime programmes. The idea was to mobilize a neighbourhood, not only so that residents feel freer to report crime but to provide them with a better quality of life and employment options. A crucial leadership role in these areas is played by local U.S. attorneys, but the process involved the local residents, social organizations as well as law agencies to create a multi-agency strategy. Another approach is a policy of 'Territorial Exclusion' where individuals are banned from certain areas, those of high crime, at the risk of being fined or returned to correctional facilities. The problem of city police adopting these more aggressive tactics is that they do infringe on civil rights of individuals, so municipalities usually need to apply to higher levels of government to provide the legal backing to adopt such policies.

12.5.1.5 Reducing Remoteness from the Public

This has been achieved by adding more foot patrols in high crime areas and by police officers making more presentations to schools and community groups to emphasize that they are devoted to helping the public and need support in preventing anti-social and criminal behaviours. In addition many police forces now use Twitter to provide and receive information from the public, which provides more contact and often valuable information about problems occurring, although the Greater Manchester Police Service site, which has been a pioneer in the use of social media, is among those who stress that information on crimes being committed should be reported via the police emergency numbers. Greater inclusiveness of police forces is also being achieved by increasing the numbers of women and ethnic groups in the forces, although the latter is still a problem in many areas, especially cities with large numbers of recent and visible minority immigrants. The Police-Public Consultative Committees that exist in many countries have also been re-organized to create more effective communication between the police and the lay members, with oversight by the latter. The public has also been encouraged to report more crime, especially via communication devices such as Email, while programmes as Crime Watch or Crime Stoppers, television reconstructions of unsolved crimes, often produce useful information for police and may provide rewards for respondents.

12.5.1.6 Greater Agency Co-operation and Information Sharing

This has already been described as an essential part of the new integrative approaches of Safe City strategies, despite the difficulties of creating co-operation between various agencies. But it is increasingly recognized that many of social agencies in particular have information, especially on youths, that may be useful in reducing the escalation of their anti-social behaviour to criminal actions, for example, by providing remedial measures to those at risk, or in providing advance warning of potential problems to police services. Access to this information is much easier now most files are computerized. However civil rights advocates warn that this may involve breeches of confidentiality when personal information from the files of various agencies is shared. This is a problem that has to be resolved.

12.5.2 Changes in the Judicial and Corrections Systems

Historically, many cities developed their own systems of justice, with courts and correction systems for those who transgressed against the laws. As larger state entities emerged, these parts of the justice system became controlled by the apparatus of the state, or some regional subdivision in federal systems. Hence the local offices that provide the delivery of justice—whether through the enactment of most laws, the court system itself, and the correctional facilities—are merely local arms of the national state or some provincial or state authority that attempts to provide one system for all its residents. However there are greater differences in the values and organization of justice systems in particular between countries, such as the adversarial system in English-speaking countries, mainly with lay juries responsible for verdicts, compared to the inquisitorial system, with judges making decisions as used in Italy. So adopting a single approach to removing existing problems in judicial systems is difficult. Yet in recent years dissatisfaction with the results of the typical top-down, remote justice system, of whatever form, has led to an increasing number of initiatives at a city and even local community level, designed to try and solve particular problems at a more local level. These often use restorative justice, rather than retribution justice approaches, although such initiatives have to get the approval of the state system before they are allowed to operate. Four in particular are proving effective.

12.5.2.1 Balanced Representation

Increasing moves have been made to create representative judges, by gender, ethnic, age balances, to decrease remoteness and increase fairness, ensuring a judgement by peers. A British House of Lords Constitution Committee (HLCC 2012) study of judicial appointments and numbers recently revealed that of the 3,694 judges in England and Wales in 2011, the majority were white males from a narrow social class. Only 22.3 % were female but this had risen from 10.3 % in 1998. The proportions of Black or Asian minority ethnic origin, was still very small, at 5.1 % compared to 1.6 % in 1998. The committee report stressed the need to increase these minority proportions of judges substantially and also to ensure that more come from under-represented parts of the city, rather than from only the richer areas or from particular social backgrounds. The practice of having lay local magistrates in the U.K., which provide the initial review of crimes and deal with low level offences, sending serious crimes to crown courts presided over by trained judges, has always tried to ensure greater local representation of the population and has a less biased pattern. More generally, in countries without strong state judicial oversight, magistrates in local courts, just like small local police services, are always prone to being unduly influenced or even corrupted by crime syndicates. In many countries with weak democratic structures the judicial system is often run by associates of the ruling party, ensuring there is little independence, leading to many injustices, an issue that has to be resolved if a fair system of justice is to occur.

12.5.2.2 Faster Justice

Courts in many cities have been encouraged to provide faster dispensation of justice, especially for minor crimes, which has occurred in some large cities in the U.S. by courts with a mandate of reviewing and sentencing within 24 h. Although effective, civil rights advocates maintain that such policies run the risk of not allowing defendants time to prepare their cases, leading to miscarriages of justice.

12.5.2.3 Specialized Reviews of Behaviour and Crime

These are being developed to deal with specific crime problems. For example, Community Courts in local areas are increasingly being used to judge young, first offenders, dispensing curfews, rehabilitation and community service in place of jail for young offenders. Such approaches, at least for minor offences, give them a chance to reflect and transform their criminal behaviour and the courts are staffed by local residents who have been trained in judicial procedures. New Drug Courts have been developed in the United States in particular to deal with those apprehended of drug charges, in which the emphasis is placed on reducing drug use through restorative measures, not jail time, which those charged have to maintain. This change is linked to the realization that addiction is a medical and not necessarily a criminal matter. In Britain the creation of Anti-Social Behaviour Orders (ASBO) in 1998 were designed to reduce incivilities and vandalism, behaviours which might not be criminalized, but cause fear among residents and if left unchecked often escalate to criminal behaviour (Flint and Nixon 2006). Behaviours, such as drunkenness, rowdiness, playing loud music at night, threatening or lewd actions do not necessarily lead to breaking any law so police interventions are limited. This led to the creation of an ASBO process which has three stages: warnings to offenders about their behaviour by a home visit from police and social agency staff; formal meetings in police stations or ASBO centres if the behaviour persists; finally the charging of persistent offenders under the new laws, which could lead to major fines or even custodial sentences for major offences. In general the policy seems to have been successful, although some youths in gangs treat the award as an ASBO as a mark of acceptance by their anti-social groups and there has been a lot of reporting of trivial offences by some neighbours. In addition there are considerable variations in the ASBO orders between various police areas, which show that there are differences in the degrees of tolerance and policing between jurisdictions.

12.5.2.4 Post-Crime Actions

One of the problems associated with the sentencing of convicted offenders is that there are disproportionate numbers of people in prisons who come from particular ethnic groups and areas. Hence there have been attempts to find alternative approaches to dealing with some of these groups, such as a return to traditional aboriginal justice methods in Canada for those in the group charged with non-violent or minor crimes. In the last decade increasing numbers of criminologists and judges have argued that that vamping up the existing repressive system is counter-productive (Greenwood 2004), as well as producing increasing marginal returns in the sense that since keeping people locked up when they are young and violent may reduce crime rates, but is not cost-effective when they become elderly and of no threat to others (Useem and Piehl 2008). Also the costs of incarceration have reached crisis levels in some countries. For example, in U.S.A. the costs of the correctional system rose from $ 9 billion in 1986 to $ 60 billion 20 years later and by 2013 was close to $ 80 billion, with 450,000 people employed in jails, a considerable increase from the 125,000 employed in in 1980. Costs may be lower in other countries but the rapid upward trend is the same. For example, in Canada the cost of the federal custodial system increased from C$ 1.6 to almost C$ 3 billion between 2005 and 2011 and each prisoner now costs C$ 113,000 per year, a substantial rise from C$ 88, 000 since 2005, whereas keeping one in the community averages C$ 29,500 (Correctional Services of Canada 2011). Similar problems have been recognised in other countries. These expenditure are leading even conservative politicians to argue that the cost explosion of prisons in a debt-ridden western society cannot be sustained, which does seem to be leading to gradual support for different sentencing laws and more rehabilitative measures to reduce the high rates of recidivism. In the summer of 2013 the U.S attorney-general announced measures to reform the automatic imposition of mandatory sentences in federal courts for some offences, giving judges the power back to make their own decisions on the evidence. This approach had already been applied in many traditional conservative states like Texas, while also providing more help for efforts to assist prisoners re-enter society after release from jail, as well as sentencing to community service rather than prison. These new policies are likely to reduce the incarceration of people convicted of minor drug offences who are non-violent and not linked to gangs, many of whom had previously been imprisoned for life because of the 'three convictions and life sentence' rule adopted in response to the spike in crime rate from the 1980s. This change, along with the related one of releasing elderly prisoners who pose little threat to anybody, is likely to reduce prison populations and should substantially reduce costs of prisons in the future.

12.6 Area-Based Changes

Two very different areal approaches to reducing crime rates in cities have been adopted. One is based on new community approaches to reduce crime. The other is upon environmental design changes to reduce crime. Both also affect the perception of crime and may improve its detection. Although they may overlap with some of the policing initiatives that have been described, such as hot-spot policing, these initiatives are carried out by different actors, or new organizations that grow out of citizen involvement to change character of areas.

12.6.1 Community Approaches for Crime Reduction

Some of these new community-based initiatives pre-date explicit Safe City agendas, but many have been re-vamped and incorporated within them.

12.6.1.1 Crime and Neighbourhood Organizations

Like the volunteer patrols described earlier, organizations such as Neighbourhood Watch provide surveillance and reporting systems by local residents on crimes or suspicious behaviour that can be used by the police. This also creates more neighbourhood co-operation and knowledge of problems, as well as providing information to the police. Block Watch Parenting programmes have also been successful especially in providing safe houses for children in distress, or those being bullied or followed by stalkers. Residents of these safe houses are carefully vetted before being allowed to participate and approved to show this sign. The problem is that when an area becomes safer these approaches often fade and disappear, and take time to be reconstituted if criminal activity increases. Also, since they depend upon neighbourly co-operation, such organizations are often missing in the areas that need them most, the high crime and poverty areas, because of the isolation of many people and their fear of residents with criminal dispositions.

12.6.1.2 Crime and Community Action Plans

These provide a new way of involving local residents in identify crime or potential problems in area, as a prelude to solving them. It has proved helpful in some Dutch cities to have a Local Facilitator who is employed to monitor anti-crime initiatives in local communities, to act as a resource person for local complaints, and is familiar enough with the various social and police agencies to put pressure on them to reduce bottlenecks or slowness in the delivery of existing anti-crime programmes. In England and Wales a new neighbourhood approach has been successfully adopted in high crime areas, where a somewhat similar key role is played by a Neighbourhood Manager who acts as a centre for information and advice and co-ordinates anti-crime activities and liaisons with the public and relevant agencies (Turley et al. 2012).

12.6.1.3 Crime and Community Delivery of Social Facilities

Policies in this category range from initiatives to provide more *social and recreational facilities* for local residents, especially the often bored teen and young adult groups who have suffered from the reduction in the number of youth and sports clubs in many areas and who may be capable of disruptive behaviours as well as crime. The addition of social facilities may reduce crime. It is quite astonishing that although many cities have a range of suburban community centres built for adult use, or for parents with small children, they have often ignored the needs of teens, apart from providing sports facilities. More specific facilities for at-risk groups have also been established in some inner city areas of high crime, such as *Mission Locales* in France that provide social and educational facilities in a safe environment, creating alternatives for those tempted by anti-social and criminal lifestyles. Far more controversial in their establishment and in their running, has been the creation of specialist '*safe drug consumption sites* in a few cities. These are essentially drop-in centres in areas of high drug use where addicts can access a range of services, from safe needles to advice for treatment. They are designed to act as safe refuges for people who would otherwise be at risk from infections or even arrest. So drug use is treated as a public health problem, not a criminal one. The evidence is still mixed about the utility of such places. Those of a conservative persuasion do not believe such sites should be funded from public funds and consider they do as much harm as good. As always, announcements of about the sites of such centres as well as the more general development of necessary half-way houses for ex-prisoners or homeless shelters, usually run into the Not In My Backyard (NIMBY) opposition from local residents, issues that have to be resolved with sensitivity. Yet these houses are needed to provide a way for former prisoners to be gradually reintroduced into society.

12.6.1.4 Gated Communities

One of the most controversial of the community policies has been the rapid development in the United States especially of gated communities, middle or wealthy areas, not just buildings, surrounded by walls and with gates monitored by guards. Obviously this is not a complete city approach since they are restricted to particular areas and the trend is almost a return to the family or high income compounds of historical cities. The obvious problem is that such developments restrict the ability of the population at large to use all of the city, creating limitation of the 'rights to the city' issues discussed in Chap. 3, as well as the possibility of vigilante justice being unfairly applied to visitors perceived as strangers. Cities in most developed countries outside the U.S.A. are still reluctant to approve these gated developments, although they have increased substantially in the wealthy suburbs of cities within less developed countries that have high crime rates.

12.6.2 Environmental Design Changes and Crime

These policies involve changing the character and form of specific areas in cities in various ways, primarily by eradicating features that may make crime easier, ensuring that the areas become more difficult for criminals to be successful. Historically, compounded houses, or in cases of upper income residences, the permanent presence of servants, reduced crime opportunities. But from the late nineteenth century the built urban fabric displayed few signs of any concerns about safety, other than those countries having concierges in apartment blocks, since houses opened straight on the street, or later had low density suburban styles with gardens in front and large windows. Pioneering investigations by Oscar Newman (1972) on crime-ridden public housing complexes in the U.S.A., led to the concept of 'defensible space', essentially the degree to which local residents are able to exert control over their surroundings, especially in preventing crime. He argued that many of these towers were crime-ridden because they had low levels of surveillance, high levels of anonymity among the residents, and multiple escape routes, all of which led to limited knowledge and common feelings about other residents, as well as ease of entry and rapid escape, all contributing to a greater incidence of crime. Coleman (1985) in Britain also observed similar concentrations of crime in these high density areas and additionally argued that the lack of maintenance and security on site, and high levels of incivility between residents also contributed to their higher crime levels. These studies, and the work of Kevin Lynch (1960) on the 'legibility of cities', stimulated research in the way in which the design of urban areas contributed to crime rates, a topic which had been ignored for too long by so many public and private developers. The result has been the identification of a series of crime-reducing features in the both the design and use of buildings and areas in a series of research articles and books (Wekerle and Whitzman 1995). These studies seek to increase both the awareness of urban environmental features that affect crime, as well as reducing the vulnerability of people in urban areas. They can be summarized in terms of nine broad categories, although some individual policies within the categories are often inter-related.

12.6.2.1 Crime and Visibility

Clear sight lines along roads, pathways and entrances should be created to reduce hiding places for criminals, while masking vegetation, especially around doors, should also be removed. In the case of routes, more careful designs should reduce the number of pathways with sharp bends, especially in pedestrian tunnels for road crossings, or if unavoidable, the bends should be provided with mirrors or metal sheets to provide vision around corners. At night it is also important to have well-lit areas, although this as much to increase perception of safety. In this context it is worth noting that lighting is often focused on roads, rather than pavements or sidewalks which remain in shadow, the lights should be adjusted to illuminate these paths, so as to give facial visibility at least 15 ft away. All street lights, especially those illuminating paths or pedestrian tunnels, should be closely monitored so that vandalised or burnt-out bulbs can be quickly replaced, since dark places increase the fear of pedestrians and may attract predators. Bus shelters and small spaces have often been used as hiding places for those with criminal intent; they should have at least two transparent sides to ensure that pedestrians are able to see who are in the shelters to avoid being taken by surprise.

12.6.2.2 Crime and Surveillance

Mixed-use areas should be encouraged, because single use areas such as commercial zones are 'dead' at night, while the addition of residences overlooking streets increases the possibility of surveillance. Although the value of the 'eyes on the street' idea has been popularized by Jane Jacobs (1961), such surveillance only provides practical results if people are prepared to assist victims of crime, to phone the police, or agree to act as witnesses. There have been many cases where people have witnessed crime but done nothing about it, meaning that the 'eyes on the street' are effectively blind. Surveillance in areas can also be increased if people in passing buses or taxis are also encouraged to report crimes or suspicious behaviours. The addition of more request bus stops, especially at night, dropping passengers closer to their residences, is also useful, so passengers have less area to traverse before reaching their buildings, thereby reducing their exposure to potential muggers in high crime areas. The addition of CCTV cameras may also be useful.

12.6.2.3 Crime and Anonymity

Anonymity between people reduces the probability of neighbours helping those in distress or subject to crimes. The creation of neighbourhood, or even local block organizations, however informal, increases the degree of social interaction and perhaps trust between people. This can lead to mutual help, or at least the reporting of crimes. Similarly, local organizations should be encouraged to develop a create sense of ownership over their public spaces, which often leads to neighbours monitoring such areas for anti-social behaviour.

12.6.2.4 Crime and Businesses or Service Activity Generators

Such facilities will create more pedestrian movements and surveillance in an area, especially if they open for longer than normal business hours. The encouragement of the so-called 'activity jewels', such as street vendors and entertainers on city pavements, and food and other kiosks in parks, also adds to the vitality of locations. All have the benefit of attracting more people to an area, thereby providing increased surveillance and comfort for people through the presence of crowds.

12.6.2.5 Crime and Design Legibility of Routes and Stops

Anxiety and fear are reduced when there are many legible signposts and directions on roads and underground parking garages, timetables at bus stops, or maps near high volume exit-entry points to transit systems. Reducing the number of exit points in high rise complexes was one of Newman's (1972) recommendations for improving safety in high rise complexes, since many exits allows criminals to escape.

The addition of security systems that only residents or workers can use is also an important safety device. A particular need is to add more security, such as regular police patrols or cameras, in designated waiting places, like bus and train stations, especially at night, thereby increasing the perception of safety.

12.6.2.6 Crime and Target Hardening

There is little doubt that reductions in crime have been helped by the increasing number of households and businesses that have alarm systems linked to security organizations that respond to break-ins and call in police help. In addition most stores and banks have developed more procedures to train staff in how to respond to robberies. Many have security guards and anti-theft devices, including the extensive use of CCTV cameras, which help to identify people after criminal acts, as well as better safes and storage facilities. The drop in car thefts is also a consequence of better anti-theft systems on cars. All these examples provide ways in which the potential targets of crime are made harder to attack.

12.6.2.7 Crime and Maintenance and Management Systems

Most believe that it is imperative that there should be gatekeepers in high density buildings, such as the concierges traditionally found in French apartment blocks. They can monitor and perhaps control access, as well as providing advice and rapid access to maintenance services when problems occur. It has been found that the presence of security systems alone, whether cameras or special locks, are often not enough to deter crimal or anti-social behaviour. One of the problems of many of the public sector high rises built in many cities from the 1950s was the absence of effective security systems or personnel, so that many quickly degenerated into crime areas. Such systems need to be added. In addition it is important to develop a pride of appearance and sense of ownership in local areas, which not only increases aesthetic appeal, but helps to deter crimes. The so-called 'broken windows concept' (Kelling and Wilson 1982) hypothesized that if broken windows, lights and other derelict features in properties were not repaired quickly they become a sign to those with criminal dispositions that an area is untended, making it more susceptible to crime. Such negative signification also applies to places where walls and buildings are defaced by graffiti, are vandalised or poorly maintained, such as places with broken doors and fences, ill-kept gardens, poor drainage systems, rotting piles of garbage and few rubbish collection bins. It has been argued that the removal of these eyesores may avoid a downward spiral of decay in an area, for neglected areas are often subject to increasingly more serious criminal behaviour, and, by being avoided and feared by the general public, create lower levels of surveillance. This led many police forces, especially New York and Los Angeles, to target such areas with a heavier police presence to deter crime and to punish those even guilty of minor infractions, such as creating graffiti, property damage, or public drunkenness. Such policies have often been seen as causing reductions of crime rates in these deteriorating areas. However there is mixed evidence about this causal connection from various studies. Sampson and Raudenbach (1999) have argued that there are intervening causal factors involved, such as the collective efficacy of the residents—their shared expectations of the social conditions of space—in wishing to reduce crime. Most studies that support the deteriorating area-higher crime hypothesis, often summarized as the 'Broken Window' concept, were of small areas, whose uniqueness may mean the positive relationship may not be scalable to large areas. However Worrall's (2002) large scale study of crime rates in all California counties between 1989 and 2000, did reveal a positive correlation between rigorous Broken Window policing and lower crime rates after controlling for several social and economic variables. Yet Worrall cautioned that the correlation did not confirm that there was a causal connection between the two. It must also be noted that the general drop in crime rates and gentrification in areas—a process that is the opposite of decay—over the period also produced contextual differences between the areas studied. So it is difficult to attribute causal connections between local crime decline and these policing methods alone. Also, there is always the problem that heavier policing may end up criminalizing more people in such deteriorating areas for minor infractions which would be ignored elsewhere. Despite these caveats, the Broken Windows-Heavier Policing policies still have a lot of support among police forces.

12.6.2.8 Crime and Urban Renewal Designs and Materials

The presence of well-designed and clean areas, especially of public spaces, may help to deter anti-social behaviour and crime. Although urban renewal was a major urban trend in 1960s, it is generally recognized that most designs, as well as the materials used in many of these projects, increased the likelihood of crime by creating uni-functional, barren concrete spaces, rarely overlooked, and devoid of people for most of the evenings. Such areas quickly became vandalised and defaced with graffiti, because of insufficient local authority control to maintain these areas. Even when such projects have been renovated, the same problems occur, so constant maintenance is a priority. A new approach to the problem has been to get private businesses with a stake in the area, or rich individuals, to assist in the financing and maintenance of such areas, with the proviso that companies with the contract to maintain the areas be given specific targets that have to be fulfilled, otherwise the contract goes elsewhere. Another policy that has achieved success in the U.S.A. is to provide tax breaks for corporations that invest in depressed or disadvantaged areas, such as the successful projects of the New York Opportunity Zones, or Philadelphia's Empowerment Areas. In general, the removal or refurbishment of slum and derelict areas can play an important part in the fight against crime in specific neighbourhoods. What is also important in these projects is the involvement of the local residents and especially youths in the process, in order to create a sense of pride and ownership over their area. This helps to deter vandals. In addition it is important to get private landlords to maintain and renovate their properties. Unfortunately it has proved very time-consuming to get owners of run-down properties to improve their buildings. Fines, licensing systems for delinquent rental owners that deny them rights of renting new properties, have all been proposed and used in various countries, although legal constraints over the civil rights of property owners have often held up the process. A real problem facing such policies in most democratic countries is that municipalities frequently lack the legal ability to implement such approaches without getting the support of national or state authorities to change existing property laws. In many inner city areas the process of gentrification has often led to the physical upgrading of inner-city areas, and because of social co-operation between the new residents there is a greater willingness to co-operate to improve an area. The downside, of course, is that the poorer people get displaced, so they may be worse-off unless adequate social housing is available.

12.6.2.9 Crime and Green Space

Chapter 4 (Green Cities) has already described how the greening of areas has many positive advantages in improving the quality of life and has shown the relationship between green areas and crime reduction (Kuo and Sullivan 2001; Branas et al. 2011). Although it has often been assumed that the presence of trees in an area reduced local visibility and helped criminals to conceal themselves, several recent studies have shown that the situation is much more nuanced. It is the low bush areas that should be removed as they can conceal lurkers, whereas areas of large trees, with maintained canopies and with their lower branches removed, improve visibility. As a result they have been shown to be correlated with low crime rates (Donovan and Prestemon 2010).

12.6.3 Crime and Behavioural Policy Solutions

The main emphasis of this third main category of policies is to try and directly reduce criminal activity, or even potential criminal activity, by changing the behaviours of people who have been, or could be engaged in such pursuits. Of course, not all will be receptive to changing their behaviour; some have a life-style of crime that is very difficult to modify, let alone eradicate. But most criminologists believe that the majority of those who are likely to commit crime in the future, would not carry out such activities if the circumstances of their upbringing and opportunities had been different, either from flaws in their developmental history, or from the social ecology of their environment, each of which contributes to their *predisposition to crime* (Hagen 1985). Crimes against persons and property are strongly associated with particular characteristics of a population, namely: poverty, unemployment, limited education, gender (with far more crime from male teenagers and young adults), single parent families, transients, as well as those with mental disabilities, addictions, such as drugs and alcohol, and the un-acculturated immigrants. This is not to imply that all people with these features are criminals. In fact most of the people in these categories are not only handicapped by one or more of these features but have higher levels of victimization and often discrimination by the rest of the population, leading to low self-esteem and poorer health. Rather the issue here is that these factors may produce a heightened risk of criminal behaviour in members of these groups, especially children brought up in environments characterised by one or many of these features, who may believe that many activities are acceptable and not really criminal, creating a contagion effect on their friends. In addition, Sampson and Groves (1989) have shown that high levels of crime also characterise areas with low 'collective efficacy', meaning there is a low social cohesion between people in an area, as well as an unwillingness to report crime out of fear of retribution.

Initially, the search for effective policies focused on attempts to reduce the incidence of these social correlates through traditional policing or areal policies involving various social services. Both are still needed. But increasingly it is recognized that such policies are not enough and attention moved first to understanding and reducing the effect of the social institutions that underlie and reinforce many of these features, such as poor schools and limited support systems from the state, and more recently to policies based on the factors involved in social cognition and social learning which predispose some individuals towards crime and delinquency. In this respect Beck's (1999) insights into what he called the primal information processing are useful, the automatic, unthinking reaction to perceived slights or insults that leads to hostility, anger and often aggression. This occurs before any rational review, or considered response, which explains why many violent acts and assault occur. The thesis is starting to form the basis of *new cognitive strategies* to reduce anger and violence by therapies that modify such behaviours, before they can turn into violence. Another trend is through similar *evidence-based* policies to those described in policing, where pioneering empirical research on the outcomes of various anti-crime and violence policies, as well as reviews of previous work, leads to the identification of the best programmes for reducing crime (Tolan and Guera 1994; Sherman et al. 1997) and violence (Elliot 1993). The Centre for the Study and Prevention of Violence, at the University of Colorado in Boulder, is especially important in this regard. Its on-going Blueprints Model Programme (BMP) has evaluated over 900 violence prevention programs. Perhaps one of the surprising conclusions of evaluations of behavioural approaches by the Boulder Centre is that many familiar programmes designed to reduce teen and young adult delinquency and future violence and crime were shown to have limited effectiveness. For example, rigorously evaluated by carefully designed studies, the following popular programmes were considered as being of limited effectiveness: 'boot camps' for young offenders, gun buy-backs, peer-counselling, summer jobs for 'at risk' teens, and house detention devices using electronic tags (which can be removed). The research centre even concluded that familiar and well established programmes such as Neighbourhood Watch were of limited effectiveness, for they are rarely established in high crime areas.

More targeted and positive behavioural policies come from new understandings of the life-course development of children that steer some to delinquency and crime. Researchers in this field have shown that two critical phases occur, one in the pre-school years, the other during adolescence, in which deficient socialization patterns often lead to the subsequent delinquent and violent behaviour that often underlies criminality. In the first phase Fonagy (2003, 2004) has argued that the older ideas that people are socialized into these anti-social and criminal behaviours by the influence of others is not the main cause of crime. Instead he maintained that most young children are 'socialised out' of these attributes by their parents, family and neighbours because of a process of 'mentalisation'. This is a process of social cognition where children learn how to make sense of the world and how to react, essentially developing self-control and empathy with others, which enables them to function within groups by losing their propensity to injure others. Those that do not undergo this development, are the ones that are subsequently disruptive in class and engage in bullying and violent behaviour at an early age. Since it often continues in later years, it has become a good prediction for subsequent delinquent and criminal activity. Some researchers (Tiihonen et al. 2014) have studied whether there is a genetic disposition to these behaviours, although this seems to be small and may be turned-off by subsequent positive parenting. The more important influences on behaviour come from the milieu in which the child is nurtured. Since this is a time of major cognitive development in the brain, a child needs high levels of attachment to, and support from, parents, family and carers, who do not allow anti-social behaviour, as well as exposure to a large number of words and positive experiences that improves cognition and expression, which Hart and Risley (2003) have shown to vary drastically between children under four brought up in welfare, working and middle class households. Environments that do not provide such support are those with high levels of people without education, or in poverty and, family breakdown, and the problems become even worse if children are abandoned, abused or are in contact with people or parents suffering from mental disabilities, addictions to drugs or alcohol, or engage in cruel and sexual practices. There are causal associations between these negative early environmental influences and crime, for they degrade the normal developmental process produced by positive parenting, producing children who are likely to stressed and anxious, with impaired cognitive development, leading to slower progress in school, health problems, emotional detachment, lack of self-esteem and often future deviancy, in which violence is used as a way out of difficulties. Chapter 13 on Healthy Cities describes how such impoverished early life characteristics are biologically embodied.

The second critical stage in the relationship between crime and personal development is found in the fact that adolescence and crime is often correlated, although this usually intensifies problems already present (Hawkins 1996). Male self-report studies in the U.S.A. have shown that almost a quarter of all males commit at least one act of serious violence before they are aged 18, with a peak of offending around 16–17 years of age and continuing for just under a decade (Elliott 1993). In addition, the need for fellowship outside of family often leads young teens into gangs in which new relationships are forged in place of parental or school associations. This often leads to a 'teen socialization' into delinquent behaviour and even crime, through imitative actions that provides initiation into a group, or a way of obtaining status and respect within it (CSJ 2009b). These groups then provide the fellowship and sense of belonging that is absent from their parents or neighbours, together with a sense of excitement and willingness to take risks, such as by engaging in criminal actions to gain money. Many of the violent acts come from excessive responses to some personal insult or some minor conflict. Most teens have learned to resolve such problems. But those without positive development histories often carry out impulsive crime and take offence at what is seen as a lack of respect, because of the type of primal processing behaviour identified by Beck (1999). So cognitive modification is needed for individuals to resolve their impulsive behaviours. Some high crime cities with large numbers of gangs, such as Glasgow, have implemented policies that target gang members with heavy surveillance which restricts their abilities to engage in crime, as well as trying to get individuals to leave their group through re-location procedures if necessary, to avoid retribution from the gang.

Many different approaches to violence prevention have been used in Safe City and other crime prevention programmes, such as those focusing upon individual offenders, their families, schools or communities. But Greenwood (2004) has argued that relatively few of the huge array of prevention programmes have been properly evaluated in a scientific sense, especially in terms of their relative cost-effectiveness and community-based programmes in particular have had few rigorous evaluations. In the case of delinquency, which is often the precursor for more serious subsequent crime, Greenwood presented a rather pessimistic view of the policies normally applied, although the opinions may be applied to the wider issue of crime.

> Today, there is a growing mismatch between delinquency prevention practice and rapidly accumulating knowledge of the life-course patterns of delinquent behaviour. Many existing prevention efforts fail to integrate accumulating knowledge about the processes of desistance from anti-social behaviour at each stage of development that might be utilized in prevention programs to hasten the termination of delinquent careers. Instead, current practice reflects a vigorous but undisciplined marketplace of competing ideas, often without sound foundations in either theory or research. (Greenwood 2004, p. 202)

Despite this pessimism, reviews of the effectiveness of various behavioural programmes are providing policy makers with a series of strategies that can improve outcomes (BMP). Three different types of policy can be recognized, sometimes described as primary, secondary and tertiary treatments: those that apply to the whole population; those that are targeted at the population that is judged to be 'at risk' of being involved in criminal activity; and those that apply to previous offenders in the hope that they will not subsequently engage in crime.

12.6.3.1 Primary Behavioural Treatments

These apply to the population as a whole, where is a pressing need for more *quality pre-schooling*, with small classes and caring teachers who enhance cognitive development and positive socialisation skills, not just traditional educational skills. Unfortunately relatively few countries have effective and universal programmes in this area, a need made more apparent now so many young families do not have a stay-at-home parent, given the dramatic increase in the rates of female participation in the workforce. Such facilities are especially in short supply in the poorest and socially disorganized parts of the city and their absence often contributes to impaired cognitive development in young children, which can subsequently lead to higher crime rates. A related example of early childhood policies can be seen in the recent initiatives seen in many Dutch and Swedish cities in encouraging immigrant families to enrol their children in pre-school centres, so they will learn the local language quickly, as well as being encouraged in recreation time to find playful positive experiences in the cold winter conditions. This means that children will not start regular school with a language handicap that many will never overcome, reducing their chances of obtaining good educational qualifications. Success has also been achieved with *mentoring programmes*, but skilled mentors may be hard to find in many areas. Ideally, school programmes would seem to be the obvious targets for intervention, since youths spend a great deal of time there. But it must be admitted that many schools have difficulty enough in dealing with their primary education role, and frequently do not have enough skilled professionals to deliver appropriate delinquency and crime prevention programmes. In any case, the daily proximity of friends and peers, often with similar problems, makes it difficult to persuade individuals to change behaviours which would mean they lose status with their friends. The biggest success is likely to be at kindergarten level, where there is a higher level of parental involvement and aberrant behaviour is more easily controlled.

Increasingly it is recognized that elementary school programmes should not only focus on the acquisition of academic skills. The PATHS programme (Promoting Alternative Thinking Strategy) seeks to *improve social cognition in lessons* and activities that develop a child's ability to understand the perspectives and actions of others and to encourage them to recognize and control their emotions, as well as improving their decision-making and risk avoidance skills. Since good parenting may already mimic the effect of these programmes, these policies may have their greatest effect in areas and families where these skills are not developed. Indeed Bradshaw and Garbarino (2004) have stated that although it is well-known that children who have been badly treated, or exposed to community violence, are prone to negative behaviours, the way this happens is not clear. They suggest that social cognition factors are involved. At a later stage in schools, programmes that develop effective information on the *dangers of drugs and alcohol* are essential and they should be reinforced by parallel programmes delivered by religious organizations and communities in areas of high crime and delinquency.

One big problem with high school programmes comes from the very size of many of these institutions, which makes counselling more difficult. There is also accumulating American evidence that part of the problem of violence and subsequent criminal behaviour may be associated with the very *size of the schools* (Devine et al. 2004). It has been shown that the biggest public schools in the U.S.A. are 6–7 times larger than the average school in E.U. countries and those with the highest rates of violent incidents are among the biggest in the country. This has led to what amounts to a climate of intimidation since these schools have heavy security systems, with guards, gates and cameras. Such situations did not exist in public schools until the 1970s and drastically impedes educational progress, especially for those who need it the most. A vigorous movement for *smaller schools* has developed, supported by philanthropic organizations like the Gates Foundation. In general, these smaller schools provide better results and less violence. However it must be noted that the creation of smaller elite schools in poor areas always run the risk of siphoning off good students, and by rejecting those already delinquent, they make the existing schools worse. More generally, some cities also have tried to eradicate some of the more obvious anti-social behaviours, such as littering and spitting, by criminalizing the behaviours and making offenders pay an 'on the spot' fine, although catching offenders is often a problem and may lead to antagonistic relationships between police and the public.

12.6.3.2 Secondary Behavioural Policies

These are targeted at people that are 'at risk', those individuals deemed to be potential recruits to anti-social and criminal behaviour, primarily, although not exclusively of people from problem families, or of low educational attainment. Unfortunately it has been easier in most cities to get money to put more police on the streets than to get enough trained nurses or counsellors to work on remedial issues. Most of these secondary programmes are delivered by various social welfare agencies and a problem has been the inadequacy of both funding and the availability of enough trained personnel to deal with these issues that need urgent attention if the interventionist programmes are to be successful. Rather than the 'hit or miss' type of approach in terms of policy implementation in the past, there is now rigorous evidence-based reviews of the effectiveness of secondary behaviour interventionist policies (Olds et al. 1997). In the U.S. Greenwood (2004, p. 207) identified ten programmes aimed at different stages in a child's development with proven success in reducing crime and delinquency in various at-risk populations compared to those who were not able to take advantage of the programmes.

In the *early years* of children the Nurse Home Visitation Programme uses specially trained nurses to regularly visit poor, high-risk mothers before pregnancy and for the following 2 years, providing advice and support. For children in kindergarten and elementary schools three programmes have been effective, although these may also be used in early teens: the Incredible Years Programme involves teachers, parents and child training services in treating behaviour problems and promoting social competence in young children; the PATHS programme (Promoting Alternative Thinking Strategy) has already been mentioned, a school-based programme designed to increase social competence among children and control of their emotions; the Bullying Prevention Programme operates in schools to promote awareness of the problem and creating rules to reduce and prevent such behaviours.

Two other programmes are also based in *classrooms*, usually but not exclusively for teens: Life Skills Training is designed to prevent drug-use and boost social resistance to pressure to use drugs and alcohol and impart general life skills; the Midwestern Prevention Project is also designed to prevent drug use but includes community as well as classroom components. Big Sisters and Brothers is a well-known and long lasting programme providing adults to mentor 'at risk' children from 6 to

18 years old, providing them with more positive experiences, advice, support and friendship.

Two successful interventionist policies involve *families*, usually with teenage children with severe behaviour problems that also involve extensive monitoring by skilled personnel. The Functional Family Therapy Programme (FFT) is mainly a short term intervention with specific techniques designed to make families and teens with truancy bullying and delinquency problems understand the problems and consequences of their actions, and to motivate them to resolve the results of negative behaviours by providing structure to their interactions with their children and how to set limits on anti-social behaviours. The Multisystemic Therapy Programme (MST) applies a number of techniques to focus on the negative conditions that exist in the environment of delinquent youths and their families in order to solve the problems before they escalate, focusing on disengagement from deviant peers and helping parents build a positive social support network

A more general programme addresses children in *foster care*. Multidimensional Treatment Foster Care (MTFC) recognizes that large fostering institutions have rarely been successful in treating children with severe behaviour and delinquency problems. So the programme places such individuals in carefully selected and trained foster families, with constant support from various trained personnel who are skilled in behaviour therapy, to create a caring and supportive environment

All these ten secondary behavioural programmes have been shown to have a very successful track record in reducing the levels of subsequent anti-social and criminal behaviour, although it must be admitted that this does not mean that all individuals enrolled respond in a positive way. Also all were cost-effective in the sense that the costs of the programmes were far less than the taxpayer costs involved in subsequent expenses for welfare, health and justice actions, although Greenwood (2004) shows the ratios do vary from programme to programme. Most rehabilitative approaches rarely have enough long term funding for these crucial programmes in most cities, while the effectiveness of many has been shown to be reduced by insufficient training for the key personnel involved in the various interventions. Different jurisdictions have developed alternatives to these general programmes. For example, in Calgary in 2010 there was a 25 % decline in youth crime over the previous 5 year average. Police attribute this change not only to their own actions but to partnerships with school boards involving two agencies in particular that are financed by the provincial Safe Communities Innovation Fund. One agency is the new Multi-Agency School Support Team that identifies negative behaviour in children 5–12 years old and gets them help before they are involved in crime. The other is the Serious Habitual Offender Programme for older children that places them under strict police supervision and only prosecutes them if they re-offend. In addition, the 2003 Provincial Youth Criminal Justice Act encouraged alternative measures for those charged with minor crimes, such as curfews and mandatory help in community projects, instead of moving offenders into juvenile detention centres where they socialise with other offenders and often learn new criminal behaviours. However tougher penalties are given for serious offences. What is clear from the general success of these programmes is that there are positive alternatives to the previous hands-off policy of intervention at early stages in the development history of 'at risk' children that often leads to criminal actions and then jailing of offenders who are caught. This often ends up by habituating adults to a life of crime, with huge costs, not only to the victims of crime, but to the taxpayers that fund the justice and prison system.

12.6.3.3 Tertiary Behavioural Treatments

Many have observed that high recidivism rates are a result of some people being criminal-minded, with lack of respect for the rights of others, concluding that only aggressive, intrusive policies seem to work in reducing crime. Yet it is increasing obvious that jails brutalize rather than rehabilitate inmates and rarely prepare those incarcerated for a life outside prison, making the situation worse for those imprisoned, although organizations, such as the John Howard Society in the U.K., have a long history of helping offenders to readjust to society. Studies of those who reoffend in the U.K. (HMSO 2002; CSJ 2009a) have shown that many have poor reasoning and thinking abilities, with low levels of literacy and numeracy, as well as few employable skills, which means they have great trouble in finding employment and housing, while in the U.S.A. a report for the Dept. of Justice revealed that 16% had mental health challenges. Such information shows that there is a great need for more rehabilitative measures, which have been limited until politicians began to realise that the escalating and heavy burden of prison costs was becoming too much for the public sector to bear. One result has been a proliferation of reports describing ways in which rehabilitative measures can help reduce rates of re-offending (HMPS 2002; Centre for Social Justice 2009). Within jails Gilligan and Lee (2004, p. 323) have emphasized that the provision of therapy and education, especially using cognitive methods for anger management and control, not punishment, in centres for human development, have produced major changes, concluding that: "if it seems utopian to dream of replacing prisons with schools, let us remind ourselves that prisons already are schools—schools in crime and violence". More recently a practical demonstration of the consequences of high recidivism rates and the effectiveness of policies to reduce recidivism was shown in a report entitled *Through the Gate* (PBE 2010), which calculated that crime by released prisoners costs the British economy £ 11 billion a year. However it was shown that the St. Giles Trust, a London-based charity, reduced the rate of re-offending in a sample of 583 prisoners by 40% after providing intensive services to criminals before, as well as after, their release from jail, in which help from reformed offenders was a crucial feature in creating trust for the prisoners. The support consisted of finding accommodation, providing drug and alcohol addiction treatment, training in various vocations, giving advice and access to role models. The study estimated that £1 spent on rehabilitation brought a £ 10 return in lower crime costs, so that a 10% reduction in re-offending rates would save £ 1 billion a year. Similar striking results have also been shown in many U.S. studies but as yet there is not enough money or staff in most countries to implement such schemes on a wider scale, though Scandinavian countries have been far more progressive than most in this regard.

12.7 Societal Policies

Although there are many situations in which policies at an urban level can reduce crime rates it must be accepted that wider societal factors are also at work, issues that can only be effected treated by action at a state level. Although this is really beyond the scope of this urban-based review some general concluding points must be made. For example, many studies have shown that crime seems to be positively correlated with unequal societies (Wilkinson 2004; Wilkinson and Pickett 2009). So the most obvious way of improving the situation is to reduce the level of inequality within a society, not in absolute income levels compared to other countries but relatively within a state. Many people have argued that inequality is usually the result of structural factors, the seemingly hidden mechanisms in society that create unequal rewards, which means some groups experience poverty, ill-health, lack of opportunity and oppression, and leads some into criminality. This is so often taken for granted and naturalised that the worse-off are assumed to be the ones at fault, creating additional marginalization through stigmas and ideological discourses which justifies repression. Scheper-Hughes (2004, p. 14) has argued on the basis of her anthropological studies of violence in many parts of the world that although it is the affluent who have the power, they often see themselves as being endangered. Hence she argues that the affluent are really the 'endangering' classes because they benefit from the unequal rewards. For most of the twentieth century the development of progressive social policies have drastically reduced inequality levels, but as Chap. 3 has shown, the last two decades has seen increasing inequalities, although the basic income safety net in most developed countries is still present, something that still needs to be applied to many countries of the developing world. However, few urban areas, except city-states, have the powers to undertake the task of reducing societal inequalities effectively, for the financing and delivery of such social welfare measures, as well as other societal based programmes, are mainly a matter of national jurisdiction, so they are mainly beyond the remit of this review. Nevertheless, as Chap. 3 has shown, there are a number of urban policies that can help reduce the level of inequalities, from housing provision by municipalities, to charity organizations such as food banks, as well as more fundamental programmes that ensure that decision-making in the city is transparent, fair to all and involves effective citizen participation.

12.8 Security Policies

Space constraints mean that this review has not dealt with additional problems that cause threats to people living in urban areas that are more usually described as relating to the security of the population. Again relatively few of these are incorporated in what are described as Safe City strategies, yet need to be identified for the sake of completeness. Some of these issues have been dealt with in other chapters, such as finding ways of making urban places more resilient against natural disasters from natural hazards (Chap. 9). Although many community and urban scale policies can reduce the impact of disasters it has been shown that their scale means that urban places alone do not have the financial or other resources to solve these problems. So massive aid, in both finance and personnel, from the national government of the affected area and from international agencies, is usually needed in the rescue and recovery phases from disasters. By contrast there are two forms of insecurities in cities that have been reduced because of urban government action. One relates to the major decline in the dangers that used to come comes from the collapse of buildings and infrastructures. In developed countries at least, improved building regulations have solved many of these problems, with regulations and safety standards that are usually enacted by cities or at least monitored by city employees. However it has often been the national state that established the legislation within which municipalities operate. Similarly the threat of fires has been drastically reduced by better construction techniques and materials as well as skilled fire-fighting bodies. Unfortunately, too many cities of the less developed world do not have adequate building regulations, effective inspections, or fire protection services. This means that the collapse of buildings, infrastructure and the destruction from fires is ever present. Yet it is important not to be too complacent about the developed world situation. In many cities the combination of an aging infrastructure and the limited financial ability of municipalities mean that there are often backlogs in the maintenance and replacement of infrastructure. Such problems need to be solved in the next decade to prevent increased levels of threat to the residents of cities. Another set of problems to the security of urban populations come from public health hazards, which are summarized in the Healthy City discussion (Chap. 13).

A more general set of security problems come from riots and external threats. Throughout history urban places have been the sites of civic unrest, from localized riots, to these that have overthrown the state and set up new governments. It is difficult to generalise about these movements, many of which are episodic, since in some cases they are allowed to take place peacefully; in others the state authorities repress them with all the military apparatus of the state and kill hundreds if not thousands of protestors. In a democracy, protests by marches or demonstrations of aggrieved people, is a legitimate form of action. Yet even originally peaceful protests have often turned to violence, sometimes escalating into riots in many areas, as seen in the 2011 English riots (Davies 2014). By the late twentieth century protests against the meetings of international leaders of the major developed countries became a regular part of the urban scene. If these demonstrations are unchecked, participants and bystanders are often injured and killed. But since part of the police role is to keep order and ensure safety in cities, the police and security services have to get involved. Policing these demonstrations is not as easy task, for attempts to monitor and control the crowds has often been over-zealous or involve deliberately provocative acts which can increase the level of violence against the demonstrators and the arrest and subsequent incarceration of citizens, often on flimsy grounds. Such problems may be exacerbated by deliberate acts of violence by police employed by repressive regimes determined to keep control, which sometimes leads to resistance and even civil war. Not only does this add to insecurity in the cities involved, but there are very considerable costs of extra policing and often the involvement of the military. Finally, the effects of major external effects such as terrorism and wars have often devastated urban places must be acknowledged. In the former case local police services are now more alert to the threats from terrorist activity, from the radical gangs in western Europe in the 1970s, to separatist movements, such as those in Basque areas of Northern East Spain, or the I.R.A. in northern Ireland—although participants in these campaigns see themselves as nationalists fighting for a cause. But again the main defence against these problems comes from the various security arms of the state, although they may alert the local forces of law and order and medical services to the impending threats, who are usually the first responders. In the case of wars cities used to have their own protections, through their walls and moats, but advances in canons by the end of the eighteenth century made such defences irrelevant, so many were torn down and converted into green spaces. This means that security from external attack is now in the hands of the military arms of the state, so urban places have little role to play, although in some countries the tradition of local militias still exists. All these examples show that the problem of ensuring the safety of the urban population remains a complex one. Policies to reduce and control crime are only part of the solutions that are needed to improve safety and security, but still a very necessary one in terms of the quality of life of citizens in urban places.

A more immediate and growing threat to cities of the developed world has been the increase in terrorist attacks in the past decade, mainly from radical Islamists who deny the verses advocating peace in their holy book in their frequent attempts to strike at the heart of what they see as a material and corrupt developed world. The major western cities have been at particular risk, as seen in the destruction of the Trade Centre towers in New York on Sept 11th 2001. Many of these risks have been reduced by new protective measures from expanded intelligence and policing services, but the need for continued vigilance and especially information and co-operation from the public, is obvious. The difference from similar terrorists in the past is that contemporary terrorist activities have the capacity to be far more destructive, given modern technology, and are also far more global in scope. They have the ability to strike anywhere, with the centres of wealth and control, such as the centres of big cities of the western world, their ports and airports, as particular targets. In all these situations municipalities have limited ability to provide security for their citizens; this is primarily the function of the security forces of national governments. The consequence of all these dangers has been that all cities, and especially vulnerable areas such as airports and port facilities, have increased their level of security programmes, with local police and emergency services geared to cope with such attacks.

12.9 Conclusions

One of the most important issues that relate to the relative liveability of places is the degree of safety from crime that creates harm to people and their possessions. This review has focused upon the issues associated with safety from crime, as opposed to security which is now the responsibility of national state actions and has only been treated briefly here. As always in dealing with changing trends, in this case, crime, there is a need to find accurate statistics. This is more difficult than in many other urban problem areas because crime is a socially constructed concept, rather than some absolute value. It varies between countries and urban places due to differences in the legal criminalization of certain behaviours and the pre-disposition of some groups to these behaviours. There are also variations in the reporting, recording and pursuit of people engaged in criminal acts, let alone the degree of success in penalizing people convicted of crime. In addition there is a wide discrepancy between crime rates as measured by statistical authorities and the perception of crime. So even though the rapid growth of crime in many western cities in the two decades from the 1970s slowed down and has been reduced back to these levels, there is still the perception that crime is still a major problem, not simply in areas of traditional high crime rates, but generally in cities. In part this is a reflection of the memory of the rapid rise in crime rates in western countries from the 1970s to the early 1990s which led to increasing concern and political actions to increase the safety of urban places. In this context the need to be more aware of the vulnerabilities of women has been highlighted in recent years by United Nations publicity and their suggested policies to improve female safety (UNW). Some of the reasons for the reduction in crime rates in the developed world are due to society-wide factors, such as the declining birth rate and fewer young males, but there can be little doubt that it is also due to the policy changes introduced by the police and justice systems. These involve new ways of anticipating and solving crime, not simply by new policies within these agencies but by more integrated and co-operative crime-fighting and prevention strategies under the Safe Cities rubic.

The emphasis in this discussion has been upon the changing attitudes and policies in the developed world. It is worth noting that although many of these new crime-fighting policies can be stand-alone programmes, there is increasing recognition of the need for more integrated approaches. This involves more than co-ordination between the various agencies that are linked to law and order issues, but also to social service organizations, as well as other stakeholders, including the general public. The result in many urban areas has been a more integrated, multi-dimensional strategy known as Safe Cities, which involves recognition of the seriousness of the problem of crime, a willingness to address it and a vision to create a viable strategy to combat it. An important part of this new process has been the increasing application of policies that have been empirically tested and shown to be effective, whether in those that relate to crime detection, areal change, or behavioural change. In addition there is greater recognition that many crimes are committed by people with serious mental disorders or with drug dependency and that some of the problems they cause should be better seen as a public health issues rather than only criminal activity. Despite the progress in so many of the fields related to crime, it must be admitted that the focus on urban places means that other types of crime, such the so-called white-collar crimes, for example, frauds or financial manipulations, are not usually part of Safe City approaches, although are still a necessary part of police activity.

A Safe City approach is based on several foundational features, especially the requirement to support policies for the long term and the provision of adequate finance as well as skilled personnel who adopt proactive measures rather than just reacting to the crime events. Many of the new policies that have been adopted under the Safe City banner are based on improvements to the existing crime detection and punishment system. The use of new technologies in laboratory work and crime scene advances and analyses of information in laboratory work are important. But so are the new computer-based procedures to map crime incidence and to quickly concentrate police forces in these areas to investigate crimes and to deter future crime acts in these areas. Making the police more visible and with greater public access and accountability is also vital. In addition organizational changes in the judicial system are also helping to create more effective systems. However there are also many ways of changing the urban environment in order to deter crime, as well as modifying criminal behaviours or predispositions to reduce crime events. In this latter case it seems important to focus on rehabilitative measures not only punitive actions, especially when first offenders are involved, and to find ways of reducing the rates of recidivism. In both categories a large number of policies have been pursued but not always with success. This is why the new trend of empirically testing the utility of particular actions is a crucial approach. It allows crime prevention and crime fighting agencies to adopt policies that have been shown to be successful elsewhere. The ability to learn from the experience of other jurisdictions, an ability that has been intensified by the speed of communication in sharing ideas, provides another example of the way that new policies adopted in cities are using the experience of others to solve, or at least mitigate, their problems—in this case crime. especially in crime-ridden cities in the developing world.

References

Alberta Justice. (2007). *Keeping communities safe: Report and recommendations*. Edmonton: Alberta Justice Communications.

BCS: British Crime Survey. (2004). http://www.crimesurvey.co.uk/previous-research.html. Accessed 10 Oct 2012.

Beck, A. T. (1999). *Prisoners of hate: The cognitive basis of anger, hostility and violence*. New York: Harper Collins.

BMP: Blueprints Model Programmes. http//www.colorado.edu/cspv/blueprints/modelprogrames. Accessed 10 Oct 2012.

Bradshaw, C., & Garbarino, J. (2004). Social cognition as a mediator of the influence of family and community violence on adolescent development. *Annals of the New York Academy of Sciences, 1036,* 85–105.

Branas, C. C., Cheney, R. A., MacDonald, J. M., Tam, V. W., Jackson, T. D., & Ten Have, T. R. (2011). A difference-in-differences analysis of health, safety, and greening vacant urban space. *American Journal of Epidemiology, 174*(11):1296–1306. doi: 10.1093/aje/kwr273, published online: November 11, 2011.

Chainey, S., & Ratcliff, J. (2006). *GIS and crime mapping*. New York: Wiley.

CMP: CompuStat. http://www.compustat.umd.edu. Accessed 10 Oct 2012.

Coleman, A. (1985). *Utopia on trial*. London: Hilary Shipman.

Correctional Services of Canada. (2011). Annual report of the office of the correctional investigator: 2010–11. http://www.oci-bec.gc.ca/cnt/rpt/annrpt/annrpt20112012-eng.aspx. Accessed 8 Oct 2012.
CSJ: Centre for Social Justice. (2009a). *Locked-up potential*. London: Centre of Social Justice.
CSJ: Centre for Social Justice. (2009b). *Dying to belong: Street gangs in Britain*. London: Centre of Social Justice.
Davies, W. K. D. (2004a). Affective dimensions of urban crime areas: The psycho-geography of urban problem areas. *Geographica Helvetica, 59*(3), 218–226.
Davies, W. K. D. (2004b). A psycho-geography of crime areas: Variations in the affective domain. In M. Pak (Ed.), *Cities in transition* (pp. 341–350). Slovenia: Dela, University of Ljubljana.
Davies, W. K. D. (2005). Towards a model of crime variations and solutions: A framework for geographic work. *Geographica Helvetica, 60*(4), 1–10.
Davies, W. K. D. (2014). Riots by a growing social periphery. Interpreting the 2011 urban riots in England, 2011. In D. P. O'Donoghue (Ed.) *Urban transformations: Centres, Peripheries and systems* (pp. 81–93). Farnham, Surrey: Ashgate.
Davies, W. K. D., & Townshend, I. J. (Eds.). (2002). *Monitoring cities: International perspectives*. Calgary and Berlin: International Geographical Union, Urban Commission.
Devine, J., Gilligan, J., Miczek, K., Shaikh, R., & Pfaff, D. (2004). Youth violence: Scientific approaches to prevention. *Annals of New York Academy of Science, 1036,* 1–417.
Donovan, G. H., & Prestemon, J. P. (2010). The effect of trees on crime in Portland, Oregon. *Environment and Behaviour, 44*(1), 3–30.
Eck, J. E., & Weisburd, D. (Eds.). (1995). *Crime and place*. Monsey: Crime Justice Press.
Elliott, D. S. (1993). Serious violent offenders: Onset, development course, and termination. *The American Society of Criminology, 32,* 1–21.
EPA: Environmental Protection Agency. http://www2.epa.gov/lead. Accessed 15 Nov 2012.
Eurostats. (2013). Statistics on police numbers per state. http//:ec.europe.eu/crime-statistics. Accessed 3 Aug 2013.
Flint, J., & Nixon, J. (2006). Governing neighbours: Anti-social behaviour orders and new forms of regulating conduct in the UK. *Urban Studies, 43*(5), 939–955.
Fonagy, P. (2003). Towards a developmental understanding of violence. *British Journal of Psychiatry, 183,* 190–192.
Fonagy, P. (2004). Early life trauma and the psychogenesis and prevention of violence. *Annals of New York Academy of Science, 1036,* 181–200.
Fonagy, P., Gergely, G., & Jurist, E. (2002). *Affect regulation, mentalisation and the development of self*. New York: Other Press.
Georges-Abeyie, D. E., & Harris, K. D. (1980). *Crime: A spatial perspective*. New York: Columbia University Press.
Gilligan, J., & Lee, B. (2004). Beyond the prison paradigm: From provoking violence to preventing it by creating 'anti-prisons': Residential colleges and therapeutic communities. *Annals of New York Academy of Science, 1036,* 300–324.
Gottfredson, M. R., & Hirschi, T. (1990). *A general theory of crime*. Stanford: Stanford University Press.
Graham, P., & Clarke, J. (2001). Dangerous places: Crime and the city. In J. Muncie & M. C. Laughlin (Eds.), *The problem of crime* (2nd ed., pp. 152–189). London: Sage Publications and Open University.
Graton, W., & Knobler, P. (1998). *Turnaround: How America's top cop reversed the crime epidemic*. New York: Random House.
Gray, E., Jackson, J., & Farrall, S. (2008). Reassessing fear of crime in England and Wales. *European Journal of Criminology, 5*(3), 363–380.
Greenwood, P. W. (2004). Cost-effective violence prevention through targeted family interventions. *Annals of New York Academy of Science, 1036,* 201–214.
Hagan, J. (1985). *Crime, criminal behaviour and its control*. New York: McGraw-Hill.
Hart, B., & Risley, T. R. (2003). *Meaningful differences in the everyday life of young American children*. New York: Brookes Publishing.
Hawkins, D. (1996). *Delinquency and crime*. Cambridge: Cambridge University Press.

Herbert, D. T. (2002). Crime and its control in urban environments. In W. K. D. Davies & I. J. Townshend (Eds.), (pp. 541–556).
HLCC-House of Lords Constitution Committee. (2012). *Report on judicial appointments*. London: HMSO.
HMPS-Her Majesty's Prison Service. (2002). *Redesigning prison reoffending*. London: HMSO.
HO: Home Office. (1999). *The Stephen Lawrence enquiry*. London: HMSO.
Hsieh, C. C., & Pugh, M. D. (1993). Poverty, income inequality and violent crime: A meta-analysis of recent aggregate data studies. *Criminal Justice Review, 18,* 182–202.
Jacobs, J. (1961). *The death and life of great American cities*. New York: Random House.
Kelling, G., & Wilson, J. Q. (1 March 1982). Broken windows. *Atlantic Monthly*. http://www.theatlantic.com/magazine/archive/1982/03/broken-windows/304465/. Accessed 12 April 2012.
Kuo, F., & Sullivan, W. (2001). Environment and crime in the inner city. *Environment and Behaviour, 33*(3), 343–367.
La Vigne, L. N., Solomon, A. L., Beckham, K., & Dedel, R. (2006). *Prisoner reentry and community policy*. Washington DC: Department of Justice, Research for Safer Communities.
Lynch, K. (1960). *Image of the city*. Cambridge: MIT Press.
Markus, D. K., Fulton, J. J., & Clarke, E. J. (2010). Lead and conduct problems: A meta analysis. *Journal of Clinical Child and Adolescent Psychology, 39*(2), 234–241.
Marple, J., & Mitchell, C. (1999). *The crime fighter: Putting the bad guys out of business*. New York: Doubleday.
Maslow, A. H. (1943). A theory of human motivation. *Psychological Review, 50,* 370–396.
McClain, P. D. (2001). Urban crime in the U.S.A. and Western Europe: A comparison. In R. Paddison (Ed.), *Handbook of urban studies* (pp. 220–240). London: Sage.
Monbiot, G. (8 June 2013). The grime behind the crime. *The Guardian*.
Mumford, L. (1961). The city in history. London: Secker and Warburg.
Nevin, R. (2000). How lead exposure relates to temporal changes in IQ, violent crime and unwed pregnancy. *Environmental Research, 83*(1), 1–23.
Newman, O. (1972). *Defensible space*. New York: Macmillan.
Olds, D. L., Eckenrode, C. R., & Henderson, C. (1997). Home visiting by paraprofessionals and by nurses: A randomized and controlled trail. *Paediatrics, 110,* 486–496.
PBE: Pro Bono Economics. (2010). *Through the gates*. London: Frontier Economics and St Giles Trust.
Pressdee, M. (2000). *Cultural criminology and the carnival of crime*. London: Routledge.
Putnam, R. (2000). *Bowling alone: Collapse and revival of American community*. New York: Simon and Schuster.
Racine, J.-B. (2002). Explaining and regulating or monitoring violence in the cities of tomorrow. In W. K. D. Davies & I. J. Townshend (Eds.), (pp. 557–598).
Radzinowicz, I., & King, J. (1977). *The growth of crime*. London: Hamish Hamilton.
Reyes, J. W. (2007). *Environmental policy as social policy? The impact of childhood lead exposure on crime*. National Bureau of Environmental Research Working Paper 13093. http//www.nber.org/papers/v130097. Accessed 10 July 2012.
Ruderman, W., & Laker, B. (2014). Busted: A tale of corruption and betrayal in the city of brotherly love. New York: Harper Collins.
Sampson, R. J., & Groves, W. B. (1989). Community structure and crime: Testing social disorganization theory. *American Journal of Sociology, 94,* 774–802.
Sampson, R. J., & Raudenbach, S. W. (1999). Systematic social observation of public spaces. *American Journal of Sociology, 105*(3), 603–651.
Scheper-Hughes, N. (2004). Dangerous and endangered youth: Social structure determining violence. *Annals of New York Academy of Science, 1036,* 13–46.
SEBP. http://www.sebp.uk. Accessed 12 July 2012.
Sherman, L. W. (1995). Hot spots of crime: Criminal careers of places. In J. E. Eck & D. Weisburd (Eds.), (pp. 35–52).
Sherman, L. W. (2007). Predicting crime. Selected works of L.W Sherman. http://crim.cam.ac.uk./people/academic-research/Lawrence-sherman/lwsDownloadPub.pdf. Accessed 19 July 2013.

Sherman, L. W., & Weisburd, D. (1995). General deterrent effects of police patrol in crime hot spots: A randomized, controlled trial. *Justice Quarterly, 12*(4), 635–648.
Sherman, L. W., Gartin, P. R., & Bueger, M. E. (1989). Hot-spots of predatory crime: Routine activity and the criminology of place. *Criminology, 27,* 27–56.
Sherman, L. W., Gottfriedson, M., MacKenzie, D., Eck, J., Reuter, P., & Bushway, S. (1997). Preventing crime: What works and what doesn't, what's promising. Report to U.S. Congress. National Institute of Justice, Washington DC.
Sherman, L. W., Farrington, D. P., Welsh, B., & Mackenzie, D. (2002). *Evidence-based crime prevention.* London: Routledge.
Taylor, C. (1986). The nature and scope of distributive justice. In F. S. Lucash and J. N. Shklar (Eds.). *Justice and equality here and now* (pp. 34–66). Ithaca, New York: Cornel University Press.
TE. (20 July 2013). The curious case of the fall in crime. *The Economist.*
Tiihonen, J., Rautiainen, M-R., Ollila, H. M. et al. (2014). Genetic background of extreme violent behaviour. *Molecular Psychiatry*, 19(10). doi:10.1038/mp.2014.130. Accessed 20 Nov 2014.
Tolan, P. H., & Guera, N. G. (1994). What works in reducing adolescent violence? Research Paper No. 001. Centre for the Study and Prevention of Violence (cspv). Institute of Behavioural Sciences, University of Colorado, Boulder.
Tomalin, C. (2011). *Charles Dickens: A life*. London: Penguin.
Turk, A. T. (1969). *Criminality and the legal order*. Chicago: Rand McNally.
Turk, A. T. (1977). Class conflict in criminalization. *Sociological Focus,* 10.
Turley, C., Ranns, H., Callanan, M., Blacknell, B., & Newburn, T. (2012). Delivering neighbourhood policing in partnerships. Home Office Report 61. London.
U. N. (2002). Gender mainstreaming: An overview. New York: United Nations.
UNW. http://www.unwomen.org/en/what-we-do/ending-violence-against-women/creating-safe-public-spaces. Accessed 10 June 2014.
Useem, B., & Piehl, A. M. (2008). *Prison state: The challenge of mass incarceration*. Cambridge: Cambridge University Press.
van den Berg, L., Pol, P. M. J., Mingardo, G., & Speller, C. J. M. (2006). *The safe city: Safety and urban development in european cities*. Aldershot: Ashgate.
Wekerle, G. R., & Whitzman, C. (1995). *Safe cities: Guidelines for planning, design and management.* New York: Van Norstrand.
Wilkinson, R. (2004). Why is violence more common where inequality is greater? *Annals of New York Academy of Science, 1036,* 1–12.
Wilkinson, R., & Pickett, K. (2009). *The spirit level.* London: Allen Lane.
Worrall, J. L. (2002). *Broken windows*. Report for California Institute for County Government in San Bernadino. http://cicg.org/publications/CIGC-Brief.Aug.2002. Accessed 11 July 2012.
WS: Weed and Seed. http://www.crimesolutions.gov. Accessed 10 July 2012.

Chapter 13
Healthy Cities: Old and New Solutions

Wayne K.D. Davies

> *Urban health problems are not the result of one single condition; they are caused by an accumulation of interrelated determinants. The coordinated improvements of health conditions requires a comprehensive vision and strategy that recognize the complex interplay between determinants.*
> Otgaar et al. 2011, p. 11

13.1 Introduction

Throughout most of history towns and cities had higher death rates than those in rural areas, creating what amounted to an urban health disadvantage. This was largely because of their densely packed and insanitary conditions, which made their residents especially prone to diseases spread by contagion and contaminated water. So the very creation of these unhealthy man-made environments was responsible for much of the increases in morbidity and mortality. Of course there were times when some of these problems were reduced. For example, from the mid-third century BCE some of the buildings in major cities in the Indus valley, such as Mohenjaro Daro and Harappa, had installed drainage systems and even running water to remove human waste in rooms we would call toilets (Possehl 2002). In subsequent centuries civilizations such as the Minoans in Crete built underground clay pipes for water supply and sewage removal. But it was Roman engineering in the centuries before the common era that endowed towns and cities in their empire with the most sophisticated water systems to date, bringing clean water from long distances to their settlements, adding water fountains and public toilets near town squares, and encouraging cleanliness through the many public baths, although their use of easily worked lead for some the water pipes did produce some poisoning. Moreover, Vitruvius's classical book on architecture (Morgan 1960, p. 21), probably written during the time of Augustus, included advice to town builders to make sure their

---

W. K.D. Davies (✉)
Department of Geography, University of Calgary, 2500 University Drive, N.W., Calgary, AB T2N 1N4, Canada
e-mail: wdavies@ucalgary.ca

W. K.D. Davies (ed.), *Theme Cities: Solutions for Urban Problems*,
GeoJournal Library 112, DOI 10.1007/978-94-017-9655-2_13

settlements were not sited near marshy ground which were known to be prone to fevers and ague, probably what we now know as malaria.

These technical advances in various early civilizations were often forgotten or ignored by subsequent urban builders. The result was especially disastrous in the rapidly expanding, but unsanitary and overcrowded large towns created in the Industrial Revolution where levels of mortality and premature death reached such critical levels that political action was finally taken to solve the problem, primarily through sanitary, water supply and building regulatory policies. These policies produced the first major changes to improve the health of urban residents over the last two centuries, at least in the countries of what became the developed world, and should be known as the first healthy city movement. The second major change in the health of people in urban places developed from the late nineteenth century onwards. Impressive reductions in mortality and disabilities occurred in part through reductions in family size, better hygiene and nutrition and healthier physical environments because of the public health interventions, but also by unprecedented increases in the medical profession's ability to cure people from many diseases and impaired bodily functions. Together with improvements in prosperity it removed the historic urban disadvantage in health, so that urban areas in the developed world had longer longevity rates than those in rural areas.

By the end of the twentieth century doubts began to be cast on the ability of existing practices alone to further improve the health of the population, especially in urban places. Three new sets of challenges can be recognized. First, new medical difficulties have arisen that threaten to turn back the gains of the past centuries, such as disease-causing bacteria that are resistant to drugs, and the increase in new chronic diseases. They not only pose problems in poor developing countries, where they are compounded by inadequate sanitary and clean water facilities and poor building structures, but in the developed world as the population ages and needs more care. Second there is a revived emphasis upon the so-called 'determinants of health' the factors that influence health and ill-health, many of which are environmental in nature and are not always effectively dealt with by individual medical care procedures and affect health over the long term. For example, one of the biggest contemporary problems is air pollution, especially from fossil burning in urban areas and in households with open fires. A recent World Health Organization report (WHO 2014a) estimates that over 7 million deaths a year occur from this cause, not directly, but by creating premature deaths from heart disease and respiratory tract cancers. The long term effect of this and other factors means there is a need to focus as much upon the reduction of these problems and the promotion of health, as upon treatments of diseases or decaying body functions. Third, various agencies, from the World Health Organization's Healthy City Programme that began in 1986, to local governments and community groups, have been advocating new ways of improving health in cities. These amount to new organizational changes and targeting of particular problems and have a strong urban basis, for this is where increasing numbers of people live. Like so many of the new urban themes that have been advocated in recent years, the approach stresses the need for a commitment to look beyond current practice in health care, as well as the need for political, community and wider stakeholder involvement as this WHO statement shows.

> Being a Healthy City depends not on current health infrastructure, rather upon, a commitment to improve a city's environs and a willingness to forge the necessary connections in political, economic, and social arenas.... It aims to: to create a health-supportive environment; to achieve a good quality of life; to provide basic sanitation and hygiene needs; to supply access to health care. (WHO 1998)

The emphasis upon factors other than the existing health care system requires a more holistic view of the various determinants that affect the health of the population in urban areas. In many ways it represents a new emphasis upon a different type of *public health* approach, one which has gradually declined from the 1880s after the success of the sanitary approach in improving the physical environment. This has has strong links with emerging ecological ideas (Ashton and Ubido 1991).

This chapter begins the review of all these changes by summarizing these new attitudes to health, not just ill-health, before reviewing the nineteenth century urban improvements that helped reverse the high mortality rates. This is not included only for historical reasons but because it provides an exemplar of policies that are still needed in many cities in developing countries today if the health of their urban residents is to be improved. The historical review is followed by sections providing a summary of the very serious new challenges to medical care, followed by descriptions of the policies that can be used to improve the various health determinants and then the initiatives of the Healthy City movement. Although many of the policies may be made and financed by national governments, there is not only an important urban component in the delivery of the services, but also the increasing realization that life in an urban habitat may be creating new health problems.

13.2 Changing Concepts of Health

Health is not an easy concept to define. Superficially it seems to relate to the idea of a population that is disease-free. However, the constitution of the World Health Organization has described health in a wider way: "*Health is a state of complete physical, mental and social well-being, and not merely the absence of illness*" (WHO 1948). The trouble with this health definition is that the use of the word 'well-being' covers much of the human condition and does not lend itself to easy measurements. Nevertheless the phrases 'well-being' and 'absence of illness' do mean a wide view of health issues is envisaged. Later this led to a more wide ranging statement about health promotion, as much as curing illness, from the European office of the WHO in its Ottawa Charter for Health Promotion in 1986.

> Health promotion is the process of enabling people to increase control over, and to improve, their health....To reach a state of complete physical mental and social wellbeing, an individual or group must be able to identify and to realize aspirations, to satisfy needs, and to change or cope with the environment. Health is a positive concept emphasizing social and personal resources, as well as physical capacities. Therefore, health promotion is not just the responsibility of the health sector, but goes beyond healthy lifestyles to wellbeing. (WHO-E 1986, p. 1)

This description emphasizes that the health sector cannot solve health problems alone. It has led to the popularization of the concept of 'wellness', which is being used to describe a more holistic approach to the improvement of health, emphasizing health promotion, as much as treatment.

> Health is not simply the absence of illness and disease, but as something we build with our families, schools, communities and workplaces, in our parks and playgrounds, the places we live, the air we breathe, the water we drink and the choices we make. (Alberta Health 20142014, p. 4)

So there is a new emphasis upon prevention, of encouraging healthy living, not just treating our way out of ill-health through existing and future medical knowledge. Yet a more healthy living is not the only benefit; the work of researchers discussed in the Knowledge City discussion (Chap. 11) has shown the link between I.Q. growth and disease reduction, which means a healthier population is likely to be a more productive and innovative one, which assists future economic and social progress (Eppig et al. 2010). These new perspectives have led to the need to pay more attention to the many factors, or health determinants, that affect health in both the long as well as the short term. They lead to interventions designed to create healthy cities that are not limited to the current health care system.

13.3 The First Healthy Cities Movement

13.3.1 Mortality Crises

It was well known throughout history that urban places were less healthy than rural areas and from the seventeenth century some individuals provided quantitative evidence of these differences. For example, John Gaunt (1662) in England showed that 1 in 30 died annually in London, whereas the rates in the country were 1 in 50, with a third of deaths from infant mortality. Plagues were particularly feared. Although always present, there were extreme epidemic plague years, usually at 10 or 20 year intervals in London but with even more severe outbreaks, such as those in 1563 and 1625 where deaths increased five to six times above normal rates (Harding 2012, p. 31). Limited medical knowledge meant that there was no clear understanding of the causes of the high mortality but most attributed it to miasma, the effects of foul air and filth in these densely populated, unsanitary urban areas. Within cities there were important variations in mortality rates as shown by Villermé's pioneering study in Paris in 1817 which showed how wealthy areas, such as the second and third arrondissements, had annual mortality rates of 1 in 62 and 1 in 60 respectively, whereas the two poorest areas, arrondissements 11 and 8, had higher rates of 1 in 43 (La Berge 1992). In recent years these historical spatial variations within cities have been investigated more thoroughly. For example, it has been demonstrated that the richer areas in the centre of seventeenth and eighteenth century London, with its more substantial houses and a better fed and clothed population, had far lower mortalities than either areas outside the wall in the inner suburbs, or the outer edges (Harding 2012). As their densities increased with more poor housing after the Industrial Revolution because of low construction standards, these areas had even higher death rates.

A major change came in the mid nineteenth century when dramatic spikes in mortality, especially associated with severe cholera outbreaks, focused attention on the problem. In Britain, Edwin Chadwick's pioneering report on the sanitation of cities (1843) showed how mortality varied with the physical and social composition of residential areas. Social commentators (Engels 1844; Mayhew et al. 1851–1862) as well as novelists such as Dickens (Tomalin 2011), described in horrific detail the squalid conditions in which the poorest lived, worked and died in London and other big cities, where many infectious diseases were rampant (Harding 2012, p. 40). But it was the maps of Dr John Snow, showing higher cholera deaths around a Soho well in London in 1854, and his demonstration of their link with its polluted water that gradually led to the acceptance that wells or rivers containing water contaminated by faecal matter and rubbish led to the growth of what we now know as the bacteria that cause cholera (Johnston 2006; Frerichs n.d.). The eventual result, although it took time, was to replace the historic belief that bad air caused disease, with one that stressed the role of contaminated water, which eventually led to the policies that built sanitary systems and the new specialism of epidemiology, of which Snow is often considered the originator.

Although there was a great deal of dissention in the early nineteenth century about the causes of death rate increases in the big cities, it is now clear that they came from a combination of eight, essentially built-environment problems. The first was the *overcrowding* in many houses and areas which increased the risk of contagion from infectious communicable diseases. The second was that urban growth overwhelmed the local ability to provide fresh water supplies, or to remove faecal material, not only from humans but from the horses that were the main form of transport and fouled the main streets, and the many animals, pigs, cows and chickens kept for fresh food. This led to well-water *contamination*, as well as piles of waste, including human faeces, that resulted in outbreaks of diseases such as typhoid and cholera in what was literally a stinking environment. A third problem came from the *lack of cleanliness* and hygiene of many in the population, making the spread of disease easier in such overcrowded areas. A fourth lay in the *limited building regulations,* as well as their adequate enforcement, either in sites chosen or construction standards. The former issue led to structures being built on wet marshy ground, or near polluted streams. The latter led to damp, cold buildings with little warmth, small rooms and with few windows. Buildings in poor areas were often built in courts to increase density, creating areas rarely penetrated by the sun, while cellars were often used as dwellings. These conditions increased the risk of respiratory diseases and also rickets because of an absence of sunlight, as well as increasing the danger from fires.

A fifth problem came from *pollution*. This was not just the stink from faecal and other rotting matter in urban places, but from the noxious fumes and toxic waste products produced by the many industrial plants that were mixed up with residences, and especially from the typical coal-fuelled household fires that may have warmed rooms, but led to indoor pollution. Outside the buildings the emissions of gases and particles from coal fires often created choking fogs, full of particulates, in damp, river valley environments subject to inversions, adding to the respiratory risks. This was related to a sixth problem, the often *dangerous working conditions* in mines and industrial plants, or even in home-based industries, which led to deaths or injuries from machinery or various industrial processes, especially those that gave off noxious fumes or by-products. A seventh problem came from the increase and concentrations of *poor people* in the cities. Their poverty was due to the limited wages and often intermittent work, which left most with limited means to buy nutritious food, warm clothing or adequate shelter. Combined with poor hygiene it made them more prone to disease. An eighth problem came from the *limited medical knowledge and access* to it at the time. This was made worse by the reluctance of many to embrace new ideas, while the poor had too little money to even get medical assistance, unless it was from charity provided by church organizations. The effect of these problems on the lower class in particular was calamitous, as seen in a description by Engels of the labouring classes who were:

> … For the most part, weak, thin and pale. Their weakened bodies are in no condition to withstand illness and whatever infection is abroad, they fall victims to it. Consequently they age prematurely and die young. (Engels 1844, pp. 118–119)

These descriptions by Engels and the more substantial works 50 years later by Charles Booth (Pfautz 1967; Davies 1978), who produced detailed maps of the degree of poverty in each street in London, district measurements of the degree of social condition in major districts (Davies 1978), and seventeen volumes on the conditions of life and work-places in the metropolis, identify many of what we now call the social determinants of health. Only a few in these slum areas were able to escape the environment within which life was lived, which created the milieu for urban ill-health and premature mortality, although conditions were better in the more prosperous areas of well-built and maintained houses. Hence, the physical build and social environment in which so many of the disadvantaged lived lay at the root of their ill-health and early mortality.

13.3.2 Changing the Built-Environment of Cities

The work of Chadwick, the descriptions of novelists and social commentators, as well as doctors such as Snow (Frerichs n.d.), in exposing these problems eventually led to increasing political debate about the need to improve these environments (Berridge and Gorsky 2012). Some of the pressure for change came from altruism, the view that we are all humans and that the poor should be helped, a particular belief among Christian religious organizations. Also important in supporting built-environment changes was the realization by many in the political classes that an unfit, weakened population needed to be improved if the country was able to assert and defend itself by healthy effective armed men. These reasons, plus what amounted to panic from the high spikes of mortality, especially during the cholera outbreaks, also meant the better-off felt threatened. It is also important to emphasize the role played by associations such as Health of Towns Association established in Britain in 1844. Branches were quickly established by business and social elites in most major centres. Their members organized public meetings to present facts on mortality variations and exerted great pressure on governments to create laws that led to the development of sanitary and clean water systems in towns, as well as Public Health officials to monitor and often agitate for change (Ashton and Ubido 1991). They helped promote the idea that there must be improvements in the physical fabric of towns and their infrastructures. Yet there were still those who believed the fault for the conditions in these areas of squalor lay in the poor themselves, either because people were too lazy or immoral to improve their condition, or they were simply inferior and prone to disease.

Four main sets of changes were crucial in transforming the urban health conditions of many settlements. The main one consisted of *improvements in the physical fabric of cities*, helped by a series of major public health acts and new technologies. For example, in Britain, the 1845 and 1875 Public Health Acts laid down enforceable building standards, not simply in construction but in minimum sizes of houses and lay-out, plus tarmac roads, as well as the requirement for small gardens and at least outside toilet facilities (Hall 1988). Critical in these changes were the additional requirements for a clean water supply and new sewer lines to connect the housing areas to plants to disperse and process the sewage, although initially many dumped it into the nearest water body, where many obtained their water. Increasingly it was the responsibility of the municipal government to provide such facilities, in addition to fire protection, garbage collection services, and the banning of animal husbandry in urban areas to reduce their waste. This led to greater powers for local governments in providing these services, replacing previously inadequate and sometimes private provision, as well as the important addition of Medical Officers of Health to monitor health conditions in local areas. Although often known as the 'sanitation phase' it may be better summarized as a *regulatory and engineering phase of improvement.* It was the engineering advances in supplying water from often distant reservoirs, and then filtering water—first through sand and later disinfection with chlorine to kill microbes—that led to clean urban water supplies, which were later helped by the invention of engines that allowed water to be pressurized for further distribution to residences. Another major change came with the development of toilets, using the new water supplies, where the flushed water containing human waste, was carried away by new sewer systems that also removed other waste water. Later, the addition of sewage processing plants to process the effluent improved the systems, instead of the earlier approach of just dumping it into the nearest water body, although far too many urban places still do. Also the new regulations included effective inspection systems that controlled and improved building quality and industrial plants, and eventually, the safe disposal of toxic by-products. Overcrowding was gradually reduced by decisions to eradicate older slum areas and build at lower densities, while the development of unions gave workers more ability to argue for workplace safety, as well as better wages.

The creation of a better physical urban environment helped reduce the spread of water and airborne communicable diseases. A second set of factors dealing with broad social changes in health improvement, were also influential. Lower fertility rates decreased family size and reduced child mortality. Increases in prosperity led not only to better nutrition from more food supplies, but better clothing and housing. More cleanliness and a knowledge of hygiene was also important. In addition there was a gradual and crucial improvement of medical knowledge and nursing capabilities.

A third trend came from greater acceptance of a *new valuation of nature* and appreciation of the utility of green space that began with the Romantic revolution in art. This stimulated the creation of public parks for leisure and escape from the noise and filth of the city, and cemeteries for the safe disposal of dead bodies (Reps 1965). These ideas later led to the ideas of adding green space to the planning of new subdivisions. Restricted at first to wealthy residents, these planning ideas gradually filtered down to middle class developments and to model towns of workers (Davies and Herbert 1993) and eventually to the Garden City ideas. They led to a more comprehensive approach, advocates for a complete town combining the best of the town and country, whose ideas influenced subsequent suburban designs, issues discussed in Chaps. 2 and 4 especially. New ideas about the restorative values of nature also led to the belief that it could contribute to better health, which in turn was assumed to develop an improved morality, unlike the social degeneration seen in cities and especially slum areas which were thought by many to breed moral, as well as physical diseases. In Britain it led to many new movements designed to encourage a more active and outdoors life-style, from the addition of sports and physical exercise in schools, to the Scout and Guides movement from 1908, with their outdoor orientation and camping trips, to the 1926 New Health Society which sought to improve nutrition, as well as to the Sunlight League in 1926, which extolled the health-giving properties of sunlight (Carter 2012). Similar movements occurred in many countries, especially in Germany where there was greater emphasis on the need for hygiene to be taught in schools and to be part of everyday life, stemming from the work of Dr Weyl's 10 volume treatise on Hygiene that began in 1893 (Schott 2012). Although many these social movements of the early twentieth century faded during the war period and its austere aftermath, the revival of the environmental movements from the 1960s and the more recent ideas of the nature deficit in children discussed in Chap. 4, can, in part, be regarded as the partial inheritors of these earlier traditions.

A fourth transforming trend was less a matter of policy or organizations than of fashion and life-style, namely the benefits of *spa and seaside towns*. These were regarded as being sited in healthier environments, allowing the opportunity to improve health by sojourns in these centres. The earliest examples came from the expansion of spa towns from the seventeenth century, places with mineral springs which were assumed to have therapeutic properties, obtained by either immersion in, or by drinking such waters (Adams 2012). Many of the early European spa towns were revivals of settlements on old Roman mineral water sites and were primarily patronised by the elite, which meant they became as much social, as medical, centres. Up to the 1930s they were patronized primarily by the elite and their therapeutic values were highly regarded. But in the U.K. many declined after the creation of the National Health Service in 1948 because this nationalized system was not prepared, except in a few cases, to subsidize patients to attend what were expensive courses of treatment. Although there has been a recent revival of some spas in Britain and the development of many new private ones, it has been concluded that they are based primarily upon methods of treatment, rather than the particular place properties of climate or water (Adams 2012). However, in continental Europe there has been a much greater continuity and use of many of the old spas, as health providers were more likely to prescribe such courses of treatment for all classes. The bracing qualities of sea air was also recognized and promoted from the late eighteenth century, providing a healthy alternative to the stink, noise and fevers found in cities. Again it was the upper classes who led the initial growth of these centres. With increasing affluence by the end of the nineteenth century, the middle classes, and eventually the working classes after the introduction of paid holiday time, also found temporary solace in these seaside places. Bodily emersion in the sea, exercise by swimming, and the apparently beneficial effects of tanning, were added to the initial atmospheric benefits, although the fear of skin cancer from over-exposure to the sun has reduced the time people spend on tanning in the last 20 years.

13.4 The Biomedical Approach and Its New Challenges

13.4.1 Medical Improvements and Dominance

By the end of the nineteenth century the changes described above began the health transformation of many urban places in developed countries that led to a rapid decline in premature deaths. Soon after, a new phase of improved medical knowledge, training, surgical techniques, the use of new technologies, and the discovery of antibacterial drugs and vaccinations for many diseases, increasingly played a major part in reducing deaths. Although advances in sanitation and environmental improvement were still being made, what became known as the biomedical model became dominant in health care. This model

> attributes morbidity and mortality to molecular level pathogens brought about by individual life-styles, hereditary biology, or genetics, and it altered public health to personal 'risk' factors such as smoking, diet and exercise. (Corburn 2009, p. 49)

Although local government continued the progress of ensuring improvements in the physical fabric, and Public Health officials monitored local health conditions, the growth and dominance of what amounted to germ theory ideas and the success of individual cure and care approach by doctors and hospitals meant it was the medical profession, not city officials, as in the engineering and regulation phase, that increasingly became the main decision-makers in health care. In addition, there was a move away from what is usually called the 'urban field view' in health terms, where residents and professionals searched for the particular qualities of place, caused by the interactions between the various elements, that caused ill-health. This was seen in the early sanitary phase, which led to more context-specific and localized policy responses. Although the move to laboratory sites and techniques produced great results, the reduced interest from what were real world, not controlled conditions, meant that the local milieus that caused or assisted diseases were often underestimated.

In addition, the new profession of planning increasingly downplayed the need to put health issues in the forefront of their concerns. Certainly many planning departments emphasized the need for a more ordered and efficient land use distribution, especially to ensure that noxious industry would be separated from residential areas, while ensuring the latter would have various effective public facilities, from roads, sewers, schools, recreational areas etc. These issues were connected to health improvements. But Corburn (2009, p. 41) has described how attempts were made by some late nineteenth century American planners (Marsh 1909), to argue that the planning profession should focus primarily on the key issues of social justice and the health of cities that had motivated the nineteenth century reformers. Such opinions failed to influence leaders of the new American Planning Association. So efficiency and also aesthetics, such as the City Beautiful ideas, became the main emphasis. Yet planners often assumed that efficient land use planning was itself a way of improving health, especially after acceptance of the Neighbourhood Unit principle, with its insistence on lots of green space, the incorporation of many facilities within the unit, and the positioning of main roads outside the unit. In addition, the removal of slum areas and their replacement with new tower blocks was also seen as a way of improving the housing conditions of the poorest classes and hence their health. However there is more than a whiff of social determinism in planning, since it is assumed that the provision of new housing in these new low density layouts and new high-rises would automatically alter behaviours and improve health. Advocates did not anticipate that the neighbourhood units would became car-suburbs as vehicle ownership increased, contributing to the lack of exercise, or that many of the high estates would become crime-ridden areas with few jobs, limited security and social isolation, creating new areas of deprivation, ill-health and often racial segregation. Hence in health terms a new set of problems emerged in these newly developed urban environments, while disadvantaged areas persisted. These were not effectively addressed by municipal planners. Also the increasing domination of the biomedical model of medicine meant that despite the sterling work of community health specialists, most attention was focused on the cure of the diseases of individuals, not on wider and often long term issues that contributed to these problems, or on the recognition that decision-making on health care issues should involve more than medical professionals.

13.4.2 New Medical Challenges for Healthy Cities

At first sight the impressive gains in medical knowledge, care, and disease prevention, and their constant improvements over the past century, would imply that the health of people in urban areas would continue to improve. However new problems threaten to reduce the life-span of people in coming generations compared to their parents, a potential general health reversal unknown in the last century, except in exceptional circumstances such as war. Among the varied problems, eight in particular, demand solutions through new medical advances and care and some of these may be influenced by the increasing urbanization trends.

13.4.2.1 The Rise of Chronic Diseases

Table 13.1 shows that the WHO estimates of the top twenty reasons for the direct causes of mortality in 2015 and 2030 is no longer dominated by the communicable diseases that so devastated the populations of cities especially in the past; most seem to have been conquered. Today the majority of the deaths come from so-called chronic diseases, linked to organ failures, such as heart attacks and strokes, to cancers of various types, and diabetes. These diseases now account for 35 million deaths a year

Table 13.1 Estimated changes in mortalities: top 20 causes (Source: Revised from tables in World Health Organization 2012a)

2015 Rank of causes of death	Numbers (%)	2030 Numbers (%)	Rank
1. Ischaemic heart disease	7.59 m (13.2 %)	9.25 m (13.2 %)	1
2. Stroke	6.70 m (11.7 %)	8.58 m (12.2 %)	2
3. Lower respiratory infections	3.22 m (5.6 %)	3.54 m (5.0 %)	4
4. Chronic pulmonary disease	3.22 m (5.6 %)	4.57 m (6.5 %)	3
5. Diarrhoeal diseases	*1.81 m (3.2 %)*	*1.62 m (2.3 %)*	*9 down*
6. HIV/AIDS	1.67 m (2.9 %)	1.79 m (2.6 %)	8
7. Trachea, bronchus, lung cancers	1.64 m (2.9 %)	2.41 m (3.4 %)	6
8. Diabetes mellitus	**1.56 m (2.7 %)**	**2.46 m (3.5 %)**	**5 up**
9. Road injury	1.42 m (2.5 %)	1.86 m (2.6 %)	7
10. Heart disease (hypertensive)	1.14 m (2.0 %)	1.46 m (2.1 %)	10
11. Preterm birth complications	*1.13 m (2.0 %)*	*0.92 m (1.3 %)*	*19 down*
12. Cirrhosis of liver	1.03 m (1.8 %)	1.2 m (1.7 %)	11
13. Tuberculosis	*0.89 m (1.5 %)*	[a]	*Down*
14 Kidney diseases	0.87 m (1.5 %)	1.15 m (1.6 %)	13
15. Self-harm	0.84 m (1.5 %)	1.01 m (1.4 %)	16
16. Liver cancer	**0.83 m (1.4 %)**	**1.19 m (1.7 %)**	**12 Up**
17. Stomach cancer	**0.80 m (1.4 %)**	**1.14 m (1.6 %)**	**14 Up**
18. *Birth asphyxia-traumas*	*0.77 m (1.3 %)*	[a]	*Down*
19. Colon, rectum cancers	**0.75 m (1.3 %)**	**1.08 m (1.5 %)**	**15 Up**
20. Falls	**0.72 m (1.2 %)**	**0.98 m (1.4 %)**	**17 Up**
Dementias (Alzheimer's etc)		**0.97 m (1.4 %)**	**18 Up**
Breast cancers		**0.81 m (1.1 %)**	**20 Up**

Italics show diseases estimated to decrease by over two rank: 9 down
Bold print. 2030 rank estimated to increase by over two grades: *18 up*
[a] Diseases that are estimated to drop out of the top 20

of a total mortality of 57 million, which has led to calls for urgent world attention to be paid to these problems (WHO 2010). Yet despite the major advances that have taken place in the treatment of such diseases, especially if the problem is diagnosed early, Table 13.1 shows they are still predicted to be the highest ranking sources of mortality in the foreseeable future. Moreover, these chronic diseases are no longer restricted to the developed world, for 80 % of the deaths from such causes now take place in low or medium income countries, places that are the least equipped to deal with such problems. Clearly there is need for more research into the cures for these problems, some of which come from long exposures to unhealthy environments or poor life-styles over many years, which means better health education and monitoring. Also more treatments for such diseases in the developing world is needed. Indeed it has been estimated that less than 3 % of the international development assistance going to developing countries is spent on non-communicative diseases (WHO 2010), showing the continued fixation with diseases spread by infections. A related issue here is the increased costs of providing care for patients with such chronic conditions, which require longer periods of care and greater expense.

13.4.2.2 Aging

These changes in the causes of mortality are also related to the general problem of coping with aging, for it is estimated that the global population of those over 65 years will almost double from the current 600 million to 1.1 billion by 2035, resulting in a change from 8 to 13 % of the population. Some of these people will be healthy and active but many will have ill-health and there is predicted to be major increases in the numbers of people suffering from various forms of dementia, such as Alzheimer's, for which there is no known cure at present, except for interventions that help to delay or reduce its impacts. Recent estimates show that 35.6 million people in the world are already suffering from this disease of cognitive impairment and loss, conditions that already leads to health care costs of $ 604 billion (WHO-ADI 2012). The numbers with dementia are predicted to almost double by 2030 and will more than triple to 115.4 million sufferers by 2050. This will require much larger numbers of senior centres and homes to cope with, and treat, the growing number of patients, as well as finding enough qualified staff. It has been known that many life-style factors are correlated with those who have the disease; for example, people who are active and socially integrated and those with higher educational attainments less likely to be at risk. In 2014 a major new study focused on the impact of a series of modifiable life-style factors on the incidence of Alzheimer's, a major type of dementia (Norton et al. 2014). It revealed that seven factors—physical inactivity, smoking, mid-life hypertension, midlife obesity, diabetes, depression and low educational attainment—were all significantly correlated with the disease and that one-third of cases could be avoided by changes in life-style. For example, those who did not engage in at least three 20 min bursts of rigorous exercise a week were 82 % more likely to develop the disease. The importance of the study is that the estimated huge increase in numbers could be drastically reduced by these life-style changes. The problem of coping with this wave of new cases will be hardest in the developing world, where there are less resources to deal with the problem, especially in cities where there are likely to be fewer family or long term friends to help with care.

Although increases in the elderly population is a growing problem in most developed and middle income countries, there are marked spatial inequalities in this effect, with aging in Japan occurring more rapidly than in other developed countries. It is creating one of the most severe current problems, but also provides a warning to other countries. The population over 65 years of age in Japan reached 23 % of the total in 2011 and is projected to be 38 % by 2050 with the population declining by a third from the current 127 million in 50 years. It is estimated that by 2035 there will be 69 persons over 65 years of age in Japan compared to 100 in the working population (25–64 years), up from 43 in 2011, whereas the world today has only 16 in the over 65 age category. Some of the effects of the aging trend may be reduced by raising fertility rates and by people working longer. But these skewed future rates will not only require more health care but will lead to potential GDP losses unless labour is replaced by capital and more immigration is allowed, which in the case of Japan, has never been a policy palatable to the majority. Since caring for this aging population is now a major problem, and many villages and small towns will soon be dominated by old people, Japan instituted a long care elderly insurance scheme. This is funded by compulsory taxes on people over 40 years of age and other taxes. Few other countries or cities facing aging populations are making such provision. Indeed in some countries, such as the U.K., government cutbacks since the 2007 financial depression has led to major cuts in the budgets of local authorities and care for the aged is falling at a time when there is greater need. However in some cities there is a gradual realization of the need to plan for this aging process. For example, in many American cities, such as New York and Cleveland, there are now Departments of Aging in the municipal organization, which are designed to prevent elder abuse and also to provide programmes to assist the elderly, especially by creating on-line websites that list the various services that seniors can access and use to obtain help. A more general approach to mitigate the effect of aging occurred with the WHO's establishment of the Global Network of Age-Friendly Cities and Communities in 2006 (GNAFCC). Initially based on 33 cities in 22 countries the network is designed to provide guidance to other cities wishing to implement policies that promote healthy and active aging by creating an active and accessible urban environment. Key components are the involvement of seniors in the discussions and policy formulation and to provide indicators to measure progress as well as evaluations of their effectiveness in subsequent years. Eight domains of city life have been identified in a guide for age-friendly cities, as well as communities, within urban places (WHO 2007). These domains are in the fields of: transportation; respect and social inclusion; outdoor spaces and buildings; social participation; housing; information and communication; community support and health services; civic partnership and employment opportunities. Historically many churches, and neighbourhood networks based on long residence, as well as families have provided fellowship and help for seniors. But in a more mobile and secular society these local support systems have decayed, making it imperative to plan for an active older population, not simply to house them. It is tragic that so many old people's homes are built in remote locations in cities, with few places to walk to, or to obtain goods, resulting in the almost warehousing and segregation of old people, rather than building premises to support active living near shopping or transit nodes and also adopting the other recommendations of the WHO's age-friendly city network (GNAFCC) issues largely ignored by New Urbanism and other themes.

13.4.2.3 Limited Care and Regulations

One of the biggest challenges to modern medicine is to create a more equitable pattern of 'cure and care'. The poorest people in developed countries and the majority in the developing world still have limited access to health care facilities, even of a basic level, with resultant high and unnecessary death rates from diseases where cures are available. One stark example can be seen in Table 13.1 which shows the persistence of high death rates from the old scourges of childbirth, such as preterm birth problems (11) and birth trauma (18). A large proportion of these 1.9 million deaths occur in rural areas, which could be cured by access to modern medical knowledge and better health care, which should occur with greater urbanization, one of the reasons why both diseases are expected to drop substantially in the next 15 years and result in lower rankings.

13.4.2.4 Control and Threats of Communicable Diseases

Much of the reduction in mortality and ill-health rates in urban places has been due to the way that the most widespread communicable disease threats, such as measles, chicken pox, polio or tetanus, have been controlled by vaccinations that provide adaptive immunity against the pathogens that cause these diseases, and by antibiotic drugs that kill bacterial infections. So only respiratory infections (no. 3 in Table 13.1), diarrhoeal diseases (no. 5) and tuberculosis (no. 13) are found in the top twenty of the major causes of death today (WHO 2012a). In total only a fifth of the annual deaths in the world now come from infectious diseases. Yet the proportions vary spatially, with levels of 40 % in the less developed world to close to only 7 % in developed countries such as the U.K. (Davies et al. 2013, p. 27). The limited availability of effective health care, antiseptic methods and the availability of vaccinations and antibiotics in the developing world, combined with limited clean water, sanitary facilities, and more airborne pollutants, means that these treatable diseases still kill far too many people, given existing medical knowledge. So there is still a lot of room for real progress in the fight against such diseases in the settlements of poorer countries. Moreover, the current control over the most infectious diseases should not lead to complacency. Some, such as smallpox (Kopolow 2003) have been eradicated, with polio well on the way until new outbreaks occurred in Afghanistan, Pakistan and Northern Nigeria, a result of war and fanatical

Islamists preventing vaccinations. Others, such as increasing cases of tuberculosis in cities such as London, seem to come from immigrants arriving with the infection from countries where the disease is still endemic. This means that more effective screening of the health background of immigrants is probably needed. Also, there is always the threat of new outbreaks of many of the diseases largely controlled, because of increasing numbers of people refusing to accept vaccinations. So vigilance to spot new outbreaks and their source, as well as mass vaccination programmes are still needed. There is also the threat of new, previously unknown diseases appearing, with the case of the acquired immunodeficiency virus (AIDS) and (Severe Acute Respiratory Syndrome (SARS) providing opposite examples of early failure and success in coping with the problems. In the former case Holt (2014) has described how the growth of the disease caused by a virus, spread through sexual contact and contaminated blood, which quickly mutates and was difficult to treat. It led to a high of 2.2 million annual deaths. Originally unrecognized from its origin in infected monkeys in Africa its spread from the 1980s was helped by governments, such as South Africa, which refused basic treatment to victims. Today antiretroviral drugs can halt its progression, allowing people to lead an active life, although they are not cured. So since 35.3 million are infected worldwide, the virus is still estimated to be the sixth most common cause of death in 2030 (Table 13.1) and remains as a major threat. By contrast, the respiratory disease, SARS that spread between November 2002 and July 2003 provides an example of the dangers that can quickly emerge in our connected world (Enserink 2013). It originated from a species jump of the virus to a Chinese farmer from its host in bats and civets and then to others in local markets and Hong Kong. In this case it was relatively quickly diagnosed when travellers from Hong Kong brought it back to North America due to vigilance of doctors in major hospitals in Canada and the United States, alerted to a new disease in southern China, and its rapid identification by Canadian researchers and The U.S. Centre for Disease Control and Prevention. Victims were isolated in special wards, with their contacts being tracked and warned (Enserink 2013). In addition, civets in the southern Chinese markets that acted as hosts were destroyed. The use of masks and other protective clothing by carers, and the careful cleaning and sanitation of all surfaces near infected people, also reduced the spread of the infections. In the longer term, the search for cures is vital. What the example also shows is the need to have large numbers of wards or places available to be turned into isolation rooms if a similar major outbreak occurs—a scale of resources which may still be rare in most urban centres, even in the developed world. In the past, separate quarantine hospitals, and even areas, were set aside in or near cities for people with infectious diseases. For example, North Brothers Island in New York fulfilled this role, especially for immigrants, although it is now abandoned and its buildings are decaying. If infectious diseases cannot be controlled by vaccines and drugs this type of separation may again have to be practiced. The history of the SARS outbreak shows the need for rapid genome typing of a suspected new disease and the ability to quickly create large numbers of isolation wards to quarantine victims. It illustrates how a multipronged and international effort is needed to prevent the spread of new communicable diseases. Yet there is little doubt that there are many diseases that are still unknown, or like the hemorrhagic fever, Ebola, kills 70% or more of infected people and does not have a cure. Until recently it had not spread beyond a few isolated areas of Africa and seemed to be containable. But in 2014 a major outbreak occurred in three countries of West Africa, which has raised the spectre of a major spread. Assistance in dealing with the outbreak is being helped by the addition of skilled international medical teams, but the burial practices of many cultures in the area, plus suspicion of western medical practices and the desire to avoid the stigma of having a family member diagnosed with the disease, handicapped containment, showing how cultural factors and very limited medical facilities impede medical progress. With the exception of HIV-Aids the world has managed to avoid pandemics for almost a hundred years since the so-called Spanish flu, a virus causing a virulent form of influenza that killed at least 50 million and infected 500 million world-wide in 1918–1920. But the new travel contacts and the increasing concentrations of people in urban places, puts this generally positive history of controlling communicable diseases in the twentieth century at greater risk, which is also linked to another emerging problem.

13.4.2.5 The Drugs Don't Work

This is the title of a provocative and chilling small book by the British Medical Officer of Health and her colleagues (Davies et al. 2013). In it she describes an emerging crisis of Antimicrobial Resistance (AMR) caused by the fact that many microorganisms (bacteria, fungus, virus or parasites) no longer respond to the drugs that have been invented to counteract their effect in creating various diseases. Although reporting is flawed, it is estimated that 25,000 people die from such problems in Europe annually from the inability of formerly effective drugs to work, more than are killed on roads. Almost similar levels of deaths occur from AMR in the U.S.A. (CDC 2014; WHO 2014b). The reason is that the bacteria have developed resistance to the drugs that killed them. It was something that Sir Alexander Fleming warned about after accepting the Nobel prize for his discovery of penicillin in 1923, namely that it is a fact of nature that bacteria and viruses mutate and develop new forms, and those new disease-causing types will be resistant to drugs that previously killed them. Hence no drug is going to be effective forever. The scale of this problem should not be underestimated, for similar conclusions have been reached by other reports on the problem in the U.S.A. (CDC 2O14) and by the WHO (2014b) in its first comprehensive survey of a number of AMR bacteria. If this mortality was linked to some visible, known specific disease, there would be public pressure for a cure. But since the deaths from these resistant microbes are scattered throughout the country, they appear to be almost hidden, so the scale of the problem has been underestimated until the last few years. The detailed tables in the WHO report (2014a) also show that there are great variations between countries in the scale of the problem, which varies with different disease-causing microorganisms. Some of the biggest worries are seen in the case of tuberculosis where 20 % of previously treated cases proved to be resistant to multidrug treatment cases in some countries, while in many places gonorrhoea cases are now being treated with the final drug that works. In the case of Cyprus 36 % of a sample of people tested proved resistant to anti *E. coli* drugs, while 68 % of sampled people in Greece were resistant to microorganisms that caused pneumonia (WHO 2014b). In addition, South East Asia is a region where new resistances to anti-malarial drugs have emerged. Certainly the sample sizes vary in each case study and cannot be used to conclude they represent the whole population. But the evidence is sufficiently worrying for the authors of the three main reports to date to use the words 'crisis' and even 'catastrophe' in descriptions of the problem. Diseases that were thought to be controlled could return and kill large numbers, while the drugs routinely used to prevent infections after surgery are proving less effective in many areas. In some areas multi-drug treatment has proved effective, but this is very expensive and involves specialised individual treatment that may be the equivalent of $ 80,000 annually, making their use out of the question for poorer people.

The problem of AMR is made worse because the major pharmaceutical companies are no longer researching new antimicrobial drugs to replace those previously effective. From 1934 to 1968 14 classes of antibiotics were created, but only five since that time and none since 1987 (Davies 2013) while currently the WHO (2014b) recognizes 27 different types of effective vaccinations. The reason for the decline in antibiotic development is the costs of innovation, perhaps over $ 1 billion in research, testing and satisfying the regulatory agencies. Companies believe the costs do not justify the potential financial return, given that many of the drugs may be needed only once or on a few occasions to cure some infection—unlike the daily drug regimes for many suffers of chronic diseases. Also, limited patent time, and the small returns from poor countries where patients cannot afford the drugs although the need is high, adds to the reluctance to invest in AMR development. So a 'market failure' has occurred, which means the inability of the market to prove a strong enough attraction to create new AMR drugs. Britain's Chief Medical Officer of Health has identified the problem starkly.

> We are losing the battle against infectious diseases. Bacteria are fighting back and are becoming resistant to modern science. In short the drugs don't work… Our response needs to be global and multifaceted to… manage and mitigate the risk of antimicrobial resistance, which is just as important and deadly as climate change and international terrorism. (Davies et al. 2013, pp. ix–xiii)

There have been many suggestions to solve the emerging crisis. One is to resolve the market failure that has reduced research. This can be done by helping companies to finance new drug research, increasing the participation of government and university researchers, as well as encouraging charitable organizations to work with companies in joint venture financing, such as the Medicines for Malaria Venture. In addition, quicker means of identifying new bacterial and viral strains are needed, which is happening through the development of genome sequencing while new serums may be effective. Linked to this is the need to have rapid communication through international health agencies about the appearance of new infections, allowing countries to take precautions, which is the role of America's AMR monitoring system set up in 1996 and the WHO's Geneva hub (CDC 2013). What is also needed is low-tech advice, such as better education to reduce the spread of infections by thorough hand-washing and sterilization on surfaces near infected patients. Also there is need for more radical cleaning methods in hospitals, such as the vaporizing technique that sterilizes everything in hospital wards (Zoutman et al. 2011). There are some signs of success in these efforts, for more careful procedures have reduced rates of a particularly troublesome bacterium called methicillin-resistant *Staphylococcus aureus* bacterium (MRSA) in Britain by 80 % since the 2008 peak, when increasing numbers were picking up infections in hospitals (Davies et al. 2013).

It has also been argued that there are other reasons for AMR. Too many antibiotics have been given out routinely, often for viral infections that they cannot solve, while many people do not take the full dose, meaning bacteria are not properly eradicated and those that resist multiply. In addition, broad spectrum drugs are often prescribed, designed to cover many diseases, making bacterial mutations more likely. More specific drugs aimed at particular diseases would help, but this requires better diagnostic tests. Also this broad spectrum approach and indiscriminate use of antibiotics are killing off some of the many millions of 'good' bacteria that exist in our bodies and are helpful for our health. So the WHO and governments at many levels are emphasizing the need to educate doctors and the public to reduce their routine drug use, especially for minor infections that do not need them, which decreases the opportunity for bacteria resistances to emerge. This over-use is especially a problem in countries where antibiotic products can be bought in shops without a prescription. It is not only a developing world issue, but one that exists in many Mediterranean countries, where resistances to five basic bacterial strains have grown, whereas the figure in the U.K. where the drugs need a doctor's prescription was only 0.4 % (WHO 2014b). The growth of illegally produced and often ineffective antibiotics is another challenge that needs to be rapidly contained. An even greater problem comes from the routine antibiotic doses given to farm animals, especially in the high density farming of chickens, pigs and cattle, without waiting for infections to emerge and isolating the infected cases. This greater exposure to the drugs again makes it more likely that resistant bacteria will emerge. Such farming practices are routine in North America but banned in Europe, resulting in Europe closing its markets to meat products from North America. To drastically reduce this drug use is going to be difficult, given the profits that pharmaceutical companies make from such practices. We may see similar problems to those experienced by attempts by governments to reduce tobacco advertising etc. which led to huge lawsuits by the tobacco firms. Action on AMR is occurring at last, but is long overdue, if what senior medical people and authorities describe as an emerging crisis is going to be solved.

13.4.2.6 Increased Dangers from Indoor Environments

Most people in the developed world now spend most of their lives in indoor environments, for time-budget surveys have shown that the average American spends only an hour a day outside. This has led to increased attention being paid to the health problems created by these indoors environments (Godish 2001). In the less developed world the continued use of charcoal or wood burning fires for cooking, with inadequate smoke extraction, leads to respiratory problems and far too many premature deaths, estimated at 4.3 million annually (WHO 2014a). In the developed world the hazards associated with the historic use of asbestos for insulation and lead in paint has become well known, and their poisonous effects have gradually been removed by eliminating them in new buildings, although older structures may still have these and other contaminants. The health and crime effects of lead in paint and from car emission fumes have already been described in the discussion on Safe Cities (Chap. 12). Although these and other poisonous materials are being identified and eradicated, a wider set of problems associated with indoor living have come from what has been described as 'the sick building syndrome'. Indeed, it was estimated in 1984 that up to 30 % of new office buildings were affected by the problem (EPA 1991). A wide range of factors contribute to this syndrome, such as: poor ventilation causing polluted air to circulate; poorly sited intakes which bring in exhaust and other fumes; biological contaminants in water ducts and tanks; various toxic chemical emissions from insulation materials, plywood and artificial fibres, as well as from the over-use of poisonous cleaning chemicals; while dust containing materials brought from outside, as well as insects, spores, dirt and skin particles, exacerbates asthmatic conditions, itself a disease that is rapidly increasing. Fortunately, the smoking bans in workplaces, restaurants and public places has helped reduced the problem. It must be acknowledged that it was often cities that began the process of making such practices illegal and subject to fines. Another problem comes from the fact that living mainly indoors in a climate-controlled environment means that the body does not have to consume so many calories to keep warm. So people with the same eating habits are more likely to become overweight by spending too much time indoors.

13.4.2.7 Sunlight Deficiencies

The increase in indoor living has also reduced exposure to sunlight and its ultraviolet radiation which creates health-giving vitamin D in the body. Without it there are sunlight deficiencies. In the densely populated slums of nineteenth century cities it led to high incidences of rickets in children—a disease of bone deformation. Traditional societies in northern climes were protected because their diet included high levels of vitamin D from oily fish. Fortunately the disease was largely eradicated in developed countries in the twentieth century by lower density layouts, and mandatory green spaces in new housing areas, as well as by dietary supplements containing the essential vitamin D, such as by the provision of free milk and the cod liver oil supplements given to children in Britain during World War II (Gille 2004). Sadly, new cases of rickets have emerged in children occupying crowded houses in parts of London (Michie 2013). A reduction in vitamin D can also lead to a depressive illness known as Seasonal Affective Disorder (SAD). Originally seen in countries with long winters and cloudy skies, such as in the deep, fiordic coastal areas of Norway in particular, it has become increasingly prevalent in those who do not venture outdoors. However, this can be cured by increasing the amount of time spent outside in sunlight, or through light therapy from intense light pads (Lurie et al. 2006) and adding vitamin D to diets.

13.4.2.8 Autoimmune Diseases

Another worrying medical trend in recent years has been the explosion in the incidence of autoimmune diseases (AI), such as Type 1 Diabetes, Multiple Sclerosis, Schizophrenia, Rheumatoid Arthritis, Crohn's disease, Coeliac disease (gluten intolerance), several forms of Cancer, and, as some researchers suggest, Autism (Kinder and Hagaman 2011; Yang et al. 2013). These diseases are given the AI label as they are caused by the body attacking its own tissues, with some researchers suggesting they currently affect 5–10% of the world population, proportions that seem certain to increase, even if they are not yet in the highest causes of mortality and ill-health (Ramagopalan et al. 2009, 2010). The exact causes are still unknown, but it has been recently found that these diseases have higher rates in the developed lands of higher latitudes, as well as in urban, compared to rural societies, and with seasonal peaks in late winter and spring (DiSanto et al. 2012). It has also been suggested that these associations seem to account for the higher levels of diseases such as MS in women in cloudy areas such as the Orkneys, in increases in black-skinned individuals who are second generation immigrants in inner city London, as well as in women in Iran since the Islamic revolution forced many women to wear all-embracing veils (Gille 2004; Michie 2013). Researchers are suggesting that the environmental effect of low sunlight exposure in both northern and, it must be stressed, *urban* places, results in vitamin D deficiencies, which may also be at least a partial cause of the explosion in the incidence of many of these non-skeletal problems. The recent increases in asthma rates also seems to have similar causes. Litonjua and Weiss (2007) have calculated that asthma now affects 300 million people worldwide and has shown a 50% increase since the 1970s, which also seems to be the result of a deficiency of vitamin D, in this case one that impairs lung development in the foetus (Hagaman et al. 2011).

More recent research has also discovered that natural levels of vitamin D are also low in mothers with children born in late winter-early spring. This is due to a lack of exposure to the vitamin D-causing properties of sunlight, again linked to more indoor living. Research reviews have implicated over 18 of these autoimmune diseases (AI) with low vitamin D levels in pregnancy so that deficits in this vitamin go beyond its role in bone metabolism that leads to rickets (Yang et al. 2013). Although the precise causes of the relationships are still under investigation, the low vitamin D effect may be exacerbated by genetic risk and perhaps mothers' diets (Weiss and Litonjua 2007; Litonjua and Weiss 2011). Researchers have suggested that it is probable that the low vitamin levels may make a pregnant mother's body create antibodies that cause particular genes to malfunction as well as damaging the brain of the foetus. These antibodies are not flushed away before birth, but remain in the body and cause certain genes in some people to malfunction in later years, resulting in the body being attacked by its own immune system. This problem in early gestation means that adding vitamin D later in life does not solve the problem. It has also been argued that this vitamin also seems to be essential in providing adequate amounts of serotonin in the brain, which have been shown to be low in autistic children who have low empathy levels, which provides an explanation for the higher incidence of the disease with males and those born in late winter (Patrick and Ames 2014).

Although AI research is still in its infancy, recommendations to combat their increase are emerging. One is the recommendation that pregnant women eat more natural sources of vitamin D in fruits and vegetables and take supplements of the vitamin. This research also suggests that the accepted wisdom since the late 1960s of reducing exposure to sunlight because of the risk of skin cancer needs to be modified, for moderate exposure amounts is needed to reduce the subsequent increase of susceptibility to these AI diseases.

> It seems likely that a gradual decrease in exposure to sun due to sun avoidance behaviours in Western societies (sunscreen, clothing, sun avoidance, increased time spent indoors) reached a critical level in the early 1970s, such that humans were not spending enough time outdoors and vitamin D levels reached acutely low levels. Vitamin D is essential to the normal functioning of the human immune system. (Litonjua and Weiss 2007, p. 747)

All these suggestions seem especially applicable to the poorest population, often in overcrowded parts of cities with few park spaces. Although attempts have been made to ensure poor mothers and their children are provided with vouchers for fruit, vegetables and free vitamins, such as in Britain's Healthy Start programme for children at risk, it has been reported that less than 10 % used this opportunity—a depressing and unnecessary result attributed to gaps in supply chains, delivery systems and poor staff training (Michie 2013).

13.4.2.9 Solutions

All these emerging problems may be amenable to further breakthroughs in medical research and the greater spread of effective care. But as the cases of the failure of some antibiotics and the aging population in particular indicate, there are huge problems developing that will need a great deal of research work to combat. Moreover, if infectious diseases cannot be controlled by antibiotics, the historic danger of diseases spreading by contagion in large urban areas will re-emerge. This risk is reinforced by the new fast transports of air and train, including cramped subway systems, which crowd people into what are effectively sealed containers and deliver them to far flung destinations where they quickly and anonymously disperse into the larger settlements and may infect populations not used to the diseases they may carry. Also, other emerging health problems, such as AI diseases, seem to have an association with our increased urban living. These negative effects of urban living mean there should be a major emphasis on providing healthier urban behaviours and environments.

13.5 Addressing the Determinants of Health

The so-called 'health field' concept promoted by a former Canadian Minister of National Health and Welfare (Lalonde 1974) is often regarded as an important marker of a renewed emphasis upon public health after its dominance faded from the 1880s once urban environmental conditions improved. The health field approach moved away from an emphasis upon medical care alone, by stressing the need to investigate all four key categories of the factors that affect health: human biology (all factors related to the body, including genetics); environment (all features outside the body that affect health, but over which individuals have little control); life-style (personal decisions made by individuals that affect health, such as drug-taking etc.); and provision of heath care facilities (their quality, availability and accessibility of health care). Other studies have provided alternative versions of the main categories of what are now known as health determinants, sometimes in diagrammatic form (Dahlgren and Whitehead 2006). These summary features have been extended by more recent research that has shown the need to look more closely at a subset of these determinants that are often overlooked, namely how various characteristics of the social realm—such as the effect of the socio-economic inequalities, or from stress or social exclusion—influence health outcomes. These factors are part of what are being described as the *social determinants of health* (Wilkinson and Marmot 1999, 2003), although the term has sometimes been used to categorize most determinants under the 'social' title (LCC 2012, p. 6).

These summaries of the various health determinants provide useful initial descriptions of many of the relevant groups of factors that influence ill-health. But it does seem worthwhile to extend and re-arrange the ideas to provide a more comprehensive view. In Fig. 13.1 eight broad domains of factors that influence health are identified together with the surroundings contexts. Within these health determinant categories, examples of the more specific factors can be identified which have deleterious effects on health. Many of these were not routinely dealt with by current health care systems, but are now being more closely investigated by epidemiologists and other researchers. These domains may be initially considered as separate categories of related factors for the sake of clarification; in reality many often combine with one another to create ill-health, especially in our complex, multi-faceted urban places. Also, many of the factors have two-way relationships with one another, including modifications on human biology through epigenetic processes linked to life-styles or environments that affect the operation of genes. Moreover, some of the features are directly linked to disease; others take more time to cause ill-health. But the influence of all these factors cannot be discounted when evaluating health in an area, since they provide the background conditions that lead the body to be weakened and prone to specific diseases. Although these factors can be looked at individually, in practice they combine in a spatial context to create differential place characteristics, not simply in physical or built-environment terms but also through the economic and social environmental characteristics and life-style choices. This ensures that each location will have very different risks and protections for health

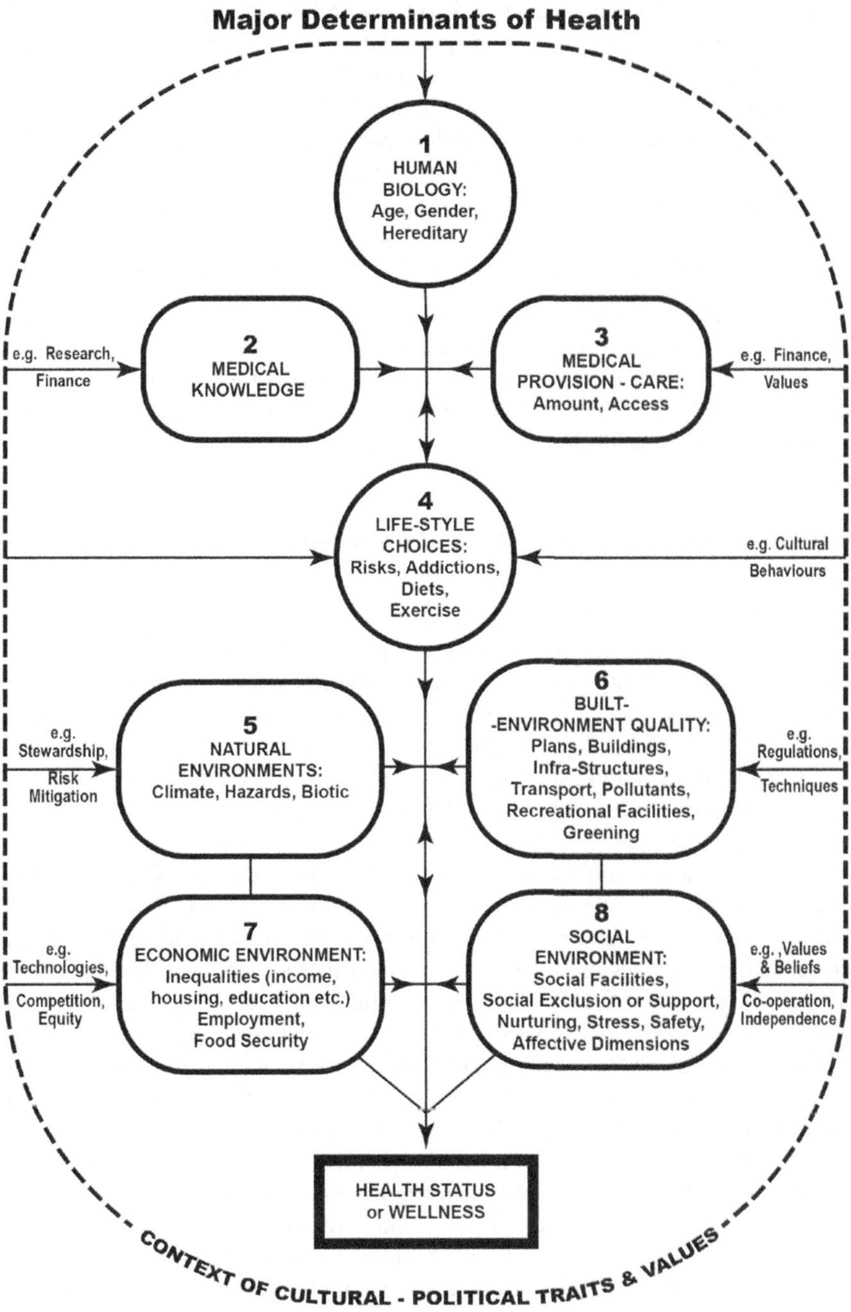

Fig. 13.1 Summary of the major determinants of health

outcomes. So a continuing challenge for a research field such as epidemiology is to get to the root causes of ill-health and premature mortality in various locations and to provide new preventative health policies that relate to all the health determinants. Unfortunately, as the authors of a text on *Unhealthy Cities* have argued, the twentieth century approach to public health has until recently ignored or at least underestimated many of the spatial variations in these determinants;

> ...In part because public health has been burdened with a tradition that overemphasizes individual-risk factors, the consequences of social and environmental conditions for health promotion and illness have been overlooked. (Fitzpatrick and La Gory 2011, p. 155)

So even though the twentieth century focus on the biomedical model has ensured major advances in the level of health and continues to do so, hopefully by solving the problem of ABR and the new diseases, there is also a pressing need to reduce the negative place-based factors that contribute to ill-health by investigating the impact of other health determinants.

13.5.1 Human Biology, Medical Knowledge and Medical Care

These are the first three health determinants shown in Fig. 13.1. The most fundamental influence on the health of persons or populations stem from their *biological constitution*, their ability to ward-off and recover from disease and disabilities. In this context features such as age, gender, life-history and the genetic make-up of people are all important. Age is really a matter of being more prone to certain problems as the body ages and experiences what amounts to wear and tear, or in the early years when a baby may not have developed the strength or immunities to fight off the dangers in the local environment. Gender is also important, as males or females may be more liable to suffer more from one type of disease or disability than others. The latter certainly live longer on average than the former, although some of the differences are due to greater risk-taking in males and their more unhealthy life-style choices. Some families may have flaws in their genetic make-up that make them more prone to certain diseases. In addition, some groups have a greater ability than others to resist certain diseases, not because of inherent racial features, but through centuries of developed immunity to problems caused by some pathogens. Others do not have these resistances. The tragic result of these differences not being understood was seen in the European Age of Exploration when the indigenous peoples of the Americas died in their tens of millions from diseases introduced by Europeans.

Table 13.1 separates *medical knowledge* from *health care* since high levels of knowledge does not mean there is 'cure and care' for all individuals in an area, especially while there are many different forms of care and differential access to it. Continued research and effective and equitable application in all of these determinants are obviously needed, especially given the need to counteract the severe challenges identified above that come from emerging medical problems. Nevertheless, there are still many spatial inequalities in the provision, availability and effectiveness of medical care, not simply within and between cities, but in urban centres in the developing world that have limited facilities and where only the affluent get adequate care. In addition there are increasing worries about accelerating health care costs, with ever more technical procedures and longer treatments for the new chronic diseases. Some indication of the escalation of costs can be seen by the fact that the budget of the National Health Service in the U.K., which provides universal coverage for all citizens, was originally close to $15 billion (£9 billion equivalent) when adjusted for inflation to provide current value in 1948 when the programme began. But the budget is now is now more than 12 times this original amount and is a growing at 4 % a year, with increasing needs from an aging population. Although more careful administration and costs savings in hospitals or in care facilities may reduce the figure, the size of the budget and its growth make it unlikely that they will do more than modify the total, meaning that new ways of funding the system, perhaps by larger tax increases will be required. These costs are more difficult to absorb in the less developed world, where access and availability to care is limited. Indeed globally, health care availability is inversely related to need (Howatt et al. 2012). This is leading to the development of so-called frugal technologies to solve medical problems at lower costs, using less sophisticated techniques, in order to increase availability in low income areas. Another problem that has emerged is the corruption and fraud in some countries. For example the U.S. system is largely a private one, although there are government agencies that provide care and subsidies for the aged (Medicare) and the poor (Medicade). Recent reports suggest fraud could be as high as $ 272 billion annually, which is around 10 % of U.S. medical spending (TE 2014). The health care fraud comes from many sources that include: suppliers charging Medicade for non-existent supplies; patients claiming benefits to which they are not entitled; doctors or health clinics charging for extra or non-essential services, and filing prescriptions for painkillers and then selling them; people stealing medical cards and using them to bill Medicare for services that they are not entitled to. Steps are being taken to reduce this enormous leakage of funds, including simplifying the claims systems and providing more vigorous checking. This problem of wasted money may be more obvious in the complicated U.S. system but certainly exists in other jurisdictions and must be reduced so that health care money can be spent more wisely and fairly.

13.5.2 Life-style Choices

The fourth determinant contributing to the health or ill-health of populations come from the life-style choices that people make, although these effects are often cumulative and take place over decades. The strong adherence to the principle that a population in democratic countries should be able to make their own choices reduces the ability to create quick solutions, especially of a coercive nature, so multi-pronged approaches are usually needed. It has already been noted that the population in the developed world in particular spends too much time indoors. There is a need to redesign areas and to provide leisure spaces where more outdoor activity is encouraged. However, the most obvious pernicious effects of life-style choices come from the increase in the use of addictive substances that drastically reduce the life-span of people using such products. For example, despite the now well-known health hazards of smoking, an estimated 1.2 billion people still use tobacco regularly and the proportions of smokers in even OECD countries is often still over 20 %. This also puts the individuals adjacent to smokers at a high risk of smoke inhalation, either first or second hand, and leads to higher rates of respiratory problems and cancers. Similarly, drug use leads to premature deaths and other crime problems. But a more general problem comes from the overuse of alcohol, which leads to liver and heart diseases, as well as contributions to traffic accidents. A recent study revealed that 25 % of Russian males die before the age of 55—compared to 7 % in the U. K.—with an average age of death for males at 64, a result of overconsumption of vodka (Zaridze and peto et al. 2014). However there is also evidence that beer and wine in small proportions has a positive effect on health (Preedy 2011) and helped reduce mortality rates in historic cities where beer with alcohol killing bacteria was safer than contaminated water.

All these problems require determined efforts to produce a reduction of addiction levels. In the case of tobacco use many initiatives have helped reduce the problem, such as: increasing taxes and banning advertising on products, adding health warnings, increasing educational programmes to avoid or reduce use, and banning their use in public spaces—but the constant litigation by tobacco companies has reduced progress. Still more effective policies are needed, which probably need a combination of persuasion and regulation. Outright prohibition of products drives the issue underground, and criminalizes the behaviour, as shown by the U.S. prohibition of alcohol in 1920–1933. Many of the anti-tobacco policies could be applied to reduce binge drinking. The same applies to drug use, where despite billions spent in attacking the sources of supply and distribution the problem is as worse as ever. Some have argued that removing criminal prosecutions and treating drug use as a medical, not criminal, problem would help reduce the addictions. But the majority opinion is still against such a change, which means that alternative methods must be found.

A more recent health problem associated with life-style relates to dietary choices. Changes in the diet of people in many western countries, with overconsumption of sugary drinks and foods rich in fat, red meat, sugar and salt has led to increased obesity, leading to a greater probability of ill-health from a number of diseases. For example, it is currently estimated that one-third of school age children in the U.S.A and two-thirds of American adults are overweight or obese, with far higher rates in Afro-American and Hispanic women (Ogden et al. 2014). Being overweight or obese increases the risk from a variety of chronic diseases, especially heart failure, stroke and diabetes which are among the top causes of mortality with numbers projected to increase substantially in the next 15 years (Table 13.1). This trend is not simply a result of increasing purchasing power—although this makes a huge contribution—but also comes from the relentless advertising by manufacturers to increase demand for their products, especially to children, creating what amounts to an obesogenic environment. In addition far larger portions of most meals and drinks are available compared to 30 years ago, which compounds the problem. Another contributing factor is the limited availability of healthy food options in low-income areas.

There is little doubt that national legislations are being enacted to address this growing problem. For example, almost a third of the population in Mexico were reported as being overweight, 14 % have diabetes and the average daily consumption of sugary drinks was half a litre. This led to Mexico's 2013 decision to increase by 10 % the tax on sweetened beverages, using major advertising campaigns to support the policy. In other countries, such as Britain, attempts are being made to create Agreement Partnerships with large retailers to reduce the number of sale prices on products with large portions of fat and sugar, as well as on alcohol products. Many also believe it is important to reduce the level of advertising on such products, especially those aimed at children. The mandatory labelling of calorie levels in tinned or packaged products is also taking place, in the hope that the public will become more aware and make more informed choices about what, and how much, they are eating. Yet only a multifaceted strategy will solve this increasing crisis as summarized in a WHO report (2000) on *Obesity: Preventing and managing the global epidemic.* Unfortunately, limited progress has been achieved since its publication. However, some cities are taking the lead in public health actions such as those relating to diet, before national regulations are developed. For example, Alcon (2012) has argued that Mayor Bloomberg's leadership in promoting healthy initiatives in New York provided one of the first examples of a municipal politician taking controversial stances in this field. This led the Health Commissioner, Thomas Farley, to note that:

> …For way too long public health departments have defined their responsibility as essentially infectious disease control, rather than the improvement of health of the population. (Alcon 2012, p. 2038)

The New York approach has been to implement policies to reduce levels of addiction and improve diets, such as by the banning of smoking in public places and businesses, plastering anti-smoking advertisements on subways, banning transfats in restaurants, and an attempted ban on large sugary drink containers that was struck down by the courts. In addition there has been greater targeting of crime problem areas, for most have high levels of ill-health, as seen in the crime measures described in the discussion on Safe Cities (Chap. 12), and in creating district public health offices in areas of low health, such as one in the Bronx.

In other countries more and more planning decisions are being made to improve the walkability of subdivisions and adding segregated bicycle paths to increase this form of transport, as described in Chaps. 4 and 6. Such policies are designed to increase exercise levels, which in turn improves health. But it is overeating the wrong types of food that is the major problem in most areas. In addition, the trend that has reduced play time, sports and exercise periods in schools is being reversed in many cities in an attempt to increase the fitness and therefore health of children, while the increasing popularity of physical health clubs for adults shows that a start has been made on the problem by at least some people. Although it is difficult to quantify the effect of exercise, recent surveys indicated that 9 % of premature mortality in non-communicative diseases is due to low levels of activity, although this varied with the type of disease or illness (Lee et al. 2012), while other researchers argued that participation in sport is associated with a 20–40 % reduction in all causes of mortality (Khan et al. 2012).

13.5.3 Natural or Physical Environments

Behind these first four domains of health factors are four categories of environmental determinants, related to the natural, built, economic and social environments. Although the causes of disability and premature mortality often relate to one or more of these factors, a WHO survey of the environmental effect on ill-health (Prüss-Üstün and Corvelán 2006) estimated that 24 % of the global disease burden (health years lost) and 23 % of the annual deaths are due to what are described as 'modifiable environmental factors', proportions that rise to closer to a third for children. These are described as the factors in the physical and built-environments that can be altered; they exclude factors such as disease vectors in water bodies, impregnated bed nets, as well as social behaviours or life-styles. The WHO report showed that the burden of environmental effects on premature mortality and ill-health is far higher in the developing countries and varies with disease type. Studies of the risk factors of different diseases show that 85 out of the 102 types covered had an environmental factor, although these varied considerably. The highest mortality proportions were seen with diarrhoea (94 %) and malaria (42 %), far higher than the 20 % from lower respiratory causes, although this rises to over 40 % in developing areas, primarily due to fumes from the use of solid fuel fires in enclosed spaces. The risk factor may be greater than previously thought, as recent work is showing that environmental effects can be passed through generations through epigenetic processes whereby genes are turned on or off (Tollefsbol 2014).

The first of these environmental influences are those from the physical conditions, features often forgotten when the focus of attention is upon cities in the developed mid-latitude world that have less extreme environments. For example, health is often affected by extreme deviations from the normal *climatic regimes*, such as in cases of extreme cold or heat and high humidity, where the elderly, young and those in ill-health, die in larger numbers in such weather conditions. Some of these problems are made worse because of the heat-island effect of all the concrete, steel and bricks in cities. For example, in July 1995 a Chicago heat wave led to temperatures reaching over 40 °C—although the effective temperatures as felt by the body in the sun and in buildings were far higher—resulting in the deaths of over 600 people (Klinenburg 2002). The size of the mortality from this event in one city is worth comparing with other statistics. In the decade after 1992, 2190 Americans died from heat-related causes compared to 880 from floods and 150 from hurricanes. Although this shows that extreme heat can result in spikes of mortality, it is important not to underestimate the effect of disasters from *natural hazards* when many thousands, even millions, of people have been killed by major storms, earthquakes, volcanic eruptions as described in Chap. 8.

The second health effect of the physical environmental comes from various *biotic hazards*, ranging from snakes to toxic plants, although these are often less of a problem in cities as they are often deliberately eradicated. Of greater significance are the diseases spread by contagion in a densely populated area by pathogens developed in, or carried by, various animals or insects. Perhaps the best known historic examples come from the various plagues, such as the Black Death plague that spread from Asia via trade routes and peaked in 1348–1350 in Europe and led to the deaths of 75–200 million people. This was between 30 and 60 % of Europe's population—although it must be noted that the century also saw the end of the medieval warming period and poor harvests which led to food shortages and a severely weakened the population (Herlithy 1997). More modern examples range from the previous descriptions of the rapid global spread of HIV-AIDS, which contrasts with the containment of SARS.

The mortality rates and ill-health effects of many tropical diseases are still far too high. Many, not all, now have cures, but these are not often available in many cities and regions. One of the biggest health problems is still associated with malaria, spread by parasites carried by certain types of female mosquitoes in tropical and subtropical areas. Although not in the top 20 causes of death in Table 13.1 it has been argued that the 0.66 million deaths estimated by the WHO represent gross underestimates, with the true figure being closer to 1.24 million, let alone the debilitating effects on many people (Murray et al. 2012). Certainly medicines such as quinine, several recent drugs, and the use of nets over beds at dusk and night when mosquitoes are flying have reduced the risk of the disease. So has the draining of marshes and standing water pools in urban areas which reduce the number of places where the insects breed. Snowden's (2006) detailed study of the effects of malaria in Italy until it was finally eradicated in 1962 showed how the disease had previously led to low productivity, premature death and economic backwardness in the country. It needed a multi-pronged state strategy from the early twentieth century to remove the scourge, one that involved medical breakthroughs, swamp draining, rural clinics and education for preventative measures. In tropical areas the hoped-for final solution by chemical spraying of marshes with DDT in the 1960s proved illusory, as the parasites have bred resistance to particular drugs and the sprays used led to the poisoning of the water which created other problems for a whole range of birds and animals. So despite determined efforts to reduce its effects, through new drugs, swamp draining and more recently through experiments with the mass inoculation of populations in some areas to reduce the number of people acting as hosts to the parasite, malaria still remains a significant cause of ill-health and premature deaths. This may get worse as some of the malaria-giving parasites are becoming resistant to the drugs used to treat them, which makes new treatment discoveries a priority.

These are only a few of the examples of death and disability caused by factors from the physical environment, especially from the hotter wetter environments where a large number of other serious diseases lurk, such as dengue, yellow fever, typhoid etc. Some people in these areas have increased tolerance to these infections, but many of these diseases have an increased ability to harm people who have travelled from other environments without the appropriate medical treatment. One of the big emerging problems is the effect of global warming. This will probably lead to lower yields from current crop varieties in an aggregate sense, leading to food shortages given world population growth, unless new varieties can be developed. In addition, the warming will allow many flies and insects, and hence the pathogens they carry, to spread further north, infecting people with little immunity. So malaria may be spread again to Europe, as West Nile disease is already becoming common in North America. The fact that diseases, such as African dengue and chikungunya fever, have recently spread to the Caribbean, probably in the blood of an infected airline passenger and then into the local mosquito population, means that these and other infectious diseases are likely to occur in more northerly locations, causing additional health problems.

13.5.4 Built-Environment and Infrastructures

The second set of environmental factors that affect health come from poor man-made or built environments, issues identified as causing high mortalities that led to the first healthy city movement. Progress in built-environment improvements, including the infrastructures such as transport systems, to reduce their negative outputs has not been halted, at least in the developed world. For example, many public sector high-rises have been razed because of high crime rates and poor construction (Coleman 1985). In terms of harmful outputs from cities, Chap. 6 has described the major improvements in the technical ability to more safely process and use the effluent in sewage processing plants, while water can be re-cycled for human use. In addition, instead of new expensive solutions to bring water from long distances, a series of new devices for reducing run-off and using this water mean that new water conservation measures are being used in some jurisdictions that will help water shortage in drier climates. However, these nineteenth century advances, let alone those of recent years, are still limited in the cities of the less developed world, for the WHO (2012b) estimates that 1.1 billion people still lack access to clean water and 2.6 billion do not have even a basic latrine, although have been major improvements in these basic sanitation and water supply facilities in the last decade. Many of these people live in rural areas. But very unhealthy conditions linked to poor water supply and limited sanitation also exist in many urban places in developing countries, especially in the informal shanty towns that are the homes to many in the towns and cities of developing countries. One result of poor sanitation is the annual death toll of 1.8 million from diarrhoea, currently fifth in the WHO table of annual deaths. Although it is estimated to fall to ninth position, it may still claim 1.6 million deaths a year in 2030, even though it is relatively easily treatable by modern medical techniques.

Health problems in cities are not simply a result of poorly constructed buildings and infrastructures; there are often health problems associated with the use of manufactured substances whose poisonous side-effects are often underestimated. Some of these were identified in Chap. 12 (Safe Cities), such as the lead in gasoline that produced more toxic car exhaust emissions that have been shown to affect brain development and subsequently I.Q levels and increased crime rates (Reyes 2007; Markus et al. 2010). More apparent are the deaths and disabilities from road accidents, which are among the most important causes of premature deaths (Davies 1997). The WHO's Global Study on Road Safety (WHO 2013a) showed that 1.4 million people die on the roads annually, projected to increase to 1.8 million by 2030, with somewhere between 20 and 50 million more suffering serious injuries today (Table 13.1). Certainly there are signs of improvement in the mortality rates in countries such as the U.S.A., with deaths now down to 34,000 a year in 2012, from levels of 54,000 in the early 1970s, which is still a rate of 1.08 deaths per 1000 people a year and the size of a small country town. Even this figure is unacceptable. In any case there is little room for complacency in world terms. After all, 91 % of the fatalities are now in low and middle income countries, places that contain only half of the world vehicles so far. This high proportion is because only 28 countries (with 7 % of the world population) have compulsory road safety standards in five critical areas: drink and driving regulations, speeding, failure to have seat belts, child-retaining devices, and the enforced use of helmets for motorcyclists and cyclists. In many countries mortality rates could be reduced if the following problems were solved: lack of driving tests; safe crossing places and pavements for pedestrians; refusal to obey traffic signals and adherence to the designation of one side of the road for traffic in one direction; erratic driving behaviour; and limited effective ambulance services that can quickly administer care and remove traffic victims to hospitals. Although mortality and injury rates have dropped in most of the developed world there is still room for progress in their cities, such as better designed roads and junctions, especially at high traffic accident areas, effective laws to reduce distracted driving, segregated bicycle lanes, more public transit, and, as some cities are showing, saving lives by reducing speed limits in residential areas to 30 km/h or below. Improved transport systems and the use of bicycles also provide more healthy options, with the latter in particular providing exercise.

The man-made urban environments also affect health through the higher noise levels and the stress that many people feel in these areas, often combined with an absence of meaningful supportive relationships, issues usually dealt with under the social determinants of heath category. More specifically, health is also affected by all the *pollutants* generated by human numbers and their activities, from faecal matter to noxious products of industrial process and in household goods, poisonous chemicals and nuclear outputs in some places, and to the fumes and particulates from burning fossil fuels (EPA 1999). These not only cause local health problems, but by altering the atmosphere are threatening to alter the climate of the planet. Many of these have already been described in the sustainable development discussion (Chap. 5). A WHO report (2014a) on pollution levels in 1600 cities of the world maintains that air pollution—both indoor due to stoves and fires and outdoor because of vehicle and industrial emissions from fossil fuel—leads to over 7 million deaths annually from these sources. Although the Indian government disputes the figures, the source shows that Delhi now has worse air quality than Beijing, as measured in small airborne particles that are less than 2.5 micrometers in size (PM25) which are associated with increased rates of bronchitis, lung cancer and heart disease.

However, not all aspects of a created environment are damaging. The addition of parks, green spaces, recreation trails and nature areas etc., have long been regarded as having a positive effects on humans as the discussion on the relationship between green space and health in the chapter on Green Cities has shown (Chap. 4). Such areas provide an antidote, in at least some cities, to the increase in urban densities and indoor living.

13.5.5 Economic Determinants

Many investigators, from the nineteenth century social commentators to the present (Adler and Ostrove 1999), have shown that people living in disadvantaged socio-economic circumstances suffer more ill-health and earlier death rates than others. However the adoption of the 'care and cure' medical approach has tended to blame some health problems on life-style choices, rather than the wider structural circumstances of their life in the socio-economic environment that negatively affect their health. These components of disadvantage are now usually labelled as the social determinants of health, with Wilkinson and Marmot (1999, 2003) using empirical research results to identify ten significant factors in causing ill-health, summarized as: the social gradient, unemployment, work conditions, food, stress, early life, social exclusion, social support, addictions and transport. But it does seem useful to separate the factors into what amount to economic and social factors, rather than having one long list, while the addictions and transport factors have already been dealt with in the previous discussions on life-style and man-made environment respectively.

One of the most powerful explanations of ill-health among the economic determinants is described by Wilkinson and Marmot (1999) as the 'social gradient'. This means that people who are in the lowest income brackets have shorter lives than people in higher income brackets, as seen in the Canadian example in Fig. 13.2, features that also apply to more ill-health. The pattern is much more pronounced in less developed countries. Less money for food, heating, adequate clothing, ability to pay for and access to medical care, knowledge of how to look after oneself, are all contributory factors, but the lack of education and hence ability to improve one's position in the social hierarchy is also a major contributing factor to this general trend. Populations that have higher levels of educational and occupational qualifications tend to have lower levels of dementia, indicating that there is some kind of cognitive reserve that counteracts the effect of this neuro-degeneration (Prince and Acosti et al. 2012). However, it was previously noted that the modification of other life-style factors, such as physical inactivity, could drastically reduce the incidence of the disease (Norton et al. 2014). Another powerful social determinant of ill-health is the *unemployment rate,* as is limited *work satisfaction* and difficult work conditions. In general, people at the lower end of the social gradient also have longer periods of poorer health in their later years due to various causes. Rates of various mental diseases are also higher in cities.

The level of *nutrition* is also critical in improving health. The increasing numbers of food additives, such as excessive salt, sugar and food preservatives found in many products in supermarkets and fast food outlets, have led to higher incidences of heart diseases and obesity in particular. Certainly some of these issues are a result of poor life-style choices by disadvantaged people, but they often lack local sources of healthy food. A good diet, with a variety of vegetables and fruit, and adequate amounts of each, is essential to maintaining health, for malnutrition leads to ill-health and also vitamin deficiencies which can lead to diseases. There is still controversy over these issues, but a meta-analysis of research findings confirmed a linear relationship between stroke risk and low fruit and vegetable consumption,

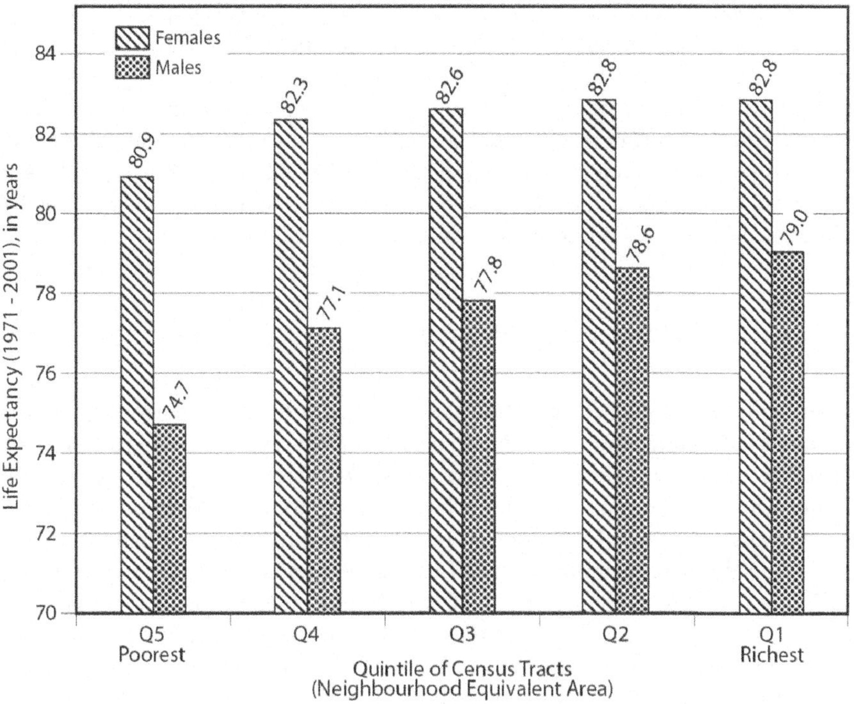

Fig. 13.2 Mortality and income levels in Canada (Source: Revised from Wilkins 2007)

with latest research suggesting that well over six portions of fruit and vegetables a day are needed to obtain the necessary vitamins (Qu et al. 2014). Nutritious food may not be available in disadvantaged areas as shown by the concept of *food deserts*, areas in poor inner city areas that are without adequate food outlets that supply healthy and fresh foods, a characteristic that was described in Chap. 4. Health education may improve the choices made, but this is rarely as effective as expected, unless it starts at an early age in school, but even then family food choices may reduce possible progress.

The presence of large social gradients or economic inequality in many countries and within urban areas also means that many people also lack the economic power to improve their shelter, or clothing, which then affects their state of health. For most of the twentieth century better wages, welfare measures and social provision led to a decline in inequality levels in developed countries. But as Chap. 3 has shown, the past 20–30 years have seen levels of inequality in income and wealth rising in developed countries, while the explosive growth of urban places in the less developed world has led to larger areas of poor squatter settlements. Progress towards healthier cities requires more attention to the provision of Basic Needs in food, shelter, education, etc., and to improve the Capabilities of the disadvantaged population to better their own condition and—as addressed by Just City initiatives (Chap. 3)—giving more people a voice in decisions that affect their areas. Some of these needs can be addressed by municipal actions, for example: in shelter by subsidized housing or community based non-profit housing agencies; in nutrition, through organizations providing food banks for those in need; or through planning decisions that take the health impact of a displaced population into account. However, the most fundamental changes are probably still made by national or state policy through taxation, welfare, social provisions and ecological regulations to remove the structural disadvantage of those at the low end of the social gradient that are created by societal not personal factors.

13.5.6 Social Determinants of Health

A series of interacting factors that may be described as the lived-experiences of people are usually called the social determinants of health. A key factor is the presence of *strong supportive networks of family or friends*, which are correlated with better health. The opposite applies to those who are subject to *social isolation*, which may be by choice, but more likely a result of the absence of friends or family, perhaps due to age, poverty, disability, poor self-esteem. Poorer health is also found in those suffering *social exclusion*, especially discrimination on ethnic or religious grounds. Indeed, loneliness and social isolation—which is different to depression—has been shown to increase the propensity to functional decline and ill-health, as well as increasing the chances of earlier mortality (Cacioppo and Patrick 2009; Perissinotto et al. 2012). In addition, there is the problem of *stress*, which comes from the workplace as well as in homes and neighbourhoods, and is now accepted as having a biological effect.

> Hormones associated with the chronic stress burden protect the body in the short run and promote adaptation (allostasis), but in the long run, the burden of chronic stress causes changes in the brain and body that can lead to disease (allostatic load and overload). Brain circuits are plastic and remodelled by stress to change the balance between anxiety, mood control, memory, and decision making. Such changes may have adaptive value in particular contexts, but their persistence and lack of reversibility can be maladaptive. (McEwen 2012, p. 1)

These social determinants create a poorer quality of life, with lower self-esteem and well-being. But others exist. For example, it is also being recognised that a healthier life is more likely when there are high levels of positive *nurturing*, such as when children are given a more supportive start to life through good education, nutrition and a caring environment with limited exposure to stress or violent behaviour (Brooks-Gunn et al. 1993). This means helping mothers as well as their children to develop in such environments. In later life, worries about *safety,* let alone actual physical violence, are also important contributory factors in ill-health. In addition, some *beliefs and values*, such as refusal to accept certain types of medical care because of religious beliefs/prejudices, are additional negative factors.

The extent to which there are *supportive facilities* in an area, from quality schools, to churches and community associations or the provision of home-help, can also create more caring social environments in an area, creating the social capital that provides support, although these may be considered as part of an area's infrastructure. Added to such problems from these essentially empirical characteristics, are those that come from the *affective dimensions*, which deal with attitudes and feelings. Areas of high crime with social and health disadvantages have at least 10 types of negative attitudinal characteristics that can negatively affect well-being. These were labelled as areas or 'terrains' of: social inadequacy; despair and limited goals; exclusion and discrimination; acceptance of decay and destruction; anxiety and fear; spontaneity of actions and emotions; indifference to others; low restraint and self-control; anti-social attitudes; and peer group (gang) allegiance (Davies 2004). All these attitudes reduce the well-being and the health of people in areas where these attitudes are found, and needed to be countered if progress is to be made in improving the lives of people affected by such features.

Many social commentators, from those of the nineteenth century to the present, have suggested that being poor and stressed seems to literally make one sick. In addition, the effects of poverty, stress, exposure to violence, poor nutrition and limited affection is also known to lead to far lower acquisitions of early language skills in children and hence subsequent intellectual progress. Also, Fonagy (2003) has argued that the usual socialization of young children 'out of aggression' that occurs in nurturing parenting is limited in disadvantaged areas, producing a rapid and irrational response to perceived threats. This type of primitive, unconscious reaction to stimuli based on previous experiences, bypassing any rational evaluation, has been identified by the cognitive psychologist Aaron Beck (1999) as the main basis of an inadequate control of emotions, which leads people to violence in subsequent life. Increasingly, it is recognized that these health and behavioural associations are not a matter of life-styles choices, although they may contribute to the situation. What amounts to a biological embedding of the effects of the early environment occurs through epigenetic processes that affect the influence of genes, while it was earlier observed (Sect. 13.2.2.8) that recent research is showing a strong relationships between vitamin D deficiency in pregnant mothers and increasing rates of autoimmune diseases in their offspring.

Stimulated by the work of Professor Elizabeth Blackburn who was awarded a Nobel prize in 2009 for her work on the way that the social environment has embedded biological effects, the mechanisms of these connections are now being explored in more detail, as seen in a special issue of Proceedings of the American Academy of Sciences (Boyce et al. 2012) and in a book edited by Tollefsbol (2014) that reviews transgenerational epigenesis effects. One of the ways in which this environmental-embodiment relationship occurs is by the shortening of telomeres, which are rather like a cap on the end of chromosomes that protects the DNA during replication (Epel et al. 2004). This reduction is associated with higher levels of ill-health and predisposition to many diseases in later life. A recent pioneering study of 9 year old Afro-American children who had suffered chronic stress from their upbringing in poverty, unstable families and who were subject to harsh parenting and maternal depression, revealed a shortening of their telomeres compared to children from more affluent and nurturing backgrounds (Mitchell et al. 2014). So these disadvantaged children are likely to experience what amounts to premature aging and morbidity through degraded physical functions that make them more prone to disease. But the effect was not uniform. This association between the social environment and telomere length (TL) was also moderated by the extent of genetic variation in the subjects within the neurotransmitter pathways of the serotonin and dopamine genes that control the extent of pleasure and excitement responses. So some of the subjects had a genetically influenced higher likelihood of sensitivity. It was shown that those with highest sensitivity levels were more affected by their environment, having the shortest TL when exposed to these stressful environments. But they had the longest TL when exposed to advantaged environments. The researchers suggest that this points to the presence of a genetic predisposition in the responses to the social environment. The children with high sensitivity levels get even worse off from the poor nurturing and environment, but have greater benefits from better conditions. Yet those with limited genetic sensitivity may not be so affected by the negative social environment, which may explain why some people survive the highly disadvantaged area and their resilience enables them to subsequently flourish. Yet the general effect is that the children have their biological structure damaged by the stressful social environment of their upbringing, leading to future health problems. These findings mean that children in these areas are not simply disadvantaged because of locally poor educational opportunities to improve their life. Their negative environment effectively ages them and makes them more liable to ill-health. This recent research provides justification for early childhood interventions providing effective nurturing, creating less stressful environments to reduce the probability of what amounts to physical damage in epigenetic processes. The research has moved the known correlations between childhood disadvantage and future ill-health to causal relationships, although not all will be affected to the same extent.

13.5.7 Cultural Contexts and Interrelations

A final domain of influential factors affecting health is more wide-ranging in the sense that it shows the cultural context-health connection, illustrating how the culture of societies—which includes political power and relationships—affects all of the other determinants. Figure 13.1 shows that cultural factors can impede or improve each of the various health determinant categories. For example, there have been many cases where a religious or sub-cultural group in countries or cities have refused to believe in particular medical treatments, while new medical advances have also been resisted by some practitioners. The adoption of malaria nets over beds in East Africa is a good example of the former. The uptake of nets supplied by international agencies in the 1990s was extremely slow at first despite overwhelming evidence showing their effectiveness. The white nets, however, were deemed too similar to the shrouds used to cover the dead, so they were not used. When the colour was changed from white to green they were quickly adopted by the local population. There is also a cultural-political background to the level of medical facilities. Governments may provide less or more money for medical care. In terms of life-style choices, regulations on access to addictive substances make it easier or harder to obtain these unhealthy substances.

The impact of each factor varies according to the various circumstances in and between urban places. Indeed all these factors combine in various proportions to make significant place differences in the health of income groups and areas, especially in an urban context, between and within diverse towns and cities. The intra-urban differences often lead to an ecology of health disadvantage within cities, creating what has been called an *urban health penalty* (Greenberg 1991). These unhealthy areas are characterised by limited health care facilities, people with poor life-style choices, substandard residences and often polluted conditions, low incomes, few economic prospects and a stressful, unsafe and socially isolated life. These conditions are made worse by feelings of hopelessness and the negative biological embedding of the harsh social environment. Increasing concern about the persistence of these pockets of ill-health has been one of the reasons why a new approach to the health care structure of many cities was developed.

13.6 The New Healthy City Movement and Organizations

Towards the end of the twentieth century, the challenges associated with improving the health of the population led many to the realization it was not appropriate to depend *only* on treating our way out of ill-health through the work of the medical profession alone. New forms of organizations and initiatives were also needed as well as a wider ecological view. It led the World Health Authority (WHO) to create the Healthy Cities programme in 1986 (Kenzer 1999). To some extent this was a broader version of The Ottawa Charter of 1986 that had established criteria for making Canadian cities healthier. Although initially restricted to the developed world, the Healthy Cities programme spread to cities in all parts of the world from 1994 and by the end of 2013 comprises a network of over a thousand cities worldwide, with both continental organizations as well as national ones, such as the 29 cities that comprise the U.K. network. The most successful Healthy Cities programmes maintain momentum from five values: a clear vision to promote health; the commitment of local community members; the ownership of policies; a wide array of stakeholders; and a process for institutionalizing the programme (WHO 1998). It is left to adopters to organize their own structures on the basis that governance varies widely between countries and locals know their own needs and priorities. The risk of this autonomy is that some cities lack the resources to develop good guidance and policies. To counteract this, membership of this network allows cities to learn from one another and to use 'best practice' ideas pioneered by other centres. The influence of this programme has also led many non-members to review their own approaches to improving health care, in which six major new trends are prominent, namely: creating health impact assessments; new political engagements; raising public awareness; adding community engagement; improving private participation; and finally, a more specific spatial targeting to improve conditions in persistent areas of ill-health.

13.6.1 Health Impact Assessments (HIA)

One way in which local governments have become more engaged in health issues can be seen in initiatives that focus on the bureaucratic deficiencies of particular departments with respect to health issues. For example, planning decisions have rarely taken public health issues into account in reviewing and approving development applications, except in zoning decisions and insuring that such issues as roads, sewers, water and other utilities are provided. However, towards the end of the twentieth century new interests emerged that stressed the need to assess the health effects of public projects on the residents of cities in particular. In part this may be seen as an extension of the increasing use of Environmental Impact Assessments. These were designed to investigate the physical and biotic impacts of new developments and were given statutory approval in many countries as relevant issues to be considered in decisions made by government planning authorities. Initial support for the Health Impact Assessment (HIA) concept came from the WHO's concerns over the effectiveness of new sanitation projects in developing countries which led the organization to produce a review of procedures for analysing the effects of new development on health (Birley 1995). A more general impetus came from the gradual realization that 'health' cannot be considered in isolation. In some regions it has become a vital consideration in *all* types of development, especially in terms of changes planned for the urban environment. This can be seen in Article 152 of the 1997 Amsterdam E.U Treaty which states:

> A high level of human health protection shall be enshrined in the definition and implementation of all Community policies and activities. (ET 1997)

In Europe the specific concept of the Health Impact Assessment (HIA) emerged from worries that public development proposals were not being analysed for health impacts (Scott-Samuel 1996). For example, concerns about noise and other health impacts over Manchester airport's second runway proposal in the 1990s led to guidelines for health assessments, Soon after, the U.K. government committed itself to using HIAs as an important strategy in development reviews as part of the need to identifying and reducing health inequalities (Scott-Samuel 1998).

The so-called Gothenburg Consensus that emerged after a meeting of the WHO in the Swedish city in 1999 attempted to bring some order into the evaluation process by identifying key features that were essential in the development of a HIA (WHO 1999). In addition to the general desire to maximize health outcomes they stressed the need for five values: democracy, so that all people can participate and help formulate policies; equity, to involve people from all walks of life; sustainable development in the long and short term; the ethical and transparent use of evidence, which recognizes that qualitative and non-scientific evidence can also be important in assessing impacts; and a comprehensive approach to health that considers the physical, mental and social well-being of all sectors of society and their involvement. Many of these values mirror the arguments used by proponents of a Just City approach to development within cities (Chap. 3) as well as those seeking to create greener, more sustainable places (Chaps. 4–6). It was also agreed in Gothenburg that a HIA should also include four major elements: reviews of the evidence that showed the effect of any new development policy or programme on the health of a population; consideration of the opinions, expectations and experience of those affected; recognition of the need for more informed understanding by both decision-makers and the public of the effects of the proposed policies; and proposals for adjustment or other options to maximize the positive effects of any policy, and to minimize their negative consequences. These elements illustrate the seriousness of the desire to take public opinion into account and to consider views from the various stakeholders.

A review of 88 HIAs from 1996 to 2004 (Davenport et al. 2005) revealed that not all were successful. It was concluded that only when certain conditions applied were they successful in influencing policy, namely: when there was institutional support for solving health problems; when key decision-makers were involved in the design of analyses; and when there was a statutory framework within government procedures for using a HIA. Without these conditions, various bureaucratic evaluation procedures just avoided the issues. Yet HIAs are now becoming as familiar as environmental reviews in some countries. A good example of the way in which health concerns raised by a community obtained crucial institutional support from a city's Public Health Department (PHD) can be seen in the Trinity Plaza proposal in inner city San Francisco in 2003, although it was not technically a HIA. The planning department supported the proposal for the redevelopment of the Trinity Plaza area near downtown in 2003, which involved plans to convert 360 rent-controlled housing units into 1700 market-rate units in several high rise towers. It ignored the fact that the development was one of the latest gentrification projects that involved a major displacement of low income people who had lived in the area for decades (Corburn 2009). The project led to a great deal of community anger and a pressure group to oppose the proposals. The Mission Anti-Displacement Coalition (MAC) created what they called a People's Plan to retain affordable family housing, preserve local jobs and asked for community oversight over land use changes. Fortunately for the community's case the Public Health Department in the city had moved away from only dealing with its traditional biomedical and sanitation focus and started to address the social determinants of health in the city, setting up workshops to learn about community concerns and providing advice about ways of improving areas. This led the PHD to intervene in the Trinity Plaza project on the side of the community, supporting its view that the displacement of the existing residents would cause financial hardship, given higher rents in other locations, and ensure the loss of community relationships and stress to the individuals who would be evicted, leading to increased levels of ill-health from these social determinants. The gradual acceptance of the health consequences of this project eventually led to discussions between the PHD, developers, planners and community activists, which led to a revised proposal in which 12 % of the project would be allocated to below market-rent housing, with existing occupants keeping their homes at current rates. In Corburn's (2009, p. 130) succinct phrase: "*the Department of Public Health had entered the world of urban development and staked-out a new space for human health*". Although there are still problems about the continued gentrification of the

Mission area, this San Francisco example provides a useful exemplar of the way in which health issues are being increasingly seen as relevant in planning reviews of new developments, irrespective of whether there is a formal HIA process in the cities.

13.6.2 New Political Engagements and Partnerships

The need for greater political engagement is another major principle of the WHO's Healthy City movement. Hopefully this would lead to the allocation of more finance and resources to improve health promotion and care. This new emphasis on local political involvement in health care, instead of only medical decision-making, has often taken time to achieve in many cities, often with limited initial results. For example, Liverpool in England moved from setting up a Healthy City Committee in 1988, to a Healthy City Team (HCT) in 1994, which produced a new plan for Liverpool in co-operation with the local Primary Care Trust (Otgaar et al. 2011). This Trust was one of the UK's 153 similar bodies under the National Health System that forms the first level of health care via family doctors and dentists, etc. Unlike the initial committee, the team had the skilled personnel to develop not only a vision statement, but a plan for achieving their goals, and consulted widely with citizens and medical professions and various other community groups. However, its plan was not simply about better health care and health promotion, but also related to how these would help to achieve greater business prosperity, social justice and environmental protection in the city. Despite its progress the HCT was wound up in 2000, in part because it seemed to need the greater involvement of the traditional medical professions. It was replaced by a new organization, essentially a new partnership, which is called Liverpool First, to plan and monitor health care in the city. This is now the joint responsibility of the Liverpool municipal council and the local medical Primary Care Trust. This organization has a wider range of stakeholders, ensuring input from many local organizations, from community groups, private businesses and various public sector services, thereby creating a more coordinated approach involving many people. This development in Liverpool has been duplicated elsewhere, which means that the planning of health care in Britain is no longer the domain of only the medical profession in those urban areas that have adopted this model. The local authority connection means that there is a more democratic basis, while the involvement of many other agencies and groups provides additional ideas and skills, to make a more comprehensive approach to improving the health care of residents in cities.

13.6.3 Creating Public Awareness of Health Issues

New members of the Healthy City network have to create a Health Profile to show their situation in comparison with other centres. This creates greater public awareness of the level of health in particular cities and frequently leads those with poor profiles to take steps to find ways of improving their record. For example, one of the reasons why Liverpool joined the Healthy City programme was its recognition that it had one of the worse health records in the U.K. where life expectancy at birth was 74 years, compared to 79 in Edinburgh and 78 years in London. Only Glasgow with a life expectancy of 71 years had a worse record among major British cities. Also, over half of Liverpool's population lived in the bottom 10% of deprived British areas, places that were classified by measures of multiple socio-economic deprivation, while a quarter of the population had long-term debilitating illnesses (LPCT 2007). Part of these problems stemmed from Liverpool's decline since the 1930s, when its population was over 800,000, and the poor health habits of many in the population. This poor health profile was not only a problem for its people but also created a negative image, which had the effect of repelling rather than attracting innovative migrants or businesses. So health promotion and participation campaigns were initiated which involved the general public as well as a range of health care and social service professionals. In some ways this decision to attract public attention is another version of the comments made earlier in the chapter about the way that the Health of Towns Association in Britain in the mid nineteenth century encouraged discussion of the mortality crises in various towns that often led to major local improvements.

The so-called Big Health Debate, which took place in Liverpool in 2006–2007, provides a good example of this process. It began by surveys to find major topics of health concern, followed by workshops involving medical professionals and representatives of the general population, then focus groups that worked on particular problems that had emerged, as well as analyses of the use of health facilities (LPCT 2007). One of the findings was that the general public felt they lacked the power to take control over many parts of their life, especially in health care, and were critical of being unable to influence decisions that affected them. So there was clearly an unmet need to involve the public in the discussions and the decisions being taken over health care.

13.6.4 Greater Community Linkages and Participation

For most of the twentieth century there were few examples of effective local community engagement in urban health planning under the older top-down approaches dominated by medical professionals. Usually the so-called 'community engagement' process involved people being informed about the decisions of medical planners after the main decisions had been taken, or were invited to attend public meetings which pretended to discuss issues of concern but were really designed to reduce resistance to changes. However the Healthy Cities goal of stressing the need for more democracy and community participation is leading to a different approach to previous practices. In the case of British cities such as Liverpool the new attitudes were encouraged by a national government white paper called Our Health, Our Care and Our Say (DH 2006), which outlined:

> The need for a wider range of community-based services, offering patients choice, convenience, fairness and a much better experience of the NHS. The White Paper also focuses on prevention and the need for the NHS to work in partnership with local authorities, voluntary agencies and local people so we are all working to build the foundations for better health in local communities. (LPCT 2007, p. ii)

The result was a model in which more community care was envisaged, given the fact that many are living longer, but often with debilitating illnesses that need local and home care, not hospital stays. Also a new type of participation was proposed to create more transparency and openness to discuss issues of public concern. It also anticipated capacity-building in the population, by improving their knowledge of the issues and encouraging their contribution to the debates, a process in which their opinions were carefully evaluated. This led to what was described in Liverpool as 'The Big Debates' about health care. They revealed that a major concern of many people was about their limited access to health care. This is a problem found in many cities, as decisions about the locations of both the offices of doctors or hospitals were usually made without effective public consultation. Indeed at a more general level it is remarkable how many hospitals in the developed world seem to be located close to wealthy areas, or have limited rapid public transport access since they are in inaccessible locations, while physician's offices are few and far between in poor areas, illustrating a failure of the market to serve all the population. Indeed few countries have adopted the Dutch model of planning which ensures that the major facilities that serve the public, or have large public use, such as hospitals, have to be located on major transit routes to maximize their accessibility.

The health surveys in Liverpool showed that people were prepared to accept the emerging trend of the increasing concentration of specialized medical facilities so long as basic facilities were locally available within a short travel time. Research results in a major report showed the locations of various health facilities and mapped the travel distances and times to these facilities from various points in the city. Geographically, it demonstrated that many residential areas had either few health care facilities or limited access to them. Incorporation of these results into discussions on how to solve these problems of access led to a new and comprehensive approach for health care in the region, with the subtitle the 'Outside of Hospital Strategy', which has created a four tiered system of health care in the metropolitan area (LPCT 2007). At the local level are the General Practices, with single or multiple doctors serving 1800–18,000 patients. Above these, 20–25 Neighbourhood Heath Centres are planned to serve populations of 20,000–25,000 and are located in places that could be reached by public transport in 15 min. These centres are anchored by doctors and health care professionals but also contain other social and community services. The third level consists of NHS Treatment Centres serving up 100–150,000 people that are located so they can be reached within 30 min by public transport. These provide primary and secondary care, such as minor surgeries, diagnostic tests, counselling and therapies, as well as outpatient clinics. The fourth level consists of the main hospitals serving 300–500,00 people that carry out major surgeries and other specialised medical procedures and care, in which some facilities concentrate on particular medical areas, such children's care or mental problems. This four-tiered system is designed to create a more integrated and community-based delivery system based on the problems of various areas, conforming to the goals of greater choice, access, fairness and convenience in a national white paper (DS 2006). Also some of the centres provided a new set of jobs within areas that previously had few medical facilities. The second and third levels of Treatment Centres made access to public transport a major location principle, showing the need to cater for people without cars, while the addition of these new, more local units was designed reduce the growing pressure on the emergency wards of major hospitals.

Liverpool also showed the way in which more effective public involvement in hospital re-planning can occur, as seen in the case of Alder Hey Children's Hospital in the West Darby suburb, which treats over 200,000 patients each year, many from other Primary Trusts outside Liverpool (Freeman 2006). Additional new hospital buildings were built and the adjacent decaying and unsafe Springfield Park was rejuvenated—with walking trails, sports pitches and greater surveillance to improve safety—in addition to creating better connections for pedestrians and cyclists to surrounding houses and retail facilities to the south. The design and location of the new buildings, both their interior and exterior plans, were discussed in public meetings, with local citizens providing ideas and modifications of initial ideas, which led to a feeling of positive involvement with the site's redevelopment. Determined efforts were also made to ensure the project would be more sustainable, such as in building construction, energy use and creating links with local transport facilities. Moreover, instead of the hospital being simply a place of 'care and cure', the hospital trained its staff to help patients and their parents adopt a healthier diet and life-style, providing greater health promotion in an area where premature deaths and disability due to poor life-style choices is far too high (Otgaar et al. 2009, pp. 49–51). These examples show how public engagement moved from the top-down, information-imparting style, to one of genuine attempts to find out the major problems that concerned patients and the public, but also to involve the local residents, as well as experts in various aspects of health care, design and accessibility, in a genuine interactive process of consultation and engagement.

In a more social context there is also evidence of increasing interest in developing local community interactions. For example, in Helsinki the city developed a series of 'neighbourhood houses' staffed by a core of city coordinators, but also by many volunteers. They organize physical and social activities, primarily, but not exclusively, for young people to provide them with new opportunities, such as sports, and in doing so guide them away from anti-social activities (Lafond et al. 2003). In some ways these 'houses', managed by the Health and Safe City Advisory Committee in the city, provide a contemporary version of older community associations found in many western Canadian cities that arose spontaneously (Davies and Townshend 1994). However in these cases there was no anti-crime agenda as in Finland; most started with sports activities for children but added additional social functions for adults, or various youth clubs, and often generated funds to build their own community hall with help from city and provincial grants. Historically, of course, even older community-based social centres and youth clubs were created by churches, where parish halls provided places for local interaction and facilities. However, the decreasing numbers of church members and rival denominations in many cities has meant that their ability to act on behalf of the local area has been drastically reduced, although many still provide support for their church members or space for other groups. There can be little doubt that there is a need for more secular local organizations like the Helsinki approach, or the contemporary YMCA and YWCA that have evolved out of their Christian origin. All these local centres provide important places for social and recreational contact, not simply for the youth, but as centres that may also alleviate the loneliness of many old people in cities.

13.6.5 Engaging the Private Sector

Although the objective of these more comprehensive health initiatives has been to involve as many stakeholders as possible, the contribution of the private business sector in health has not been as widespread as it could be, although many of the schemes discussed above do have representatives from private businesses. Part of the problem is that some firms do provide extra medical benefits for their workers, and feel that this is enough. Others, especially in Europe, provide some employees, usually at managerial levels, with stays in spas, presumably to recover from the stress of decision-making!. Hence such firms are reluctant to also contribute to general health care schemes for which the benefits seem to be beyond their immediate interests. In some cities there are signs that private firms see the advantages in participating in schemes to promote better health, especially when there are new business opportunities. This is certainly the case in Helsinki where the promotion of new technology, a focus on health and new organizations to promote healthy living, has led to many initiatives (Otgaar et al. 2011). For example, Forum Virium (FV) was established with support from major national organizations and companies to promote and create new digital services in several areas, including health care. This led to a photo and diary service for day-care centres to provide parents with images of their children's activities, a service that has subsequently spread to many countries, creating a new business venture. In addition, FV helped create the Healthy Helsinki programme that initiated several ventures to improve health through citizen empowerment, such as a Mobile Health project. Set up with help from telecom companies, it enabled participants to monitor their leisure and exercise levels, getting positive feedback through mobile devices if they increased their levels in attempts to change their behaviour. Elsewhere the development of computerized Personal Health Records, together with interactive mobile devices, has not only enabled users to store their own data but provide access to training, life-style and nutrition advice while in a neighbouring town it led to the monitoring of the growth in children, which identified abnormalities. More generally there are increasing numbers of smart applications for health-related issues being provided for use by smart or mobile phones, such as regular personal contacts to reduce isolation. Also, organizations such as the Toronto Rehabilitation Institute (TRI) are not only treating patients with long term disabilities from injury or aging, but actively pursue commercialization of their new technical products, such as their SlingSerter and RoboNurse for lifting patients (TRI). They are also in the forefront of research in providing assistance to people with cognitive impairment through smart alert systems to remind people to turn-off stoves and taps, and how and when to complete many regular routines, through sensing devices; in addition, like the Japanese, they are also building robots to help with these tasks. Such devices also provide help and relief to carers, but enable people to live longer in their homes rather than being forced into homes for the aged or hospitals, which dramatically increase costs of care.

Although there are clear business opportunities for companies to provide the hardware and software for such new monitoring systems, there is always the danger of intruding on personal privacy. However, municipal bureaucrats are often wary of getting involved since they have their functions to perform and the public benefits of some programmes may not be easy to evaluate. Nevertheless, the new electronic technologies can be used to monitor health more effectively than in the past and more private company involvement is not only possible, but needed in Healthy City initiatives so long as personal privacy is not compromised.

13.6.6 Attacking Spatial Health Inequalities

In the last decade of the twentieth century national governments of developed countries, often stimulated by the WHO Healthy Cities programme, started to emphasize policies that compel, not simply advise, all levels of government and their agencies to help reduce health inequalities, especially in cities. For example, in London it led to the creation of a London Health Commission composed of representatives from the various boroughs and organizations such as the NHS, Transport and the local Development Agency (LHC 2008). This commission was designed to ensure that health concerns are involved in all the tasks of the Greater London Authority and the 33 boroughs within it that have statutory authority for a wide range of services and functions. Previously, local governments and the Primary Care Trusts of the NHS did not consider the health impacts of their various planning and development projects. This change meant there was now a statutory obligation for the mayor of London to act to reduce the inequalities. The Health Commission developed a strategy called *Living Well in London* that ensures that investments in housing and public spaces are made to create an environment that helps people make more healthy choices and promotes links between the many agencies—from housing, to police, social services and organizations dealing with drug and alcohol abuse and homelessness—that may help to improve the determinants of health. In some deprived areas new multi-functional facilities have been developed, such as The Hub in Canning Town, a large building that houses a health centre with doctors and nurses as well as a pharmacy, with many other services indirectly linked to health and community improvement, such as a training centre, an internet café, community meeting rooms, and a safety group etc. In other boroughs of London local strategic partnerships have been established to improve the quality of life and health of residents. The board of these organizations include representatives from the public sector, including the NHS, local government and police, and the private sector, composed of people from citizen groups and various businesses. Although they have no powers to create actions, they are influential in ensuring that health considerations are given a priority in the various plans and actions of the partner organizations. They also encourage the establishment of thematic groups to focus on specific health improvements, including encouraging local residents to identify and solve health related problems. In addition to these partnerships, a 'targeted site' approach has identified areas of extreme deprivation and health problems as places for special programmes to improve the local conditions. The *Well London Alliance* programme created 14 projects in 20 of the most deprived parts of the city in 2007, dealing with quite small areas of 1500–2000 people (WLA 2007; Otgaar et al. 2011). These are not policies of planning or housing departments alone, but involve multiple agencies—including the YMCA, two charities, a technology transfer agency, the Arts Council, a local university and major hospital trust—programmes which are funded by grants from the National Lottery. The various agencies contribute their own expertise to the projects. However, in an attempt to get more local input, the residents affected were consulted about the projects through site visits, action workshops and discussions in the community cafes that were established. Surveys revealed a lack of co-ordination among the existing social services and revealed that the people in the areas were not simply deprived economically, but lacked hope and ability to improve their conditions, typical problems that affect the development of grass roots organizations (Davies and Herbert 1993, Chap. 6). An important focus of these targeted Well London programmes has been to emphasize better nutrition, create exercise opportunities, and develop a series of other programmes to improve the negative various social determinants of health problems, through a series of so-called 'heart of the community' projects. These are designed to reduce loneliness and improve community relationships, such as involving the youth in the design of projects, creating access to healthy facilities in the area, from encouraging new shops, to finding safe walking routes. They also focus on reducing crime, developing personal improvement programmes to improve the mental well-being of people in the area and try to create a sense of ownership over local facilities and projects. By having control group areas where no intervention has occurred, the effectiveness of the various initiatives were measured in the 5 year project period from 2007 in order to create evidence-based policy evaluation, in the hope that the programmes that work can be copied successfully in other areas. However, despite these targeted initiatives there is a pressing need for more social housing in London. The influx of capital from overseas rich people, combined with the strength of the financial services sector in particular, means that land and house prices continue to rise, with the result that essential workers, such as nurses, social workers, police and teachers, are priced out of the market.

13.6.7 Other Targeting Measures

The type of comprehensive, multi-agency approach described above is only one of the ways in which attempts are being made to improve the health of disadvantaged people. Another is the targeting of poor people by giving income supplements on condition the children attend school and keep their vaccinations up to date, an approach pioneered by the Brazilian *Bolsa Familia* programme which has reduced the poverty level of over 44 million people. More specific health care access for the poorest in American cities can be seen by the addition of new *community clinics,* either free standing or mobile. They provide health care in disadvantaged areas, where there are few local physicians or pharmacies, creating centres that provide a range of services, such as physicians, dentists, as well as counselling and health promotion units. In impoverished parts of American cities, *school health education and provision* is an increasing trend, which often includes the addition of clinics, especially aimed at targeting adolescents and their problems (Fitzpatrick and LaGory 2011). A National Assembly of School-Based Health Care acts as a co-ordinating and advice centre for these school-based clinics which have grown from 20 in 1980 to currently over 2000 in 44 states. Research has shown that schools with clinics have fewer disciplinary problems, course failures, and school absentee rates, as well as better health (Smith 2013).

There is also a long history of the work of particular individuals acting as the catalysts for change in particular deprived areas. For example, a Venezuelan physician, America Bracho, was influential in establishing Latino Access, a non-profit organization in the poor, mainly Spanish speaking area of Santa Ana, an area of a third of a million people in Los Angeles, one of the toughest, crime-ridden, deprived and unhealthy areas in the country (LHA 2009). Also churches have long had a role in trying to cater for the poor and the disabled. In recent years more active approaches have developed, such as the Communities of Shalom (CS) set up by the United Methodist Churches. Initiated by the Rev Joseph Sprague it has now created hundreds of shalom zones in impoverished areas from its beginning in 1992, training local residents to promote health information and social capital. In many American cities some churches have partnered with public health departments in local universities to create facilities and programmes to improve the health of people in poor neighbourhoods, using trained local volunteers and specialists, such as the Congregations for Public Health in Birmingham, Alabama (CPH). All these examples represent the contemporary equivalents of the settlement houses of nineteenth century cities, designed improve the educational, health and economic prospects of people in impoverished areas, although these were mainly secular, as seen in work of Hull House in Chicago and many other cities (Adams 1910). Another policy has tried to improve to areas of high disadvantage and ill-health by seeking to *reduce concentrations* of poor families, such as by scattering, instead of creating, public housing clustering, and by removing families to other locations where there are better facilities and more social mixture (Rubinovitz and Rosenbaum 2000; Comey et al. 2008). Although there have been some successful resettlement schemes the scale of the areas of impoverishment means that it cannot be a solution for all areas.

Space constraints mean that the examples discussed have only focused on developed countries to show that even in areas where there are extensive heath care programmes there is room for major improvements in targeting the most unhealthy places. A far, far bigger problem exists in the developing world where at least of the third of their population, over 860 million in total, live in informal shanty towns. Many of these places desperately need updated and modern versions of the types of services, from clean water supply to sewage and other utilities, that were created in the European urban improvements of the nineteenth century, deficiencies which lead to poor hygiene and debilitating illnesses. These informal settlements are characterized by high levels of deprivation, due to poverty, with flimsy informal housing—whose rents often take up a third of the non-food expenditure—unemployment, limited skills or education, low environmental quality, often with high crime, and with many residents, especially the elderly, having few social contacts and high levels of loneliness. Also, many of these areas lack proper property rights. This situation is often made worse by the presence of toxic outputs from nearly factories that ignore any environmental regulations even where they exist. But it is often argued that these slums, not all populated initially by rural migrants, should be welcomed as they provide the opportunity for improvement, unlike rural areas. Some have shown improvement, but the idea of the slum being a temporary way-station on the way to prosperity is increasingly being challenged. A large survey of slum dwellers (Marx et al. 2013) has confirmed that many people get trapped in these areas, leading to generational poverty and also illness, despite the opportunities that enable some to escape. Even in the cities of middle income countries there are still many slum areas, for example in Buenos Aires about 10 % of the population lives in the 56 shanty towns called *villas miserias*. Most of them lack schools, medical facilities, basic utilities and have poor drainage, so heavy rain causes flooding, and have limited employment and higher levels of ill-health. In 2011 the city created the Secretariat of Habitat and Inclusion to coordinate the efforts to incorporate these areas into the city and upgrade their facilities, but its efforts are handicapped by a small budget. One of its schemes is a so-called *urban acupuncture* programme that adds sports facilities, plazas and community centres to the deprived areas to act as focal and service points. However the programme has not been as effective as promised because the Secretariat is underfunded and is seen as a top-down agency. This often ignores the grass root organizations that have developed in these areas which emphasize that their real need is also for more adequate social housing and basic services, including health care. Elsewhere attempts are being made to provide more community facilities and to connect slum areas, which are often isolated on hillsides with few roads, with other parts of the city, such as by the gondolas in Bogota. In many other slum areas, such as in Rio de Janeiro, the problem of safety is paramount because of their control by local gangs. This is being tackled by what amounts to a blanket invasion by security services, followed up by stimulation of grass roots organizations, the provision of local community and a school facilities, and a permanent police presence to guarantee safety.

Despite occasional attempts to improve the condition of people in these informal settlements there is still an enormous amount of work to do, given the scale of these problems and the limited funds and trained personnel of the city administrations. Indeed, at base the older developed world solutions to create healthier, man-made environments are still needed. Although modern medicine, especially for infectious disease has kept mortality rates low so far, the constant exposure to unhealthy conditions and stress in these slums, combined with the new growth of antimicrobial resistance mean that these areas could be the breeding grounds for future epidemics. This fear should speed up efforts to improve these often squalid environments.

13.7 Conclusions

Improving the health of people in cities is part of their basic needs and a matter of social justice. It is one of the most important of the new policies contributing to liveable urban places. The old urban health disadvantages were solved by the nineteenth century engineering and regulatory approaches, and by the enormous advances in medical knowledge and care. Yet too many urban places still need basic sanitation, safe water supplies and adequate numbers of health care professionals and facilities. In addition new concerns have arisen about gaps in the delivery of health care, even in the developed world, with a need to address the societal structural conditions that lead to high inequalities and poor health. The new Healthy City programmes are creating positive changes through greater political involvement and community engagement, with policies designed to improve access to health care, such as spatial targeting and the use of Health Impact Assessments. The concept of `wellness' stresses the need for active health promotion and research into the impact of the many health determinants, especially the often downplayed environmental, economic and social domains.

Although these new approaches are undoubtedly improving the health of people in many cities, they are not enough. It has been shown that urban living creates greater risks from such problems as traffic and pollution, but also from less time spent outdoors. One problem is that some pregnant mothers who have limited exposure to sunlight may be inhibiting the development of their foetuses, which seems to account for the recent increases in many autoimmune diseases. Another comes from an enormous recent rise in levels of myopia in young urban Chinese and other Asian children in particular, with 70% of children affected in some areas. Studies have discounted the general effect of genetic factors and shown that the children affected are not spending enough time outdoors, with too much time studying indoors and exposure to electronic devices. Control groups of Australian children and rural Chinese did not have this problem, which seems to be a consequence of a lengthening of the eyeball along its main axis due to a lack of exposure to sunlight (Lougheed 2014, Rose and Saw 2013).

Increasing challenges also come from the failures of the major antibiotic drugs that counter many diseases, which could reverse many medical advances. There are also increased risks from new communicable diseases in densely populated and interconnected urban places, as shown by the case of HIV. But a more recent example is seen by the rapid spread of Ebola in West Africa, with over 5,000 deaths and at least 14,000 cases by the end of October 2014. 40 As Rene Dubois (1959) observed in many books, nature in the form of some microbe will always strike back at human bodies in some unpredictable way, for new viruses constantly emerge and existing ones mutate and become resistant to drugs. Ebola's rapid spread was due to its ability to mutate rapidly and suppress human immune systems. But it was also a consequence of several environmental and cultural factors: the inability to recognize its initial appearance, probably from cross-species infection from the virus-carrying fruit bats; its spread in areas of high density, poor sanitation, high mobility levels, and very limited health care systems in personnel and facilities; a lack of trust in authorities after years of war; and where traditional healing and funerary practices involved a lot of touching, which led to greater spread since bodies are especially infectious by extruding the body fluids that spread the virus. The heroic efforts of the charity Médicine sans Frontières helped many people, but did not have the resources to contain the outbreak, with an often 70% death rate among infected people. Unfortunately it took over six months for the WHO and other countries to start a massive effort to try and contain the disease by providing: supplies of soap, alcohol swabs and disinfections to help reduce Ebola's spread; thousands of trained doctors and nurses to attend the sick; enough protective clothing; building isolation hospitals to treat patients; as well as tracing the contacts of infected people, which was impeded by the absence of detailed local maps. This outbreak showed we cannot be complacent about the threat of new diseases. There is pressing need to have enough of the needed facilities available to be sent quickly to areas where new diseases are detected and a necessity to boost research to find vaccines and drugs to combat such new diseases and older ones like malaria that are still endemic.

Many of these problems go beyond the ability of cities alone to solve them, although their density and connections make them more vulnerable, especially those with poor sanitation. Cities rarely have the power or finance to improve health care on their own, but they can initiate new programmes to improve their facilities. Yet their very size means that their politicians and citizen groups should be able to mount successful lobbying of national governments to take the more active, comprehensive and multifaceted measures to improve health in urban settlements.

References

Adams, J. (1910). *Twenty years at Hull House*. New York: Macmillan.

Adams, J. M. (2012). Healthy places and healthy regimens: British spas 1918–50. In V. Berridge & M. Gorsky (Eds.), *Environment and health* (pp. 113–132). Basingstoke: Palgrave- Macmillan.

Adler, N. E., & Ostrove, J. M. (1999). Socioeconomic status and health: What we know and what we don't. *Annals New York Academy Science, 896,* 3–15.

Alberta Health. (2014). Alberta's strategic approach to wellness: Health for all, wellness for life. http://www.health.alberta.ca/documents/Strategic-Approach-Wellness-2013.pdf. Accessed 2 April 2014.
Alcon, J. (2012). Redefining public health in New York. *The Lancet, 379,* 2037–2039.
Ashton, J., & Ubido, J. (1991). The healthy city and the ecological idea. *Journal of the Society for the Social History of Medicine, 4*(1), 173–181. Accessed 8 Nov 2013.
Beck, A. T. (1999). *Prisoners of hate: The cognitive basis of anger, hostility and violence*. New York: Harper Collins.
Berridge, V., & Gorsky, M. (Eds.). (2012). *Environment and health*. Basingstoke: Palgrave Macmillan.
Birley, M. H. (1995). *The health impact assessment of development projects*. London: HMSO.
Boyce, W. T., Sokolowski, M. B., & Robinson, G. E. (2012). Towards a new biology of social adversity. *Proceedings of the National Academy of Sciences, 109* (supplement 2), 17143–17148.
Brooks-Gunn, J., Duncan, G. J., Klebanof, P. K., & Sealand, N. (1993). Do neighbourhoods influence child and adolescent development. *American Journal of Sociology, 62,* 353–395.
Cacioppo, J. C. T., & Patrick, W. (2009). *Loneliness: Human nature and the need for social connection*. New York: W. W. Norton.
Carter, S. (2012). Leagues of sunshine; sunlight, health and the environment. In V. Berridge & M. Gorsky (Eds.), *Environment and health* (pp. 94–112) Basingstoke: Palgrave- Macmillan.
CDC. (2013). Antimicrobial resistance threats in the United States, 2103. Centre for Disease Control and Protection, Atlanta. http://www.cdc.gov/drugresistance/threat-report-2013/index.html. Accessed 3 April 2014.
Chadwick, E. (1843). In V. Kiernan (Ed.), *Report on the sanitary condition of the labouring population of Great Britain (1965)* (New edition). London: Clowes and Co. for HMSO, London: Penguin.
Coleman, M. (1985). *Utopia on trial*. London: Hilary Shipman.
Comey, J., de Suza Briggs, X., & Weismann, S. (2008). *Struggling to stay out of high poverty neighbourhoods*. Washington D.C.: The Urban Institute.
Corburn, J. (2009). *Toward the healthy city*. Cambridge: MIT Press.
CPH: Congregations of Public Health. http://www.soph.uab.edu/csch/community/cph.
Dahlgren, G., & Whitehead, M. (2006). *Levelling up: European strategies for tackling social inequalities in health*. Copenhagen: WHO Regional Office for Europe (discussion paper).
Davenport, C., Mathers, J., & Parry, J. (2005). Use of health impact assessment in incorporating health considerations in decision making. *Journal of Epidemiology and Community Health, 60,* 196–201.
Davies, W. K. D. (1978). Charles Booth and the measurement of urban social character. *Area, 10*(4) 290–296.
Davies, A. (1997). *Road transport and health*. London: British Medical Association.
Davies, W. K. D. (2004). Affective dimensions of urban crime areas. *Geographica Helvetica, 59*(3), 218–226.
Davies, S., Grant, J., & Catchpole, M. (2013). *The drugs don't work*. London: Penguin.
Davies, W. K. D., & Herbert, D. T. (1993). *Communities within cities*. London/New York: Belhaven/Halstead-Wiley.
Davies, W. K. D., & Townshend, I. J. (1994). How community associations vary. *Urban Studies, 31*(10), 1739–1761.
DH. (2006). *Our health, our care and our say*. London: U.K. Department of Health.
Disanto, G., Morahan, J. M., Giovannoni, G., Hyppönen, E., Ebers, G. E., & Ramagopalan, S. V. (2012). Month of birth, vitamin D and risk of immune-mediated disease: A case control study. *BMC Medicine, 2012*(10), 69–76. http://www.biomedcentral.com/1741-7015/10/69. Accessed 10 Feb 2014.
Dubois, R. (1959). *Mirage of health: Utopias, progress, and biological change*. New York: Harper Brothers.
EPA (1991). Sick building syndrome fact sheet: Air facts No. 4. http://www.epa.gov.iaq/sick-building-factsheet.pdf. Accessed 20 Feb 2014.

EPA (1999). *Why you should be concerned about the quality of the air you breath?* Washington D.C.: U.S. Environmental Protection Agency. http://www.epa.gov/iaq/. Accessed 14 Jan 2014.
Eppig, C., Fincher, C. L., & Thornhill, B. (2010). Disease and intelligence. Proceedings of Royal Society B (June). http://www.royalsocietypublishing.org/content/eob/2010/06/29/rspb2010.0973. Accessed 12 July 2012.
ET: Eurotreaties. http://www.eurotreaties.com/amsterdamtreaty.pdf. Accessed 10 Nov 2013.
Engels, F. (1844). *The condition of the working class in England*. London: Panther (reprint 1969).
Enserink, M. (2013). SARS: Chronology of the epidemic. *Science, 339,* 1266–1271.
Epel, E. S., Blackburn, E. H., Lin, J., Dhabhar, F. S., Adler, N. E., Morrow, J. D., & Crawford, R. M. (2004). *Accelerated telomere shortening in response to life stress. Proceedings of National Academy of Sciences of the USA, 101*(49), 17312–17315. doi:10.1073/pnas.0407162101.
Freeman, N. (2006). Rapid HIA of the Royal Liverpool Children's NHS Trust's Outline Business Case for the modernisation of children's services. http://www.apho.org.uk/resource/item.aspx?RID=118186. Accessed 2 Feb 2014.
Fitzpatrick, K., & LaGory, M. (2011). *Unhealthy cities: Poverty, race and place in America*. New York: Routledge.
Fonagy, P. (2003). Towards a developmental understanding of violence. *British Journal of Psychiatry, 183,* 190–192.
Frerichs, R. R. (n.d.) http://www.ph.ucla.edu/epi/snow.htm. Accessed 4 April 2013.
Gaunt, J. (1662). *Natural and political observations mentioned in the following index and made upon the bills of mortality*. London: Roycroft.
Gille, O. (2004). Sunlight robbery: Health benefits of sunlight are denied by current public health policy in the U.K. Health Forum Occasional Reports, No. 1. http://www.healthresearchforum.org.uk/. Accessed 10 Nov 2012.
GNAFCC: Global network of age-friendly cities and communities. http://www.who.int/ageing/age_friendly_cities_network/en/. Accessed 12 Dec 2013.
Godish, T. (2001). *Indoor environmental quality*. New York: CRC Press.
Greenburg, M. (1991). American cities: Good and bad news about public health. *Bulletin of the New York Academy of Medicine, 67,* 17–21.
Hagaman, J. T., Panos, R. J., McCormack, F. X., Charuhas, V. T., Shipley, R. T., & Kinder, B. W. (2011). Vitamin D deficiency and reduced lung function in connective tissue and associated interstitial lung diseases. *Chest, 139*(2), 353–360. doi:10.1378/chest.10-0968.
Hall, P. (1988). *Cities of tomorrow*. Oxford: Blackwell.
Harding, V. (2012). Housing and health in early modern London. In V. Berridge & M. Gorsky (Eds.), *Environment and health* (pp. 23–44) Basingstoke: Palgrave- Macmillan.
Herlithy, D. (1997). *Black death and the transformation of Europe*. Cambridge: Harvard University Press.
Holt, N. (2014). *Cured*. New York: Penguin.
Johnston, S. (2006). *The ghost map*. London: Penguin.
Kenzer, M. (1999). Healthy cities: A guide to the literature. *Environment and Urbanization, 11*(1), 201–220.
Khan, K. M., Thompson, A. M., Blair, S. N., Sallis, J., Powell, K. E., Bull, F. C., & Bauman, A. E. (2012). Sports and exercise as contributors to the health of nations. *The Lancet, 380,* 9836, 59–64.
Kinder, B., & Hagaman, J. T. (2011). Could combating vitamin D deficiency reduce the incidence of autoimmune disease? *Expert Review Clinical Immunology, 7*(3), 255–257.
Klinenburg, E. (2002). *Heat wave: A social autopsy of disaster in Chicago*. Chicago: Chicago University Press.
La Berge, A. (1992). *Mission and method: The early C19 French public health movement*. Cambridge: Cambridge University Press.
Lafond, L. J., Heritage, Z., Farrington, J. L., & Tsouros, A. D. (2003). *National healthy cities networks: A powerful force for health and sustainable development in Europe*. Geneva: World Health Organization.

Lalonde, M. (1974). *A new perspective on the health of Canadians*. Ottawa: Dept. of National Health and Welfare.
LCC. (2012). *Liverpool health and wellbeing strategy, 2012–2015: Laying the foundations*. Liverpool: Liverpool City Council and Liverpool Primary Care Trust.
Lee, I. M., Shiroma, E., Lobela, F., Puska, P., Blair, S., & Katzmarzyl, P. (2012). Effect of physical inactivity on major non-communicable diseases worldwide. *The Lancet, 380*(9838), 219–229.
LHA. (2009). http://wwwlatinohealthaccess.org/index.shtml. Accessed 20 Feb 2014.
LHC. (2008). *London and Londoners: Making the links for health*. London: London Health Commission.
Litonjua, A. A., & Weiss, S. T. (2007). Is vitamin D deficiency to blame for the asthma epidemic? *Journal Allergy Clinical Immunology, 120*(5), 1031–1035.
Litonjua, A. A., & Weiss, S. T. (2011). The *in utero* effects of maternal vitamin D deficiency. *American Journal of Respiratory and Critical Care Medicine, 183*(10), 1286–1287.
Lougheed, T. (2014). Myopia: The evidence for environmental factors. *Environmental Health Perspectives, 122* (1), A12–19. doi:10.1289/ehp.122-A12.
LPCT. (2007). *A new health strategy for Liverpool: The outside of hospital strategy*. Liverpool: Liverpool Primary Care Trust.
Lurie, S. J., Gawinski, B., Pierce, D., & Robinson, S. J. (2006). Seasonal affective disorder. *American Family Physician, 74,* 1521–1524.
McEwen, B. S. (2012). Brain on stress: How the social environment gets under the skin. *Proceedings of the National Academy of Science, 109*(supplement 2), 17180–17185. doi:10.1073/pnas.1121254109.
Markus, D. K., Fulton, J. J., & Clarke, E. J. (2010). Lead and conduct problems: A meta analysis. *Journal of Clinical Child and Adolescent Psychology,* 39(2), 234–241.
Marsh, B. (1909). *An introduction to city planning: Democracy's challenge to the American city*. New York: Committee on Congestion of Population in New York.
Marx, B., Stoker, T., & Suri, T. (2013). Economics of slums in the developing world. *Journal of Economic Perspectives, 27*(4), 187–210.
Mayhew, H., Tuckniss W., & Beeard, R. (1851–1862). *London labour and the London poor: Volumes 1–4. Reprinted (1968)*. London: Dover Publications.
Mitchell, C., Hobcroft, J., McLanahan, S. S., Siegal, S. R., Berg, A., Brooks-Gunn, J., Garfinkel, I., & Notterman, D. (2014). Social disadvantage, genetic sensitivity, and children's telomere length. *Proceedings of National Academy of Sciences, 111*(18). doi:10.1073/pnas.1404293111.
Michie, C. (2013). Childhood rickets in back while vitamin D sits in warehouses. *The Guardian,* 25 October 2013.
Morgan, M. H. (Ed.). (1960). *Vitruvius: The ten books of architecture* (trans: M. H. Morgan). New York: Dover.
Murray, C., Rosenfeld, L. C., & Lin, C. C. (2012). Global measures of malaria: 1980–2010. *The Lancet, 379,* 413–431.
Norton, S., Matthews, F. E., Barnes, D. E., Yaffe, K., & Brayne, C. (2014). Potential for primary prevention of Alzheimer's disease: An analysis of population-based data. *The Lancet Neurology, 13,* 788–794. doi:10.1016/S1474-4422(14)70136-X.
Ogden C. L., Carroll, M. D., Kit, B. K., & Flegal K. M. (2014). Prevalence of childhood and adult obesity in the United States, 2011–2012. *Journal of the American Medical Association, 311*(8), 806–814.
Otgaar, A., Klijs, J., & van den Berg, L. (2011). *Towards healthy cities: Comparing conditions for change*. Farnham: Ashgate.
Patrick, P. R., & Ames, B. N. (2014). Vitamin D hormone regulates serotonin synthesis: Relevance for autism. *Federation of American Society for Experimental Biology, 28*(6), 2398–2413.
Perissinotto, C. M., Stijacic, I., Cenzer, K. E., & Covinsky, A. (2012). Loneliness in older persons: A predictor of functional decline and death. *Archives of Internal Medicine, 172*(14), 1078–1084.
Pfautz, H. W. (1967). *Charles Booth on the city: Physical patterns and social structures*. Chicago: Phoenix Books.

Possehl, G. (2002). *The Indus civilization: A contemporary perspective*. New York: AltaMira Press.
Preedy, V. R. (Ed.). (2011). *Beer: Health and disease prevention*. London: Academic Press.
Prince, M., Acosti, D., et al. (2012). Dementia incidence and mortality in middle income countries. *The Lancet, 380*(9836), 50–58.
Prüss-Üstün, A., & Corvelán, C. (2006). *Preventing disease through healthy environments*. Geneva: WHO.
Qu, Y., Hu, D., Huang, J., Wang, Y., & Zhang, D. (2014). Fruits and vegetables consumption and risk of stroke: A meta-analysis of prospective cohort studies. *Stroke, 45,* 1613–1619.
Ramajopalan, S. V., Maugan, N., Knight, J. C., et al. (2009). Expression of the multiple sclerosis-associated MHC class II allele HCA-DCBI*1501. *PLOS Genetics, 5,* e1000369. doi:10.1371/journal pgen.100036. Accessed 4 April 2014.
Ramagopalan, S. V., Heger, A., Berlanga, A. J., Maugeri, N. J., Lincoln, M. R., Burrel, l. A., Handunnetthi, L., Handel, A. E., Disanto, G., Orton, S., Watson, C. T., Morahan J. M., Giovannoni G., Ponting C. P., Ebers G. C., & Knight J. C. (2010). A ChIP-seq-defined genome-wide map of vitamin D receptor binding: Associations with disease and evolution. *Genome Research, 20,* 1352–1360. doi:10.1101/gr.107920.110. Accessed 4 April 2014.
Reps, J. (1965). *The making of urban America*. New Jersey: Princeton University Press.
Reyes, J. W. (2007). Environmental policy as social policy? The impact of childhood lead exposure on crime. National Bureau of Environmental Research Working Paper 13093. http//www.nber.org/papers/v130097. Accessed 10 Nov 2013.
Rose, K. A., Saw, S. M. et al. (2008). Myopia, lifestyle, and schooling in students of Chinese ethnicity in Singapore and Sydney. *Archives of Ophthalmology, 126* (4), 527–530. http://dx.doi.org/10.1001/archopht.126.4.527.
Rubinovitz, L. S., & Rosenbaum, J. E. (2000). *Crossing the class and colour lines*. Chicago: University of Chicago Press.
Schott, D. (2012). The Hanbuch der Hygiene: A manual of proto-environmental science in Germany. In V. Berridge & M. Gorsky (Eds.), *Environment and Health* (pp. 69–93). Basingstoke: Palgrave-Macmillan.
Scott-Samuel, A. (1996). Health impact assessment. *British Medical Journal, 313*(1), 183–184.
Scott-Samuel, A. (1998). Health impact assessment: Theory into practice. *Journal of Epidemiology and Community Health, 52,* 704–75.
Smith, B. L. (2013). Expanding school-based care. *American Psychological Association, 44*(8). http://www.apa.org/monitor/2013/09/school-care.aspx. Accessed 3 April 2014.
Snowden, F. (2006). *The conquest of malaria 1900–62*. New Haven: Yale University Press.
TE. (2014). The $ 272 billion swindle. *The Economist,* May 11, 26–27.
Tollefsbol, T. (Ed.). (2014). *Transgenerational epigenetics*. London: Academic Books.
Tomalin, C. (2011). *Charles Dickens: A life*. London: Penguin.
TRI: Toronto Rehabilitation Institute. http://www.uhn.ca/TorontoRehab/Research/Commercialization/Pages/default.aspx. Accessed 20 April 2014.
UMC: United Methodist Churches. http://www.centerfortransformingcommunities.org/Communities-of-Shalom.html. Accessed 20 Jan 2014.
Weiss, S. T., & Litonjua, A. A. (2007). Maternal diet vs lack of exposure to sunlight as the cause of the epidemic of the asthma, allergies and other autoimmune diseases. *Thorax, 62*(9), 746–748. doi:10.1136/thx.2007.079707.
WHO. (1948). *Constitution*. Geneva: World Health Organization.
WHO. (1998). Types of healthy settings. http://www.who.int/healthy_settings/types/cities/en/. Accessed 10 Dec 2013.
WHO. (1999). *Health impact assessments: Main concepts and significant applications*. Brussels: World Health Organization Regional Office.
WHO. (2000). *Obesity: Preventing and managing the global epidemic*. Geneva: World Health Organization.
WHO. (2007). Age-friendly cities: A guide. Geneva: World Health Organization. http://whqlibdoc.who.int/publications/2007/9789241547307eng.pdf?ua=1.

WHO. (2010). Global status report on non-communicable diseases. Geneva: World Health Organization.
WHO. (2012a). Projections of mortality and causes of death: 2015–2030. Geneva: World Health Organization. http://www.who.int/healthinfo/mortality_data/en/. Accessed 15 March 2014.
WHO. (2012b). *Global analysis of sanitation and drinking water*. Geneva: World Health Organization.
WHO. (2013). *Global study on road safety*. Geneva: World Health Organization.
WHO. (2014a). *Ambient air quality and health*. Geneva: World Health Organization.
WHO. (2014b). *Antimicrobial resistance: Global report of surveillance*. Geneva: World Health Organization.
WHO-ADI. (2012). *Dementia: A public health priority*. Geneva: World Health Organization and Alzheimer's Disease Institute.
WHO-E. (1986). *Ottawa charter for health promotion*. Copenhagen: European Office of WHO. http://www.euro.who.int/__data/assets/pdf_file/0004/129532/Ottawa_Charter.pdf?ua=1. Accessed 10 Jan 2014.
WHO-E. (1998). Health 21: Health for all in the 21st century. http://www.euro.who.int/__data/assets/pdf_file/00074/109759/EHFA5-E.pdf.
Wilkins, R. (2007). *Mortality by neighbourhood income in urban Canada, 1971–2001*. Ottawa: Statistics Canada Health Measurement Group.
Wilkinson, R., & Marmot, M. (Eds.). (1999). *Social determinants of health*. Oxford: Oxford University Press.
Wilkinson, R., & Marmot, M. (2003). *Social determinants of health: The solid facts*. Copenhagen: WHO Regional Office for Europe.
WLA. (2007). Well London strategy, delivering Well London. http://www.welllondon.org.uk/83. Accessed 20 Jan 2014.
Yang, C. Y., Leung, P. S., Adamopoulos, I. E., & Gershwin, M. E. (2013). The implication of vitamin D and autoimmunity: A comprehensive review. *Clinical Review of Allergy Immunology, 45*(2), 217–226. doi:10.1007/s12016-013-8361-3.
Zaridze, D., Peto, R., et al. (2014). Alcohol and mortality in Russia: Prospective observational study of 151,000 adults. *The Lancet, 383*(9927), 1465–1473.
Zoutman, D., Shannon, M., & Mandel, A. (2011). Effectiveness of a novel ozone-based system for the rapid high-level disinfection of health care spaces and surfaces. *American Journal of Infection Control, 39*(10), 873–875.

Chapter 14
Festive Cities: Multi-Dimensional Perspectives

Wayne K.D. Davies

> *The success of festivals and their impact on society depends on the way themes, symbols, identities, values, time and traditional social habits are organized*
> Friedrich 2000, p. 4.

14.1 Introduction

A curious characteristic of the many contemporary attempts to resolve the problems of the modernist city has been the relatively limited attention paid by most urban scholars to features of the city that relate to sensory not material attributes, to consumption not production, pleasure not work, and to episodic rather than permanent characteristics. Festive events in cities combine these four traits. These are activities consciously planned to last a limited time to mark special occasions, or to promote specific events, often using spectacles and parades designed to create excitement and joy, as well as achieving cultural, social and economic goals. Some of these may also, in Friedrich's (2000, p. xiv) words, "touch deep human instincts which relate to origin and community" in addition to the impacts described in the above quotation that primarily relates to historic events. An important early impetus for the study of festivals came from the work of cultural historians, stimulated by a three volume seminal work on Renaissance fêtes (Jacquot 1956–75). They have been increasingly investigated by other historians, anthropologists, sociologists and geographers, in both traditional and modern societies (Strong 1992; Bélanger and Côte 2005; Di Meo 2001; Gotham 2005; Gravari-Barbas 2005, 2007). More recently the economic impact of festivals in particular has attracted tourism and special events specialists (Allen et al. 2005; Picard and Robinson 2006). Some of this increasing interest in festivals in contemporary cities comes from the growth of a secular and leisure society with its members constantly seeking new experiences and pleasure, in pursuit of what they consider to be the good or at least more

W. K.D. Davies (✉)
Department of Geography, University of Calgary,
2500 University Drive, N.W., Calgary, AB T2N 1N4, Canada
e-mail: wdavies@ucalgary.ca

W. K.D. Davies (ed.), *Theme Cities: Solutions for Urban Problems*,
GeoJournal Library 112, DOI 10.1007/978-94-017-9655-2_14

exciting life, part of an emerging postmodern urbanism in which the festival and spectacle, with ephemeral displays become the focus, albeit transitory, of urban life (Harvey 1991, p. 270). But a more general intellectual stimulus to the interest in festivals comes from post-modernist concepts, not attributes, in which emphasis is placed on the study of *difference* not similarity, *individuality* over conformity, as well as the *immaterial* and the *symbolic* in cities in their search for explanations beyond the empirical (Davies 1996). The result is an increasingly rich literature illustrating the impact and varied roles that festive events have played in cities, towns and even small places, both in the past and in contemporary settlements.

Festivals have shown a remarkable growth in recent decades and now make a substantial contribution to urban vitality and distinctiveness. Some annual festivals of a secular nature associated with a single place are now so big and vital that they almost define the city in which they are located, as can be seen by such examples as the Carneval in Rio de Janeiro, the Mardi Gras in New Orleans, the Oktoberfest in Munich, the Snow Festival in Sapporo, or the Stampede in Calgary. These festivals also provide a city branding that often goes beyond the role that distinctive urban architectural features play in urban imagery and tourism and have significant economic and cultural impacts on the city in which they are located. Indeed, many participants and observers consider the most important festivals in their city to be the climax of their year, which is often true for regular festivals of religious significance, such as the annual Deepavalli or Divali Festival in India, although these are not city-specific. Other festivals are held in cycles of more than a year; probably the biggest is the Hindu Maha Kumbh Mela held every 12 years and estimated to have attracted 80 million people to Allahabad (India) at various times in the 55 day event from early February 2013 (Bismas 2013). This is really a spiritual pilgrimage for most participants, not simply a secular event. However, these world-scale festive events in a limited number of centres should not disguise the fact that few settlements today—even small towns, and communities within cities—are without some festival or festive activities that go beyond those of a religious nature. In addition, many leisure activities and commercial outlets have increasingly added festive events to their offerings. Hence it seems essential to understand the range of functions that festivals perform. The term 'festive cities' highlights a new or rather expanded theme in the emerging nature of cities that needs to be investigated in order to understand the multi-dimensional character of festivals and their varied consequences.

The association of festivals and urban places is far from new. Festive events have been characteristic of cities since urban life began (Mumford 1961; Gibson 2005). But they have experienced a tremendous increase in numbers, size and variety in the last thirty years, a change that Manning (1983) claims has been unmatched in history. Hence it has been argued that the old system of urban festivals in cities, typically composed of temporary, limited-time events, has been replaced by a situation in which many cities have an increasingly permanent festive nature because of their variety of festivals, creating what Gravari-Barbas (2005) described as 'festive cities', those distinctive centres where a number of types of festivals occur and occupy different types of space in the city, although this is a rather more restrictive use of the term than the one used here. There is no doubt that these events have become part of our increasingly 'hyper-festive society' (Murray 1999) and form an essential element of what Debord (1994) viewed as our new 'society of the spectacle'. Some indication of the change can be seen in the fact that a city such as Montreal now boasts on its website that it is home to over a hundred different festivals. Even Calgary's website claims a total of 69 festive events, dominated by the ten day annual Stampede which energizes the city for ten days in July. Over 1.2 million people now attend the Stampede grounds during the event to attend the rodeo, fun fair, agricultural exhibits, and many other attractions, with allied festive events using cowboy themes taking place in many parts of the city, organized by private firms and communities. Yet we must be cautious about claiming the contemporary uniqueness of festivals, either in size or number, at least for bigger cities. It has been suggested that the history of the city of Rome was embedded in festivals for there were regularly 42 major events per year as shown in Miller's (2010, p. 173) survey. Some were held on a specific day(s) annually, others organised at a time decided by the city magistrates, while others were episodic, and often linked to the army's victories, or were designed to placate the gods after various emergencies (Scullard 1981; Miller 2010). In the early modern period Curcio-Nagy (2004) has shown how colonial Mexico City had five major festivals and a hundred other events annually, but without the range of different rationales for the festivals found in contemporary cities. Although each festive event may be transitory, their increasing importance means that it is surely necessary for geographers and urban specialists in general to include more of the celebratory, people-centred, participatory and emotional aspects of life in their studies to make them more comprehensive, as well as meaningful, in order to improve the richness of life of urban residents. This review is not designed to study one festival, or even a sample of festivals. Instead, it is a conceptual overview of the ways that festivals in urban places can be studied and assessed, demonstrating their origins, characteristics and multi-dimensional nature, in which each dimension provides an insight into part of their functioning, an approach which is also a basis for evaluating the utility of various urban festive events.

14.2 Characteristics of Festivals

Festivals are distinctive because they take people outside their normal behaviours in time and space. They provide unusual activities and evoke feelings and emotions that are very different to the regular and material routines of the workday. Traditionally, most took place in spaces that are either normally used for other activities, such as roads, or were empty spaces, but later become the exclusive sites for the period of the event—sites usually temporarily transformed by decorations and events that add to the sense of occasion and the separateness of the experiences gained in that space. Throughout history, the major festivals were undoubtedly unusual spectacles of sensory experiences, creating a kaleidoscope of colour and sounds that transcended the typical urban day, creating fun, excitement, joy, even euphoria and awe. Many festivals include parades as a focal ingredient, with participants displaying rich, ornamented and sometimes outlandish costumes, performing various types of music and dancing, and were also associated with sporting and competitive events and high levels of consumption. They included intricate displays designed with deliberate symbolism to encapsulate the purpose of the event—displays that frequently included special effects, such as mechanical aids and later fireworks. These dramatic events were rarely spontaneous; they were usually choreographed and organized in detail, taking weeks, sometime months, to prepare. As a result festivals have always attracted skilled artists as performers, as well as the technical creators of the various elements of the event, which often include building temporary structures that have been the prototypes for more permanent buildings. As a result they transformed the arts and cultural life of places, for the festivals needed the best creators of mechanical devices, poetry, music, choreography and dance. Yet historic festivals were not simply spectacles; most were carefully designed to provide symbolic representations that celebrated or reinforced particular beliefs or to create social, political or religious solidarity within the groups involved. For example the major Hindu Deepavali festival is also celebrated by other religious groups in India for five days in late October or early November, the dates depending on the Hindu calendar. It is an event often called the Festival of Lights since lamps are lit to guide Lord Rama back from his defeat of the forces of evil. So the event celebrates the defeat of darkness and evil, with lights to bring back hope for humanity's future.

The presence of large numbers of people at these festivals usually attracted traders, providers of food and drink, in addition to entertainers of many types, who moved from one festival to another, meaning that many festivals—other than those linked to common religious practices, such as the Christian Lent—were deliberately spaced in time within regions. Hence the initial religious festival often became the stimulus for a commercial fair which outgrew its original function. Many evolved into events with a particular specialism, such as the goose fair in Nottingham, or the gingerbread event in Birmingham. The large ones, such as Stourbridge's festival in the West Midlands of England, attracted traders with specialist goods from many parts of Europe in addition to those from Britain. It led the diarist Defoe (1774) in his description of travels in England to claim, dubiously, that it was the largest in the world, not simply the nation. This association of festivals with commercial activity mean that they played important economic roles within the larger region, in which the limited demand for various goods was satisfied through the periodicity of the festive period, rather than by the type of permanent establishments we find today in the western world.

The spectacles of the festival and the associated feasting ensure that they also became periods of conspicuous consumption, sometimes with food and drink provided by the organisers of the event to reinforce their role as patrons and providers, thereby rewarding and entertaining their subjects. Indeed the organisers and patrons of historic festivals often believed that the more lavish the show, the greater the prestige and value of the event in its main purpose. However, the anticipation, excitement, pleasure and often awe that the festive spectacles produced among spectators was not always positive. Some members of the crowds attracted to these events turned to rule-breaking, even violence, as inhibitions were reduced by an excess of drink or drugs, or by criminal or wonton behaviours that went beyond the normal mores of the society concerned. For example, the origin of St Bartholomew's Fair in London lies in a charter given in 1133 to hold the event to help fund a priory. It grew from a three day event to one of two weeks that became a major trading centre of international importance characterized by large number of stalls, feasting, and entertainment of various forms, and by the early modern period as the site of many musical and dramatic events by prominent artists. But it also became notorious for its excessive merry-making, debauchery and rioting, associations that led to its closure by the London authorities in 1855, showing that disruptive behaviour in some modern festive events is not a product of our times. Indeed, the problem of maintaining public order led many historic festivals to create special festival courts to issue fines and resolve disputes before the behaviours turned to violence and disrupted the event. Yet in some societies this transgression from the normal rules of behaviour at festival time was often condoned for the period of the festival. This was regarded as providing a necessary, though temporary escape value from the build-up of the pressures and conformance to rules that constrained and limited many lives, especially in historic cities where many lived in poverty, and faced various disasters, from pestilence to privation and war. One example where such excessive behaviours were tolerated was the Roman festival of Saturnalia. Originally a winter solstice celebration of the ending of what was initially considered the death of the sun and increasing days of light, which in the Roman region marked the end of the season of planting and a time for a festival, originally to honour Saturnus, the god of seeding and sowing. But in Rome it escaped its agricultural origins and became a festive event, usually between 17 and 23 December. This end of calendar period was seen as a time outside time, a phase of chaos before the New Year, in which public gambling was allowed and where drunkenness, wonton practices and sexual licence flourished. Many of the normal rules of behaviour were overthrown, for it included a day in which masters served slaves, although the New Year brought a return to normal life. The romantic poet Catallus thought it was 'the best of days' (Mueller 2010, p. 221) while the politician and essayist Seneca (Epistles 18, 3) saw it as a period in which 'the whole mob has let itself go in pleasures', but a dangerous period in which the political order could be threatened. Similarly, the Venice carnival in the medieval and early modern city was originally a pre-Lenten Spring event to mark a farewell (vale) to meat (carne), fruit and eggs, the beginning of a period of fasting. But it turned into a riotous period of celebration, parties and masked balls in which people engaged in various excesses, where sexual licence was essentially condoned, for if it occurred during the festival no legal recourse could be taken for the misbehaviour. In other words the festival time was like a temporary release from normal life, although monarchs and city administrations were always wary of the possibility that these festivals would lead to permanent insurrection and took steps to prevent this possibility. By the early twentieth century the festival had atrophied because of increasingly strict policing and the economic decline of the city. But in 1981, Venice recreated the old aristocratic carnival as a city-wide cultural event to boost its economy, which has now become a festival of major international significance.

Not all of these features are typical of all contemporary festivals, especially those whose main purpose is now to deliver some cultural events. Yet some of the elements described above can still be seen in the older historic festivals that have been revived in recent years, although without all the excesses. However, their original rationale has often been lost to many spectators, except in cases where indigenous groups, with older belief systems linked to the original functions of the festival, participate. Hence, it is now the spectacles themselves, the associated party atmosphere, and the commercial opportunities that become the reason for their continued existence, not their original symbolic rationale. Form may have superseded function in these cases. This has led some researchers to claim that many festivals are inauthentic folkloric events (Magliocco 2001). However others have suggested that different meanings are derived by various people. So the revival or even creation of local cultural events, costumes and dances may help re-invigorate local traditions, albeit in different form, providing an antidote to national and more recently global homogenization, thereby fostering a new or revived community cohesion based on the feeling of shared origin and difference (Araza and Crouch 2006). But whatever the meanings, festivals still provide unique experiences in towns and cities, are significant generators of economic and cultural activities, and are often historically and socially referential, as the following review of their variety will demonstrate.

14.3 Understanding the Variety of Festivals

Given the variety, uniqueness and complexity of festivals it is not surprising that most investigators have been content to use case studies to illustrate their character and meanings in particular times and spaces. However, the singular nature of such works limits our understanding of the roles that they play in towns and cities in general. This makes it important to summarize the range of festivals that exist by creating a festival typology. It is proposed that there are six major categories of festivals based on the functions they perform. Yet it must be admitted that many festivals cannot be simply be pigeon-holed into one category, as some festivals have two or more types of function, while others may have originated with one function but are now primarily associated with another. It is also worth noting that most historic festivals fall into the first four of the categories, namely temporal, religious, political or related to traditions and heritage, while most of the new festivals are related to either some form of cultural association, or life-styles.

14.3.1 Temporal or Cyclical Events

Some festivals can be traced to pre-urban origins, events that acknowledge and celebrate nature's rhythms and mysteries. Inevitably, the timing and types of these festivals vary, usually according to the climates involved. Many were linked to the seasons, especially to the solstices that mark a major change in the year, or to planting times, often dictated by the particular seasonality of rainfall in an area creating what amounts to a cyclical festival type.

14.3.2 Religious Beliefs

Other festivals are based on some observance to gods, other than those of nature. Some of these religious festivals have grown out of the cyclical naturalistic festivals, which transformed old rituals into new religious themes, and may often be intimately linked with an associated political control. These festivals were linked to a celebration of spiritual beliefs that confirmed and reinforced allegiance to these convictions through the practices and rituals of the festivals, frequently using a major parade or ceremony as the focus of the festival where major religious figures and icons can be seen or displayed. In some Christian countries individual towns and cities in a region celebrated their own saint's day, resulting in a succession of festivals on different days in any region. Religions differ in the degree to which festivals are used as important reinforcement mechanisms for faith. Of all the main religions Hinduism is the one with the most celebrations and festivals, used not only mark events in its mythical past but to give thanks to their many Gods in its spiritual pantheon (Chanchreek and Jan 2007). Since many are focused on urban places, these centres in Hindu countries have traditionally had proportionally more festivals.

14.3.3 Political

Historically, monarchs or secular leaders have always had the central role in this type, in which the displays, floats, dramas and music were devoted to praising the elite and giving thanks for the protection they provided. These attempted to create political solidarity or allegiance to the regime, or in more recent years of celebrating the political state and its past. The value of political festivals in this respect has long been acknowledged, for even Machiavelli (1624, 1625) recommended that princes should always promote festivals to provide enjoyment and support for the monarch as well as distracting attention from grievances. A variation of the politically motivated festival can be seen in events that are not designed to reinforce solidarity with the dominant regime or the state, but are designed to demonstrate their opposition against existing conditions. Some of the most obvious examples are the protests of workers now associated with May Day, although this was originally a spring fertility festival, so was a cyclical festival in origin. Others include the more recent Techno-festivals held in Paris and the Essence festival in New Orleans. All have a very serious intent, but are often associated with festive behaviour to attract attention from onlookers and reinforce in-group solidarity. These examples of what amount to protest festivals are really a variation of politically motivated festivals, that are increasingly used by groups who wish to promote alternative opinions and assert their rights in multi-polar democracies.

14.3.4 Traditions and Heritages

A fourth type of festival is based upon on some local traditions, other than those based on seasonal, religious or political beliefs. There are many examples of city festivals based on some previous tradition as their primary rationale, perhaps using some reference back to a earlier period in the city or region's history that possessed distinctive ways of life, creating a heritage. Some of these are difficult to separate from re-interpretations of older religious or folkloric festivals. However the theme of basing a new festival upon some heritage, even a myth or distorted one, has been a popular one among festivals created in the last century. Another variation of this type of festival can be seen in *local place festivals*, which provide some celebration of the place in which the festival occurs. Some are based on real historical events that occurred in the area, others on local food products or cuisines, such as the biennial artesan cheese festival in the small market town of Bra (near Turin) that has attracted over 15,000 people to the event. Others, however, do not have a distinctive focus; they may just be a local celebration. All these types are designed to create a distinctive image that celebrates the place's difference, which usually has the additional function of assisting the local businesses who help sponsor the events.

A variant of the festival based on tradition that has become more prominent in our multi-cultural cities is associated with groups who celebrate their traditions or uniqueness through their distinctive *ethnicity or culture*. These ethnic festivals do not only attract members of their ethnic group; they often draw many spectators from the rest of the city or beyond. They flock to experience the colours, dances and sounds of the celebrating group that are so divergent from the typical rhythms of the city life, and who often return to the ethnic stores and restaurants of these areas. Western cities with large Chinese populations usually have annual festivals to celebrate the Chinese New Year, but many other ethnic groups have also created their own festivals to celebrate their traditions and difference.

14.3.5 New Cultural Event or Events

Obviously, most of the previous categories of festivals have a cultural component since they are based on distinctive activities and attributes that stem from religion, ethnicity or a way-of-life. But it does seem important to separate out a fifth type, a cultural festival whose function is primarily based on a new *cultural event or events*, especially the showcasing of some artistic endeavour, whether various types of music, books, films, sports or leisure, and even laughter in the case of one of Montreal's major festivals. Music festivals or arts festivals have shown a major growth in the last few decades. In addition, there are festivals based on bizarre events that represent some local cultural peculiarity, using culture in its broadest sense as some 'way of life', not only high art. One curious example, the festival associated with a wife-carrying competition began in Sonkaäjarven (Finland) in 1997, and claims, like so many unusual festivities, to be based on some local historical event. Another is the 'Elvis Presley Look-alike' Festival in Porthcawl in Wales, although these types may be stretching the normal connotation of 'culture'.

14.3.6 Life-Styles

Most of the examples of this sixth major type of festival are of recent origin, such as the gay parades and festivals which celebrate the life-styles of people once marginalized or illegal. These life-style festivals enable participants to publically acknowledge their distinctiveness and values and portray their pride in their difference, frequently through outrageous costumes. Thirty years ago such parades caused consternation to people who did not share the views expressed. Today they are generally accepted, and frequently solicited by cities and businesses that have recognized the spending ability of gay groups, at least in most developed world towns and cities. Other examples are the recent *Raves* associated with young adults and teens, although they have elements of the protest and musical types of festive event. These are still mainly late or all-night festive events that have not progressed to the formality of a festival. They are usually organized in a clandestine fashion, in an unoccupied or derelict location with information about the event circulated by mobile phone or social media contacts.

14.4 Growth Through Time

Examples of all types of festivals can be found in most countries of the world, although it is probable that temporal or seasonal festivals were the first type, followed by religious and then political festivals which often incorporated aspects of the older forms to bridge the gap between old and new ideas. In Europe most festivals from the Middle Ages to the eighteenth century were organised and controlled by the church or the rulers and their court. From the French revolution onwards, festivals were increasingly organized to buttress the role and legitimacy of the state, and the population became full participants in the spectacles, not just observers—participants who were emotionally involved. Perhaps the apex of this trend occurred in Nazi Germany in the 1930s where festivals were deliberately designed to be vehicles of political persuasion, to convince the observers and participants of the rise of a new and confident state, but with its dark and vicious side concealed. Yet new religions and political regimes often re-interpreted older rituals and symbols in festive activities to conform to the new beliefs, although the more explicit sacrificial and sexual practices of previous centuries were removed by more prudish societies, such as those of Jewish-Christian-Muslim beliefs, whose morals forbade such practices. However, the ribald and obscene behaviour that still surfaced in medieval festivals in Christian Europe were always a concern to church authorities.

In the early modern period in northern and western Europe the rise of Puritanism from the sixteenth century, with its intolerance of merrymaking, imagery and serious purpose, led to the decline of many older festivals. Later, the growth of an industrial society that created new settlements with an emphasis upon work and limited holiday time, together with the growth of more secular values and modernization of life, provided another reason for festival decline, if not the abandonment of many, which were seen as relicts of increasingly abandoned social or religious values. The growth of new national states, or political ideologies within them, from the late eighteenth century onwards, often led authorities to implement new festive events to celebrate their regimes in an attempt to create social solidarity.

The last half of the twentieth century saw a significant festival growth in cities. Some, like the Edinburgh Festival, originated as relatively small events in the late 1940s, designed to promote music, theatre and art after the deprivations of the war-time years. Although this high culture tradition has grown, the so-called Fringe festival, composed of a wide variety of formal and informal activities has become even larger, one that grew out of the situation where artists turned up in the city and found performance space, despite not being approved by the organizing committee. The explosive growth of these activities, both traditional, radical and experimental, means that the Fringe is now larger than the formal festival, and has become the mecca for young artists to showcase their talent (Jarman 2007). From the late 1950s more and more festive events have been added to urban places, aided by the increase in leisure time and discretionary income, and lubricated by the greater mobility of a population able to travel to these new events. In addition, the greater tolerance in societies, plus the growth of civil rights legislation, has enabled people of difference, whether ethnicity or life style, to add festive events to celebrate their traditions and beliefs—or in the case of alternative music festivals, to throw over the old conventions. The changes have added vibrant new events, full of colour, dance, and music to the streets of many cities, adding to their heterogeneity and interest. In the late 1960s increasing numbers of festivals sprang up in many small towns, helped by the increasing mobility of population in the developed world. Some were revivals of older events, others were new creations. Whatever their origin, the small town festivals were seen not simply as events to stimulate culture and excitement, but also as deliberate attempts to improve the local images, and economies of towns that had often suffered decay as more mobile consumers shopped in large centres. In some countries governments deliberately encouraged the development of such events. For example, the federal Community Arts Board within a revived Australian Council, wished not simply to recognize the emerging cultural growth and pluralism of the nation, but to find ways of integrating the artists into the local community as local interpreters of history, folklore and culture (Gibson and Connell 2011). This has helped stimulate festivals in many small towns, as well as street festivals in the bigger cities. Although these trends continued into the last two decades of the twentieth century, this period also saw the initiation or enhancement of older celebrations into what can only be described as mega-event festivals, usually seen as those that attract over a million people and have budgets over $ 500 million (Allen et al. 2005).

14.5 The Multi-dimensional Nature of Festivals

Although festivals can be described in terms of the main reasons or functions that lie behind the events, this is not the only way in which they differ from one another and how they may be interpreted. Unfortunately, most studies of festivals are rather one-sided. In a methodological sense most focus on the uniqueness character of a festival, the local particulars that make them distinctive, rather than searching for the generalities that underlie them, in order to uncover theoretical structures. Yet there have been attempts made to identify such structures. For example, in terms of community effects, festivals have been viewed as consolidating collective consciousness (Durkheim 1976), while others stress that they represent the contestation of different groups competing for power (Bakthin 1984). In cultural terms, it has been noted above that some view modern festivals in essentially negative terms, as inauthentic events in an historical sense, only loosely related to real traditions; others see them as re-energizing older local cultures and opportunities. In socio-economic terms a similar difference of opinion occurs between those who see festivals as important engines of economic growth and attracting international visitors to cities in particular (Allen et al. 2005), while some view them as version of the older Roman idea to provide 'bread and circuses' to feed and entertain the masses that might otherwise revolt over their condition. These examples demonstrate that individual festivals have been interpreted in different ways. But to continue such polarized views does seem counter-productive, for most festivals contain elements of all these characteristics, although some may be more important than others in particular cases. Indeed, as a study of New Orleans argued, festivals, like spectacles in general, are often conflicted, contradictory and should be seen as multi-faceted events with multi-dimensional consequences (Gotham 2005). Yet viewing festivals in this way poses two key questions that cannot be answered by the study of a single festival or just a small sample. *If festivals are multi-dimensional, what are the various dimensions, or sources of variation, by which festivals can be understood and assessed? What alternative and perhaps opposed consequences can occur within each dimension?*

Figure 14.1 provides an answer to these questions by suggesting that the character of individual festivals can be summarized in terms of eight major dimensions of variation. This provides a preliminary multidimensional festival framework for exploring the nature and impact of festive events in cities and perhaps a way of theorizing their character, although further research is likely to produce additional dimensions. In addition, it will be shown that there are possible alternative consequences within most dimensions. The result is that their effects cannot be seen as being unidirectional; instead festivals produce contested and conflicted results.

14.5.1 Ephemeral Character

The most obvious characteristic of festive events is that they are ephemeral, occupying only part of time and space in the city. Temporally, they are events outside

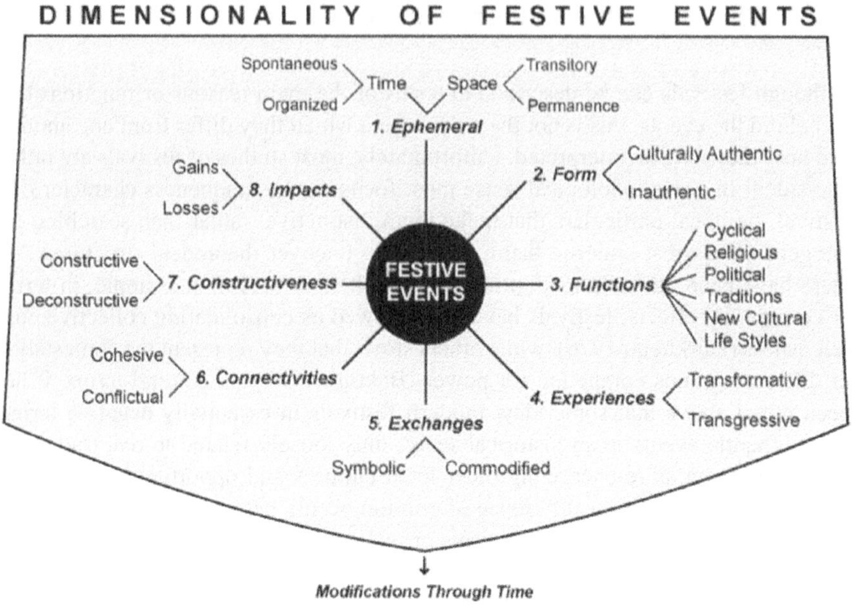

Fig. 14.1 A model of the dimensionality of festive events

the rhythms of normal life, which for most people are dominated by the regular and material routines of the workday, providing an escape from regular routines. Their time-limitation creates a heightened sense of anticipation before the event that permanent features cannot provide, for the latter will always be there after the festive event is over, as well as an occasional experiential opportunity to view or participate in these events. Their temporality allows the provision of events and functions that are too complex, expensive, or lack sufficient demand, to provide on a permanent basis. Moreover festivals may differ in how time is viewed. It has been suggested that pre-modern and contemporary festivals differ in the way that they use time (Lefebvre 1971). In the past, events were not measured in time; time was qualitative, integrated into the regular routines of life, so the events were often open-ended; today festivals run like clockwork, with organized timed events, possessing routines little different from regular life.

Spatially, most festivals are also ephemeral. They take place in areas that are temporarily abstracted from their normal usage, often a street, square or vacant space. Such areas are usually transformed for the duration of the event by decorations or by features needed for the event, thereby defining its separate existence. Gravari-Barbas (2005, 2007) has shown that different types of festivals in Paris are often found in specific and different spaces. This temporary use may cause conflict with the more permanent occupiers of the spaces within the city. More generally in countries or areas where one group has conquered or replaced others, there may also be an important symbolic meaning in the re-occupation of the particular festival space by the marginalised group for the limited time of the event. For example, the annual

Mayan Spring festival of Cha'an (enjoy) that marks the end of the planting season takes place on empty ground adjacent to and in the town square in Chemax (Yucatan), a site that became the centre of the Spanish colonial town and now functions as its commercial node with people of mainly non-Mayan, Mexican heritage for most of the year (Brown 2007). Similarly, the annual recurring clashes in Northern Ireland over the routes of the Orange Parade marches, show the differences between the celebrations of the Protestants at their ascendancy, because of King William of Orange's victory in the battle of the Boyne in 1690 that vanquished the attempt of the Catholic King James to retake the English, Scottish and Irish thrones, and the resentment of the Roman Catholics over this annual display of triumphalism.

The temporary occupation of space is not the only trend involved, for over time special spaces for festive events have been built in cities. Many of the central piazzas in Italian towns in which festivals are held were originally common areas outside the town that were absorbed by the town as it grew, but remained the site of various festivals, as seen by the word campos used to describe many central areas. This type of trend also accounts for the inner city location of many fairgrounds and agricultural exhibition sites that are found in North American cities that were on the edge of cities in the early twentieth century when they were first created. The spaces occupied by the various festivals are now complemented by additional business functions, so the area is not only occupied for the limited festival time.

Some festivals have grown so big that special areas have been built to accommodate them, whilst others have become permanent spaces because of their need for buildings to contain them, or because they use expensive and difficult to move facilities or machinery. For example, the Sambadrome is now the focal point of the Rio Carneval, which houses the spectators and the judges of the spectacular floats, costumes and dance parades that provide a climax to the Lenten festival, activities that take the various neighbourhood samba schools much of the year to build and to choreograph. This venue is a purpose-built six block series of stands, imitating the street form through which the parades used to pass in the centre of the city. A more general example can be seen in the Darling Basin area of inner Sydney, in which the need for a major site for the massive Australian Bicentennial celebrations of 1988 provided the stimulus for the renewal of the area's derelict railway and harbour lands. Today, it is a new zone of leisure and entertainment for the city, the site for many festive events during the year. Although each of the multiple festive events are temporary, they succeed one another in this permanent space for festivals and exhibitions. This may be seen as spatially crystallizing the extended number of festive events in cities. So the loss of a spatial ephemeralness through festive events occupying different spaces, is replaced by a temporal sequential festivity in the same place, through the addition of more temporary festive events in these permanent locations. It also must be noted that some festivals may move around a country or region in different years, such as the annual Eisteddfod, a celebration of Welsh culture in Wales, or the Saint-Vincent-Tourante wine festival in the Cotes d'Or in Burgundy that moves sequentially between the regional settlements, with the most famous wine villages hosting it every twenty or thirty years on January 22–23rd. This was a new event created in the 1930s to promote regional wines as a reaction against the decline in the trade in a world depression. Finally, the more successful festivals often spin-off related activities to different parts of the city, as seen in the many local events that take place during the time of the Mardi Gras in New Orleans. Many Fringe events of the Edinburgh Festival are held in a myriad of locations in the city; events originally informal, are now organized, and in total are far larger than the basically highbrow cultural Arts festival that was initiated in the late 1940s (Jarman 2007).

14.5.2 Festival Functions

The outward forms of contemporary festivals may be viewed empirically, simply as a set of events, performances, or spectacles. Many are just designed to provide new cultural events for the town and to attract tourists. But historically, most festivals had a deeper symbolic meaning, since they were designed as representations of the particular *functions* that the festive event was created for, relating to one or more of the six major types already described: celebrating temporal or cyclical (often seasonal) events, religious beliefs, political control, traditions, cultural attributes, and life-styles. The complexity of some festive events means that there can be overlap between these functions, with some festive events having more than one of these features, or having morphed from one function to another through time. However, a modern empirical interpretation of a festival based on contemporary beliefs misses the point that its rituals and forms were carefully designed to provide some symbolic representation that celebrated or reinforced the particular reason for the event, frequently to create social, political or religious solidarity within the groups involved, to celebrate some tradition, culture or life-style. For example, parades involving dignities have usually been rigidly organized with firm hierarchical principles in the placement of various groups. The spiritual or political leaders occupied positions close to the main symbols of the event, with each lower status group located further away, with the rank and file of the population, both rural and urban as observers. In colonial countries these festivals had a two-fold hegemonic function. The political leaders or their representatives were paraded before the populace as visible objects of the regime, frequently accompanied by lavish displays and military prowess to show their wealth and strength. By providing a spectacle, often based on a formal holiday, they were also one of the many ways of socializing subject peoples to celebrate the regime—an approach that was less coercive than other measures of subjection, such as restrictive laws, but essentially provided the same normalizing function.

14.5.3 Festival Forms

The third dimension relates to the various *forms* of the events, which can be thought of as a series of artistic empirical attributes that involve one or more features, providing an expression of the unique costumes, music etc, of the culture that created the event. At the most basic level the form of the festivity may be based simply on one activity, one ritual or one artistic endeavour, such as a type of music that dominates many small town festivals. At the most complicated level the larger festivals involve many activities with lavish spectacles, frequently with a parade as the main event, with participants displaying rich, ornamented and sometimes outlandish costumes. Many use vestiges of the past, such as religious icons or traditional costumes and musical instruments, to give them distinctiveness. Most are complemented by music and dancing that add to the range of sights, sounds, and activities in the urban places. The result is that places with these festive activities have more complex and varied cultural characters than those without them. They add to their vitality, prestige, often to their uniqueness, increasing the national and international profile of the host towns and cities. The festivals become a focal part of their tourist trade and with many towns deliberately choosing particular themes, or accepting the form pioneered by some local group, often of quite a bizarre character, to be distinctive from their neighbours. However the uniqueness of the forms may be questioned. Some festivals, such as music festivals, are copied by other towns, or are derived from similar origins, such as the Lenten festivals. In addition, as the events become attractive to outside visitors, they often add successful elements from other festivals to enhance their own offerings to create new experiences for their spectators, which have the danger of reducing their uniqueness.

In recent years there has been a *questioning of the cultural authenticity* of many of the forms and activities seen in contemporary festive events that are derived from older traditions. There seems little doubt that many of the forms of the new or revived festive events have been invented, or are seriously modified versions of older traditions. Yet one must remember that all so-called 'authentic traditions' get created at some point and are modified through time, so 'authentic' is always a contested word. Indeed, many forms, that seem authentic to the contemporary observer are nothing more than invented traditions of the late eighteenth century, based on a romantic interpretation of older forms, as seen in a classic book (Hobsbawn and Ranger 1983). Nevertheless, it does appear that many contemporary festivals now function more as spectacles attempting to provide unique experiences, rather than as events with symbolic meanings that can be linked to some prior rationale or function, whether seasonal or religious. But one must be careful in assuming a singular result. Some researchers have claimed that many folkloric festivals are inauthentic in the sense that they have invented or distorted the local historic traditions that were part of the older cultures. Magliocco (2001) argued this point in relation to La Cavalcata Sarda in Sardinia, a parade of local traditions that welcomed the King of Italy to the island in 1899 and was revived as a tourist promotion for the Italian Rotary Club conference in 1951. Others have suggested that this festival has fostered a revival of community cohesion in recent years, based on the feelings of shared origin and cultural difference by participants that are manifested in the festive events, providing an antidote to global homogenization (Araza and Crouch 2006). Other folkloric festivals have had a similar history and result, helping to re-invigorate or re-invent local cultures, even in highly modified forms.

14.5.4 Festival Experiences

Festive events usually produce many new personal or group experiences for both participants and spectators, although they can either be of *transformative* or *transgressive natures*. In the former case they frequently alter people's feelings, creating emotions such as pleasure, excitement, even euphoria or awe at the performances involved. New attitudes may emerge, such as pride in the displays or in the city in which they occur—even new or revived interest or insights into the culture of the forms and their underlying functions. Throughout history, the major festivals were undoubtedly unusual spectacles of sensory experiences, given the profusion of colour, sights and sounds that transcended the escape from the 'real', the typical urban day. Hence they contributed substantially to the range and character of human life within towns and cities, providing experiences often unmatched by more permanent features within the settlements. From an 'insider' perspective there can be little doubt that these festivals have become major contributors to the revival of local or regional feelings in towns and cities in a globalized world, providing examples of how the celebration of cultural difference, not similarity, is promoted by festive events. From an 'outsider' perspective, especially those of long distant travellers, they also reinforce the feeling of having observed something unique during the festive events, creating a belief that the town or city involved is somehow unique, even though the life-styles of the residents for the rest of the year are little different to their own.

The experiences found in festive events are not always positive for the society involved; they can also be *transgressive,* in the sense that festive behaviours often challenge or break existing societal mores. Without tight controls, which in themselves can reduce the positive experiences for spectators, some members of the crowds attracted to festive events turn to rule-breaking, even violence, especially as inhibitions are reduced by an excess of drink or drugs, or even to criminal or wanton behaviours that go beyond the norms of the society. In some cases these transgressions can be considered to be liberating for those who enjoy overthrowing conventional behaviours, even if others are hurt. But this provides a problem of governance within the jurisdiction of the festive events. However, city authorities have always been alert to the threat of these behaviours becoming too extreme and providing a stimulus to the overthrow of the established social and political order. Indeed, without appropriate regulation and policies to cope with crowds, any festive event can get out of control and inflict serious damage upon other spectators and the local environment, whilst in extreme case riots may develop. Historically, the problem of maintaining public order led many festivals to create special temporary courts during their operation, to issue fines and resolve disputes before the behaviours turned to violence and disrupted the event. It has already been observed that during the Protestant Reformation in England from the seventeenth century many historic festivals were closed down by Puritan authorities because of their blatant debauchery and flouting of more moral societal standards. Yet it can be argued that by allowing these annual outlets to take place, the negative effects of the temporary disorders were demonstrated, which often led to support for the existing societal system. However, city authorities have always been alert to the threat of these behaviours becoming too extreme and providing a stimulus to overthrowing the established social and political order. Indeed, without appropriate festival regulation and policies to cope with crowds, any festive event can get out of control and inflict serious damage upon other spectators and the local environment, whilst in extreme case riots may develop. For example, the early post World War II urban ethnic festivals, such as Notting Hill in London, celebrating the heritage of West Indian immigrants, often led to major unrest and police action, although subsequently community consultation and careful monitoring by the police and local authorities has led to more peaceful events during the last few decades. Pioneering music festivals, such as Beaulieu in southern England in the early 1960s, were abandoned after a few years because of the damage produced by anti-social elements. The same happened to the annual German Love Festival that was held in Berlin for many years, then moved to Dusseldorf and finally to Bochum in 2010 largely because of overcrowding in poor facilities and disruptive behaviour that has led to rioting and deaths during the event.

14.5.5 Festival Exchanges

Festivals also normally involve exchanges, although these can be of very different types. Historically, there were often *symbolic exchanges* in festive events, where humans or animals were sacrificed as part of the rituals designed to provide a scapegoat for the assumed transgressions of the people. Others were designed to ensure *prosperity*, such as success in harvest or war, or propitiating the Gods. The sanitized form of some of these exchanges may still be seen in modern festive events, but their original bloodletting and meaning has been largely lost. Other festivals were dominated either by *reciprocity*, through exchanges of social capital, or by *redistribution*, in the sense that goods were given or exchanged as part of the celebrations. But the exchange can also be symbolic, in the form of spectators pledging allegiance to, or at least acceptance of, the secular or religious leaders during the festivals in return for their protection and governance.

The exchanges, or limited versions of them, still occur in some modern versions of historic festivals, but most goods and services in our contemporary society have been commodified. This means that *market exchange mechanisms* increasingly dominate the festivals. Festivals, especially the largest ones, are seen as money generators for the organisers and the associated businesses provided during the festive events. Certainly they may provide cultural or other functions, such as the music or other performances in arts festivals in exchange for money provided from ticket sales to the various events and activities. But increasingly the exchange value of the major festivals is seen as primarily of economic benefit, rather than their original functions.

14.5.6 Festival Connectivities

Festivals influence the connectivities between the local communities, or the people who are involved in, and affected by, the events. There is not necessarily one outcome;

the linkages may be positive and provide *cohesive ties,* or may be negative if they create *conflictual relations*. Many festive events create community solidarity, such as by: providing identity to the people involved through reinforcing existing beliefs or political hegemonic structures; revitalizing local interactions and communities; reconstituting differences between the participants and the host population. These consequences may be considered positive from the viewpoint of the festival rationale.

Festivals may also be contested or conflictual, in the sense that the festive events may lead to greater differences of opinions by the people affected by the festival. One source of potential conflict comes from alternative views of why the festival is created, or because of rivalry over the way the event is organized or how people are positioned in the festivities. Others may come from criticism from people who are not involved in the festivals and who deplore or even oppose the very reason for the festivals, perhaps on the basis of different beliefs. Frequently, complaints are made that the festive events disrupt the life-styles and daily regimes of some people, while the excessive behaviour of some, especially that of a sexual nature, is deplored by those with traditional family values, a particular problem in the Mardi Gras in New Orleans. Of course, these alternative views of the utility of the festive events may be held by different people in the festive town or city at the same time. This means that either one group is ignored, or some reconciliation strategy must be employed between the groups, for there are situations where a minority gain control and disrupt, abolish or transform the festive event into some other form or even function.

14.5.7 Festival Constructions

Most festive events are not spontaneous, even though they seem to suddenly appear in a time and space. They have to be *constructed*, which usually involves major investments in time, effort and resources, but also may contain *deconstructive* tendencies, especially in a societal sense. In the first case, historic festivals, like many today, have always attracted skilled artists, as performers, or creators of the various elements of the event and of the temporary structures. Both can add to the utility of the festive events by adding to the cultural attributes of an urban place. Many festivals have dramatic or spectacular features; they are choreographed and organized in detail and take weeks, sometime months, to prepare. In recent years a new field of management has grown up to explore the mechanisms by which large events, now called Special Event Festivals, are created and can be successful (Allen et al. 2005). The construction of these events can be summarized as a series of stages, namely: Conceptualising and Planning; Finance and Sponsorship; Marketing; Budgeting; Risk Assessment; Employee Acquisition; Event Control and Management; Aftermath Impacts and Evaluation. The personnel involved in these stages certainly add to the capabilities and payrolls of the festive city or place. Of course, similar sequences were always needed in the historic organization and implementation of large festive events in cities, and even in the local ones. But the processes by which they evolved were passed on verbally through the generations, so our knowledge of their construction is limited, although they were, perhaps, made easier to implement in the past by dominant religious or political leaders with the control and wealth to create such events.

An often forgotten part of the constructive process of successful festive events is the ability to cope with problems that crop up before, during, and after the events, especially the difficulties in dealing with the crowds that attend festive events. In most countries and cities, permission has to be granted by the appropriate authorities for the festive events, which may not always be forthcoming, certainly in totalitarian regimes. It is also necessary to create festival coping mechanisms to deal with the crowds attracted to the events, which involve a range of issues, such as: effective policing, the provision of waste disposal and sanitation facilities, as well as ambulance and health services. Since all of these have to be on a temporary basis they involve provision and extra costs. Many of the music festivals from the 1960s were woefully unprepared in these areas and the resultant problems, especially those of uncontrolled anti-social behaviour, even riots and deaths caused by overcrowding, led many to be abandoned.

It has already been noted that the relaxation of many societal mores during the time period of these events can lead to alternative festival behaviours that would not be tolerated normally. But these openings can be a *societal deconstructive effect* by providing the opportunity for a serious critique of the existing regime, controls, or even injustices, such as to a particular group. Their communicative effect is magnified by the large crowds often attracted to such events. One approach is seen in the gay parades in which members openly flaunt their preferred sexuality with the object of gaining support for their cause and especially of changes in discriminatory laws. Other methods involve the use of controversial plays that might normally be banned, or from the deliberate creation of demonstrations, workshops, topics that deal with ways in which particular injustices in society exist and can be addressed. One of the best examples is provided by the Essence festival in New Orleans which started off as a celebration of Afro-American heritage but has increasingly become an annual critique of the continuing injustices that still afflict many members of this group, as well as providing workshops and exhibitions that help foster pride in their heritage and show ways of resolving their problems. Such examples illustrate how a festival can be used as a demonstration to publicize the existing problems that affect the life of these minority groups in cities, with the object of breaking the taboos or even laws that discriminate against them. This may be a deconstructive societal function yet is one that leads to social solidarity within the group and may help fight injustice.

14.5.8 Festival Impacts

The final dimension shown in Fig. 14.1 concerns the societal impacts of festivals upon the host town or city. Like so many of the other dimensions, it can result in alternative and often contradictory features, in this case, gains or losses. There is an extensive literature on *festival economic impacts* for the various events, especially from economists or special events specialists (Allen et al. 2005). They have promoted festivals as part of urban cultural redevelopment, using the leisure consumption of the festival as the new driver of change and new profit opportunities. They have calculated the direct monetary effects of festivals through measuring local and outside visitor numbers and spending, capital expenditure on the facilities needed for the festival, and the wages and salaries of people involved in the events, from performers to organisers and sponsors. But festivals also have spill-over effects, filling the local hotels, restaurants and stores, thereby boosting the economy of the centre. There are also multiplier effects which accrue from the way that money spent directly on the event circulates in the local economy, substantially magnifying the direct effect of the festive event. Not surprisingly these impacts vary from festival to festival but some idea of the importance can be seen from two examples. A recent study of the economic impact of the Mardi Gras in New Orleans in 2011 showed that the direct festival economic impact on the city alone generated $ 144 million, with an additional $ 156 million in indirect effects, which together accounted for 1.5 % of the city's gross domestic product; in addition the study revealed that for every dollar spent on costs such as policing, waste disposal and medical services etc., the city gained $ 8.5 in revenue return, illustrating its positive economic value (NO 2012).

The impact of the 10 day Calgary Stampede is also positive, for Calgary is often known as the Stampede city, even though its role as the headquarters of oil and gas companies operating in Canada is its major economic function and accounts for much of its growth to 1.2 million people. It has been estimated that the Stampede provided $ 189 million to the city economy, with another $ 256 million to the province of Alberta in 2012 (CS 2013). However a large number of other activities, from trade shows to exhibitions, are now carried out in the 2033 acre Stampede grounds throughout the year, which means that 3.6 million people visit the grounds during the year, a much larger number than the 1.2 million who attend the Stampede itself, with its rodeo, chuck-wagon races, stage-show as well as its funfair, agricultural exhibitions and many concerts. This shows how the festival has spun-off many activities on its site—a similar trend to many festivals with permanent sites. It has also affected the city, as seen in the way that businesses and offices decorate their premises with western themes, people wear cowboy-cowgirl inspired clothes and hundreds of pancake breakfast and barbecues are held throughout the city to celebrate the event.

Festivals also have longer term effects, for the financial flows usually stimulate new investments that create the renewal or expansion of the existing urban fabric, such as new hotels or restaurants to serve the festivals, providing new business opportunities and employment growth. Yet not all festivals are financially positive; some make a loss and have to be subsidized by public funds, which might be limited and could lead to the eventual closure of the event. Also, individual entrepreneurs who produce or sell goods and services for the festivals may over-estimate their sales, leading to losses and bankruptcies. These examples indicate that there are risks involved in the economic exchanges involved in the creation and running of festive events.

These generally positive economic effects are often criticized as being one-sided in their views. Three key *negative effects of festivals* can be recognized. Critical theorists, such as Harvey (1992), see festivals as the new masks of postmodern exploitation that obscure the societal losses that occur from ignoring their impact on the rest of the life of the city, especially the poor and disadvantaged. Certainly such groups may take part in the festival and may be enthused by it. However the spaces that they have occupied as dwellings or leisure areas are frequently taken over by the growth of the festival grounds justified by arguments that relate to maximizing the profit from sites that become the new spaces of festivity. Meanwhile, their badly required social needs—of affordable housing or various social services—are either poorly financed, or ignored. Instead, public tax money often goes to support the development of the festival, encouraging the conspicuous consumption designed for out-of-town tourists, or better still in the eyes of the business elite involved in such events, in high spending international visitors, rather than to help disadvantaged locals.

A second negative effect comes from the increasing corporate control of festivals, often to the disadvantage of small, local businesses. The larger the festivals get, the more major national and international corporations seek exclusive sales in particular activities during the event—at least in the areas controlled by the festival—in food, drink and other goods. In return, the festival gets financial support or sponsorship from the companies involved. Increasingly, these payments help to finance many of the events in the festivals, so their organizers become dependent on such support. Local businesses often get squeezed out, unless there is a strong, deliberate emphasis upon local or regional products in the function of the festival. Third, when the emphasis is placed on the festival as a tourist event something local is frequently lost. Gibson (2005) has described how the local elite encouraged the transformation of the Lent festival in New Orleans into a national and international secular event, building upon the uniqueness of the city as the home of jazz and the blues and its historical French flair. Combined with its ethnic composition these roots produced life-styles more tolerant than most American cities, which contributed to the development of a party atmosphere and the extension of the boundaries of accepted morality. But many in the city deplore these excesses and the internationalization of the event means some of its original localism is lost, while its high prices exclude many residents from participating.

14.6 Utility of Festivals

The creation of so many new types of festivals in the past thirty years and the revival or re-orientation of many old ones provide many points of value to cities, that complement more permanent places of leisure and entertainment in cities, from theatres and bars to parks. They ensure that festive events should be seen as essential features of contemporary urbanity, perhaps re-creating their importance in pre-Reformation and Enlightenment days, although with very different emphases and roles. There seems little doubt that festivals of whatever type have always provided an unique identity for the celebrating city, town or community, creating not only a sense of civic pride in the distinctiveness of the events, but community interaction in the process of creating the events. They also provide a showcase for a display of cultural talents, both local and imported, adding to the existing set of cultural and leisure activities. Many festivals provide a link with the past, or with particular beliefs, revitalised traditions and create the possibilities of new identities in a homogenizing world, even though few believe in the original meanings and rationale for the events. Also, there can be few doubts that these events improve the overall profile and prestige of the settlement in which they are located, providing an important competitive advantage over other centres that do not have such high profile events. They can be used to showcase other economic and cultural aspects of the city, which may lead to additional growth. In a human context it must not be forgotten that they provide an antidote to the conformity and pressures of everyday life. Many bring excitement, pleasure, a celebration of difference, and in the case of some festivals, a sense of fun to participants and spectators; all are human experiences and even needs that are often forgotten in the focus upon the material and the workplace in urban studies.

Despite these varied advantages there is little doubt that the rationale on which many festivals are based today is firmly based in economic considerations. The examples of Calgary and New Orleans described above show the direct and indirect effects that boost the economy of these centres, filling the local hotels and restaurants, generating money through ticket sales to the various events and activities. Successful festivals also stimulate new investments that create renewal or expansion of the existing urban fabric, providing new business opportunities and employment growth. However there are also social benefits beyond the festival itself, for many community groups, firms, and individuals host their own events, usually linked to the festival theme to foster social co-operation and to profile themselves in the case of companies. Many cities have deliberately created festivals to improve their economic prospects and image, such as Venice's revival of their old Carnival, or Lyon's decision to create a Festival of Lights in December to remove, or at least downplay, the city's early winter image as a foggy, dark place.

All these features show how festivals can be seen as important value-added features to the character of urban life in very different dimensions, which explain their growth and rehabilitation within cities, even if many of the original temporal, religious and political reasons for the older festivals have been lost. Nevertheless, it is essential not to be carried away with these positive values. Table 14.1 outlines ten of the most important advantages of festivals but shows that each has a parallel downside, features that need to be resolved or at least managed if festivals are not to be burdened by negative effects.

Some of the problems of festivals, such as the disturbance of local life-styles and the damage to local environments by improving the infra-structure needed for festivals may be mitigated by community consultation and more careful physical planning. But there is little doubt that the impact of the crowds and dealing with their exuberance needs a great deal of money and investment in skilled management and trained personnel. This includes coping with the large amounts of waste

Table 14.1 Major positive and negative features of festivals

Advantages	Disadvantages to be resolved, managed
Increased local pride and possible international prestige	Annoyance at disturbances from event
Additional crowds providing:	Noise, congestion, crowd control. safety. housing
Economic gain	Waste disposal, sanitation, and medical needs
Environmental improvements near festival sites	Physical damage through vandalism
Excitement and pleasure	Anti-social behaviour, substance abuse
Community participation/cohesion	Intra-urban rivalry and alienation
Revitalisation of traditions	Loss of authenticity
Improving culture & expectations	Loss, downgrading of old cultures
New investment and renewal	Loss through event failure; unrealised projections
Business opportunities, job growth	Increased costs, price inflation in hotels. houses
Government and corporate support	Possible loss of local consultation and control

and ensuring its removal, installing portable toilet facilities, providing temporary accommodation and transport links, to sensitive policing that does not alienate the spectators. Another vital need is the provision of medical care via paramedical staff and ambulances for people who fall ill or may be overcome by the event. Many early and often spontaneous music festivals, or those organized in remote locations, failed to realize the need for these facilities and controls, which created huge problems, mainly associated with the requirement to suddenly accommodate the effects of such large numbers. A similar effect is seen in seaside or inland resorts where the population may grow rapidly in the summer months, which puts great strains on things like sewage facilities which have to built to the seasonal population capacity. The problems of coping with the enormous increase in the population of a centre can be seen in the largest festival in the world. The Maha Kumbh Mela festival—literally the 'largest pitcher gathering or fair'—occurs in Allahabad at the junction of the Ganges and Vamuha rivers and the myhthical Saraswati. It is the largest of the three yearly Hindu Kumbh festivals held in one of four locations every 12 years to celebrate the places where Hindus believe a container of immortality was spilled by Gods fighting over it. The festival involves an estimated 80 million people attending, although not all of the celebrants stay for the whole time. A focus of the festival is the dawn bathing in the Ganges where areas for different sectors within Hinduism are set aside. It is believed that the immersion will cleanse sins and help people find a path to salvation, an eternal life outside the cycles of earthly birth and death, meaning that the event is as much a pilgrimage as a festive event. In 2013 the most auspicious day within the whole timespan of the festival was February 14th where it is estimated that 30 million people bathed in one of 18 locations along the Ganges (Bismas 2013), the biggest daily gathering on earth. A major panic in the crowds in the 1950s resulted in hundreds of pilgrims being crushed. It led the state authority to create what amounts to a temporary or 'pop-up' mega-city for tens of millions of pilgrims on the flood plains of the rivers. Every four years a large area is laid out with accommodation tents, meeting places for the celebrants to hear their leaders, eating areas, cooking sites and sanitation places, the major parts being separated by wide roads. This deliberate planning has eradicated many of the old problems of crowd congestion, sanitation and waste disposal in particular, although the ritual bathing of so many people in an already polluted river still causes huge health concerns. Similar planning to accommodate the temporary crowds drawn to festivals has been a vital need in many other festival sites, although not on this vast scale.

Another set of problems may occur with crowd behaviour. Crowds may be initially exuberant but can quickly turn negative when overcrowding or coercive monitoring occurs. Some people take the opportunity of their anonymity in crowds to engage in anti-social behaviour; others use the occasion to indulge in excessive drinking or drugs to heighten their experiences. So festival policing is important; trouble-makers need to be calmed down or perhaps removed, discretely, before they affect the behaviour of others. These examples show that without appropriate regulation and policies to cope with crowds, the event can get out of control and inflict serious damage upon people and the local environment. In extreme case riots may develop, as seen in Berlin's Love Festivals, or the early Notting Hill Festivals—although careful monitoring has now led to this latter event being largely peaceful.

Another series of possible problems stem from the extent of community participation in festivals. Frequently the term 'participation' is used almost as a motherhood concept, with the assumption that co-operation and nurturing will occur within the area if a festival is created. In practice, different people in the community may have divergent views about how the event should be organised, about its content, or wish to exert their own control, leading to rivalry and dissention, as seen in examples already described. It is possible that acceptance of some overall community or group benefit will outweigh other factors and provoke cohesion, but this is far from guaranteed. So some attempt to find benefits that appeal to as wide a population as possible is needed for the event to be successful. Leadership is a key ingredient of the initial stages of a festival's conception; many festivals have been developed through the persistence and skills of one individual. Similarly there is no assurance that a festival will be successful. Historically, perhaps, fear of not giving thanks for seasonal changes, participating in religious rituals or the celebratory effect of political success, was enough to help a festival. Yet for the participants the release from normal workdays by providing holidays for the festival and the anticipation of socialization plus indulging in goods in short supply certainly helped success. In the contemporary world there is little doubt that a great deal of work needs to be done on striving for excellence in the events, in advertising their merits, especially if the festival is to be more than something that is for local participants. But there is always something entrepreneurial involved and not all people hoping to benefit from the event will make the right choice, leaving unwanted goods or services and financial failures for some companies. The experience of the Millennium Dome in London, a centre point of the Britain's 2000 festivities, was one that has failed to find a subsequent rationale. This stands in stark contrast to successes such as the Eiffel Tower, built to mark the entrance to a World Fair site in 1889 that celebrated

France's republication centennial. It not only attracted over 7 million visitors in 2011 but has become an internationally recognized symbol for the city. A related point is that festivals often drive up local prices for the period of the event, a simple matter of demand for goods and resources being greater than supply, which can seriously affect the poor and those on limited incomes. They also suffer when redevelopment takes away many of the cheaper, older properties in which they have lived. This is a problem that is often ignored in the places with big international festivals, although in the past, religious and political leaders were known to provide at least some special facilities for the poor as part of their charitable acts.

In a cultural context new modernist or economic rationales may have led to the loss of authenticity in local traditions in festivals, which is always a danger that needs to be anticipated and dealt with. It is noticeable how the largely indigenous towns south of Managua, the capital of Nicaragua, have revived festivals based on their own heritages—often linked to local saints days which were imposed by the Catholic missionaries on older traditions—rather than simply criticizing the superficial, modernised versions in the capital city. This means they have provided an outlet for their own artists rather than losing the old craft and theatrical skills to modernization and imported goods. Finally, it is worth noting how easy it is for a festival to lose its local roots and influence when it increases in size and becomes dependent on government and especially corporate support. Many of the largest events have signed contracts with major corporate giants, so that only one type of beverage or food supplier is involved in return for corporate donations. Certainly the money is used for bigger prizes for the events or shows, but there is a concomitant loss of choice and local input which impacts upon the authenticity of the events; their spectacles and shows often hide the capitalist monopolies that lie behind them.

All these issues mean that although there can be many positive features associated with the creation, perpetuation and enlargement of festivals in cities, or in settlements in general, there are a set of negative features that can counteract the value of the event. In the former case it has already been noted that a new field of management has grown up around what are usually now called Festival Special Events which involve a series of stages, from Conceptualising and Planning to Aftermath Impacts and Evaluation (Allen et al. 2005). Yet such stages were always needed in the historic organization and implementation of large festive events in cities, and even in the local ones. They were rarely conceptualised and were perhaps made easier to implement by religious or political leaders with the control and wealth to deal with the events. However the historical record shows that while the festive events were used to support existing beliefs and regimes, the very presence of large crowds often led to apprehension that they would form the base for a regime overthrow, a fear that sometimes turned out to be true. Yet one must not forget that festivals promote new emotions and sensations, vital stimulants of a healthy humanity and urbanity that was often forgotten in the nineteenth and twentieth century focus upon work and rationality. The satisfaction and frequent feelings of exuberance generated by historic festivals, with their provision of relaxation, pleasure, indulgences, and sometimes euphoria, provided a temporary yet welcome release from the routines, daily grind and problems of normal urban life. It is also possible to view festivals as escape valves which resolve and diffuse the general tensions and desires that are always present in local communities and broader societies. It has already been noted that festival days often led some to abandon usual life-styles, leading to indulgence or even chaos, behaviours which were temporarily tolerated before normal order and mores were restored. These features can be still seen as some of the basic tenets of many festivals today. However, whatever exists behind these outward forms there can be no doubt that historical festivals were designed to reinforce beliefs or allegiance and create solidarity.

14.7 Conclusions

The creation of so many new types of festivals in the past thirty years, as well as the revival or re-orientation of old ones, has added substantially to the vibrancy and differentiation of urban places. This makes it possible to argue that they should be considered as one of the important new ways in the transformation of contemporary urbanization, perhaps part of the culture-led urban regeneration that Miles and Paddison (2005) have suggested as playing a pivotal position in new urban developments through entrepreneurial activity. Ponzini and Rossi (2010, p. 1039) are more specific in describing this urban policy culturalisation as

> a new stage of entrepreneurial urbanism centred on the dynamic of the immaterial (atmosphere, vibrancy, creativity) and material factors (regeneration areas, monuments etc.).

Festivals, therefore, provide a way of expressing the revival of these interests that have too long been forgotten in the mundane, work-orientated approach to urban development.

The specific impact of festivals is especially important in those centres—not only big centres—that have a number of sequential festive events in the year, resulting in an almost continuous air of festivity to parts of some settlements (Gravari-Barbas 2007). Yet even a single important festive event effectively brands a community, town or city, and provides an important addition to their range of experiences and life-styles. So the increasing importance of these events, in experiential as much as economic terms, should be recognized as providing an additional and valuable component to the new urban themes of the past few decades described in other chapters, such as New Urbanisms, Green, Safe or more Sustainable cities etc., and also expand the people-centred approaches to urban places promoted by some writers (Gehl 2010). However there is a danger that festivals become dominated by the elite, resulting in extensions of exploitation and corporate control, ignoring the needs of the disadvantaged in urban places. Certainly festive events were an important part of the pre-industrial city, but their role diminished, first with the growth of a dour, fun-less and moralistic nonconformity in the western world which led to the loss of many older festivals, second with an industrialization that focused on work, leaving little time for leisure, and third with the decline of religion and the secularization of society which led to a modernity that resulted in many people abandoning traditional ways of life. The revival of the role of festive events in cities marks another demonstration of the various ways in which contemporary urbanization is moving away from the vestiges of the squalor and work orientation of the old industrial city, or the rigidity and depersonalization of so much post World War II urbanism. This makes it imperative that we seek to understand the nature of festivals and their varied impacts on our towns and cities. They are not simply of direct economic value to local businesses, but often create a new image for a place focused on the festive event which can have even greater commercial benefit in the increasingly competitive international world. Perhaps the meanings of the original symbolism of many traditional festive events have been lost; but they still provide celebrations, spectacles, or perhaps new urban cultural experiences, that in some cases may still create solidarity and identity as described in Singapore (Foley 2007). By adding to the range of pleasures and excitements in urban places they provide an important complement to the more permanent places of leisure and entertainment in cities. However, festive events need to be viewed through the lenses of the eight different dimensions of their character, many of which can produce quite contradictory results. Not all of these are positive, so these negative effects need to solved, or at least managed, if the full benefit from these events is achieved.

Acknowledgements I must acknowledge the value of many conversations on this topic with my Calgary colleague, Dr Denise Brown, who has extensive experience of festive events in Mexico in particular, as well as to Dr Maria Gravari-Barbas of the Sorbonne, whose stimulating work on Festive Cities in France helped initiate this chapter.

References

Aitcheson, C. A., & Pritchard, A. (2007). *Festivals and events: culture and identity in leisure, sport and tourism*. Eastbourne: University of Brighton, Leisure Studies Association.

Allen, J., O'Toole, W., Harris, R., & McDonnell, I. (2005). *Festival and special event management* (3rd ed.). Milton Queensland: Wiley.

Araza, I., & Crouch, D. (2006). *La cavalcata Sarda: performing identities in a contemporary Sardinian festival*. In D. Picard & M. Robinson (Eds.), (pp. 32–45). Festivals, tourism and social change. Clevedon: Channel View Publications.

Bakthin, M. (1984). *Rabelais and his world* (trans: H. Iswolsky). Bloomington: Indiana University Press.

Barcsay, T. (2000). The 1896 millennial festivals in Hungary. In K. Friedrich (Ed.), (pp 187–211).

Bélanger, A., & Côte, J. F. (2005). Le spectacle des villes. *Sociologie et Société: Numéro Spécial de la Revue, 37*(1), 266.

Bismas, S. (2013). The Kumbh Mela festival holds its most auspicious day. http://www.bbc.co.uk/news/world-asia-india-21395425. Accessed 9 March 2013.

Brown, D. F. (2007). The Cah: Place and the identity of the Chemax Maya. In H. Kraay (Ed.), *Negotiating identities in modern Latin America*. Calgary: University of Calgary Press. (pp. 118–129).

Chanchreek, K. L., & Jain, M. (2007). *Encyclopaedia of great festivals*. New Delhi: Shree.

Cooper, J. C. (1995). *The dictionary of festivals*. London: Harper-Collins.

CS: Calgary Stampede. (2013). http://www.calgarystampede.com/Media-Resources/Media-Guide-2013. Accessed 26 Oct 2013

Curcio-Nagy, L. (2004). *The great festivals of colonial Mexico City*. Albuquerque: University of New Mexico Press.
Davies, W. K. D. (1996). Post modernism and cities. In R. J. Davies (Ed.), *Contemporary city structuring*. Cape Town: Society of South African Geographers and I.G.U. Urban Commission. (pp. 1–17).
Debord, G. (1994). *The society of the spectacle* (trans: D. Nicholson-Smith). New York: Zone Books.
Defoe, D. A. (1774). *Tour through the whole island of Great Britain*. London: Reprinted Penguin. (1971).
Demossier, M. (2000). Wine festivals in contemporary France: Reshaping power through time in Burgandy. In K. Friedrich (Ed.), (pp. 107–121).
Di Meo, G. (2001). *La géographie en fêtes*. Paris: Géophrys.
Durkheim, E. (1976). *The elementary forms of the religious life*. London: Allen and Unwin.
Florida, R. (2002). *The rise of the creative class*. New York: Basic Books.
Foley, M., McPherson, G., & Matheson, C. (2007). Cultural identity and festivity: Generating Singapore through citizenship and enterprise in events activity. In C. A. Aitcheson & A. Pritchard (Eds.), (pp. 9–18).
Friedrich, K. (2000). *Festive culture in Germany and Europe from the sixteenth to the twentieth century*. Lewiston: The Edward Mellen Press.
Gehl, J. (2010). *Cities are for people*. Washington D.C: Island Press.
Gibson, T. (2005). La ville et le spectacle: Commentaries sur l'utilisation dans la sociologie urbaine contemporaine. *Sociologie et Sociétés, 37,* 170–195.
Gibson, C., & Connell, J. (2011). *Festival places: Revitalizing rural Australia*. Bristol: Channel View Publications.
Gotham, K. (2002). Marketing Mardi Gras: Commodification, spectacle and the political economy of tourism in New Orleans. *Urban Studies, 39*(10), 1735–1756.
Gotham, K. (2005). Theorizing urban spectacles, festivals and tourism and the transformation of urban space. *City, 9*(2), 225–246.
Gravari-Barbas, M. (2005). *La ville festive: espaces, expressions, acteurs*. Angers: Universite d'Angers.
Gravari-Barbas, M. (2007). De la fête dans la ville Ã la ville festive: Les faits et les espaces festifs, object géographique emergent. In A. Da Cunha & L. Matthey (Eds.), *La ville et l'urbain: Des savoirs émergents*. Lausanne: Presses Polytechniques et Universitairies Romandes. (pp. 387–415).
Harvey, D. (1992). *The condition of postmodernity*. New York: Free Press.
Hobsbawn, E., & Ranger, T. (Eds.). (1983). *The invention of tradition*. Cambridge: Cambridge University Press.
Jacquot, J. (1956–75). *Les fêtes de la Renaissance* : Vols.1–3. Paris: Editions du CNRS.
Jarman, D. (2007). Mirror of the nation: the Edinburgh festivals and Scottish national identity. In C. Aitchison & A. Pritchard (Eds.), (pp. 1–8)
Lefebvre, H. (1971). *Everyday life in the modern world*. New York: Harper and Row. (Reprint 1984).
Lenfant, C. (2003). Lyon: Lumières de décembre. *Urbanisme, 331,* 64–65.
Machiavelli, N. (1624). *The Prince: Chapter 21* (Translated and edited by R. M. Adams (1977). New York: Norton.
Machiavelli, N. (1625) History of Florence: Book 7. In A. Gilbert (Ed.), (1989), *The Chief Works of Machiavelli*. Durham North Carolina: Duke University Press, 84, 1352.
Magliocco, S. (2001). Coordinates of power and performance: Festivals as sites of (re) presentation and reclamation in Sardinia. *Ethologies, 23*(1), 167–188.
Manning, F. E. (Ed.). (1983). *The celebration of society*. Bowling Green: Bowling Green University Press.
Mile, S., & Paddison, R. (2005). The rise of culture-led urban regeneration. *Urban Studies, 42*(5/6), 833–839.

Miller, J. F. (2010). Festivals: Roman. In M. Gagarin M (Ed.), *The Oxford encyclopaedia of ancient Greece and Rome* (Vol. 3, pp. 171–174). Oxford: Oxford University Press.

Mueller, H. F. (2010). Saturnalia. In M. Gagarin (Ed.), *The Oxford encyclopaedia of ancient Greece and Rome* (Vol. 6, pp. 221–222). Oxford: Oxford University Press.

Mumford, L. (1961). *The city in history*. London: Secker and Warburg.

Murray, P. (1999). *Apres l'histoire*. Paris: Les Belles Letters.

NO: New Orleans. (2012). http://www.neworleanscvb.com/articles/index.cfm?action=viewarticle ID=6560menuID=1603. Accessed 25 Oct 2013

Picard, D., & Robinson, D. M. (2006). *Festivals, tourism and social change: Remaking worlds*. Clevedon: Channel View Publications.

Ponzini, D., & Rossi, U. (2010). Becoming a creative city: The entrepreneurial mayor, network politics and the promises of an urban renaissance. *Urban Studies, 47*(10), 1–21.

Scullard, H. H. (1981). *Festivals and ceremonies of the Roman republic*. Ithaca: Cornell University Press.

Strong, R. (1992). *Les fêtes de la Renaissance*. Paris: Solin.

Chapter 15
Slow Cities

Susan Ball

Slow living ... is not an escape from global culture into an ossified past but rather it is part of contemporary arguments about how we are to live now and in the future. The idea of slow living represents a contemporary interpretation of the past of places and communities and a mobilization of their traditions, principles and values in order to critique the present and provide alternatives for the future.
Parkins and Craig 2006, p. 78

15.1 Introduction

One of the eternally fascinating features of urban places is the fact that they are so marvellously different, with such a variety of historical forms, sights, sounds and ways of life. Yet the rapid homogenizing effects of contemporary globalization have been threatening to make many urban places similar to one another, which some fear will rapidly eradicate centuries of historic and cultural differences, in morphologies and life-styles especially. This will destroy much of the pleasure and delight that many find in the particular character of urban places and reduce our range of sensory experiences in cities that classic urban commentators (Mumford 1961) saw as an enhancement of our human condition. Critics of this homogenizing trend, such as Radstrom (2011), have argued that towns and cities need to counter these pressures and develop 'place-sustaining' policies to preserve and expand the uniqueness of urban places, not simply for themselves, but within the wider context of contemporary forces and the homogenizing forces of globalization. The recently developing Cittaslow or Slow City movement, composed of towns and cities from around the world who stress the importance of slow living and sustainable development,

S. Ball (✉)
Département d'études des pays anglophones, University of Paris 8,
2 rue de la Liberté, 93526 St Denis, France
e-mail: s.ball@orange.fr

W. K.D. Davies (ed.), *Theme Cities: Solutions for Urban Problems*,
GeoJournal Library 112, DOI 10.1007/978-94-017-9655-2_15

provides one example of the ways that such objectives can be attained. The result is a new process of 're-territorialisation' in a globalization context (Parkins and Craig 2006), through 'the politics of topographies' (Massey 2005, p. 172). In some ways this new emphasis upon appreciating the character of local places and regions can be seen as a partial revival of the work of early twentieth century ideas of French possibilist geographers, such as Paul Vidal de la Blache, who focused on describing the distinctiveness of the local 'pays' that had such varied characters because their history and unique 'terroir' created such individual and authentic 'genres de vie' (ways of life) and cuisines in these regions and their market towns (Buttimer 1971). However, this tradition in geographic work was often criticized for ignoring the forces of the emerging modern, interconnected and industrial world that were diminishing the distinctiveness of these areas and has faded as a major trend in geographical study. Today, this new emphasis upon the local and the particular in the Cittaslow movement is not based simply on local distinctiveness; it is providing an alternative to global forces, not ignoring them. This trend seems to support Massey's views that in the contemporary world "*most struggles around globalization are inevitably 'local' in some sense or other*" (2005, p. 181). So the challenge that Cittaslow members face is to develop a politics that can draw on relations from both within, and beyond, the towns that comprise the movement to promote and enhance their ideas, not to simply attempt to recreate some fossilized version of the past.

One of the ways in which these new, or perhaps revived, concepts about localism have been centred is through the spectacular diffusion of the attribute 'slow', which has been applied to many sectors of human life (Honoré 2004), from its initial application in slow food, to the labelling of cities that accept certain principles developed by Cittaslow International (pronounced 'cheeta-slow'). The Cittaslow organization came into existence in 1999, and although it has not attracted a very wide publicity, the Cittaslow membership increased rapidly to include 147 towns across four continents by mid June 2011.These members share a philosophy that recasts the term 'slow' as a positive attribute, not a negative one in the increasingly fast-paced and interconnected world. It is one that draws on a city's traditional assets and ways of life, with the goal of obtaining a qualitatively enhanced way of life and standard of living. This review outlines the development of Cittaslow as a movement, network and organization, before providing an outlook for its future. At the outset it should be noted that there are inconsistencies in information on, and by, the Cittaslow movement, which is matched by variations in the spelling of the name, namely: Cittàslow (Pink 2008); CittàSlow (Miele 2008); Cittaslow (most of the English-speaking members and the website of Cittaslow International); Città Slow (Beatley 2004); Città Lente (Mayer and Knox 2006). In this chapter, outside of quotations, the organisation is referred to as Cittaslow, which is the version usually used by most of the English-speaking members and the website of Cittaslow International. In addition, it is worth noting that the word 'city' is usually used to describe these places, but in the English-speaking world 'town' would be a more appropriate term, given their restriction to small centres.

15.2 The Origins of Cittaslow

Cittaslow is widely viewed as part of the recent emergence of what amounts to an international movement that emphasizes slow living. Rather than being a specific organization, this movement is comprised of many different approaches and individuals who are reacting to what Harvey (1990) has described as the 'shortening of time', or 'time poverty', brought about by the rapid changes and intense contacts of globalization. Given the pressures of this fast, pressured and constantly interconnected pace of life, with little time to focus on social activities, increasing numbers of people are advocating a cultural shift towards a slower pace of life. Yet this critical approach to 'fast', and a re-appraisal of the value of 'slow', cannot be viewed as only a recent trend; it dates at least as far back as the classical adage *festina lente*, meaning 'make haste slowly' or 'hurry slowly', which is a translation by Sventonio from the Greek: σπευδε βραδεωσ. However, *festina lente* re-emerged during the European Renaissance when, for example, it was the motto of the publisher Aldus Manutius. Later, the Medici Dukes of Florence (1531–1569) used the idea as their personal emblem, by a turtle with a snail on its back (Werness 2006, p. 416). This symbolized the concept of finding a proper balance between urgency and diligence, which the oxymoron *festina lente* encapsulates. The concept has also made its presence felt during more recent social movements, including the emergence of Romanticism in the context of the Industrial Revolution in Britain in the late eighteenth century with its reaction against industrialization and an emphasis upon nature and older forms of life (Honoré 2004). In the last decades of the twentieth century renewed interest in 'slow' as a positive attribute re-emerged, as reflected in various movements that stressed slow food, slow money, slow reading, slow travel, slow fashion, slow science, slow design, slow schools, slow cities and the values of taking time to appreciate things, therefore leading a slower and more contemplative lifestyle. These ideas have been summarized in the Slow Planet website (SP) and in Honoré's comprehensive book, *In Praise of Slow*, which has been translated into over thirty languages, while specific *Slow Guides* exist on how to live such a lifestyle in such cities as Melbourne. London, Sydney and Dublin (see, for example, Howard 2009). These various attempts to improve the quality of life, both today and in the future, view 'slow' as a characteristic to be enhanced, rather than overcome—perhaps paralleling the Winter City advocates (Chap. 8) that stress the need to embrace, not fear winter. Collectively these ideas have been referred to as the 'slow movement' or 'slow living'. Many reasons account for the recent growth in the popularity of this movement, including a release from the pressures of daily life, and providing an alternative to the homogenizing processes of globalization (May and Thrift 2001; Parkins and Craig 2006). Yet rather than seeking to stop the clock, or returning to the past, the Slow Movement reflects on the present and the past to find alternative and better futures to the current trends, so as to improve life-styles and the liveability of places.

The origins of the Slow Food movement can be traced to the Northern Italian counterculture of the 1970s, in which young socialists in Piedmont began to embrace the traditional music, agriculture and food of the region. Together they formed a loosely knit organization called Arcigola, which was the gastronomic wing of the *Associazione Ricreativa Culturale Italiana (ARCI)*, the Italian Cultural Recreational Association (Andrews 2008; Petrini 2001a). Arcigola was led by Carlo Petrini, an Italian food and wine journalist, who had become increasingly concerned about the threat posed to local foods, wines and cultures in Italy because of competition from global fast food chains and the rapid growth of industrialized farming that was seen as destroying local, age-old practices enshrined in permaculture ideas that had sustained the land and people for centuries (Mollison 1988; Holmgren 2003). Outraged by the opening of the first McDonald's restaurant in Italy in the Piazza di Spagna, one of Rome's iconic places, Petrini led a protest against the chain, a protest that involved eating plates of a local dish, *penne*, outside the fast food outlet. Three years later, the Slow Food International movement was officially inaugurated on 10th December 1989 when delegates from twenty countries met in Paris and endorsed the movement's manifesto (Petrini 2001b). Since this time the movement has grown to include over 100,000 members with 'convivia' (local groups of members) in 132 countries.

Andrews (2008) has provided a comprehensive overview of the rise and politics of the Slow Food movement. Its philosophy is based on 'eco-gastronomy', which refers to how food is produced, distributed and consumed. The core principle of the movement is that food should be 'good, clean and fair'. 'Good' means that food should give pleasure; 'clean' means that it should harm neither human welfare nor the environment; and 'fair' refers to the insistence that food producers should receive adequate compensation for their labour. Petrini has written extensively about the philosophy and aims of the movement (2001, 2006, 2007 and 2009). In 1997 the Slow Food Congress was held in the historic city of Orvieto—a market town of 21 thousand people in Umbria (Italy) dramatically located on a volcanic outcrop set in a wide valley that was a historic route between Rome and Florence. At this meeting Petrini first introduced the idea of extending and applying the principles of Slow Food to urban design and living. Two years later the Cittaslow (literally, Slow City) movement was formally established when the mayors of the four founding cities (Bra, Greve, Orvieto and Positano) signed the Cittaslow Charter, along with Petrini and the Slow Food movement. The Charter had been drawn up at the initiative of the movement's founder Paolo Saturnini, then mayor of Greve in the Chianti area of Tuscany, a small historic market town of 14 thousand inhabitants. The other two founding towns are also unique places: Bra, in Piedmont, is a market town of 30,000 people, 50 km south of Turin, the home of Petrini, and is now the site of the first University of Gastronomy. Positano is a former fishing village in the middle of the Amalfi coast south of Naples, which has been transformed into a chic tourist resort since the 1950s. Figure 15.1 shows that these founding centres are in different parts of Italy with varied regional and historical traditions. It also shows the distinctive Cittaslow logo which depicts buildings on the back of a snail, clearly a link back to part of the Medici logo and motto.

Fig. 15.1 Cittaslow logo and location of founding towns

15.3 The Cittaslow Charter

Although Cittaslow has clearly evolved out of the Slow Food movement it goes further than the organization that spawned it by creating a process by which other centres can join the association which is based on the requirement to support a number of policies that are designed to improve the quality of life and uniqueness in individual towns. In the words of Cittaslow U.K. this means that "The Cittaslow philosophy is meant to be embedded in the municipality" (CSN). Attachment C in the Cittaslow Charter identifies the core ideas of the movement which are described as the 'requirements for excellence' and are shown in Table 15.1. These consist of a total of 54 policy goals grouped into seven categories, dealing with the *environment, infrastructure, facilities and technologies, autochthonous production, hospitality, awareness promotion,* and what is called an *extraordinary* category. In addition the Charter maintains it is important to accept seven additional projects that are fundamental to Slow Food activities. It must be emphasized that almost half of these policies (25) in the seven categories are considered obligatory features for towns wishing to become members.

The various policy goals in the latest version of the Charter are shown in Table 15.1 which summarizes and rephrases the ideas. The *Environment* category lists a familiar set of sustainability policies, such as those to reduce pollution, waste, and the promotion of renewable energy. The list of polices to improve *Infrastructure* are an eclectic group, but follow the sustainability approach in stressing the importance of greening policies and providing alternative transport options to the use of cars but also emphasize the upgrading of places and the reclamation of historical areas. This is an indication of the conservation part of this movement that has rarely been mentioned in New Urbanist ideas or the other new urban themes of the past thirty years that have been described in previous chapters.

It is worth noting that the *Facilities and Technologies for Urban Quality* category does not simply show the need to improve facilities and reduce both the problems of noise and what probably refers to the visual pollution produced by the unco-ordinated kaleidoscope of colours in the signs found in most contemporary cities. It is also forward-looking in stressing policies that add new technologies to the towns, such as fibre optic cables, while the desire to promote bio-architecture, creates another link with green ideas. The policies listed under the *Hospitality* group clearly demonstrate an awareness of the importance of improving the experiences of visitors, showing that the movement wishes to attract tourists to the centres, since they are essential to maintain and stimulate the local economy. The policies in the *Awareness* and *Extraordinary* categories recognize that it is also important to promote the Cittaslow approach and to convince local citizens of its value. However, the most original category that identifies one of the major differentiating features of Cittaslow deals with the *Safeguarding of Autochthonous Features* in the city and its region. They range from production methods, traditions in crafts, cultural events, to protection for plant and animal species in the area that are under threat due to increasing use of a small number of types for food and wine production—a problem that threatens to reduce the seed pool and create a threat to ecological diversity. This trend, if allowed to continue, will produce a loss of our heritage in cultivation and also has the potential danger of creating a vulnerability to diseases that may attack specific plant species that dominate particular food crops. The need to support *slow food policies* also draws attention to the importance of educating children in nutrition and taste, so they can appreciate the value and diversity of foods in the development of a more healthy diet. This combination of what amounts to worries about both environmental and traditional cultural losses, and the importance of more healthy living, lies at the heart of the Cittaslow movement and is a key reason why it is unique among the various new ideas that have developed to improve the quality

Table 15.1 The Cittaslow charter requirements. (Source: Rephrased and re-organized from the latest Cittaslow International Charter, CSIC)

A. Environmental Policies: Obligatory
(1) A municipal plan for saving energy, particularly using renewable sources (2) A ban on the use of genetically modified organisms (3) Add systems and programs for controlling light pollution (4) Adoption of systems of environmental management (EMAS and ECOLABEL or ISO9001;ISO 14000 SABOOO) and participation in Agenda 21 protects (from 1992 Rio Sustainable Development Conference)
Seven additional criteria
(i) Verify air, water and soil quality in accordance with law (ii) & (iii) promote and disseminate advice on both differentiated refuse collection and industrial domestic composting (iv) add a purification plant for urban sewage (v) plan for the regulation of commercial art and traffic signs (vi) & (vii) create systems for controlling pollution from noise and electro-magnetic sources
B. Infrastructure Policies: Obligatory
(1) Plan to favour alternative mobility systems over private transportation, integrate traffic with public means of transport and pedestrian areas, including adding external car parks with park-ride systems, etc (2) Verify that the disabled have access to public places, and also remove barriers to access new technologies (3) Promote programs to facilitate: family life, local activities to create bonds between school and family, home assistance for the old and sick, provision of social centres, and regulations for business hours of municipalities and public toilets (4) Create urban restyling and upgrading programmes (5) Integrate public information functions with Slow City information windows
Eight additional criteria
(i) Plan to improve and reclaim historical centres and works of cultural-histoncal value (ii) create safe mobility and traffic plans (iii) add bicycle tracks to connect schools and public buildings (iv) add a medical assistance centre (v) provide quality and interconnected green areas and service infrastructures (vi) create centres for the distribution of natural products (vii) develop agreement with shopkeepers to create 'friendly shops' to assist people in trouble (viii) redevelop deteriorating urban areas for urban re-utilization
C. Facilities and Technologies for Urban Quality: Obligatory
(1) Create and promote bio-architecture by training people to work on such projects
Eight additional criteria
(i) Provide optical fibre and wireless systems in the city (ii) adopt systems to monitor electromagnetic fields (iii) provide refuse containers, to blend with environment, and ensure they are emptied at regular times (iv) add programs to plant environmentally suitable and local plants (v) plan to provide municipal services via a civic internet and to educate citizens in its use (vi) promote use of the internet in teleworking (vii) develop plans to control noise in the loudest areas (viii) make plans to review colour schemes

Table 15.1 (continued)

D. Ways of Safeguarding Autochthonous Production: Obligatory
(1) Develop organic farming (2) Certify artisan products and artistic crafts (3) Create programs to safeguard artisan and artistic products facing extinction (4) Safeguard traditional work methods and professions at risk of extinction (5) Use organic and local products; preserve old traditions in restaurants, schools, etc (6) Create nutrition and taste education programmes in co-operation with Slow Food ideas (7) Favour local wine and food species, as well as preparations facing extinction, via Slow Food Presidia (8) Develop a census of the territory's products; support their commercialisation and marketing (9) Preserve and promote local cultural events
Two additional criteria
(i) Carry out audits of trees in the city noting age and size; (ii) promote local and traditional cultures in urban and school gardens
E. Hospitality Features: Obligatory
(1) Provide training courses to deliver quality hospitality experiences and quality tourist information (2) Create guided itineraries, and use international signage in historical centres, to assist tourists
Three additional criteria
(i) Help visitors by providing information on: scheduled events and municipal facilities, such as opening hours, parking and tourist offices (ii) prepare 'slow' guides to the city, through brochures and websites (iii) make guides and shopkeepers aware that they need to make their charges clear and to display these outside their premises
F. Awareness Policies: Obligatory
(1) Develop a campaign to inform citizens of the intention of city officials to become a Slow City and outline the movement's objectives (2) Initiate programmes to show how the community can acquire the Slow City philosophy and its projects, including educational green areas, libraries and the development of a seed and species bank (3) Spread knowledge of Slow City and Slow Food activities
One additional criterion
(i) Support the Slow Food Presidia, which promotes a range of economic, environmental, social and cultural objectives
G. Extraordinary Requirements: Obligatory
(1) Members are expected to promote the aims of the Stow City movement. They should also display the Cittaslow trademark as part of their municipal letterhead and add the 'slow' philosophy ideas to their website
H. Support to Slow food Activities & Projects
(1) Establish local slow food convivium (a local chapter of the Slow Food network) which meet for various events, from evening meals to international conferences (2) Add adult educational programmes in taste and nutrition, which should also be compulsory in higher education establishments, in cooperation with the Slow Food movement (3) Create school vegetable gardens, assisted by Slow Food members (4) Implement one or more Slow Food or Area centres for those species or production processes at risk of extinction

Table 15.1 (continued)

(5) Use the products from the local area that are safeguarded by Slow Food and the maintenance of nutritional traditions in collective food service, in protected structures, school canteens, together with programs for food education
(6) Support the typical products of the local area through implementation of the 'Mercati della Terra' (a market for locally grown and produced products) in cooperation with Slow Food members
(7) Support the Terra Madre (TM) project and food committees through joint twinning TM is a world market first held in Turin in 2004 (Petrini 2009)

EMAS: Eco-Management and Audit Schemes are management instruments developed by the European Union which help companies and other organisations to assess their environmental performance. *ECOLABEL*: The voluntary labelling of food and consumer products which provide consumers with a sustainability measurement. ISO9001: A standard produced by the International Organisation of Standardisation (ISO) which covers quality management systems. ISOI4000: An ISO standard addressing environmental management. SA800: An international set of auditable standards concerning the social aspects of workplaces. Agenda 21 projects are aimed at sustainable development, which implement the Agenda 21 ideas adopted in 1992 by the United Nations Conference on Environment and Planning, Rio de Janeiro.

and liveability of urban places in the last three decades. Many of these polices give an added meaning to the word 'slow' which goes beyond speed, since they include the time spent in appreciating and achieving a qualitatively different lifestyle.

Clearly some of these goals overlap with some of the ideas expressed in the Transition Town and Ecocities movements as well as in settlements attempting to be more sustainable (Chap. 7). But the Cittaslow policies are distinctive in focusing on several issues that are rarely found in the recently developed new themes influencing urban policies. One is the importance of *maintaining traditions* in many aspects of life, as well as in urban morphology, features derived from a local heritage—not one imported from elsewhere, as in the copying of older forms found in neo-traditional urbanism developments in North America. Related to this is the focus on *autochthonous developments* in the town and its region. Three other unique features are the emphases on *quality, hospitality* and the *promotion* of the ideas of slow living, with its privileging of diversity rather than standardization. Finally, at the heart of the approach is the distinctive focus on *food and taste*, issues that also form the basis of the Slow Food movement.

15.4 Membership of Cittaslow

Cittaslow membership is open to towns with a population of less than 50,000 that are not regional capitals—showing the emphasis is upon smaller centres, most of which are the historic market towns of their region. A mixed committee comprising members of Cittaslow International and representatives of the national Cittaslow Coordination Committee undertakes the work of 'translating' the goals that are expressed in its charter (Table 15.1) to the cultural context of the country in question (Miele 2008, p. 150). Cittaslow U.K., for example, works from a list of 55 goals.

The Charter does not give information on the self-assessment other than in its Appendix E which refers to 'cards'—presumably based on the goals or polices of the Charter, which are each assessed on a scale of 1 to 3. The applicant towns assess themselves against this scale, attributing a high score to those goals they are already addressing, a medium score to those they have identified and are in the early stages of addressing and a low score to those they are not pursuing (Cittaslow U K. 2011). Applications to join the network must be submitted by a recognized agency in each town, such as the local council, an established community group or business association. In order to be accredited with Cittaslow status a town must score at least 50 % of the set of weighted goals in a self-assessment exercise that is overseen by a certifying body, usually in *loco*. Certification confirms that the town applying to be a member of the movement has already put in place policies which accord with the movement's philosophy. With reference to the certification procedure, Mayer and Knox (2006, p. 327) have remarked that the process favours applicants that have already developed programs that fit into the Cittaslow criteria.

In the case of the U.K., most towns scored between 50 and 60 % overall against the 55 goals when they first joined the movement (Cittaslow U.K. 2011). An application fee of £ 500 is charged to cover the costs of the initial assessment exercise. On being granted membership an annual membership fee is levied in accordance with the size of the town. The annual membership fee is payable to Cittaslow International and the national coordination committee in the cases of Italy, the U.K., Germany, Netherlands, Nordic countries (Norway, Denmark and Sweden), South Korea, Poland, Spain and the U.S.A. Current membership fees range from £ 600 for towns with less than 1,000 residents to £ 3,500 for towns with more than 30,000 residents (Cittslow U.K. 2011). As a means of confirming that the movement's philosophy will be 'embedded in the municipality', on receipt of its Cittaslow status the local government body concerned is required to include its logo in its letterhead.

Obtaining Cittaslow accreditation is confirmation that a town has adopted its philosophy and has achieved many of the goals *before* becoming a member. It also implies a continuing commitment to developing this philosophy *after* official recognition, through those policies, projects, activities and events which have been formally stated in the Charter and assessment exercise. Members are re-evaluated every four years by means of a self-assessment report addressing how and whether they have fulfilled their previously stated goals and added programs to conform to the Charter's policy objectives. A joint committee comprising members of the local government body and representatives of Cittaslow reviews the self-assessment exercise, and advises on the setting of goals for the following four year period. The stringency of this exercise serves a dual purpose: not only does it aim to ensure that the member towns are continuing to develop the Cittaslow philosophy, but also that they do not become members of the movement for short-term marketing benefits. However, there is some doubt as to the regularity of this exercise in some centres, with different sources having claimed that it is conducted once every 3 or 4 years, while others have claimed that the assessment occurs 'periodically' (Miele 2008, p. 140).

15.5 Cittaslow: A Varied and Fluid Movement

Given that scope exists for (re)defining and applying the criteria for membership, it may not be surprising that a 'typical' Cittaslow defies definition. Indeed, earlier case studies of Cittaslow members point to the movement being varied and fluid. Miele claimed that the members are "ordinary towns and each one differs from the others" (2008, p. 141). These 'ordinary towns' come from a range of national and regional contexts, have a variety of resources to draw on, and vary considerably in size. This can be from 1,500 people in Casalbeltrame (Piedmont), up to the self-imposed size limit of 50,000. Yet in an urban context they are all small towns, many are the historic market centres of distinctive regions that had shown relatively little growth in recent decades due to the expanding influence of larger cities in the area, evidenced by longer commuting and shopping trips on new road networks. The policies and projects they pursue are correspondingly varied. Many towns have promoted the following Cittaslow policies: shared meals, the provision of information on the range of local shopping facilities, food festivals, the setting up of food cooperatives, farmers' markets, initiatives for taste education, annual censuses of typical local products, training young people in 'traditional' skills, campaigns for the use of low energy light bulbs, the installation of sewage treatment systems, car-free Sundays, free alternative transport services and projects for the development of local sites of land or the refurbishment of buildings (Parkins and Craig 2006; Mayer and Knox 2006; Miele 2008; Pink 2008, 2009). However, the Cittaslow towns vary in the degree to which these policies are advanced.

Such a policy variety, both in terms of cultural contexts and approaches to developing an alternative development strategy, could lead to the charge that Cittaslow is too heterogeneous to comprise a single movement. However, it is from this heterogeneity that Cittaslow finds some unity, primarily in two features. First, the fact that each member sets out to protect the *unique* quality of its town is seen as a way of bringing the movement together. Second, the individual members each use a variety of means to create alternative development strategies. Indeed, Beatley (2004, p. 335) argued that although the towns may be pursuing a variety of goals they are united by the desire to protect their unique and distinctive communities.

So the use of a range of alternative development strategies by each of the Cittaslow members may also be considered—somewhat paradoxically—as a unifying theme. Rather than setting out to endorse a single alternative plan, membership gives official recognition to Cittaslow diverse approaches, both within a town—where a range of strategies are brought together and recognized in the process of applying for and renewing membership—and across the movement, when alternative developments are compared. By officially recognizing diverse approaches between and within towns, Cittaslow makes it a defining characteristic of the movement as a whole and of the individual members.

Since the Cittaslow movement does not apply a rigid set of criteria for membership, but rather sets out policy goals to be translated in different local contexts, the movement can also be seen to exhibit *fluidity* as seen by the following summary.

> In order to work and to reproduce a new version of slowness in each new and diverse/distant locality there is always change and adaptation to local conditions and contingencies. But this suggests that both the qualification of the slow objectives, practices and spaces, and the variable procedures for joining the CittàSlow network, may be understood as fluid technologies that create mutable mobiles. (Miele 2008, p. 135)

Miele even goes as far as claiming that the quality of fluidity and slowness accounts for the success of the movement. Parkins and Craig also take up these ideas as defining characteristics, specifically with reference to the movement's fluidity over time.

> Cittàslow … does not seek to promote 'static' cultures defined through their stubborn opposition to the 'monolithic fluidity' (if one can use such a phrase) of modern global culture but is itself a 'fluid' organization, defining itself through its on-going negotiation of emerging cultural change and traditional ways of living. (2006, pp. 82–83)

From this viewpoint, Cittaslow towns are seen as taking an *active* role in choosing between new influences, rather than *passively* conforming to the influence of the current form of globalization, as well as cherishing traditional ideas.

Being identified as a fluid movement is important to Cittaslow. It can help to counter claims that the movement, and each of its members, are preservationist in nature, seeking to preserve towns in the past, rather than moving forward with the times. Indeed, in the English-language town planning literature an important distinction is made between the terms 'conservation' and 'preservation'. The latter is usually applied to individual buildings within a town, which are viewed as being worthy of preservation on architectural grounds, while 'conservation' refers to the means by which the preservation of such buildings may be brought into their current urban context, perhaps by using them for different functions, in order to give coherence and new life to the end result (Allison et al. 1996; Cohen 2001). Miele indicated that this idea of preserving things from the past within the context of an evolving environment is basic to the Cittaslow conservation approach:

> the translation of Slowness for CittàSlow can take different directions and it does not limit itself to the preservation of old Slow objects, practices and spaces … making Slowness means also engaging in a process of qualification of what is Slow and different from 'fast', in the new objects, technologies, practices and spaces. (2008, p. 147)

In her study of Italian Cittaslow centres, Radstrom refers to conservation as a 'flexible dynamic' which 'may be central to place-sustaining' (2011, p. 109). She claimed that Cittaslow provides an important example of the implementation of a 'place-sustaining' approach in which towns seek to conserve their unique qualities and sense of place by means of drawing on their cultural traditions. So the 'place-making' strategies, used by increasing numbers of cities seeking to improve their competitive position, risk failure in the absence of a more holistic approach to achieving a 'place sustaining' goal.

> The true risk lies in the fact that while a focus on place-making continues, existing places may be lost because of the lack of a cohesive approach to place-sustaining. When a place that already exists is endangered, this does not need to be "made" but rather, it needs to be sustained.... The danger is that by not focusing on place-sustaining, planners will be doomed to a continual cycle of place-making, which will always aim, but may not succeed, to re-create the original sense of place. (Radstrom 2011, pp. 109–110).

Radstrom's research was conducted mainly on Cittaslow towns in Italy, where urban and regional identities are often based on a long tradition of historic and often competitive city-states which were only absorbed into the new state of Italy in the mid-nineteenth century. Outside of Italy this strong regional identity and continuing city-state feeling from their historic heritage does not occur to the same extent; or if it did, existing regional cultures may not have survived modernization, so their conservation forms a minor part of the holistic approach to sustainable urban development. However, even in Italy conservation is a minor concern in some towns. For example, in San Vincenzo very few of the old spaces, traditions and crafts have survived. Hence, this town's Cittaslow goals are aimed at devising environmentally sound versions of *new* practices, such as beach tourism (Miele 2008). Although conservation may only play a minor part in the overall strategies of some members, there is no doubt that the Cittaslow movement has become associated with this wider conservation trend. Knox has pointed to a possible negative effect of such policies, which may act contrary to the overall goal of producing 'slowness' in a town.

> Paradoxically, Slow City designation becomes a form of brand recognition within the heritage industry. Because they are small … the charming attraction of Slow Cities could all too easily be overwhelmed by tourism. So the more they flaunt their gentle-paced life, the faster they may end up changing. In this scenario prices will rise … cafés will lose their spilled-drink, smoky, messy, authenticity … affluent outsiders will choose to make their second homes in them … and the poor and the young will be pushed out. (Knox 2005, pp. 7–8)

Some of these problems, of course, have surfaced in many attractive small towns in western Europe, in particular where second home owners account for a large part of the local housing stock, leading to accusations of rural and small settlement gentrification.

One of the crucial features of the official recognition of Cittaslow status is the impetus it gives to a town's alternative development strategies. But at the same time the town needs to be aware that the acquisition of this status carries the risk that the intensity with which the town is used by others may increase. At an early stage this may bring about a welcome increase in tourism. But over time it may also lead to the displacement of low-income residents and businesses as property prices increase and only new incoming middle class households can afford properties. It would be incorrect, however, to view any such displacement as a simple outcome of Cittaslow status. Increases in property prices in historic or distinctive towns are part of a wider trend in which the poor are displaced (Smith and Williams 1986). So while Cittaslow status may play a part in this movement, alone it does not account for it. Also, while both variety and fluidity are important in terms of defining the Cittaslow movement, they do not set the movement's members apart from other towns pursuing alternative development strategies. Pink (2009) has suggested that what makes the movement distinctive is that it enrols local government in slow living, while Radstrom believes it should be viewed as an organization rather than a movement.

> Both Slow Food and Cittaslow have previously been described as movements. In fact, Cittaslow is more accurately defined as an organization of small cities which have voluntarily signed on to be members. … This is a unique organisation of cities, moving forward with a common interest that spans borders, cultures and city identities. Whereas Slow Food can be seen as operating as a grassroots movement, Cittaslow is more formally organized and operates on a city policy level. (2011, p. 94)

15.6 Cittaslow: An International Network

The organization refers to itself as an international 'Cittaslow network' of towns. Since 1999 it has grown to include 147 members distributed across four continents, although the vast majority, namely 128, are located in Europe of which 68 are in Italy". Figure 15.2 shows the growth in Cittaslow membership since 1999, obtained by personal correspondence between the author and Cittaslow International. Data on the year that cities joined the movement was provided for 105 towns in the network; unfortunately information was not available for 25 towns in Italy, 1 town in New Zealand and 4 towns in Poland. Given that records on the date of membership for 25 of the 68 member towns in Italy are missing, the peaks shown in 2007 and 2008 can only be estimates. As towns in Italy were the first members of the movement, it is nevertheless probably correct to assume that there has been an annual

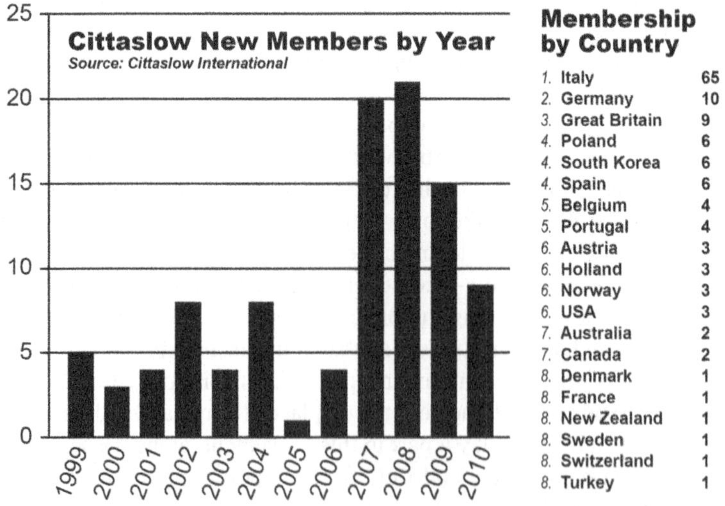

Fig. 15.2 Growth and location of Cittaslow members. (Source: Compiled from survey by author in 2011)

decline in the number of new members since 2008. For example, a comparison of the list of members for 2010 and July 2011 indicates that although there was an overall increase in the number of members, six cities left while 15 centres joined the movement.

In contrast to Slow Food, which began as an international movement with a manifesto endorsed by delegates from fifteen countries at its inaugural meeting in Paris, Cittaslow was originally conceived as a national network of towns. However, the Charter's goals were drawn up with Italian cities in mind, especially cities of the late-medieval and Renaissance in Italy, in which their piazzas were still the core of many social activities. Within four years of its launching, however, the Italian concept of Cittaslow had developed European ambitions. Pink (2009, p. 455) described an interview conducted in 2003 with two members of the Slow Food Press Office, in which Stefano Cimicchi, the President of the movement at the time, stated:

> We want the association to become a player at European level to make sure that the Constitution currently being drawn up (by Strasbourg) takes into account the reality of small towns and cities.

Between 2002 and 2006 eight European towns from outside Italy joined the movement (3 from Germany, 2 from Norway and 3 from the U.K.). This led to a series of meetings for the purposes of reflecting on the movement's early history, and identifying the problems to be encountered when applying the concept of slowness outside of Italy (Miele 2008). In 2007 the first members from outside Europe joined the movement (4 towns in South Korea and 2 in Australia), and by 2011 towns in 24 different countries had become members of Cittaslow thereby attesting to the transnational nature of the movement. However it is worth noting that French towns, especially those with similar historic traditions and local cuisines, have been slow to join the movement. The first joined in 2010, four more were accepted in 2011 and 3 more in 2013 (Roux 2013).

In addition to developing the network in terms of its geographic spread, Cittaslow has also started to spread its influence further up the urban hierarchy. Although membership is limited to towns with a population no greater than 50,000, in 2002 the mayor of Paris, Bertrand Delanoë, was presented with the annual award in the non Cittaslow administrators section. A *Cittaslow Supporter* category was created in 2005 for towns and local authorities with a population exceeding 50,000, making it clear that the movement has broadened from its earlier small-scale focus. Moreover, the movement has started to encourage its members to conform with international assessment criteria established outside Cittaslow—such as standards set for environmental management by the International Organization for Standardization (ISO series 14000) and Eco-Management and Audit Schemes (EMAS)—as well as trying to influence the policy of supra-national governmental organisations. Hence the network is developing a more active international role.

The creation of the network of towns that make up Cittaslow is also important for the transfer of ideas. This may take place within the network of accredited towns, but ideas may also be exchanged between a member and one or more towns preparing for membership, as noted by Miele (2008, p. 146).

> San Vincenzo's attentiveness to the conservation of natural resources and the implementation of these innovative environmental policies has also become an example for other towns, whose main economic activity is beach tourism, that want to joint CittàSlow as in the case of the first three Portuguese towns that form the Portuguese network.

As the international network continues to develop, and as ideas are transferred and adapted, a greater *transnational* character of Cittaslow may develop. Pink holds out some hope for this when she observed that its national and international connections are shaping the organization.

> For example, in 2005 representatives of the Cittàslow national networks signed the 'Skondal Agreement' at the first International Congress of Cittàslow in Norway, 'pledging to work together to promote sustainable energy projects' (Cittàslow UK Statutes) and Cittàslow proposes to create a Co-ordination Committee which will have a representative from 'Every country with a city in the Network. (Pink 2009, p. 455)

The Co-ordination Committee was convened on 9th April 2011 in Pollica, Salerno (Italy). This is the town of which Angello Vassallo—a former Vice-President of Cittaslow—was Mayor, prior to his murder by the Camorra on 5th September 2010, because of his anti-crime activities. It is hoped that this Committee will play a role in helping to translate 'slow practices' between towns, which may themselves change by integrating diverse ways of slow-living into a new dynamic culture. If this is achieved, not only will Cittaslow cease to be viewed as an Italian movement, it may also progress from being seen as an international network of towns to being defined as a transnational movement capable of translating and assembling slow ideas and practices from different cultures.

A further development towards international networking can be seen in the commitment made by Cittaslow International in 2009 to promote adhesion among its members to the Covenant of Mayors. This initiative, launched by the European Commission in 2008, brings together on a voluntary basis local and regional authorities agreeing to implement Sustainable Energy Action Plans with the aim of reducing carbon dioxide emissions by at least 20 % by 2020. By becoming a Covenant Supporter, Cittaslow International agreed to promote the Covenant of Mayors among its members, to guide them through the process of becoming signatories, and to facilitate exchanges between those who joined.

15.7 Governance

As a Covenant Supporter, Cittaslow International participates in what has been 'portrayed by European institutions as an exceptional model of multi-level government' (Covenant of Mayors, undated). The headquarters of Cittaslow International, which comprises a small, full-time secretariat and director are based in the Palace of Taste situated in the Convent of San Giovanni, Orvieto. At the global level the organization is governed by the Cittaslow International Co-ordinating Committee. The members of this committee are elected for a three-year term of office at the International Assembly, to which all member towns are invited and have full voting rights. The fourth International Assembly was held in Poland in June 2011, when Gian Luca Marconi (Mayor of Castelnovo ne' Monti in Italy) was re-elected President. In addition to the President, the International Co-ordinating Committee comprises nine Vice-Presidents and 16 members elected from member towns around the world. Although the election of officers gives the national networks the possibility of being represented at the international level, only 14 of the 24 countries in which there are Cittaslow members are represented directly on the International Co-ordinating Committee. So Italy still dominates the movement, accounting for half of its 26 places, with each of the other 13 countries having a representative each. This seeming 'over-representation' of members from Italy on the International Co-ordinating committee is nevertheless in line with the fact that around half of the membership of Cittaslow comprises towns from Italy.

Members of the International Co-ordinating Committee meet two or three times per year in different member towns. These meetings are held in order to identify the initiatives that are of interest to the network as a whole and to approve their financing, to review membership criteria, and to monitor the progress of international projects. Sub-committees are also established to consider specific topics as and when required, such as marketing and ways of strengthening links between member towns. The experts comprising the Cittaslow Scientific Committee also assist the International Co-ordinating Committee on cultural and scientific matters. There are 14 members of the Scientific Committee, providing expertise in such fields as: architecture, planning and economic development, law, food culture, and tourism.

Below the level of international committees, there are a number of national networks. These come into existence once a country has three member towns. At this point, Cittaslow International invites these towns to form a new national network with the purpose of working together to promote the movement and recruit new members in their country. By November 2011 there were ten national Cittaslow networks, with each one adopting its own form of organization. In the case of Cittaslow U.K. the national network takes the form of a Community Interest Company (CIC) with a Board of Directors appointed from each of the member towns, together with a representative of Slow Food U.K. CICs were introduced by the British government in 2005 and are designed for social enterprises which want to benefit the community rather than private shareholders. The Board works by holding meetings three or four times a year and uses Skype to keep in contact. These meetings are also regularly attended by the Director and other members of Cittaslow International. Cittaslow U.K. does not have a single head office to carry out administrative functions, rather it shares responsibilities and activities between the volunteer members. For example, the town of Alysham in Norfolk provides Cittaslow U.K. with its secretariat, while Mold in North Wales, looks after finances as well as co-ordinating the work of members of the British network.

These national and international networks are made up of the recognized agencies (usually the town councils) who are the members of the Cittaslow movement. Given the variety of ways in which town councils in different countries throughout the movement are internally organized, there is no single method of incorporating Cittaslow into any municipal organization. In the U.K. the movement has normally been represented by a formally constituted sub-committee of the town council. The members of this committee would typically include the town councillors supporting the Cittaslow membership, together with representatives of key public, commercial and voluntary organisations. Pink provides an outline of the Cittaslow Committee in Aylsham, U.K.

> (It is) Chaired by … a Town Councillor … and a leader in the Cittàslow process, its membership consisted of representatives of a wide range of agencies in the town, including the Traders' Association, the Felllowship of Churches, the Aylsham Partnership, the County Council, Slow Food Aylsham and the Aylsham Guides. The Committee also divided some aspects of its work into two further sub-committees, the Carnival Committee and the Arts Committee, later (in April 2006) amalgamated into one committee, the Cittàslow Community Events Committee. (Pink 2008, p. 173)

As a legally constituted element of local government, members of Cittaslow Committees act in order to initiate policies in line with the goals they have agreed with the International Cittaslow movement.

Cittaslow's formalized position within the structure of local government led Pink to observe that it provides a further example of the trend identified by Mayer (2006, p. 203), namely closer relationships between urban social movements and the state (Pink 2009, p. 457). This relation may also be considered as a form of corporatism (Simmie 1981), in which social and environmental interests align with small businesses and the state at the local level. Although Cittaslow's incorporation into local government may point to a closer relation between the local state and social movements, it is nevertheless a delicate relation that may be broken. In April 2011, Ludlow Town Council—the first member of Cittaslow in the UK—took the decision not to re-apply for membership of the organization in 2011/2012. The minutes of the meeting of Ludlow Town Council on 18th April 2011 stated:

> Members discussed the Town Council's continuing membership of Cittaslow. Each member expressed their views regarding the benefits to the town and how controversial the issue had become within the town and Council. Resolved: That Ludlow Town Council does not apply for membership of Cittalow when the CIC (Community Interest Company) ends. (Ludlow City Council 2011)

Furthermore, it is worth noting that the longevity of a town's relationship is often based upon the electoral fortunes of the persons supporting Cittaslow within each town. This may be a weakness, since if the members supporting the organization lose their seats it should not be assumed that new councillors will continue to support membership of the movement. However, in some cases this problem may be resolved when a town council's Cittaslow sub-committee is comprised of councillors and non-councillors. For example, Cittaslow U.K.'s website notes that Ludlow, a former member, provided a more open form of participation in local democracy:

> Ludlow operates as a part of the Town Council but includes non-councillors in the democratic process. Indeed, non-councillors have so far been in the majority in Cittaslow Ludlow meetings, which are attended by representatives of a wide range of Ludlow organisations… and representatives of South Shropshire District Council. Individuals who represent a particular concern or interest are also welcomed. …Participation in local democracy is widened, and the Cittaslow Goals provide a set of ambitions that local organisations find it easy to sign up to, and work towards. (Cittaslow Ludlow 2006)

This type of more diverse governance is also typical of some of the New Regionalism initiatives (Chap. 2). But claims for open participation in the workings of Cittaslow sub-committees should be treated with a degree of caution on two grounds. First, the commitment of committee members to their own town's local uniqueness may reduce their ability or willingness to see beyond 'the local'. However, Cittaslow members do take part in international and national networks, and the appeal of the movement's "principles have travelled beyond the 'local' towns where its model is applied" (Pink 2009, p. 457). However, committee members may view neighbouring towns as a potential threat to their own local authenticity and economy, or may attach less importance to the democratic claims that derive from the adjacent towns and regions *beyond* the locality. Second, a member's emphasis on local uniqueness skirts the potential for conflicts *within* the locality. In most cases, as Pink noted (2008, p. 186), the populations within Cittaslow member towns are often largely unaware of the movement, let alone committed to its goals. Indeed, Pink goes on to advise against referring analytically to these towns as 'communities' as this will obscure how their policies are shaped by a small number of local actors. Mayer and Knox (2006, p. 332) have also noted that although local organisations involved in Cittaslow initiatives continue to make efforts to involve local citizens in their activities as part of their commitment to the Charter, their Cittaslow agendas do not seem to have obtained grassroots citizen support. Pink (2009, p. 456) concluded that at the local level this continues to be a top-down organization, formed by a few broadly middle class, committed activists. Although these middle class alliances of small businesses, traders, artisans and eco-activists in Cittaslow towns in the U.K. may be representative of the range of the main political parties, they are nevertheless homogeneous in their commitment to what has been referred to as the "politics of lifestyle—based around issues of eating, shopping and travelling" (Furedi 2011). These activists may set out to promote open participation in their meetings and through events and initiatives. But it remains to be seen if this will be taken up by the population at large, or if Cittaslow supporters will use their legalized position in local government to impose their own 'lifestyle politics' on the wider community.

15.8 Conclusions and Prospects

In many parts of the world small historic towns that used to thrive as centres of their local region, and were shown by possibilist geographers in France in particular as having distinctive ways of life and traditions, suffered from two changes. One was the ability of larger urban nodes to increase their functional importance, as well as extending their spheres of influence which reduced the number of functions in these smaller centres. Another came from the pressures to adopt the seemingly inexorable standardization of modernist life-styles, which made smaller centres seem old-fashioned and without opportunities or dynamism. In many ways the Cittaslow movement has been a reaction against these pressures by stressing the importance of alternatives to these trends, by emphasizing the role of small historic towns in creating more fulfilling, pleasurable, relaxing and even sensory patterns of life in urban places that are part of the movement, especially those linked to traditions. However they also include many aspects of the current sustainability and greening movements, in addition to recognizing the need to promote themselves to visitors and ensure they have positive experiences. Yet what is also clearly essential to the movement is a specific set of ideas in its charter about what a 'community' is supposed to be, unlike the vague almost motherhood references to the concept that are never clarified in the New Urbanist mantra as seen in Chap. 2.

> The development of local community is based among others, on the ability to share and recognize their specific traits, of regaining their own identity, visible from the outside and deeply lived within. (Cittaslow Charter, p. 21)

Since its creation in 1999, Cittaslow has gone beyond its roots in Italy to develop an organizational structure extending to four continents by 2011. During the same period, other individual towns have started to work towards developing the means to conserve and improve their 'unique' qualities, while also responding to the challenges posed by the increasing competition from regional and national centres as well as globalization. Indeed, by improving their unique qualities they can counter one of these challenges by attracting visitors from long distances, thereby turning the greater spatial interaction possibilities of the contemporary world into an advantage for their economies.

Although membership has increased during the last decade, the Cittaslow movement still remains small. But more towns can be anticipated to join the network in the future since a number of centres are currently going through the Cittaslow self-assessment exercise required for membership. By doing so they will be able to capitalize on their assets and draw upon the experiences of their fellow members in the network. This may help them prosper and avoid the stagnation of so many other small centres that have been left behind by economic change. Yet it must be admitted that not all the threatened small towns in the world have the types of long historic tradition and distinctive regional ecology seen in Italy that led to unique foods and ways of life, which means that the movement may have limitations in its potential for spread. Also an unintended consequence of their focus upon uniqueness, which has been promoted by the assertive minority of their residents' that support the ideas and are overwhelmingly middle class, may be an increase in the intensity with which their town is used, with more people coming to shop, dine and visit the centre. This may also result in a greater interest in living in these towns, leading property prices to rise, which is likely to alter the social mix of the town as local low-income residents may be displaced by new, incoming middle-class residents. However, since some of the main goals of the movement are to improve hospitality and visitor experience, as well as to promote their ideas within the population of the member towns, the participating towns seem well aware of these potential problems.

Cittaslow's future influence is also likely to be felt in other centres which adopt some of the movement's ideas, even though they are not formal members, or have left the network. This is increasingly likely to be the case given the spread of other aspects of the slow living movement that Cittaslow sprang from, as two examples show: slow money is being used to support local enterprises, especially farming, by encouraging people to invest locally (Busse 2011). Also, Whole Foods Market Inc., the largest chain of natural and organic food supermarkets in the U.S.A., has recently introduced an animal welfare labelling programme which tells consumers exactly how the animals they are consuming were raised (Leeder 2011), attracting those who deplore industrial farming.

Cittaslow's diverse origins and the focus upon historic roots have resulted in a movement that is accredited with 'variety' and 'fluidity', while legalized local activism is also organized into a system of both national and international networks. What Parkins and Craig (2006, p. 79) have described as examples of 'explicitly bureaucratic creations', are united by a shared Cittaslow philosophy that focuses on taking the time to achieve and to appreciate—in a unique local setting—a different lifestyle that favours quality and traditions. Yet the creation of the international and national Cittaslow networks, the aim to promote open-participation, and the involvement of the organization in external multi-level governmental associations, are all moves to go beyond the local. Yet a note of caution about the future needs to be introduced. There is a tendency of some members of Cittaslow to romanticize their individual 'uniqueness', which may limit their ability to enter into a wider politics. This is a challenge that the individual Cittaslow local agencies will have to face and one that illustrates Massey's (2005, p. 182) point that:

> will demand far more of the agents of local struggle in the construction of both identity and politics than there is room for, in that topography where identity seemingly emerges from the local soil.

If the Cittaslow movement does not address the struggles that participation in a wider politics implies, then membership of the movement may provide little more than a badge of approval for use in promotional activities. Alternatively, if Cittaslow members actively engage in a wider politics, not only will they recreate and reshape their own identities, they will also help to reconfigure the movement as a whole. This will create a more significant impact upon the general urban scene, but not only by improving the sustainability and liveability of our towns that other new urban themes are also persuing. They will also emphasize the often forgotten need for historic conservation, as well as their unique focus on quality, taste, traditions and hospitality. These are all important benefits from slow living that rarely appear in the other new urban strategies that have been developed over the last three decades.

References

Allison, G., Ball, S. Cheshire, P., Evans, A., & Stabler, M. (1996). *The value of conservation? A literature review of the economic and social value of the cultural built heritage*. London: English Heritage.

Andrews, G. (2008). *The slow food story: Politics and pleasure*. Montreal: McGill-Queen's University Press.

Beatley, T. (2004). *Native to nowhere: Sustaining home and community in a global age*. Washington D.C.: Island Press.

Busse, P. (2 October 2011). Slow money and Wisconsin farming, 3.0. Isthmus: *The Daily Page*. http://www.thedailypage.com/isthmus/article.php?article=32197. Accessed 30 Oct 2012.

Buttimer, A. (1971). *Society and milieu in the French Geographic tradition, Monograph Series of the Association of American Geographers*. Chicago: Rand McNally.
Cittaslow International website (CIW). http://www.cittaslow.org/. Accessed 15 Nov 2013.
Cittaslow, UK. (2011). *An introduction to Cittaslow*. Cittaslow.
CL: Cittaslow Ludlow. (2006). Case study: Catalysing partnerships. http://www.cittaslow.org.uk/page.php?Pid=8PLv=1. Accessed 24 July 2011, (withdrawn by 30 Oct. 2013).
CM: Covenant of Mayors. http://www.eumayors.eu/about/covenant-of-mayors_en.html. Accessed 30 Oct 2013.
Cohen, N. (2001). *Urban planning conservation and preservation*. New York: McGraw-Hill.
CSIC: Cittaslow International Charter. http://www.cittaslow.org/download/DocumentsiUfficiali/2009/newcharter%5B1%5D.pdf. Accessed 31 Oct 2013.
CSN: Cittaslow Network. http://www.Cittaslownews.files.wordpress.com/2009/08/newcharter.pdf. Accessed 29 Nov 2011.
Furedi, F. (11 May 2011). Hating Tesco: A passion shared by the PC and the BNP. *Spiked*. http://www.spiked-online.com/index.php/site/article/10503/. Accessed 15 Nov 2013.
Harvey, D. (1990). *The condition of postmodernity*. Oxford: Blackwell.
Holmgren, D. (2003). *Permaculture: Principles and pathways to sustainability*. Australia: Holmgren Design Press.
Holmgren, D. (2005). The end of suburbia and the beginning of mainstream permaculture. *Permaculture Magazine, 46*, 7–9.
Honoré, C. (2004). *In praise of slow: How a worldwide movement is challenging the cult of speed*. San Francisco: Harper.
Howard, A. (2009). *Slow guide to Dublin*. South Melbourne: Affrim Press.
Knox, P. L. (2005). Creating ordinary places: Slow cities in a fast world. *Journal of Urban Design, 10*(1), 1–10.
Leeder, J. (8 August 2011). Grocer's animal welfare initiative rolling out to Canadian stores. *The Globe and Mail*.
LTC: Ludlow Town Council. (2011). Agenda and minutes. http://www.ludlow.gov.uk/Contents/Text/Index.asp?SiteID=817SiteExtra=1562671TopNavId=825NavSideId=14480. Accessed 30 Oct 2013.
Massey, D. (2005). *For Space*. London: Sage.
May, J., & Thrift, N. (Eds.). (2001). *Time-Space: Geographies of temporality*. London: Routledge.
Mayer, M. (2006). Manuel Castells, the city and the grassroots. *International Journal of Urban and Regional Research, 30*(1), 202–206.
Mayer, H., & Knox, P. L. (2006). Slow cities: Sustainable places in a fast world? *Journal of Urban Affairs, 28*(4), 321–334.
Miele, M. (2008). CittàSlow: Producing slowness against the fast life. *Space and Polity, 12*(1), 135–156.
Mollison, B. (1988). *Permaculture: A designer's manual*. New South Wales: Tagari Publications.
Mumford, L. (1961). *The city in history*. New York: Harcourt Brace.
Parkins, W., & Craig, G. (2006). *Slow living*. Oxford: Berg.
Petrini, C. (2001a). *Slow food: The case for taste*. New York: Columbia University Press.
Petrini, C. (Ed.). (2001b). *Slow food: Collected thoughts on taste, tradition and the honest pleasures of food*. Vermont: Chelsea Green Publishing.
Petrini, C. (2006). *Slow food revolution: A new culture for eating and living*. New York: Rizzoli International Publications.
Petrini, C. (2007). *Slow food nation: Why our food should be good, clean and fair*. New York: Rizzoli International Publications.
Petrini, C. (2009). *Terra Madre: Forging a new global network of sustainable food communities*. Vermont: Chelsea Green Publishing.
Pink, S. (2007). Sensing cittàslow: Slow living and the constitution of the sensory city. *Senses and Society, 2*(1), 59–77.
Pink, S. (2008). Rethinking contemporary activism: From community to emplaced sociality. *Ethnos, 73*(2), 163–188.

Pink, S. (2009). Urban social movements and small places: Slow cities as sites of activism. *City, 13*(4), 451–465.

Radstrom, S. (2011). A place-sustaining framework for local urban identity: An introduction and history of Cittaslow. *International Journal of Italian Planning Practice, 1*(1), 90–113.

Roux, A. (24 June 2013). Laisser du temps au temps avec la label cittaslow. *La Gazette*.

Simmie, J. (1981). *Power, property and corporatism*. London: Macmillan.

Smith, N., & Williams, P. (Eds.). (1986). *Gentrification of the city*. London: Unwin Hyman.

SP: Slow Planet. http://www.slowplanet.com. Accessed 15 Nov 2013.

TLS: Tourism and Leisure Solutions. (2005). *A study to explore the potential to brand Perth as a food town*. Scotland: Scottish Enterprise.

Werness, H. B. (2006). *The continuum encyclopaedia of animal symbolism in Art*. New York: Continuum.

Chapter 16
Conclusions: Creating More Liveable and Sustainable Urban Places

Wayne K.D. Davies

The decades around the beginning of the twentieth first century have seen an explosion of new ideas to improve urban places. These have been designed not only to solve the deficiencies of current forms of growth but also to resolve a series of new problems associated with urban places, as well as the effect they are having on their regions and environment. Many of the most prominent of these new ideas add an adjectival descriptor to the words urban or city. These shorthand summaries have been interpreted as a series of city or urban themes, or as a series of clusters of urban concepts focusing on solutions to particular urban problems. They have been reviewed in the various chapters, ranging from themes such as New Urbanism ideas (Chap. 2) to Slow Cities (Chaps. 14 and 15). These themes have been described in terms of their origins and key features, the problems they are designed to solve, together with a critical review of their utilities. However, as noted in the Introduction there other themes that could not be covered except in abbreviated form because of space constraints, although these were viewed as being more restricted to particular types of urban places, or are still emerging in the sense that they are in early stages of development. Yet despite their scope, not all of current urban problems have been addressed by these new ideas.

All of these themes have primarily focused upon one major idea or sets of ideas to create better cities but inevitably there are overlaps between some of them. For example, exponents of Slow Cities also advocate some sustainable initiatives, while the policies that are described as Green Cities are often incorporated into Sustainable City agendas. Yet these overlaps do not detract from the fact that each of the themes is primarily focused upon a specific content area or approach, in which a number of distinctive policies have been advocated to improve urban development. In advocating an individual theme there is always a danger of it being viewed as being the only one worth studying, or using to develop policies. Certainly in many cases there may be one major problem—whether sprawl, economic downturns,

W. K.D. Davies (✉)
Department of Geography, University of Calgary, 2500 University Drive, N.W., Calgary, AB T2N 1N4, Canada
e-mail: wdavies@ucalgary.ca

W. K.D. Davies (ed.), *Theme Cities: Solutions for Urban Problems*,
GeoJournal Library 112, DOI 10.1007/978-94-017-9655-2_16

safety, or sustainability etc.—that demands greater attention at a particular point in time. However in many cases the condition of most towns and cities today requires attention from several of these themes in order to develop policies of very different types to improve the quality of urban life, ecology and economy.

Although the themes are relatively distinct in terms of their content and policy derivatives it can be argued that all relate to one of two wider categories, those linked to the desire to create more liveable and sustainable urban places. These terms now have a wide variety of connotations which often makes them difficult to pin down. For example, in terms of *liveability* many of the designs and policies of New Urbanism were developed as an antidote to the faceless, auto-based sprawl of many cities and to make new subdivisions more walkable, attractive and community–orientated. Festive Cities illustrate the varied origins and range of festivals that can often provide historical connections, but in all cases add episodic excitement to the life of urban residents and attract visitors to boost a town's economy. Slow City ideas seek to reduce the pressures of contemporary life and provide more enjoyable and human-based experiences, frequently based on autochthonous cultural practices. Just City ideas may be a unique and much needed new approach as inequalities rise within society. But since they are based on the value of 'fairness' for those living in cities—giving all residents greater equality in service provision, opportunities, and to participate and influence decision-making processes—they may be considered as part of the attempt to make urban places more open to all. This will improve the liveability of places for *all* residents, not just the wealthy or social elite.

In terms of *sustainability* the obvious negative environmental effects created by cities are increasingly being addressed by more attempts to improve the natural ecology of places through creating more green spaces and rehabilitating the natural environment, by Green City ideas, although this also adds to their liveability. Similarly, actions to reduce such features as fossil fuel consumption and their negative externalities, waste, and excessive resource use, form the basis of Sustainable City and Transition Town initiatives. In a physical context the specific problems posed by life in settlements in cold climates are discussed in Winter Cities. Resilient Cities relate to the ability of settlements, and people within them, to mitigate the threats posed by extreme, episodic and often unpredictable natural processes, thereby ensuring their continued existence and ability to recover from devastations. In economic terms a different type of sustainability is involved. Many cities search for innovative ways of ensuring their economic prosperity, given the new challenges that have come from the loss of their older locational advantages for particular industrial processes, and the effect of new spatial imperatives based on the new economy and swifter transport links and electronic communication devices. It has led some to search for ways for keeping and attracting one class, namely more creative workers (Creative Cities); others have targeted a wider approach through focusing on new knowledge opportunities as a basis for future growth, especially as manufacturing has moved to low cost locations away from the developed world (Knowledge Cities). Sustainability can also be seen in a social sense through the Safe City policies that are designed to ensure that people in towns and cities are as free from crime as possible, which in the developed world has helped to counteract the rise in crime rates seen from the 1970s. Rising concerns about the need to improve health and care availability and to continue the progress made in removing the historical urban disadvantage in mortality rates has led to new types of Healthy City policies, which can be seen as another form of the social sustainability of urban populations. Yet the very fact of making places more sustainable, whether in environmental, economic or social terms, also makes places more liveable in the long term. For example, reducing crime rates and removing persistent health problems in cities makes them more attractive to entrepreneurs who wish to develop new economic activities. Hence the broad concepts of liveability and sustainability should be seen as two sides of a more general approach to improving the quality and maintenance of life in urban places that are expressed in the various foci of each chapter.

However, unlike the progress in developing physical sustainability indices, as seen in the EIU's Green Index described in Chapter 5, attempts to develop liveability scales and city rankings on these scales have been less successful. Perhaps the best known at a global scale are the rating schemes developed by the Economist Intelligence Unit (EUI) and Mercer International (MI). Since the indicators used vary, with 30 in the former case and 39 in the other, it is not surprising that their final ranking of cities varies, with only three of the same centres present in the top ten cities of each list. Some of the individual indicators used are similar in concept but may be interpreted in a contrary fashion, while 20% are quite different. In addition the indicators used are scaled in different ways, on a five point scale in the case of the EUI and a ten point one for Mercer, with varying weights also applied to categories of the indicators in the former case. However since the scaling is based on subjective evaluations by a very limited number of people, mainly local correspondents and head office personnel, there is clearly room for divergence. But a more general problem for their general utility is the fact that these indexes are primarily used by companies and governments to provide salary compensations for employees transferred or working in various countries. Hence the indicators reflect the needs of middle-upper income people with good qualifications, not the concerns of the vast majority of people in the various cities. So judged on the basis of these liveability and quality of life scales for high earning expats, the ranking scales of the cities may be useful, but are certainly far from adequate for the general concept of liveability and quality of life for the whole resident population in the cities surveyed.

Given the complexity of many of these themes and their policy derivatives it is not surprising that advocates of these various ideas have focused on developing one of these approaches, ignoring the importance of other themes. Indeed, cities are often confronted by a problem that seems to take precedence over others, which leads to the policies designed to solve the problem, whether adopting Resilience City mitigations, creating Knowledge City opportunities to improve economic prosperity, or using Safe City policies to reduce crime. However, a truly comprehensive approach to create more liveable and sustainable cities usually requires the application of more than one of these themes. To focus only on one of a few themes runs the risk of creating another phase of what might be called '*silo development*'. This means that planning effort is too often concentrated on a single theme while forgetting the need for policies that resolve other problems, or as in too many urban renewal schemes, a failure to thoroughly investigate and resolve, to the satisfaction of local residents, the negative effects of some new project. Perhaps the most obvious case is the way that New Urbanism developments in particular have largely ignored the severe problems of inequality and poverty pockets in cities that has led to the Just City literature. Another is the way that smart growth polices of increasing density will certainly help reduce their carbon emissions but often forget that green spaces are necessary, not only as places for relaxation but by providing health-giving properties. Obviously, not all of these themes are relevant to all cities. Some, such as Winter City ideas, are only applicable to places facing extreme cold, although a case can be made for places having extreme heat or rain to develop and share a new cluster of ideas to help cope with such conditions. In other cases, such as the problems posed by natural hazards, some cities are at real immediate risk from such hazards as earthquakes or floods, or have forgotten past disasters caused by such problems and have been too complacent to adopt policies to reduce future risks of such events. Others have far less levels of jeopardy from extreme natural processes, although the changing nature of world climate through anthropogenic warming does mean that many more urban places need to measure the extent to which they may be threatened by the effects of climatic change and plan accordingly. Similarly economic prosperity cannot be taken as being for all time; the rapidity of economic change means that urban places need to develop the capability to be alert to changes that can affect their future and act to encourage the development of alternative activities.

Despite the occasional overlap, each of these themes is distinctive enough to deal with either a particular content area, or in the case of Transition Towns with a new community-based approach to improve a town's sustainability and independence, from which various policies have been generated to solve problems. Hence the themes can be viewed not simply as identifying a problem or problems, but as ideas also containing clusters of policies designed to resolve them, whether improving safety or health, or providing more ecological diversity, or reducing pollution. Increasingly many of the policies being adopted are being subject to evidence-based reviews to determine their effectiveness relative to other policies, which should also involve monitoring through time to determine their long term utility. Yet these new clusters of urban concepts should not be viewed only as providing empirical contents. In many ways they have a variety of distinguishing features, other than their co-incidence in time around the beginning of the new millennium, that identify a distinctive character of the urban themes in the history of urbanism. Five in particular stand out.

First, they have originated from a *larger variety of sources* than the mainly top-down national or state government actions after World War II, that created New Towns, Green Belts or Industrial Controls on industry in large urban places, or the creation of Enterprise Zones and Urban Development Corporations for decaying inner cities after the stagnation of the period after the 1970s. Certainly many municipalities have been involved in creating or implementing various policies within their political jurisdictions, or combining with surrounding areas to form what amount to New Regionalist governance structures, or joining networks to seek out new ideas. But the majority of the new themes have come from wider sources than local or national governments, ranging from citizen initiatives in the case of Transition Towns and Slow Cities, from planners and architects promoting Neo-Urban ideas, to social activists and academics developing Just Cities concepts and practices, and, in the case of recent Resilient City initiatives, with major contributions from international organizations from the World Bank to various NGOs, such as Red Cross, as well as from research from academics studying natural hazards. Hence governments have been less important in the creation of most of these themes but have used many of the policies and, of course, need to be involved in later stages if the ideas are to be implemented and really flourish.

Second, advocates of most of the themes emphasize the need to *reduce policy and management fragmentation*. They see the importance of breaking away from what amounts to the specialized roles of various departments in government or agencies that traditionally deal with particular urban service provision or problems, such as dependence only upon the police in ensuring safety, medical professionals in terms of health care organization, or planners in development decisions. Most of the policies associated with the various themes emphasize the need for wider stakeholder engagement and involvement. This goes beyond the often top-down information-imparting meetings of city officials with individual citizens, community organizations, businesses and non-profit groups, replacing them by actively working with these groups in a true communicative fashion. It also involves finding ways of breaking down the often rigid barriers between various government departments or agencies to provide more co-operation, scrutiny and greater flexibility in creating and applying new policies, as well as counteracting the tendency of existing administrations to stick to their traditional practices. Indeed new agencies that involve all representatives of levels of government are often needed to ensure progress, since these control many different functions, from land through specialised services to finance. As can be seen in the case of the Safe and Knowledge City themes this also usually means acceptance by decision-makers and the public of the importance of solving particular problems and the willingness to create a vision, strategies, and then individual policies to solve the difficulties identified through joint initiatives, to circumvent the silos of separate practices. However it must be recognized that there may be trade-offs or compromises required between alternative policies. For example, increasing the density of urban places does reduce carbon emissions since it makes it possible to substitute public transit for cars, but this could reduce the amount of accessible green space which is increasingly recognized as essential for health.

Third, many of the themes have a much wider *scope* than previous concepts and policies designed to improve the condition of urban places, most of which in the past relate to the provision of various services, or regulations over development. For example, the whole environmental sustainability movement is not just based on improving the local ecological conditions, important though this may be for general relaxation and health. It also recognizes the many negative externalities of cities upon their regions, and increasingly through greenhouse gas emissions, upon the composition of the atmosphere. Hence the effects are not simply local or regional; they contribute to a global warming that is projected to have many different effects on urban places that include: sea level rises which put coastal cities in more danger; increases in temperatures which will lead to higher mortalities in some places; and also to the greater frequency and magnitude of various storms, drought and fire risks in various areas, potentially leading to greater urban damage. The increased scope is also seen in the way that interest in improving the quality of urban life has been extended to increasing the number of festivals and adopting slow living in various towns.

Fourth, advocates of all of the themes have created *networks* of various types between cities, or people within them who are interested in similar urban ideas. The networks promulgate and co-ordinate the policies associated with various themes, or as in the case of the C40 group of major metropolitan centres implement policies—such as emission reductions—that make up for the absence of progress from their national governments. Many urban municipalities are likely to be involved in one or more of these networks, providing them with far better access to new ideas developed in other cities and experience of best practices, as well as ways of resolving the barriers to the adoption of new policies. In some cases, like the charters for New Urbanism or Cittaslow, they also provide foundational sets of principles, or ways of increasing interest in sustainability, as the Transition Towns network has done through citizen interest and involvement.

An important feature of these networks is that they increasingly bind member cities into horizontal ties, linking towns and cities with similar problems or interests, such as improving resilience against natural hazards, or finding new ways of coping with severe winters. So urban places are not only connected in vertical relationships, downwards with small dependent places in their region, and upwards through the size hierarchy to larger places. Obviously there were always horizontal links between places of similar size in a country. But these new networks and specific city to city (C2C) relationships are increasingly more significant than their vertical links in terms of policy knowledge acquisition, especially as national governments in big cities or capitals have become less activist in creating urban initiatives. These horizontal urban linkages, often of continental and global reach, expand the interests and horizons of cities. They have made many municipalities more independent in outlook and more open to innovative ideas, although most are still circumscribed by governance rules that restrict them to their own defined territory through state or national laws. Moreover the participants in the linkages are not simply municipal politicians or officials but also citizens, local businessmen and academics who also visit other cities, either to obtain first-hand experience of problems and solutions, or to participate in various meetings and conferences that deal with new approaches to solving and managing urban problems. However the need to provide justification for these visits and the subsequent sharing of information is paramount, otherwise there is a danger of the exchanges being seen by the public-at-large as just perks or junkets at taxpayer expense. These exchanges sometimes involve the deliberate transfer of expertise to other settlements in the developing world, by advanced cities sending their specialists in particular problems to cities in the developing world to provide advice over ways of solving particular problems. Examples include Bonn's links to a set of cities round the world to help build resilience, or the way that conservation experts from Lille helped officials in the historic Vietnamese capital of

Hué to map, conserve and provide advice on the identification, legal protection and adaptive re-use of heritage buildings, not only by expert-led visits to the south east Asian city but by return visits by Vietnamese to study practices and management of the problems in the host country.

Although the number, variety and objectives of these networks have exploded in recent years, helped by the more rapid electronic communication systems, it is worth remembering that there have been networks of towns and cities with mutual interest in the past. Historically, the best known is the Hanseatic League of commercial centres in medieval North Europe, towns with common commercial interests. They gave each other trading privileges and asserted their rights with feudal lords or local states, as well as providing co-ordination over protecting trade routes from pirates. But even the nineteenth century there were networks concerned about improving health, such as the British Health of Towns Association from the 1840s that agitated for improved sanitary and water facilities to reduce the horrendous mortality rates consequent upon industrialization and unregulated urban growth. In the 1860s European cities had regular conventions of their mayors. During the early twentieth century there were many networks devoted to particular issues, such as the British Committee for the Study of Municipal Institutions in 1904, as well as many funded tours of other cites to investigate their successes in particular functions (Ewan and Hebbert 2007), while European cities with socialist ideals also established their own network (Dogliani 2002). Since World War II the number of these networks has grown and moved away from being primarily leaders or groups of bureaucrats from urban places discussing common problems or co-ordinating professional management qualifications. Far more specialist forums involving experts in many fields, many international in scope and dealing with the types of themes reviewed in the various chapters, have developed, assisted by the ease of travel and instant electronic linkages.

Fifth, the growth of networks and their use in providing more personal contacts and knowledge acquisition has led to an emerging interest in the whole *city-learning process*, namely of how cities gather and apply knowledge to improve their conditions. This could easily be considered another emerging urban theme of the early twenty-first century. This changes the focus from collecting and applying the empirical results of particular information and policy ideas, such as those associated with the various themes, to the *mechanisms* by which this new knowledge is acquired and used. If these mechanisms of learning are deliberately studied and applied, they take a city and its citizens away from the accidental, random or individual acquisition of either policy information in particular fields, or of problem identification. They offer the prospect of creating a more explicit and formalized process of what amounts to learning, not simply in one content area, but more generally across many areas. So towns and cities *may* be able to create or use their networks and contacts to become smarter and more capable of adapting to changes, which given the complexity and changeability of cities is always difficult. This gives another meaning to the term Smart Cities, which has previously been applied to new ways of 'doing things', such as smart growth in urban development, or in the way that new technologies, such as Smart Grids, provide more effective solutions to energy reductions. Instead, this third type of 'smartness' comes from the deliberate use of the mechanisms of learning to ensure progress. However the conditional word 'may' used in the preceding statement must be stressed. Relatively few cities have availed themselves of this type of opportunity. Certainly some cities have developed new innovative procedures from their own administrative departments, such as: Portland's innovative sustainability programmes; Bilboa's impressive regeneration; or in middle income countries, Curitiba's planning initiatives since the late 1960s or Medellín's social urban projects. Others have brought in experts from private companies that specialise in a variety of development projects, from mass transit to new management procedures. But if the emphasis is placed on the extent to which a collective learning process in a city is developed there is more scope for cities to create future new policies in very different content areas, since it is the *process of knowledge creation* that is being developed, not a particular '*way of doing*', issues discussed in the Knowledge City (Chap. 10).

The big problem is how to translate the knowledge into new urban actions. Campbell (2012) identified four crucial elements that can be viewed as improving city-learning. One is the *knowledge-gathering* activity. This is how new knowledge about such features as improving urban life, functioning and sustainability is obtained from internal and external sources. The improvement possibilities also deal with finding ways in which blockages to the application of ideas can be resolved. Another element consists of whether urban places have an *agency or office for managing this process* and also provide a repository for the information obtained, one that is effective, valued and can be used by the decision-makers in the city. A third element is described as the '*institutional process*', procedures for ensuring that the knowledge obtained is disseminated and discussed with appropriate officials in the various organs of government and also with other stakeholders in the city, making it easier to create support for new policy options. Often this takes the form of a type of superagency in the city to initiate and coordinate all or major developments, such as the IPPUC established in Curitiba in 1965 (Campbell 2012) or Bilboa's Metropoli-30. A fourth element is probably the most difficult to achieve, for it involves creating a *milieu of trust* among the not only the various decision-makers but also the stakeholders in the city. Within all places there are always rivalries and differences of opinion between competing government departments, businesses and other interests. If a common goal to improve the liveability and sustainability of urban place then there must be discussion and resolution of differences in pursuit of a common aim. This is probably the most difficult part to achieve, for trusting relationships are needed to ensure the exchange of information that is not codified, but is internalized and tacit in character, information that is only effectively shared by personal contacts. These issues have already been discussed in the learning process of Knowledge Cities (Chap. 11). Frequently the personal contacts and trust come from participation in the conferences and study tours linked to the networks associated with the various themes. Or they may emerge simply from C2C linkages that have been developed, and even in a more general way from the more limited goals of the city business or trade associations that promoted their own agendas and often pressured their municipalities to follow policies that supported their aims.

Finally, it is worth adding an additional two-pronged element to Campbell's list, which might best be called a *problem recognition and willingness to act* element. This means that that the leaders in a city, from local politicians, businesses and various organizations, not only accept that their city or urban place has some major problem that needs to be solved, but also develop a willingness to act to solve the difficulty. Hopefully this leads to the cooperative process to develop some vision and subsequently policies to solve the issue, which has been seen especially in the citizen-led Transition Town movement, and in cities that have pursued Safe, Knowledge and Slow City agendas in particular. This recognition of both the need and also the necessity to resolve the perceived problem through deliberate management strategies and supported by major stakeholders is often the first step in the learning process that leads to a problem resolution, and with its success the identification of other issues and problems that reduce the liveability and sustainability of cities. In democracies it either takes some far-sighted and often charismatic leadership, as displayed by some mayors, to achieve such progress, or some catastrophic event to stimulate consensus to embark on such action. In many cases, however, there may be little consensus about particular strategies. What is then needed are a series of initially small steps that will produce improvements and advantages that all can see and get used to, thereby building support for more widespread progress. This is the type of step-by-step actions recommended for expanding pedestrianization (Gehl 2010) or San Francisco's waste management strategy. It is not new. The idea can be traced back to Alinsky's action approach to community development in seriously disadvantaged inner city neighbourhoods that he began practicing in the late 1930s but only formalized decades later (Alinsky 1972). However it must be recognized that not all actions have the capability of beginning with small steps.

In reality few cities in the world have developed the capacity to learn by adopting all these elements. Campbell (2012, 56) created a city-learning classification, identifying five types based on the degree to which a city organizes its effort to finding and using information to make it a better place, or to solve particular problems. He then provided examples of cities that fall into these categories. His first category consists of the most active cities that deliberately organize learning missions to a range of other places, managed by a dedicated agency that encourages the building of social capital among those travelling to other places, and subsequently tries to find ways of building upon this information to improve aspects of their own city. A second type consists of cities that engage in episodic visits or exchanges to other centres, best illustrated by the Sister Cities network where one-on-one linkages occur, although some of these cities have more than one sister city or twin town. But the type rarely has any agencies to control the learning process. A third group consists of those centres that are part of clusters of cities in a network that shares common program objectives or campaigns, such as the World Heritage group that exchange ideas about historical conservation, or Agenda 21 centres that promote sustainable principles. In this type a central role is played by the organization and its secretariat, such as UNESCO in the Heritage City case, rather than any deliberate learning by the city itself from other cities. The fourth type describes those involved in networks of common interest. The objective is not learning itself but the information that can be derived from others in the network, such as in the case of Healthy Cities, where there are usually some rules for joining and in which common problems and policies are discussed and which may be stimulated by some outside agency, such as the WHO in this example. Finally there are the cities that are part of networks but are essentially passive participants in conferences or networks, such as national or regional conferences of City Mayors in many countries, where information on new practices or problems may be discussed. But no deliberate process of accumulation of knowledge gained or implementation occurs, although ideas may subsequently be shared and acted upon by the city.

Campbell (2012) admits that these types are not mutually exclusive, for a city may at different times use one or more of these approaches. But the classification does provide a way of showing a gradient of learning effort, from the most active cities to the most passive. He maintains that one of the best cases of the active learning city can be seen by the case of Seattle. It experienced an economic downturn in the 1980s, which led to many layoffs in its major industry, the Boeing aircraft company. This led many in the city to the conclusion that it would be useful to study how other cities coped with economic change. The city began its study visits in 1992 with groups of city officials and businessmen visiting Stuttgart, Amsterdam and Rotterdam. Initially organized by its Chamber of Commerce, the city and its region subsequently created a Seattle Trade Development Alliance to manage the programme, which in subsequent years went to Vietnam and later to Japan and China and other European cities. This began the process of deliberate learning from other cities through study missions, annually sending as many as a hundred civic and business leaders to other cities to observe how various problems were solved and how new initiatives were developed and implemented. Although some people may go on successive visits, there is usually an annual sixty per cent turnover, so that the mission consists of people who have been on a visit and also new ones. This rolling membership is designed to expose new people in leadership or technical positions to development patterns in other cities that may provide useful new policies in Seattle, as well as exchanging views with groups in other cities on a variety of topics that go beyond business links. The study missions also include the opportunity for the participants to build relationships with fellow travellers, helping to build the social capital between the participants to create the milieu of trust among decision-makers in Seattle and region. However there is no doubt that the missions also enable participants to promote the business opportunities available in Seattle to potential overseas customers, who may import their products, invest in the city, and to establish personal links between companies that might be subsequently engaged in business relationships. Although these study missions form the major thrust of Seattle's learning process, other groups, from its port to various industries and educational institutions have their own overseas visits, although these are more tightly confined to their own businesses and functions, rather than being part of the city's more general and varied learning agenda.

In Europe the former iron and steel manufacturing city of Bilboa has also engaged in a deliberate process of learning to alter its economy from the declining and polluting industrial base to new ventures and structures, but also with innovative local inputs. This was not carried out by regular study visits in the Seattle manner, a city with an enviable economic base. Rather it created a highly organized agency, Metropoli-30, with a staff and a defined budget to develop and help implement as strategic mandate. However the agency helped bring in experts from other cities to provide guidance on its plans to transform the city from its declining former heritage and provide ways of gauging progress (Campbell 2012). These plans led not only to the creation of a major new attraction, the Guggenheim museum, that also acts as a focal point for many conferences. The associated Bilboa Rio 2000 agency, a non-profit, public partnership has also proved crucial (Plöger 2008). It has overseen the modernization of its port, rapid transit systems, as well as rehabilitation of old industrial sites, in addition to re-integrating the city within its surrounding region rather than being an export-orientated industrial centre. The restructuring has led to new technology parks, successor companies and new services that in total employ as many people as the older industrial developments. Since different governments control land planning, finance and land ownership etc., the agency is designed to have involvement from all levels of government to effect meaningful change.

These developments have meant that some urban places around the world have become more proactive in obtaining and applying ideas from the new themes and networks, not only those reviewed in this book. In many ways it is creating a new phase of municipal endeavour, rather like the ways in which urban places developed links and new services in late nineteenth century Europe, although much of this was assisted by national government leadership in setting out legal frameworks to be applied in cities. Most of these early networks subsequently atrophied because of two world wars and the economic depression in between. Yet one must be cautious; cities vary drastically in the extent to which they are engaging in new practices. The eternal barriers of the limited powers and financial base of most urban places provide real constraints for many cities, for few national or state governments are prepared to delegate more powers or tax rights. Yet by their sheer scale and voting power they can influence higher levels of government, and many have been pursuing new policies, especially in sustainability, because of the failure of higher levels of government to act.

What has been the result of all these new themes and the networks that often sustain them? It is obviously too soon to produce a clear evaluation, for many have only been operating for a few years. Some have developed comprehensive new strategies within which individual policies have been applied; others have been content with implementing individual polices without any coherent strategy. In both cases there are many examples of successful actions. The trouble is that too few places are actively pursuing even one of the various approaches. Indeed the saddest interpretation to the changes being promoted by advocates of these new themes is that where there is a need for technical transformations, such as reducing the use of fossil energy, we do not lack the means to implement change. What is holding up greater progress is a lack of the political will and leadership to implement change or develop innovative new solutions to existing problems. Certainly some of the solutions are more costly, but not all are, and some cities have pioneered the way for progress, as three examples show. So if most cities adopted the carbon-neutral policies of Copenhagen, the metro planning of Portland, the waste management policies of Edmonton and San Francisco, the future would be far brighter. These examples are only a few of the range of approaches that have been successfully generated by the various themes that could improve our urban future. Adopting more of the policies being advocated would be particularly useful in providing better development goals for building the new cities, or the extensions to existing places, that will be needed to house the extra 2.4 billion people anticipated by 2050, although there are few signs of this happening on a large enough scale. This intense growth pressure, plus many emerging problems—especially those from greater global warming, and the failure to solve inequalities—persuade many that the future is not being effectively addressed by our current actions. Certainly there are positive changes where cities have adopted policies from the range of new themes that have been shown to improve our settlements. But far too many of the adoptions are small, incremental changes, that are not effective enough, especially in the developing countries. In their settlements there is a pressing need for many changes, such as making settlements more resilient against natural hazards, improving housing stock, creating enough employment, and also solving the problems of providing so many necessary services, especially the disposal of waste. Moreover there is almost certainly a need to develop new themes to solve the particular problems of urban places in the developing world. In other cases, such as the state policy of urban development in China, the rate of progress in building new high rises, factories, roads and high speed rail links is unprecedented in history. But the negative side effects of the urban transformation in most of the developing world can be seen in such features as high pollution levels, and relatively few parks or green spaces in the new, dense urban areas. The former may be solved by the new state policies, but it will take a decade or more; the advantages of the latter seem to have been forgotten as the space has been built-over. In general, therefore, more of our urban places, in the developed and less developed world, should take advantage of the policy improvements suggested by the various themes, either with individual examples or better still with more co-ordinated and sustained efforts to achieve their various goals, and with the creation of new themes to solve other pressing problems. They would not only improve existing settlements but could be used to provide better alternatives for the massive growth of urban places expected in the next thirty to forty years. Although most of the themes described in the various chapters have been based on the developed world experience, it may be hoped that the cities of the developing world will be able to identify and learn from the mistakes of the former, in order to create a far better future for their residents. Yet this involves not simply knowledge of the empirically tested solutions and concepts described here in the various themes, but the political and managerial necessity to develop more of the structural capacities to implement these programmes in settlements. Unfortunately, however, the type of urbanization, and the often squalid conditions suffered by far too many that has been taking place in many settlements of the developing world in the past two decades, and the problems that still face many cities in advanced economies should not lead us to be optimistic, despite progress in some countries. Yet these conditions could be modified by applying more of the urban themes reviewed in the various chapters. But even these have objectives that are far more limited than Mumford's rhetorical flights of over fifty years ago that led him to conclude in his masterful *City in History* that the major role of cities should be "the magnification of all the dimensions of life" (Mumford 1961, 576). Such a goal seems implausible in the light of most recent urban experiences. But by using more of the new themes that have emerged in recent years it may yet be possible for us to at least substantially improve current conditions and create more attractive, satisfying, liveable, sustainable and just cities, through processes that do not further harm our environment that has nurtured us.

References

Alinsky, S. (1972). *Rules for radicals*. New York: Random House.
Campbell, T. (2012). *Beyond smart cities: how cities network, learn and innovate*. Abingdon: Earthscan.
Dogliani, P. (2002). European municipalism in the first half of the twentieth century: The socialist network. *Contemporary European History, 11*, 573–596.
Ewan, S., & Hebbert, M. (2007). European cities in a networked world during the long twentieth century. *Environment and Planning C: Government and Policy, 25*, 327–340.
Gehl, J. (2010). *Cities are for people*. Washington DC: Island Press.
Mumford, L. (1961). *The city in history*. London: Secker and Warburg.
Plöger, J. (2008). Bilbao city report. http://eprints.lse.ac.uk/3624/1/Bilbao. Accessed 26 March 2014.
EIU: http://www.eiu.com/liveability2012. Accessed 30 June 2014.
MI: http://www.mercer.com/newsroom/2014-quality-of-living-survey.html. Accessed 20 June 2014.

Index

W. K.D. Davies (ed.), *Theme Cities: Solutions for Urban Problems*,
GeoJournal Library 112, DOI 10.1007/978-94-017-9655-2

Q

R

T

The manufacturer's authorised representative in the EU is Springer Nature Customer Service Centre GmbH, Europaplatz 3, 69115 Heidelberg, Germany. If you have any concerns regarding our products, please contact ProductSafety@springernature.com

Printed and bound by CPI Group (UK) Ltd, Croydon, CR0 4YY
30/09/2026
02225436-0015